MW00844540

AMPHIBIANS *and* REPTILES *of* BAJA CALIFORNIA

Organisms and Environments

Harry W. Greene, Consulting Editor

1. *The View from Bald Hill: Thirty Years in an Arizona Grassland,* by Carl E. Bock and Jane H. Bock
2. *Tupai: A Field Study of Bornean Treeshrews,* by Louise H. Emmons
3. *Singing the Turtles to Sea: The Comcáac (Seri) Art and Science of Reptiles,* by Gary Paul Nabhan
4. *Amphibians and Reptiles of Baja California, Including Its Pacific Islands and the Islands in the Sea of Cortés,* by L. Lee Grismer
5. *Lizards: Windows to the Evolution of Diversity,* by Eric R. Pianka and Laurie J. Vitt

AMPHIBIANS *and* REPTILES *of* BAJA CALIFORNIA

INCLUDING ITS PACIFIC ISLANDS AND THE ISLANDS IN THE SEA OF CORTÉS

L. LEE GRISMER

WITH A FOREWORD BY HARRY W. GREENE

UNIVERSITY OF CALIFORNIA PRESS
Berkeley Los Angeles London

THE PUBLISHER GRATEFULLY ACKNOWLEDGES THE GENEROUS CONTRIBUTION TO THIS BOOK PROVIDED BY THE GENERAL ENDOWMENT FUND OF THE UNIVERSITY OF CALIFORNIA PRESS ASSOCIATES.

University of California Press
Berkeley and Los Angeles, California

University of California Press, Ltd.
London, England

The photograph facing the title page is of *Elgaria multicarinata* (Southern Alligator Lizard) from near Tijuana, BC; that facing the table of contents is of *Uma notata* (Colorado Desert Fringe-Toed Lizard) from near Mexicali, BC. All photographs are by L. Lee Grismer unless credited otherwise.

Library of Congress Cataloging-in-Publication Data

Grismer, L. Lee, 1955–
Amphibians and reptiles of Baja California, including its Pacific islands and the islands in the Sea of Cortés / L. Lee Grismer ; with a foreword by Harry W. Greene.
p. cm.–(Organisms and environments ; 4)
Includes bibliographical references (p.) and index.
ISBN 0-520-22417-5 (cloth : alk. paper)
1. Reptiles–Mexico–Baja California (Peninsula).
2. Amphibians–Mexico–Baja California (Peninsula).
3. Reptiles–Mexico–California, Gulf of. 4. Amphibians–Mexico–California, Gulf of. I. Title. II. Series.
QL655.G75 2002
597.9'0972'2–dc21 2001005193

Manufactured in Singapore

11 10 09 08 07 06 05 04 03 02
10 9 8 7 6 5 4 3 2 1

The paper used in this publication meets the minimum requirements of ANSI/NISO Z39.48–1992 (R 1997) *(Permanence of Paper)*. ♾

TO MY FAMILY, WHO SPENT MANY VACATIONS, WEEKENDS, AND BIRTHDAYS WITHOUT ME. AND LACY, I AM SORRY I WAS NOT THERE FOR YOUR FIRST DAY OF KINDERGARTEN.

A TODOS LOS RANCHEROS Y PANGUEROS DE BAJA CALIFORNIA QUIENES ABRIERON SUS CASAS Y SUS CORAZONES PARA BRINDARME COMIDA, HOSPEDAJE Y AFECTO Y QUE TAN AMABLEMENTE HAN ATENDIDO A MIS HERIDAS.

CONTENTS

FOREWORD

Harry W. Greene

Amphibians and Reptiles of Baja California is the fourth volume in a new University of California Press series on organisms and environments. Our themes are the diversity of plants and animals, the ways in which they interact with each other and with their surroundings, and the broader implications of those relationships for science and society. We seek books that promote unusual, even unexpected connections among seemingly disparate topics, and we want to encourage projects that are special by virtue of the unique perspectives and talents of their authors. Our first volume, by Carl E. Bock and Jane H. Bock, concerned the ecology of Arizona grasslands. This was followed by a field study of Bornean treeshrews, by Louise H. Emmons. Forthcoming books focus on Seri ethnoherpetology, by Gary Paul Nabhan; the American bison, by Dale F. Lott; and lizard biology, by Eric R. Pianka and Laurie J. Vitt.

L. Lee Grismer's volume comprehensively summarizes the herpetology of a remarkable and in some ways deceptively simple chunk of the earth. Seen from space, the world's second largest peninsula appears as a narrow extension of western North America, separated from mainland Mexico to the east by the Sea of Cortés and seeming altogether distinct from the country to which it belongs. In fact, several million years ago this spectacular landscape was cobbled together when a sizable piece of continent broke free, drifted west into the Pacific Ocean, and then fused with a smaller peninsula from the north. Seen from the other extreme, for a foot traveler stranded without water, some parts of Baja California would be exceptionally foreboding desert, seemingly unable to support life. However harsh, though, the peninsula and its offshore islands are far from desolate, having long been hotbeds of ecological and evolutionary processes. Flanked by rich marine environments, their terrestrial habitats include windblown sand dunes, precipitous rocky slopes, semitropical oases, and montane forests. Under scrutiny, especially after a good rain, these places reveal astonishing biological diversity and natural beauty. Among Baja California's more than 160 species of amphibians and reptiles, many of them endemic and highly distinctive, are incomparably vivid blue and yellow lizards, rattlesnakes without rattles, and other such living treasures.

We often care more about what we understand, and this makes the growth and spread of knowledge critically important in fostering conservation. Detailed natural history, however, requires years of work by many people, particularly for arid regions where so many organisms are usually out of sight. Although the first summaries of the Baja California herpetofauna were made during the late nineteenth century, this volume documents far more species than were known to those early scientists, and future studies will no doubt yield further surprising discoveries. But the book goes beyond the description of new forms of life, for our view of that diverse biota as a mixture of once geographically disjunct elements also reflects a changing interpretive context. Barely forty years ago the earth's landmasses were thought to be constant in location, and Baja California biogeography was paradoxical: the closest relatives of some species found at the tip of the peninsula are in southwestern Mexico, rather than farther north on the same landmass. That would imply that their ancestors had marched up around the Sea of Cortés and down to the cape region yet left no fossils or relict populations along the way. By 1980, though, the stage was set for reinterpretation of faunal distribution in the light of continental drift, and we can now more profoundly appreciate Baja California for its wonderful exposition of geological history, ecology, and evolution.

Amphibians and Reptiles of Baja California not only summarizes the findings of many previous biologists but also presents the results of a tremendous amount of painstaking, sometimes dangerous fieldwork by the author and his associates. The book is enriched throughout by that firsthand experience, the consequence of a lifelong passion for natural history. A self-confessed "born herpetologist," fascinated by Baja California since early childhood, L. Lee Grismer grew up to become a professional evolutionary biologist and biogeographer. Indeed, Grismer has been an active player in the recent conceptual advances in those fields, and his personal perspectives, especially regarding the problem of which populations to recognize as separate species, pervade this book. Grismer's skills as a world-class photographer also are abundantly displayed here, perhaps most dramatically in those images that capture a species' place in nature, taking us into a landscape with an animal in the foreground.

In a country with phenomenal natural riches, Baja California and its offshore islands stand out as a priceless setting for studying the factors that shape ecological communities, and they are replete with stunningly beautiful locales for recreation and education. Scholars, students, and lay naturalists will find *Amphibians and Reptiles of Baja California* an invaluable sourcebook for the present; in an age of global environmental change, this volume also provides a critical benchmark for the future. For everyone, residents and visitors alike, who wants to study and enjoy nature in this famous landscape, and for all of us who want to preserve such spectacular diversity in the face of ever-growing human impact, Grismer's book will be an essential resource.

PREFACE

The Baja California peninsula is a narrow, jagged, strip of land extending some 1,300 km southeast from the southern border of California to Cabo San Lucas. Its long, irregular, desert coastline of nearly 3,300 km is corrugated with thousands of isolated coves and embayments and adorned with several uninhabited islands of varying sizes, shapes, and habitats. The peaceful union of such contrasting environmental elements as desert and sea embellishes this region with a mystical beauty found nowhere else in North America. Its often violent and complex geological evolution has in turn contributed to the evolution of forms of life found nowhere else on earth. This uniqueness, coupled with the rugged and unforgiving terrain, leaves Baja California a region steeped in mystery and intrigue, one of the last biological frontiers in North America.

Soon after the fall of Mexico in 1521 at the hands of Hernán Cortés, rumors began to circulate of a large Pacific island called California lying off the west coast of Mexico, some ten days' travel from the capital. With its promises of pearls, gold, mermaids, and Amazons, this island became an obsession for Cortés, and in 1533 he dispatched the exploration vessel *Concepción* from the Isthmus of Tehuántepec in search of it. Cortés placed the *Concepción* under the command of a harsh, arrogant captain named Diego Becerra de Mendoza. After sailing for several days without sighting land, the crew's initial contempt and disrespect for Becerra began to escalate with their anxiety and sense of isolation. Eventually, the ship's pilot, Fortún Jiménez, murdered Becerra in his sleep and took control of the ship. A short time later, the *Concepción* reached the southern end of Baja California and anchored in what is now known as Bahía de la Paz. Here the European explorers encountered a group of Indians known as the Guaycura. Owing to their initial hospitality, Jiménez and his crew remained in La Paz for a short time. Eventually, the Guaycura came to distrust Jiménez, and he and twenty-two others were murdered in an ambush while refilling their water barrels. The remaining crew managed to escape and return to mainland Mexico.

The difficulties of exploring Baja California persisted for hundreds of years. Before the completion of Mexican Highway 1 in 1973, few *norteamericanos,* other than the most adventurous, had ever experienced the peninsula's interior, much less ventured to any of the uninhabited desert islands in the Sea of Cortés. Today, however, a journey into Baja California is much easier. A road trip to the tip of the peninsula, once a grueling two- to six-week journey over rutted, unpaved roads, can now be accomplished in two to three days in a motor home. The highway has opened up Baja California's long-mysterious interior and its Gulf waters to sport fishermen, sun worshippers, hikers, and off-roaders.

Still, however, Baja California possesses a wild, intangible allure. An innate sense of adventure, heightened by a hint of danger and the unknown, beckons many explorers off the highway into Baja's harsh, poorly understood interior. Baja California's jagged, snow-covered peaks, volcanic badlands, parched deserts with relentless summer temperatures, and arid, uninhabited desert islands have been reluctant to give up their secrets. There is still much knowledge to be attained and many personal challenges to be met.

An inescapable passion for herpetology was something I was born with. My first memories are of backyard expeditions in Corona del Mar, California, trying to catch Western Fence Lizards on rocks alongside my swimming pool. A few years later, I was old enough to understand the amazing stories about sea turtles, iguanas, and rattlesnakes my father told from his travels as a bush pilot in Baja California and a bizarre place he called the Sea of Cortés. In 1959, at the age of four, I flew down to Baja with my father for the first time. I remember sitting in the back seat (next to a rather large, nauseated woman) in his Cessna 180, peering out of the window as we headed south along the east coast of the peninsula. To this day I have no idea where we went and eventually landed. However, I vividly remember the intense feelings of intrigue and excitement that overcame me when I first saw the Gulf islands—feelings that have remained with me ever since. The islands struck me as golden parcels of desert surrounded by a turquoise sea. I knew they had to harbor some of the most fascinating species of reptiles in the world, and all I could think about for the next forty years was finding out what they were.

The idea of becoming an expert and writing a book on this herpetofauna came much later. My focus in the beginning was just to get into the area to hike and take pictures. Later, as my interests matured in college, I wanted to follow in the footsteps of researchers such as Ted J. Case and Robert W. Murphy, who continue to publish great papers on this herpetofauna. Using their early works as a foundation to evaluate my observations in Baja, I slowly began to accumulate publications on the herpetofauna. It was not until the early 1990s that, at the persistent urging of a number of people, I decided to write this book. Fortunately I had kept good field notes from even my earliest visits. On reading through these I believed I had a worthwhile story to tell and overcame my initial reluctance. But this book is not the final word, nor am I the only authority. Rather, this is a progress report based on the efforts of many scientists who have published their findings elsewhere. I hope they will be pleased with the way I tell our story.

ACKNOWLEDGMENTS

A book of this nature is rarely the work of a single individual. Rather, it reflects the collaborative effort of many people over a long period, and only one of these people has the honor of putting his or her name on it.

For scholarly guidance throughout various aspects of this project I thank R. Bezy, P. Buchiem, R. Carter, G. Casas, R. Etheridge, D. Frost, J. Galusha, E. Gergus, T. Goodwin, J. Guzman, L. Harris Jr., B. Hollingsworth, A. Kluge, C. Lowe Jr., J. McGuire, E. Mellink, G. Pregill, A. Ramírez, A. Reséndez, B. Reséndez, F. Reynoso, H. Smith, and J. Wright. For field assistance I am indebted to D. Barrios, K. Beaman, G. Cota, M. Cryder, B. Dawson, P. Galina-T., M. Galván, J. Giordenango, F. Grismer, J. Grismer, L. Grismer, M. Grismer, H. Lawler, C. Mahrdt, C. Mattison, S. Messenger, R. Murphy, G. Polis, G. Pregill, D. Rumbaugh, E. Ryder, N. Scott, R. Sosa, T. Van Devender, V. Velázquez, H. Wong, and N. Wong.

For the loan of specimens I thank C. W. Myers (American Museum of Natural History), H. Marx and H. Voris (Field Museum of Natural History), J. Wright and R. Bezy (Los Angeles County Museum of Natural History), A. Leviton and J. Vindum (California Academy of Sciences), G. Pregill (San Diego Natural History Museum), W. Tanner and J. Sites (Brigham Young University), C. Lowe and G. Bradley (University of Arizona), D. Wake and D. Good (Museum of Vertebrate Zoology, University of California, Berkeley), P. Alberch and J. Rosado (Museum of Comparative Zoology, Harvard University), G. Zug, R. McDiarmid and R. Crombie (National Museum of Natural History), Francisco Reynoso (Universidad Autónoma de Baja California Sur), and H. Smith (University of Colorado).

For financial support I thank the Department of Biology of La Sierra University, the Department of Natural Sciences of Loma Linda University, and the San Diego Herpetological Society. I am indebted to the Herpetologists' League for partial funding to publish the color plates and to my father, Larry Grismer, who has flown me into remote areas of Mexico for nearly thirty years. I thank José Luis Genel García, Jorge Reyes of the Secretaría de Desarrollo Urbano y Ecología, and the Instituto Nacional de Ecología for the issuance of Mexican collecting permits and many others who listed me on their permits throughout the years. I am grateful to S. J. Beaupre, R. L. Bezy, J. R. Dixon, K. de Queiroz, S. Eckert, D. R. Frost, E. A. A. Gergus, D. A. Good, B. D. Hollingsworth, J. E. Lovich, R. W. McDiarmid, J. A. McGuire, R. R. Montanucci, T. J. Papenfuss, J. P. Rebman, J. A. Rodríguez-Robles, D. A. Rossman, J. A. Seminoff, H. M. Smith, D. E. Spiteri, J. M. Walker, D. B. Wake, J. J. Wiens, L. D. Wilson, and H. Wong for reviewing species accounts.

Last, I thank a person who gave me the support that only a best friend can give. In the field he worked as hard on my projects, day and night, as I did. He ate the same spoiled food, was harassed by the same local and federal police, was stranded on the same islands, nearly capsized in the same hurricanes, and climbed to the top of many of the same mountains and islands under unforgiving circumstances. Even after he broke the front windshield of my car with his forehead during a head-on collision, he elected to continue our fieldwork. His intellect and insight into systematics and biogeography, coupled with his knowledge of the natural history of the organisms with which he works, aided me when I faced difficult decisions I could no longer view objectively. I look forward to future collaborations with Jimmy A. McGuire.

GENERAL INTRODUCTION

Baja California is the second longest and the most geographically isolated peninsula in the world. Over the last four to five million years, it has undergone a uniquely complex tectonic origin and ecological transformation. What we see today as the Baja California peninsula was originally connected to the west coast of mainland Mexico but was torn away by differential movements of the Pacific and North American plates. Since then, it has been carried approximately three hundred kilometers to the northwest along what has become known as the San Andreas Fault. This separation occurred in various stages of uplift, submergence, and geographical fragmentation. Concurrent with these tectonic upheavals were climatic changes that transformed the peninsula from a generally cool and more mesic region to one of extreme aridity. This change was gradual at first but has rapidly accelerated over the last eight to ten thousand years.

Because of its latitudinal extent, its complex topography, and its location between two very dissimilar bodies of water, Baja California accommodates a wide range of climates. This range in turn supports a striking degree of environmental diversity, with habitats ranging from the extremely hot and arid Lower Colorado Valley Desert in the northeast to the humid, tropical deciduous forests of the south. Additionally, a thick coniferous forest is found in its northern mountains and an endemic pine-oak woodland in the Sierra la Laguna of the Cape Region. Central Baja California is covered with an extensive network of volcanic badlands interrupted by palm-lined oases, and its north-central region is one of only three fog deserts in the world. Yet despite its ecological diversity and unique environmental history, Baja California remains one of the most poorly studied regions in North America.

Exploration and Research

Exploration and research in Baja California have not yet enjoyed the same attention or success as investigations in the United States or mainland Mexico. This is not a situation that has developed by chance alone but rather one that has emerged through design. Baja California, on the whole, is extremely arid and rugged. The vast majority of its interior is accessible only by foot or on horseback, and thus a researcher's field time is limited by the amount of fresh water and other supplies that can be carried. The scarcity of permanent sources of fresh water in central Baja California has allowed development only in isolated regions near oases or semipermanent riparian areas in the mountainous regions. Even today, simply traveling from one rancho to the next often requires a moderate degree of isolation and travel through unforgiving terrain.

The logistic difficulties impeding research were made clear in two prominent episodes of Middle American exploration that resulted in a noticeable lack of publications dealing with Baja California's natural history. The first involved the series Biología Centrali-Americana—a 67-volume compilation dedicated to the advancement of knowledge of the flora and fauna of Mexico and Central America. This publication, which spanned the period from 1879 to 1915, expressly excluded Baja California. The second period encompassed the Hobart M. Smith and Edward H. Taylor expeditions in Mexico, during which nearly fifty thousand specimens of amphibians and reptiles were col-

OPPOSITE *Masticophis fuliginosus* (Baja California Coachwhip) from Ancón, BCS

lected between 1932 and 1941. Although this effort culminated in the publication of three major annotated checklists by Smith and Taylor (an unparalleled advancement in our knowledge of Mexico's herpetofauna), not a single specimen was collected from Baja California. And still, in spite of the information presented in this book, a gap remains in our knowledge of the herpetofauna of Baja California and the Gulf of California (the Sea of Cortés). Virtually no detailed ecological or natural history studies have been conducted on any species in Baja California, and there is uncertainty as to the exact distribution of others. The harshness and isolation of the islands in the Gulf of California have dissuaded detailed ecological studies on all but a few species. Additionally, many isolated areas such as the Vizcaíno Peninsula, the Sierra los Cucapás, and the upper elevations of the Sierra la Asamblea, Sierra San Borja, Sierra la Libertad, Sierra Guadalupe, and Sierra la Giganta probably contain species that have not been reported and unknown species that have yet to be described. Baja California and the Sea of Cortés are long overdue for a comprehensive treatment of their herpetofauna.

The Environment

A thorough understanding of any faunal element requires a sound knowledge of the historical and contemporary ecology of the region in which it occurs. This is especially true of the herpetofauna of Baja California. Unlike continental regions, Baja California has undergone complex tectonic and ecological transformations as a result of uplift, submergence, isolation, and desertification, and these transformations have contributed to its environmental diversity. Thus an understanding of the dynamic relationships and interactions among these abiotic components, both past and present, is paramount in understanding the relationships, distribution, and geographic variation of its herpetofauna.

Paleoenvironmental History

FORMATION OF THE BAJA CALIFORNIA PENINSULA

The most significant factor contributing to the endemism and the biotic diversity seen today in Baja California and the Sea of Cortés is the complex geological origin and evolution of these regions. Between 8 and 13 million years ago (mya), most of Baja California lay submerged beneath the Pacific Ocean and nestled against the northwest coast of mainland Mexico, placing what is today the city of San Felipe against the west coast of Isla Tiburón and placing Cabo San Lucas just north of Puerto Vallarta, Jalisco (Gastil et al. 1983). A shallow epicontinental seaway, the proto–Gulf of California, inundated western Mexico, extending north to at least Isla Tiburón and perhaps into southern California by 6 mya (Stock and Hodges 1989; Winker and Kidwell 1986). With the onset of several complex tectonic events associated with differential movements between the Pacific and North American plates, along what would eventually develop into the San Andreas Fault, Baja California was being torn away from the west coast of Mexico and began to migrate northwest (Lonsdale 1989). At the same time, the peninsular ranges were being uplifted, and depositional filling began forming the Vizcaíno Desert and the Magdalena Plain. During its migration, Baja California underwent many changes in contour and topography. Parts of the peninsula may have been influenced by marine inundations, giving it the appearance of an archipelago, while other portions were giving rise to mountain ranges and were besieged with volcanic activity. For a short period, the Cape Region became separated from the rest of the peninsula and existed as an island, while the Gulf of California remained in the Coachella Valley of southern California at least as far north as Whitewater Canyon in Riverside County (Winker and Kidwell 1986). In fact, it may have been the recent uplift of coastal mountains in southern California that contributed to the Gulf's final regression to its present location.

Overlaid on the geomorphological and physiographic evolution of the Baja California peninsula was a series of climatic changes. These changes were the result of an overall drying trend in North America, which actually began during the Eocene—before the peninsula began to separate from mainland Mexico. The intermittent glacial periods of the Pleistocene and the formidable rainshadows caused by the uplift of the Peninsular Ranges brought severe localized drying trends to Baja California that persist today. In the past, however, the central portion of Baja California was a cool, mesic, volcanic, wooded grassland whose upper elevations supported large stands of oaks and other trees much like those seen today in some portions of the Sierra Madre Occidental of Sonora, Mexico. Horses and giant tortoises ranged through these forests. The Cape Region of Baja California was tropical and supported populations of crocodiles, green iguanas, and boa constrictors, as well as semiaquatic elephants, giant hares, and large cats (Miller 1980). Eventually Baja California became the long, topographically complex, generally arid peninsula that we recognize today. This is not to say that its dynamic past has become dormant. To the contrary, the continued tectonic activity along the San Andreas Fault and its associated fault systems attest to the peninsula's unrest and continued northwestward movement.

It has been this intricate evolution of Baja California's

physiography, coupled with its complex historical ecological transformation and currently diverse climatic regimes, that has resulted in organisms becoming isolated in particular regions at various points in time and permitted their invasions and reinvasions into other areas during other periods (Grismer 1994a,b). The overall effect of these events has been the evolution of unique flora and fauna and distinct biotic provinces. We must keep in mind, however, that we are merely observing the evolution of Baja California and its inhabitants at a particular moment (the present). Just as conditions and organisms have changed and evolved in the past, they will continue to do so in the future.

FORMATION OF ISLANDS

The islands associated with Baja California vary in age, origin, and geological composition. In the Sea of Cortés, there are three principal types of islands, defined by the manner of their origin: oceanic, continental, and landbridge. Oceanic islands have never been connected to mainland Mexico or to Baja California but originated in the Sea of Cortés. Their origin may be due to uplift, as a result of extensional rotation or compression of the underlying tectonic plates, or to volcanic deposition from a series of underwater eruptions. Both are phenomena associated with the crustal extension and the northwesterly movement of the peninsula as it continues to drift away from mainland Mexico. Isla Tortuga, off the coast of Santa Rosalía, is an example of an oceanic island formed by volcanic deposition.

Continental islands were once connected to mainland Mexico or the Baja California peninsula but became separated as a result of tectonic displacements along coastal fault zones. In the Gulf of California, these are usually islands that broke off the trailing edge of the peninsula as it moved northwest. An example is Isla Ángel de la Guarda.

By far the most common type of island in the Sea of Cortés, and along all of Baja California's Pacific coast (with the exception of the Islas San Benito), is the landbridge island. Landbridge islands are relatively young islands (no more than fifteen thousand years old) that were once connected to the mainland or the peninsula. Many of them are simply the emergent peaks of nearby coastal ranges. For the most part, these peaks became isolated because of a rise in sea level. Occasionally landbridge islands are formed from coastal submergence where the earth's crust is thinned as it is stretched. This thinning causes the ground to sink and water to enter from a nearby sea. Such an event may have been the first step in the formation of the proto–Gulf of California. Some landbridge islands have been formed by erosion, which can cut off prominent points of coastal areas and leave them isolated. Isla Espíritu Santo, off the coast of La Paz, may have been created by coastal erosion. There are a few other landbridge islands—such as Isla Willard, which forms part of the northern end of Bahía San Luis Gonzaga, and Isla Requesón, in Bahía Concepción—that are connected to the peninsula by a sandy isthmus at low tide but separated at high tide.

Physical Characteristics

An in-depth knowledge of the physiography (shapes and contours) of Baja California is a prerequisite for understanding the geographical orientation of its various phytogeographic regions. Because the distribution of many species is correlated with phytogeography, understanding the physiographic nature of Baja California is essential to comprehending why much of the herpetofauna occurs where it does. Additionally, this knowledge provides insight into the various climatic regimes of Baja California, which reciprocally dictate the distribution of the phytogeographic regions. Thus, physiographic insight is fundamental to understanding the geographical interactions between the herpetofauna and the environment. Major geographic features and towns mentioned below are shown in map 1.

Today Baja California is a thin northwest to southeast–tending peninsula nearly 1,300 km long. It is situated between 32° 30' N latitude and 117° W longitude at its northwestern corner and 23° N and 110° W at its southern tip. Its width ranges from approximately 240 km along the U.S.-Mexican border to less than 30 km at the Isthmus of La Paz. It is separated from the state of Sonora by the Río Colorado in the north and from the rest of Sonora and mainland Mexico by the Gulf of California, approximately 160 km wide. The area of Baja California is approximately 143,400 km^2, and its coastline is approximately 3,300 km long. Associated with the coastline are forty-five major islands, each at least 1.3 km^2 in area. Several smaller islands are also associated with Baja California, and an additional ten or so major islands are principally associated with the Mexican states of Sonora and Sinaloa (maps 2 and 3).

The most distinctive feature of Baja California's dramatically sculpted topography is a series of mountain ranges, known collectively as the Peninsular Ranges, that run nearly uninterrupted from its northern border to the Isthmus of La Paz. This massive fault-block system is tilted westward toward the Pacific Ocean, and its crest lies slightly to the east of the peninsula's central axis. On the west side the mountains slope gradually toward the coast, whereas the eastern face rises abruptly out of the desert floor, often presenting a precipitous escarpment. Many passes and

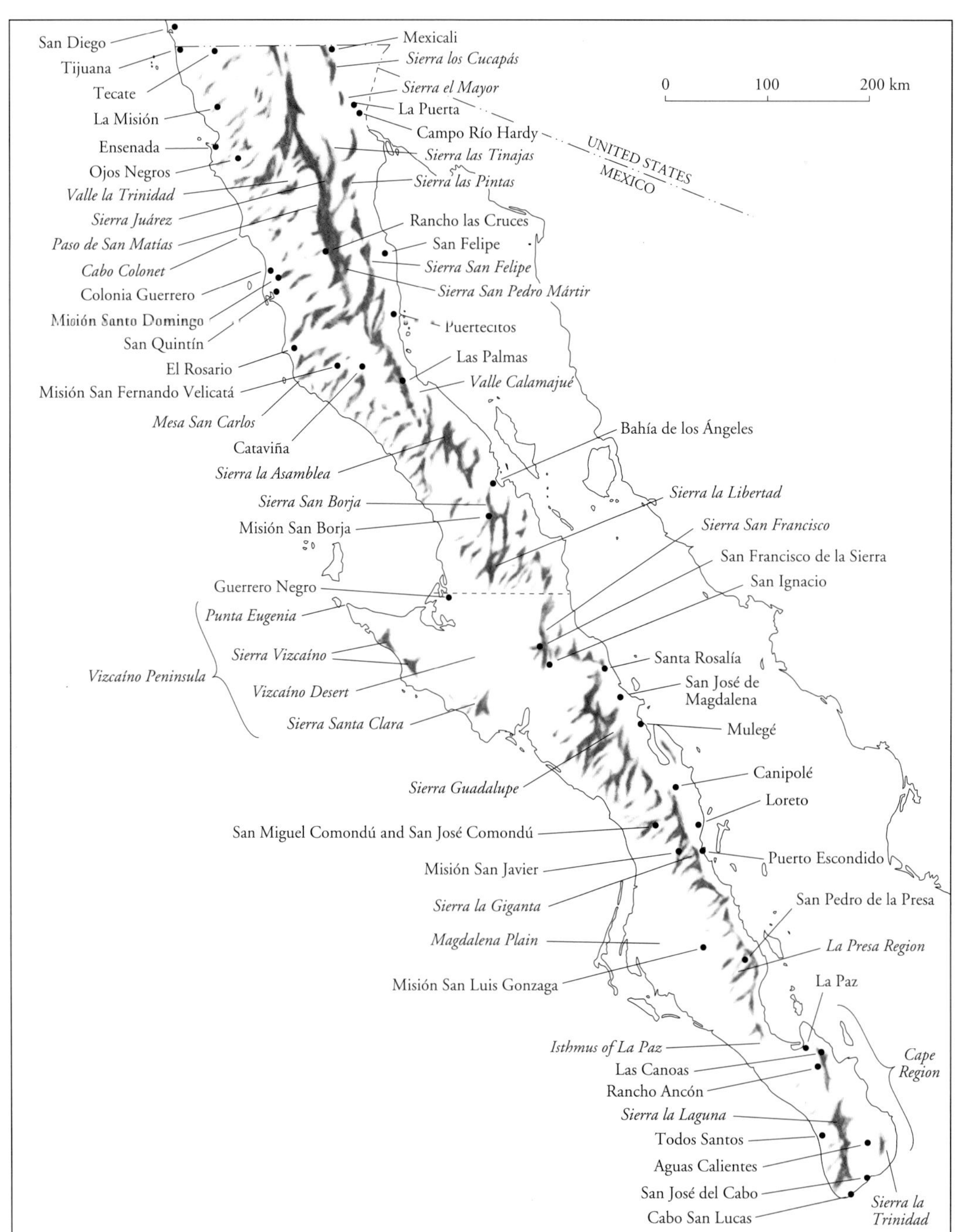

MAP 1 Geographic features and towns of Baja California

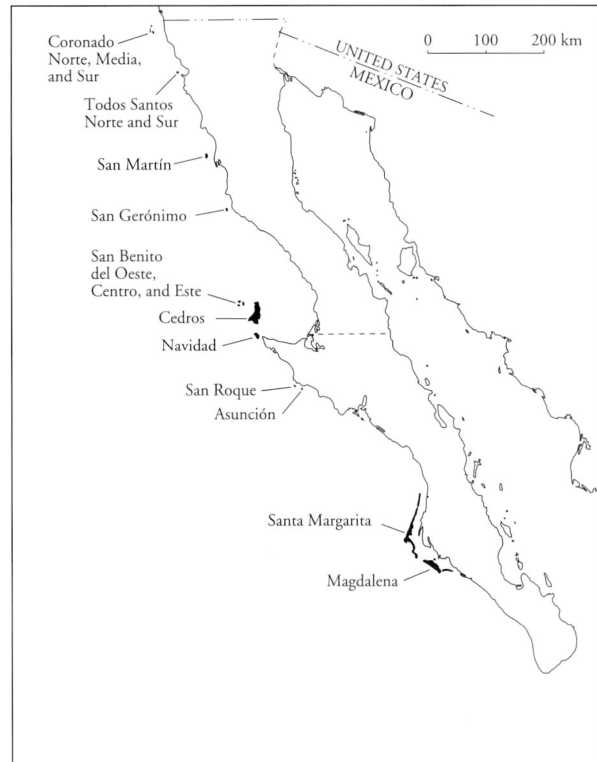

MAP 2 Pacific islands of Baja California

MAP 3 Gulf islands of Baja California

minor depressions have been cut into the main crest of the Peninsular Ranges, and other, minor fault systems have given rise to additional ranges that radiate outward from the main massif at various angles. The result is a rugged and complex topography that allows the Peninsular Ranges to be conveniently divided into a series of less extensive ranges.

Dominating northern Baja California are two major mountain ranges known as the Sierra Juárez and the Sierra San Pedro Mártir. The Sierra Juárez, which reaches an elevation of nearly 1,410 m, is part of a mountain range that extends from southern California. At its upper elevations, the terrain is flat and supports an ephemeral body of water known as Laguna Hanson. Its eastern face is rocky and drops sharply into a dry-lake basin known as Laguna Salada. The Sierra San Pedro Mártir is a much more dominant feature. Its highest peak, Picacho del Diablo, reaches 3,096 m, and from this elevation one can see both coastlines of Baja California as well as that of Sonora. Situated just west of the crest of the Sierra San Pedro Mártir are three large meadows: La Grulla, La Encantada, and Vallecitos. These meadows, which lie between 1,800 and 2,400 m elevation, drain a large segment of the western face of the Sierra San Pedro Mártir, in a stair-step fashion, toward the Pacific Ocean. Delimiting the southern end of the Sierra Juárez and the northern end of the Sierra San Pedro Mártir is Paso de San Matías. The most significant mountain pass in the Peninsular Ranges, it is situated at the eastern end of Valle la Trinidad, which extends 100 km from southeast to northwest. It lies between 800 m at its northwestern end just east of Ensenada and 975 m at its southeastern terminus at Paso de San Matías. The valley is constricted at the town of Valle la Trinidad by a small, northeast to southwest–tending range known as the Sierra Warner. Northwest of Valle la Trinidad and north of Ensenada is the prominent Valle Guadalupe.

East of the Sierra Juárez lies a series of smaller mountains. The Sierra los Cucapás, Sierra el Mayor, and most of the Sierra las Pintas are separated from the Sierra Juárez by the Laguna Salada basin. The Sierra de los Cucapás and Sierra el Mayor, the more northerly of these ranges, are a contiguous, isolated mass extending south approximately 80 km below the U.S.-Mexican border. These ranges support only scant vegetation and a depauperate herpetofauna. Their highest peak reaches 1,000 m, and the entire range is surrounded by the ancient Río Colorado flood plain. The Sierra los Cucapás is a relatively narrow range, tending southeast and composed primarily of granitic rock. Its southern end, which is marked by a narrow, low pass known as Cañón David, supports broad expanses of basaltic desert pavement. South of this pass lies the Sierra el Mayor, which is more volcanic in composition and somewhat more spread out, angling off in a more southerly direction. It is estimated that as little as ten thousand years ago these mountains existed as an island or islands, and this isolation prompted the evolution of some of their endemic flora and fauna. Currently, the Laguna Salada basin borders their western foothills. Not long ago it was little more than an ephemeral dry lake that filled only during seasons of above-average rainfall. Lately, water from the Río Colorado has been channeled in through the "no-man's-land" separating the Sierra el Mayor from the more southerly Sierra las Pintas. In the early 1980s the level of this lake rose so much that it stretched the entire length of the mountain ranges and wrapped eastward around the southern terminus of the Sierra el Mayor. Mexican Highway 5 had to be elevated to avoid flooding, and culverts were constructed to allow water to pass beneath the highway.

Sixty kilometers south of the Sierra el Mayor and across the southwestern margin of the Laguna Salada basin lies the northern end of the Sierra las Pintas, a contorted, jagged, colorful volcanic range whose peaks protrude through the southern end of the dry lakebed. The Sierra las Pintas and Sierra Juárez are linked at their southern bases by a wide set of rocky, volcanic hills that come together in the north to form the Sierra las Tinajas. This range continues north into the Laguna Salada basin as an isolated finger of land between the Sierra Juárez and Sierra las Pintas. The southern connection of the Sierra Juárez, Sierra las Pintas, and Sierra las Tinajas may account, in part, for the greater floral and faunal diversity of the latter two as compared to that of the more isolated Sierra los Cucapás and Sierra el Mayor.

Immediately south of the Sierra las Pintas lies the Sierra San Felipe, which reaches 1,332 m in elevation and is separated from the Sierra San Pedro Mártir to the west by the Laguna Diablo basin in Valle San Felipe. This is an ephemeral dry lake filled primarily by runoff from the eastern slopes of the Sierra San Pedro Mártir during bouts of heavy summer precipitation. Currently, the Mexican government is sponsoring agricultural development of Valle San Felipe, with the primary crop being Jojoba *(Simmondsia chinensis).* The Sierra San Felipe angles away from the Sierra San Pedro Mártir to the southeast and generally runs uninterrupted past Bahía San Felipe. At this point it grades into a small series of extremely rugged and broken volcanic mesas and canyons known as the Sierra Santa Isabel and Sierra Santa Rosa. These ranges reach 1,200 m in elevation and slope gradually toward the Gulf of California.

The cismontane areas west of the Sierra Juárez and Sierra San Pedro Mártir, from the U.S.-Mexican border to just north of Cabo Colonet, approximately 200 km south of Tijuana, consist of foothills, plains, and mesas of varying sizes. These features are occasionally interrupted by wide canyons and deep arroyos. This region descends gently toward the Pacific Coast from the crest of the Peninsular Ranges and in some places terminates in precipitous coastal bluffs. The dominant geographic feature of this region lies inland from Punta Santo Tomás, approximately 110 km south of the U.S.-Mexican border. This is a small range of three narrowly connected mountains known from north to south as the Sierra Peralta, Sierra Warner, and Sierra San Miguel. The foothills of these ranges stop short of the Pacific and give way to a series of narrow, coastal plains known collectively as the San Quintín Plain. This plain extends south to Rancho Socorro, where it gives rise to an even narrower plain extending nearly to El Rosario. The Sierra Warner and the Sierra Peralta form a portion of the northwestern border of Valle la Trinidad. Between the Sierra San Miguel in the west and Sierra San Pedro Mártir in the east lie a series of narrow, arid valleys known as Valle San José. These valleys are buffered from cool Pacific breezes by these western ranges and consequently heat up sharply during the summer. As the Sierra San Miguel extend southward beyond El Rosario, they become a jumbled mass of low, often volcanic foothills that blend imperceptibly into the southern end of the Sierra San Pedro Mártir and the Sierra Santa Isabel.

Cismontane Baja California south of El Rosario is a rugged, complex mass of canyons, mountains, low hills, volcanic mesas, ephemeral dry lakes, and deep, sinuous arroyos that collectively make up the Sierra Columbia. The southern sides of the Sierra Columbia abruptly merge into the Vizcaíno Desert just north of the 28th parallel. The dominant features of this area are Mesa San Carlos and Mesa la Sepultura. Mesa San Carlos is a wide, flat, volcanic structure roughly 500 m high and 15 km long on the west coast, approximately 60 km south of El Rosario. Mesa la Sepultura is situated approximately 27 km southeast of El Rosario inland from Mesa San Carlos.

From Mesa San Carlos south to Laguna Manuela, just north of Guerrero Negro, the Sierra Columbia comes within a few kilometers of the coast. Here it constitutes the lower western slopes of the elevated central portion of north-central Baja California, leaving a flat, low-lying, wind-blown, fog-drenched, narrow coastal strip between them and the Pacific Ocean. Inland from the Sierra Columbia and west of the Sierra Santa Isabel, Sierra Calamajué, and Sierra la Asamblea of the Gulf coast, the peninsula opens up into a long, narrow, relatively flat region crisscrossed by a series of low hills and valleys. At the center of this area is a rugged and extensive region composed of granitic uplifted blocks and boulders, through which course palm-lined arroyos supporting sources of semipermanent freshwater such as those at Cataviña and Misión Santa María. The most intriguing feature of this region, El Pedregoso, is a gigantic, isolated pile of granitic boulders stacked some 200 m above the floor of Valle Gato. This entire rocky region is drained primarily by the picturesque arroyos La Bocana and Santa María, which intricately negotiate the southern end of the Sierra Santa Isabel down to the narrow plain along the Gulf coast. Along their courses they support various sources of semipermanent surface water. South of the boulder region the peninsula gives way to a series of large, ephemeral lakes. The first of these, Laguna Chapala Seca, is located approximately 60 km south of Cataviña, just west of the Sierra Calamajué. Two other much larger dry lakes, Valle Agua Marga and Valle Laguna Seca, occur northwest of Bahía de los Ángeles.

Along the Gulf side of the peninsula, between Bahía San Luis Gonzaga in the north and Bahía de los Ángeles in the south, lie the Sierra Calamajué and, to the south, the much more extensive Sierra la Asamblea. The latter is a prominent granitic uplift reaching nearly 1,700 m in elevation and maintaining two boulder-lined meadows fringed with Mexican Blue Fan Palms *(Brahea armata)* and Piñon Pine trees *(Pinus monophylla)* in its upper elevations. The Sierra la Asamblea blends into the Sierra San Borja to the south through a series of small, rugged, volcanic hills in the vicinity of Bahía de los Ángeles. The Sierra San Borja and the more southerly Sierra la Libertad coalesce as a mixture of granitic and volcanic ranges and mesas reaching 1,940 m in elevation. They extend south of Bahía de los Ángeles and grade into the volcanic foothills of the Sierra San Francisco at the northern edge of the Magdalena Region and eastern edge of the Vizcaíno Desert. Consequently, the Sierra San Borja and Sierra la Libertad lie inland from the Gulf coast and form the western edge of Valle San Rafael, which extends from Bahía de los Ángeles to below Bahía San Francisquito.

Just north of Guerrero Negro on the Pacific coast the Vizcaíno Desert begins. This is a low-lying, flat, wedge-shaped area extending 150 km between the volcanic foothills and badlands of the Sierra San Francisco and the Sierra Guadalupe to the east and the Vizcaíno Peninsula to the west. The southern tip of this triangular desert reaches as far as Bahía San Juanico, where it grades into the northern end of the Magdalena Plain. Together the Magdalena Plain and Vizcaíno Desert make up a distinctive cool coastal desert. The dune region of the Vizcaíno Desert is unique in North America: its dunes appear as long, parallel furrows that have formed under the influence of continuous light, steady breezes (Wiggins 1980). The southern portion of the Vizcaíno Desert, south of Laguna San Ignacio, is a broad, flat, sandy plain precipitously edged to the east by the volcanic foothills of the Sierra Guadalupe.

The Vizcaíno Peninsula is a prominent point of land jutting northwest from the center of Baja California. It is composed of the Sierra Vizcaíno in the north and the Sierra Santa Clara at its base. The Sierra Vizcaíno thrusts northwest into the Pacific Ocean, terminating at Punta Eugenia and emerging offshore to form Isla Natividad and Isla Cedros. The southwestern slopes of the Sierra Vizcaíno give rise to a narrow coastal plain that frequently terminates in precipitous ocean bluffs. Punta Eugenia and, in part, Islas Natividad and Cedros cradle the southern portion of the extensive Bahía Sebastián Vizcaíno along the west coast of central Baja California. Where the head of this bay inundates the shallow plain of the northern Vizcaíno Desert, one of the largest lagoon systems in North America has developed. Originally one system, it has now been transformed into three separate lagoons: Laguna Manuela in the north, Laguna Guerrero Negro in the center, and the extensive Laguna Ojo de Liebre (Scammon's Lagoon) in the south. On the southwestern margin of the Vizcaíno Desert is the Sierra Santa Clara, which lies immediately north of a large Pacific inundation known as Laguna San Ignacio. The Sierra Santa Clara forms a series of seven closely proximate, heavily eroded, ancient volcanic peaks and remains one of the least explored regions in Baja California.

Northeast of the Vizcaíno Desert lies the Sierra San Francisco. This range is a dominant feature of central Baja California, with a majestic expanse of volcanic mesas and peaks reaching over 2,100 m. This system is embellished with a complex network of deep, sheer, palm-lined canyons containing permanent fresh water. Within these canyons are found some of the most elaborate cave paintings in Baja California. The Sierra San Francisco gives way in the southeast to a low-lying, broad expanse of volcanic fields and cinder cones through which courses the Río San Ignacio before it empties into Laguna San Ignacio. Just east of the town of San Ignacio, the Peninsular Ranges continue as the Sierra Guadalupe in the north and Sierra de la Giganta in the south. The northernmost portion of the Sierra Guadalupe begins majestically with Volcán las Tres Vírgenes.

These are three volcanic peaks that arise from a common base approximately 30 km north of Santa Rosalía. The dominant peak attains an elevation of nearly 2,000 m, and in some years snow can be seen on its north-facing slopes. These are active volcanoes that occasionally spew vapors from their lower slopes. The area surrounding their base is crisscrossed with lava flows that may be as recent as 1746.

The Sierra Guadalupe extends south to near the southern end of Bahía Concepción and has peaks reaching as high as 1,800 m just west of Mulegé. In the north, these mountains begin as two ranges separated by a wide valley wherein Misión Guadalupe is situated. These ranges, characterized by precipitous cliffs, provide some of the most spectacular and dramatic landscapes in all of Baja California. They come together southwest of Mulegé and eventually give way to the Sierra la Giganta, which reaches 1,767 m in elevation at Cerro la Giganta northeast of San José Comondú. From here the range continues irregularly south to the vicinity of the Gulf coast village of San Evaristo. From San Evaristo south, the Sierra la Giganta gradually decreases in elevation, blending smoothly and nearly imperceptibly into the Magdalena Plain at the Isthmus of La Paz. Viewed from the Gulf of California, the eastern face of the Sierra la Giganta appears to rise out of the sea, presenting a mural of strange geologic formations and sinuous bands of color. This face is an abrupt clifflike escarpment dissected by numerous arroyos and sculpted with jagged foothills and volcanic flows that wind their way to the water's edge.

The volcanic badlands immediately west of the Sierra Guadalupe and Sierra la Giganta consist of mesas and deep, palm-lined arroyos created by thousands of years of erosion from streams draining the western slopes of the mountains. Many of these arroyos have large supplies of permanent fresh water that have supplied villages such as La Purísima, San Ysidoro, San José Comondú, and San Miguel Comondú.

The western foothills and badlands of the Sierra la Giganta slope toward the Pacific Ocean and grade into the Magdalena Plain. This is a 55-km wide, flat, sandy area east of Bahía Magdalena, extending south some 300 km, where it tapers to a point near the town of Todos Santos in the Cape Region. The Magdalena Plain was formed from thousands of years of fluvial deposits from the western slopes of the Sierra la Giganta. Currently, its fertile soil supports extensive agriculture. Unfortunately, however, as groundwater for irrigation is being pumped out faster than it is being replaced, seawater incursions from the adjacent coastline are contaminating groundwater supplies. At the southern terminus of the Sierra la Giganta, the Magdalena Plain angles east and fans out across the Isthmus of La Paz, contacting the Gulf of California on the northwestern shore of Bahía de la Paz. Along its central western edge, the Magdalena Plain is bordered by Bahía Magdalena. This bay runs parallel to the west coast for approximately 50 km, where it joins Laguna Santo Domingo in the north. Bahía Magdalena is bordered in the west by Isla Magdalena (shaped like a crooked finger), the mountainous Isla Santa Margarita, and the sandbar Isla Creciente. Laguna Santo Domingo consists of a narrow body of water flowing out of Bahía Magdalena and running parallel to the coast for approximately 145 km. It is flanked to the west by the thin, sandy, northern portion of Isla Magdalena. This western border is interrupted in three locations that form inlets into the bay.

The southernmost portion of Baja California, south of the Isthmus of La Paz, is known as the Cape Region. It is likely that this region had an origin related to but separate from that of the remainder of the peninsula, and throughout its evolution it has made intermittent land-positive connections with the northern areas. The Cape Region is dominated by a range of mountains whose axis begins at Punta Coyote in the north, just east of La Paz, and runs nearly due south, diagonally across the Cape Region, to Cabo San Lucas. Collectively, these mountains are known as the Sierra la Laguna, but the southwestern portion occasionally goes by the name of Sierra la Victoria. Jutting northward from the central portion of the Sierra la Laguna near the town of San Bartolo and reaching the Gulf coast at Ensenada de los Muertos is the Sierra del Álamo. This range forms the southern border of a large, alluvial plain known as San Juan los Planes. Its northern border is formed by the northern section of the Sierra la Laguna, often referred to as the Sierra la Pintada and Sierra la Palmillosa. The highest peak of the Sierra la Laguna, El Picacho, is a pinnacle of rock lying east of Miraflores that rises to nearly 2,200 m. The northern portion of the Sierra la Laguna, south of San Bartolo, supports a large meadow called La Laguna, lying at roughly 1,600 m. This meadow once supported a shallow lake; hence the name. The Sierra la Laguna is a single fault-block, but, unlike the mountains to the north, it tilts to the east, so that the most precipitous slopes face the Pacific Ocean. There are a few large arroyos that cut through the crest of the Sierra la Laguna, the most prominent of which are Arroyo Santiago and Cañón San Bernardo.

A smaller, disjunct series of mountains, the Sierra la Trinidad, fringe the southeastern margin of the Cape Region and border the Gulf coast. The highest peak, Cerro del Venado, reaches about 1,000 m. Approximately ten thousand years ago the Sierra la Laguna and Sierra la Trinidad were separated by a shallow seaway extending through Valle San José, which separates them today.

The islands off the Pacific coast of Baja California are all landbridge in origin, except for the Islas San Benito, which are oceanic (see map 2). The largest and most environmentally diverse Pacific island is Isla Cedros, which reaches nearly 1,200 m in elevation. The remaining islands are generally low, small, and rocky, with the notable exception of Isla Creciente, the long, narrow sandbar enclosing the southern end of Bahía Magdalena. The islands in the Gulf of California are extremely variable in size, habitat, physiography, and geological origin (see map 3). Generally speaking, all are arid and rocky with ridges running from north to south, which on Ángel de la Guarda form an extensive system of mountains. These ranges are granitic or volcanic. Some islands, such as Isla Coronados, Isla Tortuga, and Isla San Luis, are merely the emergent tops of oceanic volcanoes.

Climate

With the exception of the northwestern and southeastern sections of the peninsula, most of Baja California and the Sea of Cortés form the southwestern portion of the Sonoran Desert (Axelrod 1979). The climate is characterized by relatively high annual mean temperatures with low precipitation. Baja California's length, complex topography, and location place it weakly under the influence of moisture from four different sources (Hastings and Turner 1965). It lies at the southern edge of winter cyclonic systems from the westerlies that affect weather patterns in the north; at the western edge of summer monsoons originating in the Gulf of Mexico that affect weather patterns of eastern Baja California and the Gulf of California; at the eastern edge of tropical storms and hurricanes that range across the eastern North Pacific in autumn and affect weather patterns of southern Baja California; and at the western limit of fall activity in the easterlies. However, the region is dominated by two major climatic regimes: winter cyclonic storms from the north and fall hurricanes from the south (Turner and Brown 1982).

Northern Baja California receives the majority of its precipitation from winter storms, which usually originate in the western Pacific and sweep southeast over the peninsula (Hastings and Turner 1965). These storm fronts weaken as they stretch farther south (Humphrey 1974), and usually they do not extend past Laguna San Ignacio. Occasionally, high-pressure cells over the western United States will push storms south into central Baja California and the Cape Region, where they are known as *equipatas* and are usually accompanied by cold temperatures. The gradual rise of the western slopes of the Sierra Juárez and the Sierra San Pedro Mártir induces a great deal of precipitation in the form of either rain or snow and relieves these passing storm fronts of the majority of their moisture. Consequently, little rain falls east of these mountains (Markham 1972; Hastings and Humphrey 1969), although their associated low-pressure systems are responsible for strong winds. Because of this rainshadow effect, the adjacent desert regions are the hottest (Meigs 1953) and driest areas in North America, receiving only about half as much annual rainfall as Death Valley, California (Markham 1972).

Freezing temperatures are rare in the western coastal areas but quite common in the northern mountains. Snow begins to fall on these ranges in late November but usually does not persist for more than a few days, except at the higher elevations in the Sierra San Pedro Mártir, where it may last considerably longer and fall as late as March. Further south, snow is rare but has been observed on Cerro Matomí, Cerro San Borja, and Volcán las Tres Vírgenes; during the unusually cold winter of 1987–1988, snow fell west of Bahía de los Ángeles and was observed on Isla Ángel de la Guarda. From March to November, the northwest portion of the peninsula is characterized by windy days and cloudy nights. Although little rain falls (Humphrey 1974), early-morning low cloud cover usually extends inland to blanket the foothills of the Peninsular Ranges and adds a significant amount to this region's precipitation (Markham 1972). Much of this cloud cover comes in the form of radiational fog, which develops as the ground ceases to radiate heat during the night. This causes air temperatures to drop, and if humidity is high enough the water vapor condenses to form fog (Logan 1968). Such fogs, which are common in the northwest during the spring and fall, usually burn off by mid-morning.

The northern winter storms generally lose their effectiveness somewhere between Bahía de los Ángeles in the northeast and Laguna San Ignacio in the southwest (Aschmann 1959). South of here, the peninsula, and more specifically the Gulf coast, receives the majority of its precipitation from two kinds of summer storms. The more common are convectional storm systems, which result from the orographic lifting and cooling of humid tropical air over the mountains (Humphrey 1974). As the northern desert regions of North America heat up during the summer, the rising air creates a low-pressure cell that draws warm, moist air north from the tropical Pacific areas. As this air mass gains momentum and surges across the Gulf of California, it picks up additional moisture. When it moves onto land, it slams into the precipitous eastern face of the Peninsular Ranges, forcing this moisture-laden air to rise as it crosses the mountains. As it rises, it quickly cools and releases its moisture in the form of heavy and often violent rainshowers with spectacular displays of thunder and lightning. When tropical moisture is present, such storms *(aguaceros)* occur almost daily in the higher elevations of the Sierra San Pedro Mártir, Sierra la Asamblea, Sierra San Borja, Sierra la Libertad, Sierra Guadalupe, Sierra la Giganta, and Sierra la Laguna; this is the case from July to September (Humphrey 1974) and occasionally into early October. Often the tops of these storms flatten and spread out, bringing rain to the surrounding lowland areas as well, but they may be extremely localized, affecting only the upper elevations of the mountains.

The region is also affected by hurricanes. These are powerful anticyclonic systems that develop over the warm waters off the west coast of southern and central Mexico from July through November (Markham 1972). Instead of turning out to sea, a hurricane may sweep up the Gulf of California, hitting the east coast of the peninsula, or cross the Pacific Ocean and hit the southwestern coastline. Hurricanes that do not come onshore, but continue up the Gulf of California, feed on the warm waters and increase in intensity as they move north, generating winds of over 200 kph. On September 28, 1976, Hurricane Kathleen hit La Paz with 208-kph winds and torrential rains; these caused a dam above the town to burst, burying two to three thousand people in its debris. On rare occasions, severe hurricanes have reached as far north as southern California and have even caused deaths at San Felipe on the east coast and El Rosario in the west. After moving off the warm Gulf waters onto land, hurricanes soon dissipate.

Chubascos or *toritos* are smaller anticyclonic systems that usually develop locally, within the Gulf of California. They too are characterized by strong winds and locally heavy rains and can cause severe localized damage when they move onto land. Although they generally last only a few hours, they are some of the most dangerous storms in the Gulf of California. They build very rapidly and have caused numerous deaths among boaters and fishermen who find themselves suddenly trapped in strong winds and rough seas.

These southern summer storms affect the southeastern portion of the peninsula much more than they do the northern and western areas. In fact, they often extend no farther north than Bahía de los Ángeles in the east and Laguna San Ignacio in the west. When both winter and summer storms of the same season falter, the central portion of Baja California may receive no rainfall at all. Occasionally, southern cismontane areas are unaffected by such storms because they lie in the rainshadow of the Sierra la Giganta (Markham 1972).

The strongest influence on the temperature of Baja California is the dissimilarity of its surrounding bodies of water. The Pacific Ocean dominates the temperature regime of western Baja California with its south-flowing, cold California Current, which is responsible for the coastal advection fog (Shreve 1951) and cloud cover that, with steady onshore breezes, keep the coastal temperatures relatively low. In consequence, the west coast of Baja California, from approximately El Rosario south to Todos Santos, is a cool coastal desert that Humphrey (1974) referred to as the Pacific Coastal Desert. In fact, the northern portion of this area, south to Punta Eugenia and inland 6 to 10 km, is one of only three fog deserts in the world (Meigs 1966). Its fogginess results from warm, moist belts of air cooling as they descend and contact the cold surface waters of the California Current. As this air cools, its moisture forms fog near the surface of the water and is carried inland by continuous onshore breezes. The California Current is submerged beneath the north-flowing, slightly warmer Davidson Current in the vicinity of Bahía Magdalena during the fall. Here, along the Magdalena Plain, the Pacific Ocean also

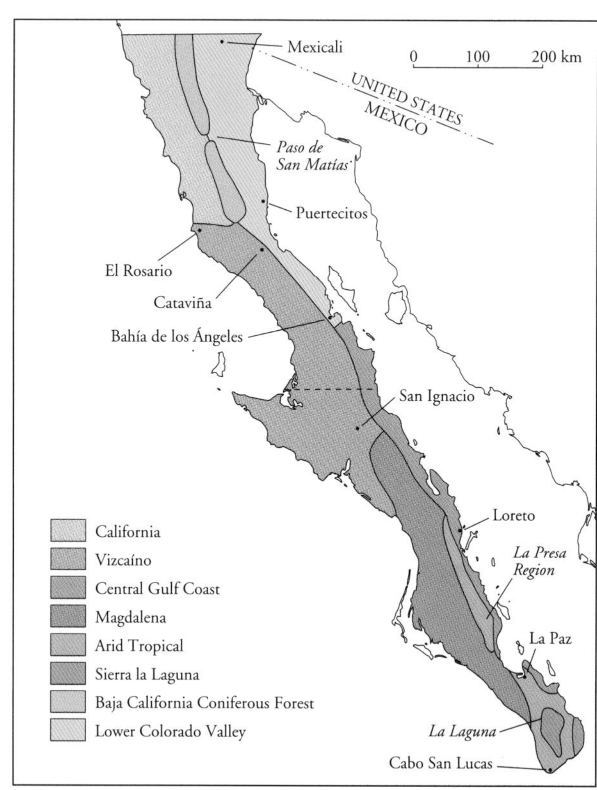

MAP 4 The phytogeographic regions of Baja California

influences the climate of the peninsula, although cloud cover is less and annual mean temperatures are slightly higher (Hastings and Turner 1969).

The temperature of the southern and eastern portions of Baja California is controlled primarily by the Sea of Cortés. A much warmer body of water than the Pacific (Robinson 1973), the Sea of Cortés offers little amelioration to the peninsula in the form of precipitation or cooling. The areas east of the Peninsular Ranges receive no cooling from onshore Pacific breezes, and consequently the Gulf coast becomes extremely hot during the summer (Markham 1972).

Phytogeographic Provinces

Baja California's wide range of climatic variables and well-sculpted topography support clearly defined phytogeographic regions (Shreve and Wiggins 1964; Wiggins 1980; Grismer 1994b). Phytogeographic regions are good indicators of natural biotic provinces because they are ecological reflections of the interactions of climate, topography, and soil. They are distinguished from one another with the notion that many of the plants therein share the same ecological and environmental history. A comparison of the distribution and geographic variation of the herpetofauna with the phytogeographic regions of Baja California (map 4) reveals that the two coincide very closely (Grismer 1994b). Therefore, it is important to understand not only the floristic composition of these regions, which sets them apart from one another, but also the environmental reasons why they exist where they do. The following characterization of the various phytogeographic regions follows Grismer 1994b and is adapted from Shreve and Wiggins 1964, Wiggins 1980, Turner and Brown 1982, and Cody et al. 1983.

CALIFORNIA REGION

The California Region is a relatively cool, mesic area occupying the cismontane areas of northern Baja California from the U.S.-Mexican border south to the vicinity of El Socorro. Its eastern border lies on the lower western slopes of the Sierra Juárez and Sierra San Pedro Mártir at the edge of the Jeffrey Pine Belt. This region, which is basically a southern extension of the coastal scrub and chaparral communities of southern California, occupies the Pacific coastal scrub and foothill chaparral areas of northwestern Baja California. The Pacific coastal scrub areas (fig. 1) occur in the sporadic, flat, low-lying western portions of this region from sea level to the edge of the chaparral belt at approximately 300 to 600 m elevation. The dominant plant species here are drought-deciduous forms such as California Sagebrush *(Artemisia californica),* White and Black Sage *(Salvia apiana* and *S. mellifera),* California Buckwheat *(Eriogonum fasciculatum),* and Coastal Agave *(Agave shawii).*

FIGURE 1 Pacific coastal scrub vegetation of the California Region at Salsipuedes, 20 km north of Ensenada, BC

FIGURE 2 Chaparral of the California Region just above Rancho San José (Melling's Ranch) in the foothills of the Sierra San Pedro Mártir, BC

The chaparral portion of the California Region occurs primarily in canyons in the western part of the region and at the higher elevations in the foothills of the Peninsular Ranges to the east. This chaparral assemblage is characterized by larger (1–3 m) evergreen shrubs (fig. 2) such as Chamise *(Adenostoma fasciculatum),* Hoary Leaf-Lilac *(Ceanothus crassifolia),* Chaparral Ash *(Fraxinus trifoliata),* Toyon *(Heteromeles arbutifolia),* Scrub Oak *(Quercus berberidifolia),* Laurel Sumac *(Malosma laurina),* Lemonadeberry *(Rhus integrifolia),* and Red Shank *(Adenostoma sparsifolium).* South and inland from Cabo Colonet, this area begins a gradual transition into the more arid southerly Vizcaíno Region.

BAJA CALIFORNIA CONIFEROUS FOREST REGION

The Baja California Coniferous Forest Region is the southernmost disjunct and depauperate section of the

FIGURE 3 Montane meadow of the Baja California Coniferous Forest Region at Laguna Hanson in the Sierra Juárez, BC

FIGURE 5 Lower Colorado Valley Region along the eastern foothills of the Sierra Juárez near Cañón Santa Isabel, BC

FIGURE 4 Baja California Coniferous Forest Region with Quaking Aspen *(Populus tremuloides)* near La Tasajera in the Sierra San Pedro Mártir, BC

broader and more inclusive Sierran Montane Conifer Forest (Pase 1982), which, in Baja California, is composed of two disjunct sections. This cool mesic area occupies the upper elevations of the northern Sierra Juárez and southern Sierra San Pedro Mártir above the chaparral belt of the California Region in the west and the creosote bush scrub of the Lower Colorado Valley Region in the east. It extends south from the U.S.-Mexican border approximately 300 km to near Cerro Matomí and is a southern extension of the coniferous forests of southern California. The Baja California Coniferous Forest Region receives more precipitation than any other area in northern Baja California. The majority of this comes from cold northern winter storms, but a significant portion also results from convectional southern summer storms. Consequently, the floristic composition is relatively diverse, composed of many large shrubs and trees but with a conspicuous lack of understory vegetation. The dominant species of this region are Piñon Pine *(Pinus quadrifolia)* and Jeffrey Pine *(P. jeffreyi),* the former being more prevalent in the Sierra Juárez (fig. 3) and the latter more prevalent in the Sierra San Pedro Mártir in the higher elevations. Other conifers, such as Lodgepole Pine *(P. contorta* subsp. *murayana),* can be found among the Jeffrey Pine, Sugar Pine *(P. lambertiana),* White Fir *(Abies concolor),* Quaking Aspen *(Populus tremuloides),* and Incense Cedar *(Calocedrus decurrens)* (fig. 4).

LOWER COLORADO VALLEY REGION

The Lower Colorado Valley Region is the largest subdivision of the Sonoran Desert, but only a thin southern extension of its western portion enters Baja California. It occupies the areas between the western coastline (or the Río Colorado in the extreme northeast) of the Gulf of California and the eastern foothills of the Peninsular Ranges, extending from the U.S.-Mexican border south to just below Bahía de los Ángeles. In the northern portion of eastern Baja California, the region is a low-lying, flat, sandy, hot, and arid area lying in the rainshadow of the Sierra Juárez and Sierra San Pedro Mártir. South of here, from Puertecitos to

FIGURE 6 Coastal fog desert of the Vizcaíno Region in the Sierra Santa Clara, BCS

FIGURE 7 Inland vegetation of the Vizcaíno Region along the base of the Sierra la Asamblea, BC

Bahía San Luis Gonzaga, and again south to Bahía de los Ángeles, the region is rugged and characterized by small, eroded volcanic terraces and mountain ranges. Owing to its temperature and aridity, the Lower Colorado Valley Region of Baja California is dominated by various small-leafed, drought-resistant plants (fig. 5). The most common of these in the flat, sandy areas to the north are Creosote Bush *(Larrea tridentata),* White Bursage *(Ambrosia dumosa),* Ocotillo *(Fouquieria splendens),* Brittlebush *(Encelia farinosa),* and Desert Agave *(Agave deserti).* In the arroyos and rugged foothill areas, Mesquite *(Prosopis glandulosa),* Smoke Tree *(Psorothamnus spinosa),* Little-Leaf Palo Verde *(Cercidium microphyllum),* Ironwood *(Olneya tesota),* and chollas (*Opuntia* spp.) are also common.

VIZCAÍNO REGION

The Vizcaíno Region is an arid region occupying the central third of western Baja California. The marine influence of the Pacific Ocean leaves this region cooler than the other desert areas, but because of its southerly and cismontane location it receives little in the way of winter or summer rainfall. Therefore it is composed mostly of desert species. The northern margin of the Vizcaíno Region begins near El Rosario and extends east to the crest of Sierra San Miguel and south to the vicinity of San Ignacio. Here it is bordered by the volcanic foothills of the more southerly Magdalena Region. From this point, it arcs southwest along the foothills of the Magdalena Region to the Pacific Coast near Bahía San Juanico. In the north, maritime influences extend inland approximately 6 to 10 km, and vegetation along this narrow coastal strip is scant. This area, extending from El Rosario to Laguna San Ignacio, forms a unique subdivision within the Vizcaíno Region that Meigs (1966) considered to be a "fog type" temperate desert. Vegetation in open areas along this strip is stunted, widely spaced, and depauperate (fig. 6) because of continuous onshore winds, which release little moisture in their passage. In areas protected from the winds but available to precipitation, floral diversity increases sharply. The dominant plants of this coastal subregion are bursages *(Ambrosia chenopodifolia* and *A. camphorata),* Coastal Agave *(Agave shawii* subsp. *goldmaniana),* Bush Sunflower *(Encelia farinosa),* Palo Adán *(Fouquieria diguetii),* and Jojoba *(Simmondsia chinensis).* Inland from this coastal subregion and above 500 m elevation, the conspicuous species are Blue Agave *(Agave cerulata),* bursages *(Ambrosia dumosa* and *A. camphorata),* Palo Adán, Cirio *(Fouquieria columnaris),* Cardón *(Pachycereus pringlei),* Elephant Tree *(Pachycormus discolor),* Pitahaya Agria *(Stenocereus gummosus),* Little-Leaf Palo Verde *(Cercidium microphyllum),* and Mesquite *(Prosopis glandulosa)* (fig. 7).

The southern portion of the Vizcaíno Region, which is much flatter, comprises the Vizcaíno Desert. The Vizcaíno Desert is a triangularly shaped area beginning just south of Punta Rosarito and extending inland to the low,

FIGURE 8 Datilillo forest in the Vizcaíno Desert of the Vizcaíno Region along the southwestern edge of the Sierra Santa Clara, BCS

FIGURE 9 Prostrate vegetation in the Vizcaíno Desert of the Vizcaíno Region just east of the Sierra Vizcaíno, BCS

rocky foothills of the Peninsular Ranges. From here it continues south along the southern and eastern margins of the foothills to near Bahía San Juanico. The western margin of the Vizcaíno Desert north of Guerrero Negro is bordered by the Pacific Ocean. South of Guerrero Negro, it is bordered by the mountainous Vizcaíno Peninsula, from Punta Eugenia south to Laguna San Ignacio. South of Laguna San Ignacio, the Vizcaíno Desert is bordered by the Pacific Ocean once again. The Vizcaíno Desert is characterized by cool, cloudy, and breezy days most of the year. As a result, the vegetation is relatively depauperate and prostrate. Much of the Vizcaíno Desert (fig. 8) is characterized by Datilillo *(Yucca valida),* Palo Adán *(Fouquieria diguetii),* Lomboy *(Jatropha cinerea),* Yerba Reuma *(Frankenia palmeri),* and salt bushes *(Atriplex julacea* and *A. polycarpa).* In the west-central portion surrounding the town of Guerrero Negro, the vegetation becomes prostrate and uniform (fig. 9). Here the dominant plants are *A. julacea* and *F. palmeri.*

CENTRAL GULF COAST REGION

The Central Gulf Coast Region is a narrow zone beginning near Bahía de los Ángeles and continuing south along the eastern foothills and escarpments of the Peninsular Ranges to the Isthmus of La Paz. At the isthmus, it is interrupted by the eastern expanse of the Magdalena Region but begins again on the coastline northeast of La Paz and continues south to Ensenada de los Muertos in the Cape Region. South of this ensenada, it is interrupted a second time by the Arid Tropical Region but persists once more on the coastline east of Sierra la Trinidad. This region is very hot and arid and receives virtually no winter rainfall. The majority of its precipitation comes in the form of runoff from southern convectional storms in the bordering Peninsular Ranges during the summer. Occasionally it receives rainfall from *chubascos* and large convectional storms that flatten and spread out at high altitudes. As a result, the Central Gulf Coast Region is composed of desert vegetation and is characterized by Leatherplant *(Jatropha cuneata),* Lomboy *(Jatropha cinerea),* Palo Adán *(Fouquieria diguetii),* Copál *(Bursera hindsiana),* Elephant Tree *(B. mi-*

FIGURE 10 Central Gulf Coast Region along the eastern face of the Sierra la Giganta west of Puerto Escondido, BCS

FIGURE 11 Vegetation along the volcanic slopes and flats of the Magdalena Region approximately 100 km northeast of Ciudad Insurgentes, BCS

crophylla), Cardón *(Pachycereus pringlei),* and Little-Leaf Palo Verde *(Cercidium microphyllum)* (fig. 10). In the Cape Region, the flora of the Central Gulf Coast Region consists of additional dominants such as Pitahaya Agria *(Stenocereus gummosus).*

MAGDALENA REGION

The Magdalena Region occupies the western desert slopes and Pacific drainages of the Sierra Guadalupe and Sierra la Giganta. The Magdalena Region begins at San Ignacio in the north and arcs southwest toward the coast, contacting the Pacific near Bahía San Juanico. Here it continues south to just north of Todos Santos. The eastern border extends south from San Ignacio along the base of the Peninsular Ranges to the Isthmus of La Paz, where it arcs east and contacts the Gulf coast. Its relatively warmer climate is influenced by the warmer Davidson Current during the fall and the colder California Current at other times of the year. However, morning fogs are common, and annual precipitation is low and unpredictable, owing to the rainshadow effect of the Sierra la Giganta and its

FIGURE 12 Palm-lined oasis in the Magdalena Region at San José de Magdalena, BCS

southerly location, which keeps it outside most winter storm tracks. In fact, this area has gone as long as five years without rain.

The Magdalena Region is topographically divisible into two sections. An eastern portion extends between San Ignacio and the Isthmus of la Paz. This area consists of a "badlands" composed of volcanic hills, fields, and mesas extending along the base of the Sierra la Giganta (fig. 11). It is edged to the west by a small southern section of the Vizcaíno Desert and the Magdalena Plain. Here the dominant species are Creosote Bush *(Larrea tridentata),* Leatherplant *(Jatropha cuneata),* Lomboy *(Jatropha cinerea),* chollas (*Opuntia* spp.), Palo Blanco *(Lysiloma candida),* Pitahaya Agria *(Stenocereus gummosus),* Organ Pipe Cactus *(Stenocereus thurberi),* Palo Adán *(Fouquieria diguetii),* Little-Leaf Palo Verde *(Cercidium microphyllum),* and Cardón *(Pachycereus pringlei).* Palm-lined oases (fig. 12) are characteristic of some of the deeper canyons and arroyos in these volcanic badlands, and there the dominant species are the Mexican Fan Palm *(Washingtonia robusta)* and Palma Palmía *(Brahea brandegeei).*

West of the Comondú area, the Magdalena Region opens up into the Magdalena Plain, which constitutes the region's western section. This section occupies nearly half the width of the southern third of Baja California. Here the volcanic fields become less complex along the fringe of the western foothills of the Sierra la Giganta and eventually give way to a low, flat, open and extensive sandy plain that borders the Pacific Ocean. The Magdalena Plain is composed primarily of fluvial deposits from the western slopes of the Sierra la Giganta and in unusually wet years has served as a large, shallow, freshwater basin. The climate of the Magdalena Plain is influenced strongly by the cool Pacific Ocean. This region is often shrouded in morning advectional fogs whose moisture supports the growth of Span-

FIGURE 13 Spanish Moss *(Ramalina reticulata)* on Magdalena Plain vegetation of the Magdalena Region at Estero Salina, BCS

FIGURE 14 Magdalena Plain of the Magdalena Region 80 km north of La Paz, BCS

ish Moss *(Ramalina reticulata)* on the vegetation (fig. 13). In fact, the fogs produce enough moisture to support the growth of bromeliads *(Tillandsia recurvata)* on the high-tension wires in the towns of Ciudad Constitución and Ciudad Insurgentes, and there are even road signs warning of a *zona de neblina* (fog zone) along the main highway. The dominant floral components of the Magdalena Plain (fig. 14) are Mesquite *(Prosopis glandulosa),* Coast Desert Thorn *(Lycium californicum),* chollas (*Opuntia* spp.), Palo Adán *(Fouquieria diguetii),* Pitahaya Agria *(Stenocereus gummosus),* Organ Pipe Cactus *(Stenocereus thurberi),* and Cardón *(Pachycereus pringlei).*

ARID TROPICAL REGION

The Arid Tropical Region is composed of two disjunct sections. The first occurs on the mountainous upper elevations of the Sierra Guadalupe and Sierra la Giganta, beginning approximately due east of Mulegé and terminating just north of the Isthmus of La Paz. This region varies in elevation from 504 to 2,088 m, and the majority of its precipitation comes between June and October, from southern convectional storms that leave between 200 and 300 mm of rain each year. Winter fogs coming off the Gulf of California sometimes reach this area as well. Because of its elevation, temperatures are cooler than along the Gulf coast, and, coupled with the relatively high amount of summer rainfall, they enable the region to support many large tropical deciduous shrubs and trees. The dominant floristic components are Palo Blanco *(Lysiloma candida),* Creosote Bush *(Larrea tridentata),* Palo Fierro *(Pithecellobium confine),* Vinorama *(Acacia brandegeana),* Palo Chino *(Acacia peninsularis),* Mesquite *(Prosopis glandulosa),* Palo Zorillo *(Cassia emarginata),* Pitahaya Dulce *(Stenocereus thurberi),* and Little-Leaf Palo Verde *(Cercidium microphyllum)* (fig. 15). Also very common in the Sierra la Giganta are several species of pincushion cactus (*Mammillaria* spp.) and chollas (*Opuntia* spp.). In the bottoms of many canyons that have water or underground streams, it is common to find thick stands of Palma Palmía *(Brahea brandegeei)* and the fig Zalate *(Ficus palmeri).*

FIGURE 15 Arid Tropical Region vegetation of Valle de Guadalupe in the Sierra Guadalupe, BCS

The southern section of the Arid Tropical Region encompasses the low-lying, flat areas of the Cape Region, beginning at the Isthmus of La Paz and continuing south to Cabo San Lucas. The Arid Tropical Region of the Cape surrounds the Sierra la Laguna. As a result, weather patterns and floristic makeup can be significantly different on opposite sides of the mountains. On the west side of the Sierra la Laguna, temperatures are slightly cooler because of the influence of the Pacific Ocean, and precipitation is generally lower because this area lies in a rainshadow and is not greatly affected by the summer storms. The western area is basically a mesquite and thorn-scrub woodland characterized by many of the same species that occur in the lower

FIGURE 16 Arid Tropical Region vegetation at La Victoria in the Sierra la Laguna of the Cape Region, BCS

elevations of the Sierra la Giganta. The Arid Tropical Region on the east side of the Sierra la Laguna receives much more precipitation and has been considered by some to be an impoverished jungle (fig. 16). Species present on the west side of the Sierra la Laguna also occur here but are not usually the dominant forms. More characteristic of the eastern flora are species such as Coralvine *(Antigonon leptopus),* Palo Fierro *(Pithecellobium confine),* Vinorama *(Acacia brandegeana),* Cardón-Barbón *(Pachycereus pecten-aboriginum),* Coral Tree *(Erythrina flabelliformis),* Ciraelo *(Cyrtocarpa edulis),* Trumpet Bush *(Tecoma stans),* Little-Leaf Palo Verde *(Cercidium microphyllum),* Palo Blanco *(Lysiloma candida),* Plumeria *(Plumeria acutifolia),* Copál *(Bursera hindsiana),* Elephant Tree *(Bursera microphylla),* and Leatherplant *(Jatropha cuneata).*

SIERRA LA LAGUNA REGION

The Sierra la Laguna Region occupies the upper elevations of the Sierra la Laguna, which runs diagonally across the Cape Region from near La Paz to Cabo San Lucas. Owing to its geographic position and altitude, this region has a variety of climates, ranging from the hot, relatively dry foothills of the north to the wet, tropical climate of its upper elevations in the south. These high elevations induce summer storms, moving north out of the southern Gulf and Pacific, to deposit much of their moisture. This area receives as much as 700 mm of rainfall each year, more than any other area in southern Baja California. The vegetation of this region corresponds to its climates. In areas below 800 m, the dominant plants are Copál *(Bursera hindsiana),* Elephant Tree *(Bursera microphylla),* Morning Glory (*Ipomoea* spp.), Palo Blanco *(Lysiloma candida),* and Mauto *(Lysiloma divaricata).* In areas lying between 800 and 1,300 m there is a transition to oak and pine-oak communities (fig. 17) dominated by Encino Negro *(Quercus devia),* Encino Roble *(Quercus tuberculata),* and a pine *(Pinus cembroides).*

FIGURE 17 Oak woodland of the Sierra la Laguna Region near La Laguna in the Sierra la Laguna, BCS

INSULAR REGIONS

Because of climatic similarities, insular flora of the Gulf of California and along the Pacific coast are generally extensions of those of the adjacent continental phytogeographic regions. Any differences are usually due to the island's relatively small size and low elevation and its being surrounded by water: for example, more moderate mean temperatures and slightly less diverse vegetational structure and composition. Generally, on islands greater than 3 km^2, however, the vegetation is essentially the same as on the adjacent mainland (Cody et al. 1983). Therefore, all islands within the Gulf of California, with the exception of the Isla Encantada Archipelago, lie within the Central Gulf Coast Region (fig. 18). The Isla Encantada Archipelago is included in the Lower Colorado Valley Region. On the Pacific coast,

FIGURE 18 Central Gulf Coast Region vegetation of Arroyo Mota, Isla Santa Catalina, BCS

FIGURE 19 Guano-covered slope of Isla Coloradito, BC, showing the characteristic lack of vegetation created by seabird rookeries

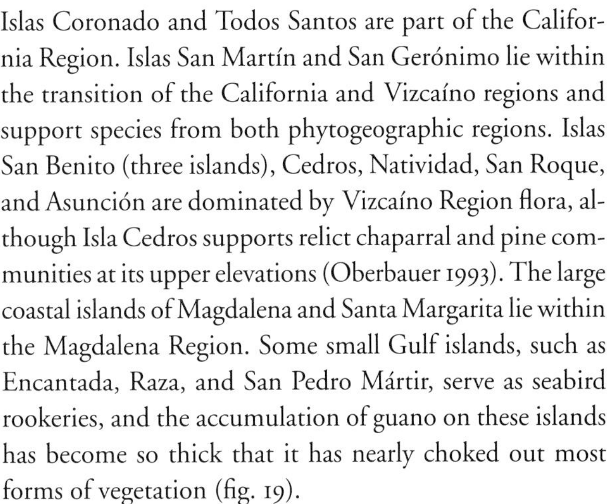

Islas Coronado and Todos Santos are part of the California Region. Islas San Martín and San Gerónimo lie within the transition of the California and Vizcaíno regions and support species from both phytogeographic regions. Islas San Benito (three islands), Cedros, Natividad, San Roque, and Asunción are dominated by Vizcaíno Region flora, although Isla Cedros supports relict chaparral and pine communities at its upper elevations (Oberbauer 1993). The large coastal islands of Magdalena and Santa Margarita lie within the Magdalena Region. Some small Gulf islands, such as Encantada, Raza, and San Pedro Mártir, serve as seabird rookeries, and the accumulation of guano on these islands has become so thick that it has nearly choked out most forms of vegetation (fig. 19).

The Herpetofauna

History of Herpetological Research

The first written report concerning amphibians and reptiles to emerge from Baja California came from the Jesuit missionary Francisco Javier Clavigero in 1789. His account of the amphibians and reptiles comes primarily from an unpublished manuscript written by the Jesuit Miguel del Barco during his tenure and travels in Baja California from 1738 until the expulsion of the Jesuit order from Mexico in 1768. Clavigero writes:

> In California there are few kinds of reptiles, to wit, large and small lizards, frogs, toads, turtles, and snakes. Among the species of large and small lizards, we do not know that there are any poisonous ones. Frogs are very scarce. Toads are abundant when it rains, but they disappear entirely when the land is dry again. Among the turtles, besides the common land variety and the fresh-water ones, there are two other species of large sea turtles. One of these is that from whose shell tortoise shell is obtained to use in prized and curious works. . . . There are two kinds of snakes, the rattlesnakes or *crotali,* as Linnaeus calls them, and those which are not. The latter are smaller than the former; but their poison is more active.

The second report came from the German naturalist F. Deppe as a result of his overland trek from La Paz to San Diego during the spring of 1837. Deppe stated, "The peninsula is poor beyond description in living forms of organic life. Neither reptiles nor insects were seen" (quoted in Lichtenstein 1839). These reports remained the most current sources of published information concerning the herpetology of Baja California for nearly half a century.

Biological exploration in Baja California and the Sea of Cortés did not really get under way until the early to mid-nineteenth century. The first expeditions were primarily concerned with the collection of plants and birds, and very few amphibians and reptiles were taken. The first significant herpetological collections to come out of Baja California were made by L. J. Xantus de Vésey in the Cape Region between 1859 and 1861. During the next twenty-five years or so, great naturalists such as Spencer F. Baird, Lyman Belding, M. P. E. Botta, Townsend S. Brandegee, Walter E. Bryant, Gustav Eisen, William M. Gabb, and others added so significantly to these collections that by 1895 John Van Denburgh was able to publish the first review of the herpetofauna of Baja California and its associated islands. We now know that Van Denburgh's 1895 review of Baja California herpetology was very incomplete, but it served as a starting point on which others could build.

During the years 1888–1892, the first major collection from central Baja California was made by Léon Diguet, a French chemical engineer employed by the Boleo Mining Company at Santa Rosalía. Based on this material, in 1899 François Mocquard published a descriptive work that significantly augmented that of Van Denburgh. In 1911, the United States Bureau of Fisheries and the American Museum of Natural History sponsored the Albatross Expedition, whose mission was a biological reconnaissance of the coast and bordering islands of Baja California. The expedition was in the charge of Charles H. Townsend, who devoted special attention to the collection of reptiles and fishes. As a result of Townsend's efforts, Mary C. Dickerson (1919) described one new genus and twenty-three new species of lizards. Shortly afterward, Charles Nelson presented a faunal analysis of the Baja California herpetofauna, and Karl P. Schmidt produced a comprehensive, descriptive work. Both were among the first to entertain hypotheses concerning the evolutionary origin of the herpetofauna as a faunal unit. John Van Denburgh's 1922 *Reptiles of Western North America* considered species from Baja California and some of the islands in the Sea of Cortés but excluded the amphibian fauna. Schmidt 1922 and Van Denburgh 1922 have remained the best descriptive accounts of the Baja California and Gulf of California herpetofauna to date.

Other works concerning large segments of Baja California's herpetofauna were mostly island checklists (Van Denburgh 1905; Van Denburgh and Slevin 1914, 1921a; Cliff 1954a,b; Soulé and Sloan 1966; Savage 1967; Bostic 1975; Wilcox 1980; Murphy and Ottley 1984, Grismer 1993, 1999b, 2001a), none of which treated simultaneously the Pacific and Gulf islands. Savage (1960) presented a dispersal-based hypothesis to account for the origin of the herpetofauna of Baja California, and later Murphy (1983a,b), Welsh (1988), and Grismer (1994a,b, 2000) discussed vicariant models.

Biogeography

FAUNAL RELATIONSHIPS

As currently understood, the native, nonmarine herpetofauna of Baja California, its associated Pacific islands, and the islands in the Sea of Cortés is composed of 161 species. Of these, 4 are salamanders, 13 are frogs, 4 are turtles, 84 are lizards, 55 are snakes, and one is an amphisbaenian. Readily apparent is the depauperate nature of the amphibians compared to the modest diversity of lizards and snakes. This is easily accounted for by the generally hot and dry climate of Baja California. In fact, of the species of amphibians present, most are restricted to the northwestern portion of the peninsula in cismontane or montane mesic areas or just marginally enter Baja California in the extreme northeast, in association with the Río Colorado and its delta. Only three species of amphibians naturally range into Baja California Sur. Three others, the Bullfrog *(Rana catesbeiana),* Río Grande Leopard Frog *(R. berlandieri),* and Forrer's Grass Frog *(R. forreri),* have been introduced.

Much of the herpetofauna of Baja California and the Sea of Cortés occurs in the United States and mainland Mexico, but a large percentage is endemic to Baja California. In this work, the term *endemic* pertains to species restricted to the region of study as a geographical and biological unit (Grismer 1994a) rather than a political one. Therefore, some species, such as the Peninsular Leaf-Toed Gecko *(Phyllodactylus xanti)* and the Baja California Rat Snake *(Bogertophis rosaliae),* which range continuously throughout the peninsula and enter southern California, are considered endemic to Baja California. In this context, the northern limit of Baja California is the northern limit of the Peninsular Ranges at Mount San Jacinto in Riverside County, California, in the east and the Los Angeles Basin in Los Angeles County, California, in the west. These areas are all part of the same massive fault-block system from which the Baja California peninsula is derived. Some taxa, however, such as the Sandstone Night Lizard *(Xantusia gracilis),* which are endemic to the Baja California peninsula as defined here, occur outside the geopolitical boundary of Baja California and are not covered in this work.

Of the 161 nonmarine species in the region of study, 23 percent are shared only with the United States, 29 percent are shared with the United States and Mexico, less than 1 percent are shared only with Mexico, and 48 percent occur only in Baja California. Of these, approximately 64 percent are believed to be more closely related to and in some cases derived from mainland Mexican lineages, 6 percent are more closely related to species in the United States, and 30 percent are related to species occurring in the United States and Mexico (Grismer 1994a,b). Thus, it is clear that the origin of this herpetofauna is more intimately associated with mainland Mexico than with the United States (table 1). These numbers do not reflect the fact that 12 percent of the species shared exclusively by Baja California and the United States only marginally enter the northern areas. At least five additional species have been introduced into the region of study.

HISTORICAL BIOGEOGRAPHY

Grismer (1994a) divided the peninsular herpetofauna into various biogeographical complexes based on hypotheses concerning their evolutionary origin. Since that time, the phylogenetic relationships of some of these species have been reevaluated, and some of them have been placed in different groups (see table 2). One group of species that is endemic to Baja California appears to have evolved in

Table 1

Faunal Relationships of the Nonmarine Herpetofauna of Baja California, Including Its Pacific Islands and the Islands in the Sea of Cortés, with That of the United States and Mainland Mexico

	Shared only with U.S.	Shared with U.S. and Mainland Mexico	Shared only with Mainland Mexico	Endemic (▲ = insular endemic)	Introduced
SALAMANDERS					
PLETHODONTIDAE: Lungless Salamanders					
Aneides lugubris (Arboreal Salamander)	•				
Batrachoseps major (Garden Slender Salamander)	•				
Ensatina (Ensatina Salamanders)					
eschscholtzii (Monterey Ensatina)	•				
klauberi (Large-Blotched Ensatina)	•				
FROGS AND TOADS					
BUFONIDAE: True Toads					
Bufo (Toads)					
alvarius (Colorado River Toad)		•			
boreas (Western Toad)	•				
californicus (California Toad)	•				
cognatus (Great Plains Toad)		•			
punctatus (Red-Spotted Toad)		•			
woodhousii (Woodhouse's Toad)		•			
HYLIDAE: Treefrogs					
Hyla (Treefrogs)					
cadaverina (California Treefrog)	•				
regilla (Pacific Treefrog)	•				
RANIDAE: True Frogs					
Rana (Pond Frogs)					
aurora (Red-Legged Frog)	•				
berlandieri (Río Grande Leopard Frog)					•
boylii (Foothill Yellow-Legged Frog)	• (?)				
catesbeiana (Bullfrog)					•
forreri (Forrer's Grass Frog)					•
yavapaiensis (Yavapai Leopard Frog)		• (?)			
PELOBATIDAE: Spadefoot Toads					
Scaphiopus couchii (Couch's Spadefoot)		•			
Spea hammondii (Western Spadefoot)	•				
TURTLES AND TORTOISES					
EMYDIDAE: Pond Turtles and Sliders					
Clemmys marmorata (Western Pond Turtle)	•				
Trachemys nebulosa (Baja California Slider)				•	
TESTUDINIDAE: Land Tortoises					
Gopherus agassizii (Desert Tortoise)		•			
TRIONYCHIDAE: Soft-Shell Turtles					
Apalone spinifera (Spiny Soft-Shell)					•
LIZARDS					
CROTAPHYTIDAE: Collared and Leopard Lizards					
Crotaphytus (Collared Lizards)					
dickersonae (Dickerson's Collared Lizard)			•		

	Shared only with U.S.	Shared with U.S. and Mainland Mexico	Shared only with Mainland Mexico	Endemic (▲ = insular endemic)	Introduced
grismeri (Sierra los Cucapás Collared Lizard)				•	
insularis (Isla Ángel de la Guarda Collared Lizard)				▲	
vestigium (Baja California Collared Lizard)	•			•	
Gambelia (Leopard Lizards)					
copeii (Baja California Leopard Lizard)				•	
wislizenii (Long-Nosed Leopard Lizard)		•			
IGUANIDAE: Iguanas and Chuckwallas					
Ctenosaura (Spiny-Tailed Iguanas)					
conspicuosa (Isla San Esteban Spiny-Tailed Iguana)				▲	
hemilopha (Cape Spiny-Tailed Iguana)				•	
nolascensis (Isla San Pedro Nolasco Spiny-Tailed Iguana)				▲	
Dipsosaurus (Desert Iguanas)					
catalinensis (Isla Santa Catalina Desert Iguana)				▲	
dorsalis (Desert Iguana)		•			
Sauromalus (Chuckwallas)					
hispidus (Spiny Chuckwalla)				▲	
klauberi (Isla Santa Catalina Chuckwalla)				▲	
obesus (Northern Chuckwalla)		•			
slevini (Slevin's Chuckwalla)				▲	
varius (Isla San Esteban Chuckwalla)				▲	
PHRYNOSOMATIDAE: Sand, Rock, Spiny, Horned, Brush, and Side-Blotched Lizards					
Callisaurus draconoides (Zebra-Tailed Lizard)		•			
Petrosaurus (Banded Rock Lizards)					
mearnsi (Banded Rock Lizard)				•	
repens (Central Baja Banded Rock Lizard)				•	
slevini (Slevin's Banded Rock Lizard)				▲	
thalassinus (San Lucan Banded Rock Lizard)				•	
Phrynosoma (Coast Horned Lizards)					
coronatum (Coast Horned Lizard)	•				
mcallii (Flat-Tailed Horned Lizard)		•			
platyrhinos (Desert Horned Lizard)		•			
solare (Regal Horned Lizard)			•		
Sceloporus (Spiny Lizards)					
angustus (Isla Santa Cruz Spiny Lizard)				▲	
clarkii (Clark's Spiny Lizard)			•		
grandaevus (Isla Cerralvo Spiny Lizard)				▲	
hunsakeri (Hunsaker's Spiny Lizard)				•	
licki (Cape Spiny Lizard)				•	
lineatulus (Isla Santa Catalina Spiny Lizard)				▲	
magister (Desert Spiny Lizard)		•			
occidentalis (Western Fence Lizard)	•				
orcutti (Granite Spiny Lizard)				•	
vandenburgianus (Southern Sagebrush Lizard)	•				
zosteromus (Baja California Spiny Lizard)				•	
Uma notata (Colorado Desert Fringe-Toed Lizard)		•			

(continued)

Table 1, *continued*

	Shared only with U.S.	Shared with U.S. and Mainland Mexico	Shared only with Mainland Mexico	Endemic (▲ = insular endemic)	Introduced
Urosaurus (Brush Lizards)					
graciosus (Long-Tailed Brush Lizard)		•			
lahtelai (Baja California Brush Lizard)				•	
nigricaudus (Black-Tailed Brush Lizard)				•	
ornatus (Tree Lizard)		•			
Uta (Side-Blotched Lizards)					
encantadae (Enchanted Side-Blotched Lizard)				▲	
lowei (Dead Side-Blotched Lizard)				▲	
nolascensis (Isla San Pedro Nolasco Side-Blotched Lizard)				▲	
palmeri (Isla San Pedro Mártir Side-Blotched Lizard)				▲	
squamata (Isla Santa Catalina Side-Blotched Lizard)				▲	
stansburiana (Side-Blotched Lizard)		•			
tumidarostra (Swollen-Nosed Side-Blotched Lizard)				▲	
EUBLEPHARIDAE: Eyelidded Geckos					
Coleonyx (Banded Geckos)					
gypsicolus (Isla San Marcos Barefoot Banded Gecko)				▲	
switaki (Barefoot Banded Gecko)				•	
variegatus (Western Banded Gecko)		•			
GEKKONIDAE: Geckos					
Gehyra mutilata (Stump-Toed Gecko)					•
Hemidactylus frenatus (Common House Gecko)					•
Phyllodactylus (Leaf-Toed Geckos)					
bugastrolepis (Isla Santa Catalina Leaf-Toed Gecko)				▲	
homolepidurus (Sonoran Leaf-Toed Gecko)			•		
partidus (Isla Partida Norte Leaf-Toed Gecko)				▲	
unctus (San Lucan Leaf-Toed Gecko)				•	
xanti (Peninsular Leaf-Toed Gecko)				•	
TEIIDAE: Whiptails					
Cnemidophorus (Whiptails)					
bacatus (Isla San Pedro Nolasco Whiptail)				▲	
canus (Isla Salsipuedes Whiptail)				▲	
carmenensis (Isla Carmen Whiptail)				▲	
catalinensis (Isla Santa Catalina Whiptail)				▲	
celeripes (Isla San José Western Whiptail)				▲	
ceralbensis (Isla Cerralvo Whiptail)				▲	
danheimae (Isla San José Whiptail)				▲	
espiritensis (Isla Espíritu Santo Whiptail)				▲	
franciscensis (Isla San Francisco Whiptail)				▲	
hyperythrus (Orange-Throated Whiptail)				•	
labialis (Baja California Whiptail)				•	
martyris (Isla San Pedro Mártir Whiptail)				▲	

	Shared only with U.S.	Shared with U.S. and Mainland Mexico	Shared only with Mainland Mexico	Endemic (▲ = insular endemic)	Introduced
pictus (Isla Monserrate Whiptail)				▲	
tigris (Western Whiptail)		•			
XANTUSIIDAE: Night Lizards					
Xantusia (Night Lizards)					
henshawi (Granite Night Lizard)	•			•	
vigilis (Desert Night Lizard)		•			
SCINCIDAE: Skinks					
Eumeces (Skinks)					
gilberti (Gilbert's Skink)	•				
lagunensis (San Lucan Skink)				•	
skiltonianus (Western Skink)	•				
ANGUIDAE: Legless and Alligator Lizards					
Anniella (Legless Lizards)					
geronimensis (Baja California Legless Lizard)				•	
pulchra (California Legless Lizard)	•				
Elgaria (Alligator Lizards)					
cedrosensis (Isla Cedros Alligator Lizard)				•	
multicarinata (Southern Alligator Lizard)	•				
nana (Islas Coronado Alligator Lizard)				▲	
paucicarinata (San Lucan Alligator Lizard)				•	
velazquezi (Central Baja Alligator Lizard)				•	
WORM LIZARDS					
BIPEDIDAE: Two-Footed Worm Lizards					
Bipes biporus (Five-Toed Worm Lizard)				•	
SNAKES					
TYPHLOPIDAE: Blind Snakes					
Rhamphotyphlops braminus (Asian Blind Snake)*					•
LEPTOTYPHLOPIDAE: Blind Snakes					
Leptotyphlops humilis (Western Blind Snake)		•			
BOIDAE: Boas and Pythons					
Lichanura trivirgata (Rosy Boa)		•			
COLUBRIDAE: Colubrid Snakes					
Arizona (Glossy Snakes)					
elegans (Glossy Snake)		•			
pacata (Baja California Glossy Snake)				•	
Bogertophis rosaliae (Baja California Rat Snake)				•	
Chilomeniscus (Sand Snakes)					
savagei (Isla Cerralvo Sand Snake)				▲	
stramineus (Sand Snake)		•			
Chionactis occipitalis (Western Shovel-Nosed Snake)		•			
Diadophis punctatus (Ringneck Snake)		•			
Eridiphas (Baja California Night Snakes)					
marcosensis (Isla San Marcos Night Snake)				▲	
slevini (Slevin's Night Snake)				•	
Hypsiglena (Night Snakes)					
gularis (Isla Partida Norte Night Snake)				▲	
torquata (Night Snake)		•			

* Known from only one specimen.

(continued)

Table 1, *continued*

	Shared only with U.S.	Shared with U.S. and Mainland Mexico	Shared only with Mainland Mexico	Endemic (▲ = insular endemic)	Introduced
Lampropeltis (Kingsnakes)					
catalinensis (Isla Santa Catalina Kingsnake)				▲	
getula (Common Kingsnake)		•			
herrerae (Islas Todos Santos Mountain Kingsnake)				▲	
zonata (California Mountain Kingsnake)	•				
Masticophis (Whipsnakes and Racers)					
aurigulus (Cape Striped Racer)				•	
barbouri (Isla Espíritu Santo Striped Racer)				▲	
bilineatus (Sonoran Whipsnake)			•		
flagellum (Coachwhip)		•			
fuliginosus (Baja California Coachwhip)				•	
lateralis (California Striped Racer)	•				
slevini (Isla San Esteban Whipsnake)				▲	
Phyllorhynchus decurtatus (Spotted Leaf-Nosed Snake)		•			
Pituophis (Bull, Gopher, and Pine Snakes)					
catenifer (Gopher Snake)	•				
insulanus (Isla Cedros Gopher Snake)				▲	
vertebralis (Baja California Gopher Snake)				•	
Rhinocheilus (Long-Nosed Snakes)					
etheridgei (Isla Cerralvo Long-Nosed Snake)				▲	
lecontei (Long-Nosed Snake)		•			
Salvadora hexalepis (Western Patch-Nosed Snake)		•			
Sonora semiannulata (Ground Snake)		•			
Tantilla planiceps (Western Black-Headed Snake)	•				
Thamnophis (Garter Snakes)					
elegans (Western Terrestrial Garter Snake)	•				
hammondii (Two-Striped Garter Snake)				•	
marcianus (Checkered Garter Snake)		•			
validus (West Coast Garter Snake)			•		
Trimorphodon biscutatus (Lyre Snake)		•			
ELAPIDAE: Coral Snakes					
Micruroides euryxanthus (Western Coral Snake)		•			
VIPERIDAE: Vipers and Pit Vipers					
Crotalus (Rattlesnakes)					
angelensis (Isla Ángel de la Guarda Rattlesnake)				▲	
atrox (Western Diamondback Rattlesnake)		•			
caliginis (Islas Coronado Rattlesnake)				▲	
catalinensis (Isla Santa Catalina Rattlesnake)				▲	
cerastes (Sidewinder)		•			
enyo (Baja California Rattlesnake)				•	
estebanensis (Isla San Esteban Rattlesnake)				▲	
lorenzoensis (Isla San Lorenzo Rattlesnake)				▲	
mitchellii (Speckled Rattlesnake)	•				
molossus (Black-Tailed Rattlesnake)			•		
muertensis (Isla El Muerto Rattlesnake)				▲	
ruber (Red Diamond Rattlesnake)				•	
tigris (Tiger Rattlesnake)		•			
tortugensis (Isla Tortuga Rattlesnake)				▲	
viridis (Western Rattlesnake)	•				

southern Baja California as the peninsula was separating from mainland Mexico during the Miocene and is referred to as the Southern Miocene Vicariant Complex. All species (or monophyletic groups of endemic species) of this complex have close relatives in southwestern Mexico. Additionally, because of the Cape Region's intermittent isolation from the more northerly portions of the peninsula, many of these forms have distinctive pattern classes, or in some cases sister species, in the southern portion of the peninsula, whose contact zones are generally situated around the Isthmus of La Paz. Among the fossil species known from the Cape Region are the large land tortoise (*Geochelone* sp.), boa constrictor (*Boa* sp.), and crocodile (*Crocodylus* sp.).

Many species were widespread in northern Baja California, southwestern United States, and northwestern Mexico when northern Baja California existed as a western extension of mainland Mexico. Eventually the distribution of these forms was bisected by the northern extension of the newly forming proto–Gulf of California in the late Miocene to early Pliocene, when it extended at least as far north as Banning Pass in Riverside County, California. This separation prompted the formation of closely related forms on either side of the northern portion of the Gulf of California. When the Gulf finally receded in the Pleistocene, these sister taxa came into contact with one another and are now parapatric near the head of the Gulf of California or occur on opposite sides of the Coachella Valley. This group is referred to as the Northern Pliocene Vicariant Complex.

Another group, composed of species that are currently distributed continuously around the head of the Gulf of California in close association with the Sonoran, Mojave, and Great Basin desert, is referred to as the Western Desert Complex. Presumably these species evolved outside Baja California in association with the formation of these deserts and entered the region from the northeast after the regression of the Sea of Cortés and the union of the southern peninsular regions. As a result, their geographic variation on the peninsula is generally weak and clinal, and their closest relatives occur in northern mainland Mexico. A portion of this group includes species that are situated at the head of the Gulf of California and only marginally enter Baja California. These species are closely associated with the low-lying, sandy regions left behind after the regression of the Gulf or the flora thereon and have not dispersed further south into Baja California. A few species with diverse origins outside Baja California are restricted to its northeast portion. These forms probably entered Baja California after the regression of the Gulf. These are strictly associated with the Río Colorado, its drainages, or its flood-plain vegetational associates and are referred to as the Río Colorado Group.

Similarly, a group of mesophilic forms occurring in northwestern Baja California appears to have entered from North America. These are referred to as the Northwestern Complex. The dispersal of these forms into Baja California may have been facilitated by the uplift of coastal mountains in southern California during the Pleistocene. This uplift may have forced the Gulf to recede to its present position and provided a mesic corridor through which these species could disperse southward. The fact that they generally dispersed no further south than El Rosario was likely the result of the simultaneous formation of the arid Vizcaíno Region, which would have served as an effective xeric barrier to the dispersal of these mesophilic forms, just as it does today. The closest relatives of these forms are found in the Pacific Northwest and central United States.

Another group of mesophilic species is restricted to cismontane and montane woodland mesic environments of northern Baja California. These species were presumably widespread throughout North America during cooler and wetter periods, but with the onset of Pleistocene drying trends became fragmented and restricted to higher elevations where cool and moist climates still prevail. The closest relatives of these species occur in the northern portions of the Sierra Madre Occidental and the mountains of central Arizona. This group is referred to as the Chaparral-Madrean Woodland Complex.

One group of species is ubiquitous throughout Baja California, the United States, and much of Mexico, along with its closest living relatives. Because these species are so widespread, it is exceedingly difficult to form viable hypotheses concerning their historical origin in Baja California. Finally, there is one other group of species whose relationships to other species remain obscure and of whose distribution our knowledge is incomplete. Before any reasonable hypotheses can be made regarding its evolutionary origin in Baja California, some of these issues must be more fully addressed.

ECOLOGICAL BIOGEOGRAPHY

Although some species of the Baja California herpetofauna are ubiquitous in distribution, many others can be placed into loosely-defined ecological groups in accordance with their general natural history and distribution (Grismer 1994b; table 3). Thus, unlike historical biogeography, ecological biogeography (or ecogeography) groups species according to contemporary environmental factors and is not concerned with their geographic or phylogenetic history.

Table 2
Historical Biogeographical Complexes of the Native Herpetofauna of the Baja California Peninsula

	Southern Miocene Vicariant Complex	Northern Pliocene Vicariant Complex	Western Desert Complex	Río Colorado Group	Northwestern Complex	Chapparral-Madrean Woodland Complex	Ubiquitous Group	Placement Uncertain
SALAMANDERS								
PLETHODONTIDAE								
Aneides lugubris					•			
Batrachoseps major					•			
Ensatina								
eschscholtzii					•			
klauberi					•			
FROGS AND TOADS								
BUFONIDAE								
Bufo								
alvarius				•				
boreas								•
californicus						•		
cognatus				•				
punctatus							•	
woodhousii				•				
HYLIDAE								
Hyla								
cadaverina					•			
regilla					•			
RANIDAE								
Rana								
aurora					•			
boylii					•			
yavapaiensis								•
PELOBATIDAE								
Scaphiopus couchii							•	
Spea hammondii					•			
TURTLES AND TORTOISES								
EMYDIDAE								
Clemmys marmorata					•			
Trachemys nebulosa	•							
TESTUDINIDAE								
Gopherus agassizii			•					
LIZARDS								
CROTAPHYTIDAE								
Crotaphytus								
grismeri		•						
vestigium		•						
Gambelia								
copeii		•						
wislizenii		•						

	Southern Miocene Vicariant Complex	Northern Pliocene Vicariant Complex	Western Desert Complex	Rio Colorado Group	Northwestern Complex	Chapparral-Madrean Woodland Complex	Ubiquitous Group	Placement Uncertain
IGUANIDAE								
Ctenosaura hemilopha	•							
Dipsosaurus dorsalis			•					
Sauromalus obesus			•					
PHRYNOSOMATIDAE								
Callisaurus draconoides			•					
Petrosaurus								
mearnsi								•
repens	•							
thalassinus	•							
Phrynosoma								
coronatum					•			
mcallii		•						
platyrhinos		•						
Sceloporus								
hunsakeri	•							
licki	•							
magister		•						
occidentalis					•			
orcutti	•							
vandenburgianus					•			
zosteromus		•						
Uma notata			•					
Urosaurus								
graciosus			•					
lahtelai	•							
nigricaudus	•							
Uta stansburiana							•	
EUBLEPHARIDAE								
Coleonyx								
switaki								•
variegatus		•						
GEKKONIDAE								
Phyllodactylus								
unctus	•							
xanti	•							
TEIIDAE								
Cnemidophorus								
hyperythrus	•							
labialis						•		
tigris								•
XANTUSIIDAE								
Xantusia								
henshawi								•
vigilis								•

(continued)

Table 2, *continued*

	Southern Miocene Vicariant Complex	Northern Pliocene Vicariant Complex	Western Desert Complex	Rio Colorado Group	Northwestern Complex	Chapparral-Madrean Woodland Complex	Ubiquitous Group	Placement Uncertain
SCINCIDAE								
Eumeces								
gilberti						•		
lagunensis	•							
skiltonianus	•							
ANGUIDAE								
Anniella								
geronimensis								•
pulchra								•
Elgaria								
multicarinata					•			
paucicarinata	•							
velazquezi	•							
WORM LIZARDS								
BIPEDIDAE								
Bipes biporus	•							
SNAKES								
LEPTOTYPHLOPIDAE								
Leptotyphlops humilis								•
BOIDAE								
Lichanura trivirgata								•
COLUBRIDAE								
Arizona								
elegans			•					
pacata			•					
Bogertophis rosaliae	•							
Chilomeniscus								
stramineus		•						
Chionactis occipitalis			•					
Diadophis punctatus						•		
Eridiphas slevini	•							
Hypsiglena torquata								•
Lampropeltis								
getula							•	
zonata						•		
Masticophis								
aurigulus								•
flagellum		•						
fuliginosus		•						
lateralis								•
Phyllorhynchus decurtatus			•					

	Southern Miocene Vicariant Complex	Northern Pliocene Vicariant Complex	Western Desert Complex	Rio Colorado Group	Northwestern Complex	Chapparral-Madrean Woodland Complex	Ubiquitous Group	Placement Uncertain
Pituophis								
catenifer					•			
vertebralis	•							
Rhinocheilus lecontei								•
Salvadora hexalepis			•					
Sonora semiannulata		•						
Tantilla planiceps	•							
Thamnophis								
elegans					•			
hammondii					•			
marcianus				•				
validus	•							
Trimorphodon biscutatus		•						
VIPERIDAE								
Crotalus								
atrox			•					
cerastes								•
enyo	•							
mitchellii								•
ruber		•						

Source: Adapted from Grismer 1994a.

Looking at the herpetofauna in terms of its ecological biogeography is useful because it serves to underscore the diversity of this region and the wide array of adaptive types that inhabit it.

There is a Northwestern Mesophilic Group containing cismontane species whose distributions closely follow or are contained within the California Region. These species avoid the more arid portions of the peninsula and thus generally range no further south than the vicinity of El Rosario. A distinctive Northern Montane Group occurs in the Baja California Coniferous Forest Region of the Sierra Juárez and Sierra San Pedro Mártir. Some of the members of this group are endemic to these mountains, and all represent the fragmented southern distribution of more widely ranging northerly species or closely related forms. A Northeastern Xerophilic Group of species occupies northeastern Baja California in the Lower Colorado Valley Region, east of the Peninsular Ranges, and generally extends no further south than Bahía San Luis Gonzaga. Most of the members of this group have various specializations for living in the open, sandy, and arid extremes of this area. Another group, whose distribution in Baja California is contained entirely within the Lower Colorado Valley Region, is the Colorado River Riparian Group. Species of this group enter Baja California along the Río Colorado and its tributaries. Some are strictly aquatic; others find shelter and food in the vegetation along the watercourses or irrigation canals. None of the members of this group would occur in Baja California if it were not for these water sources. A Northern Transpeninsular Mesophilic Group contains species that are widespread in the mesic California and Baja California Coniferous Forest regions of the north; but as they extend south into the more arid regions of the peninsula, their distributions become highly fragmented and restricted to mesic refugia such as springs, oases, and mountaintops (Grismer 1988a, 1990a; Grismer and McGuire 1993; Grismer and Mellink 1994; Grismer et al. 1994; McGuire and Grismer 1993). Those that extend into the Cape Region become more widespread in distribution once again in the more mesic Arid Tropical and Sierra la Laguna regions. A similar but more widely distributed Transpeninsular Xerophilic Group ranges throughout the arid Lower Colorado Valley, Vizcaíno, Central Gulf Coast, and Magdalena regions of the peninsula. Members of this group

Table 3

Ecological Biogeographical Groups of the Native Herpetofauna of the Baja California Peninsula

	Northwestern Mesophilic Group	Northern Montane Group	Northeastern Xerophilic Group	Río Colorado Riparian Group	Northern Transpeninsular Mesophilic Group	Transpeninsular Xerophilic Group	Saxicolous Group	Southern Mesophilic Group	Cape Region Group	Southern Xerophilic Group	Ubiquitous Group	Placement Uncertain
SALAMANDERS												
PLETHODONTIDAE												
Aneides lugubris	•											
Ensatina												
eschscholtzii	•											
klauberi		•										
FROGS AND TOADS												
BUFONIDAE												
Bufo												
alvarius				•								
boreas	•											
californicus	•											
cognatus				•								
punctatus						•						
woodhousii				•								
HYLIDAE												
Hyla												
cadaverina	•											
regilla					•							
RANIDAE												
Rana												
aurora	•											
boylii		• (?)										
yavapaiensis				• (?)								
PELOBATIDAE												
Scaphiopus couchii						•						
Spea hammondii	•											
TURTLES AND TORTOISES												
EMYDIDAE												
Clemmys marmorata	•											
Trachemys nebulosa								•				

	Northwestern Mesophilic Group	Northern Montane Group	Northeastern Xerophilic Group	Río Colorado Riparian Group	Northern Transpeninsular Mesophilic Group	Transpeninsular Xerophilic Group	Saxicolous Group	Southern Mesophilic Group	Cape Region Group	Southern Xerophilic Group	Ubiquitous Group	Placement Uncertain
LIZARDS												
CROTAPHYTIDAE												
Crotaphytus												
grismeri			•									
vestigium							•					
Gambelia												
copeii										•		
wislizenii			•									
IGUANIDAE												
Ctenosaura												
hemilopha										•		
Dipsosaurus												
dorsalis						•						
Sauromalus obesus							•					
PHRYNOSOMATIDAE												
Callisaurus												
draconoides						•						
Petrosaurus							•					
Phrynosoma												
coronatum												•
mcallii			•									
platyrhinos			•									
Sceloporus												
hunsakeri									•			
licki									•			
magister			•									
occidentalis	•											
orcutti							•					
vandenburgianus		•										
zosteromus										•		
Uma notata			•									

(continued)

Table 3, *continued*

	Northwestern Mesophilic Group	Northern Montane Group	Northeastern Xerophilic Group	Río Colorado Riparian Group	Northern Transpeninsular Mesophilic Group	Transpeninsular Xerophilic Group	Saxicolous Group	Southern Mesophilic Group	Cape Region Group	Southern Xerophilic Group	Ubiquitous Group	Placement Uncertain
Urosaurus												
graciosus			•									
lahtelai							•					
nigricaudus							•					
Uta stansburiana											•	
EUBLEPHARIDAE												
Coleonyx												
switaki							•					
variegatus											•	
GEKKONIDAE												
Phyllodactylus												
unctus									•			
xanti							•					
TEIIDAE												
Cnemidophorus												
hyperythrus											•	
labialis	•											
tigris											•	
XANTUSIIDAE												
Xantusia												
henshawi	•											
vigilis						•						
SCINCIDAE												
Eumeces												
gilberti	•											
lagunensis								•				
skiltonianus	•											
ANGUIDAE												
Anniella												
geronimensis	•											
pulchra	•											

	Northwestern Mesophilic Group	Northern Montane Group	Northeastern Xerophilic Group	Río Colorado Riparian Group	Northern Transpeninsular Mesophilic Group	Transpeninsular Xerophilic Group	Saxicolous Group	Southern Mesophilic Group	Cape Region Group	Southern Xerophilic Group	Ubiquitous Group	Placement Uncertain
Elgaria												
cedrosensis	• (?)											
multicarinata	•											
paucicarinata									•			
velazquezi								•				
WORM LIZARDS												
BIPEDIDAE												
Bipes biporus								•				
SNAKES												
LEPTOTYPHLOPIDAE												
Leptotyphlops												
humilis											•	
BOIDAE												
Lichanura trivirgata							•					
COLUBRIDAE												
Arizona												
elegans			•									
pacata										•		
Bogertophis rosaliae							•					
Chilomeniscus												
stramineus						•						
Chionactis occipitalis			•									
Diadophis punctatus	•											
Eridiphas slevini							•					
Hypsiglena torquata											•	
Lampropeltis												
getula											•	
zonata		•										
Masticophis												
aurigulus									•			
flagellum			•									
fuliginosus											•	
lateralis					•							

(continued)

Table 3, *continued*

	Northwestern Mesophilic Group	Northern Montane Group	Northeastern Xerophilic Group	Río Colorado Riparian Group	Northern Transpeninsular Mesophilic Group	Transpeninsular Xerophilic Group	Saxicolous Group	Southern Mesophilic Group	Cape Region Group	Southern Xerophilic Group	Ubiquitous Group	Placement Uncertain
Phyllorhynchus												
decurtatus						•						
Pituophis												
catenifer	•											
vertebralis										•		
Rhinocheilus lecontei			•									
Salvadora hexalepis											•	
Sonora semiannulata						•						
Tantilla planiceps					•							
Thamnophis												
elegans		•										
hammondii					•							
marcianus				•								
validus									•			
Trimorphodon												
biscutatus							•					
VIPERIDAE												
Crotalus												
atrox			•									
cerastes			•									
enyo										•		
mitchellii							•					
ruber											•	
viridis	•											

Source: Adapted from Grismer 1994b.

are conspicuously absent from the mesic California and Baja California Coniferous Forest regions of the northwest and, to a large extent, the Sierra la Laguna Region of the Cape. There is also a Saxicolous Group that is generally restricted to the more arid, rocky areas of the Peninsular Ranges. Members of this group usually range throughout many phytogeographic regions and show little geographic variation. The limiting factor of their distribution seems to be the presence of rock, and the species of this group are usually endemic to Baja California. There is a Southern Xerophilic Group that is also composed entirely of endemic species. Its members are widespread in the more southern, xeric regions of the peninsula but do not enter the Lower Colorado Valley Region of the northeast or the northern mesic areas. These species generally do not range further north than El Rosario in the west and Bahía de los Ángeles in the east. A Southern Mesophilic Group, which is similar in ecology to the Northern Transpeninsular Mesophilic Group, is composed of endemic species that are widespread in the mesic Arid Tropical and Sierra la Laguna regions of the Cape Region. As these species extend north into the more arid central regions of the peninsula, their distributions become fragmented and restricted to springs, oases, and mountaintops. Last, there is a Cape Region Group composed of endemic species that are confined to the mesic Arid Tropical and the Sierra la Laguna regions south of the Isthmus of La Paz.

DISPERSAL CORRIDORS, DISPERSAL BARRIERS, AND HABITAT ISLANDS

The distribution limits of any reproductively cohesive group of organisms or any population result from the interaction between the population's genetic limitations and the ecology of the region within which it occurs. For example, organisms that are not genetically adapted for living in deserts usually do not do so, and the distribution limits of the population to which they belong often track the fringes of adjacent desert environments. Such relationships between populations and their environments are most obvious in areas where geographically proximate environmental extremes result from interactions between major physiographic features and climate. This principle is eloquently demonstrated in the general ecogeography of the herpetofauna of Baja California and even more so in general trends involving dispersal barriers, dispersal corridors, and habitat islands.

Dispersal corridors are generally localized regions whose physiography and ecology allow certain groups of species access into regions where they would not occur without such corridors. Five such corridors occur in northern Baja California (see map 1). The first is an east to west–tending depression in the Peninsular Ranges known as Paso de San Matías. This is a narrow, low-elevation pass, approximately 0.5 km wide, 980 m in elevation, and 6.5 km long. This pass separates the Sierra Juárez in the north, which reaches an elevation of 1,800 m, from the Sierra San Pedro Mártir in the south, which reaches 3,007 m. Paso de San Matías provides a corridor through which some desert species extend west into the coastal California Region (Grismer 1994a, 1997; Grismer and McGuire 1996; Welsh and Bury 1984), whereas, in other parts of their distribution, they are unable to cross the Peninsular Ranges. Species that are generally restricted to arid regions but are filtering through Paso de San Matías into the cooler, more mesic coastal areas of Valle la Trinidad, immediately to the west of Paso de San Matías, include the Desert Iguana *(Dipsosaurus dorsalis),* Zebra-Tailed Lizard *(Callisaurus draconoides),* Long-Nosed Leopard Lizard *(Gambelia wislizenii),* Banded Rock Lizard *(Petrosaurus mearnsi),* Desert Spiny Lizard *(Sceloporus magister),* Desert Night Lizard *(Xantusia vigilis),* Shovel-Nosed Snake *(Chionactis occipitalis),* and Ground Snake *(Sonora semiannulata).* Interestingly, however, there appear to be no coastal species extending east onto the desert floor. Apparently desert environments exceed the physiological capabilities of most cismontane species, although coastal environments do not exceed those of some desert species.

Another dispersal corridor, known as Valle San José, lies between the Sierra San Miguel to the west and the Sierra San Pedro Mártir to the east. This is a northwest to southeast–tending valley approximately 47 km long. Its northern end is confluent with Valle la Trinidad, and its southern end opens onto the Vizcaíno Region near the southern terminus of the Sierra San Pedro Mártir. The westerly Sierra San Miguel blocks the cooling maritime influence of the Pacific Ocean, leaving the valley hot and arid. This allows a northerly invasion of the arid-adapted Vizcaíno Region flora (Shreve 1936). It also enables arid-adapted species of reptiles common to the Vizcaíno Region to extend northward, west of the Peninsular Ranges into the California Region (Welsh 1988; Grismer 1997). Such species include the Baja California Leopard Lizard *(Gambelia copeii),* Baja California Collared Lizard *(Crotaphytus vestigium),* Zebra-Tailed Lizard *(Callisaurus draconoides),* Desert Night Lizard *(Xantusia vigilis),* Sand Snake *(Chilomeniscus stramineus),* and Baja California Gopher Snake *(Pituophis vertebralis).* Some of these species extend north all the way into Valle la Trinidad (e.g., *C. draconoides, G. copeii,* and *P. vertebralis*), and one continues into the United States through similar habitat *(G. copeii).*

Three other major dispersal corridors in northern Baja California are particularly significant to saxicolous (rock-dwelling) species. The first is the Sierra San Felipe. This large, rocky mountain range, extending some 70 km from the northern end of Laguna Diablo in the north to the Sierra Santa Isabel to the south, reaches over 1,200 m elevation. The Sierra San Felipe lies 18 km to the east of the Sierra San Pedro Mártir, the main block of the Peninsular Ranges, across the dry lake known as Laguna Diablo. A connection with the southern end of the Sierra San Pedro Mártir, by way of a complex network of rocky canyons and low mountain ranges generally referred to as the Sierra Santa Isabel, provides a dispersal corridor through which saxicolous Peninsular Ranges species can extend their distribution eastward into the Lower Colorado Valley Region. Species utilizing this corridor are the Baja California Collared Lizard *(Crotaphytus vestigium),* Northern Chuckwalla *(Sauromalus obesus),* Banded Rock Lizard *(Petrosaurus mearnsi),* Granite Spiny Lizard *(Sceloporus orcutti),* Peninsular Leaf-Toed Gecko *(Phyllodactylus xanti),* Rosy Boa *(Lichanura trivirgata),* Lyre Snake *(Trimorphodon biscutatus),* and Speckled Rattlesnake *(Crotalus mitchellii).*

The remaining two dispersal corridors in northern Baja California are the Sierra las Tinajas and the Sierra las Pintas. Both ranges extend from a common base on the southern end of the Sierra Juárez, just north of Paso de San Matías, and fan out to the northeast like two large fingers. The Sierra las Pintas is the larger and more easterly of the two. Like the Sierra San Felipe, these rocky, volcanic, mountain ranges extend east into the Lower Colorado Valley Region and allow certain saxicolous species to extend their distributions. Saxicolous species known from these ranges are the Baja California Collared Lizard *(Crotaphytus vestigium),* Northern Chuckwalla *(Sauromalus obesus),* Banded Rock Lizard *(Petrosaurus mearnsi),* Barefoot Banded Gecko *(Coleonyx switaki),* Peninsular Leaf-Toed Gecko *(Phyllodactylus xanti),* Rosy Boa *(Lichanura trivirgata),* and Speckled Rattlesnake *(Crotalus mitchellii).* One can see that these ranges are depauperate relative to the Sierra San Felipe, and potential colonizers still occur in the Sierra Juárez. Noticeably absent are the Granite Spiny Lizard *(Sceloporus orcutti)* and Lyre Snake *(Trimorphodon biscutatus).*

Dispersal barriers prevent populations from extending their distributions from one locale to an adjacent one. They are usually the result of major physiographic features such as mountain ranges, but they may also be zones of ecological transition from one environmental region into another as a result of climate. An example of the former would be the northern Peninsular Ranges, which prevent species from dispersing from the Lower Colorado Valley Region into the California Region simply because they are unable to cross the mountains. An example of an environmental barrier created by an ecological transition occurs in the cismontane region between San Quintín and El Rosario. Here, although there are no large physiographic features that would prevent the southern dispersal of many of the more northerly-occurring species of northwestern Baja California, a number of these species range no further south than El Rosario. This is because the degree of aridity in this region begins to exceed their physiological capabilities and has altered the ecology of the environment. There are also some less obvious physiographic dispersal barriers within climatically homogeneous regions that play a significant role in the local distribution of many species. One such barrier is a large volcanic flow in the Lower Colorado Valley Region between Puertecitos and El Huerfanito. This flow, which is approximately 28 km wide and of relatively recent origin, bisects what would otherwise be a generally flat, narrow, sandy coastal desert plain extending south from San Felipe to Bahía San Luis Gonzaga. Its presence limits the southern distribution of some widely distributed species of the Lower Colorado Valley Region to the vicinity of Puertecitos. These include the Long-Nosed Leopard Lizard *(Gambelia wislizenii),* Desert Spiny Lizard *(Sceloporus magister),* Glossy Snake *(Arizona elegans),* and Western Shovel-Nosed Snake *(Chionactis occipitalis).* It also limits the northern distribution of southern species along the east coast of the peninsula to the vicinity of El Huerfanito (e.g., *Gambelia copeii*).

Similarly, the continuity and volcanic nature of the southern extent of the Sierra Columbia, the Sierra la Libertad, the Sierra Guadalupe, and the Sierra la Giganta in central Baja California may limit many lowland species to areas west of the Peninsular Ranges despite the existence of suitable habitat along the east coast of the peninsula. The Sierra Columbia, which contacts the west coast of the peninsula at approximately 17 km north of Jesús María, marks the northern boundary of the Vizcaíno Desert. From here, this range arcs southwest toward the interior of the peninsula, contacting the Sierra la Libertad and forming the eastern boundary of the Vizcaíno Desert. The Sierra la Libertad continues south as a generally unbroken, rocky barrier that merges into the Sierra Guadalupe just southeast of San Ignacio. The Sierra Guadalupe, in turn, grades into the Sierra la Giganta, which continues south to the northern edge of the Isthmus of La Paz. This mountainous barrier delimits the southern edge of the Vizcaíno Desert and the northern and eastern boundaries of the Magdalena Plain. At the Isthmus of La Paz, the Sierra la Giganta gently fades into the Magdalena Plain, which then extends east across the peninsula, contacting the Gulf coast at Bahía de la Paz. These mountains may act as a dispersal barrier to certain species, denying them access to the low-lying Central Gulf Coast Region to the east. Species whose distributions are generally restricted to the Vizcaíno Desert and Magdalena Plain and only reach the Gulf coast through the Isthmus of La Paz are the Five-Toed Worm Lizard *(Bipes biporus),* the Baja California Leopard Lizard *(Gambelia copeii),* and the Baja California Glossy Snake *(Arizona pacata).*

Ever since the legendary works of Alfred Russel Wallace and Charles Darwin, islands have held a special place of reverence for evolutionary biologists. Their isolation and general inaccessibility often preserves them as natural laboratories for evolutionary studies. When we think of islands we generally envision some landmass surrounded by water, comprising a fortuitous array of species that happened to disperse over water from an adjacent continental source. This is a limited conception, however, because there are other types of islands that are *not* surrounded by water and may be far more isolated than those that are. These are known as habitat islands. A habitat island is a localized region of habitat within, and surrounded by, a much broader area of a differing habitat. Examples include permanent water holes in the desert that maintain a streamside flora and an array of aquatic fauna. Unlike the case of an island surrounded by water, where colonization through overwater dispersal is a *passive,* ongoing (albeit slow) phenomenon, in habitat islands many species lack dispersal opportunities and are truly isolated. The reason is simple. For organisms to reach a habitat island they would have to actively disperse across uninhabitable terrain, something of which they are not biologically capable. It is difficult to envision a small frog hopping across a wide expanse of desert to reach an isolated water source. However, being washed from the side of a riverbank during a storm and floating out to an island at sea on vegetation is more plausible. Therefore, habitat islands differ from other islands in the way they maintain their species diversity. Islands surrounded by water have been shown repeatedly to be dynamic and evolving communities regulated by the colonization and extinction of species (MacArthur and Wilson 1967). Habitat islands, in contrast, are more like museums whose exhibits (species) are relics of bygone environments. In other words, many of the species occupying habitat islands were always there, but the habitat has shrunk around them. Habitat islands are extremely important to evolutionary biologists and paleoecologists because they serve as living windows through which we can look into the past and glimpse the organisms that lived in an area during a previous ecological episode.

There are three major types of habitat island in Baja California. The first is formed by rocky mountain ranges surrounded by broad, flat expanses of low-lying, sandy desert. These ranges offer refuge for saxicolous species, and the surrounding desert prevents immigration and colonization from adjacent rocky areas. There are three such mountain ranges in Baja California. In the northeastern portion of the peninsula, just south of the U.S.-Mexican border, lie the Sierra los Cucapás and Sierra el Mayor. These form an extensive range of mountains, approximately 80 km long, bordered by the Río Colorado flood plain on the east and Laguna Salada on the west. This poorly explored mountain range harbors a depauperate, saxicolous herpetofauna containing the endemic Sierra los Cucapás Collared Lizard *(Crotaphytus grismeri),* Northern Chuckwalla *(Sauromalus obesus),* Speckled Rattlesnake *(Crotalus mitchellii),* and perhaps others (McGuire 1994). In west-central Baja California, two closely situated mountain ranges, the Sierra Vizcaíno and the Sierra Santa Clara, define the Vizcaíno Peninsula. The Sierra Vizcaíno lies northwest of the Sierra Santa Clara and extends southeast from near Punta Eugenia approximately 115 km into the Vizcaíno Desert. This mountain range is very poorly explored, and the Northern Chuckwalla *(Sauromalus obesus)* is the only saxicolous species currently known to occur there (Grismer et al. 1994). The Sierra Santa Clara, located approximately 50 km to the southeast, consists of seven closely spaced and eroding volcanic peaks. This system is known to harbor the Northern Chuckwalla *(Sauromalus obesus),* Night Snake *(Hypsiglena torquata),* and Lyre Snake *(Trimorphodon biscutatus).* Conspicuously absent from all three of these mountain ranges are the common and widespread Central Baja California Banded Rock Lizard *(Petrosaurus repens),* Granite Spiny Lizard *(Sceloporus orcutti),* and Peninsular Leaf-Toed Gecko *(Phyllodactylus xanti).* Their absence may be explained by the fact that these ranges never had a rocky connection with the adjacent Peninsular Ranges. It is likely that the area was colonized by saxicolous species through chance overwater dispersals during a period when they were surrounded by water. The former three species just happened not to disperse (Grismer et al. 1994; McGuire 1994).

The second form of habitat island harbors isolated populations of species on high mountaintops. In northern Baja California, examples include the upper elevations of the Sierra Juárez and Sierra San Pedro Mártir. These mountains extend approximately 160 km south of the U.S.-Mexican border, and the Sierra San Pedro Mártir reaches over 3,000 m elevation. Both are dominated by Baja California Coniferous Forest Region flora and offer refuge to some interesting montane species at the southernmost extent of their distributions. Both the Sierra Juárez and Sierra San Pedro Mártir maintain allopatric (isolated) populations of the California Mountain Kingsnake *(Lampropeltis zonata).* The Sierra San Pedro Mártir also maintains populations of the Large-Blotched Ensatina Salamander *(Ensatina klauberi),* Foothill Yellow-Legged Frog (*Rana boylii;* possibly extinct), Southern Sagebrush Lizard (*Sceloporus vandenburgianus*), and Western Terrestrial Garter Snake *(Thamnophis elegans)* (Mahrdt et al. 1999; Welsh 1988). The Sierra la Asamblea is another isolated mountaintop habitat island. This mountain range rises precipitously out of the arid Vizcaíno Region just north of Bahía de los Ángeles and reaches over 1,400 m elevation. Its upper elevations support a relict chaparral, manzanita, and piñon pine woodland in which a population of the Southern Alligator Lizard *(Elgaria multicarinata)* was discovered (Grismer and Mellink 1994). This mountaintop is accessible only by climbing and has been visited only twice by herpetologists. It is likely that additional relict cismontane species will be found during future visits. Last, the upper elevations of the Sierra la Laguna in the Cape Region of Baja California serve as a habitat island. The Sierra la Laguna extends generally from La Paz to Cabo San Lucas, reaches over 2,100 m, and maintains an endemic pine-oak woodland at its upper elevations. Endemic to this region is a population of the small, secretive, and peculiar Desert Night Lizard *(Xantusia vigilis).*

The last major type of habitat island in Baja California consists of widely scattered oases that extend fragmentedly for approximately 700 km from Misión San Fernando Velicatá in the north to at least the La Presa region in the south (Grismer and McGuire 1993). Up to fifteen thousand years before the present, Baja California was much more mesic than it is today, and many mesophilic species had transpeninsular or nearly transpeninsular distributions (Grismer 1994b; Welsh 1988), ranging widely throughout central Baja California. With the rapid onset of extreme aridification between fifteen and eight thousand years before the present, Baja California was transformed into an arid peninsula, and the only mesic habitat that remained was that near permanent sources of water. This reduced the transpeninsular distribution of many mesophilic species to being widespread and continuous only in northwestern Baja California and the Cape Region, where aridity is not a factor. In central Baja California, their distributions became fragmented and restricted to the mesic refugia of these relict oases. Today the oases range in size from extensive bodies of water, such as those at La Purísima, to small, deep pools no more than two meters across, such as that found at Las Palmas (Grismer and McGuire 1993). Nonetheless, all maintain relict populations of various combinations of species (Grismer and McGuire 1993). Species commonly found in the oases are the California Treefrog *(Hyla cadaverina),* Pacific Treefrog *(Hyla regilla),* Baja California Slider *(Trachemys nebulosa),* and Two-Striped Garter Snake *(Thamnophis hammondii).*

Conservation and Commercialization

Conservation issues are a rapidly growing concern for government agencies in Mexico. Several biosphere reserves have been created in key peninsular areas, and all the islands within the Gulf of California and along Baja California's Pacific coast are now protected. Because of the high degree of both peninsular and insular endemism and the monetary value that unfortunately accompanies such phenomena, the commercial market for this region's herpetofauna has grown at a staggering rate over the last few years. Even though this commercialization was started by just a few notorious reptile collectors and dealers trafficking in illegal wildlife, it has now become a serious threat to many species in the region of study.

Mellink (1995) notes some of the negative environmental effects that illegal reptile collecting has had in Baja California. For example, certain areas of the Sierra Juárez and Sierra San Pedro Mártir have been heavily degraded by collectors breaking apart rock piles in search of California Mountain Kingsnakes *(Lampropeltis zonata).* Similar devastation can be seen along Mexican Highway 1 as it passes through rocky areas in northern Baja California (such as Cataviña and Jaraguay), where collectors are looking primarily for Rosy Boas *(Lichanura trivirgata)* and banded rock lizards *(Petrosaurus).* The sheer numbers of specimens that have been removed from some areas are also alarming. I was told by two Mexican nationals from La Paz that they helped one American reptile dealer collect over one thousand San Lucan Banded Rock Lizards *(Petrosaurus thalassinus)* in the Cape Region for over two years. Many of these were sold, and others provided the parental stock for illegal captive propagation.

Commercialization of this herpetofauna threatens insular endemics most of all. For example, the Isla Todos Santos Mountain Kingsnake *(Lampropeltis herrerae)* is confined to a single, tiny island off the coast of Ensenada. Mellink (1995) reported finding snake traps on the island baited with live mice, and Anglos with pillowcases turning over rocks. I know of one collector who took a gravid female off the island. The long-term potential problems of removing gravid individuals from populations of species with extremely limited gene pools is obvious. The same problems affect islands in the Gulf of California. For example, when I saw several species of endemic rattlesnakes appearing on reptile price lists from Florida, I became very concerned. I talked to a Mexican national who participated in the collection efforts of the American who shipped those reptiles to Florida, and he told me that nearly one hundred snakes were collected. I have found abandoned pitfall traps on several islands filled with dead animals because the person who set them never came back to close them up. Mellink provides many more examples.

Despite the sanctimonious claims of many dealers, breeders, and collectors, unregulated captive propagation of illegally obtained animals has done nothing, and will do nothing, to protect against the potential loss or local extirpation of Baja California or Gulf of California amphibians and reptiles. Additionally, as Mellink notes, many individuals from different populations are crossed to produce more commercially valuable color phases, thus creating individuals with genetic constitutions and color patterns that do not even occur in the wild. Furthermore, the commercialization of this herpetofauna not only has created its own market but also has resulted in the environmental degradation of many areas (Mellink 1995). When people see the outrageous prices for which some species sell and realize that they are only a few hours or days away from areas where these species can be collected, some develop an urge to be adventurous and go catch their own.

Besides the fact that overcollecting, destroying microhabitats, and smuggling reptiles across the border into the United States is considered an illegal act in both countries, it is environmentally and ethically unconscionable, and it constitutes an act of aggression against our ecosystems. So the next time you see a price list with protected species on it or attend a reptile expo where Baja California and Gulf of California herps are being paraded for sale in hundreds of little plastic boxes, remember that the original populations from which these specimens came were most likely illegally taken (Mellink 1995). Then ask yourself if you want to be a part of, or cater to, that level of our society that abuses wildlife for profit.

Crotaphytus insularis (Isla Ángel de la Guarda Collared Lizard) from Puerto Refugio, Isla Ángel de la Guarda, BC

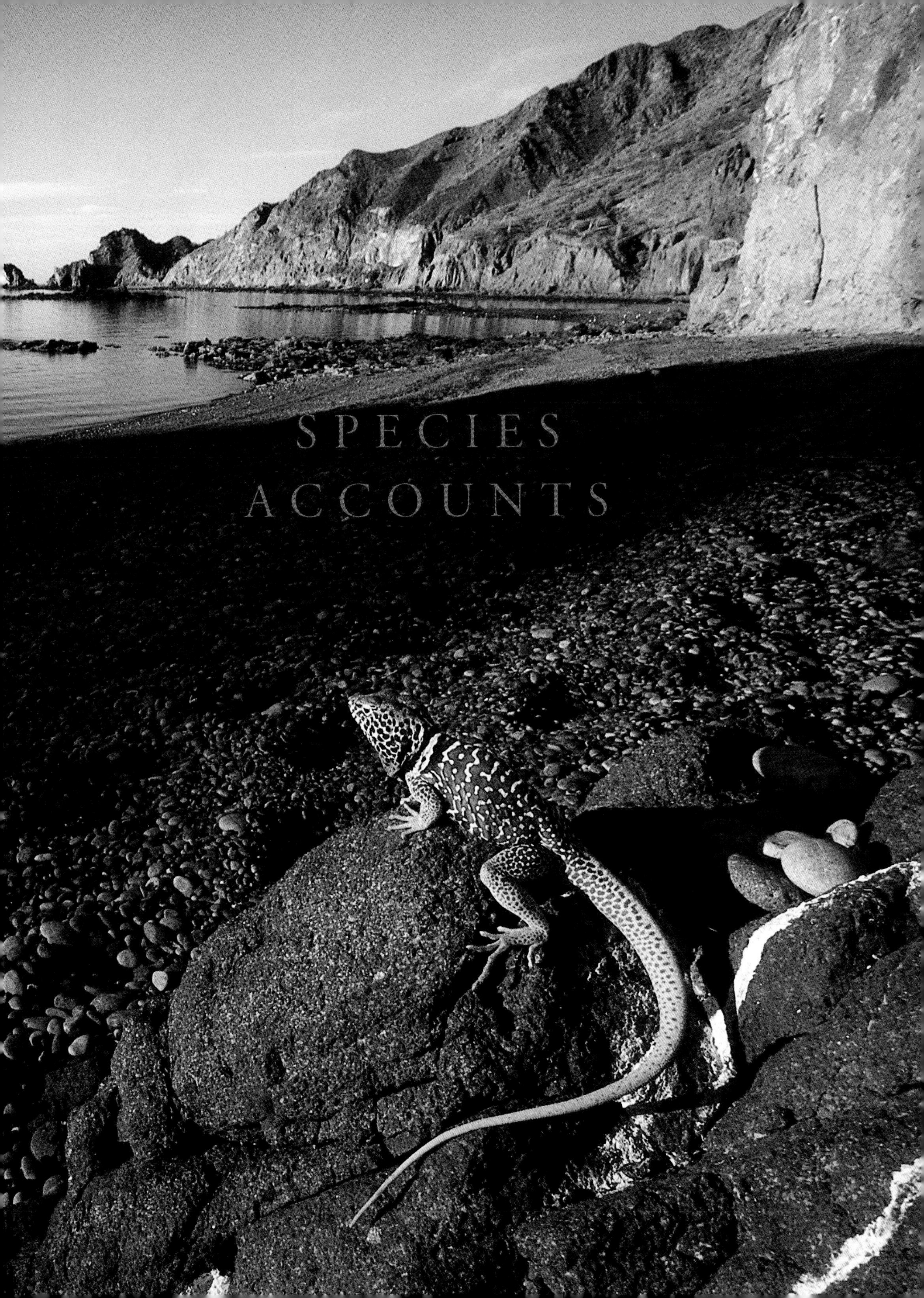

SPECIES ACCOUNTS

INTRODUCTION

Arrangement and Use of Species Accounts

The primary intent of this book is to serve as a guide to the identification, distribution, natural history, and taxonomy of each species of amphibian and reptile in Baja California, its associated Pacific islands, and the islands in the Gulf of California (hereafter referred to as the region of study). Throughout this book "Baja California" refers to the landmass of the Baja California peninsula from the U.S.-Mexican border south to Cabo San Lucas, and not to the geopolitical state of Baja California, which extends from the U.S.-Mexican border south to the 28th parallel. The state of Baja California is here referred to as BC, and the state of Baja California Sur, which encompasses the remainder of the peninsula south of the 28th parallel, is referred to as BCS.

This book is not intended to provide an exhaustive descriptive account or taxonomic history of each species, although it does provide information on the type description and type locality of every species. Museum acronyms follow Leviton et al. 1985. The taxonomic keys are written to allow the user to identify all species of amphibians, turtles, and snakes. The keys differentiate adult males of all species of lizards and the females and juveniles of nearly all species. Identification of females and juveniles can be refined by reading the descriptions in the species accounts. The scientific name, standardized common English name (Liner 1994, in part), and local Mexican common name are given for every species. In some cases the standardized common names are omitted because they make no sense. For example, the whiptail lizard *Cnemidophorus carmenensis* from Isla Carmen is referred to by Liner as the Carmen Island Orange-Throated Whiptail, even though its throat is blue. In such cases I provide an alternative common name. The local Mexican common names were determined from interviews with residents and indigenous people of Baja California and Sonora, and they vary from region to region. They do not follow the standardized Mexican names listed by Liner because such names are of little utility when making an inquiry of a Mexican layperson about species in the area. This is not to say that Liner's names are not useful, only that I favor the utility of the local vernacular. All genera are treated alphabetically under their respective headings (Frogs, Lizards, etc.). To reduce redundancy in the written descriptions, the species are treated in phylogenetic order where possible.

Included in every species account are sections titled Identification, Relationships and Taxonomy, Distribution, Description, and Natural History. The identification section provides a list of characters for each species, the combination of which allows the reader to distinguish a particular species from all other species in the region of study. The character list is not intended to provide a complete taxonomic identification or to serve in differentiating species from other species outside the region of study.

The distribution section provides a written description of the entire range of each species on the basis of museum records and personal observations. The descriptions for areas outside the region of study are general. A much more detailed description is given for distributions within the region of study and is referenced to a distribution map. Species and pattern classes are keyed to colors on the maps, with colored circles used for populations on the smaller islands as well as for isolated peninsular populations. Colored question marks denote unconfirmed populations; colored crosses denote extirpated populations.

The relationships and taxonomy section discusses what is currently hypothesized about the phylogenetic relationships and taxonomy of each species and provides references to the most recent systematic treatments in which older citations can be found.

The description section provides a description of each species, including standard measurements and counts commonly used in taxonomic studies, as well as more explicit morphological and color pattern characteristics used to diagnose each particular species in the region of study.

The natural history section focuses on the natural history of each species only as it pertains to the region of study. It is not intended to reproduce information already published in many fine field guides (e.g., Stebbins 1985) or research papers concerned with populations outside this book's regional coverage, unless stated otherwise. It is the natural history of this herpetofauna that is the least understood, and although much of the information presented here is anecdotal, I hope it will provide avenues for future research. (I would be delighted if I could say with accuracy that this area of study is in its infancy; but, excepting only a handful of species, it has not yet even been born.) Topics discussed include habitat preference, seasonal and daily activity periods, general behavior, diet, and reproductive biology. The majority of this information comes from personal field observations between 1975 and 2000 and scattered reports throughout the literature. Considerably more information about the diet and reproductive biology of this herpetofauna could be obtained by the examination of museum material. Unfortunately, such an endeavor exceeded the constraints of completing this book in a timely fashion. I strongly encourage others, however, to pick up where this book leaves off.

When appropriate, species accounts include additional sections. A section on geographic variation discusses notable geographic trends of more widely ranging species and species with insular populations. The Baja California peninsula extends 1,300 km and over ten degrees of latitude; the extensive geographic variation in this region's herpetofauna underscores the dynamic relationship between the environment and its inhabitants. When possible, these geographic trends are correlated with environmental parameters such as climate, physiography, and phytogeography.

A remarks section discusses miscellaneous topics believed to be pertinent and interesting, such as conservation, commercial exploitation, and anomalous distribution patterns.

A section on folklore and human use concerns the aboriginal and current myths surrounding particular species as well as a particular species' effect on the lives of local residents. Much of this information was gathered from interviews with the local inhabitants.

Taxonomy

The taxonomy (use and construction of scientific names) of the herpetofauna on islands in the Gulf of California and islands along the Pacific coast of Baja California, as applied by Grismer (1999b, 2001a) and used here, follows an evolutionary species concept (Wiley 1978; Frost and Hillis 1990; Grismer 2001a), and therefore no subspecies are formally recognized. This is not to say that subspecies designations are uninformative: in many cases they delimit the concordant geographic distribution of suites of morphological and color pattern characteristics within a widely distributed species, and they are often correlated with environmental parameters. Therefore, where useful in this context, the subspecific epithets are used as pattern classes here by capitalizing the first letter, removing them from italics, and using them singularly (*sensu* Frost 1995). A continental pattern class, therefore, represents a group of individuals from a defined geographic area who share one or more characteristics that allow us to differentiate them from other individuals of the same species from different geographic areas. Gene flow between individuals of different continental pattern classes occurs where their individuals approach one another. In such areas, individuals usually appear intermediate. It is this gene flow that differentiates a pattern class from a named, independent, evolutionary lineage such as a species. Additionally, insular subspecies that show trends in character variation that tend to differentiate them to a certain degree from their peninsular counterparts (although not differentiating them discretely; i.e., the range of character variation overlaps between the insular and peninsular populations) are treated as insular pattern classes. Discrete diagnosability is taken as the cutoff level of lineage identity, under the assumption that it is unequivocal evidence of an absence of gene flow (Davis and Nixon 1992; see Grismer 1999b and 2001a for application). Recognizing pattern classes in this way is an attempt to avoid losing the information contained within the subspecies designations but also to avoid confusing these populations with populations that are demonstrable, genetically isolated, evolutionary lineages.

Taxonomic Keys

The majority of the anatomical terms used in the following keys are identified in the accompanying line drawings. A Spanish version of the keys is given in appendix B.

SALAMANDERS

1. Body very elongate and wormlike; limbs greatly reduced .... *Batrachoseps major*
1. Body proportions not as above 2
2. Tail constricted at base; digits slender, not square, webbing absent 3
2. Tail not constricted at base; digits short and square, webbing present .... *Aneides lugubris*
3. Body covered with large orange blotches .... *Ensatina klauberi*
3. Body lacking large orange blotches .... *Ensatina eschscholtzii*

FROGS AND TOADS

1. A single conspicuous black spade on heel immediately posterior to first toe; pupils vertical 2
1. Heels lacking single spades, or, two enlarged spades or tubercles present on heel, one on the inside and one on the outside; inside spade or tubercle largest; pupils not vertical 3
2. Spades rectangular .... *Scaphiopus couchii*
2. Spades semicircular .... *Spea hammondii*
3. Hind limbs very rough and warty 4
3. Hind limbs smooth 9
4. Top of hind limbs with distinct, large, isolated warts; a series of one to four distinct white warts at corner of mouth .... *Bufo alvarius*
4. Warts on top of hind limbs much smaller; no distinctive white warts at corners of mouth 5
5. Very distinct, thin, well-defined white line running down the center of back 6
5. Back not marked as above 7
6. Cranial crests present .... *Bufo woodhousii*
6. Cranial crests absent .... *Bufo boreas*
7. Snout laterally compressed, almost pointed; back covered with red tubercles or warts ...... *Bufo punctatus*
7. Snout not laterally compressed; back not as above ... 8
8. Dark, paravertebral markings on back paired and symmetrical; parotoid glands generally unicolor .... *Bufo cognatus*
8. Dark markings on back not symmetrical; anterior ¾ of parotoid glands pale-colored and posterior ¼ blotched .... *Bufo californicus*

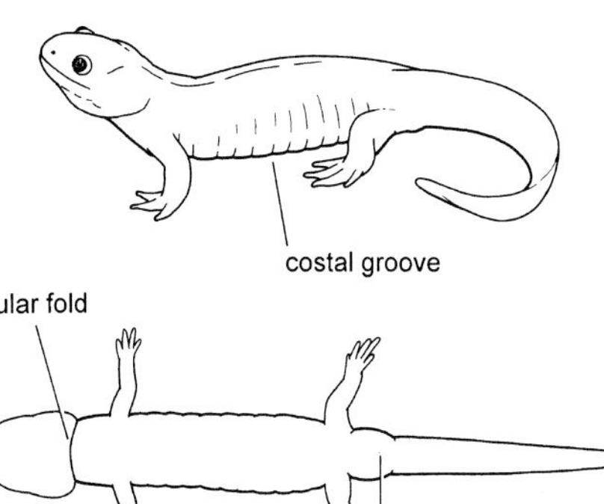

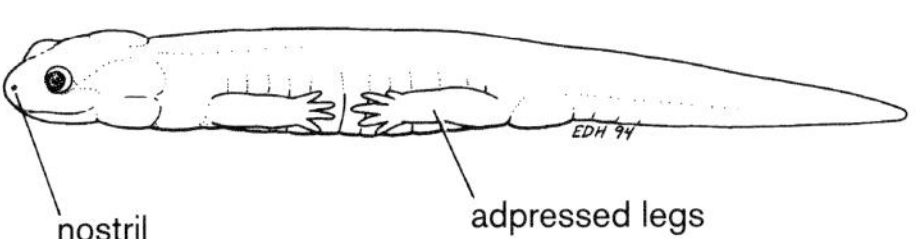

Common diagnostic characteristics of salamanders. (Illustration by Errol D. Hooper, Jr.)

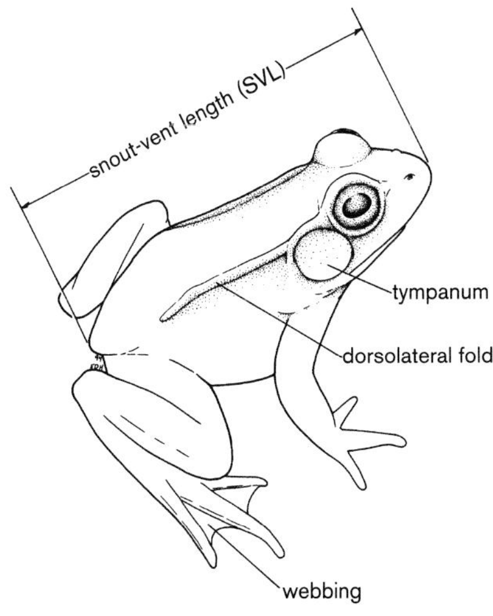

Common diagnostic characteristics of frogs and toads. (Illustration by Errol D. Hooper, Jr.)

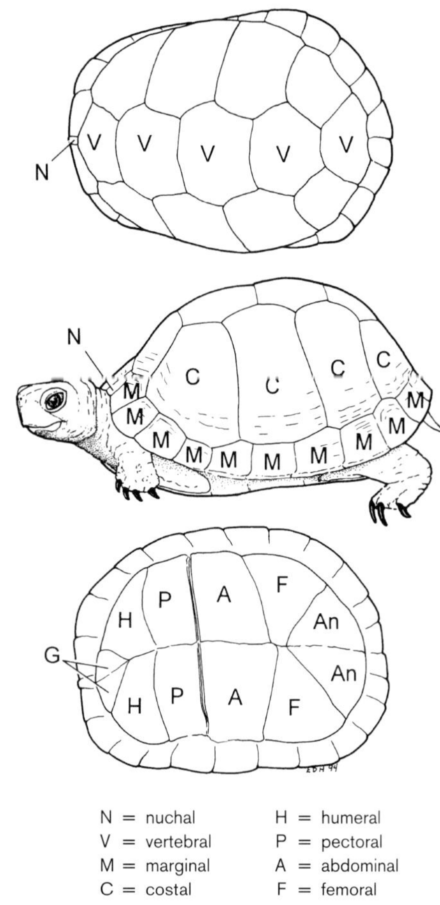

Shell laminae of turtles and tortoises. (Illustration by Errol D. Hooper, Jr.)

9. Tips of digits expanded and disk-shaped 10
9. Expanded digital disks absent 11
10. Dark stripe running along side of head through eye . . . . *Hyla regilla*
10. Dark stripe through eye absent . . . . *Hyla cadaverina*
11. Dorsolateral folds absent; large, semicircular fold or crease behind tympanum . . . . *Rana catesbeiana*
11. Dorsolateral folds present; no fold or crease behind tympanum 12
12. Dorsolateral folds distinctive . . . . *Rana forreri*
12. Dorsolateral folds poorly developed, present only posteriorly . . . . *Rana aurora*

TURTLES AND TORTOISES

1. Hand flattened or oarlike; one or two claws present 2
1. Hand less flattened, never oarlike; five claws present 6
2. Shell leathery with seven longitudinal ridges on carapace . . . . *Dermochelys coriacea*
2. Shell not as above 3
3. Four costal laminae present 4
3. Five costal laminae present 5
4. Single pair of prefrontals; laminae not imbricate . . . . *Chelonia mydas*
4. Two pairs of prefrontals; laminae usually imbricate . . . . *Eretmochelys imbricata*
5. Four inframarginals on bridge; carapace broad; a single pair of abdominal laminae . . . . *Lepidochelys olivacea*
5. Rarely more than three inframarginals on bridge; carapace not noticeably broad . . . . *Caretta caretta*
6. No laminae on shell; snout elongate and tubular . . . . *Apalone spinifera*
6. Laminae on shell; snout not elongate 7
7. Hind limbs elephantine and stumplike . . . . *Gopherus agassizii*
7. Hind limbs not as above 8
8. Light-colored post-tympanic patch on side of head . . . . *Trachemys nebulosa*
8. Post-tympanic patch absent . . . . . *Clemmys marmorata*

LIZARDS

KEY TO THE GENERA

1. Eyelids absent 2
1. Eyelids present 6
2. Four legs present 3
2. No legs present . . . . *Anniella*
3. Subdigital lamellae transversely expanded throughout length of digit or just at tip 4
3. Subdigital lamellae not expanded . . . . *Xantusia*

4. Subdigital lamellae expanded throughout length of digit 5
4. Subdigital lamellae expanded only at tip . . . . . *Phyllodactylus*
5. Claw absent from digit 1 of forefoot . . . . . *Gehyra mutilata*
5. Claw present on digit 1 of forefoot . . . . . *Hemidactylus frenatus*
6. No trace of gular or postgular folds, even on side of neck 7
6. Gular or postgular folds present at least on side of neck, but gular fold usually complete 9
7. Body scales small and granular . . . . . *Coleonyx*
7. Body scales imbricate 8
8. Body scales keeled . . . . . *Elgaria*
8. Body scales smooth . . . . . *Eumeces*
9. Body and head covered with large spines . . . . . *Phrynosoma*
9. Body and head not as above 10
10. Single, raised, enlarged middorsal scale row forming a weak to moderate vertebral crest on back 11
10. Single, middorsal scale row not as above 12
11. Caudal scales arranged in whorls of distinct spines . . . . . *Ctenosaura*
11. Caudal scales rectangular; tail with distinct, middorsal crest . . . . . *Dipsosaurus*
12. Axillary spot present (sometimes very faint) . . . . . *Uta*
12. Axillary spot absent 13
13. All dorsal body scales keeled . . . . . *Sceloporus*
13. Dorsal body scales not keeled or only vertebral rows weakly keeled 14
14. Head large, distinctly and abruptly wider than neck; elongate eyelid fringe scales on lower eyelid only . . . . . *Crotaphytus*
14. Head not as above; elongate eyelid fringe scales absent or on upper and lower eyelids 15
15. Head scales greatly enlarged, platelike, and reduced in number 16
15. Head scales numerous and more uniform in size 18
16. Ventral body scales enlarged, rectangular, and imbricate; dorsal body scales granular . . . . . *Cnemidophorus*
16. Ventral body scales small and not rectangular 17
17. Single, complete, black, dorsal body band present immediately anterior to forelimbs . . . . . *Petrosaurus*
17. Complete, black, dorsal body band absent anterior to forelimbs . . . . . *Urosaurus*
18. Supralabials strongly keeled and obliquely directed; lower jaw slightly countersunk into upper jaw 19
18. Supralabials not keeled; lower jaw not countersunk 20

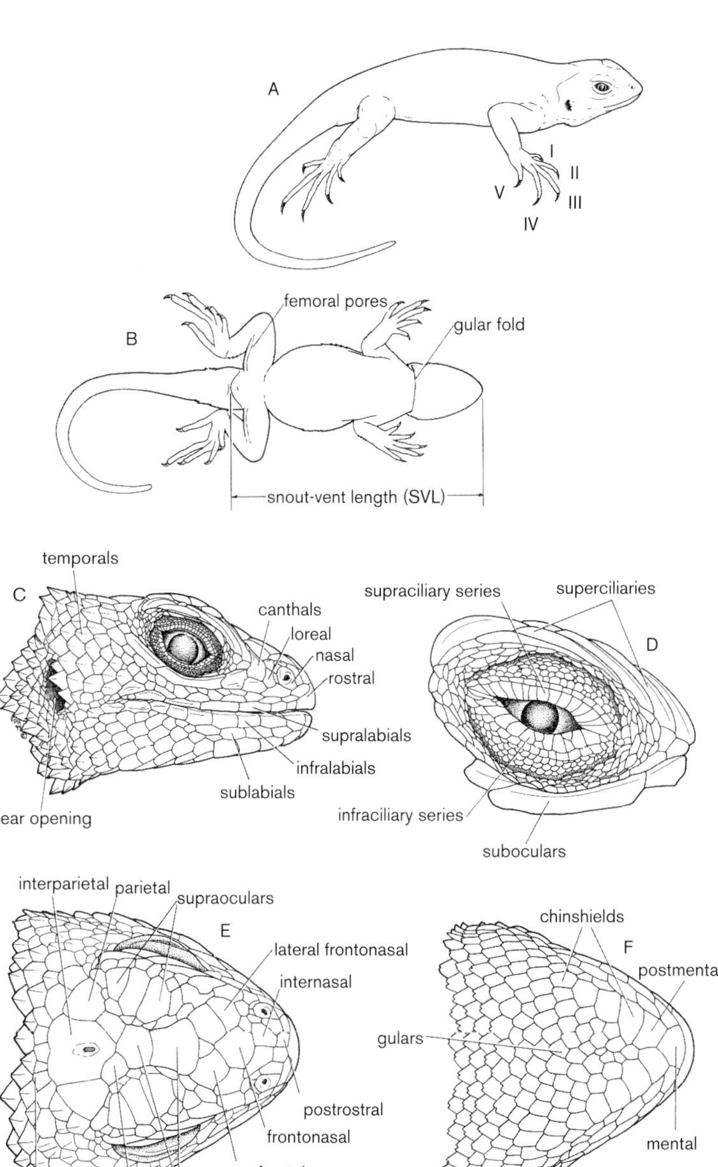

Common diagnostic characteristics of lizards. (Illustrations by Errol D. Hooper, Jr.)

(A) Lateral view of a standardized lizard showing digital sequence; (B) ventral view, showing major anatomical features; (C) lateral view of head, showing scale morphology and nomenclature; (D) lateral view of eye, showing surrounding scales; (E) dorsal view of head, showing scale morphology and nomenclature; (F) ventral view of head, showing scale morphology and nomenclature.

19. Single, dark spot on side of belly equidistant between forelimbs and hind limbs; body very wide and flat ..... *Uma notata*
19. Two to three dark markings on side of body usually closer to forelimbs; body not wide .... *Callisaurus draconoides*
20. Body flat and obese with large folds of loose skin on sides; greatly enlarged scales on anterior edge of ear opening .... *Sauromalus*
20. Body not flat; no enlarged or only weakly enlarged scales anterior to ear opening .... *Gambelia*

KEY TO THE SPECIES

ANNIELLA (PP. 241–244)

1. Rostrum sharply pointed in lateral profile; fourth supralabial largest .... *geronimensis*
1. Rostrum rounded in lateral profile; second supralabial largest .... *pulchra*

CNEMIDOPHORUS (PP. 211–233)

1. Frontoparietal single 2
1. Frontoparietal divided 8
2. Light crossbars present in lateral fields .... *ceralbensis*
2. Light crossbars absent 3
3. Ventrum bluish; mesoptychial scales small 5
3. Ventrum grayish or orange; mesoptychial scales large 4
4. Belly orange in adult males .... *hyperythrus*
4. Belly grayish in adult males .... *espiritensis*
5. No stripes on tail 6
5. Stripes present on tail 7
6. Dorsal stripes extremely faded or absent .... *pictus*
6. Dorsal stripes present .... *carmenensis*
7. Light markings on hind limbs present ... *franciscensis*
7. Light markings on hind limbs absent ...... *danheimae*
8. Mesoptychials small .... *labialis*
8. Mesoptychials large 9
9. Ventrum gray with black mottling 10
9. Ventrum black 11
10. Dorsal stripes vivid in adults; dorsal ground color nearly black .... *celeripes*
10. Dorsal stripes present in adults but not exceptionally vivid or wide; light, dorsal pattern often appearing reticulated .... *tigris*
11. Dorsal pattern nearly unicolor tan with faint, fine reticulations .... *canus*
11. Dorsal pattern dark with fine reticulations or light-colored spots 12
12. Dorsal pattern with light-colored spots 13
12. Dorsal pattern dark with fine, reticulated pattern that is slightly lighter than ground color .... *martyris*
13. Dorsal spots on anterior of body .... *catalinensis*
13. Dorsal spots not on anterior of body .... *bacatus*

COLEONYX (PP. 195–202)

1. Tubercles on body 2
1. Tubercles absent .... *variegatus*
2. Eyelid fringe scales 42 or more .... *gypsicolus*
2. Eyelid fringe scales 41 or fewer .... *switaki*

CROTAPHYTUS (PP. 108–113)

1. Posterior black collar distinctly wider than anterior black collar .... *dickersonae*
1. Posterior black collar as wide or slightly narrower than anterior black collar 2
2. Dorsal pattern containing conspicuous, white, transverse bars .... *vestigium*
2. Dorsal pattern composed of spots or fused spots forming short wavy lines 3
3. Posterior black collar nearly complete across dorsum; forelimbs immaculate .... *grismeri*
3. Posterior black collar absent or very short; forelimbs mottled .... *insularis*

CTENOSAURA (PP. 117–121)

1. Light bands on limbs and post-thoracic region of body retained throughout life .... *hemilopha*
1. Post-thoracic region of adults unicolor or weakly spotted 2
2. Rounded, dark marks on ventral surfaces of hind limbs of juveniles retained into adulthood ...... *conspicuosa*
2. Ventral surfaces of hind limbs in adults with dark flecks but no rounded markings .... *nolascensis*

DIPSOSAURUS (PP. 121–125)

1. Gular region light with dark longitudinal streaks .... *dorsalis*
1. Gular region nearly solid chocolate brown .... *catalinensis*

ELGARIA (PP. 244–251)

1. Scales of body, limbs, and tail strongly keeled; no conspicuous black and white labial markings 2
1. Keeling weak; black and white labial markings conspicuous 3
2. Adult SVL less than 115 mm; from the Islas Coronado .... *nana*
2. Adult SVL more than 116 mm .... *multicarinata*
3. 14 longitudinal dorsal scale rows .... *cedrosensis*
3. 16 longitudinal dorsal scale rows 4

4. Dorsal color pattern never with light-colored transverse bands in hatchlings or adults .... *paucicarinata*
4. Dorsal color pattern with light-colored transverse bands always in hatchlings and nearly always in adults .... *velazquezi*

EUMECES (PP. 237–241)

1. Tail usually dull red; body unicolored .... *gilberti*
1. Body striped 2
2. Tail reddish or pale to bright blue .... *skiltonianus*
2. Tail bright pink or purple 3
3. Last supralabial in broad contact with secondary upper supratemporal .... *lagunensis*
3. Last supralabial not in broad contact with secondary upper supratemporal .... *gilberti*

GAMBELIA (PP. 113–117)

1. Dorsal ground color pale with numerous asymmetrically arranged punctations on the head and body .... *wislizenii*
1. Dorsal ground color dark with large paired paravertebral markings on the dorsum .... *copeii*

PETROSAURUS (PP. 143–150)

1. Dorsal scales of tail approximately same size as dorsal scales of body and not distinctly keeled or spinose 2
1. Dorsal scales of tail much larger than dorsal body scales and distinctly keeled and spinose 3
2. Thin, distinct, postorbital stripe present; one scale between nasal scale and supralabials .... *repens*
2. Postorbital stripe absent; two scales between nasal scale and supralabials .... *thalassinus*
3. Widely separated round white spots in gular region .. .. *mearnsi*
3. Gular region overlaid with a dense, light-colored reticulum .... *slevini*

PHRYNOSOMA (PP. 150–157)

1. Bases of occipital spines connected, forming a crown .... *solare*
1. Bases of occipital spines separate 2
2. Tympanum thin and normal in appearance; six to eight greatly enlarged longitudinal rows of gular scales .... *coronatum*
2. Tympanum thick and covered with scales, sometimes not differentiated from surrounding scales and difficult to see; gular scales small, only one or two enlarged longitudinal rows present 3
3. Thin black vertebral line on dorsum .... *mcallii*
3. No black vertebral line on dorsum .... *platyrhinos*

PHYLLODACTYLUS (PP. 204–210)

1. Body tubercles present 2
1. Body tubercles absent .... *unctus*
2. Scales around snout at third supralabial more numerous than interorbital scales 3
2. Scales around snout at third supralabial less numerous than interorbital scales 4
3. Dorsal coloration dark brown .... *partidus*
3. Dorsal coloration grayish .... *homolepidurus*
4. Belly scales large .... *bugastrolepis*
4. Belly scales not large .... *xanti*

SAUROMALUS (PP. 125–135)

1. Color pattern consisting of large, irregular, reddish-brown to black blotches on a yellowish background .... *varius*
1. Color pattern banded or unicolor 2
2. Scales of neck, dorsum, limbs, and tail very spiny .... *hispidus*
2. Scales of neck spiny or not, those of back, limbs, and tail generally smooth 3
3. Body bands absent 4
3. Body bands usually present .... *obesus*
4. Nuchal scales homogeneous .... *slevini*
4. Nuchal scales heterogeneous, with smaller scales interspersed among larger scales .... *klauberi*

SCELOPORUS (PP. 157–175)

1. Lateral body scales much smaller in size and sharply differentiated from dorsal body scales 2
1. Lateral body scales nearly equal in size to dorsal body scales 3
2. Lateral body scales sharply differentiated from dorsal body scales .... *grandaevus*
2. Lateral body scales grade smoothly into dorsal body scales .... *angustus*
3. Dorsal scales relatively small with posterior spinose margin not upturned, making back appear smooth in lateral view; scales of tail nearly twice as large as scales on back .... *vandenburgianus*
3. Dorsal scales larger with posterior spinose margin upturned, giving the back a spiny appearance in lateral view; scales of tail the same size or only slightly larger than those on back 4
4. Forelimbs with conspicuous black and white bands ... . *clarkii*
4. Forelimbs lacking conspicuous black and white bands 5

5. Distinctive black shoulder patch absent ... *occidentalis*
5. Distinctive black shoulder patch present 6
6. Dorsal body scales wide, flat, and weakly keeled with short, posteriorly projecting spine; no light-colored border behind black shoulder patch .... *orcutti*
6. Dorsal body scales conspicuously keeled and more elongate, resulting from a prominent, posteriorly projecting spine; light-colored border present behind black shoulder patch 7
7. First sublabial in contact with mental usually on both sides, always on at least one side 8
7. First sublabial not contacting mental 10
8. Thin, postorbital black lines present .... *magister*
8. Thin, postorbital black lines absent 9
9. Dorsal pattern uniformly golden tan ...... *lineatulus*
9. Dorsal pattern not as above .... *zosteromus*
10. Posterior supraocular in contact with supraciliary ... . *hunsakeri*
10. Posterior supraocular separated from supraciliary by small supranumerary scales .... *licki*

UROSAURUS **(PP. 177–185)**

1. Dorsolateral fold on body containing enlarged scales or tubercles; frontal scale usually divided 2
1. Dorsolateral fold on body not containing enlarged scales or tubercles 3
2. Supraocular scales strongly convex .... *ornatus*
2. Supraocular scales weakly convex or flat .... *graciosus*
3. Long, thin, wavy longitudinal black, paravertebral lines present .... *lahtelai*
3. Paravertebral lines absent .... *nigricaudus*

UTA **(PP. 185–195)**

1. Ventral scales 75 or more (counted midventrally between gular fold and cloacal opening) .... *palmeri*
1. Ventral scales 74 or fewer 2
2. Dorsum nearly uniformly green; ventrum bluish .... *nolascensis*
2. Dorsum not uniformly green 3
3. Turquoise spots on top of head and rostrum .... *squamata*
3. Top of head and rostrum lacking turquoise spots ... 4
4. Prefrontal region flat .... *stansburiana*
4. Prefrontal region conspicuously convex in lateral profile 5
5. Two prefrontal scales between frontonasal and anterior frontal, at least on one side .... *tumidarostra*
5. One prefrontal scale between frontonasal and anterior frontal 6
6. Frontoparietals not or only slightly contacting medially .... *lowei*
6. Frontoparietals in broad contact medially .... *encantadae*

XANTUSIA **(PP. 233–237)**

1. Body very flat; limbs not short; dorsum covered with large, dark, irregularly shaped blotches .... *henshawi*
1. Body cylindrical; limbs short; no large dark blotches on dorsum .... *vigilis*

SNAKES AND WORM LIZARDS

KEY TO THE GENERA

1. Two forelimbs present .... *Bipes biporus*
1. No limbs present 2
2. Body pattern of red-orange, black, and white-yellow bands with red-orange and white-yellow in contact .... *Micruroides euryxanthus*
2. Body pattern not as above 3
3. Tail laterally compressed with a yellow and black reticulated color pattern .... *Pelamis platurus*
3. Tail cylindrical in cross-section 4
4. Ventral scales same size and shape as dorsal scales .... *Leptotyphlops humilis*
4. Ventral scales transversely elongate 5
5. Deep, conspicuous loreal pit present between eye and nostril .... *Crotalus*
5. Loreal pit absent 6
6. Large head plates lacking; ventral scales small .... *Lichanura trivirgata*
6. Large head plates present; ventral scales large 7
7. Anal plate entire 8
7. Anal plate divided 14
8. Some or all of dorsal scales strongly keeled 9
8. All dorsal scales smooth or weakly keeled vertebral series 10
9. 27 or more dorsal scale rows at midbody; pattern consisting of dark blotches .... *Pituophis*
9. Fewer than 27 dorsal scale rows at midbody; pattern lineate at least on sides of body .... *Thamnophis*
10. Rostral scale greatly enlarged, triangular, and extending a great distance posteriorly into the prefrontal region; front of head squarish in lateral view .... *Phyllorhynchus decurtatus*
10. Rostral scale not as above; head normal in lateral view 11
11. Eyes large and protuberant; head triangular, much wider than neck; pupils elliptical .... *Trimorphodon biscutatus*
11. Eyes normal; head not triangular; pupils round 12

12. Rostral scale enlarged and projected anteriorly; 50 percent or fewer of subcaudals divided ...... *Rhinocheilus*
12. Rostral scale normal; 50 to 100 percent of subcaudals divided 13
13. Dark bar across head connecting eyes; dorsal pattern blotched; ventrum lacking dark markings ... *Arizona*
13. No dark bar across head; pattern banded or striped; ventrum with dark markings .... *Lampropeltis*
14. Pupils elliptical 15
14. Pupils round 17
15. Prominent tripartate nuchal collar; dark stripe along side of head extends through eye .... *Hypsiglena*
15. Nuchal collar reduced or absent; no stripe along side of head extending through eye 16
16. Light-colored, lyre-shaped marking on top of head behind eyes .... *Trimorphodon biscutatus*
16. Head speckled .... *Eridiphas*
17. Rostral scale greatly enlarged; middorsal stripe present .... *Salvadora hexalepis*
17. Rostral scale not enlarged 18
18. Suboculars present; body red-brown to tan .... *Bogertophis rosaliae*
18. Suboculars absent 19
19. Loreal absent 20
19. Loreal present 21
20. Lower jaw countersunk; anterior ½ of head not black .... *Chilomeniscus*
20. Lower jaw not countersunk; entire head black .... *Tantilla planiceps*
21. Snout flattened, appearing spadelike in lateral view; lower jaw countersunk .... *Chionactis occipitalis*
21. Snout normally shaped; lower jaw not countersunk 22
22. Orangish ring on neck .... *Diadophis punctatus*
22. No ring on neck 23
23. Head entirely black .... *Thamnophis*
23. Head not black 24
24. Distinct canthal ridge present .... *Masticophis*
24. Canthal ridge absent .... *Sonora semiannulata*

KEY TO THE SPECIES

ARIZONA (PP. 263–265)

1. 51 to 83 short, transverse body blotches ...... *elegans*
1. 36 to 41 elongate, subcircular body blotches ... *pacata*

CHILOMENISCUS (PP. 267–270)

1. Prefrontal scales contact on midline .... *stramineus*
1. Prefrontal scales do not contact .... *savagei*

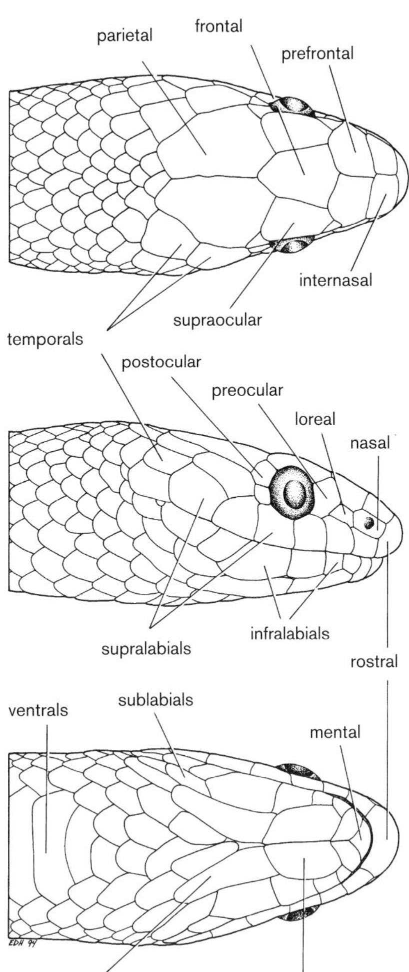

Head scales of snakes. (Illustration by Errol D. Hooper, Jr.)

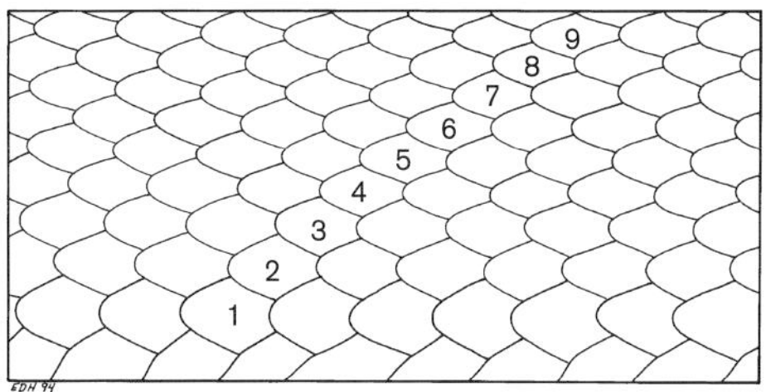

Sequence of dorsal scale row numbers in snakes. (Illustration by Errol D. Hooper, Jr.)

CROTALUS (PP. 319–341)

1. Rattle matrix greatly shrunken, no loose rattle segments (or only one, rarely two) .... *catalinensis*
1. Rattle matrix normal or only moderately shrunken, at least one loose segment past juvenile button stage ... 2
2. Dorsal ground color reddish; first pair of infralabials transversely divided; distinctive black and white caudal rings 3
2. Dorsal ground color ashy or light brown; first pair of infralabials not transversely divided; black and white caudal rings with fuzzy margins 4
3. Black caudal rings slightly less wide than white caudal rings or of nearly equal width .... *ruber*
3. Black caudal rings much thinner than white caudal rings .... *lorenzoensis*
4. Lateral edges of supraoculars conspicuously elevated 5
4. Lateral edges of supraoculars not elevated 6
5. Lateral edges of supraoculars form an upward-pointing, hornlike process .... *cerastes*
5. Lateral edges of supraoculars lack hornlike processes ... . *enyo*
6. Numerous dark flecks and punctations in dorsal color pattern, giving a speckled appearance; prenasals usually separated from rostral 9
6. Dorsal pattern not speckled; prenasals usually contacting rostral 7
7. Adult SVL less than 637 mm; from Isla El Muerto ... . *muertensis*
7. Adult SVL greater than 636 mm; not from Isla El Muerto 8
8. Adult SVL greater than 1,114 mm; from Isla Ángel de la Guarda .... *angelensis*
8. Adult SVL less than 1,113 mm; not from Isla Ángel de la Guarda .... *mitchellii*
9. Usually 3 or more internasals 10
9. Usually 2 internasals 11
10. Adult SVL less than 675 mm; from Isla Coronado Sur .... *caliginis*
10. Adult SVL usually greater than 674 mm ...... *viridis*
11. Markedly contrasting, alternating black and white caudal rings 12
11. Caudal rings not as above 13
12. Upper loreal scale present .... *atrox*
12. Upper loreal scale absent .... *tortugensis*
13. Tail solid black or brown 14
13. Tail not as above .... *tigris*
14. Prefrontal region very dark .... *molossus*
14. Prefrontal region faintly darkened ...... *estebanensis*

ERIDIPHAS (PP. 273–275)

1. More than 196 ventral scales .... *marcosensis*
1. Fewer than 197 ventral scales .... *slevini*

HYPSIGLENA (PP. 275–279)

1. One or more pairs of gular scales nearly equal to chinshields in size .... *gularis*
1. Gular scales smaller than chinshields .... *torquata*

LAMPROPELTIS (PP. 279–284)

1. Prominent white band on posterior portion of head in contact with parietal scales 2
1. Posterior head band absent 3
2. Red bands present; red on rostrum .... *zonata*
2. Red bands absent; no red on rostrum .... *herrerae*
3. Dorsal pattern of brown-black and white-yellow alternating body bands or varying degrees of striping .... *getula*
3. Dorsal pattern not as above; sides of body covered with small yellow spots; dorsal ground color purplish to black .... *catalinensis*

MASTICOPHIS (PP. 284–292)

1. Conspicuous dorsolateral stripe present on dorsal body scale rows 3 and 4; striping may or may not continue the length of the body 2
1. Striping on scale rows 3 and 4 absent 6
2. Anterior ⅓ of stripe even in width; chin and throat with dark blotches .... *lateralis*
2. Anterior ⅓ of stripe uneven and swollen at regular intervals, every 2 to 7 scale rows; chin and throat usually immaculate 4
3. Dark blotches on first 10 to 15 ventral scales .... *bilineatus*
3. Dark blotches on at least first 40 ventral scales .... *slevini*
4. Dorsolateral stripes continue the length of the body 5
4. Dorsolateral stripes fade posteriorly .... *aurigulus*
5. Anterior portions of dorsolateral stripes not expanded; no crossbands on nape of neck .... *lateralis*
5. Anterior portions of dorsolateral stripes expanded at regular intervals; cross bands present on nape of neck .... *barbouri*
6. Body with zigzag pattern of thin, dark bands following the edges of the scales; or head and body entirely black .... *fuliginosus*
6. Body color not as above .... *flagellum*

PITUOPHIS (PP. 294–300)

1. Black subcaudal stripe present 2
1. Black subcaudal stripe absent .... *catenifer*
2. More than 52 dark body blotches .... *insulanus*
2. Fewer than 53 dark body blotches .... *vertebralis*

RHINOCHEILUS (PP. 300–303)

1. Loreal scale square; dark dorsal body blotches not contacting ventral scales .... *etheridgei*
1. Loreal scale rectangular; dark dorsal body blotches contacting ventral scales .... *lecontei*

THAMNOPHIS (PP. 308–313)

1. Vertebral stripe present 2
1. Vertebral stripe absent 3
2. Dorsal stripe two to three scale rows wide; no conspicuous dark blotches on side of body .... *elegans*
2. Dorsal stripe one scale wide; conspicuous dark blotches on side of body .... *marcianus*
3. Dorsal ground color olive-drab; head not black .... *hammondii*
3. Dorsal ground color black; head black .... *validus*

1

SALAMANDERS

BECAUSE OF THE ARID NATURE OF MOST OF BAJA CALIFORNIA, ITS SALAMANDER FAUNA IS DEPAUPERATE, CONSISTING OF ONLY THREE GENERA AND FOUR OR FIVE SPECIES, ALL OF WHICH BELONG TO THE FAMILY PLETHODONTIDAE. ANOTHER LINEAGE, THE SALAMANDRIDAE, CLOSELY APPROACHES BAJA CALIFORNIA, WITH *TARICHA TOROSA* RANGING SOUTH INTO MONTANE AREAS OF SOUTHERN SAN DIEGO COUNTY, CALIFORNIA, BUT IT HAS NOT BEEN REPORTED IN SIMILAR, CONTIGUOUS HABITATS IN NORTHERN BAJA CALIFORNIA. ALL THE KNOWN SPECIES OF SALAMANDERS IN BAJA CALIFORNIA OCCUR IN THE RELATIVELY COOLER AND MORE MESIC CALIFORNIA OR BAJA CALIFORNIA CONIFEROUS FOREST REGION (WITH ONE POSSIBLE EXCEPTION; SEE *BATRACHOSEPS MAJOR* BELOW), AND THEIR SURFACE ACTIVITY PATTERNS ARE GENERALLY RESTRICTED TO WINTER EVENINGS DURING PERIODS OF PRECIPITATION.

FIGURE 1.1 *Ensatina klauberi* (Large-Blotched Ensatina) from Sierra San Pedro Mártir, BC

Plethodontidae

LUNGLESS SALAMANDERS

The Plethodontidae are the world's largest group of salamanders, containing approximately 27 genera and 320 species (Frost 1985; Duellman 1993), and the only group of salamanders represented in Baja California. Plethodontids can be distinguished from all other salamanders by the presence of nasolabial grooves and parasphenoid tooth patches, and the absence of lungs. Lacking lungs, they rely entirely on cutaneous respiration. With the exception of several species from France and Italy, the Plethodontidae are confined to the New World, ranging from southern Canada to central South America. Plethodontids inhabit a wide range of habitats and occupy a variety of terrestrial, fossorial, aquatic, and arboreal niches. Three genera of plethodontids—*Aneides, Batrachoseps,* and *Ensatina*—occur in Baja California. With the exception of *Batrachoseps,* they only marginally enter the more mesic northwestern portion of the peninsula.

Aneides Baird

ARBOREAL SALAMANDERS

The genus *Aneides* consists of a group of small to moderately sized scansorial salamanders occurring along the Pacific coast of North America, the Sacramento Mountains and neighboring mountains of New Mexico, and the Appalachian Mountains of the eastern United States. The genus contains seven species (Jackman and Wake 1998), one of which, *Aneides lugubris,* ranges into northwestern Baja California and occurs on some of the adjacent Pacific islands.

FIG. 1.2 **_Aneides lugubris_**

ARBOREAL SALAMANDER; SALAMANDRA

Salamandra lugubris Hallowell, 1849:126. Type locality: "Monterey, California," USA. Holotype: ANSP 1257

IDENTIFICATION *Aneides lugubris* differs from all other salamanders in the region of study by having an enlarged head; 13 to 14 costal grooves; square, flat, and expanded digits; a prehensile tail that is not constricted at the base; and a small to moderate amount of yellow speckling on its sides.

RELATIONSHIPS AND TAXONOMY *Aneides lugubris* may be the sister species of a natural group containing *Aneides vagrans* and *A. ferreus* (Jackman and Wake 1998) of the Pacific Northwest or possibly the sister species of a natural group containing *A. ferreus, A vagrans,* and *A. flavipunctatus* of the Pacific Northwest (Larson et al. 1981).

DISTRIBUTION *Aneides lugubris* occurs from sea level to near 1,270 m in elevation and ranges from Eureka in Humboldt County, California, south through the Coast Ranges, ter-

FIGURE 1.2 *Aneides lugubris* (Arboreal Salamander) from Arroyo Santo Tomás, BC

minating near Valle Santo Tomás in northwestern Baja California (Lynch and Wake 1974). Populations also occur in the foothills of the central Sierra Nevada of California, South Farallon Island off the coast of San Francisco, many islands within San Francisco Bay, Catalina Island off the coast of San Pedro, and Isla Coronado Norte off the coast of Rosarito, BC, Mexico (map 1.1). Although the southern record for this species in Baja California is near Santo Tomás, it is likely that it ranges at least as far south as San Vicente because of the continuity of similar habitat, and it undoubtedly ranges farther east into the foothills of the northern Peninsular Ranges.

DESCRIPTION Relatively large, with adults reaching 75 mm standard length (SL); head square, enlarged, much wider than body at angle of jaws because of enlarged mandibular musculature; males have heart-shaped mental gland and well-developed hedonic glands; snout flat anteriorly; prominent nasolabial groove; premaxillary, maxillary, and dentary teeth enlarged and flattened, giving the appearance of an overbite; eyes large and protuberant; gular fold well-developed, extending dorsally onto side of body; 13 to 14 costal grooves; limbs large and robust with well-developed, flat, wide digits; adpressed forelimbs cover 7 to 8 costal grooves; adpressed hind limbs cover 8 to 9 costal grooves; tail moderately sized, prehensile, not constricted at base, tapering and becoming laterally compressed posteriorly, and composing up to nearly 50 percent of total length. **COLORATION** Varies from light brown to chocolate brown above and from dull white to pale yellow or grayish below; abrupt transition between dorsal and ventral coloration occurs on sides of body; dorsum and sides usually covered with pale yellow speckling, varying considerably in size and number of spots; speckling usually absent from limbs; juveniles dark ventrally with patchy network of small, brassy, light blue iridophores on dorsum.

GEOGRAPHIC VARIATION At least two distinct populations of *A. lugubris* exist in Baja California: one from the peninsula and another from Isla Coronado Norte. The variation between these two populations is not great, but some notable differences exist (Grismer 1993). Although nonspeckled individuals occur in the peninsular populations from La Misión just north of Ensenada, they are relatively rare. However, on Isla Coronado Norte, 20 to 30 percent of the individuals lack any notable lateral speckling on the body. Also, peninsular populations of *A. lugubris* have a large amount of interdigital webbing, whereas salamanders from Isla Coronado Norte have noticeably less. Bishop (1943), Stebbins (1962), and Lynch and Wake (1974) state that *A. lugubris* rarely has 14 costal grooves and usually 15. These counts are based on specimens from farther north in California. Specimens from Baja California have 14 costal grooves and rarely 13.

NATURAL HISTORY Although no natural history data have been reported for *A. lugubris* from Baja California, it is likely that the natural history differs little, if at all, from populations in southern San Diego County, California. *Aneides lugubris* is most commonly observed within areas of thick chaparral and wooded, riparian canyons with oaks and sycamores in the California Region and may range into the lower reaches of the Baja California Coniferous Forest Region, although none have yet been reported (Welsh 1988). If moisture conditions are right, it can commonly be found beneath rocks, logs, and boards and inside decaying logs and stumps, wood rat nests, and rodent burrows. The population on Isla Coronado Norte occurs in a somewhat different habitat. This northernmost island of the Islas Coronado archipelago is approximately 1.5 km long and 300 m wide. It is a narrow, rocky island covered with steep sloping hills rising to 110 m. This island, like the others, has no trees and only a few small shrubs but supports abundant herbaceous vegetation. This is rather atypical habitat for continental populations of *A. lugubris*. Therefore, the Islas Coronado populations may be influenced by somewhat different ecological pressures, which may have selected for their slightly differing morphology. Similar ecological conditions exist on South Farallon and Año Nuevo islands off the coast of California, whose populations of *A. lugubris* differ little from their continental counterparts (Storer 1925; Morafka and Banta 1976).

Aneides lugubris is unique among Baja California salamanders for its arboreal proclivities. As many as 35 individuals have been found in the cavities of oak trees as much as 10 m above the ground (Stebbins 1962). The species is nocturnal and active during periods of precipitation from fall to spring. It is commonly found on the trunks of oak trees in the irregularities of the bark. During the day, this salamander remains hidden in moist microhabitats, and it is inactive above ground during summer. It eats small arthropods, worms, and occasionally slender salamanders *(Batrachoseps)*. Mating takes place during the spring, and eggs are usually laid during the summer. Hatchlings are usually not observed until the first fall rains.

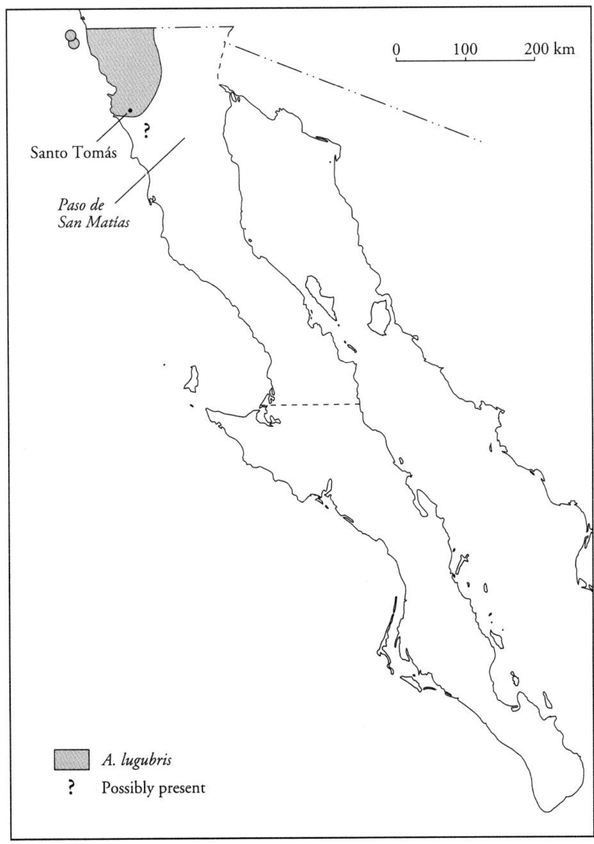

MAP 1.1 Distribution of *Aneides lugubris* (Arboreal Salamander)

Batrachoseps Bonaparte

SLENDER SALAMANDERS

The genus *Batrachoseps* comprises a group of small, short-limbed, elongate, secretive, fossorial salamanders that range disjunctly throughout the Pacific Northwest southward to southwestern Mexico. The genus contains 10 to 18 species (Jockusch et al. 1998; Wake 1996; Wake and Jockusch 2000), one of which *(B. major)* ranges into northwestern cismontane and montane Baja California and some adjacent Pacific islands.

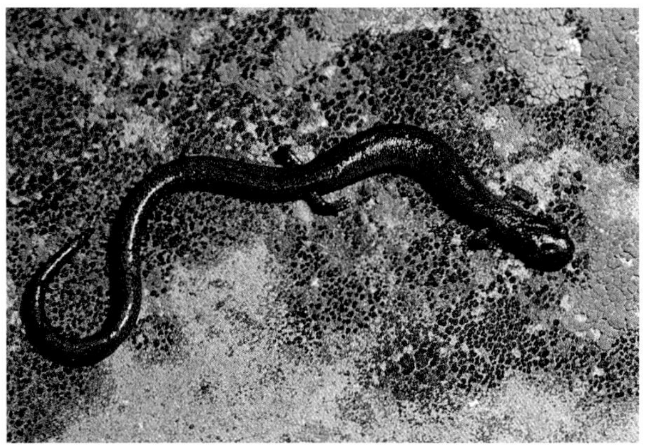

FIGURE 1.3 *Batrachoseps major* (Garden Slender Salamander) from Isla Coronado Sur, BC

FIG. 1.3 **Batrachoseps major**

GARDEN SLENDER SALAMANDER; SALAMANDRA PEQUEÑA

Batrachoseps major Camp, 1915:327. Type locality: "Sierra Madre (the town), 1,000 feet altitude, Los Angeles County, California," USA. Holotype: MVZ 611

IDENTIFICATION *Batrachoseps major* can be differentiated from all other salamanders in the region of study by its greatly elongate, wormlike body; small limbs; 4 as opposed to 5 toes on the hind foot; and 17 to 20 costal grooves between limb insertions.

RELATIONSHIPS AND TAXONOMY Wake (1996) places *Batrachoseps major* in his attenuate clade, which includes all other *Batrachoseps* except the more robust species *B. wrighti* and *B. campi.* Wake and Jockusch (2000) consider *B. major* and *B. pacificus* of the northern Channel Islands in California as sister species. Marlow et al. (1979), Yanev (1980), Wake (1996), and Wake and Jockusch (2000) discuss taxonomy.

DISTRIBUTION As currently constituted, *B. major* ranges from Los Angeles County south to the vicinity of El Rosario in northwestern Baja California. In Baja California, *B. major* is restricted to the northwestern portion of the peninsula, ranging as far south as Arroyo Grande, approximately 24 km southeast of El Rosario (Grismer 1982; map 1.2). It ranges to at least 1,785 m elevation in the Sierra San Pedro Mártir (Welsh 1988). It occurs on the Pacific islands Coronado Norte, Media, and Sur, Todos Santos Sur, and perhaps San Martín (Grismer 2001a).

DESCRIPTION Small, with adults reaching 50 mm SL; head flat and oval in dorsal profile, widest at angle of jaw, and equal in width to or only slightly wider than body; snout obliquely truncate in lateral profile; eyes large, protuberant, and elliptical in shape; deep postorbital groove present; well-defined, middorsal furrow running length of body and tail; 17 to 20 costal grooves between limb insertions; costal grooves weakly encircle body ventrally; 6 to 11 costal grooves present between adpressed limbs; limbs very small, both with four digits; digit 1 extremely reduced and vestigial; tail long, wormlike, and annulated by lateral grooves.

COLORATION Coloration is variable between and within populations. Generally, ground color of dorsum dark, composed of dense network of dark iridophores arranged in a reticulate pattern; dorsum overlaid by brassy and pale blue iridophores forming inconspicuous to conspicuous dorsolateral stripe; less dense bluish white speckling ventrally, making sides appear lighter; speckling of pale blue iridophores often gives sides a bluish cast; dark coloration extends toward midline of belly; dark iridophores in mental, gular, pectoral, and pelvic regions; midventral region light; coppery iridophores present on snout and tops of eyes; iris black with brassy iridophores; limbs lighter in color than dorsum, lacking pigment ventrally.

GEOGRAPHIC VARIATION The peninsular populations of *B. major* and that of the Islas Todos Santos are long, slender, and very similar in general body stature. However, the insular populations from the Islas Coronado are much more robust in body stature, having a wider head and shorter tail (Grismer 1993). The belly also tends to be lighter in color, and some individuals have copper-colored blotches on the body. In peninsular and Islas Todos Santos salamanders, the tail tapers smoothly to a point and is slightly narrower than the body. In specimens from the Islas Coronado, the tail is as wide or wider than the body nearly all the way to the tip, where it tapers abruptly. Also, in the Islas Coronado salamanders, the V-shaped groove on the top of the head is less obvious, and the gular fold is much more distinct. This morphology, however, may be associated with their larger size. Individuals from the adjacent peninsula that approach these sizes tend to be very similar in morphology.

Grismer (1982) reported that the southernmost population of *B. major* from Arroyo Grande, approximately 24 km southeast of El Rosario, is morphologically closer to *B. major* of the Sierra San Pedro Mártir than to the adjacent coastal populations in the vicinity of San Quintín. Individuals from the Sierra San Pedro Mártir population tend to be less robust and darker in coloration than individuals from populations of northwestern Baja California. They also tend to have a slightly lower number of costal grooves between adpressed limbs (6 to 9 rather than 7 to 11), a slightly broader head, and shorter limbs.

NATURAL HISTORY *Batrachoseps major* is a semifossorial species that commonly occurs in cismontane and montane habitats of northern Baja California. It is most frequently found in dense chaparral, foothill forests, and riparian corridors within the California Region. Here it resides beneath rocks, logs, and other surface litter. It is by no means restricted to such situations, however, and has been found in locally moist areas in more arid regions farther south. On the Islas Coronado, salamanders are common beneath rocks and logs near the water's edge. Along the coastal plain from just north of Colonet south to El Consuelo and again in the vicinity of El Rosario, *B. major* occurs among Vizcaíno Region vegetation and has been reported beneath decaying Desert Agave (Grismer 1982). In the Sierra San Pedro Mártir, *B. major* is most frequently found in riparian habitats and meadows (Welsh 1988). Here, it occurs beneath rocks, logs, large leaves, and occasionally within rotten logs (Welsh 1988).

Batrachoseps major is nocturnal and active above ground during periods of precipitation from fall to spring. It remains inactive below ground during the drier months. Although fossorial, it is not a burrower and must use the burrows of other animals or cracks in expansive soils to gain access to its subterranean retreats. Welsh (1988) reported a gravid specimen in late June from Yerba Buena Spring at 2,420 m in the Sierra San Pedro Mártir. I have observed gravid females in Arroyo Grande in early April. It is likely that the lowland populations have an earlier reproductive season than those of the Sierra San Pedro Mártir.

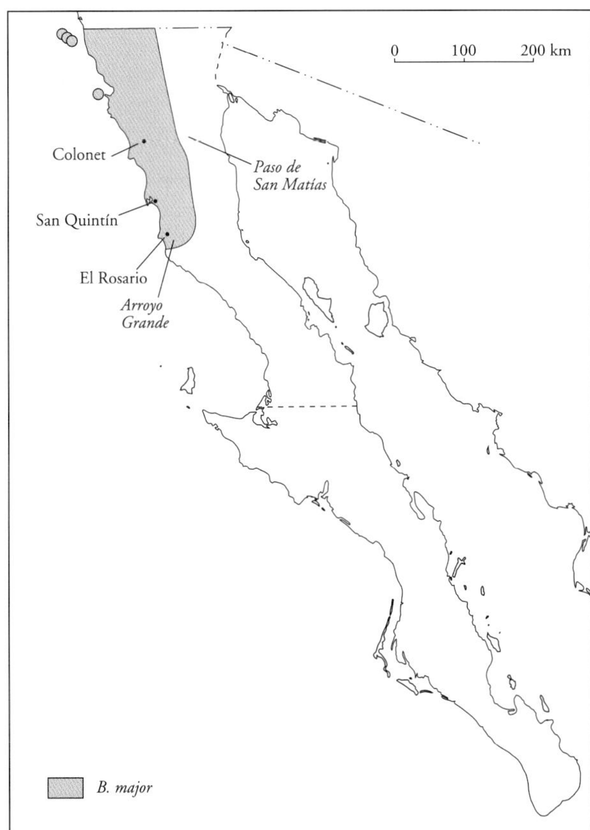

MAP 1.2 Distribution of *Batrachoseps major* (Garden Slender Salamander)

REMARKS Lockington (1880) reported on two specimens of *Batrachoseps* collected from La Paz, in the Cape Region, which Wake (1966) considered to be valid specimens. The central portion of the Cape Region is dominated by the Sierra la Laguna, with abundant oak and pine-oak woodlands. The upper elevations remain cool and moist throughout the year and would provide suitable habitat for *Batrachoseps.* It is likely that slender salamanders will be rediscovered in the Cape Region.

Ensatina Gray

ENSATINA SALAMANDERS

The genus *Ensatina* contains a group of small to moderately sized terrestrial salamanders occurring in mesic regions of the Pacific Northwest from southern Canada to northern Baja California. Ensatinas are nocturnal salamanders that are most active during periods of precipitation and are often locally abundant. The taxonomic composition of *Ensatina* remains controversial (see Highton 1998; Wake and Schneider 1998), and two species *(E. eschscholtzii* and *E. klauberi),* both of which range into northwestern Baja California, are recognized here.

FIG. 1.4 ### *Ensatina eschscholtzii*

MONTEREY ENSATINA; SALAMANDRA

Ensatina eschscholtzii Gray, 1850:48. Type locality: "California," restricted by Boulenger (1882:54) to "Monterey, California," USA. Holotype: none designated, but one specimen exists in the British Museum: BM 8000.

IDENTIFICATION *Ensatina eschscholtzii* differs from all other species of salamanders in the region of study by having relatively long digits; 11 to 12 costal grooves; a conspicuously constricted tail base; and orange coloration covering the entire upper limb region.

RELATIONSHIPS AND TAXONOMY Recent taxonomic treatments of *Ensatina* by Moritz et al. (1992) and Jackman and Wake (1995) indicate that it may contain more than one species (Highton 1998). This issue, which is controversial, is discussed in depth by Highton (1998) and Wake and Schneider (1998). Larson et al. (1981) hypothesized that *Ensatina* is the sister taxon to the genera *Plethodon* and *Aneides.*

FIGURE 1.4 *Ensatina eschscholtzii* (Monterey Ensatina) from La Misión, BC

DISTRIBUTION *Ensatina eschscholtzii* ranges along the Pacific coast of North America from southwestern British Columbia south to at least La Misión, north of Ensenada in northwestern Baja California (Stebbins 1985; Mahrdt 1975; map 1.3).

DESCRIPTION Medium-sized, with adults reaching 61 mm SL; head ovoid, slightly wider than body; eyes protuberant; gular fold moderately developed; 11 to 12 costal grooves; limbs moderately sized; digits cylindrical in cross section, narrow, long, and lacking interdigital webbing; palmar tubercles present; adpressed forelimbs cover 8 to 11 costal grooves; adpressed hind limbs cover 8 to 11 costal grooves; tail moderately sized, making up more than 50 percent of total length, conspicuously constricted at base, longer and slimmer in males, annulated with distinct grooves, glandular dorsally. **COLORATION** Dorsal ground color varies from red-brown to dark brown; dorsal surface of eyes have few iridophores, rendering skin above eyes translucent and greenish; proximal portions of limbs orange, distal portions of limbs mottled with orange and brown; small to moderately sized orange blotches occurring on dorsal surface of limbs and tail; ventral surfaces generally immaculate; occasional dark peppering on forelimbs, chest, and tail; dorsal and ventral ground colors unite abruptly at level of limb insertions, forming a jagged, emarginate border.

NATURAL HISTORY In Baja California, *E. eschscholtzii* is known only from the California Region in the vicinity of La Misión. Specimens were taken from a steep, boulder-strewn hillside in an area of chaparral vegetation (Mahrdt 1975). La Misión is situated in Arroyo San Miguel, which opens toward the Pacific Ocean and is kept cool and moist by continuous onshore breezes and frequent fogs. *Ensatina eschscholtzii* is usually nocturnal and most active during periods of precipitation from fall to spring. However, if surface conditions are favorable, salamanders may be active during the day (Mahrdt 1975; Stebbins 1962). In Arroyo San Miguel, *E. eschscholtzii* have been observed during the day from fall to spring beneath surface objects if the moisture content is sufficient. Salamanders are also abroad shortly after or during periods of precipitation (Mahrdt 1975). During the summer, when the ground is dry, surface activity ceases, and moist, subterranean shelters within the rocky areas are sought out. *Ensatina eschscholtzii* eats various small arthropods and earthworms. Courting pairs have been observed at La Misión from February through April (C. Mahrdt, personal communication, 1996).

REMARKS It appears that *E. eschscholtzii* marginally enters northwestern Baja California and does not range any further south than La Misión. Because of habitat continuity, however, it may range as far south as Valle Santo Tomás and perhaps as far south as San Vicente. It is likely that *E. eschscholtzii* occurs further inland in the chaparral foothills of the lower elevations of the Peninsular Ranges, as it does in San Diego County, California.

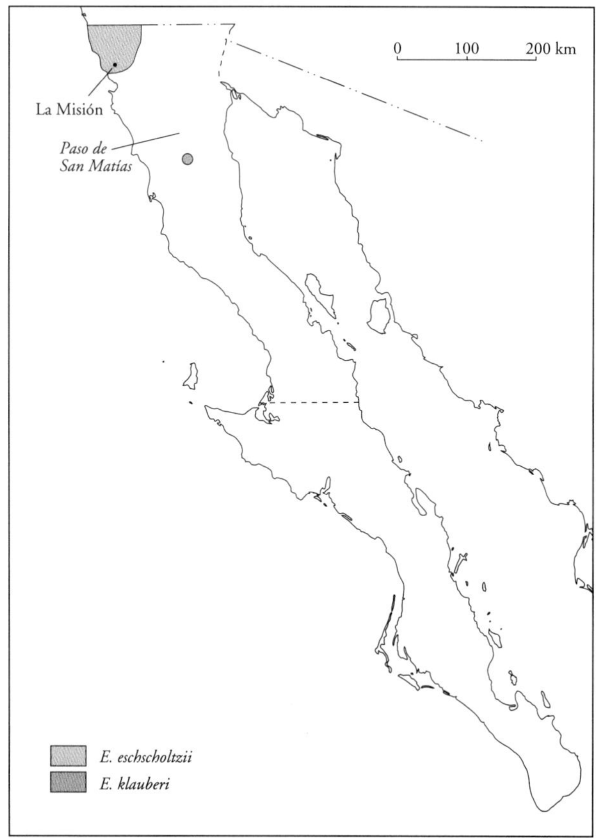

MAP 1.3 Distribution of *Ensatina* (Ensatina salamanders)

FIGURE 1.5 *Ensatina klauberi* (Large-Blotched Ensatina) from La Tasajera, Sierra San Pedro Mártir, BC

FIG. 1.5 ***Ensatina klauberi***

LARGE-BLOTCHED ENSATINA; SALAMANDRA

Ensatina klauberi Dunn 1929:1. Type locality: "Descanso, San Diego County, California." Holotype: USNM 74337

IDENTIFICATION *Ensatina klauberi* can be distinguished from all other species of salamanders in the region of study by its conspicuously constricted tail base and large orange blotches on a nearly black ground color.

RELATIONSHIPS AND TAXONOMY See *Ensatina eschscholtzii.*

DISTRIBUTION *Ensatina klauberi* ranges disjunctly along the mountain ranges of southern California from San Jacinto, San Bernardino County, south to the Sierra San Pedro Mártir (Jackman and Wake 1995; Mahrdt et al. 1998; see map 1.3). A population also exists in the Sierra Juárez.

DESCRIPTION Same as *E. eschscholtzii* except that the body is less gracile and somewhat stocky. **COLORATION** Dorsal ground color nearly black, grading to gray ventrally; orange blotches in parotoid area may extend to eyelids and postorbital regions, may or may not connect on midline forming a U-shaped marking; large orange dorsal body blotches variable in size and arrangement, tending to be on either side of midline; proximal portions of limbs orange, light orange patches in metatarsal and metacarpal regions; orange markings on tail tend to form bands; ventral surfaces generally immaculate.

NATURAL HISTORY Three specimens of *E. klauberi* were found in the vicinity of La Tasajera beneath rocks or within decaying logs on a southwest-facing talus slope adjoining a sparsely vegetated meadow (Mahrdt et al. 1998). All were found during early August, a period when this species is not usually encountered in the northern portion of its distribution. At this time of year in the Sierra San Pedro Mártir, however, afternoon thunderstorms are common and presumably provide sufficient soil moisture to allow surface activity.

REMARKS The presence of *E. klauberi* in Baja California has been confirmed only in the Sierra San Pedro Mártir (Mahrdt et al. 1998).

Lockington (1880) reported the Large-Blotched Salamander, *E. klauberi,* from 75 miles southeast of San Diego, which Dunn (1926) suggested was the Sierra San Pedro Mártir. Klauber (1927) believed this locale was near Laguna Hanson in the Sierra Juárez. Although recent authors (e.g. Brown 1974; Jackman and Wake 1995; Moritz et al. 1992; Stebbins 1985) have not reported it from the Sierra Juárez, unconfirmed sightings have been reported by commercial snake collectors for this species from the vicinity of Laguna Hanson. Unfortunately the disposition of Lockington's specimen is unknown.

Cope (1889) reported having examined a specimen of *E. klauberi* from Cabo San Lucas collected by John Xantus. Van Denburgh (1916) suggests that this record is probably erroneous and that the specimen came from Fort Tejon, California, where Xantus lived for many months and did some collecting. I agree that this record is probably erroneous, but it is interesting to note that the local inhabitants of the region surrounding San José Comondú speak of a brightly colored *salamandra* that comes out during the summer rains. When shown pictures of various lizards and salamanders, they emphatically point to *E. klauberi.*

2

FROGS AND TOADS

THE FROG AND TOAD FAUNA OF BAJA CALIFORNIA IS REPRESENTED BY ONLY FOUR FAMILIES AND FIVE GENERA. OF THE 13 SPECIES PRESENT, ONLY 3 *(BUFO PUNCTATUS, HYLA REGILLA,* AND *SCAPHIOPUS COUCHII)* RANGE FURTHER SOUTH THAN BAHÍA DE LOS ÁNGELES. THE MAJORITY OF FROGS AND TOADS OCCUR IN THE COOLER AND MORE MESIC CISMONTANE AND MONTANE AREAS OF NORTHERN BAJA CALIFORNIA, AND A DISTINCTIVE GROUP OCCURS IN THE VICINITY OF THE RÍO COLORADO AND ITS TRIBUTARIES IN THE ARID NORTHEASTERN REGION. TWO OR THREE SPECIES *(RANA BERLANDIERI* [?], *R. CATESBEIANA,* AND *R. FORRERI)* ARE INTRODUCED, AND THREE OTHERS *(BUFO ALVARIUS, RANA BOYLII,* AND *R. YAVAPAIENSIS)* MAY RECENTLY HAVE BECOME EXTINCT IN BAJA CALIFORNIA.

FIGURE 2.1 Mating pair of *Hyla regilla* (Pacific Treefrog) from Arroyo Calamajué, BC

Bufonidae

TRUE TOADS

The family Bufonidae contains approximately 33 genera and 400 living species. Its distribution is worldwide with the exception of Australia, Madagascar, oceanic islands, and very cold regions. Within its vast distribution, it occupies a tremendous variety of ecological niches. Bufonids for the most part are short, squat, and warty, although there are some very slender and smooth-skinned forms. They are distinguished from all other frogs and toads by the retention of a Bidder's organ in adult males, the absence of teeth, the presence of the "otic element," and unique characteristics of their musculature (Ford and Cannatella 1993). The Bufonidae are represented in the region of study by the genus *Bufo.*

FIGURE 2.2 *Bufo alvarius* (Colorado River Toad) from San Carlos, Sonora

Bufo Laurenti

TOADS

The genus *Bufo* consists of a very large group of terrestrial toads varying considerably in size. With the exception of Arctic regions, New Guinea, Australia, and adjacent islands, *Bufo* is cosmopolitan in distribution and occurs in a great variety of habitats. Six species of *Bufo* occur in the region of study, one of which is found on islands in the Gulf of California and is known from some of the Pacific islands of Baja California.

FIG. 2.2 ***Bufo alvarius***

COLORADO RIVER TOAD; SAPO

Bufo alvarius Girard, 1859:26. Type locality: "Valley of Gila and Colorado," restricted to "Colorado River bottomlands below Yuma, Arizona" (Schmidt 1953:61), modified to "Fort Yuma, Imperial County, California," USA (Fouquette 1968:71). Lectotype (cotype, Cochran 1961): USNM 2572

IDENTIFICATION *Bufo alvarius* can be differentiated from all other frogs and toads in the region of study by having hind limbs with distinct, large, isolated warts and a series of one to four distinct white warts behind the corners of the mouth.

RELATIONSHIPS AND TAXONOMY The relationships of *Bufo alvarius* to other species remain somewhat obscure. In a survey of selected North American taxa, Graybeal (1997) considered *B. alvarius* basal to a natural group containing *B. boreas, B. debilis, B. americanus, B. microscaphus, B. speciosus,* and *B. cognatus.*

DISTRIBUTION Fouquette (1970) indicates that *Bufo alvarius* ranges throughout southern Arizona and extreme southeastern California south throughout most of Sonora and northern coastal Sinaloa (map 2.1). Jennings and Hayes (1994), however, note that its last sighting in southern California was in 1955. *Bufo alvarius* has been reported in northeastern Baja California approximately 75 km south of Mexicali (Brattstrom 1951).

DESCRIPTION Large, attaining maximum length of 187 mm; limbs relatively short; skin smooth, covered with low, rounded, scattered tubercles; head wide and flat; cranial crests distinct; conspicuous canthus rostralis gives rostrum a somewhat laterally compressed appearance in dorsal profile; series of one to four distinct white warts posterior to corners of mouth; vocal sac absent or vestigial; parotoid glands large, divergent, and kidney-shaped; tympanum teardrop-shaped and approximately half the size of parotoids; forelimbs smooth with small, round tubercles proximally; tips of digits bulbous; large, isolated warts on hind limbs; two enlarged metatarsal tubercles; inner tubercle somewhat spadelike and slightly larger; moderate degree of interdigital webbing on foot; ventral surfaces granular in texture. **COLORATION** Ground color of dorsum dull green to light green, overlaid with small, pale orange to orange-brown spots encircling tubercles; spots whitish and bordered in black in juveniles; ventrum whitish to cream-colored, immaculate in adults; juveniles with occasional spotting in gular and pectoral regions; iris bronze to rust-colored in life. **TADPOLES** Larvae reach a total length of near 56 mm; eyes dorsal; body broadest between eye and spiracle and tapers posteriorly; spiracle on left side of body near middle; tail fins narrow and widest along midpoint of tail; dorsal and ventral fins similar in height; oral papillae present only laterally; labial tooth formula 2/3. Larvae appear gray-brown above and transparent below, with no iridescence; dorsal fin weakly spotted. **MATING CALL** Weak, low-pitched whistling approximately 0.5 second in duration (Sullivan and Malmos 1994).

NATURAL HISTORY *Bufo alvarius* occurs in a wide range of habitats outside the region of study (Stebbins 1985). In northern Baja California it was reported in association with flood-plain vegetation along the Río Hardy (Brattstrom 1951) but has not been reported there since, although I heard a calling male at Campo Río Hardy during late July 1991. Nothing is known of the natural history of *B. alvarius* from northeastern Baja California. In southern Arizona, it is nocturnal and a spring and summer breeder whose activity is stimulated by but not dependent on rainfall. Its general activity period in Sonora is from May to October. Although it is most active at night during periods of precipitation, toads can be found at other times near permanent sources of water. *Bufo alvarius* is an opportunistic feeder and is known to eat arthropods and small lizards (King 1932; Bogert and Oliver 1945; Ortenburger and Ortenburger 1926). Females lay 7,500 to 8,000 eggs in ropelike strands in shallow pools or streams (Stebbins 1962).

REMARKS The only specimen of *B. alvarius* reported from Baja California (Brattstrom 1951) was found nearly 75 km south of Mexicali along the Río Hardy drainage, which suggests that the original occurrence of this species in northeastern Baja California was not marginally along the international border. *Bufo alvarius,* however, may have suffered the same fate as *Rana yavapaiensis;* it has been extirpated in northeastern Baja California following the introduction of *R. catesbeiana* and possibly *R. berlandieri* (Platz et al. 1990).

The secretions of this toad are known to be highly toxic (Hanson and Vial 1956). They have caused paralysis and death among dogs and should be handled with caution.

Malkin (1962) reported *B. alvarius* from Isla Tiburón, although R. Crombie (personal communication, 1995) believes Malkin mistook *B. punctatus* for *B. alvarius.* Grismer (1999a) followed Crombie.

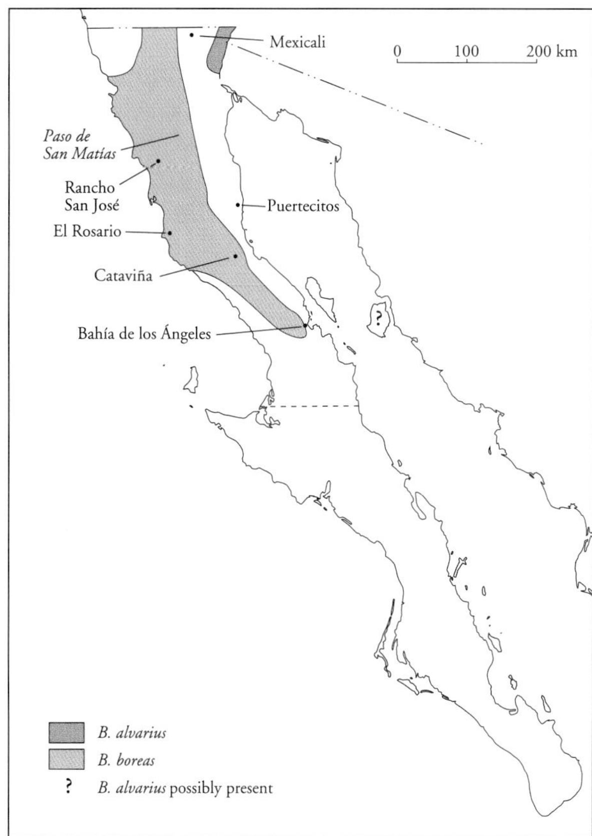

MAP 2.1 Distribution of *Bufo alvarius* (Colorado River Toad) and *B. boreas* (Western Toad)

FIG. 2.3 ***Bufo boreas***

WESTERN TOAD; SAPO

Bufo boreas Baird and Girard, 1852c:174. Type locality: Columbia River, Puget Sound, Washington; restricted to "vicinity of Puget Sound," Washington, USA, by Schmidt (1953:61). Syntypes: USNM 15467–70

IDENTIFICATION *Bufo boreas* can be differentiated from all other frogs and toads in the region of study by the combination of a distinct, thin white line running from the tip of its snout to its vent and its lack of cranial crests.

RELATIONSHIPS AND TAXONOMY *Bufo boreas* is a member of the *Bufo boreas* group, which is a natural group comprising *B. boreas* and the relict species *B. canorus* of the central Sierra Nevada, *B. exsul* of Inyo County, California, and *B. nelsoni* of Nye County, Nevada (Blair 1972).

DISTRIBUTION *Bufo boreas* is distributed widely throughout western North America from southern Alaska south through western Canada and the United States to southern Colorado, Utah, Nevada, and northern Baja California (Stebbins 1985). From here it continues south, west of the deserts, into northern Baja California. In Baja California it extends south through the Peninsular Ranges and cismontane areas to at least Bahía de los Ángeles (see map 2.1). Occasionally individuals are found on the western edge of the Lower Colorado Valley Region, where desert habitats abut the eastern slopes of the Peninsular Ranges. It is believed that they reach these arid habitats through riparian corridors that drain the eastern escarpment of the Peninsular Ranges.

DESCRIPTION Moderately sized, attaining a maximum length of 130 mm; limbs relatively short; skin rough; variously sized tubercles on body and hind limbs; tubercles usually absent on forelimbs; head broad, flat, and lacking distinct cranial crests; snout subelliptical in dorsal profile; eyes large and protuberant; parotoid glands oval, pitted, and nearly same size as upper eyelids; series of tubercles above body fold extend posteroventrally from parotoids across side of body; tympanum small, round, and approximately one-third the

FIGURE 2.3 *Bufo boreas* (Western Toad) from the base of La Rumarosa grade at the eastern foot of the Sierra Juárez, BC

size of the parotoids; vocal sac small and inconspicuous; digits on forelimb have little or no webbing; expanded toe pads absent; forearm smooth, upper arm warty; large tubercles on hind limbs; membranous fold of skin present on inner surface of tarsus; well-developed interdigital webbing; two large metatarsal tubercles present; inner tubercle slightly larger and more elongate; ventral surface smooth to wrinkled. **COLORATION** Dorsal ground color gray to greenish; thin, conspicuous cream-colored vertebral stripe running from snout to vent; stripe occasionally broken by warts; conspicuous, well-defined white patch below lower eyelid; iris coppery to greenish-yellow; tubercles of dorsum light and most often encircled by black blotches; limbs with larger dark blotches sometimes merging to form bands; ventral surfaces of hands and feet yellow in young; other ventral surfaces white to yellowish with varying amounts of spotting; ventral spotting heaviest on belly of young individuals. **TADPOLES** Larvae reach just over 55 mm in length; eyes nearer lateral margin of body than midline; body broadest between eye and spiracle, tapering posteriorly; spiracle on left side of body near middle; tail fins narrow and widest along midpoint of tail; dorsal and ventral fins similar in height; oral papillae present only laterally; labial tooth formula 2/3. Larvae appear solid black above and lighter below with no iridescence; tail fins lack spots and are clouded with minute speckling of melanophores. **MATING CALL** A soft, faint, slow, melodic trill reminiscent of a peeping chick. The call lasts approximately one second.

GEOGRAPHIC VARIATION *Bufo boreas* from the Sierra Juárez and Sierra San Pedro Mártir and their eastern foothills have a pale, light green to dark brown ground color that contrasts vividly with the dark-colored dorsal blotches. This same ground color is often observed in individuals in the more xeric regions of this species' distribution, commencing in the vicinity of Rancho San José and continuing south to Bahía de los Ángeles. In xeric regions the tubercles also become more red in color, and *B. boreas* is often mistaken for *B. punctatus*. *Bufo boreas* from lower areas in the California Region are usually dark green to gray in dorsal coloration.

NATURAL HISTORY *Bufo boreas* is a common inhabitant of the California and Baja California Coniferous Forest regions of the northwest, where it occupies many types of habitat, from sandy stream beds and chaparral hillsides to mountain meadows. It is most commonly found in the vicinity of some form of permanent water but is by no means confined to such places. In the more arid regions of its distribution it is found almost exclusively in the vicinity of locally mesic areas. *Bufo boreas* is most active from January to October in Baja California provided there is sufficient rainfall. At lower elevations, it is nocturnal and abundant during periods of winter precipitation. At higher elevations, it is active during the day, except when temperatures are unseasonably warm. During the day (or night if diurnal), toads may seek refuge by burrowing in the soft soils of rodent burrows or beneath surface objects such as logs, boards, and rocks. *Bufo boreas* breeds from January to June at low elevations and from March to July at higher elevations in the northern Peninsular Ranges (Welsh 1988). Up to 17,000 eggs may be laid in two to three long strings at the margins of a water source, usually entwined around vegetation. Tadpoles take between 30 and 50 days to transform. During mid to late summer, hundreds of thousands of newly metamorphosed *B. boreas* often crowd the Laguna Hanson valley in the Sierra Juárez in the vicinity of the lake.

REMARKS *Bufo boreas* is usually considered to be an inhabitant of mesic cismontane areas west of the Peninsular Ranges and was previously reported no further south in Baja California than Rancho Socorro. Its occurrence in xeric localities such as Bahía de los Ángeles and even Cataviña is somewhat surprising, given that the former is located at the southern end of the Lower Colorado Valley Region, one of the driest deserts in North America.

FIG. 2.4

Bufo californicus

CALIFORNIA TOAD; SAPO

Bufo cognatus californicus Camp 1915:331. Type locality: "Santa Paula, 800 feet elevation, Ventura County, Calif.," USA. Holotype: MVZ 4364

IDENTIFICATION *Bufo californicus* can be separated from all other frogs and toads in the region of study by having bicolored parotoid glands that are dark posteriorly and light-colored anteriorly, and the usual occurrence of a light patch on the sacral humps.

RELATIONSHIPS AND TAXONOMY *Bufo californicus* is most closely related to *Bufo microscaphus* of Arizona, Nevada, and New Mexico (Gergus 1998). Gergus (1994) discusses the taxonomy and relationships of the *Bufo americanus* group, of which *B. californicus* is a member.

DISTRIBUTION *Bufo californicus* is generally distributed west of the deserts from near Santa Margarita in San Luis Obispo County, California, southward through cismontane California into northwestern Baja California (Price and Sullivan 1988). In Baja California it ranges west of the Sierra Juárez and Sierra San Pedro Mártir (Welsh 1988) at least as far south as Arroyo San Simón, just south of San Quintín (Gergus et al. 1997; map 2.2).

DESCRIPTION Small to medium-sized, attaining length of approximately 75 mm; skin very rough, with several small prominent tubercles of varying sizes; head relatively wide and flat, with low cranial crests; crests occasionally absent from interorbital region; rostrum very short, and widely rounded in dorsal profile; vocal sac spherical; parotoid

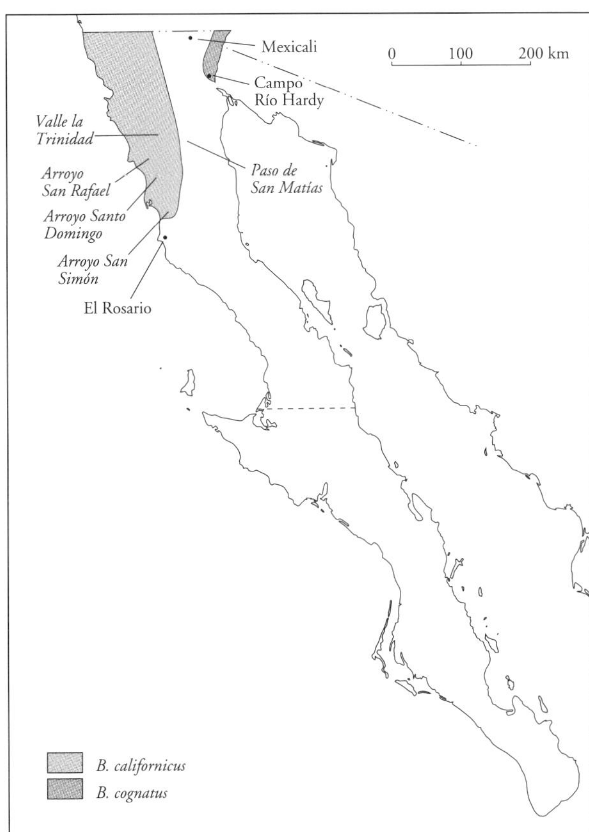

MAP 2.2 Distribution of *Bufo californicus* (California Toad) and *B. cognatus* (Great Plains Toad)

FIGURE 2.4 *Bufo californicus* (California Toad) from Arroyo Santo Domingo, BC

glands elongate, oval, and 1 to 1.5 times as long as eyes; tympanum teardrop-shaped; forelimbs not particularly short; hind limbs robust; femoral portion occasionally concealed within large fold of skin; dorsal surface of limbs covered with small, spinose tubercles; two enlarged tubercles on heel, outer tubercle half the size of inner; inner tubercle well-developed with rounded cutting surface on free, outer edge; ventral surfaces granular. **COLORATION** Dorsal ground color dusky yellowish or light olive drab to dark green; ash white in juveniles; dorsal surfaces occasionally punctate with yellow- to yellow-orange-tipped tubercles; back covered with moderately sized, irregularly shaped blotches encompassing larger tubercles; blotches much more conspicuous in juveniles; light patch usually centered on back near sacral humps; conspicuous light-colored, V-shaped marking on head; parotoid glands bicolored, light anteriorly and dark posteriorly; forelimbs weakly banded; hind limbs generally spotted or blotched; throat of adult males pinkish; ventrum whitish to light yellow with occasional encroachment of dark dorsal pigmentation; ventral coloration brighter in juveniles. **TADPOLES** Larvae reach approximately 40 mm in length; labial tooth formula 2/3; oral papillae along sides of mouth; dorsal fin of tail weakly arched; ventral fin nearly equal in height throughout length; spiracle sinistral. Larvae olive-gray to tan dorsally and nearly white ventrally; body and tail commonly mottled with moderately sized, irregularly shaped dark blotches. **MATING CALL** The call begins with a slurred introduction up to a higher note that is delivered as a clear prolonged musical trill lasting usually 3.3 to 10.6 seconds (Gergus et al. 1997; Sullivan 1992).

GEOGRAPHIC VARIATION Although intrapopulational variation in coloration can be moderate, there is a general geographic trend in dorsal pattern. Individuals from Valle la

Trinidad and the upper elevations of the Sierra San Pedro Mártir are much darker in overall dorsal ground color, and their blotching pattern is much less evident. This dark coloration is especially noticeable ventrally, where it encroaches heavily onto the pectoral region, the bottoms of the hands and feet, and around the lateral gular margins. A very distinct line of separation between the dorsal and ventral ground color appears on the limbs and body. Specimens from lower elevations to the west, and especially to the south, have a much lighter ground color. Their dorsal blotches are much more vivid, and there is no dark pigmentation on any of the ventral surfaces. Additionally, their dorsal and ventral ground colors blend smoothly into one another.

NATURAL HISTORY *Bufo californicus* is found in almost all riparian habitats within northwestern Baja California, from the coastal sage scrub and woodlands of the California Region to the coniferous forests of the Baja California Coniferous Forest Region. Toads are most abundant in the riparian habitats coursing through coastal sage scrub in areas with swiftly flowing water along open, sandy arroyos (Welsh 1988). *Bufo californicus* is active from March to September and rarely observed during other months. Unlike other toads that congregate in large breeding choruses, *B. californicus* males often space themselves along the arroyo bottoms at wide intervals when producing advertisement calls. In Arroyo Santo Domingo and Arroyo San Simón, calling *B. californicus* occur in both the swiftly moving shallow portions of the stream and along the edges of quiet backwaters. In contrast, Welsh (1988) noted that *B. californicus* occurred in only the swiftly moving portions of streams in the Sierra San Pedro Mártir. When disturbed, toads swim to the bottom of the watercourse and remain motionless. *Bufo californicus* does not seem to begin calling as early in the evening as *Hyla regilla, H. cadaverina,* or *Spea hammondii,* with which it is sympatric. In fact, in Arroyo San Rafael, *B. californicus* is most common in areas along the watercourse where densities of the other species are much lower.

The breeding season of *B. californicus* in Baja California seems to run from mid-March to June. I have heard chorusing males in Arroyo San Simón during mid-March, in Arroyo Santo Domingo during early April, and in Arroyo San Rafael during May and June. During the early part of the reproductive season, however, toads remain inactive if it is too cold or windy. Females lay several thousand eggs in tangled strands one to three eggs wide, which come to rest on the bottom margins of the watercourse. Juvenile toads first begin to appear around early June and can be found as late as mid-August. *Bufo californicus* is known to feed on several different types of small arthropods, the most common of which are crickets, beetles, ants, bees, and moths (Stebbins 1985). They occasionally eat snails, and even fragments of plant material have been found in their stomachs.

REMARKS It is likely that *B. californicus* ranges further south in Baja California than is reported here. Although it is generally considered to be a mesophilic species, a population from the Mojave River from the desert area of San Bernardino County, California, attests to its xerophilic proclivities. Some large arroyos with permanent water exist below Arroyo San Simón, and it is likely that with further collecting efforts this species will be discovered as far south as Arroyo Grande, east of El Rosario.

FIG. 2.5

Bufo cognatus

GREAT PLAINS TOAD; SAPO

Bufo cognatus Say in James 1823:190. Type locality: "The alluvial fans of the [Arkansas] River," Powers County, Colorado, USA. Holotype originally deposited in Philadelphia Museum (Baird and Girard 1853) but may have been destroyed by one of two fires (Krupa 1990).

IDENTIFICATION *Bufo cognatus* can be distinguished from all other frogs and toads in the region of study by the combination of cranial crests and paired, symmetrical, dark paravertebral markings.

RELATIONSHIPS AND TAXONOMY *Bufo cognatus* is a member of a natural group consisting of *Bufo compactilis* of the southern portion of the Central Plateau of Mexico and *B. speciosus* of the northern portion of the Central Plateau and southwestern United States (Blair 1972; Rogers 1972).

DISTRIBUTION *Bufo cognatus* is widely distributed across the Great Plains from southern Canada southward to San Luis Potosí, Mexico, and west through Texas to southeastern California and northeastern Baja California, ranging from sea level to near 2,500 m (Krupa 1990). In northeastern Baja California, it has not been found any further south than Campo Río Hardy in the extreme northeast, in association with the Río Colorado and its tributaries (see map 2.2).

DESCRIPTION Moderately sized, attaining a maximum length of 115 mm; limbs relatively short; skin rough, covered with small tubercles of nearly equal size; head relatively small, broadly triangular in dorsal view, with well-developed cranial crests uniting in interorbital region, forming a prominent cranial boss; parotoid glands ovoid in shape, approximately same size as eyes, widely separated, and diverging posteriorly; slightly enlarged series of tubercles pos-

FIGURE 2.5 *Bufo cognatus* (Great Plains Toad) from Mexicali, BC

terior to parotoids, extending ventrally toward hind limb insertion; tympanum teardrop-shaped; vocal sac sausage-shaped when inflated; moderate degree of interdigital webbing; two conspicuous, dark, sharp-edged metatarsal tubercles; outer tubercle smaller than inner; ventral surfaces of limbs and body smooth; small tubercles on seat patch. **COLORATION** Dorsal ground color generally yellowish brown to gray, with moderate to large darker paravertebral blotches; faint vertebral stripe occasionally present; dorsal blotches occasionally paired and linearly arranged; blotches generally irregular in outline and edged with color lighter than ground color; juveniles occasionally covered with small reddish spots; ventrum immaculate, cream-yellow to white with off-white to pink seat patch; dark central blotch on upper eyelids; limbs blotched. **TADPOLES** Tadpoles reach 26 mm in length; tooth row formula 2/3; oral papillae present laterally; dorsal fin highly arched; ventral fin similarly shaped but less arched; spiracle sinistral. Dorsum much darker than ventrum; tail fins weakly pigmented and usually translucent; body with network of silvery and gold iridophores. **MATING CALL** A harsh, abrasive, prolonged trill lasting 5 to 50 seconds. When one is in close proximity to several of these toads calling, the noise is uncomfortably loud.

NATURAL HISTORY *Bufo cognatus* occurs in the Lower Colorado Valley Region of northeastern Baja California in association with riparian areas along the Río Colorado and its tributaries, as well as on irrigated cropland. Nothing has been reported on the natural history of *B. cognatus* from Baja California, and, given its marginal distribution, it is unlikely that it varies significantly from populations in the adjacent southwestern United States. In the southwestern United States, *B. cognatus* spends long periods underground in self-constructed burrows and may be active above ground for only a few weeks during the summer monsoon season, when it breeds. *Bufo cognatus* is usually most active at night during periods of precipitation. In areas with permanent water, however, toads may be found at all times during the day, especially in overcast or rainy weather.

The breeding season in Baja California usually begins in late summer, with the onset of the first rains, but may occur earlier if the rains arrive sooner. *Bufo cognatus* prefers breeding ponds consisting of clear shallow water, such as flooded fields, and generally avoids deeper ponds, bodies of water with swiftly flowing currents, or those that are excessively muddy. Although a male and female may enter into amplexus at night, the female may not lay her eggs until the following afternoon. Approximately 20,000 eggs are laid in two long strands wound about each other in the space of a couple of feet. The eggs are usually anchored to nearby debris. After hatching, tadpoles can metamorphose in less than two weeks.

REMARKS *Bufo cognatus* is naturally distributed into Baja California along the Río Colorado drainage and its smaller tributaries, such as the Río Hardy. Recently, however, its range has expanded within the Río Colorado delta (see Vitt and Ohmart 1978) because of many newly built aqueducts. Also, during the breeding season, many of the fields are flooded with water, which provides ideal breeding ponds.

FIG. 2.6

Bufo punctatus

RED-SPOTTED TOAD; SAPO

Bufo punctatus Baird and Girard 1852c:173. Type locality: "Rio San Pedro [Devil's River], Val Verde County, Texas." Syntypes: USNM 2618 (three specimens), apparently lost.

IDENTIFICATION *Bufo punctatus* can be distinguished from all other frogs and toads in the region of study by the combination of lacking a vertebral stripe; having red tubercles on its back; the top of the snout being laterally compressed and somewhat pointed; and having small round parotoid glands.

RELATIONSHIPS AND TAXONOMY Probably basal to a natural group containing the *Bufo debilis* group (Ferguson and Lowe 1969), *B. cognatus,* and the *B. americanus* group (Graybeal 1997).

DISTRIBUTION *Bufo punctatus* ranges from southwestern Kansas and central Texas south to Hidalgo, Mexico, and west to southeastern California and Baja California, where it ranges throughout the peninsula except for the mesic northwestern section. On the western side of the Peninsular Ranges, it extends only as far north as El Rosario (map 2.3). *Bufo punctatus* is also known from the Gulf islands of

FIGURE 2.6 *Bufo punctatus* (Red-Spotted Toad) from La Victoria in the Sierra la Laguna, BCS

Cerralvo, Espíritu Santo, Partida Sur, and Tiburón and the Pacific island Santa Margarita.

DESCRIPTION Medium-sized, with adults reaching approximately 80 mm SVL; skin rough, covered with small, sometimes weakly spinose tubercles; head wide and flat, with shallow depression between eyes; cranial crests not well-developed; upper eyelids with small spinose warts; conspicuous, laterally directed wart above tympanum; conspicuous canthus rostralis; rostrum laterally compressed, appearing pointed in dorsal profile; vocal sac spherical; parotoid glands minutely tuberculate, nearly spherical in shape, and roughly same size as upper eyelid; tympanum round to teardrop-shaped; limbs short, with small spinose warts; hind limbs with two slightly enlarged metatarsal tubercles; inner tubercle somewhat spadelike and slightly larger than outer; moderate degree of interdigital webbing; ventral surfaces granular with widely spaced, small warts in femoral region only. **COLORATION** Variable; usually covered dorsally with reddish tubercles, commonly set in small black blotches on light background, producing a spotted appearance; males generally darker than females; tubercles redder and more conspicuous in juveniles; ground color light gray in center of back and darker laterally; eyelids light anteriorly and dark posteriorly; upper labial region lighter than ground color; forelimbs somewhat banded; reddish warts on hind limbs; gular and abdominal region whitish to cream-colored; pectoral region sometimes weakly spotted; ventral portions of limbs may be orangish distally; iris copper-colored in life. **TADPOLES** Larvae attain a length of nearly 40 mm; labial tooth row formula 2/3; oral papillae continuing ventrally for short distance; tail approximately same height as greatest depth of body; spiracle sinistral, lying somewhat near midline; body usually black with metallic flecking ventrally; some large individuals weakly mottled; tail fins transparent with evenly spaced, dark-colored dots; iris bronze to copper-colored in life. **MATING CALL** A prolonged crisp musical trill lasting 6 to 10 seconds. The pitch is usually constant but may begin slurring up to the trill and dropping off slightly at the end of the call. If several toads are calling in close proximity, the noise is loud enough to be uncomfortable.

GEOGRAPHIC VARIATION Geographic variation in *Bufo punctatus* in Baja California is primarily associated with color pattern and the number and size of its tubercles. Some of this variation shows clinal geographic trends. Cape Region populations generally have slightly larger tubercles on the dorsum. These are dispersed among more numerous, smaller, reddish tubercles, with as many as five highlighted within a single dark blotch. North of the Cape Region, especially in the area between Loreto and San Ignacio, the tubercles are more numerous and spinose but smaller. In populations north of Cataviña, the tubercles are numerous but inconspicuous because they are not always set off in a dark blotch. Cape Region populations also have a lateral ground color that is dark, making them appear to possess a very wide, light-colored vertebral stripe. This lateral darkening decreases in intensity from south to north and is usually not

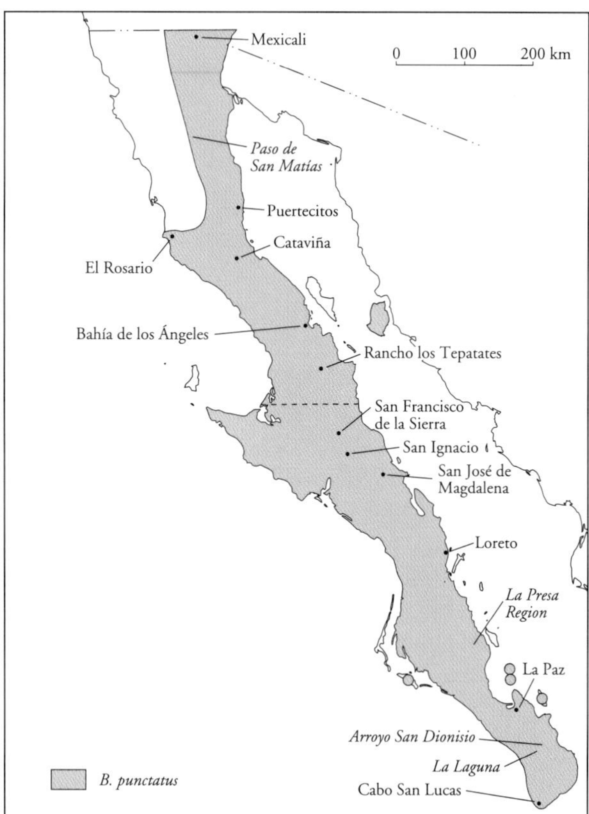

MAP 2.3 Distribution of *Bufo punctatus* (Red-Spotted Toad)

found in populations north of Loreto. *Bufo punctatus* exhibits localized color patterns as a function of substrate matching. Specimens from the thickly vegetated foothills and arroyos of the Sierra la Laguna, particularly La Laguna and to some extent La Burrera, are much darker in overall ground color than toads of the nearby sandy, more open coastal plain. Toads from the open and sparsely vegetated Vizcaíno Peninsula in west-central Baja California may have a unicolored tan dorsum (Grismer et al. 1994). In the Sierra la Libertad of central Baja California, toads tend to have little in the way of black blotches but have well-developed reddish tubercles, except in the Los Tepatates area, where they live in granitic arroyos. Here they lack nearly all red markings as well.

NATURAL HISTORY *Bufo punctatus* occurs in a wide variety of habitats ranging from the tropical deciduous forests of the Arid Tropical Region in the south to the creosote bush scrub communities of the Lower Colorado Valley Region in the northeast. Within its varied habitats, *B. punctatus* is usually terrestrial, even though it can be a remarkable climber. I have found individuals 1 to 2 m above the ground on the side of a nearly vertical rock face along an arroyo in the Sierra la Giganta, near Loreto. I even observed one toad in the Cape Region hopping up a nearly vertical tree trunk to perch on a branch 3 m above the ground.

In desert areas, *B. punctatus* is most commonly found near rocky streams or springs, where it is able to reproduce. Stebbins (1962) stated that its distribution in such areas is spotty and coincides with the occurrence of water. In Baja California, however, I have observed many specimens from areas where no such springs or seepages occur for several kilometers. In areas lacking permanent or even seasonal sources of water, *B. punctatus* must rely on temporary rain pools in which to reproduce. On Isla Tiburón, *B. punctatus* is known to breed in the El Sauzal water hole (Felger and Moser 1985), and larvae have been collected from Caracól and Chalate during late September. On the southwest end of Isla Cerralvo, the steep rocky foothills meet the low coastal dunes along the back edges of the beaches, forming a long, narrow basin in which rainwater and runoff collect and provide breeding pools. *Bufo punctatus* also breeds in a small, palm-lined canyon with semipermanent water at the northeastern end of the island. The breeding periods for *B. punctatus* are episodic, somewhat aseasonal, and localized, coinciding largely with the occurrence of storm tracks. Overall, the breeding season extends from late February through November. This is considerably longer than those of other toads in Baja California, and individuals have been observed abroad at San José de Magdalena and in the Cape Region during early January. During mid-March, just north of Cataviña, I observed a gravid female and an adult male sitting side by side next to a small, water-filled depression in a granite rock. Several toads were heard calling and found in amplexus in a small well in the town of San Francisco de la Sierra during late July. Although it had not rained, many toads were observed on two consecutive nights at least 100 m from the well, heading in the direction of the calls. McClanahan et al. (1994) demonstrated that *B. punctatus* resides in moist microhabitats during the summer months, and at night travels relatively long distances to water. I have found tadpoles and nearly metamorphosed individuals in the Cape Region during early October. Banks (1962) reports collecting tadpoles in a small spring in an arroyo on the northeast side of Isla Cerralvo during late October.

Bufo punctatus is chiefly nocturnal and most active during periods of precipitation, when hundreds of individuals seem to emerge from nowhere and are heard calling throughout the desert. During nighttime rainstorms in the Cape Region, toads cover the roads, making it impossible to drive without running them over. Following a June downpour in the southern Sierra la Giganta, I observed a *B. punctatus* floating downstream in an arroyo I was attempting to cross. On warm, dry days, toads remain underground in burrows or beneath rocks and surface debris if the substrate is sufficiently moist. At Rancho los Tepatates in the Sierra la Libertad, I have found toads residing in the cracks of large granite boulders near ponds in the arroyo bottom. It is common, however, to find *B. punctatus* out at night from June through October even if it has not rained. McClanahan et al. (1994) demonstrated that by residing in moist microhabitats during the day and being able to tolerate a 40 percent loss in body water, toads can forage at night during the dry season.

Bufo punctatus lays black and white eggs in short strands or sometimes singularly along the bottom of a breeding pool. Their gelatinous covering, which is very sticky, in most cases serves to anchor them to the substrate or each other. In temporary rainpools, eggs may hatch within three days, and the tadpoles may transform in as little as one week (McClanahan 1967). Because of such a rapid transformation, a newly transformed froglet may measure less than 15 mm in total length.

Bufo punctatus eats a variety of small arthropods including bugs, beetles, bees, and ants. I have observed large individuals eating newly transformed froglets in Arroyo San Dionisio in the Cape Region. One night during mid-June in a canyon of the San Francisco de la Sierra, several *B. punctatus* were drawn to our lanterns to eat the flying insects attracted by the light.

REMARKS A locality record for *B. punctatus* from 9.2 miles south of Rosarito, BC (Stebbins 1985), is approximately 300 km north of the next northernmost record for this species in northwestern Baja California and would be the first and only true record of this species within the California Region. However, this record more likely refers to El Rosarito, 39 km south of Punta Prieta, well within the recorded range of this species.

FOLKLORE AND HUMAN USE In the Cape Region, many of the rancheros believe that if a person is bitten by a rattlesnake, its venom can be drawn out by cutting open the belly of *B. punctatus* and applying its viscera to the wound. It is also believed in some areas that if this toad is stepped on, its red warts will sting and cause the foot to become infected.

FIG. 2.7 ***Bufo woodhousii***

WOODHOUSE'S TOAD; SAPO

Bufo woodhousii Girard, 1854:86. Substitute name for *Bufo dorsalis* Hallowell 1852 because of preoccupation (Kellogg 1932:32). Type locality: "[Territory of] New Mexico, having so far been found in the province of Sonora, and in the San Francisco Mts." Restricted to "San Francisco Mountain, Coconino County," Arizona, USA, by Smith and Taylor (1948:40). Holotype: USNM 2531

IDENTIFICATION *Bufo woodhousii* can be differentiated from all other frogs and toads in the region of study by a faint, light-colored, thin vertebral stripe beginning at the base of the cranial crests and terminating at the vent and by well-developed cranial crests.

RELATIONSHIPS AND TAXONOMY *Bufo woodhousii* is a member of the *Bufo americanus* group and most closely related to a North American group of toads comprising *B. hemiophrys, B. fowleri, B. houstonensis,* and *B. americanus* (Gergus 1994). Gergus (1994) discusses taxonomy.

DISTRIBUTION *Bufo woodhousii* ranges widely throughout the central United States south to Durango, Mexico (Stebbins 1985). It is also known from numerous isolated populations in the western United States and northern Mexico. In Baja California, it occurs only in the extreme northeast (Linsdale 1932) in association with the water systems of the Río Colorado delta (map 2.4).

DESCRIPTION Moderately large, adults reaching 135 mm SVL; skin rough and covered with numerous, small, spinose tubercles; larger tubercles tipped with more than one spine; head relatively small, rounded in dorsal profile; cranial crests well-developed, interorbital portions parallel; preorbital crests poorly developed; vocal sac rounded; spinose tubercles on upper eyelids; parotoid glands elongate, 1.5 to 2 times longer than upper eyelids, and with spinose tubercles; tympanum small and vertically oval; limbs short and robust, covered with small, spinose tubercles; upper portion of hind limb hidden within fold of skin on body; two enlarged, dark-colored, metatarsal tubercles present; inner tubercle larger than outer and possessing a spadelike cutting edge; interdigital webbing well-developed; ventral surfaces of body and limbs weakly tuberculate. **COLORATION** General ground color dusky yellow to olive above with tubercles tipped in dark reddish-brown; females usually lighter in overall ground color; tubercles on margin of lower jaw in males set off by lighter ground color; parotoid glands may be blotched posteriorly; upper eyelids usually darkly marked centrally; most larger warts on body set in small, darker-colored blotch; pale, thin, vertebral stripe beginning at level of cranial crests and terminating at vent; sides of body weakly mottled; limbs weakly banded dorsally; posterior margin of upper hind limbs mottled; ground color of ventrum beige to dusky cream and unmarked to weakly spotted; males occasionally have gular stippling; chest spot common; iris brass-colored in life. **TADPOLES** Larvae attain a maximum length of 25 mm; labial tooth row formula 2/3; oral papillae present; ventral fin not arched and of similar height throughout length; spiracle sinistral and directed dorsoposteriorly; body dark drown to black; dorsal portion of tail musculature usually slightly lighter than ventral portion of tail musculature; dorsal fin slightly pigmented with a few scattered spots; ventral fin usually immaculate; eyes black with metallic flecking dorsally. **MATING CALL** The call of *Bufo woodhousii* is a nasal *wh-a-a-a-a* that begins suddenly and usually drops off after one to three seconds (Sullivan et al. 1996).

FIGURE 2.7 *Bufo woodhousii* (Woodhouse's Toad) from Mexicali, BC

NATURAL HISTORY *Bufo woodhousii* is most commonly found in the open, flat flood plains of the Río Colorado delta in close proximity to a water source. In historical times, these areas were marshy grasslands supporting, in some places, thick stands of Macdougal Cottonwood *(Populus macdougalii)* and willows *(Salix gooddingi* and *S. hindsiana)*. Much of the native vegetation has been cleared, and some of the more arid portions of the delta, away from the river, have been converted to agricultural areas. *Bufo woodhousii* is most abundant here and can be found in roadside ponds or very small, restricted, marshy areas, where males can be heard calling from the reeds. In the southwestern United States and presumably in extreme northeastern Baja California, *B. woodhousii* breeds from February to July but may resume breeding later in the year if the rains are heavy enough. Up to 25,000 eggs are laid in a stringy, tangled mass that is anchored to debris or vegetation. *Bufo woodhousii* is usually nocturnal; it is most abundant during periods of summer precipitation but can be heard calling in Baja California in early April. It can occasionally be found abroad during the day. *Bufo woodhousii* is equipped with well-developed metatarsal tubercles with which it constructs burrows in soft soils for use during periods of inactivity. It feeds on a variety of arthropods (Stebbins 1985). In Baja California, it feeds primarily on pestilent species that occur in association with crops or arthropods associated with some of the original riparian systems of the Río Colorado delta region.

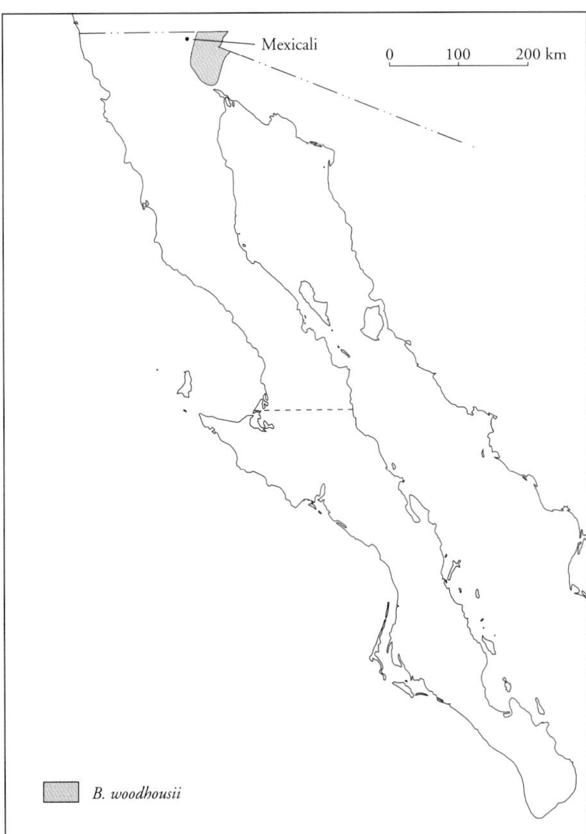

MAP 2.4 Distribution of *Bufo woodhousii* (Woodhouse's Toad)

REMARKS *Bufo woodhousii* originally entered Baja California along the Río Colorado and its tributaries, but its distribution on the delta has been further enhanced by the construction of irrigation canals and the practice of flooding agricultural fields.

Hylidae

TREEFROGS

The family Hylidae contains approximately 38 genera and 760 species (Pough et al. 2001). This large group of frogs is represented throughout the New World, the Australo-Papuan region, temperate Eurasia, northern Africa, and the Japanese Archipelago. Its greatest diversity occurs in the New World tropics. Hylids are generally small to moderately sized arboreal frogs with large eyes, long legs, and expanded toe disks, although some hylids are much more terrestrial. Hylids are distinguished from nearly all other frogs and toads by their possession of claw-shaped terminal phalanges (Ford and Cannatella 1993) and the combination of an incompletely fused pectoral girdle, dilated sacral diapophyses of the pelvic girdle, incomplete fusion of the astragalus and calcaneum bones of the ankle, and small segments of extra cartilage known as intercalary elements that occur between the last and second-to-last phalanges and set off each toe disk from the rest of the digit. In Baja California, the Hylidae are represented by the genus *Hyla*.

Hyla Laurenti

TREEFROGS

The genus *Hyla* is a very large group containing over five hundred species of small to moderately sized frogs (Pough et al. 2001). This genus exhibits a tremendous diversity of life styles and occurs in a broad array of habitats, essentially covering the range of that found in the Hylidae. *Hyla* ranges throughout central and southern Europe, eastern Asia, northwestern Africa, North, Central, and South America, and the Greater Antilles in the Caribbean. Two species of *Hyla* occur in Baja California, and one of these is known from an adjacent Pacific island.

FIG. 2.8 ***Hyla cadaverina***

CALIFORNIA TREEFROG; RANITA

Hyla cadaverina Hallowell, 1854:96 (nec. *Hyla nebulosa* Spix 1824). Type locality: "Tejon Pass," Los Angeles County, California, USA. Syntypes: ANSP 1987–88

FIGURE 2.8 *Hyla cadaverina* (California Treefrog) from Cañón Santa Isabel along the east face of the Sierra Juárez, BC

IDENTIFICATION *Hyla cadaverina* can be distinguished from all other frogs and toads in the region of study by having expanded toe tips, no webbing on the hand, and warty skin on the dorsum, and by lacking a distinct postorbital stripe.

RELATIONSHIPS AND TAXONOMY *Hyla cadaverina* is most closely related to *H. regilla* of the western United States and northern Mexico (Cocroft 1994; Hedges 1986).

DISTRIBUTION *Hyla cadaverina* ranges from eastern San Luis Obispo County, California, south to Las Palmas in Baja California (Gaudin 1979; map 2.5). It has recently been extirpated farther south, at Bahía de los Ángeles, because of habitat conversion by humans (Grismer and McGuire 1993).

DESCRIPTION Small, with adults reaching approximately 45 mm SVL; skin of dorsum weakly to moderately tuberculate; distinct fold of skin running across chest; head wide, flat, and slightly broader than body; cranial crests absent; canthus rostralis round; nostrils slightly elevated; vocal sac spherical when inflated; parotoid glands absent; fold of skin extending from posterior corner of eye to angle of mouth; eyes large and protuberant; tympanum round; skin on ventral surfaces granular; forelimbs relatively short and not particularly robust; digits long and slender, bearing truncated discs at tips; interdigital webbing present; hind limbs slender and over twice the length of forelimbs; weak tarsal fold present; small, ovoid metatarsal tubercle present at base of first digit; approximately 75 percent of length of digits webbed; toes relatively long, slender, and bearing small truncate disks at tips. **COLORATION** Extremely variable; ground color of dorsum dull grayish-brown, with or without dorsal spots of varying size and intensity; usually narrow, irregular, light-colored labial stripe present; varying degrees of gray mottling between larger dorsal spots on body; hind limbs banded; ventral surfaces yellow to beige; gular region whitish in females and darker in males; iris bronze with dark reticulations in life. **TADPOLES** Tadpoles attain a maximum length of approximately 44 mm; labial tooth row formula 2/3; oral papillae encircle mouth except dorsally; spiracle sinistral, directed posteriorly; dorsal and ventral fins nearly same height; tadpoles generally dark brown in color above and tan below; dorsal surface of body speckled with a combination of dark and brassy flecks matching color of substrate; dark flecking in caudal fins. **MATING CALL** A series of short, low-pitched, monotonic, gosling-like, paired squawks starting and ending abruptly and lasting no longer than half a second.

GEOGRAPHIC VARIATION This species is very good at substrate matching, and thus its color pattern varies from arroyo to arroyo and is influenced by the color of the rocks along the watercourse it inhabits. Some general trends, however, are noticeable. Frogs from desert or arid regions tend to have much more contrast between their light ground color and their dark dorsal blotching. Also, the dorsal blotches tend to be smaller. This tendency is most common in populations inhabiting arroyos of the eastern foothills of the Peninsular Ranges. Farther south, this trend starts to develop in the vicinity of El Rosario at the northern reaches of the more

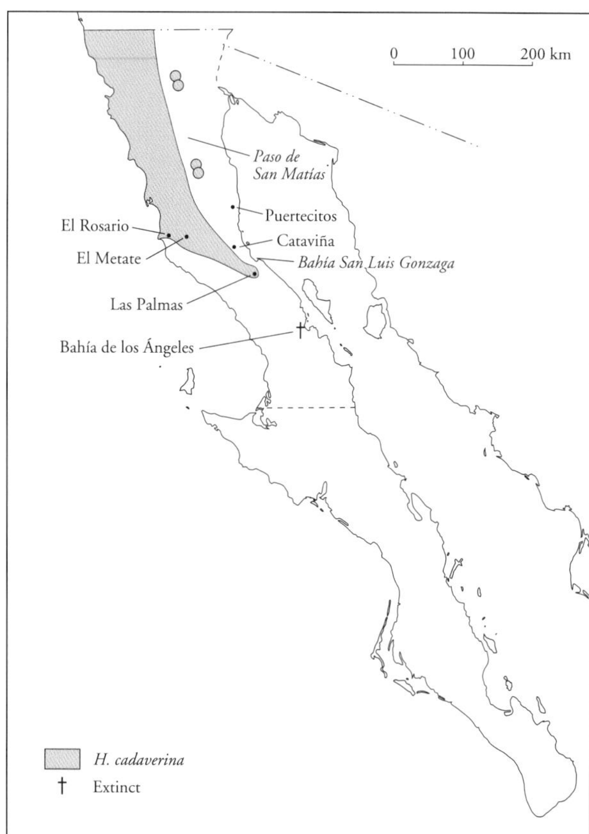

MAP 2.5 Distribution of *Hyla cadaverina* (California Treefrog)

arid Vizcaíno Region. It is even more obvious a short distance to the south at El Metate. *Hyla cadaverina* from cismontane environments are generally more brown than gray and tend to have a darker overall coloration that is more mottled, obscuring the blotching pattern to some extent. Some individuals from a small population in Arroyo Grande, approximately 18 km east of El Rosario, lack dorsal blotches entirely, and their dorsal ground color is brick red. An explanation for this coloration is difficult to find because this arroyo lacks red rocks, and the frogs are quite noticeable when perched on the water-worn gray granite boulders.

The southernmost extant population of *H. cadaverina* occurs at Las Palmas, approximately 28 km south of Bahía San Luis Gonzaga (Grismer and McGuire 1993). This population occurs in a small seepage surrounded by Fan Palms (*Washingtonia robusta*) that has been converted into a well by local ranchers, who have lined it with gray cement blocks. The frogs of this population have a highly variable color pattern. When basking on the cement blocks they have large, dark, irregular blotches that match the mottled coloration of the cement. When basking on the palm fronds, they may be unicolored dull white, white with very fine black peppering, or covered with small blotches. It is not known whether an individual frog can manifest each color pattern.

NATURAL HISTORY *Hyla cadaverina* primarily inhabits rocky canyons, where it is found in the vicinity of pools or swiftly moving streams, although near El Rosario in Arroyo Rosario, frogs can be found calling along the muddy banks in the silty flood plain east of Mexican Highway 1. It is most common in the deep rocky arroyos that drain the western and eastern slopes of the Peninsular Ranges. In such areas, *H. cadaverina* may congregate in great numbers. It is common to see individuals sitting on granitic boulders near the edge of the water in direct sunlight, and some have even been observed foraging during the day at great distances from the water (McClanahan et al. 1994). *Hyla cadaverina* generally breeds from March through July, but I have heard males calling as late as early October at Cataviña. Females deposit eggs in the quiet parts of the watercourse and anchor them to rocks, sticks, leaves, or floating debris. Eggs are deposited in a single gelatinous envelope, which may contain several hundred. According to Stebbins (1962), transformation from tadpole to froglet takes 40 to 75 days. *Hyla cadaverina* feeds primarily on the small arthropods found along the margins of the streams, including water bugs, ants, spiders, and centipedes.

REMARKS It is puzzling that *H. cadaverina* does not occur further south, as there are plenty of suitable rocky watercourses throughout Baja California. Its discovery in small seeps or springs at great distances from major watercourses is always surprising; the population at Las Palmas, mentioned above, is a particularly good example. Originally this spring consisted of a small pool less than 1 m in diameter, and even then frogs were present. The local ranchers told me that "the frogs were always there, they are frogs. Our ancestors found this spring by hearing their [the frogs'] sad songs during the night." When asked why their songs were sad, the ranchers replied, "Because there is so little water." The nearest major water source containing *H. cadaverina* is in Arroyo Santa María (Grismer and McGuire 1993), approximately 15 km to the north. Amazingly, this population may have evolved a color pattern that matches that of the cement blocks. Prior to the enlargement of this seep, the frogs resembled the surrounding rocks.

FIG. 2.9

Hyla regilla

PACIFIC TREEFROG; RANITA

Hyla regilla Baird and Girard, 1852c:174. Type locality: "Sacramento River in Oregon and Puget Sound." Restricted to "Sacramento County, California," by Schmidt (1953:720). Restricted further to Fort Vancouver, Washington, USA, by lectotype designation of Jameson, Mackey, and Richmond (1966:553). Syntypes: USNM 9182 (Puget Sound); USNM 15409 (Sacramento River); USNM 9182 designated lectotype by Jameson, Mackey, and Richmond (1966:553).

IDENTIFICATION *Hyla regilla* can be differentiated from all other frogs and toads in the region of study by having expanded toe disks and a wide, dark line running from the tip of the snout posteriorly through the eye to the forelimb insertion.

RELATIONSHIPS AND TAXONOMY See *Hyla cadaverina.* Jameson et al. (1966) and Duellman (1970) discuss the taxonomy of *H. regilla.*

DISTRIBUTION *Hyla regilla* ranges widely throughout western North America from southern British Columbia, Canada, and western Montana, USA, south to the tip of Baja California. In northern Baja California, *H. regilla* ranges fairly continuously throughout the montane and cismontane areas of the northwest. South of El Rosario, *H. regilla* occurs in several isolated oases and small pools (Grismer and McGuire 1993), continuing south into the less arid Cape Region, where its distribution becomes more continuous. It also occurs in many isolated canyons, backyards, and gardens on Isla Cedros, off the west coast of central Baja California (Grismer and Mellink 1994; map 2.6). There is a specimen (USNM 23861) catalogued as being from Isla Natividad, between Isla Cedros and the penin-

FIGURE 2.9 *Hyla regilla* (Pacific Treefrog) from Arroyo San Pablo, Sierra San Francisco, BCS

sula. There are, however, no known sources of fresh water on this island, and the record is considered questionable (Grismer 2001a).

DESCRIPTION Small, with adults reaching 44 mm SVL; skin of dorsum smooth to weakly pustulate; fold of skin extending from slightly above forelimb insertion to dorsal margin of hind limb insertion; skin granular ventrally; single median, spherical vocal sac; prominent fold of skin extending across chest; head equal in width to body; nostrils and eyes protuberant; fold of skin extending from posterior margin of eye across top of tympanum to forelimb; tympanum small; forelimbs short, not robust; dorsal fold of skin across wrist; digital discs expanded; webbing on hands absent; hind limbs long, slender to moderately robust, with weak tarsal fold; transverse fold extending across heel; toe disks smaller than those of fingers; conical tubercle medial to metatarsal tubercle; digits webbed for half to two-thirds of their length. **COLORATION** Ground color variable, ranging from dark gray, tan, or copper to bright lime green; lateral margins of body with varying degrees of dark blotches and flecking; large markings occasionally present on dorsum; posteriorly directed, triangularly shaped marking between eyes; wide dark stripe running posteriorly from tip of snout through eye to slightly past forelimb insertion; stripe set off by thin white dorsal and ventral outlining; forelimbs and hind limbs unicolored or banded; ventrum dull yellow to off-white. **TADPOLES** Tadpoles reach approximately 30 mm in total length; labial tooth row formula 2/3; spiracle sinistral, directed posteriorly; caudal musculature moderately slender; dorsal and ventral fins 1.5 times as deep as caudal musculature; oral papillae encircling mouth except dorsally; ground color dark brown with scattered golden chromatophores and guanophores dorsally and laterally; caudal musculature and dorsal fin darkly flecked; caudal fin nearly transparent. **MATING CALL** Consists of a series of short diphasic notes at a rate of 30 notes per minute (Duellman 1970). The call consists of two parts, *rib-it,* at a rate of about one per second. The second syllable rises in inflection.

GEOGRAPHIC VARIATION *Hyla regilla* in northern Baja California, north of San Ignacio, shows tendencies toward a more pustulate dorsum, thinner and longer hind limbs, and slightly less webbing on the foot. These observations prompted Jameson et al. (1966) and Duellman (1970) to consider the populations from Baja California Sur as *H. r. curta* and those from Baja California as *H. r. hypochondriaca.* Duellman (1970) remarked that specimens from the southernmost part of Baja California are slightly larger and have proportionately larger tympanums. I have noticed that the dorsal color pattern of treefrogs from central Baja California, in the vicinity of San Ignacio, is less diffuse than that of individuals from the Cape Region. Specimens in the Cape Region also tend to have a more distinct series of tubercles on the ventrolateral margins of their forearms.

Deep within the canyons of the Sierra San Francisco, the color pattern variation in *H. regilla* is remarkable. The ground colors range from shades of green, yellow, brown, and copper to tan and brick red with many different overlying dorsal patterns, all of which can be found in the same small pond. I know of no other population in Baja California that shows this degree and range of variation.

NATURAL HISTORY In northern Baja California, *H. regilla* occurs west of the crest of the Peninsular Ranges from sea level to 2,750 m in the Sierra San Pedro Mártir (Welsh 1988). It is a common inhabitant of streamsides and meadows. In the Vizcaíno Region, *H. regilla* is restricted to palm-lined arroyos with permanent water and is usually sympatric with *H. cadaverina* where rocks are present. In Baja California Sur, *H. regilla* occurs in disjunct oases throughout the Vizcaíno, Magdalena, Arid Tropical, and Central Gulf Coast regions (Grismer and McGuire 1993). Such habitats are usually dominated by typical streamside vegetation and occur in areas ranging from open plains to deep, sheer-sided volcanic canyons. At one locale in the Sierra Vizcaíno, the frogs occur in a hand-dug well, a small natural pool measuring less than 2 m across, situated beneath a rock, and in the rock cracks of a small seepage. *Hyla regilla* can also be found in cattle watering tanks.

Hyla regilla is usually nocturnal and most active during times of precipitation from November to April in the north and from mid-July to early November in the south. During the day, frogs are commonly found beneath rocks, logs, and other surface litter and at the base of vegetation next to watered areas. Frogs may take refuge within the ribs of palm tree leaves, and their silhouettes are easily seen from beneath. During dry periods, frogs remain underground, taking refuge beneath rocks and logs or in burrows and rock cracks. I found 13 frogs beneath a small rock (75 cm) in the bottom of a dry, sandy arroyo in the Sierra la Libertad at Los Corrales. Adjacent to the rock was a small, hand-dug pit into which water seeped. The nearest permanent water source was approximately 8 km upstream at Tres Palmas, at the foot of the mountain. The ranchers at Los Corrales say that during heavy rains, the arroyo floods and fills with frogs. These frogs are presumably washed down from the permanent sources upstream.

Although *H. regilla* is a treefrog and has expanded toe disks, it is equally common in terrestrial microhabitats. Compared to other species of *Hyla,* the toe disks of *H. regilla* are greatly reduced, and this frog spends the days and drier parts of the year underground or underneath fallen vegetation. During daylight hours in Arroyo San Jorge in the Cape Region, I have observed this species within the confines of rock crevices and actually syntopic with the Peninsular Leaf-Toed Gecko (*Phyllodactylus xanti*). Densities of *H. regilla* can be very high. Within the canyons of the Sierra San Francisco, frogs occur on the sandy arroyo bottoms, on rocks, in the water, on brush, and in trees, as high as 3 m above the ground. On 24 June 1995, I estimated an average of 10 frogs per square meter along the sides of and within the streams. The local ranchers told me that there are even more frogs after it rains. I observed an even greater density of *H. regilla* at La Huerta, further north in the Sierra la Libertad, where I estimated 40 frogs per square meter.

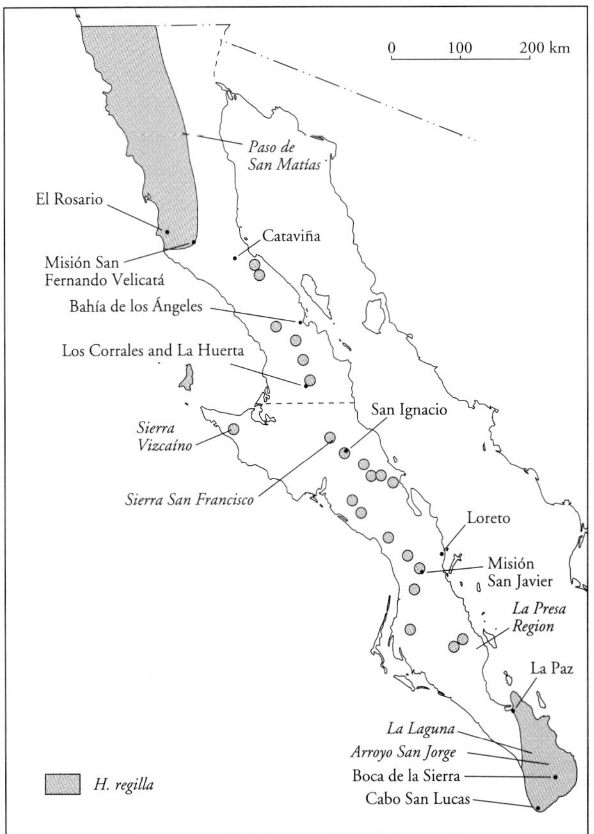

MAP 2.6 Distribution of *Hyla regilla* (Pacific Treefrog)

Hyla regilla breeds primarily during the wetter months of its activity period. I have seen tadpoles in mid-March on Isla Cedros and in the Sierra la Libertad. I have heard calling males from San Ignacio and Boca de la Sierra during April and seen tadpoles during late May at San Ignacio. In the Cape Region, I have heard males calling and observed amplexus from January through October and have found females with eggs in late March. In mid-March at Los Tepatates, I watched two chorusing males wrestle for possession of a particular calling site for nearly 15 minutes until one of them was displaced. In early January, chorusing *H. regilla* were observed from Misión San Fernando Velicatá in the north, southward throughout their range in central Baja California to Misión San Javier. Chorusing males were heard in Arroyos San Simón and Rosario during mid-March as well. It appears that in places with permanent water, such as areas north of El Rosario, the Cape Region, and some of the oases in central Baja California, *H. regilla* can be heard calling much later in the year, from March through July. These calls, however, may be territorial rather than advertisement for mates. In areas with semipermanent water and unpredictable rainfall, as in the oases throughout the Sierra la Libertad, *H. regilla* does most of its breeding from November through January. By March, little calling is heard in these areas, and the ponds contain tadpoles that are two to four weeks old. Egg clusters are attached to sticks, vegetation, or floating debris and may consist of as many as 70 eggs per cluster. Larvae may hatch within a week; the complete metamorphosis may take as long as 75 days. *Hyla regilla* eats a variety of small arthropods, the majority of which appear to be nonaquatic.

REMARKS North of the Isthmus of La Paz, *H. regilla* is known only from isolated mesic refugia where there are areas of permanent water. It is likely that this species was widespread and continuously distributed in Baja California during wetter periods of the Pleistocene, but with the formation of the peninsular deserts became restricted to mesic outposts (Grismer and McGuire 1993; Grismer et al. 1994). It is certain that there are other isolated populations of this species in the many unexplored canyons of the Sierra la Giganta that harbor hidden permanent springs. In some of the water systems west of San Ignacio, there is evidence that *H. regilla* is being displaced by the Bullfrog, *Rana catesbeiana* (Grismer and McGuire 1993).

Ranidae

TRUE FROGS

The family Ranidae is an unnatural assemblage of frogs containing approximately 47 genera and over 700 living species (Ford and Cannatella 1993). Except for southern South America, most of Australia, and Antarctica, ranids are cosmopolitan in distribution. Most ranids are moderately sized, smooth-skinned, aquatic frogs, although some genera are stout, squat warty burrowers that are more toadlike in appearance and lifestyle. In Baja California, the Ranidae are represented by the genus *Rana*.

Rana Linnaeus

POND FROGS

The genus *Rana* is a large group of frogs containing over 450 species. *Rana* is widespread on every continent except in southern South America, Antarctica, and most of Australia. Species are generally amphibious, are moderate to large in size, and have smooth skin, long legs, pointed snouts, and distinct dorsolateral folds. Three species of *Rana* naturally occurred in Baja California—*R. aurora, R. boylii,* and *R. yavapaiensis*—although the latter two have probably been extirpated. Two other species, *R. catesbeiana* and *R. forreri,* and also possibly *R. berlandieri,* have been introduced.

FIG. 2.10 **_Rana aurora_**

RED-LEGGED FROG; RANA

Rana aurora Baird and Girard, 1852c:174. Type locality: "Puget Sound," Washington, USA. Syntypes: USNM 11711 (four specimens)

IDENTIFICATION *Rana aurora* can be distinguished from all other frogs and toads in the region of study by having a pointed snout and poorly developed dorsolateral folds that are discernible only posteriorly, and lacking large, dark dorsal spots.

RELATIONSHIPS AND TAXONOMY Green (1986a) proposed that *Rana aurora* is most closely related to *R. cascadae* of the Pacific Northwest. The "subspecies" *R. aurora draytonii,* which occurs in Baja California, is most likely a distinct species. *Rana a. aurora* and *R. a. draytonii* are very different from one another in allozyme frequencies, karyology, morphology, and behavior (Green 1985a,b, 1986a,b; Hayes and Miyamoto 1984) and show no signs of intergradation where their ranges come into contact (Green 1986a).

FIGURE 2.10 *Rana aurora* (Red-Legged Frog) from the lower portion of Arroyo San Rafael near Colonet, BC

DISTRIBUTION *Rana aurora* ranges from the Cascade-Sierra crest west from southwestern British Columbia and Vancouver Island and south to Arroyo Santo Domingo in northwestern Baja California (Altig and Dumas 1972; map 2.7). In Baja California, *R. aurora* appears to be restricted to the west side of the Peninsular Ranges, ranging from sea level to 2,200 m in the Sierra San Pedro Mártir (Welsh 1988). It is not known to occur in the Sierra Juárez or its foothills.

DESCRIPTION Moderately large, with adults reaching over 120 mm SVL; skin smooth, dorsolateral fold weak; head no wider than body; snout pointed; eyes weakly protuberant; cranial crests absent; nostrils flush to slightly elevated; upper lip slightly swollen and elevated above preorbital region; vocal sacs paired; tympanum round; forelimbs short and robust; interdigital webbing in hand absent; hind limbs robust and of moderate length; one elongate inner metatarsal tubercle; low rounded ridge of skin extends posteriorly to metatarsal tubercle; digits webbed for nearly entire length in males and slightly less so in females. **COLORATION** Ground color dark brown to olive drab; juveniles slightly brighter, males tending to be more spotted; a distinct to weak dark brown eye mask extending from nostril to angle of jaw and usually bordered below by dull-white band; dorsum covered with poorly defined small dark spots, with occasional lighter centers; dorsolateral fold occasionally set off in slightly lighter color; poorly defined bands on limbs; groin distinctly mottled in olive green or black and white; lower surfaces of upper portions of hind limbs dull yellow, foreleg with reddish tint; ventral region slightly speckled; individuals capable of color change from dark to light. **TADPOLES** Reach a length of nearly 30 mm; spiracle sinistral, directed posteriorly, centered between mouth and base of tail; dorsal fin slightly higher and more arched than ventral fin; labial tooth row formula 2/3; oral papillae not continuous dorsally; dorsal and lateral surfaces of body dark brown with small, irregularly shaped, poorly defined darker spots; belly lighter; tail usually lighter in color than body; faint mottling pattern on caudal fins; irregular dark lines bordering caudal myomeres. **MATING CALL** A stutter or grating sound usually followed by a low-pitched growl. The entire call lasts an average of three seconds, and the pitch expresses little fluctuation. This frog may emit a squawking sound as it jumps into the water.

NATURAL HISTORY *Rana aurora* occurs in suitable riparian systems of the California and Baja California Coniferous Forest Regions. It is an amphibious species that resides in streams and ponds with deep permanent pools, where it is seldom found more than one leap away from the water. It tends to avoid fast-moving water in rocky streams. During the day, *R. aurora* basks and feeds at the edge of the watercourse and escapes predators through crypsis or by jumping into the water and burrowing into the mud. At higher elevations it will remain burrowed in the mud through the winter. The overall breeding season lasts from late January to late July, with populations at lower elevations probably reproducing earlier than those in the Sierra San Pedro Mártir. Welsh (1988) noted larvae with well-developed legs in early August in Arroyo San Rafael at 1,525 m. Around 4,000 eggs are laid in relatively shallow water and occur in flat masses that are usually attached to vegetation. After the tadpoles hatch, they may take as long as seven months to transform. At higher elevations, they typically overwinter as tadpoles because of the short growing season. *Rana aurora* feeds on all small arthropods that inhabit the stream or pondside in which it resides. Isopods, beetles, and caterpillars appear to be the most common food sources.

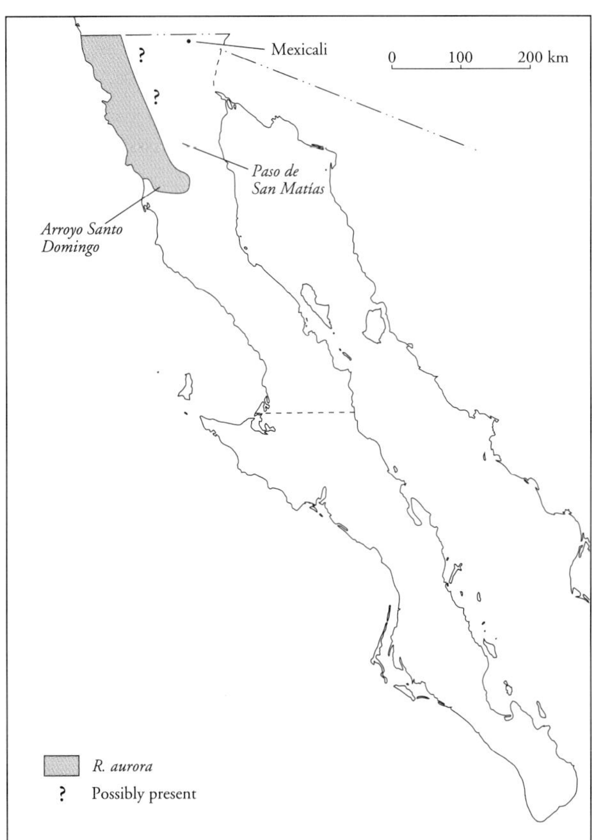

MAP 2.7 Distribution of *Rana aurora* (Red-Legged Frog)

REMARKS The range of *R. aurora* in Baja California may extend further south than is reported here, especially in the inland areas of the southwestern foothills of the Sierra San Pedro Mártir. This region has some suitable habitat, but its inaccessibility leaves it rather unexplored. Northern Baja California may be one of the last refuges for the southern population of *R. aurora* (recognized by some as *R. a. draytonii*) owing to the destruction of its habitat, the introduction of bullfrogs (*R. catesbeiana*) into southern California, and its initial commercial exploitation during the late 1800s and early 1900s (Jennings and Hayes 1985).

FIG. 2.11

Rana catesbeiana

BULLFROG; RANA DEL TORO

Rana catesbeiana Shaw, 1802:106. Type locality: "North America"; restricted to "vicinity of Charleston, South Carolina," USA, by Schmidt (1953:79). Holotype: not known to occur.

IDENTIFICATION *Rana catesbeiana* can be distinguished from all other frogs in the region of study by its smooth skin and an enlarged, prominent fold running from the posterior margin of the eye to the anterior portion of the forelimb insertion. This fold of skin forms the dorsoposterior border of the tympanum.

RELATIONSHIPS AND TAXONOMY Part of the *Rana catesbeiana* group (Frost 1985).

DISTRIBUTION *Rana catesbeiana* naturally ranges from the Atlantic coast to eastern Colorado and New Mexico. West of the Rocky Mountains, it usually occurs in nondesert areas from southern British Columbia south to northwestern Baja California, to at least Colonet. It is also known from the Río Colorado drainage in northeastern Baja California and areas east of Tijuana in the northwest. Disjunct populations occur in San Ignacio, Mulegé, San José de Magdalena, and again in the Cape Region at Santiago and Aguas Calientes (Grismer and McGuire 1993; map 2.8).

DESCRIPTION Large, with adults reaching 200 mm SVL; skin on body smooth; head wide and flat; snout rounded;

FIGURE 2.11 *Rana catesbeiana* (Bullfrog) from San Ignacio, BCS

eyes weakly protuberant; cranial crests absent; nostrils slightly raised; vocal sacs paired; tympanum round, nearly twice as large as eye; distinct fold of skin extending from posterior corner of eye to tympanum, forming its dorsoposterior border, then angling ventrally nearly 90 degrees and continuing to anterior margin of forelimb insertion; forelimbs short and robust; transverse fold of skin extending across top of wrist; interdigital webbing on hand absent; hind limbs robust and of moderate length; elongate, round, inner metatarsal tubercle present; outer metatarsal tubercle absent to weakly evident; webbing between digits 3 and 4 and 4 and 5 on foot deeply emarginate. **COLORATION** Ground color dark green to brown with poorly defined spots anteriorly that are much darker posteriorly and grade into a well-defined blotching pattern on upper surfaces of hind limbs; ventral surfaces dull white to cream-yellow with varying degrees of dark reticulate markings on throat and body, forming a faint but dense network. **TADPOLES** Reach approximately 140 mm in length; spiracle sinistral, directed dorsoposteriorly and located midway on body; dorsal fin arched and greater in depth than ventral fin at midtail; labial tooth row formula 2/3 or 3/3; oral papillae not continuous dorsally, occur in a single row ventrally and double rows laterally; dorsal surface of body olive green with scattered black spots of varying density; tail mottled with cream and dull green; ventrum cream-colored, sometimes with a weak reticulum of darker colors; chin grayish. **MATING CALL** A well-developed, deeply pitched bellow that suggests a *h-ooo-ow* or *h-aa-rum,* which increases in intensity and then tapers off toward the end. The call lasts from one to two seconds. This frog may make a squawking sound as it leaps into the water.

NATURAL HISTORY *Rana catesbeiana* is remarkably adaptable and may occur anywhere with permanent water. In cismontane northern Baja California, *R. catesbeiana* is found in pools along river bottoms, lakes, marshes, and reservoirs, in both shallow and deep water systems. In northeastern Baja California, *R. catesbeiana* is common in irrigation canals and along the banks of the Río Colorado (Clarkson and de Vos 1986). In central Baja California at San Ignacio, bullfrogs are abundant in the large ponds along the river course in areas of thick streamside vegetation. In Mulegé and San José de Magdalena, *R. catesbeiana* is surprisingly abundant in the small pools along the arroyos that are filled with cow urine and excrement. In the Cape Region at Santiago, bullfrogs can be heard calling from deep within the reed-covered marshes at the northwestern end of town. Near Aguas Calientes, I visited a narrow cave with flowing water that was filled with *R. catesbeiana,* and I even found specimens in the hot-water spring.

Rana catesbeiana is a semiaquatic frog. During periods of rain or even very humid weather, bullfrogs may leave a pond in search of a new one. *Rana catesbeiana* has been found as much as 2 km from the nearest water source, and I have observed dead specimens on Mexican Highway 3, just south of La Puerta in northeastern Baja California—even farther from water. Males may establish calling territories within a particular pond. These territories are usually 6 to

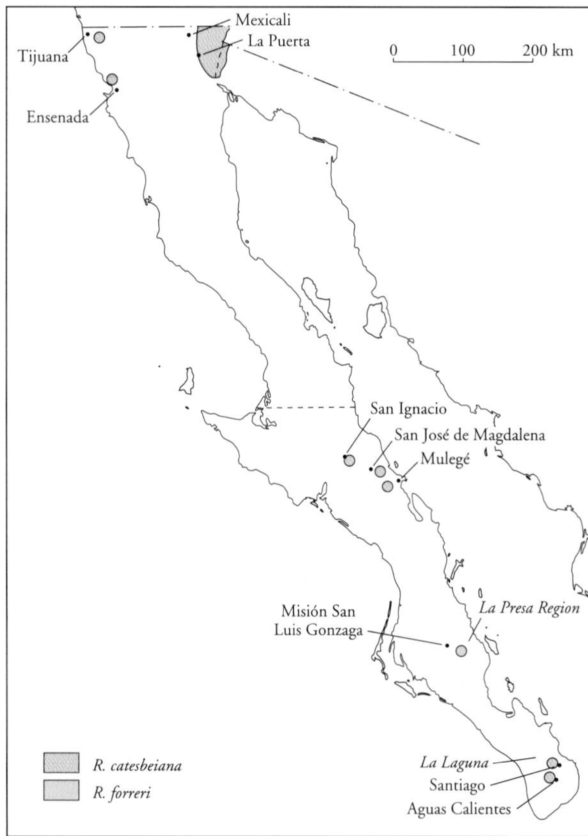

MAP 2.8 Distribution of *Rana catesbeiana* (Bullfrog) and *R. forreri* (Forrer's Grass Frog)

7 m from the shore where females are most prevalent. During daylight hours, males usually stay along the shore under the cover of vegetation. During the evening, when not calling, they are found in open areas not more than one leap's distance from the water. In northern Baja California, the breeding season usually extends from February to May, but I have heard males calling throughout the peninsula as late as August. As many as 15,000 eggs are deposited in flat masses nearly a meter across and may be free-floating or attached to aquatic vegetation. Eggs usually hatch within a week. In northern Baja California, tadpoles do not transform before winter but complete their metamorphosis the following spring and summer. This probably is not the case in central and southern Baja California, where temperatures remain relatively high year-round. *Rana catesbeiana* will eat just about anything that will fit into its mouth. The usual food items are aquatic pond or stream-associated arthropods; however, in the Lower Colorado River, Crayfish *(Procambarus clarki)* and wolf spiders (Lycosidae) are the most common prey. I have seen bullfrogs eating small Two-Striped Garter Snakes *(Thamnophis hammondii),* lizards, mice, mosquito fish, and even baby birds. Allan Blank of El Chorro at Aguas Calientes informed me that he observed a bullfrog knock down and attempt to eat a Xantus's Hummingbird *(Hylocharis xantusi)* that was feeding at the edge of a pond. Clarkson and de Vos (1986) report that *R. catesbeiana* in the Lower Colorado River feed on numerous species of arthropods, and they even found such diverse items as a young Muskrat *(Ondatra zebethicus),* a Western Diamondback Rattlesnake *(Crotalus atrox),* a Common Kingsnake *(Lampropeltis getula),* a Spiny Soft-Shell Turtle *(Apalone spinifera),* an Asiatic Clam *(Corbicula fluminea),* and a scorpion *(Hadrurus arizonicus)* in the stomachs they examined.

REMARKS *Rana catesbeiana* occurs naturally east of the Rocky Mountains and was introduced into Contra Costa County, California, in 1896 (Jennings and Hayes 1985). Since then, its range has expanded considerably to the north and south through additional introductions and natural dispersal. The occurrence of this species in the oases at San Ignacio, Mulegé, San José de Magdalena, and in the Cape Region is the result of unnatural introduction. It is not known, however, when this introduction occurred, and it is clear that this species affects the ecology of these systems (Grismer and McGuire 1993). An apparent population decline in *Thamnophis hammondii* and possibly *Hyla regilla* may be occurring in San Ignacio. In other such oases lacking *R. catesbeiana, T. hammondii* and *H. regilla* are much more common (Grismer and McGuire 1993).

FIG. 2.12 *Rana forreri*

FORRER'S GRASS FROG; RANA

Rana forreri Boulenger, 1883:343. Type locality: "Presidio, Sinaloa, Mexico." Holotype: BM 1882.12.5.7

IDENTIFICATION *Rana forreri* can be distinguished from all other frogs and toads in the region of study by having a pointed snout, strongly developed dorsolateral folds extending from the eye to the base of the hind limb, and dark dorsal spots.

RELATIONSHIPS AND TAXONOMY *Rana forreri* is the sister species of *R. berlandieri.* Frost (1985) discusses taxonomy.

DISTRIBUTION *Rana forreri* ranges from southern Sonora south along the Pacific coast to Jalisco, Mexico, and perhaps to northern Costa Rica (Frost 1985). In Baja California, it is known only from the La Presa region at Rancho San Juanito (see map 2.8) and Misión San Luis Gonzaga. It is likely that it also occurs in other nearby areas.

DESCRIPTION Moderately large, with adults reaching over 120 mm SVL; skin smooth, dorsolateral fold distinct; head no wider than body; snout pointed; eyes weakly protuberant; cranial crests absent; nostrils flush to slightly elevated; upper lip usually flush with preorbital region; vocal sacs paired; tympanum round, bordered dorsally by dorsolateral fold; forelimbs short and robust; no interdigital webbing in hand; hind limbs robust and of moderate length; prominent, elongate inner metatarsal tubercle; low, rounded ridge of skin extends posteriorly to metatarsal tubercle; digits webbed for nearly three-quarters of their length in males and females. **COLORATION** Ground color dark green to tan; eye mask absent; dorsum covered with large brownish blotches; dorsolateral fold conspicuous, set off in lighter color; well-defined dark bands on hind limbs; groin distinctly mottled in olive green or black and white; lower surfaces of upper portions of hind limbs dull yellow and occasionally speckled. **TADPOLES** Not seen from region of study. **MATING CALL** A low soft growling trill lasting 1.5 to 2 seconds that is best represented as a *w-a-a-a-a-a.*

NATURAL HISTORY In Baja California, *Rana forreri* occurs in ponds of the La Presa region, approximately 100 km north of La Paz. During the evening, individuals sit at the water's edge or along the shore up to 2 m from the water. I have observed calling males and gravid females during mid-July. Juveniles have been observed in January (E. Gergus, personal communication, 1998).

REMARKS *Rana forreri* is known only from the water systems near Rancho San Juanito in the La Presa region. I discovered this population in July 1995. The owner of the

FIGURE 2.12 *Rana forreri* (Forrer's Grass Frog) from Rancho San Juanito in the La Presa region, BCS. (Photograph by Chris Mattison)

ranch said that they have been in the ponds for a couple of years; therefore, they were probably introduced between 1991 and 1993. Also introduced into this water system were a single specimen of *Kinosternon* sp., a species of large freshwater shrimp, various aquarium fish, and a species of bass.

OTHER *RANA* THAT MAY OCCUR IN BAJA CALIFORNIA

Rana berlandieri
RÍO GRANDE LEOPARD FROG

Rana berlandieri has never been reported from Baja California but is very common in the Lower Colorado River and Gila River valleys near Yuma, California (Platz et al. 1990). Apparently this species was accidentally introduced during the mid 1960s to 1970s but was not observed by Vitt and Ohmart (1978). Platz et al. (1990) stated that *R. berlandieri* would likely expand its range west into the area of Mexicali, BC. I observed a large leopard frog in April 1999 at the confluence of the Río Hardy and Río Colorado, which I was unable to catch but I believe was *R. berlandieri.*

Rana boylii
FOOTHILL YELLOW-LEGGED FROG

Loomis (1965) reported two specimens of *Rana boylii* (LBSC 1080–81) from the lower end of the La Grulla Meadow in the Sierra San Pedro Mártir of northern Baja California. The specimens were lost in shipment, but not before they were identified by R. C. Stebbins and R. G. Zweifel. Loomis suggested that the inaccessibility of the collection site precluded the possibility of these frogs being introduced. Welsh (1988) searched unsuccessfully for more specimens and concluded that *R. boylii* may be extirpated because of the abundance of *R. aurora.* To date, no additional specimens have been found. *Rana boylii* was probably a naturally occurring species in the Sierra San Pedro Mártir. Interestingly, the northern Peninsular Ranges appear to be the southern extent of the distribution of many other more northerly distributed, disjunct species or their relatives, such as *Sceloporus vandenburghianus, Lampropeltis zonata,* and *Thamnophis elegans.*

Rana yavapaiensis
YAVAPAI LEOPARD FROG

Although specimens of *Rana yavapaiensis* from the Río Colorado delta region of northeastern Baja California exist in museum collections, and two specimens were observed as late 1981 (Clarkson and de Vos 1986), this species is believed to have been extirpated from the entire lower Colorado River (Vitt and Ohmart 1978; Platz and Frost 1984). No additional specimens have been collected from the Colorado River or its drainages in 25 years. Its probable extirpation is likely due to the introduction of *R. catesbeiana* (Clarkson and de Vos 1986; Mellink and Ferreira-B. 2000).

Pelobatidae
SPADEFOOT TOADS

The family Pelobatidae contains 2–3 genera and 11 living species (Pough et al. 2001), which collectively range throughout Europe, western Asia, extreme northwestern Africa, southwestern Canada, and the United States south into southern Mexico. Pelobatids are short, squat, terrestrial toads with relatively smooth skin, large, catlike eyes with vertical pupils, and a single, cornified, black inner metatarsal tubercle or spade. Pelobatids are distinguished from all other frogs and toads by the fusion of the sacrum and the coccyx, exostosed frontoparietal bones, and the presence of a metatarsal spade supported by a well-ossified prehallux (Ford and Cannatella 1993). The Pelobatidae are represented in Baja California by the two genera *Scaphiopus* and *Spea.*

Scaphiopus Harlen
SOUTHERN SPADEFOOTS

The genus *Scaphiopus* is composed of two species of nocturnal, burrowing toads and ranges across the southern and northeastern United States south to northern Nayarit, Zacatecas, San Luis Potosí, and northern Veracruz, Mexico. One species, *S. couchii,* occurs in Baja California and on four islands in the Gulf of California.

FIG. 2.13 *Scaphiopus couchii*
COUCH'S SPADEFOOT; SAPO CAVADOR

Scaphiopus couchii Baird, 1854:62. Type locality: "Coahuila and Tamaulipas"; restricted to "Matamoros, Tamaulipas," Mexico, by Smith and Taylor (1950:345). Probable syntypes: USNM 3713–15

IDENTIFICATION *Scaphiopus couchii* can be distinguished from all other frogs and toads in the region of study by having vertical pupils and a single rectangularly shaped, black cornified spade on its heel.

RELATIONSHIPS AND TAXONOMY *Scaphiopus couchii* is the sister species of *S. hurterii* of the eastern United States, and together they form the sister group to the genus *Spea,* the other North American pelobatids (Cannatella 1985).

DISTRIBUTION *Scaphiopus couchii* ranges from southwestern Oklahoma west through Texas, southern New Mexico, and Arizona to southeastern California. From here, it extends south into mainland Mexico as far as San Luis Potosí and Nayarit, on opposite sides of the Sierra Madre Occidental. In Baja California, *S. couchii* occurs east of the Peninsular Ranges from the U.S.-Mexican border south to approximately Bahía de los Ángeles. From here, it ranges southwest across the peninsula toward Guerrero Negro and south into the Cape Region (Wasserman 1970; map 2.9). *Scaphiopus couchii* occurs on the Gulf islands of Cerralvo, Espíritu Santo, Partida Sur, and Tiburón.

DESCRIPTION Medium-sized, reaching 90 mm total length; skin on body loose, with moderately sized tubercles; tubercles largest laterally; head no wider than body; snout rounded in dorsal profile, nearly square in lateral profile; external nares elevated, giving snout an upward tilt; cranial crests absent; eyes large, protuberant, and widely spaced, with vertical pupils; vocal sac spherical; parotoid glands absent; tympanum rounded, heart-shaped, or oblong; forelimbs robust; digits weakly webbed; hind limbs short and robust, upper portion sometimes hidden by posterolateral fold of skin on body; skin on legs smooth; enlarged, darkened, rectangular inner metatarsal tubercle present; outer metatarsal tubercle absent; thick interdigital webbing present. **COLORATION** Ground color of dorsum dull white, light green, or brown with an irregular, dense to sparse network of darker lines and spots; dark pattern of females usually composed of linear markings; males usually with poorly developed pattern and nearly unicolored light green; majority of larger tubercles on sides of body white; limbs blotched to weakly banded; ventrum dull white to cream and immaculate. **TADPOLES** Reach 20 to 25 mm in length; spiracle sinistral, aperture located approximately midway between mouth and vent, directed posteriorly with slight upward tilt; labial tooth row formula 2/4, 3/4, 4/4, 5/4, to 5/5, usually 4/4; oral papillae encircling mouth, discontinuous dorsally; caudal musculature curving upward posteriorly, giving tail an upward tilt; caudal fins equal in height at midtail and slightly wider than caudal musculature; ground color dark iridescent bronze, sometimes covered with reticulum of yellowish or light green spots; abdominal region translucent; upper half of caudal musculature dark and usually heavily spotted, with lower half and tip generally having much less spotting; caudal fins transparent and finely stippled with black. **MATING CALL** Reminiscent of the bleat of sheep. It lasts about 0.5 to 1.5 seconds and consists of a *me-oow* that diminishes in pitch and intensity.

FIGURE 2.13 *Scaphiopus couchii* (Couch's Spadefoot) from 15 km south of Ciudad Constitución, BCS

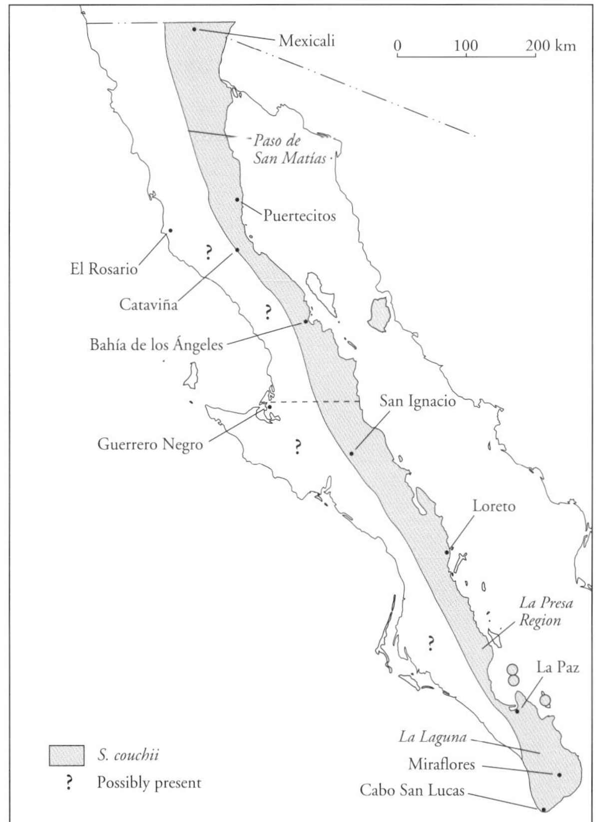

MAP 2.9 Distribution of *Scaphiopus couchii* (Couch's Spadefoot)

GEOGRAPHIC VARIATION Despite its nearly ubiquitous distribution in Baja California, *S. couchii* varies little. The most notable variation occurs in the dorsal coloration and pattern. Most females from the vicinity of Loreto tend to have a much thinner, darker reticulum on the dorsum than those of more northern and southern areas. *Scaphiopus couchii* from the upper elevations of the Sierra Guadalupe and the Sierra la Giganta tend to be darker overall than toads from other areas. The single specimen from Isla Cerralvo is lighter in overall coloration than those of the adjacent Cape Region (Etheridge 1961).

NATURAL HISTORY *Scaphiopus couchii* is found in a number of habitats in Baja California, ranging from the Arid Tropical Region of the Cape Region to the arid creosote bush scrub community of the Lower Colorado Valley Region in the extreme northeast. Wherever *S. couchii* occurs, it resides in the vicinity of soft, sandy soils, where it constructs its burrows. It is most prevalent at the edges of arroyos or in open areas at the base of vegetation. *Scaphiopus couchii* is generally nocturnal but will emerge and call during the day following heavy rainstorms. During the fall, toads occasionally can be found on humid nights following periods of precipitation. In the Cape Region, *S. couchii* is commonly seen at night from mid-July through October during the rainy season. At night, during and after rainstorms, there may be so many individuals on the road that it is impossible to drive without running them over. In northeastern Baja California, however, *S. couchii* may remain underground in self-constructed burrows for as long as two years and has a number of physiological adaptations allowing it to do so (McClanahan et al. 1994). Some toads have been found as deep as 61 cm below the surface.

The sound of summer rains hitting the desert floor causes *S. couchii* to emerge from burrows in great numbers and begin calling from the edges of and within temporary rain pools. I estimated 350 toads to be present in a single roadside pool measuring 1.5 by 3 m in the Cape Region, just south of Miraflores, during mid-August. Females lay approximately 10 to 30 eggs, and clutches from various females may be clumped together. All eggs tend to be fastened to a stationary object just below the surface. The eggs usually hatch within one to two days but have been known to hatch in as little as nine hours (McClanahan 1967). The speed at which the tadpoles transform depends to a large extent on the rate of evaporation of the pond. As the pool evaporates, the rate of metamorphosis increases, and tadpoles have been known to metamorphose in as little as 10 days (McClanahan 1967). *Scaphiopus couchii* eats small arthropods in the vicinity of the breeding pool. Tadpoles become cannibalistic as soon as the tail begins to be reabsorbed. *Scaphiopus couchii* is known to breed in the El Sauzal water hole on Isla Tiburón (Felger and Moser 1985). On the southwest end of Isla Cerralvo, the steep, rocky foothills meet the low sand dune areas along the back edges of the beaches, forming a long, narrow basin in which rainwater and runoff become trapped. This basin provides breeding pools for *S. couchii.*

Spea Cope

WESTERN SPADEFOOTS

The genus *Spea* is composed of five species of terrestrial, burrowing toads, ranging from south-central Canada south to the southern edge of the Central Plateau of Mexico (Wiens and Titus 1991). One species, *Spea hammondii,* ranges into northwestern Baja California.

FIG. 2.14 ### *Spea hammondii*

WESTERN SPADEFOOT; SAPO

Scaphiopus hammondii Baird, 1859a:12. Type locality: "Fort Reading," California, USA. Holotype: USNM 3695.

IDENTIFICATION *Spea hammondii* can be distinguished from all other frogs and toads in the region of study by having a single semicircular black spade on its heel and vertical pupils.

RELATIONSHIPS AND TAXONOMY *Spea hammondii* is most closely related to *Spea intermontana* of the northwestern United States (which might comprise more than one species; Wiens and Titus 1991) and *S. bombifrons* of the central United States (Wiens and Titus 1991).

DISTRIBUTION *Spea hammondii* ranges throughout the Central Valley of California south through the Coast Ranges into northwestern Baja California (Stebbins 1985). In Baja California, it extends as far south as Mesa de San Carlos, west of the Peninsular Ranges (map 2.10), but may occur even farther south in some of the larger arroyos.

DESCRIPTION Medium-sized, with adults reaching 65 mm SVL; skin of body loose with small vertebral tubercles; head as wide as body; snout rounded to weakly pointed in dorsal profile; external nares elevated, giving snout an upward tilt; canthus smoothly rounded; preorbital region flat; cranial crests absent; eyes large and protuberant, with vertical pupils; vocal sac single and spherical; parotoid glands small and indistinct; fold of skin extending from lower eyelid to forelimb insertion; tympanum small and rounded to oblong; forelimbs relatively short and stout; slight interdigital webbing on hand; hind limbs short and robust; foreleg tuberculate dorsally; enlarged, darkened, semicircular or wedge-shaped, cornified, inner metatarsal tubercle present; outer metatarsal tubercle absent; thick interdigital webbing on foot. **COLORATION** Dorsal ground color light green to gray with scattered blotches of darker coloration; juveniles usually lighter in color with a more vivid pattern; pair of medially curved, irregularly outlined, broad, light-colored paravertebral stripes extending posteriorly from eyes usually present; body tubercles orange or reddish, occasionally set off with a dark blotch on a whitish background; dorsal surfaces of limbs irregularly blotched; ventrum whitish to dull cream-yellow; two light-colored, dorsolateral spots usually bordering anus. **TADPOLES** Reach over 70 mm; spiracle sinistral, directed posteriorly, turned slightly upward, and located approximately midway between mouth and vent; labial tooth row formula 2/3, 2/4, 3/3, 3/4, 4/4, or 5/4, but in most cases 3/4; oral papillae encircling mouth, lacking large lobes, and may be interrupted dorsally; caudal fins nearly equal in depth and slightly wider than caudal musculature; ground color of dorsum light gray to dark greenish-brown; ventrum cream-colored with occasional light reddish iridescence; intestines visible posteriorly; tail generally transparent. **MATING CALL** The call consists of a loud, well-developed, horsey snore lasting 0.5 to 1 seconds,

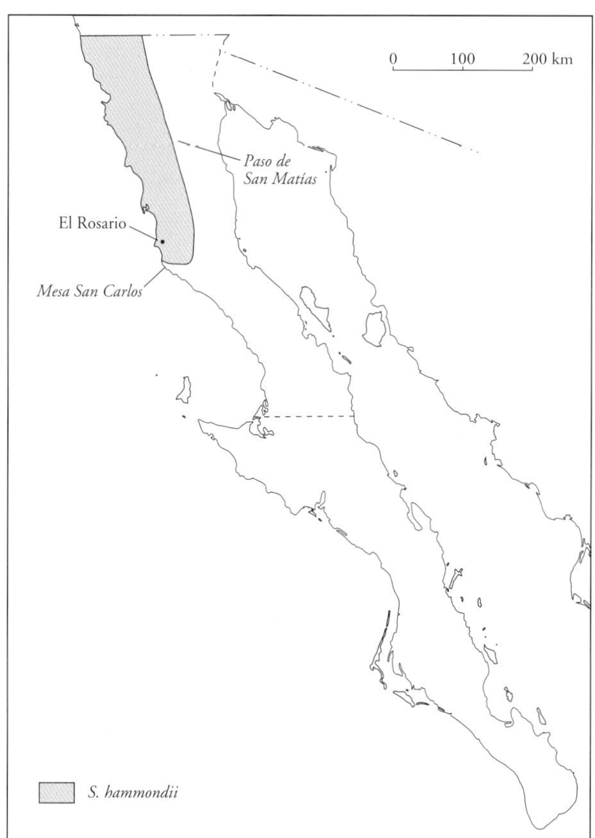

MAP 2.10 Distribution of *Spea hammondii* (Western Spadefoot)

FIGURE 2.14 *Spea hammondii* (Western Spadefoot) from Tecate, BC

with little fluctuation in pitch or intensity. It is best represented as a *w-a-a-a* or *r-a-a-a-w.*

GEOGRAPHIC VARIATION *Spea hammondii* varies little in coloration and pattern. Adults from the relatively arid Vizcaíno Region tend to have a lighter and more vivid color pattern than do adults from more northerly locales in the California Region. It appears that the former retain the juvenile coloration of the more northern populations into adulthood.

NATURAL HISTORY *Spea hammondii* occurs in sandy areas or regions with an abundance of loose soil and is usually found in arroyos, fields, and open plains in the California Region. It is principally nocturnal but may be heard calling during the day after winter and spring rains. The voice of this species is ventriloquial, making individual callers exceedingly difficult to locate. Males congregate at temporary breeding pools and call for females from late December to mid-May. Outside the breeding season, *S. hammondii* is secretive and may spend months at a time underground in self-constructed burrows. A female will lay upward of 600 eggs, which she usually attaches to vegetation. The eggs may hatch in less than five days. The larvae grow rapidly and have the curious habit of hanging vertically in the water and gulping air or feeding on surface material. *Spea hammondii* eats a variety of arthropods. Those most commonly taken are species found within the vicinity of the breeding pools. Larvae are highly carnivorous and will eat tadpoles of other species as well as those of conspecifics.

3

TURTLES AND TORTOISES

AS WITH THE AMPHIBIAN FAUNA OF BAJA CALIFORNIA, NONMARINE TURTLE AND TORTOISE DIVERSITY IS QUITE LOW. ONLY FOUR SPECIES ARE REPRESENTED ON THE PENINSULA, AND EACH IS RESTRICTED IN ITS DISTRIBUTION. THE THREE SEMIAQUATIC TURTLES ARE CONFINED TO THE MORE MESIC REGIONS OF BAJA CALIFORNIA OR OASES IN THE MORE ARID PORTIONS OF THE PENINSULA. THE ONLY TORTOISE IN THE REGION OF STUDY OCCURS ON ISLA TIBURÓN, PERHAPS ON ISLA DÁTIL, AND AS AN INTRODUCED POPULATION IN THE CAPE REGION.

THE MARINE TURTLES OF THE PACIFIC COAST OF BAJA CALIFORNIA AND THE SEA OF CORTÉS ARE REPRESENTED BY FIVE GENERA AND SPECIES, REPRESENTING BOTH THE MAJOR LIVING FAMILIES CHELONIIDAE AND DERMOCHELYIDAE. ALTHOUGH EGG-LAYING ALONG THE BEACHES OF BAJA CALIFORNIA MAY HAVE BEEN COMMON IN THE PAST, IT IS CURRENTLY RARE AND GENERALLY RESTRICTED TO ISOLATED STRETCHES OF BEACH AND THE MORE REMOTE ISLANDS IN THE GULF OF CALIFORNIA.

FIGURE 3.1 *Clemmys marmorata* (Western Pond Turtle) from Arroyo Hondo, BC. The steel tanks of the gravel pit factory along this stream demonstrate the problem of habitat destruction.

Emydidae

POND TURTLES AND SLIDERS

The Emydidae contain approximately 33 genera and 91 species (Ernst and Barbour 1989) and are the largest and most diverse lineage of turtles. They are distributed nearly worldwide, with most representatives occurring throughout the nondesert areas of North and Central America and the West Indies south to southwestern South America (Pough et al. 2001). In the Old World, emydids range throughout Europe, northwestern Africa, and Asia as far south and east as Sulawesi in the Indo-Australian Archipelago. Emydids are absent only from the continents of Australia and Antarctica.

The majority of the species of this lineage are semiaquatic omnivores with webbed feet and hydrodynamically shaped shells. Only a relatively small radiation is principally terrestrial. Emydids can be distinguished from other turtles and tortoises by their generally small heads, short tails, and broad connections of the carapace and plastron, and their tendency to form a secondary palate. In Baja California, the Emydidae are represented by two genera and two species.

Clemmys Ritgen

POND TURTLES

The genus *Clemmys* contains four species of small to medium-sized turtles represented by freshwater semiaquatic and terrestrial forms. Three species collectively range throughout the eastern United States (Iverson 1992), and the remaining species, *C. marmorata,* ranges along the west coast of North America from southern Washington to northwestern Baja California. Recent analyses indicate that *Clemmys* may be a nonnatural group (Bickham et al. 1996).

FIG. 3.2 **_Clemmys marmorata_**

WESTERN POND TURTLE; TORTUGUITA

Emys marmorata Baird and Girard, 1852:177. Type locality: "Puget Sound," Washington, USA. Syntypes: USNM 7594–96 (formerly USNM 7593)

IDENTIFICATION *Clemmys marmorata* can be distinguished from all other turtles and tortoises in the region of study by the combination of having five claws on the forelimbs, epidermal shields on the shell, and nonelephantine hind limbs, and lacking a broad, light-colored postorbital patch.

RELATIONSHIPS AND TAXONOMY Based on electrophoretic data, Merkle (1975) indicated that *Clemmys marmorata* was most closely related to *C. muhlenbergii* of the eastern United States. Lovich et al. (1991), on the basis of shell morphology, placed *C. marmorata* closest to *C. guttata* and *C. insculpta* of the eastern United States. Based on nucleotide sequence data, Bickham et al. (1996) indicated that *C. marmorata* was actually more closely related to *Emydoidea blandingi* of the eastern United States than to other species of *Clemmys.* Seeliger (1945), Carr (1952), Bury (1970), Smith and Smith (1979), and Holland (1992) discuss intraspecific taxonomy. On the basis of morphology, Seeliger (1945) noted that *C. marmorata* from Baja California was so distinct that he could not assign it to either *C. m. pallida* of central and southern California or *C. m. marmorata* from central California to Washington. This view was supported by Janzen et al. (1997) who, on the basis of cytochrome *b* data, noted that the Baja California population was so divergent that it might even represent a different species.

FIGURE 3.2 *Clemmys marmorata* (Western Pond Turtle) from the lower portion of Arroyo San Telmo, BC

DISTRIBUTION *Clemmys marmorata* ranges from extreme southern Washington southward, west of the Sierra Nevada and Peninsular Ranges, to northwestern Baja California (Bury 1970). A disjunct population occurs in western Nevada, just east of Lake Tahoe, and in the Mojave River as far east as Afton Canyon. In Baja California, *C. marmorata* occurs in riparian ecosystems along the Pacific drainage from at least Río San Carlos (Roberts 1982), 13 km south of Ensenada, to Arroyo Grande, east of El Rosario (map 3.1). It may extend further south in Baja California, but no extant naturally occurring populations have been found.

DESCRIPTION Medium-sized, with adults reaching 200 mm in carapace length; head large, triangular, and covered with smooth skin; snout sharply pointed in dorsal profile; nostrils situated high on snout; upper jaw hooked and usually notched at symphysis; tomial surfaces smooth; eyes large; tympanum elliptical; skin of neck and gular region granular with several loose folds; carapace smoothly rounded anteriorly and posteriorly in dorsal profile, wider than long; prominent vertebral ridge in small individuals; posterolateral marginals flared laterally in adults; nuchal plate small; 12 marginals on each side; 4 pleurals on each side; 5 vertebrals; plastron extending anteriorly as far as carapace and nearly as far posteriorly; pectorals and abdominals contact marginals 4 through 7, forming a wide bridge; anals notched; concavity usually present posteriorly in adult males; axillary plates small; inguinal plates absent; forelimbs short, containing five clawed and webbed digits; claw 5 sometimes vestigial; hind limbs short with five webbed digits, first four clawed; adult males tend to have wider tail bases than adult females. **COLORATION** Variable; top of head dark and covered with reticulum of light-colored spots; ground color of chin and throat cream with varying degrees of dark longitudinal striping or spotting; upper surface of limbs dark with weak indications of lateromedial striping; underside of limbs black to cream with various intermediate patterns; tail dark with black vertebral stripe; carapace dark brown to olive green, with or without darker markings; many specimens have dark lines radiating out from plate centers; plastron olive green to cream; dark markings on plastron generally confined to edges of plates; entire central region of plastron dark in small individuals.

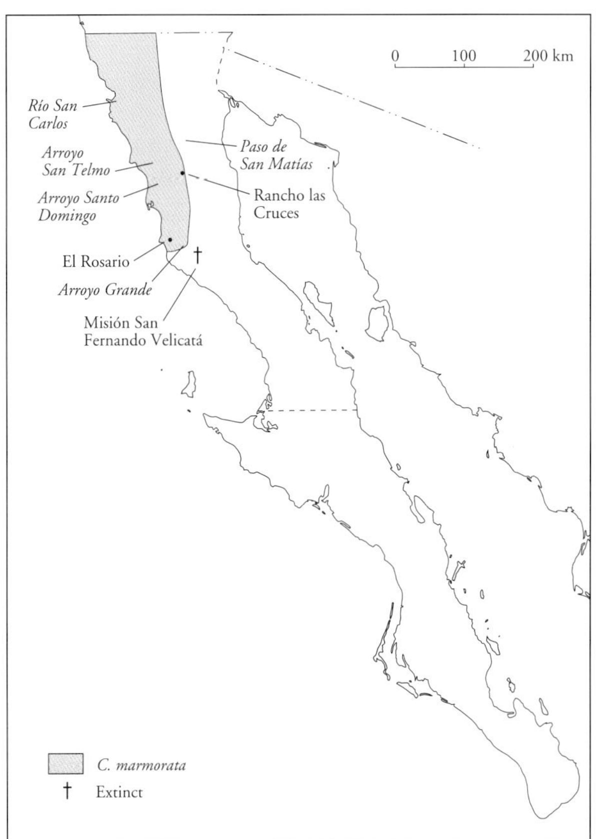

MAP 3.1 Distribution of *Clemmys marmorata* (Western Pond Turtle)

GEOGRAPHIC VARIATION Little variation exists among populations of *C. marmorata* within Baja California, although they differ in a number of morphological trends from populations outside Baja California (Seeliger 1945). What variation does occur involves the contrasts in overall ground color. *Clemmys marmorata* from low-elevation coastal localities such as Arroyo San Telmo, Arroyo Santo Domingo, and Arroyo Grande are dark in overall ground color, whereas individuals from populations at higher elevations have a lighter ground color. In these populations, the patterns of head spots are accentuated, and the light ground color highlights the dark patterns on the limbs and tail. This is especially noticeable in specimens from the Rancho las Cruces Spring in the Sierra San Pedro Mártir.

NATURAL HISTORY Nothing of consequence has been published on the natural history of *C. marmorata* from Baja California, but considerable information has been presented about this species elsewhere in its range (see references in Ernst et al. 1994; Holland 1994; Rathbun et al. 1992). *Clemmys marmorata* ranges west of the crest of the Peninsular Ranges in the California Region and occurs in riparian habitats with permanent or ephemeral water sources. Where it exists in perennial watercourses, as in the western ends of Arroyos San Telmo, Santo Domingo, and Grande near or within the northern Vizcaíno Region, it is restricted to deeper pools that are present year-round. *Clemmys marmorata* is semiaquatic and diurnal and can be observed basking on aquatic vegetation and floating logs or at the water's edge. I have observed active turtles from February to early December throughout their range in Baja California. Stebbins (1985) reports that 3 to 11 eggs are laid, and Klauber (1934) noted that in San Diego County of southern California, nesting takes place in July. *Clemmys marmorata* is extremely wary and usually very difficult to approach. At the first sight of danger, turtles enter the water and may bury themselves in the mud or crawl under overhanging stream banks. In the shallow and narrow portion of Arroyo San Telmo, just west of the town of San Telmo, they are especially adept at hiding underwater beneath streamside ledges that are covered with vegetation. In this arroyo, I have observed *C. marmorata* feeding on aquatic vegetation and decaying rodents floating in the water.

Clemmys marmorata is commonly reported to feed on fish, insects, and earthworms.

REMARKS *Clemmys marmorata* was formerly found in the ponds below Misión San Fernando Velicatá, approximately 120 km south of its present distribution, which was its known southern extent. However, in 1978 the ponds silted up following a large hurricane (Roberts 1982). Recent searches for specimens in this area have been unsuccessful, and local ranchers living along the watercourse state that they cannot remember seeing turtles for several years.

The lower ponds of Arroyo San Telmo were also filled in with silt during the winter of 1992–1993, and I have been unable to locate turtles in these areas since that time. Welsh (1988) reports the introduction of this species to the ponds of Arroyo Grande at Rancho El Metate, approximately 50 km east of El Rosario. On 26 April 1991, I was given a *C. marmorata* by a small boy at the gas station in El Rosario, who had collected it in Arroyo Grande, approximately 15 km east of town. His father told me that he occasionally sees turtles in the arroyo, but only after large rainstorms.

Trachemys Agassiz

SLIDERS

The genus *Trachemys* is composed of a group of medium to large semiaquatic freshwater turtles. It contains at least six species (Ernst and Barbour 1989) which collectively range from the central United States to central South America. *Trachemys* also occurs on several West Indian islands, making it one of the most widely distributed turtle genera. One species is endemic to southern Baja California.

FIG. 3.3 ***Trachemys nebulosa***

BAJA CALIFORNIA SLIDER; TORTUGUITA, TORTUGUITA DEL RÍO

Chrysemys nebulosa Van Denburgh, 1895:84. Type locality: "[at] Los Dolores [Misión de Dolores], L. C. [Lower California]," on "Mainland abreast of San Jose Island," Baja California Sur, Mexico. Holotype: CAS 2244

IDENTIFICATION *Trachemys nebulosa* can be differentiated from all other turtles and tortoises in the region of study by its clawed digits, nonelephantine hind limbs, and wide, light-colored postorbital patch.

RELATIONSHIPS AND TAXONOMY Van Denburgh (1895, 1922) and Stejneger and Barbour (1943) considered *Trachemys nebulosa* to be a full species. Most recent authors, however, (see Smith and Smith 1976 and Legler 1990 and references therein) consider it a subspecies of *T. scripta.* All authors, however, agree that it is distinctive and consistently diagnosable on the basis of having a yellow or orange rather than a red postocular stripe, and lacking dark rings on the carapace and black ocelli on the pleurals (Carr 1952). Therefore, the taxonomy of Van Denburgh (1895, 1922) and Stejneger and Barbour (1943) is followed here, and *T. nebulosa* is considered a distinct species.

FIGURE 3.3 *Trachemys nebulosa* (Baja California Slider) from El Bosque in the La Presa region, BCS

DISTRIBUTION *Trachemys nebulosa* ranges throughout the southern half of Baja California, where it occurs disjunctly in all large, isolated, permanent bodies of fresh water from San Ignacio south to San José del Cabo (Grismer and McGuire 1993; map 3.2). Grismer and McGuire (1993) believe that much of its distribution north of the Cape Region is due to human introduction.

DESCRIPTION Moderate to large, with adults reaching at least 370 mm in carapace length; head large, triangular, covered with smooth skin; snout protuberant in dorsal profile; nostrils situated high on snout; upper jaw protruding past lower jaw, not hooked and usually not notched at symphasis; tomial surfaces smooth; eyes large and protuberant; tympanum round to elliptical; skin of neck and gular region granular, containing several loose folds; carapace smoothly rounded posteriorly and straight to slightly emarginate anteriorly, longer than wide, low in lateral profile, slight vertebral ridge usually present; posterolateral marginals flared laterally; nuchal plate small; 12 to 13 marginals on each side, 4 to 5 pleurals on each side, 5 to 6 vertebrals; plastron extending anteriorly as far as carapace and nearly as far posteriorly; plastral plates weakly roughened by longitudinal furrows; pectorals and abdominals contact marginals 4 or 5 through 7, forming wide bridge; posterior margins of anals notched; slight concavity present in adult males posteriorly; forelimbs short, flat, containing five clawed digits; series of enlarged scales forming crest on dorsolateral margin of forelimb; hind limbs short, robust, flattened, with five fully webbed digits, of which the first four have claws; tail moderately long, scales arranged in caudal whorls, adult males with wider tail base and longer tails. **COLORATION** Top of head olive drab with indistinct pale lines; chin and throat lighter, with central yellow marking; pair of yellow lateral stripes in gular region extend anteriorly across lower jaw onto upper jaw; prominent yellow postorbital stripe extends below tympanum onto shoulder; usually small, thin, dorsal extension of stripe extends onto tympanum; two distinctive pale markings run through eye; large, elliptical, longitudinally oriented, light marking in temporal region; shoulder and neck with one to three distinctive, wide yellowish longitudinal lines or reticulations; yellow stripe extends from shoulder to base of digit 2 on forelimbs; hind limbs lack distinctive markings; two wide yellow lines extend from base of foot to base of tail; pair of light dorsolateral lines on tail; plastron yellow; paired, symmetrical black markings present on plastral plates; carapace dark brown to olive green; marginal plates with ovoid black marks bordered by light line; upper portions of carapace with numerous, irregularly arranged, black marks on a faint reticulum of pale lines, most obvious in juveniles; large adult males nearly solid black above.

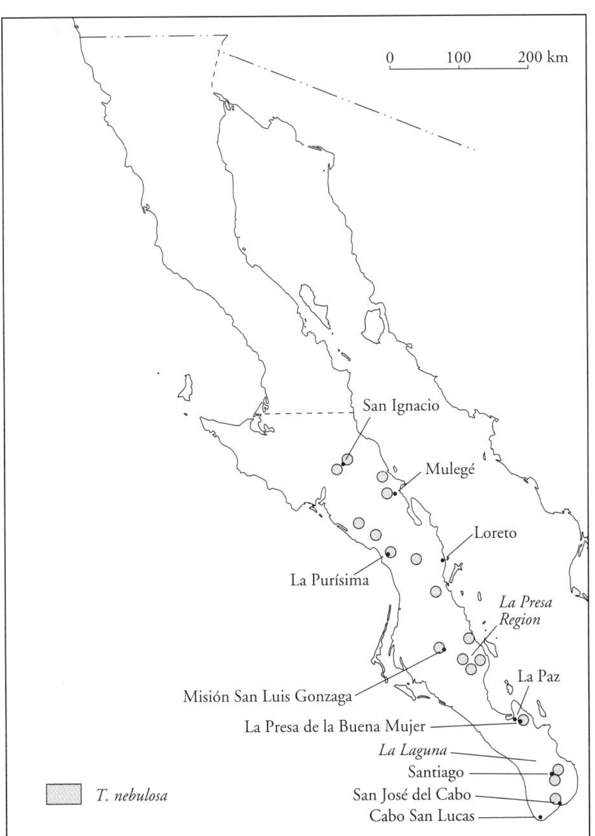

MAP 3.2 Distribution of *Trachemys nebulosa* (Baja California Slider)

GEOGRAPHIC VARIATION Geographic variation of color pattern in *Trachemys nebulosa* is slight, and what is not obscured by ontogenetic changes associated with overall darkening appears to be related to substrate matching. Specimens from La Purísima are lighter in overall coloration and have a less obscured carapace pattern, whereas turtles from San Ignacio and Mulegé are much darker, and the carapace pattern is nearly unicolored in adults. La Purísima lies at the western edge of the volcanic badlands of the Sierra la Giganta in the low, flat, more sandy regions of the northern Magdalena Plain. Ríos San Ignacio and Mulegé, on the other hand, are situated in areas with extensive regions of dark, andesitic rock. Waering (1943, in Carr 1952:265) noted the darkness of the San Ignacio specimens as compared to Cape Region forms. He also noted differences in the color of the temporal spot, which I cannot confirm. Specimens from other areas are intermediate in coloration.

NATURAL HISTORY *Trachemys nebulosa* is restricted to localities with year-round fresh water. Most such areas are large, palm-lined, artesian oases in the volcanic arroyos of the Magdalena and Arid Tropical regions. *Trachemys nebulosa* seems to prefer large bodies of water with muddy bottoms but has been found in high densities in shrinking ponds that form along the bends in drying arroyos (Grismer and McGuire 1993). This species occurs in nearly all the major water systems in southern Baja California (Grismer and McGuire 1993), as evidenced by their remains in household middens and by the reports of local inhabitants. *Trachemys nebulosa* can be seen at a distance basking along the shore or on emergent rocks and floating logs. In some of the ponds southwest of San Ignacio, as many as 20 turtles can be seen at one time, and population estimates of nearly 100 individuals for a single pond have been made (Roberts 1982). *Trachemys nebulosa* is extremely wary and very difficult to approach. During the winter, the local ranchers near San Ignacio say this species buries itself in the mud at the bottom of the ponds but may occasionally come out to sun itself on unseasonably warm mornings. Such observations have yet to be confirmed. Turtles are most active from mid-March to late October. Little has been reported on the reproductive biology of this species. Carr (1952), quoting from a letter written to him by Eric Waering of Petroleos Mexicanos, states that in June 1943 a female was observed about 35 feet from the shore of a lagoon near San Ignacio feet first in a hole, laying eggs. He also noted that the hole was presumably filled with urine. I have seen hatchlings from July through October in the La Presa

Region, which suggests that this turtle has a spring-summer nesting period.

REMARKS Conant (1969) and Murphy (1983a) suggested that *T. nebulosa* may have been introduced into Baja California by native Indians or Jesuit priests (Murphy 1983a) as a food source. But if this species was introduced by Jesuits from mainland Mexico, it is reasonable to assume that its consumption would have been reported in their literature, as were all other foods they imported. However, no mention is made in the Jesuit literature of eating nonmarine turtles. Even Johann Jakob Baegert, who lived with the Guaycura Indians at Misión San Luis Gonzaga and wrote extensively on their eating habits (Baegert 1772), made no mention of nonmarine turtles being eaten. Furthermore, no mention of turtles was ever entered into the supply manifests of the ships sailing to replenish the mission food inventories, although many other food sources were listed (Crosby 1994). Additionally, there is weak evidence at best that mainland Mexican Indians migrated into southern Baja California, and no evidence of permanent residency (Aschmann 1959; Laylander 1987). Transgulf migrations of Indians did occur in the midriff region of the Gulf of California, far to the north, where there are no populations of *T. nebulosa* on the adjacent peninsula. In contrast, Clavigero (1789) indicated that the presence of *T. nebulosa* on the peninsula is natural. He discussed the turtles of Baja California in *Storia della California* and stated that "among the turtles, besides the common land variety and the freshwater ones, there are two other species of large sea turtles." This statement implies not only that testudinids occurred in Baja California before 1789 (see Crumly and Grismer 1994) but also that freshwater species (emydids) were present. Although Clavigero was never in Baja California, he had access to all classified documents, many of which are no longer extant, and corresponded firsthand with many of the Jesuits living on the peninsula. Clavigero was heralded as a literary genius who wrote accurately and impartially (Gray 1937), and it is unlikely that his mention of freshwater and terrestrial turtles is a fabrication. Therefore, *T. nebulosa* probably had a natural origin in Baja California, one related to the formation of the Baja California peninsula (Grismer 1994a), and was probably transported throughout southern Baja California as a food source in more recent times (Grismer and McGuire 1993). This would explain the presence of this species at populated places such as San Ignacio and its absence at San Francisco de la Sierra, an extensive freshwater system less than 40 km to the north. As the water systems of the latter are far from the nearest community and accessed only by a day's journey down exceedingly steep cliffs, it would be impractical to stock them with a food source.

In the Cape Region, Belding (1887) reports this species from Río San José, Mocquard (1899) reports it from Todos Santos [Todos Santos Santiago], and Van Denburgh (1922) reports it from Santiago and Miraflores. Roberts (1982) states that this species no longer exists in the Cape Region, but I observed a specimen at Miraflores in 1987 and was informed by the people of Santiago that it still occurs in the ponds at the center of town. I have observed additional specimens at Aguas Calientes and La Presa de la Buena Mujer, just southeast of La Paz. The latter were introduced by local ranchers as a food source.

An excellent place to view this often hard-to-see species is from the overlook along the edge of Río Mulegé, just behind the mission.

FOLKLORE AND HUMAN USE In more remote areas of southern Baja California, such as the La Presa region and some of the ranchos southwest of San Ignacio, *T. nebulosa* are a source of food, and their shells are conspicuous components of the garbage scattered around the ranchos. The turtles are collected with a large net or by hand. One resident near San Ignacio captures turtles by wearing a diving mask and following them underwater until they bury themselves in the mud. He then extracts them from the mud, taking care not to be bitten. If the turtles are not eaten right away, they are tethered to a tree or hung from the roof with a string running through a hole placed in the posterior marginals. The meat of the limbs is fried or boiled and somewhat gamy in flavor. It is usually mixed with chilis and makes for a peculiar-tasting burrito.

Testudinidae

LAND TORTOISES

The Testudinidae are a strictly terrestrial group containing approximately 11 genera and 40 species (Pough et al. 2001). In the Old World, this lineage generally ranges throughout non-Saharan Africa and southern Europe east through western Asia to southeast Asia. It is also known from Madagascar and the Aldabra Islands. In the New World, testudinids occur sporadically throughout the southern United States and northern Mexico but occupy nearly all of South America east of the Andes Mountains as well as the Galápagos Islands. The Testudinidae are characterized by a domed carapace, short, broad, unwebbed feet with no more than two phalanges in each digit, and elephantine hind limbs. Testudinids have a good fossil record from Baja California (Miller 1978, 1980; Crumly and Grismer 1994), but the only living Baja California representative, *Gopherus agassizii* (Ottley and Velazquez 1989), may be a recent introduction into the Cape Region (Crumly and Grismer 1994), although it has long been known from Isla Tiburón, Sonora, in the Gulf of California.

Gopherus Rafinesque

GOPHER TORTOISES

The genus *Gopherus* is composed of a group of small to medium-sized terrestrial burrowing tortoises that frequent desert, semiarid, and mesic habitats. The four species that make up this genus (Crumly 1994; Crumly and Grismer 1994) collectively range disjunctly across the southern United States and northern Mexico. One species *(G. agassizii)* is known to occur on Isla Tiburón, Sonora, in the Gulf of California and probably has been introduced into the Cape Region of Baja California.

FIG. 3.4 **_Gopherus agassizii_**

DESERT TORTOISE; TORTUGA DEL DESIERTO

Xerobates agassizii Cooper, 1863:120. Type locality: "mountains of California, near Fort Mojave," USA. Syntypes: originally three specimens, USNM 7888; two may have been destroyed at CAS in 1906 fire (Auffenberg and Franz 1978). Surviving syntype from "Utah Basin, Mojave River." USNM catalog reads "Solado Valley, California," USA (Cochran 1961)

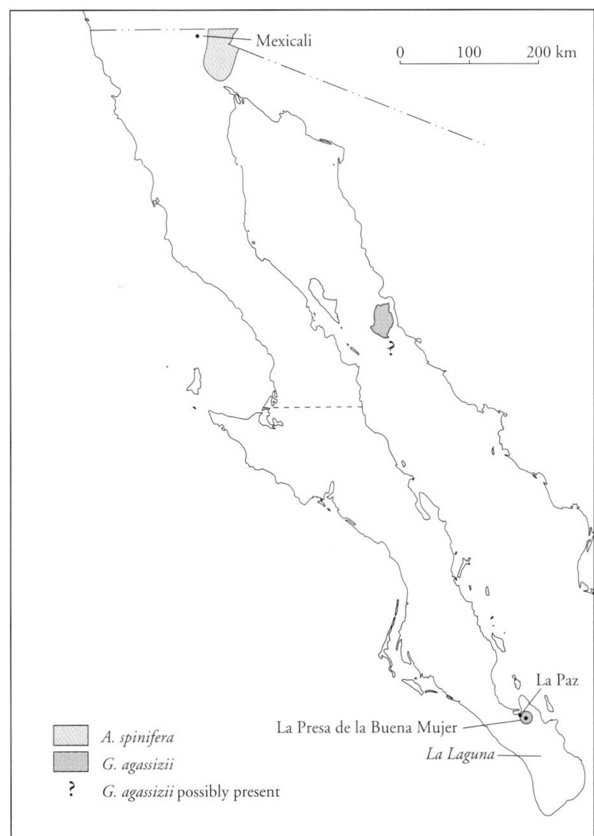

MAP 3.3 Distribution of *Gopherus agassizii* (Desert Tortoise) and *Apalone spinifera* (Spiny Soft-Shell)

FIGURE 3.4 *Gopherus agassizii* (Desert Tortoise) from Ensenada de los Perros, Isla Tiburón, Sonora

IDENTIFICATION *Gopherus agassizii* can be distinguished from all other turtles in the region of study by its elephantine hind limbs.

RELATIONSHIPS AND TAXONOMY Crumly (1994) indicates that *Gopherus agassizii* may be one of the most basal species of *Gopherus* and is currently part of a trichotomy that also includes *G. berlandieri* of southern Texas and northeastern Mexico, the sister species *G. flavomarginatus* of north-central Mexico, and *G. polyphemus* of southeastern United States. Lamb et al. (1989) discuss taxonomy.

A single whole specimen and partial skeleton of *G. agassizii* from the Cape Region was described as the new species *Xerobates lepidocephalus* by Ottley and Velazquez (1989). Crumly and Grismer (1994) noted that the characteristics considered diagnostic for this species are variable in populations of *G. agassizii* from Isla Tiburón. A desert tortoise in the Cape Region is understandable: the past climatic regime of central Baja California (Axelrod 1979) would favor such a distribution. Fossil testudinids have been reported from the Cape Region, near Santa Rita, near San Miguel Comondú from south-central Baja California (Miller 1978, 1980), and just west of Bahía de los Ángeles in northern Baja California (Crumly and Grismer 1994).

DISTRIBUTION *Gopherus agassizii* extends north from northwestern Sinaloa, Mexico, through most of Sonora to southwestern Arizona, extreme southwestern Utah, southern Nevada, and the Mojave Desert of southeastern California, USA (Auffenberg and Franz 1978; Fritts and Jennings 1994). It is also known from Isla Tiburón and possibly Isla Dátil (see Grismer 1999a) in the Gulf of California, Sonora, and the Cape Region of Baja California (Crumly and Grismer 1994; Ottley and Velazquez 1989; map 3.3).

DESCRIPTION Medium to large, with adults reaching at least 400 mm in carapace length; head triangular, covered with scales of varying size, sometimes platelike; scales of parietal region occasionally raised; snout obtuse; nostrils orientated anteriorly on end of snout; upper jaw protrudes past lower jaw, not hooked or notched at symphysis; tomial surfaces strongly denticulate; tympanum round to elliptical; skin of gular region granular, that of neck less so and containing several loose folds; carapace oval to subrectangular in dorsal profile, moderately flat in lateral profile; posterolateral marginals flared laterally; surface of carapace undulate; well-defined concentric rings present within plates; cervical plate small; 11 or 12 marginals on each side with single supracaudal plate; 4 pleurals on each side; 5 vertebrals; plastron extends anteriorly past carapace but not as far posteriorly; free edges of all plates smooth; gular projection often notched at midline; plastral plates smooth to weakly roughened by concentric furrows; plastron contains two gular plates followed sequentially by pairs of humerals, pectorals, abdominals, femorals, and anals; pectorals and abdominals contact marginals 4 to 6 or 7, forming a wide bridge; posterior margins of anals deeply notched; posterior concavity prominent in males; two pairs of axillary and inguinal plates present; forelimbs short, stout, containing five clawed, unwebbed digits and large, juxtaposed to weakly imbricate scales; hind limbs elephantine, containing five clawed, unwebbed digits; claws short and wide; hind limbs covered with small, granular, juxtaposed scales; tail thick and short, scales not arranged in caudal whorls; adult males with wider tail base, longer tail, and concave plastron. **COLORATION** Skin dark gray to dark brown, slightly lighter in juveniles; carapace brown to very dark brown with central portions of plates somewhat lighter; carapace of juveniles lighter and color contrast slightly accentuated; plastron slightly lighter and mottled with dull yellow.

GEOGRAPHIC VARIATION No major patterns of geographic variation have been observed in the Isla Tiburón population. There are reports, however, of the tortoises of this island being much larger than their mainland counterparts (Felger et al. 1981) and Reyes-Osario and Bury (1982) report on a specimen with a carapace length of 371 mm.

NATURAL HISTORY On Isla Tiburón, *G. agassizii* is most common in creosote bush and mixed desert scrub and subtropical thornscrub within the Central Gulf Coast Region (Bury et al. 1978). This vegetative structure provides dense cover for hiding and resting sites. The maximum elevation of *G. agassizii* on Isla Tiburón is unknown, although there are reports of it on some of the lower slopes of the more central mountain ranges (Reyes-Osario and Bury 1982), and fecal material has been observed as high as 300 m above the valley floor on the slopes of Cerro Kunkkak (N. Scott, personal communication, 1990) toward the central portion of the island. In the Cape Region of Baja California, *G. agassizii* occurs in the Arid Tropical Region in the vicinity of La Presa de la Buena Mujer, just southeast of La Paz (Ottley and Velazquez 1989). This is a steep, rocky area with pronounced wet and dry seasons.

Tortoises on Isla Tiburón are most active from September to November, with peak activity occurring in October (Reyes-Osario and Bury 1982), following rainstorms. Occasional specimens may be observed in December (Bury et al. 1978), but few are seen at other times of the year. Shallow burrows are constructed along the sides of arroyos and at the base of vegetation. I observed one individual at night just outside a shallow burrow on a rocky hillside. Burrows may be occupied by as many as four individuals (Reyes-Osario and Bury 1982). It is believed that tortoises here may not spend as much time constructing burrows as those in the more northerly portion of their distribution because of the mild winters on this island. The distribution of *G. agassizii* on Isla Tiburón is not well known, and all reported observations are from the eastern side of the island. The western side of the island is predominantly steep and rocky and possibly poor habitat.

Virtually nothing is known of the natural history of *G. agassizii* in the Cape Region. The only known living specimen was collected by a rancher who happened to see the rear half of the tortoise protruding from a rock fissure. The natural history data reported by Ottley and Velazquez (1989) for this population are contradictory and unsubstantiated. Locals of the area where the living tortoise was found have told me that tortoises were once common and widespread before goats were allowed to roam the hillsides. They believe that tortoises are currently restricted to one small hill. They say they occasionally see tortoises out feeding on the Yucca *(Merremia aurea)* following summer rains.

REMARKS Some authors have mentioned the possibility of *G. agassizii* occurring in northern Baja California (Schmidt 1922). Others have noted its occurrence but provided neither localities nor specimens (Ottley and Velazquez 1989; Smith and Taylor 1950; Terrón 1920, 1921) to support their recordings. The presence of land turtles in Baja California was first mentioned by Clavigero (1789; see remarks on *T. nebulosa*). The current distribution of *G. agassizii* closely approaches the northeastern corner of Baja California, and there is little doubt that this species once occurred there. However, with the continued drying trend of the southwest, the range of this species has been naturally shrinking (Smith and Smith 1976), and with the agricultural development of the Imperial Valley on both sides of the U.S.-Mexican border (Roberts 1982), it is likely the reduction in its distribution has been accelerated. Felger et al. (1981) report *G. agassizii* as occurring on Isla Dátil, but this observation is unsubstantiated (see Grismer 1999a).

The mainland populations of this species in North American deserts have seriously declined in the past twenty years, primarily because of habitat disruption and exploitation (see Bury and Corn 1995; Corn 1994 and references therein). The population density of *G. agassizii* on Isla Tiburón is the highest outside southern California (Reyes-Osario and Bury 1982). Its presence here is important because Isla Tiburón provides a protected, unaltered habitat in which this species can survive naturally and be studied. It is currently the largest isolated preserve for this species.

FOLKLORE AND HUMAN USE *Gopherus agassizii* and its eggs have long been part of the diet of the Seri Indians and other local inhabitants of Sonora (Felger et al. 1985), although this practice appears to be declining (Nabhan, forthcoming). The plastron is usually sawed off, and the liver and stomach are removed, roasted, and eaten. The Seri used to cook the meat by placing a heated stone in the carapacial cavity and by building a ground fire next to the tortoise's body (Felger et al. 1985). When cooked, the legs and the rest of the meat are twisted off and eaten. The Seri Indians were well aware of the desert tortoises' existence and abundance on Isla Tiburón and have extensive knowledge of their natural history (Malkin 1962; Nabhan, forthcoming). Felger et al. (1981) noted that *G. agassizii* was prominent in Seri folklore. The Seri referred to the tortoise as *ziix heht cöquiij* (thing bushes what-sits-in) and stated that it does not walk in the direction of the sun. To regain good eyesight, a person would break the plastron open with a sharp rock, replace the viscera with a hot stone, and hold his or her face over the rising steam. If a cactus fruit looked ripe but turned out to be bitter, the Seri believed that it was the result of trickery by the star Aldebaran to fool the tortoise. If a woman gave birth only to girls, she was said to have eaten the gonads of a female *G. agassizii.* If only boys were born, it was believed that, as a small child, the woman had been struck in the small of her back by the gonads of a male tortoise thrown playfully at her by a girlfriend. Many other beliefs are related in Nabhan (forthcoming).

Cheloniidae

SEA TURTLES

The Cheloniidae include five living marine genera, represented by perhaps as many as seven or eight species (Iverson 1992; Pritchard 1997) which collectively range throughout the warm waters of the world. Some species are occasionally found in northern temperate waters. The forelimbs of cheloniids are large and paddlelike and are the principal locomotor elements. In the Gulf of California, the Cheloniidae are represented by four genera and four species. The Gulf of California and the areas along the west coast of the Baja California peninsula have become some of the more important feeding areas of cheloniid turtles in the eastern Pacific because they offer a high diversity of nearshore habitats, which maintain robust assemblages of marine algae, sea grass, and invertebrates. Although there are few nesting sites in this region, recent data support the belief that this area is an extremely important feeding ground in need of protection, and community-based research efforts are now under way in Bahía Magdalena (Nichols et al. 2000). Much of the information in the natural history accounts was generously provided to me by Jeffrey A. Seminoff.

Caretta Rafinesque

LOGGERHEAD

The monotypic genus *Caretta* contains a large marine turtle that has been reported off the coasts of all the world's continents except Antarctica, as well as the coasts of the Hawaiian Islands, the West Indies, and several islands in the Atlantic and Indian Oceans. It is known to occur along both coastlines of the Baja California peninsula and along the east coast of the Gulf of California.

FIG. 3.5 ***Caretta caretta***

LOGGERHEAD; CAGUAMA JABALINA, CAGUAMA PÉRICO, AMARILLA

Testudo caretta Linnaeus, 1758:197. Type locality: "insulas Americanas [American Islands]"; restricted to "the Bermuda Islands" by Smith and Taylor (1950:315); restricted to "Bimini, British Bahamas" by Schmidt (1953:107). Holotype: none designated

IDENTIFICATION *Caretta caretta* can be distinguished from all other marine turtles surrounding the Baja California peninsula by the combination of five pairs of lateral scutes, a red-brown to brown color, and a disproportionately large head.

RELATIONSHIPS AND TAXONOMY Limpus et al. (1988) place *Caretta caretta* as the sister species to the Hawksbill Sea Turtle *(Eretmochelys imbricata).* Zangerl et al. (1988), however, con-

FIGURE 3.5 *Caretta caretta* (Loggerhead) from near Bahía de los Ángeles, BC

sider it more closely related to Ridley sea turtles, *Lepidochelys*. Groombridge (1982), Cogger et al. (1983), Pritchard and Trebbau (1984), and Dodd (1988, 1990) review taxonomy.

DISTRIBUTION *Caretta caretta* has been found in all subtropical and tropical seas, but only juveniles have been observed in the eastern Pacific.

DESCRIPTION Very large, attaining 130 cm in carapace length; head triangular, not distinct from neck in adults, covered with smooth skin laterally and large scales dorsally; large external nares; upper jaw with single cusp; outer and tomial surfaces smooth; tympanum obscure; head scalation variable but with at least 2 pairs of prefrontals, often partially separated by one or more irregular median prefrontals; frontal occasionally fused anteriorly to frontoparietal in adults, 2 supraoculars, followed by 2 to 3 parietals, 1 to 3 small interparietals, 2 small temporals, 3 to 4 postoculars; skin of neck and gular region granular with several loose folds; carapace moderately broad but usually longer than wide; 11 to 15 marginals; posterolateral marginals serrate and flared laterally; 5 pairs of pleurals; 5 vertebrals; first costal contacts first vertebral; plastron extends anteriorly as far as carapace, is shorter posteriorly, and free edges of all plates are smooth; two prominent, paravertebral, longitudinal ridges extending the length of plastron in juveniles; occasionally small intergular present followed sequentially by pairs of gulars, humerals, pectorals, abdominals, femorals, and anals; 3 to 6 pairs of inframarginals contacting abdominals and forming bridge; small axillary plates present; forelimbs long and oarlike with moderately enlarged scales on anterior edges; hind limbs small, flat, and nearly rounded in shape; outer edges surrounded by enlarged scales; forelimbs and hind limbs have two claws; tail extends beyond hind limbs in males and not in females.

COLORATION Variable regionally and individually but generally rusty red in eastern Pacific specimens; head scales reddish-brown centrally with cream-colored margins; lower jaw yellow-brown; neck and limbs cream to brown, usually darker dorsally, with enlarged scales of limbs highlighted with lighter-colored outer margins; carapace reddish-brown and plastron dusky cream to yellowish-brown; neck uniformly dark or light with dark streaks.

NATURAL HISTORY *Caretta caretta* is a wide-ranging turtle inhabiting both the continental shelf and the open ocean; it is one of the most common species along the Pacific coast of Baja California. Along the Pacific coast, *C. caretta* is referred to as *amarilla* because of the often bright yellow coloration of its skin. On calm days, *C. caretta* up to 75 cm in carapace length can be seen basking at the water's surface, floating for hours, often with birds perched on their carapaces. *Caretta caretta* is generally a mollusk eater and enters salt marshes, estuaries, bays, lagoons, and even the mouths of rivers in search of crabs, fish, tunicates, sponges, and jellyfish. In the eastern Pacific, it feeds heavily on pelagic red crabs *(Pleuroncodes),* which may account for its reddish coloration. Turtles may dive to depths of 45 m along the peninsular shelf off the west coast of Baja California as far as 30 km out to sea. *Caretta caretta* is not as common as some other species of sea turtles in the eastern Pacific, and no records are known from any of the mainland Mexican states south of the Gulf of California (Smith and Smith 1979). Despite its rarity in the Gulf of California, it has been observed as far north as Bahía Ometepec near the mouth of the Río Colorado (Shaw 1947). Off the west coast of Baja California, it is known from as far north as the Islas Coronado (Caldwell 1962a) and Isla Guadalupe (Smith and Smith 1979) but is most common south of the Vizcaíno Peninsula. Recent work has shown that *C. caretta* is capable of transpacific migrations (Bowen, Kamezaki, et al. 1994; Bowen, Abreu-Groboise, et al. 1995; Nichols et al. 2000; Resendiz, Nichols, et al. 1998; Resendiz, Resendiz, et al. 1998).

Brattstrom (1955) reported the capture of a Clarion Island Whipsnake *(Masticophis anthonyi)* from Isla Clarión of the Revillagigedo Archipelago with what was a hatchling of either *Lepidochelys olivacea* or *C. caretta* (see Frazier 1985) in its stomach, which indicates that nesting may occur on that island.

REMARKS Frazier (1985) notes errors in the literature stemming from confusing juvenile *C. caretta* with *L. olivacea.*

On a global scale, populations of *C. caretta* are under severe pressure from human exploitation. The meat is highly sought after, its eggs are harvested from the nests, and individuals are often accidentally caught in shrimp trawls and fishing nets. In Baja California, this species has been susceptible to incidental bycatch in net fisheries. Lobster fisheries in southern Baja California have also been implicated in mass strandings of *C. caretta.* Currently, one of the most severe threats is from long-line fisheries in the Pacific high seas (Wetherall 1989).

FOLKLORE AND HUMAN USE The Seri Indians of Sonora have two different names for this turtle (Felger and Moser 1985). Populations north of Bahía Kino are known as *xpeyo,* while the populations found from Bahía Kino southward are referred to as *moosni ilitcoj caacol,* which means "large-headed sea turtle." The latter is differentiated from *xpeyo* by its larger head and greener color but may be confused with *L. olivacea* (although the Seri do have knowledge of this species as well), which used to be prevalent in the central and southern waters of the Gulf of California (Caldwell 1962a). For some reason, the Seri do not like the meat of *C. caretta,* and it is usually not harpooned (Felger and Moser 1985).

Chelonia Brongniart

GREEN SEA TURTLE

The genus *Chelonia* comprises two or three large marine turtles and has been reported off the coasts of all the world's continents as well as the coasts of many Pacific islands, the West Indies, and several islands in the Atlantic and Indian Oceans. *Chelonia* is known to occur along both coastlines of the Baja California peninsula.

FIG. 3.6 ### *Chelonia mydas*

MEXICAN GREEN SEA TURTLE, EAST PACIFIC GREEN SEA TURTLE, BLACK SEA TURTLE; CAGUAMA PRIETA, CAGUAMA COMÚN, TORTUGA NEGRA, MESTIZA

Testudo mydas Linnaeus, 1758:197. Type locality: "insulam Adscensionis [Ascension Island]"; restricted to "Insel Ascension" by Mertens and Müller (1928:23). Syntypes: Possibly three specimens: NRM 19, 26, and 231. See Iverson (1992) for discussion.

IDENTIFICATION *Chelonia mydas* can be distinguished from all other marine turtles surrounding the Baja California peninsula by its lack of a leathery shell containing longitudinal ridges; the possession of four nonimbricate lateral laminae; a single pair of prefrontals; and upper and lower jaws with serrated incisiform "teeth" on the beak.

TAXONOMY AND RELATIONSHIPS *Chelonia mydas* forms the sister group to either the remainder of the Cheloniidae (Dutton et al. 1996; Limpus et al. 1988) or the Hawksbill Sea Turtle, *Eretmochelys imbricata* (Zangerl et al. 1988), or may be part of a basal trichotomy involving the Flatback

FIGURE 3.6 *Chelonia mydas* (Green Sea Turtle) from near Bahía de los Ángeles, BC. (Photograph by Jeffrey A. Seminoff)

Turtle, *Natator depressus,* and a clade composed of *Eretmochelys, Caretta,* and *Lepidochelys* (Bowen and Karl 1997). Pritchard (1983, 1997) and Pritchard and Trebbau (1984) discuss taxonomy and suggest that *Chelonia agassizii* is distinct from *C. mydas* because they are sympatric in western Mexico, Galápagos, and Papua New Guinea. However, this taxonomy has not been followed by subsequent authors (e.g., Bowen and Karl 1997; Kamezaki and Matsui 1995; Zug 1996). Caldwell (1962a) discusses the taxonomy of the populations surrounding Baja California. Bowen, Meylan, et al. (1992) and Karl et al. (1992) discuss global phylogeography and genetic structure, and their taxonomic implications.

DISTRIBUTION *Chelonia mydas* is pantropical (Hirth 1980). Occasionally it is found as far north as Massachusetts, USA, in the Atlantic (Bleakney 1965), and Prince William Sound, Alaska, in the Pacific. It has also been reported from as far south as southern Chile in the Pacific (Iverson 1992).

DESCRIPTION Much of this description is adapted from Caldwell (1962b). Adults large, attaining 60 to 110 cm carapace length; head triangular, not distinct from neck, covered with smooth skin laterally and large scales dorsally; snout pointed in dorsal profile; slight evidence of beak; external nares open on anterior margin of snout; upper jaw vertically grooved, tomial surface weak to coarsely denticulate, and alveolar surface with 2 prominent denticulate ridges; tomium of lower jaw strongly denticulate; eyes moderately sized; 1 pair of prefrontal scales followed by single frontal partially separating prefrontals posteriorly; 1 pair of supraoculars; frontoparietal followed by pair of parietals; 2 or 3 temporals; skin of neck and gular region granular with several loose folds; 2 to 5 postocular scales; 2 or 3 large mandibular scales; carapace slightly wider than long; carapace smoothly rounded anteriorly and weakly pointed posteriorly in dorsal profile, more tapered posteriorly in males; nuchal rectangular; 11 to 13 marginals; 4 to 6 pairs of pleurals; 5 to 9 vertebrals; plastron extends anteriorly as far as carapace but is shorter posteriorly, all free edges smooth; two central plastral ridges present and most pronounced in juveniles; gulars paired, followed sequentially by pairs of humerals and pectorals, 2 pairs of abdominals, 1 pair of femorals and anals; interanal sometimes present; 3 to 5 pairs of inframarginals forming a wide bridge; ventral surface of plastron weakly concave; one inguinal plate and a series of irregularly shaped, axillary plates; forelimbs long and oarlike with at least one claw; anterior and posterior edges have enlarged and irregularly shaped scales; hind limbs smaller than forelimbs, more rounded in dorsal profile, outer edges covered with enlarged scales; occasionally one claw present on hind limb; tail prehensile in vertical plane, extending far beyond hind limbs in males and just reaching past carapace in females. **COLORATION** Generally dark but variable; carapace and dorsal surfaces of head, neck, and limbs slate gray to black; scale edges of head and limbs edged in lighter coloration; often a radiating pattern of wide brown, olive, or yellow lines at posterior margin of carapacial plates, most commonly occurring in juveniles; plastron bluish to dark gray, plastral ridges lighter; plastron yellow in young; large, dark brown spot on hands and feet in young.

NATURAL HISTORY Pacific populations of *C. mydas* are common in lagoons and bays in the vicinity of mangroves, beds of eelgrass, or seaweed, and less common in open waters. Smaller turtles are commonly found in pelagic environments off the Pacific coast of Baja California in close association with drifting sargassum and *Macrocystis* mats. The coastal lagoons along the Pacific side of Baja California provide developmental habitats. On average the turtles in Ojo de Liebre, San Ignacio, and Bahía Magdalena are much smaller than turtles in the Gulf. These Pacific populations feed primarily on *Zostera* and marine algae. *Chelonia mydas* is by far the most common sea turtle in the Gulf of California, although its numbers have declined considerably over the years. It is ubiquitous in the Gulf and has even been found a few miles south of the Río Hardy in freshwater tributary canals of the Río Colorado (Smith and Smith 1979). In historical times it was known to occur as far north as El Mayor, 80 km north of the mouth of the Río Colorado (Cliffton et al. 1982).

Apparently *C. mydas* remains in a common area and shows a high degree of site fixity in nesting. Nesting colonies are known to be well established in Michoacán and to a lesser extent in the Revillagigedo Islands, Galápagos, and Costa Rica. No confirmed nesting colonies of this species occur on the Baja California peninsula. Caldwell (1962a) reports an unconfirmed nesting beach along the Pacific coast in Bahía Tortugas, and a breeding population reported along the west coast in the vicinity of Todos Santos (Pritchard 1979) is highly questionable. Older unconfirmed reports (Dawson 1944; McGee 1896; Townsend 1916) state that nesting occurs near the mouth of the Colorado River and along the Sonoran coast opposite Isla Tiburón. No insular nesting sites in the Gulf of California have been confirmed, although anecdotal accounts of one or two individuals nesting on Gulf islands, most notably Monserrate, do exist.

Chelonia mydas has overwintering sites in the northern Gulf of California, in Canal del Infernillo between Isla Tiburón and the Sonoran coastline. Here individuals aggregate in large numbers and lie partially buried in the mud, approximately 3 to 5 m apart, in fields of eelgrass (Felger et al. 1976). Turtles may remain buried for one to three months (Cliffton et al. 1982). The accidental discovery of this behavior by Mexican lobster hunters resulted in the near-extirpation of *C. mydas* from these areas. In 1975, five boats of fishermen with diving equipment caught as many turtles as one boat of Seri Indians with harpoons caught during the entire decade of the 1960s (Cliffton et al. 1982). An additional overwintering site has recently been discovered along the east coast of Baja California along rocky habitats of insular areas within Bahía de los Ángeles (J. A. Seminoff, personal communication, 2000). Much as in Canal del Infernillo, torpid turtles are taken by divers using compressed air.

Chelonia mydas is primarily herbivorous and feeds on eelgrass, red algae, and occasional marine invertebrates such as sponges, sea hares, and sea pens (Seminoff et al. 1998, 1999, 2000). It also has the curious habit of pulling itself to shore to bask on remote beaches.

REMARKS Recent genetic studies (Dutton et al. 1996) indicate that there are two populations of *C. mydas* in the Gulf of California that have weakly differentiated yet distinctive mitochondrial DNA haplotypes. One of these haplotype groups appears to be most closely related to Hawaiian populations, whereas the other seems to be unique to the eastern Pacific. Aside from their genetic differences, the populations also show differences in skull morphology (Kamezaki and Matsui 1995), nesting behavior, number of eggs per nest, number of nests per season, having geographically separate nesting areas, and differences in coloration and shell shape. Bowen and Karl

(1997), Kamezaki and Matsui (1995), and Zug (1996) do not believe that this population should be accorded separate specific status.

Many turtles have conspicuous discolored round marks on their shells. These are actually the turtle barnacle *Chelonibia testudinaria.* Another barnacle, *Tubicinella cheloniae,* burrows into the shell. Both barnacles are endemic to sea turtles and are found on Loggerheads *(Caretta caretta)* as well.

FOLKLORE AND HUMAN USE As the single most important food source for most of the Seri Indian tribes, *C. mydas* was known by eight different names, each of which corresponded to a different geographic group or growth stage (Felger and Moser 1985). The most common name appears to be *moosni* (Nabhan, forthcoming). Most of the groups recognized by the Seri had biological significance, and the Seri were often much more knowledgeable about this species' natural history than were the scientists studying them. Unfortunately, most of the "folk taxa" became extinct by the mid-twentieth century as a result of overharvesting (Cliffton et al. 1982). In addition to providing food, *C. mydas* was an important resource in Seri material and spiritual culture (Nabhan, forthcoming). The stomachs were dried and used as water bags and containers, the skin was used to make sandals, and the carapaces were used as large multipurpose containers (Felger and Moser 1985). See Nabhan (forthcoming) for many additional uses.

This species continues to be illegally harvested in large numbers. It is consumed locally in coastal villages and shipped to interior regions of Mexico. A large, well-organized black market continues today in Baja California. The majority of turtles are poached in Bahía Magdalena, Laguna San Ignacio, Loreto, and the vicinity of Bahía de los Ángeles and Bahías las Ánimas and San Rafael. In 1992 I visited a restaurant in Bahía de los Ángeles that had turtle steaks on its menu and sea turtle conservation posters on its walls. Recent estimates suggest that as many as 7,000 turtles die each year from illegal harvesting, incidental capture in set net fisheries, and trawling activities (J. A. Seminoff, personal communication, 2000).

Eretmochelys Fitzinger

HAWKSBILL SEA TURTLE

The monotypic genus *Eretmochelys* contains a small to medium-sized marine turtle that has been reported off the coasts of all the world's continents except Antarctica as well as off many Pacific islands, the West Indies, and several islands in the Atlantic and Indian Oceans. It is known to occur along both coastlines of the Baja California peninsula.

FIGURE 3.7 *Eretmochelys imbricata* (Hawksbill) from near La Paz, BCS. (Photograph by Jeffrey A. Seminoff)

FIG. 3.7 *Eretmochelys imbricata*

HAWKSBILL; CAREY, CAGUAMA

Testudo imbricata Linnaeus, 1766:350. Type locality: "Mari Americano, Asiatico [America and Asiatic seas]"; restricted to "the Bermuda Islands" (Smith and Taylor 1950:315); restricted to "Belize, British Honduras" (Schmidt 1953:106). Holotype: possibly ZIUS 130 (Smith and Smith 1979)

IDENTIFICATION *Eretmochelys imbricata* can be distinguished from all other marine turtles surrounding the Baja California peninsula by strongly imbricate laminae on the carapace and four prefrontal scales.

RELATIONSHIPS AND TAXONOMY The relationships of *Eretmochelys imbricata* to other marine turtles are controversial. Limpus et al. (1988) suggest that *E. imbricata* is the sister species to *Caretta caretta,* whereas Zangerl et al. (1988) hypothesize that it is most closely related to *Chelonia.*

DISTRIBUTION *Eretmochelys imbricata* occurs in nearly all tropical and occasionally subtropical seas around the world. It is not known much further north than Isla Cedros, off the west coast of Baja California in the eastern Pacific, or further north than Bahía Kino, Sonora (Smith and Smith 1979), and Bahía de los Ángeles in the Gulf of California (Caldwell 1962a).

DESCRIPTION Small to medium-sized, with adults reaching 90 cm carapace length; head triangular in dorsal profile, not distinct from neck, and covered with smooth skin laterally; snout pointed with hooked uncusped beak; external nares positioned far anteriorly on snout; edges of jaws smooth, tomium of upper jaw smooth; lower jaw not denticulate; eyes moderately sized; 2 pairs of prefrontals followed by a pair of frontals partially dividing prefrontals posteriorly; single frontoparietal and supraocular followed posteriorly by a temporal; 1 pair of parietals; interparietals occasionally present; 3 or 4 postoculars; single large supralabial at angle of jaw; large temporal ventral to uppermost temporal; 2 postoculars; 2 large mandibular scales; skin of neck and gular region granular with several loose folds; dorsal profile of carapace teardrop-shaped in adults and more heart-shaped in juveniles; vertebral keel and lateral keels in juveniles, occurring posteriorly in adults; plates of carapace strongly imbricate in young and small adults; plates juxtaposed in juveniles and very old adults; posterior periphery of carapace distinctly serrate in adults due to lateral flaring of posterolateral marginals; nuchal plate followed by 5 vertebrals; 11 marginals, and 4 pleurals on each side; plastron extends anteriorly as far as carapace, is much shorter posteriorly and with smooth free edges; plastron rounded anteriorly with prominent lateral keel on each side; intergular present followed sequentially by a pair of gulars, pectorals, 2 pairs of abdominals, femorals, and a pair of anals; small interanal occasionally present; 4 poreless inframarginals, 2 to 4 small axillaries, and a single, minute inguinal; forelimbs long and oarlike with moderately enlarged scales along anterior and posterior edges; hind limbs much shorter than forelimbs, flat, and more rounded; a series of enlarged scales anteriorly and posteriorly; both forelimbs and hind limbs have two claws; tail much longer in males than in females; tail not extending past carapace in females.

COLORATION Overall color of carapace yellowish-tan to greenish or reddish-brown with faint to moderate, radiating, light-colored marbled pattern in each plate on adults; dorsal surface of head and limbs deep, glossy black with occasional light margins; scales on sides of head prominently edged in light orange; horny sheath of upper jaw mostly black anteriorly and laterally, but with vertical, light bar extending ventrally behind nostril; gular region dusky yellow-white with a few dark-centered scales; plastron yellow with a central, dark spot in most anterior plates.

GEOGRAPHIC VARIATION Individuals in the northern portions of the Gulf of California are usually smaller than those of the southern Gulf.

NATURAL HISTORY *Eretmochelys imbricata* is principally a warm-water species. It occasionally ranges as far north as Japan, but south of the equator it generally does not venture into colder waters and thus does not come into contact with Atlantic populations around the tips of Africa and South America. In the Gulf of California, *E. imbricata* is common in the southern waters. Carr (1952) noted that it is a frequent inhabitant of shallow mangrove-bordered bays, lagoons, and estuaries with or without submerged vegetation. *Eretmochelys imbricata* is also known to enter the mouths of shallow creeks, and local residents at Mulegé report that they occasionally see *carey* in Río Mulegé. *Eretmochelys imbricata* is primarily a spongivore but will consume a wide variety of other foods, such as the fruits and leaves of mangroves, marine vegetation, tree bark, crustaceans, coelenterates, mollusks, sea urchins, and fish. Nesting information for this species in the eastern Pacific is scarce, and it is not known to form nesting colonies along any portion of the Pacific coast of Mexico. Nesting probably takes place from August to November, and females lay anywhere from 50 to over 220 eggs per clutch.

FOLKLORE AND HUMAN USE The plates of the carapace (known commercially as tortoiseshell) of the Hawksbill have been items of trade since Spanish colonial times. The Seri Indians made ornamental jewelry from them. The Seri rarely ate the meat because they believed it was inferior to that of *Chelonia mydas*. Their name for the Hawksbill is *moosni quipaacalc* (sea-turtle-that-which-overlaps [Felger and Moser 1985]), a name that refers to the imbricate nature of the carapacial plates. They also have a *moosni sipoj* (sea-turtle osprey), which has a more pointed snout than *moosni quipaacalc*, but the Seri say it has become very rare since the 1960s (Felger and Moser 1985).

Currently Japan is trying to persuade Cuba to reopen its Hawksbill fishery to supply Japanese traditional artisans with tortoiseshell. In 1997 Cuba nearly received permission through CITES to do so. The Cubans are still trying at the time of this writing (2001).

Lepidochelys Fitzinger

RIDLEY SEA TURTLES

The genus *Lepidochelys* is composed of two medium-sized marine turtles that collectively range along the east and west coasts of North, Central, and South America, the mid-Atlantic, the west coast of Europe, throughout the Mediterranean, and the coasts of Africa and India. One species, *L. olivacea*, is known to occur along both coastlines of the Baja California peninsula.

FIG. 3.8 *Lepidochelys olivacea*

OLIVE RIDLEY, PACIFIC RIDLEY; CAGUAMA MESTIZA, GOLFINA

Chelonia olivacea Eschscholtz, 1829:3. Type locality: "Bai von Manilla [Manila Bay, Philippines]." Holotype: possibly in the Museum of Tartu (formerly Dorpat), Estonia (Smith and Smith 1979)

FIGURE 3.8 *Lepidochelys olivacea* (Pacific Ridley) from La Barra, Oaxaca. (Photograph by Jeffrey A. Seminoff)

IDENTIFICATION *Lepidochelys olivacea* can be distinguished from all other marine turtles surrounding the Baja California peninsula by having six or more pleural scutes; a high, flat-topped carapace; and pores in the bridge scales.

RELATIONSHIPS AND TAXONOMY *Lepidochelys olivacea* is the sister species of *L. kempi* (Limpus et al. 1988; Zangerl et al. 1988) of the Atlantic Ocean (Iverson 1992). The relationships of *Lepidochelys* to other marine turtles, however, is controversial. Limpus et al. (1988) suggest that *Lepidochelys* is the sister group of the genus *Natator*, whereas Zangerl et al. (1988) hypothesize that it is most closely related to *Caretta.*

DISTRIBUTION *Lepidochelys olivacea* is known from the warmer regions of the Indian and Pacific Oceans as well as from the southern Atlantic. It is occasionally seen in the southern Gulf of California but seldom sighted in its northern waters, although it has been found at San Felipe (Caldwell 1962a). *Lepidochelys olivacea* ranges along the entire west coast of Baja California and extends as far north as Washington State, USA (Iverson 1992; Zug et al. 1998).

DESCRIPTION Moderate-sized, with adults attaining 80 cm carapace length; head relatively large, markedly compressed laterally, distinct from neck, triangular, pointed in dorsal profile, and covered with large scales laterally and dorsally; external nares situated far forward; beak moderately hooked; upper jaw large, alveolar surface with occasional flat swelling running parallel to smooth tomium; eyes large; tympanum obscure; 2 pairs of prefrontals partially separated posteriorly by median frontal; single supraocular; single frontoparietal; 1 or 2 temporals; parietals paired; 3 or 4 postoculars; single enlarged mandibular scale; skin of neck and gular region granular, containing several loose folds; dorsal profile of carapace in adults broad, slightly narrower than long, smoothly rounded anteriorly, and weakly pointed posteriorly; prominent middorsal keel and two prominent lateral keels in juveniles; carapace flattened along vertebral plates and lateral keels scarcely evident in adults; carapace slightly serrated posteriorly in juveniles and moderately domed in adults; 5 vertebral humps present on carapace in juveniles; nuchal plate thin; 12 or 13 marginals on each side; posterolateral marginals flared, producing a weakly serrate margin; 5 to 9 pairs of pleurals; 5 vertebrals; plastron extends anteriorly nearly as far as carapace, is much shorter than carapace posteriorly, and is more concave in males than in females; gulars paired, followed sequentially by a pair of humerals, abdominals, femorals, and anals; pair of interanals and 1 or 2 intergulars occasionally present; 4 inframarginals, each with a deep pore; axillary plates present; forelimbs oarlike, with enlarged scales along anterior and posterior edges; one claw in adults, but a very small additional distal claw present in juveniles; hind limbs much shorter than forelimbs, with a series of enlarged scales anteriorly and posteriorly; one claw on each hind limb; tail much longer in males, not extending to end of carapace in females. **COLORATION** Carapace in adults generally uniformly olive to gray and a uniform gray-black in juveniles; head slightly lighter in color and scales edged in cream; plastron yellow to greenish-yellow in adults, uniformly gray-black in hatchlings, turning to white in juveniles, except for keels, which remain lighter; limbs and neck olive above and lighter below.

NATURAL HISTORY *Lepidochelys olivacea* is the most prevalent marine turtle in the world, and it is commonly seen basking at the surface with the tips of the forelimbs extending above the water. This species commonly basks during July and August in the central portions of the Gulf of California. Large numbers have been observed in the open ocean off the Mexican coast (Oliver 1946), but it is usually observed in clear, shallow waters in bays, lagoons, and estuaries. *Lepidochelys olivacea* feeds primarily on crustaceans, mollusks, and other invertebrates. Unlike some other marine turtles, it does not consume plant material. Nesting and egg-laying in the eastern Pacific populations take place from mid-August through November (Carr 1952), and roughly 30 to 120 eggs per clutch are laid on isolated beaches. Fritts, Stinson, and Márquez-M. (1982) noted that nesting is common along the southern Pacific coast of Baja California from Punta Cornejo southward to near Todos Santos during August and September and may extend from July to October. Based on interviews with local residents,

Márquez et al. (1976) estimated that there are one to five thousand nests (equaling approximately one thousand females) in Baja California. Fritts, Stinson, and Márquez-M. (1982) believe these numbers may be high, noting that many nests excavated by *L. olivacea* collapse before the eggs can be laid, perhaps because the sand is too dry. They note that summer rainstorms may play a role in providing the beaches with sufficient moisture for successful nest excavation. Brattstrom (1955) reported the capture of a whipsnake *(Masticophis anthonyi)* from Isla Clarión of the Revillagigedo Archipelago with what was a hatchling of either *L. olivacea* or *Caretta caretta* (see Frazier 1985) in its stomach, which indicates that one of these species may nest on that island.

FOLKLORE AND HUMAN USE The Seri Indians referred to *L. olivacea* as *moosni otac* (sea-turtle toad), referring to its wide carapace. They did not favor its meat over that of *Chelonia mydas* (Felger and Moser 1985). Some Seri used to keep hatchling *L. olivacea* as pets (Nabhan, forthcoming).

Lepidochelys olivacea is not poached as often as *C. mydas* in the Gulf of California or along the Pacific coast of Baja California, although some poaching does occur. This species is prized for its skin because of its leatherlike qualities. Large numbers of individuals are illegally harvested from southern Mexican waters each year.

Dermochelyidae

LEATHERBACK SEA TURTLES

The Dermochelyidae are a monotypic lineage containing the largest and most distinctive of all living turtles, *Dermochelys coriacea*. This species is characterized by scaleless, leathery skin on its back, with seven longitudinal ridges formed by enlarged dermal platelets running its length. Five similar ridges occur on the belly. Individuals may reach nearly 700 kg in weight. The limbs are clawless, and the hind limbs may be broadly connected to the tail by a web. *Dermochelys coriacea* is pelagic and worldwide in distribution; it is more commonly observed in temperate and subarctic waters than in warmer regions. This species is rare in the Gulf of California.

Dermochelys Blainville

LEATHERBACK

The monotypic genus *Dermochelys* is the largest living turtle in the world. It is widely distributed throughout the waters of the Atlantic, Pacific, and Indian Oceans from Labrador, Iceland, the British Isles, Norway, Alaska, and Japan south to Argentina, Chile, Australia, and the Cape of Good Hope, South Africa (Pritchard 1980). It is infrequently found along both coastlines of the Baja California peninsula.

FIGURE 3.9 *Dermochelys coriacea* (Leatherback) from Michoacán. (Photograph by Jeffrey A. Seminoff)

FIG. 3.9 ***Dermochelys coriacea***

LEATHERBACK; LAÚD, CAGUAMA SIETE FILOS, CAGUAMA ALTURA

Testudo coriacea sive *mercurii* Rondeletius, 1554 (*fide* Agassiz 1857). Type locality: Mediterranean Sea. Holotype: unknown

IDENTIFICATION *Dermochelys coriacea* can be distinguished from all other marine turtles surrounding the Baja California peninsula by its leathery carapace and the presence of seven longitudinal ridges on its dorsum.

RELATIONSHIPS AND TAXONOMY *Dermochelys coriacea* is the sister species to all other cheloniids (Gaffney and Meylan 1988). Pritchard (1971, 1980) and Pritchard and Trebbau (1984) conclude that previously described subspecies (see Smith and Smith 1979) are poorly differentiated.

DISTRIBUTION *Dermochelys coriacea* is pelagic, worldwide in distribution, and more commonly observed in temperate and subarctic waters than in warmer regions. In the eastern Pacific, *D. coriacea* ranges from Canada to southern Chile. In the Gulf of California, where it is considered rare, *D. coriacea* has been found as far north as Puerto Peñasco, Sonora, Mexico (Smith and Smith 1979).

DESCRIPTION *Dermochelys coriacea* is the largest of all living turtles, with adults reaching 244 cm in carapace length; head large, not distinct from neck, and covered with smooth, leathery skin dorsally and laterally in adults; head covered with several small fragmented scales including a rostral, nasals, two pairs of prefrontals, supraoculars, postoculars, supratemporals, and a large frontoparietal in hatchlings or very small (less than 20 cm) juveniles; snout smoothly rounded in dorsal profile, and beak distinctly bicuspid with deep median notch; edges of jaws denticulate; eyes moderately sized; tympanum obscure; skin of neck and gular region wrinkled; carapace covered with smooth, leathery skin; carapace lower in lateral profile in adult males than in females and elongate in dorsal profile; carapace with two anterior projections resulting from protuberant extensions of lateral keels; posterior end of carapace greatly protracted; seven longitudinal ridges on carapace; carapace covered with several small scales in hatchlings; plastron extending slightly farther anteriorly than carapace but much shorter posteriorly; plastron covered with smooth, leathery skin with five longitudinal ridges, ending with a distinct osteoderm in males; plastron covered with small scales in hatchlings; forelimbs long, oarlike, lacking claws, and covered with smooth skin; forelimbs of hatchlings covered with small scales with a slightly enlarged row on anterior border; hind limbs much shorter, with undulate posterior margin; hind limbs covered with smooth, leathery skin in adults and small scales in hatchlings; tail longer in males than females. **COLORATION** Head black and faintly blotched with a pineal spot; jaws lighter in color; five longitudinal rows of light spots on neck; carapace slate gray with three or four longitudinal rows of small white spots between each pair of ridges; hatchlings bluish dorsally, with anterior margins of forelimbs and carapacial ridges edged in white; plastral ridges white.

NATURAL HISTORY *Dermochelys coriacea* is a wide-ranging, pelagic turtle that is only occasionally found in the Gulf of California but is more common in open waters along the west coast of Baja California. Very little is known about its natural history. This species is reported to be a powerful swimmer with an aggressive temperament. Owing to its large size and the fact that it has countercurrent heat-exchanging capillary systems in its extremities, it is able to withstand colder waters. Some specimens have been known to raise their body temperature to 17°C above that of the surrounding water.

Nesting occurs in the vicinity of Todos Santos on the Pacific coast from November through February, although individual females only lay eggs once every two to three years. Sixty-five to 80 viable eggs per clutch are laid, and a female may nest up to nine times in a single season, every two to four years. Juveniles may grow as fast as 50 cm per year. Fritts, Stinson, and Márquez-M. (1982) report that *D. coriacea* nests along the beaches of southwestern Baja California from October through March. *Dermochelys coriacea* feeds almost exclusively on jellyfish.

FOLKLORE AND HUMAN USE *Dermochelys coriacea* features prominently in Seri Indian rituals. The Leatherback, the Boojum Tree, and the Teddy Bear Cholla are considered some of the original kinds of humans (Nabhan, forthcoming). According to Felger and Moser (1985), the Seri believe that in the beginning, when the earth was just finished and before anyone had died, the Leatherback, *moosnipol* (sea-turtle its blackness), was the female member of a family that also included a sailfish and a large moth. The first person ever to die was a sibling of *moosnipol,* and so the turtle's face became spotted from weeping. When a *moosnipol* is caught by the Seri, a fiesta is given to make her happy so that nothing bad will happen to the person who caught her. The fiesta may last as long as four days, during which the turtle is kept in a specially built shelter to shade it. The turtle is released after the fiesta (if it is still alive). It is said that when a man finds and harpoons a Leatherback, she speaks to him and leads him ashore. The harpooner is required to supply *moosnipol* with drinking water as soon as she is beached. The Seri also believe that *moosnipol* understands Seri, so people whisper in her presence. Occasionally the meat is eaten, and it is believed that one will be struck blind if one does not close one's eyes on eating the meat for the first time. See Nabhan (forthcoming) for additional spiritual references.

Dermochelys coriacea is now a species of great concern and the most threatened of all sea turtles. The majority of its decline has been at the hands of man (Eckert 1991; Eckert and Sarti 1997; Sarti et al. 1996). Mexico and Costa Rica have witnessed a hundredfold decrease in nesting over the last twenty years. Satellite telemetry shows that most nesting Leatherbacks in these regions move south to the cold, nutrient-rich waters off the coast of Chile, where they have been caught in large numbers in the long lines of the swordfish boats. Informants now say that during peak times of the year in this fishery, 10 to 15 Leatherbacks per day per boat were being killed. Additionally, autopsies on dead individuals found floating in the water and washed up on beaches around the world reveal that many turtles have choked to death on plastic bags. Apparently clear plastic bags floating in the water resemble some of the pelagic species of jellyfish on which this species feeds. Along the coastline of Michoacán, Guerrero, and Oaxaca in southwestern Mexico, where *D. coriacea* comes to lay eggs, poachers have been reported to capture and kill females and remove the eggs from their bodies before they have even been laid. *Dermochelys coriacea* eggs, which are claimed to increase energy and virility, are worth about $10 each in some countries.

Trionychidae

SOFT-SHELL TURTLES

The Trionychidae are a peculiar-looking group of turtles containing 14 genera and approximately 25 species (Pough et al.

2001). The family ranges from sub-Saharan and eastern Africa through southern and eastern Asia into the Indo-Australian Archipelago. In the New World, trionychids generally occur only in the central, southern, and southeastern United States and northeastern Mexico. This group is characterized by a flat, round, flexible shell lacking epidermal plates, and nostrils at the end of a proboscislike snout. The feet are webbed and paddlelike, and each forefoot bears three claws. In Baja California, trionychids are represented by the single genus *Apalone*.

Apalone Rafinesque

NORTH AMERICAN SOFT-SHELLS

The genus *Apalone* is composed of three small to large freshwater turtles that collectively range widely across the United States, except for the Pacific Northwest, and just enter northern Mexico. One species, *A. spinifera*, is known to occur in the Río Colorado and associated drainages of northeastern Baja California, having been introduced by humans.

FIG. 3.10 **_Apalone spinifera_**

SPINY SOFT-SHELL; TORTUGA BLANCA

Trionyx spiniferus LeSueur, 1827:258. Type locality: "New-harmony, sur le Wabash [New Harmony, Wabash River, Posey County, Indiana, USA]." Syntypes: MNHN 1949, 6957, 8807–12; MNHN 8808, designated lectotype by Webb (1962:491). MNHN 1949 and 6957 may be *Trionyx ocellatus* (see Iverson 1992:304)

FIGURE 3.10 *Apalone spinifera* (Spiny Soft-Shell) from Mexicali. (Photograph by Laurie J. Vitt)

IDENTIFICATION *Apalone spinifera* can be distinguished from all other freshwater turtles in Baja California by a flat, rounded, soft shell lacking epidermal scutes and nostrils terminating in a proboscislike snout.

RELATIONSHIPS AND TAXONOMY *Apalone spinifera* is the sister species of *A. mutica* of the eastern United States (Meylan 1987). Webb (1962) discusses taxonomy.

DISTRIBUTION *Apalone spinifera* ranges from southern Quebec and Ontario, Canada, south to the Río Grande systems and central Tamaulipas, Mexico, and east to the Atlantic seaboard north of peninsular Florida. In the west, it is widespread in the Colorado and Gila River drainages from central Arizona and southwestern Utah to northeastern Baja California (Iverson 1992). In Baja California, it is restricted to the Río Colorado and its drainages, including numerous irrigation channels in the delta region (see map 3.3).

DESCRIPTION Moderately large, with adults reaching approximately 450 mm carapace length; head smooth and moderately sized; snout thin, flexible, and protuberant, with terminal nares; margins of maxilla and mandible smooth anteriorly and denticulate posteriorly; eyes moderately sized; tympanum obscure; skin of neck and gular region smooth and wrinkled; carapace flat, scaleless, and covered with leathery skin; carapace ovoid in dorsal profile and wider posteriorly; posterior portion flexible and with several dorsal knobs of low profile, anterior margin sometimes tuberculate; plastron smooth, scaleless; forelimbs stout, flat, and with three large claws; digits 4 and 5 embedded in webbing of flat, wide hand; hind limbs stout, slightly larger than forelimbs; claws of digits 1 to 3 prominent; digits 4 and 5 embedded in webbing of flat, wide foot; tail longer in males than females. **COLORATION** Head grayish to brownish; rostrum with pale, slightly curved light stripes extending from corner of eyes onto snout; stripe bordered laterally with dark stripe; occasional thin, dark, transversely oriented stripes above eyes; interrupted postorbital stripe extending onto neck from back of eye; dark markings on head much more prominent in juveniles; variable pattern of dark dorsal spots overlaying grayish to brownish ground color on soft parts of body; tail with wide, dark dorsal stripe; juveniles have more prominent dark markings on limbs; carapace grayish to brownish, lacking white dots on anterior half; whitish dots confined to posterior one-third and usually lacking in juveniles; occasional small black dots on carapacial margin; carapace with pale rim bordered medially by black line in juveniles; underside of posterior portion of carapace with occasional dark, poorly defined blotches; plastron whitish to grayish with occasional dark, symmetrical discolorations on bridge.

NATURAL HISTORY *Apalone spinifera* occurs in slow-moving or still permanent streams, ponds, irrigation ditches, and canals with muddy to gravelly bottoms. Large amounts of streamside vegetation are not necessary, although barren shorelines are generally avoided. Little has been reported on the natural history of *A. spinifera* from Baja California (Smith and Smith 1979). In the delta region, *A. spinifera* is occasionally seen sunning itself in the morning along the shores or on floating logs and vegetation in the dirty, often stagnant, irrigation canals. It is generally very wary and difficult to approach and is most often seen near the shore with only its nose and eyes exposed. *Apalone spinifera* is active from February through November, depending on the weather. Its occurrence in Baja California is the result of a southerly migration from its region of human introduction in the upper reaches of the Gila River, Arizona, during the early 1900s (Miller 1946).

FOLKLORE AND HUMAN USE *Apalone spinifera* is found on the menus of some Chinese restaurants in Mexicali. The limbs are fried and have a sweet gamy flavor. It is captured with a hook and line or net from the various irrigation canals. It is also collected from some of the more polluted and stagnant ponds.

LIZARDS

Lizards are the most conspicuous and speciose group of reptiles in Baja California and its associated islands. Nine major groups are represented, and only one, the Xantusiidae (night lizards), lacks an insular representative. The lizards of Baja California and the Sea of Cortés show a broad range of adaptive types, ranging from fossorial legless lizards *(Anniella),* arenicolous sand lizards (e.g., *Uma notata, Phrynosoma mcallii*), and saxicolous scansorial rock-dwellers (e.g., *Petrosaurus, Phyllodactylus, Sauromalus*) to arboreal *(Urosaurus graciosus, U. ornatus)* and vegetative *(Xantusia vigilis)* specialists. At the present count, the lizard fauna contains 21 genera, one of which *(Petrosaurus)* is endemic; and 84 species, of which 57 (69 percent) are endemic. Thus a significant portion of the lizard fauna of this region occurs nowhere else in the world.

Lizards are well represented on both Pacific and Gulf islands. The majority of these species are most closely related to the nearest peninsular or mainland populations, although there are indications that some transgulf colonizations have occurred (Grismer 1994a). Insular endemism in the Gulf of California is high. Of the 61 species recorded from islands, 32 (52 percent) are endemic.

Crotaphytidae

COLLARED AND LEOPARD LIZARDS

The Crotaphytidae (Frost and Etheridge 1989) is a New World group of lizards containing the two genera *Crotaphytus* and *Gambelia,* which collectively contain 12 species (McGuire 1996). The Crotaphytidae range from eastern Oregon south into northern Mexico and east to the Mississippi River. They are diurnal predators, feeding largely on other lizards and insects, which they often ambush and crush with their powerful jaws. Although these two genera coexist throughout a large portion of their range, they probably compete only minimally. *Crotaphytus* are generally restricted to rocky foothills and alluvia, whereas *Gambelia* are more commonly associated with open flatlands. The Crotaphytidae are distinguished from other iguanian lineages by several skeletal characteristics, including the presence of a posterolaterally projecting jugal-ectopterygoid tubercle; the parietal and frontal broadly overlapping the medial process of the postorbital; the supratemporal lying in a groove along the parietal; contact of the prefrontal and jugal along the anterolateral border of the orbit; palatine teeth; a posteromedially curving tympanic crest of the retroarticular process of the mandible; and the presence of posterior coracoid fenestrae (McGuire 1996).

Crotaphytus Holbrook

COLLARED LIZARDS

The genus *Crotaphytus* is a group of medium-sized, terrestrial to saxicolous lizards containing nine living species. *Crotaphytus* is renowned for its strikingly colorful adult males and for being voracious, aggressive predators. Collectively, *Crotaphytus* ranges from the western and central United States south throughout most of Baja California and northern Mexico. Two species, *C. grismeri* and *C. vestigium,* occur in Baja California, and two others, *C. dickersonae* and *C. insularis,* occur on islands in the Gulf of California.

FIG. 4.2 **_Crotaphytus dickersonae_**

DICKERSON'S COLLARED LIZARD; ESCORPIÓN DE LA PIEDRA

Crotaphytus dickersonae Schmidt 1922:638. Type locality: "Tiburon Island," Sonora, Mexico. Holotype: USNM 64451

IDENTIFICATION *Crotaphytus dickersonae* can be distinguished from all other lizards in the region of study by two black collars bordering a central white collar; a posterior black collar wider than the anterior black collar, the anterior collar being complete ventrally; black oral melanin; and granular body scales.

RELATIONSHIPS AND TAXONOMY *Crotaphytus dickersonae* is the sister species to a monophyletic group of collared lizards containing *C. bicinctores, C. grismeri, C. insularis,* and

FIGURE 4.2 Adult male *Crotaphytus dickersonae* (Dickerson's Collared Lizard) from Ensenada de los Perros, Isla Tiburón, Sonora

C. vestigium of the southwestern United States and northern Mexico (McGuire 1996).

DISTRIBUTION *Crotaphytus dickersonae* ranges along the west coast of Sonora, Mexico, from El Desemboque south to just north of Guaymas (McGuire 1996). In the Gulf of California, it is known from Isla Tiburón, Sonora (map 4.1).

DESCRIPTION Medium-sized, with conspicuously enlarged head usually twice the width of neck; adults reaching 116 mm SVL; dorsal head scales small, those of temporal region granular; enlarged interparietal with conspicuous parietal eye; 12 to 17 supralabials; fringe scales of eyelids elongate and triangular; 2 postmentals; gulars granular; gular fold extending into antehumeral and scapular region; 10 to 16 infralabials; dorsal body scales granular, grading ventrally into slightly larger, flat, juxtaposed ventrals; forelimbs moderately sized; hind limbs long and muscular, approximately equal in length to SVL; 16 to 21 dark, well-developed femoral pores in adult males; 17 to 21 subdigital lamellae on fourth toe; deep postfemoral mite pockets present; tail approximately twice as long as body; tail laterally compressed anteriorly in adult males, less so in females and juveniles; dorsal and lateral caudal scales occurring in fairly uniform whorls; no enlarged postanal scales.

COLORATION Depends on sex and SVL. ADULT MALES: Dorsal ground color aquamarine to cobalt blue; top of head and limbs brownish; top of head immaculate; dark blotches on side of head, occasionally uniting to form indistinct postorbital stripe; central gular region slate gray to black, contacting anterior black collar; two black collars present, anterior collar complete ventrally; prominent white collar between black collars, posterior black collar wider than anterior black collar and sometimes complete across dorsum; dorsal body pattern usually consisting of reticulum of small white spots; dorsal surface of forelimbs may or may not be overlaid with small white spots; dorsal surface of hind limbs usually covered with circular blotches of ground color; light-colored, caudal vertebral stripe; black inguinal patches develop ontogenetically. FEMALES AND JUVENILES: Dorsal ground color brownish to dull yellow, hind limbs and tail occasionally lemon yellow; side of head and gular region weakly blotched; gular region nearly immaculate in some individuals; first black collar not extending ventrally across throat; in adult females, spots and white collar may become suffused with orange during breeding season; orange also present in juvenile males.

NATURAL HISTORY The only natural history observations reported to date on *C. dickersonae* come from McGuire 1996. On Isla Tiburón, this species occurs within the Central Gulf Coast Region and is found primarily on south-facing hillsides with scattered rocks of various sizes, in areas where the vegetation is not excessively thick. Lizards may also be found on rocky flats, especially at the borders of arroyos. I have observed *C. dickersonae* to be active from mid-March to early November. During the winter adults probably hibernate, and only minimal activity occurs in juveniles. *Crotaphytus dickersonae* is diurnal and emerges in the morning to bask on the tops of rocks and watch for potential prey. McGuire (1996) suggests that juveniles may be found more frequently around rocky outcrops on the summits of small hills than among the scattered rocks on the slopes. *Crotaphytus dickersonae* is a fast, agile predator capable of running across irregular ground and over rocks bipedally, using its tail as a balance organ. On capture, prey is crushed with its powerful jaws and swallowed whole. Food consists largely of other lizards, but insects also compose a major portion of its diet. Schmidt (1922) noted that an adult female collected in April had a beetle, a wasp, and a "lizard tail" in her stomach. I have observed grasshoppers, seeds, leaves, and the flowers of Thorn Bush *(Lycium andersonii)* in the stomachs of preserved specimens collected in late March. I found a Side-Blotched Lizard *(Uta stansburiana)* measuring 51 mm SVL in the stomach of a subadult male measuring 77 mm SVL.

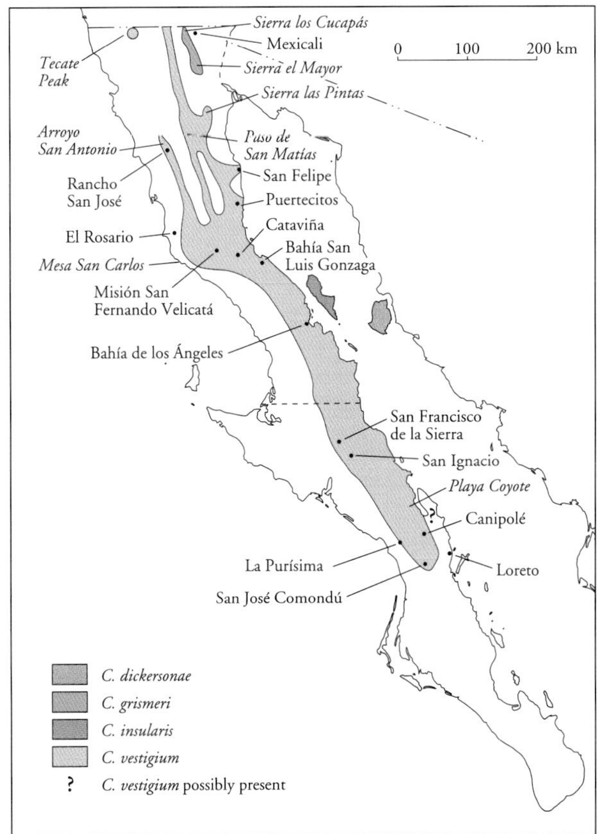

MAP 4.1 Distribution of *Crotaphytus* (Collared Lizards).

Crotaphytus dickersonae seems to have a typical spring breeding and summer egg-laying season. I have seen females with gravid coloration in late March, and McGuire (1996) reports seeing females with gravid coloration during mid-April, presumably near the start of the breeding season. Schmidt (1922) reports a gravid female (SVL 95 mm) collected in mid-April, and three eggs were found in a female (SVL 79 mm) collected in early June. Two adult males (SVL 104 and 106 mm) collected in mid-June had enlarged testes. In late March I have observed hatchlings from the previous year to be active, but not neonates. Hatching probably begins in mid-July.

REMARKS The taxonomic history of *C. dickersonae* has been somewhat controversial (Schmidt 1922; Allen 1933; Smith and Tanner 1972; Montanucci et al. 1975). A recent revision of the Crotaphytidae (McGuire 1996) has revealed that *C. dickersonae* is a valid species.

FOLKLORE AND HUMAN USE This species is believed by local inhabitants to be venomous.

FIG. 4.3 ***Crotaphytus grismeri***

SIERRA LOS CUCAPÁS COLLARED LIZARD; ESCORPIÓN

Crotaphytus grismeri McGuire 1994: 440. Type locality: "east face of the Sierra de los Cucapás, approximately 50 km south of Mexicali, Baja California, Mexico." Holotype: Centro Ecológico de Sonora (CES) 067–629

IDENTIFICATION *Crotaphytus grismeri* can be distinguished from all other lizards in the region of study by a white collar between two black collars; posterior black collar nearly complete dorsally; anterior collar complete ventrally; forelimbs lacking white markings; absence of black oral mel-

FIGURE 4.3 Adult male *Crotaphytus grismeri* (Sierra los Cucapás Collared Lizard) from Cañón David, Sierra los Cucapás, BC

anin; granular body scales; and small adult size (SVL less than 100 mm).

RELATIONSHIPS AND TAXONOMY *Crotaphytus grismeri* is part of a natural group containing the other Southwestern collared lizards *C. bicinctores, C. insularis,* and *C. vestigium* (McGuire 1996).

DISTRIBUTION *Crotaphytus grismeri* is known only from the Sierra los Cucapás and Sierra el Mayor in northeastern Baja California (see map 4.1).

DESCRIPTION Same as *C. dickersonae,* except for the following: adults reaching 99 mm SVL; 13 to 20 supralabials; 13 to 18 infralabials; 19 to 23 dark femoral pores; suprahumeral mite pocket present; 17 to 20 subdigital lamellae on fourth toe; posterior caudals keeled; scales posterior to vent usually enlarged in males. **COLORATION** Coloration and pattern depend on sex and SVL. ADULT MALES: Dorsal ground color brown; top of head tan, immaculate; white reticulate pattern on side of head; central gular region blue-gray to black, contacting anterior black collar; brownish blotches on side of head; two black collars present; prominent white collar between them suffused with light green; posterior black collar same width as anterior black collar and nearly complete dorsally; anterior collar complete ventrally; dorsal body pattern consisting of white spots; upper forelimbs tan with yellow blotching; ground color of thighs brown, overlaid with white spots; forelegs yellow and immaculate; light-colored, vertebral stripe on tail; black, inguinal patches develop ontogenetically. FEMALES AND JUVENILES: Dorsal ground color generally tan; hind limbs and tail dull orange in subadult females; central gular region pale; first black collar not extending ventrally across throat; dorsal body pattern consisting of white spots; orange transverse body bars develop during breeding season, orange also present in juvenile males; neonates and juveniles have 2 or 3 large ventrolateral black spots on sides of body outlined in white.

NATURAL HISTORY *Crotaphytus grismeri* is a saxicolous species occurring in the Lower Colorado Valley Region. It is common on rocky hillsides at certain locales and is found at all levels, ranging from the rocky rubble at the base of the hills to the crests, 100 to 200 m above the surrounding desert floor. Adult *C. grismeri* have been observed from early March through mid-September (McGuire 1996) and juveniles in early December. This species emerges in the early morning to bask on the tops of rocks and boulders. From these vantage points it is able to see its prey, which consists primarily of arthropods and Side-Blotched Lizards (*Uta stansburiana*). Like other collared lizards, *C. grismeri* is adept at running bipedally across rocks in pursuit of prey or to avoid predators. Females with gravid coloration have been observed in early to mid-May, and gravid females have been found as late as early September (McGuire 1996). By early September, the majority of active individuals appear to be juveniles, and individuals as small as 45 mm have been found. These observations suggest a laying season of at least May through September, but McGuire (1996) suggests that two clutches per year may be laid. Three large eggs were felt within a female measuring 82 mm SVL, and gravid coloration has been observed on females as small as 79 mm SVL.

REMARKS The Sierra los Cucapás and Sierra el Mayor are isolated from the Sierra Juárez by Laguna Salada, a nearly abiotic, ephemeral dry lake approximately 25 km across. Apparently the Sierra los Cucapás has never been in contact with the Sierra Juárez (Barnard 1970), and it existed as a large island during the Pliocene (Gastil et al. 1983). This island may have become colonized as the result of an overwater dispersal by the ancestor of *C. grismeri* (McGuire 1994). A more recent overland dispersal across Laguna Salada would be extremely unlikely because of the complete absence of appropriate habitat.

FIG. 4.4

Crotaphytus insularis

ISLA ÁNGEL DE LA GUARDA COLLARED LIZARD; ESCORPIÓN

Crotaphytus insularis Van Denburgh and Slevin 1921b:96. Type locality: "east coast of Angel de la Guarda Island seven miles north of Pond Island, Gulf of California [Baja California], Mexico." Holotype: CAS 49151

IDENTIFICATION *Crotaphytus insularis* can be distinguished from all other lizards in the region of study by an anterior

FIGURE 4.4 Adult male *Crotaphytus insularis* (Isla Ángel de la Guarda Collared Lizard) from Puerto Refugio, Isla Ángel de la Guarda, BC

black collar that is incomplete dorsally and complete ventrally in both sexes; irregular white dorsal spots forming a weak, sinuous banding pattern; absence of black oral melanin; and granular body scales.

RELATIONSHIPS AND TAXONOMY *Crotaphytus insularis* is part of a natural group containing *C. bicinctores, C. grismeri,* and *C. vestigium* and is the sister species of *C. vestigium* (McGuire 1996).

DISTRIBUTION *Crotaphytus insularis* is endemic to Isla Ángel de la Guarda, in the Gulf of California, BC (see map 4.1).

DESCRIPTION Same as *C. dickersonae,* except for the following: adults reach 120 mm SVL; 13 to 19 supralabials; 11 to 17 infralabials; 18 to 23 light-colored, well-developed femoral pores in adult males; suprahumeral mite pocket present; 19 to 24 subdigital lamellae on fourth toe; shallow postfemoral mite pockets; posterior caudal scales keeled; scales posterior to vent somewhat enlarged in males. **COLORATION** Coloration and pattern depend on sex and SVL. **ADULT MALES:** Dorsal ground color brown; top of head immaculate beige to brown and darkening with increasing SVL; dorsal ground color of limbs and tail dull yellow to brown, darkening with increasing SVL; well-defined dark blotches on white background on side of head; blotching continues ventrally onto white ground color of lateral gular region; central gular region with narrow black stripe extending anteriorly; rest of gular region olive drab and black centrally; single black collar anterior to forelimbs, continuous ventrally and incomplete dorsally, with prominent white collar posterior to black collar in both sexes; second, posterior, black collar short and inconspicuous or absent; dorsal body pattern consisting of reticulum of small white spots arranged transversely and connected to form a thin, sinuous banding pattern; dorsal surface of forelimbs spotted with ground color; dorsal surface of hind limbs covered with reticulum of light markings; reticulum more distinct in adult males than in females and juveniles; vertebral region of tail immaculate; ventrum white; dorsal color pattern changes abruptly to a uniform dull olive green in lateral regions of belly, with large, well-defined black spots in region of contact; black inguinal patches develop ontogenetically in males but never contact medially. **FEMALES AND JUVENILES:** Blotching on side of head less pronounced; gular region whitish; anterior, black collar only weakly present laterally; spots and white collar may become suffused with orange in juvenile males and gravid females; dull olive green lateral belly coloration absent but large dark spots present, increasing in intensity with SVL.

NATURAL HISTORY *Crotaphytus insularis* occurs in the Central Gulf Coast Region on Isla Ángel de la Guarda. It is usually found on rocky hillsides and flats bordering arroyos where the vegetation is not excessively thick, although I have occasionally seen individuals basking on fallen Cardón Cactus *(Pachycereus pringlei)* in bajadas. Here lizards commonly use the larger boulders as lookouts. Juveniles are most common on talus slopes composed of small rocks (< 120 cm diameter). Generally, *C. insularis* is active from mid-March through early to mid-November. It is a diurnal species, spending much of its day on rocky perches basking and looking for prey. Like other *Crotaphytus, C. insularis* is quick and agile and runs bipedally through its habitat in pursuit of prey or to escape from predators. Its long, thick tail is used to counterbalance the body while running. Prey consists primarily of lizards and insects; however, plant material is also probably eaten. Bugs (Hemiptera) were removed from an adult male and female collected in June, and bugs, grasshoppers, and gravel were removed from the stomach of an adult female collected in September. The reproductive season of *C. insularis* probably parallels that of *C. vestigium* from the adjacent peninsula (see below), which extends from April through late June or early July. Females in breeding coloration have been observed in late June, but I have never seen gravid females in mid-March. Adult males collected from early May to mid-July had enlarged testes. Males collected in early September had regressed testes, which suggests that the breeding season peaks in June and tapers off through July and September.

FOLKLORE AND HUMAN USE This species is believed to be venomous by the Mexican fishermen who occasionally camp on Isla Ángel de la Guarda.

FIG. 4.5

Crotaphytus vestigium

BAJA CALIFORNIA COLLARED LIZARD; ESCORPIÓN

Crotaphytus insularis vestigium Smith and Tanner 1972:29. Type locality: "Guadelupe Canyon, Juarez Mountains, Baja California," Mexico. Holotype: BYU 23338

FIGURE 4.5 Adult male *Crotaphytus vestigium* (Baja California Collared Lizard) from 33 km west of San Ignacio, BCS

IDENTIFICATION *Crotaphytus vestigium* can be distinguished from all other lizards in the region of study by two black collars bordering a central white collar; anterior collar complete ventrally; thin, sinuous, transverse white body bands; lack of black oral melanin; and granular body scales.

RELATIONSHIPS AND TAXONOMY See *C. insularis.*

DISTRIBUTION *Crotaphytus vestigium* ranges along the east face of the Peninsular Ranges from Palm Springs, California, south to approximately 28 km south (by road) of San José Comondú (McGuire 1991a, 1996; see map 4.1). McGuire (1991a) noted that its distribution probably extends slightly farther to the southwest, following a particular type of volcanic rock. In northern Baja California, *C. vestigium* extends west around the southern end of the Sierra San Pedro Mártir and continues north along the semiarid western foothills of the Peninsular Ranges to at least Arroyo San Antonio, just west of the town of Valle la Trinidad (Welsh 1988). An apparently isolated population exists as far west as Tecate Peak on the U.S.-Mexican border. *Crotaphytus vestigium* also occurs on the west coast of Baja California at Mesa San Carlos, just south of El Rosario (Bostic 1971). I have observed specimens at Misión San Fernando Velicatá between Mesa San Carlos and the Peninsular Ranges, which indicates that the former population is not an isolated one (McGuire 1996). *Crotaphytus vestigium* does not occur on the mountaintops of the Sierra Juárez or Sierra San Pedro Mártir. It does range to the tops of all mountains farther south. It is conspicuously absent from the Gulf coast south of Bahía Concepción, despite the presence of seemingly suitable habitat (McGuire 1996).

DESCRIPTION Same as *C. dickersonae,* except for the following: adults reaching 125 mm SVL; 10 to 19 supralabials; 10 to 17 infralabials; 15 to 25 light-colored femoral pores; 15 to 25 subdigital lamellae on fourth toe. COLORATION Coloration and pattern variable, depending on sex and SVL. ADULT MALES: Dorsal body ground color brown; top of head usually light, with some degree of dark mottling; posteromedial gular region black; rest of gular region olive drab and abruptly contacting lateral blotching; usually two black collars bordering a single white collar; anterior collar complete ventrally; dorsal body pattern usually consisting of a faint reticulum of small white spots arranged transversely and connected to form thin, sinuous bands; white spots occur between bands; dorsal color pattern changes abruptly to a uniform dull light green or orange in lateral regions of belly; lateral coloration of belly region abruptly contacts white midventral region; black inguinal patches develop ontogenetically and occasionally contact medially. FEMALES AND JUVENILES: White banding pattern pronounced in juveniles; hatchlings and small juveniles tend to have a weak banding pattern of linearly arranged, large, circular dark markings alternating with light bands; large dark spots usually present in females on sides of body, increasing in intensity with SVL; gravid females and juvenile males have crimson orange bars on the body and head.

GEOGRAPHIC VARIATION Geographic variation within *Crotaphytus vestigium* is most obvious in color pattern and to some extent SVL, with the largest lizards occurring in the north. Color pattern takes on two variations, one local and the other clinal. Local variation is primarily restricted to substrate matching of the ground color, or at least approximating the coloration of the parent rock. In areas of dark volcanic outcroppings, such as San Ignacio, ground colors tend to be dark shades of brown. On granitic hillsides, as in the Sierra Juárez and Sierra San Pedro Mártir, ground color is generally lighter. Often these differences are subtle. There appears to be a clinal shift in color pattern between Bahía de los Ángeles and San Francisco de la Sierra, involving a series of light spots between the dorsal body bands. From Bahía de los Ángeles north, these spots are generally irregular in shape and randomly distributed in the interspace. From San Francisco de la Sierra south, the spots are more circular and linearly arranged transversely. Similar geographic trends are found in the gular and ventral pattern characteristics. From Bahía de los Ángeles north, the large infralabial and lateral gular blotches are dark brown and very well-defined in males. In males from San Ignacio south, the lateral gular blotches are much lighter and less well-defined. Dark lateral blotching or mottling on an olive green belly is common in adult males and most females from Bahía San Luis Gonzaga north, but generally absent from populations from Bahía de los Ángeles south (McGuire 1996). Lizards south of Bahía San Luis Gonzaga have orange lateral belly regions whose color is more intense among adults in populations farther south. This coloration is most intense in males from the vicinity of San Ignacio south. These characteristics are in rough geographic concordance with the gradual transition of the Vizcaíno Region into the Magdalena Region, and Grismer (1994c) notes similar patterns of geographic variation in other species ranging through this phytogeographic ecotone.

A cismontane population of *C. vestigium* from Rancho San José is known from a single hatchling and adult male. Both have a very well-defined dorsal banding pattern composed of straight transverse bands. This pattern is fairly common in hatchlings and juveniles of other populations but never maintains this degree of definition into adulthood.

NATURAL HISTORY *Crotaphytus vestigium* occurs in a broad range of arid habitats encompassing the California, Lower Colorado Valley, Vizcaíno, Magdalena, and Central Gulf Coast regions. Within these regions, however, it is strictly limited in distribution by the presence of rock. It is most abundant on rocky hillsides and less common in bajadas, unless there is a rocky arroyo nearby. It may be locally abundant on rocky volcanic flats with low rolling hills, even on the west coast of Baja California north of La Purísima, where the climate is cool and damp and the rocks are cloaked with lichens. *Crotaphytus vestigium* is known to occur on many types of rock (including sandstone) of varying sizes. It is infrequently found on extremely large boulders and most abundant on hillsides with moderately sized rocks (up to 2 m in diameter) interspersed with sparse vegetation. Such rocks serve as lookouts and basking sites.

Because of the latitudinal extent of its distribution, *C. vestigium* has a broad range of annual activity periods. I have observed adult males basking as early as mid-March in the foothills 30 km south of San Felipe and as late as early November near Bahía San Luis Gonzaga. At other times of year, this species usually hibernates beneath rocks. I found a hibernating female in a shallow depression beneath a small rock approximately 30 km south of San Felipe in early February. Most lizards probably emerge from hibernation in March and early April, depending on the temperature. Shortly after emergence, females take on an orange breeding coloration, indicating the start of the reproductive season. This coloration persists late into the year in the southern portion of this species' distribution, as evidenced by a female with gravid coloration observed during early October just east of Canipolé. Three to eight eggs are laid during the summer (Stebbins 1985). Hatchlings begin to emerge around late July in the more northerly portions of this species' range and around mid-August from approximately San Ignacio south.

Crotaphytus vestigium is diurnal and can be observed basking on the tops of rocks throughout the day. One specimen near Playa Coyote, just south of Mulegé, was observed in a small hole in a rock approximately 1.5 m above the ground. These basking rocks, and others in the vicinity, are also used as prey and predator lookouts. Prey is captured by ambush and crushed with strong jaws. Food usually consists of insects, other lizards, and plant material. I observed an immature male in the Sierra las Pintas approximately 1.5 m above the ground in a small Palo Verde tree *(Cercidium microphyllum),* presumably grazing on new blossoms. *Crotaphytus vestigium* is a quick and agile predator capable of running bipedally across rocks. Its agility also aids in escape from predators. While attempting to capture an adult male basking on top of a large rock just south of Mulegé, I watched it leap nearly half a meter away from the rock at a 45-degree angle into the air, land squarely on all fours, and escape. Often when approached, however, lizards will flatten themselves against the rock.

FOLKLORE AND HUMAN USE Many Mexicans believe that the saliva of *C. vestigium* is toxic and will cause large red welts to appear at the site of a bite.

Gambelia Baird

LEOPARD LIZARDS

The genus *Gambelia* contains three medium-sized species of terrestrial lizards. All are active, diurnal predators that generally inhabit flat, open, arid to semiarid regions and can be seen lying at the bases of bushes or rocks waiting to ambush prey. *Gambelia* ranges throughout the southwestern United States and northern Mexico, with two species occurring in Baja California: *G. copeii,* which is endemic and occurs on three Pacific islands, and *G. wislizenii,* which ranges widely throughout the American Southwest, northwestern Mexico, and northeastern Baja California. It also occurs on Isla Tiburón, Sonora, in the Gulf of California.

FIG. 4.6 ***Gambelia copeii***

BAJA CALIFORNIA LEOPARD LIZARD; CACHORA

Crotaphytus copeii Yarrow 1882b:441. Type locality: "La Paz, Cal.," Baja California Sur, Mexico. Holotype: USNM 12663

FIGURE 4.6 *Gambelia copeii* (Baja California Leopard Lizard) from 42 km southeast of Guerrero Negro, BCS.

IDENTIFICATION *Gambelia copeii* can be distinguished from all other lizards in the region of study by its lack of a distinct collar; granular, dorsal body scales; eyelids; smooth supralabial scales; a long, slender tail at least twice the length of the body; a generally dark dorsal coloration with large paired paravertebral spots; or a nearly uniform tan dorsal coloration with obscured paravertebral blotches; and a lack of enlarged vertebral crest scales.

RELATIONSHIPS AND TAXONOMY *Gambelia copeii* is most closely related to *G. wislizenii.* McGuire (1996) discusses taxonomy.

DISTRIBUTION *Gambelia copeii* is endemic to Baja California and ranges from extreme southern San Diego County, California, south to at least Todos Santos on the west coast of the Cape Region. In northwestern Baja California, *G. copeii* is restricted to the easternmost foothills of the California Region and does not contact the Pacific coast until at least Rancho Socorro. *Gambelia copeii* ranges along the east coast of the peninsula, from El Huerfanito in the north to at least Mulegé in the south. It is apparently absent from the Gulf coast south of Mulegé. It may be restricted from the Gulf coast by the Sierra Guadalupe and Sierra la Giganta. It is abundant in the Vizcaíno Desert and south through the Magdalena Plain into the western portion of the Cape Region. *Gambelia copeii* is known from the Pacific islands of Cedros, Santa Margarita, and Magdalena (McGuire 1996; map 4.2). On Isla Cedros, it occupies all the low-lying areas and arroyos surrounding the central mountains.

DESCRIPTION Moderate to large, with adults reaching 126 mm SVL; head relatively large, triangular, and wider than neck; snout somewhat elongate; scales of head slightly enlarged and juxtaposed; parietal scale slightly enlarged and containing a parietal eye; parietal scales grade smoothly into dorsal scales; 13 to 17 supralabials; 2 postmentals; gulars flat to convex and beadlike, often separated from each other by interstitial granules; 12 to 17 infralabials; gular and lateral neck folds present; dorsal scales small, granular, and juxtaposed, grading into larger, flat ventral scales; lateral folds of skin extend from axilla to groin; forelimbs moderately sized and robust; hind limbs robust and longer than forelimbs; deep postfemoral mite pocket; 20 to 31 well-developed femoral pores in adult males; 20 to 24 subdigital lamellae on fourth toe; tail long and cylindrical, at least twice SVL; caudal scales small, with a paired median row of subcaudals; enlarged postanal scales usually found in males. **COLORATION** Dorsal ground color dark gray to light golden brown, capable of considerable lightening with an increase in body temperature; ground color usually overlaid with large paired dark paravertebral spots separated by broad cream-colored transverse bars on body and tail; large spots on hind limbs; sides of body and forelimbs with dark flecks but no spots; dense reticulum of white punctations on dorsum; light gray longitudinal lines in gular region.

GEOGRAPHIC VARIATION Northern populations of *G. copeii* from cismontane northwestern Baja California, Cataviña, Bahía San Luis Gonzaga, and the vicinity of Bahía de los Ángeles have a relatively dark overall ground color with well-defined, relatively large dark paravertebral spots that are conspicuously encircled by distinct small, white dots. However, this general pattern is often absent in lizards from the Sierra San Borja south of Bahía de los Ángeles. In the more southerly populations, from the Vizcaíno Desert south, lizards are much lighter in overall coloration, and the ground color is a light golden brown. Some individuals also have extensive amounts of yellow on the posterior portions of their flanks, the hind limbs, and the base of the tail. The paravertebral dorsal spots generally break up in adults and are nearly absent anteriorly. The transverse cream-colored bars are generally absent or indistinguishable from the ground color. The darker lateral flecks on the sides of the body and forelimbs are absent, and these regions are generally unicolored. The white punctations, although not as distinctive, tend to dominate the overall dorsal color pattern with a fine speckling. This trend is somewhat more obvious in populations from Isla Cedros, where adults from the coastal flats on the southeastern portion of the island have a generally unicolored dorsal body pattern. Lizards from the rocky arroyos in the central portion of the island are less uniform in dorsal pattern, and the paravertebral blotching is more obvious. Hatchlings and juveniles from Isla Cedros have distinctive red dorsal spots with broad cream-colored transverse bars.

NATURAL HISTORY *Gambelia copeii* is a common inhabitant of arroyo bottoms, bajadas, and flat, open, sandy areas throughout an environmentally diverse array of phytogeographic regions. It ranges from the eastern foothills of the California Region to at least 14 km west of Ojos Negros, south to the Arid Tropical Region of the Cape Region. I have observed specimens on the edge of the piñon pine belt east of Rancho Melling in the Sierra San Pedro Mártir and on the tops of the high, cool, windswept coastal dunes of Rancho Socorro on the Pacific coast. *Gambelia copeii* generally prefers relatively open spaces without too much vegetation or rock, although at Rancho San Juan de Dios east of El Rosario I found two specimens, approximately 100 m up on a rocky hillside and in dense chaparral in Valle la Trinidad.

The seasonal activity period of *G. copeii* probably varies with latitude. In northern Baja California, populations are not active until at least mid-March, and adults remain active at least through September. Juveniles may remain active into early November. In the southernmost portion of its range, it is probably active year-round, with minimal

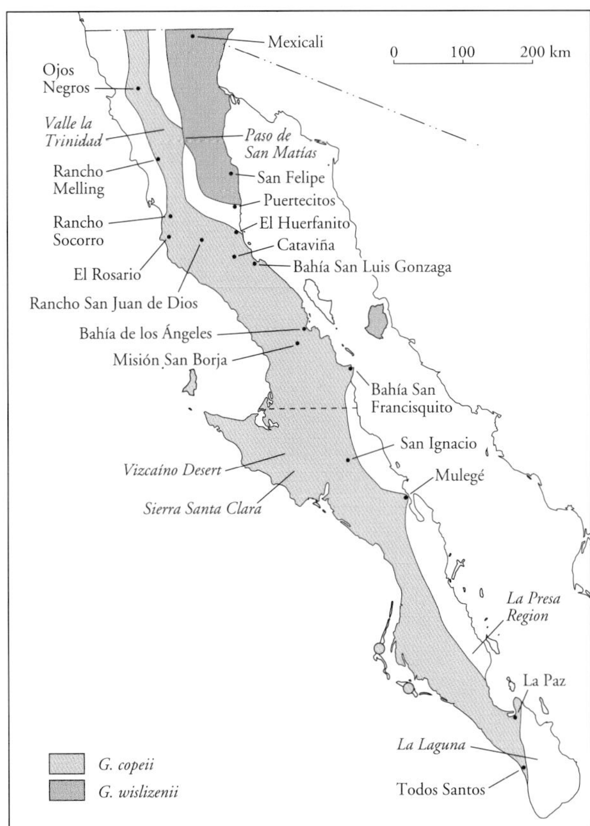

MAP 4.2 Distribution of *Gambelia* (Leopard Lizards)

activity during the winter. Like most diurnal species, *G. copeii* emerges from burrows or from beneath rocks during the early morning to bask. It is commonly seen sitting on the tops of small rocks, gaining a vantage on its surroundings. *Gambelia copeii* is principally an ambush predator and can be observed at the edge of vegetation waiting for prey, which consists primarily of lizards and arthropods. I found one individual near Cataviña emerging from its burrow in the early morning with the feet and tail of a Zebra-Tailed Lizard *(Callisaurus draconoides)* protruding from its mouth. It had obviously begun to eat the lizard the day before but could not fit it all in its stomach at one time.

The breeding season for *G. copeii* in the northern portion of its range lasts from April until at least early or mid-July. I found a dead female with gravid coloration on the road just east of Bahía de los Ángeles in mid-April and observed a gravid female during early May at Paso de San Matías. Two females with breeding coloration observed in late April in the Vizcaíno Desert had ova measuring 2 and 3 mm (Murray 1955). I observed a gravid female near Misión San Borja during late June. I have seen females with gravid coloration in the Sierra Santa Clara during early July, and on Isla Cedros I observed gravid females during July and hatchlings in late July. An emaciated female that appeared to have deposited eggs recently was observed in the Sierra Santa Clara during early July (McGuire 1996). Fitch (1970) reported on two gravid females from Baja California that were collected in March. He did not give a locality, and so it is not certain which species was involved. In the Cape Region, the breeding season may extend much later into the year, as evidenced by a female with breeding coloration found near Todos Santos during late August.

When approached, instead of running *G. copeii* may flatten itself against the substrate and rely on its cryptic coloration for camouflage. Under these circumstances, it can be approached to within a few feet, and in the early morning it can occasionally be picked up by hand. If on a rock, this species usually slinks down the back side and runs a short distance to cover.

REMARKS Banta and Tanner (1968) considered *G. copeii* a subspecies of *G. wislizenii*, indicating that the two forms intergraded in the vicinity of San Felipe. McGuire (1996) noted that *G. copeii* and *G. wislizenii* do not contact one another along the east coast of the peninsula (see above) but were observed to be syntopic for a few hundred meters in Paso de San Matías in northern Baja California, where they show no signs of intergradation. McGuire thus considered them separate species.

Montanucci (1978) considered the Isla Tiburón population of *G. wislizenii* to be conspecific with *G. copeii* from Baja California. McGuire (1996), however, demonstrated that although these populations are similar in certain aspects of color pattern, the Isla Tiburón population has spots continuing onto the dorsal surface of the head, a characteristic that does not occur in *G. copeii* and is diagnostic for *G. wislizenii.*

FIG. 4.7 **_Gambelia wislizenii_**

LONG-NOSED LEOPARD LIZARD; CACHORA

Crotaphytus wislizenii Baird and Girard 1852a:69. Type locality: "near Santa Fe," New Mexico, USA. Holotype: USNM 2270

IDENTIFICATION *Gambelia wislizenii* can be distinguished from all other lizards in the region of study by its lack of a distinct collar; granular dorsal body scales; eyelids; smooth supralabial scales; no enlarged vertebral crest scales; a long slender tail at least twice the length of the body; and a pale dorsal pattern with numerous punctations asymmetrically arranged on the head and body.

RELATIONSHIPS AND TAXONOMY See *G. copeii.*

DISTRIBUTION *Gambelia wislizenii* ranges from central Idaho and eastern Oregon south throughout the Great Basin to western Colorado and western Texas, and through the eastern deserts of California. It extends south to western Coahuila, northern Zacatecas, eastern and central Chihuahua, central Sonora, and northeastern Baja California, Mexico. The southernmost specimen of this species in Baja California is from Arroyo Matomí, just north of Puertecitos (McGuire 1996; see map 4.2). It is believed that *G. wislizenii* is prevented from extending any farther south along the east coast of the peninsula because of a volcanic expanse approximately 30 km wide between Puertecitos and El Huerfanito. Similarly, *G. copeii* extends north along the east coast to the southern edge of this expanse, near El Huerfanito, but does not cross it. Thus, these two species do not come into contact along the east coast of Baja California, notwithstanding the range map of Banta and Tanner 1968. *Gambelia wislizenii* is also known from Isla Tiburón, Sonora, in the Gulf of California.

DESCRIPTION Same as *G. copeii,* except for the following: adults reaching 144 mm SVL; supralabials 12 to 17; 15 to 24 femoral pores; 18 to 25 subdigital lamellae on fourth toe. **COLORATION** Dorsal ground color off-white to pale tan; ground color overlaid with network of moderately sized, dark reddish-brown spots on body, hind limbs, and tail, generally largest toward the midline of body; smaller spots on head; small, lighter reddish spots on forelimbs; spots on

FIGURE 4.7 *Gambelia wislizenii* (Long-Nosed Leopard Lizard) from Mexicali, BC

body and hind limbs usually encircled by a single row of small white spots; inconspicuous to conspicuous, transversely oriented, thin white bands on body and proximal portion of tail; gular region with light to dark gray longitudinal lines; belly usually immaculate and whitish; weak banding pattern on ventral surface of tail; gravid coloration consisting of transversely oriented red or orange spots on head or neck, two rows of elongate spots on sides, occasionally on thighs and forelegs, and on ventral surface of tail.

GEOGRAPHIC VARIATION Geographic variation is slight and limited to an overall darkening of the dorsal ground color of individuals from the vicinity of Paso de San Matías. Here, the population extends west, with a rise in elevation as it ranges into a Lower Colorado Valley–California Region ecotone. This darkening serves to highlight the white transverse bars, which are often inconspicuous among the eastern populations. Populations from Isla Tiburón differ slightly from those of northeastern Baja California in that they tend to lack the white transverse dorsal body bars as adults. Also, the small white punctations on the dorsum are more conspicuous.

NATURAL HISTORY *Gambelia wislizenii* generally occurs in the creosote bush scrub community of the Lower Colorado Valley Region of northeastern Baja California. Here it is commonly found in the open, sandy areas of bajadas and arroyos where the vegetation is not too dense and the substrate not too rocky. *Gambelia wislizenii* is active from early spring to late fall. It is diurnal, emerging from burrows or from beneath rocks in the morning to bask. It is often seen sitting on small rocks in the open or lying in ambush at the base of vegetation. When observed by a potential predator, *G. wislizenii* may remain motionless, flattening itself on the substrate to remain inconspicuous; it can occasionally be picked up by hand. More often, lizards run to the base of a nearby bush and flatten themselves on the substrate. *Gambelia wislizenii* feeds mostly on lizards and arthropods but will occasionally eat small flower buds and fruits. I coaxed an adult female from Isla Tiburón, measuring 110 mm SVL, to regurgitate a *Cnemidophorus tigris* (Western Whiptail) measuring 107 mm SVL. Females develop crimson-orange spots during the breeding season, from April to early June. After they have laid their eggs, the orange coloration begins to fade. I have observed hatchlings during mid-July near the Sierra San Felipe. Hatchlings also have the crimson-orange coloration, which may be associated with thwarting male aggression.

FOLKLORE AND HUMAN USE The Seri told Malkin (1962) that the bite of *G. wislizenii* is fatal only to Seri.

Iguanidae

IGUANAS AND CHUCKWALLAS

The Iguanidae (Frost and Etheridge 1989) are a group of moderate-sized to large lizards containing eight genera and approximately 36 species (de Queiroz 1987, 1995; Hollingsworth 1998). Collectively, iguanids range from the southwestern United States south to southern Brazil and Paraguay, and they are represented by endemic genera in the West Indies, Galápagos, Fiji, and Tongan islands. Iguanids are commonly referred to as marine iguanas *(Amblyrhynchus),* Fijian iguanas *(Brachylophus),* Galapagos land iguanas *(Conolophus),* spiny-tailed iguanas *(Ctenosaura),* West Indian iguanas *(Cyclura),* desert iguanas *(Dipsosaurus),* green iguanas *(Iguana),* and chuckwallas *(Sauromalus).* Chuckwallas, desert iguanas, and spiny-tailed iguanas range widely throughout the region of study.

Iguanids are notable for their herbivorous habits and generally large size. In fact, as a result of the latter, many species have served as food for humans for thousands of years. Iguanids are a successful group and manifest a wide array of adaptive types, ranging from arboreal tropical forest forms, saxicolous forms, and open desert species to marine algal feeders. The Iguanidae are a well-defined group, diagnosed by having caudal vertebrae with two pairs of transverse processes; valves in the descending colon; multicusped, compressed, and flared posterior marginal teeth; a more medially oriented supratemporal bone on the parietal; and a herbivorous diet (de Queiroz 1987).

Ctenosaura Wiegmann

SPINY-TAILED IGUANAS

The genus *Ctenosaura* comprises a group of medium-sized to large scansorial, omnivorous lizards. It contains 14 species that collectively range from northern Mexico, exclusive of the Central Plateau, south to central Panama (Buckley and Axtell 1997; Grismer 1999b; Smith 1972). The large size of most species of *Ctenosaura* makes them an impressive sight when viewed basking on a large boulder or the top of a Cardón Cactus *(Pachycereus pringlei).* They are wary and difficult to approach. *Ctenosaura* occurs on various offshore islands in the eastern Pacific and western Caribbean. One species, *C. hemilopha,* occurs in Baja California and the adjacent Isla Cerralvo in the Gulf of California, and two other species, *C. conspicuosa* and *C. nolascensis,* occur on Islas San Esteban and San Pedro Nolasco, respectively, in the Gulf of California.

FIG. 4.8 ***Ctenosaura hemilopha***

CAPE SPINY-TAILED IGUANA; IGUANA DE PALO

Cyclura (Ctenosaura) hemilopha Cope 1863:105. Type localities of syntypes: "Cape St. Lucas [Cabo San Lucas]"; "near Soria Ranch, Cape San Lucas, Baja California [Sur], Mexico" [USNM 5295; three specimens], and "San Nicolás, between Cape San Lucas and La Paz, Baja California [Sur], Mexico" [USNM 69489; one specimen; Cochran 1961]. See de Queiroz (1995) for a discussion of the discrepancies in the literature concerning the type material.

IDENTIFICATION *Ctenosaura hemilopha* can be distinguished from all other lizards in the region of study by a single, slightly enlarged, middorsal scale row forming a weak crest; caudal scales arranged in whorls of distinct spines; a nearly uniform green color pattern in hatchlings; mottled hind limbs in adults, with only weak evidence of banding; and a post-thoracic banding pattern in adults.

RELATIONSHIPS AND TAXONOMY *Ctenosaura hemilopha* is a member of a natural group containing *C. conspicuosa, C. macrolopha,* and *C. nolascensis* (Cryder 1999; Hollingsworth 1998). Cryder (1999)and Grismer (1999b) discuss taxonomy.

DISTRIBUTION *Ctenosaura hemilopha* ranges from near Loreto south along the Sierra la Giganta to the west coast near Arroyo Seco and throughout the Cape Region (Smith 1972). In the Gulf of California, *C. hemilopha* is known only from Isla Cerralvo (map 4.3). Ranchers from the western foothills of the Sierra Guadalupe speak of this species' presence there, but no specimens have been found.

DESCRIPTION Very large, with adults reaching 308 mm SVL; head and body stout; head long and triangular, more so in adult males; distinctly enlarged jowls in adult males; head scales small, interparietal not greatly enlarged; parietal eye small; parietals abruptly contact small, flat dorsal body scales; 9 to 14 supralabials; 11 to 15 infralabials; 4 postmentals; gular scales small and flat; gular fold extending up over forearm insertion; dorsal scales grade ventrally into slightly

FIGURE 4.8 *Ctenosaura hemilopha* (Cape Spiny-Tailed Iguana) from Arroyo Viejos, Isla Cerralvo, BSC

larger ventrals; dorsal crest of enlarged scales extending posteriorly one-quarter to nearly four-fifths of the way between axilla and groin; anterior scales of crest greatly enlarged in adult males; forelimbs relatively short and robust; hind limbs robust, longer than forelimbs; 9 to 15 enlarged femoral pores in adult males; 30 to 38 subdigital lamellae on fourth toe; tail approximately 1.5 to 2.0 times length of body, and rounded with enlarged upturned spines dorsally and laterally; spines arranged in caudal whorls; anterior whorls separated by two to three secondary rows of scales; enlarged postanal scales absent in males. **COLORATION** Ground color of dorsum light to dark gray in adults, hatchlings usually gray, turning to a uniform green; three large black blotches anteriorly in adults, third blotch encompassing forelimbs in adult males; occasionally wide, short dark line laterally on neck confluent with anteriormost blotch and most prominent in juveniles; dorsal surface of head unicolored; two moderate to faint darker postorbital stripes; weak speckling and banding pattern on body; several small black spots on sides; tail banded posteriorly; forelimbs black in large adult males, mottled in females and juveniles; hind limbs mottled and weakly to moderately banded below knee; tail banded posteriorly; ground color of ventrum pale yellow to light gray; gular region, chest, and ventral surface of forelimbs black in large adult males; gular region lighter and lineate in females and juveniles; chest and forelimbs spotted; widely separated, moderately sized spots on belly and hind limbs, with slight transverse orientation on belly.

GEOGRAPHIC VARIATION There is little variation among peninsular populations of *C. hemilopha,* although some slight differences exist between peninsular populations and that of Isla Cerralvo. Peninsular populations have 9 to 13 femoral pores, whereas the Isla Cerralvo population has 11 to 15. An enlarged series of dorsal scales anterior to first enlarged caudal whorl is always present in the Isla Cerralvo population but variably so in populations from Baja California. The hind limbs of the Isla Cerralvo population are often more banded and less mottled than the hind limbs of lizards from Baja California. The centers of the bands are somewhat lightened. The ventra of lizards from the Isla Cerralvo population are slightly more spotted, occasional reticulations are found in the gular pattern of females and juveniles, the spines on the tail are not as sharp or upturned, and the dorsal crest may extend farther posteriorly. This last observation is contrary to Smith (1972), who states that the dorsal crest in the Isla Cerralvo population terminates near mid-abdomen. The variation reported here is not discrete: it occurs in both Isla Cerralvo and peninsular populations to varying degrees. Thus they are considered conspecific (Grismer 1999b).

NATURAL HISTORY *Ctenosaura hemilopha* is generally restricted to the Arid Tropical and the Central Gulf Coast regions of the Sierra la Giganta, the Cape Region, and Isla Cerralvo and ranges no higher than 1,000 m elevation (Alvarez et al. 1988). Adults are commonly found in the vicinity of large rocks on cliffsides, in bajadas, and within arroyos, whereas juveniles are more often observed on vegetation farther from rocks. Spiny-Tailed Iguanas are abundant on moderate to large trees such as Mesquites (*Prosopis* sp.) and Cardón Cactus *(Pachycereus pringlei).* In the low-lying areas of the Isthmus of La Paz, Blázquez et al. (1997) and Blázquez and Rodríguez-Estrella (1997) reported *C. hemilopha* commonly basking up to 6 m above the ground on highly branched Cardón Cacti with crevices and woodpecker holes that could be used for retreats. I have observed the same near San Antonio de la Sierra, where mesquites are used. Banks and Farmer (1963) observed much the same on Isla Cerralvo. Ranchers in the La Presa region state that *iguana de palo* lives in tree hollows. In the town of San Bartolo, *C. hemilopha* are commonly seen on cement walls, rock walls, and fence posts among human habitations. On Isla Cerralvo, *C. hemilopha* are abundant on small rock piles along the edge of the beach (Banks and Farmer 1963) as well as inland on the rocky hillsides bordering the arroyos. Blázquez et al. (1997) report that lizards of the Isla Cerralvo population use a much greater range of basking sites, including open areas on the ground, than the adjacent peninsular populations. They suggest that this behavior is due to lower predation pressure on the island than on the peninsula.

Ctenosaura hemilopha is active year-round in the Cape Region. Adults can be seen basking from November to January, although peak activity occurs from April to October. *Ctenosaura hemilopha* is diurnal, and during the summer and early fall, it begins basking as soon as the sun rises. From its basking sites, *C. hemilopha* keeps a wary watch over its surroundings and usually flees into rocky crevices at the slightest hint of danger. Often, when inside a crevice, iguanas will orient their spiny tails toward the opening, presenting an imposing obstacle to anything or anyone attempting extraction. (A caudal spine under a fingernail is very painful!) On large outcroppings, *C. hemilopha* often occurs in pairs, or trios of one adult male and two females. In Arroyo Viejos, along the southern end of Isla Cerralvo, I have observed a colony of three to six lizards that has lived on the same rock pile for at least ten years.

Ctenosaura hemilopha is primarily a herbivore but will eat small arthropods and carrion. In Arroyo San Dionisio, I observed a juvenile dragging a dead hatchling Northern Cardinal *(Cardinalis cardinalis)* across the trail. On Isla Cerralvo, Banks and Farmer (1963) observed *C. hemilopha* feeding on the flowers of Cardón Cacti and saw one individual with a dragonfly in its mouth. A rancher in San Bartolo informed me that *C. hemilopha* come down off the boulders and raid his garden. He put up a fence to keep them out but said the lizards climbed over it and burrowed under it.

Very little is known about the reproductive biology of this species except that it is oviparous; hatchlings appear in late summer through late fall, and breeding occurs before late summer (Asplund 1967). Asplund speculates that there may be a spring hatching as well, although I have seen no evidence of this.

REMARKS Smith (1972) gives the range of *C. hemilopha* as extending north of Santa Rosalía. This assertion, however, was based on the distribution of Savage's (1960) San Lucan herpetofaunal area. No specimens have been confirmed north of Loreto.

FOLKLORE AND HUMAN USE *Ctenosaura hemilopha* is used as a food source in some of the more remote ranchos, and its meat is also considered a cure for whooping cough (Alvarez et al. 1988).

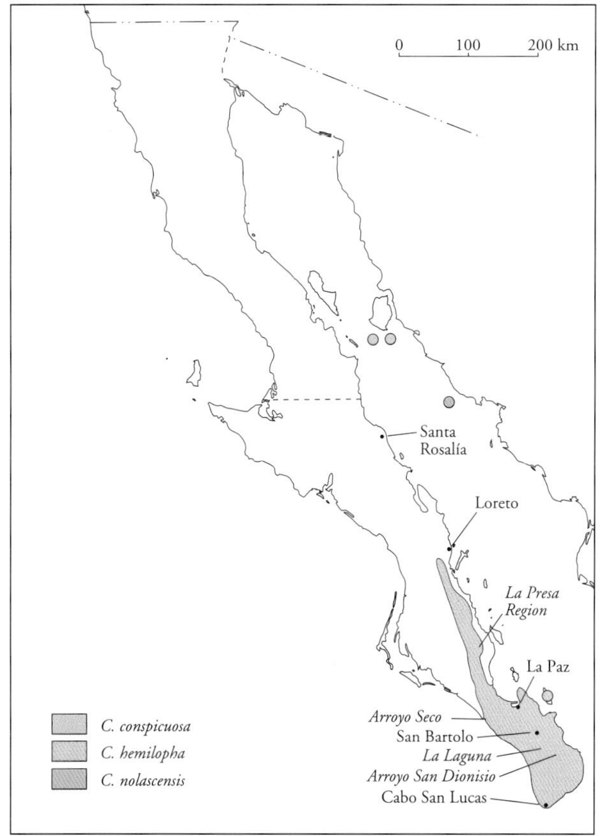

MAP 4.3 Distribution of *Ctenosaura* (Spiny-Tailed Iguanas)

FIG. 4.9

Ctenosaura conspicuosa

ISLA SAN ESTEBAN SPINY-TAILED IGUANA; IGUANA

Ctenosaura conspicuosa Dickerson 1919:461. Type locality: "San Esteban Island [Sonora], Gulf of California, Mexico." Holotype: USNM 64440

IDENTIFICATION *Ctenosaura conspicuosa* can be distinguished from all other lizards in the region of study by a single, slightly enlarged middorsal scale row forming a weak crest; caudal scales arranged in whorls of distinct spines; distinctly banded hind limbs in adults; round dark markings on the ventral surfaces of the limbs and body; the absence of a post-thoracic banding pattern in adults; and grayish and banded hatchlings.

RELATIONSHIPS AND TAXONOMY See *C. hemilopha.*

DISTRIBUTION *Ctenosaura conspicuosa* is known only from Islas San Esteban and Cholludo, Sonora, from the Gulf of California (see map 4.3).

DESCRIPTION Same as *C. hemilopha,* except for the following: adults reach 304 mm SVL; 8 to 12 supralabials, posteriormost supralabial usually elongate; 11 to 14 infralabials; 4 to 6 postmentals; dorsal crest extending posteriorly one-third to four-fifths of the way between the axilla and groin; 7 to 13 femoral pores; 29 to 34 subdigital lamellae on fourth toe; enlarged midcaudal series of scales variably extends anteriorly to first enlarged caudal whorl; caudal spines not upturned. **COLORATION** Same as *C. hemilopha,* except for the following: ground color of dorsum light gray (almost white) to light tan; occasionally two faint postorbital stripes; dorsum weakly speckled; moderate degree

FIGURE 4.9 *Ctenosaura conspicuosa* (Isla San Esteban Spiny-Tailed Iguana) from Arroyo Limantur, Isla San Esteban, Sonora

of black in pectoral region of adult females; wide, faint dark bands between hind limb insertions in adults; ground color in juveniles grayish, with prominent dark body bands; hind limbs mottled in juveniles, with very little banding; faint to absent lineate pattern on side of neck in hatchlings only; dense lineate pattern in gular region in juveniles; ventral spots large and circular with distinct transverse orientation.

NATURAL HISTORY Isla San Esteban is a rocky, mountainous island within the Central Gulf Coast Region, crisscrossed by deep arroyos and steep hillsides. *Ctenosaura conspicuosa* is more common in the arroyos than on the hillsides. It is distinctly arboreal, and adults are often seen perching 10 to 15 m off the ground on the tops of Cardón Cactus *(Pachycereus pringlei)* and large rocks. However, adults take shelter on the ground in the shade of trees on excessively hot summer afternoons. Juveniles more commonly occur on the branches of the smaller trees in the arroyo bottoms. *Ctenosaura conspicuosa* is less commonly observed on rocky hillsides above 300 m but is present at lower elevations as long as vegetation is abundant. Adults are often observed in trees in groups of two to three. The groups usually contain only a single male. *Ctenosaura conspicuosa* is active year-round, with a minimum of activity during December and January. It is a diurnal species, and during the summer it comes out to bask shortly after the sun rises. It is less wary than *C. hemilopha* and easier to approach.

Ctenosaura conspicuosa is omnivorous, with plants and arthropods forming the majority of its diet. Scales of *Cnemidophorus tigris,* however, have been found in scats of *C. conspicuosa,* and adults have been observed feeding on juvenile *Sauromalus varius* (Sylber 1988). A bat was found in the scat of a specimen on Isla Cholludo (C. Sylber, personal communication, 1995). Hatchlings have been observed from August through early October, which suggests that breeding takes place during the spring and summer. Case (1982) has observed *C. conspicuosa* and *S. varius* sharing the same overnight burrow.

REMARKS Felger (1966) reported *C. conspicuosa* as being common on Isla Cholludo, and Etheridge (1982) noted that its occurrence was possible there. In March 1991 I observed a large adult male that was in poor health. It had an infected jaw, and I removed four to five 3 cm cholla cactus *(Opuntia)* spines that were projecting into its abdomen. In October 1995, I found a juvenile within a hollow Cardón, which indicates that this population is reproductively viable. In late March 1997, I saw several adults on the rocky southwestern portion of the island.

FOLKLORE AND HUMAN USE Felger and Moser (1985) document the use of *C. conspicuosa* by the Seri Indians as a food source. After the Seri of Isla San Esteban were slaughtered by the Mexican government in the mid-nineteenth century (Cornejo 1987), mainland Seri still traveled to Isla San Esteban to gather food. Among many things they brought back from their journeys were *C. conspicuosa (heepni).* This species was particularly important to the Seri because it could be transported easily and kept alive for days or weeks. For some reason, the bones of this species were also eaten, but those of the San Esteban Island Chuckwalla, *Sauromalus varius,* were not. The transport of *Ctenosaura* by the Seri Indians (see below) and its absence from Isla Tiburón and adjacent coastal Sonora led Grismer (1994b) to speculate that its occurrence on Isla Cholludo and Isla San Esteban was the result of aboriginal introduction. The Seri told Nabhan (forthcoming) that their forefathers moved this species to Isla Cholludo to establish a food supply there in case fishermen became stranded on the island during rough weather. This hypothesis was further supported by Cryder (1999), who found that *C. conspicuosa* of Islas San Esteban and Cholludo were sister populations. Additionally, *C. conspicuosa* is more closely related to the Isla San Pedro Nolasco Spiny-Tailed Iguana, *C. nolascensis,* of Isla San Pedro Nolasco (an island also visited by the Seri), than to mainland Sonoran populations of *C. macrolopha.* Nabhan reports that the Seri also hunted *C. nolascensis* and brought them back to Isla San Esteban.

FIG. 4.10 *Ctenosaura nolascensis*

ISLA SAN PEDRO NOLASCO SPINY-TAILED IGUANA; IGUANA

Ctenosaura hemilopha nolascensis Smith 1972:107. Type locality: "Isla San Pedro Nolasco, Sonora," Mexico. Holotype: UCM 26391

FIGURE 4.10 *Ctenosaura nolascensis* (Isla San Pedro Nolasco Spiny-Tailed Iguana) from Isla San Pedro Nolasco, Sonora

IDENTIFICATION *Ctenosaura nolascensis* can be distinguished from all other lizards in the region of study by a single enlarged middorsal scale row forming a crest; caudal scales arranged in whorls of distinct spines; hatchlings grayish to green, with a distinct banding pattern; ventral surface of the hind limbs covered with black flecks in adults; and the absence of a post-thoracic banding pattern in adults.

RELATIONSHIPS AND TAXONOMY See *C. hemilopha.*

DISTRIBUTION *Ctenosaura nolascensis* is endemic to the Sonoran island of San Pedro Nolasco in the Gulf of California (see map 4.3).

DESCRIPTION Same as *C. hemilopha,* except for the following: adults reach at least 285 mm SVL; 11 to 13 supralabials; 13 to 15 infralabials; 13 to 16 femoral pores; 35 to 38 subdigital lamellae on fourth toe; 5 or 6 postmentals; dorsal crest extending posteriorly one-third to one-half the way down back; enlarged, midcaudal series of scales does not extend anterior to first enlarged caudal whorl; caudal spines not upturned. **COLORATION** Ground color of dorsum light to dark gray to tan or nearly white in adults; hatchlings grayish with dark transverse bands, lacking any shade of greenish coloration; juveniles orangish middorsally; wide, short, dark longitudinal line on side of neck occasionally present in juveniles; dorsum nearly unicolored in adults, sides of body weakly speckled; tail distinctly banded along entire length; inguinal region and ventral surface of hind limbs heavily suffused with black in adult males, spotting absent; gular region boldly marked with somewhat lineate pattern in females and juveniles.

NATURAL HISTORY Isla San Pedro Nolasco is a small, steep-sided, rocky island covered with fairly dense arid tropical scrub. *Ctenosaura nolascensis* is ubiquitous, occurring on the ground, among rocks, and in vegetation from the shoreline to the crest of the island. Adults prefer to bask in areas with large boulders and along cliff faces, regardless of whether vegetation is present. In such areas, a large male is often in the presence of one to three adult females. *Ctenosaura nolascensis* is probably active year-round, with peak activity lasting from March to October. In the early morning, adults usually bask in the vicinity of rocks, in which they take refuge if approached too closely. During the summer, juveniles emerge at their basking sites before sunrise. I have seen adults as high as 3 m above the ground in Organ Pipe Cactus *(Stenocereus thurberi),* feeding on ripening fruit. The fruit of the cholla cactus *(Opuntia)* and the leaves and flowers of various other species are also eaten. The majority of lizards I have observed had cactus spines in various parts of their bodies, likely from their habit of climbing inside the plants to feed. Many of the spines in the temporal region were impaled as deep as 2 cm, and I believe most were actually exiting the head through the roof of the mouth. Nearly all lizards had spines in their tongues, and I pulled one out of the eyeball of a hatchling. I have seen juveniles feeding on grasshoppers as well as plant material in early spring. I have observed juveniles in late March and early July and hatchlings during early October. This pattern suggests an early summer breeding season, with a late summer hatching period that coincides with the late summer rainy season.

REMARKS One of the most remarkable aspects of this species is its lack of fear of humans. All other *Ctenosaura* in the region of study that I have observed are wary and difficult to approach. *Ctenosaura nolascensis,* on the other hand, can be approached to within a few meters before it moves away, and often a lizard can be touched before it flees. This behavior suggests that these lizards have no large predators on the island, a circumstance that may account for their larger size and seemingly higher density. Such a hypothesis has also been suggested for the Isla Cerralvo population of *C. hemilopha* (Blázquez et al. 1997).

FOLKLORE AND HUMAN USE See *C. conspicuosa.*

Dipsosaurus Hallowell

DESERT IGUANAS

The genus *Dipsosaurus* is composed of a group of medium-sized, terrestrial, arid-adapted, omnivorous lizards containing two living species. *Dipsosaurus* are conspicuous lizards on the desert flats and are often seen climbing in low brush to feed on succulent leaves and flower blossoms. One species, *D. dorsalis,* ranges throughout the combined distribution of the Sonoran and Mojave Deserts of the southwestern United States and northern Mexico, and the other, *D. catalinensis,* is endemic to Isla Santa Catalina in the Gulf of California.

FIG. 4.11 **_Dipsosaurus dorsalis_**

DESERT IGUANA; CACHORA, CACHORÓN GÜERO

Crotaphytus dorsalis Baird and Girard 1852d:126. Type locality: "Desert of Colorado, Cal. [California]"; restricted to "Winterhaven [Fort Yuma], Imperial County," California, USA (Smith and Taylor 1950:355). Holotype: USNM 2699

IDENTIFICATION *Dipsosaurus dorsalis* can be distinguished from all other lizards in the region of study by a single, slightly enlarged middorsal scale row forming a weak crest that continues from the neck along the length of the tail; a rounded body in cross section; small, rectangular caudal scales; and a streaked to nearly immaculate gular region.

RELATIONSHIPS AND TAXONOMY *Dipsosaurus dorsalis* is most closely related to *D. catalinensis* of Isla Santa Catalina. Together these species form a basal lineage within the Iguanidae (de Queiroz 1987; Sites et al. 1996). Hulse (1992) recognizes two subspecies of *D. dorsalis* on the Baja California peninsula: *D. d. dorsalis* in the north and *D. d. lucasensis* in the south. Grismer et al. (1994) demonstrated that the diagnostic characters reported to separate these two taxa were variable and subtle and are not even useful in forming pattern classes. For similar reasons, Grismer (1999b) did not recognize *D. d. carmenensis* of Isla Carmen as a valid taxon.

DISTRIBUTION *Dipsosaurus dorsalis* ranges throughout the Mojave and Sonoran Deserts of the southwestern United States and northwestern Mexico. In Baja California, *D. dorsalis* is transpeninsular, ranging throughout most parts of the arid regions and into the Cape Region. One population extends west through Paso de San Matías and inhabits much of the cismontane areas in Valle la Trinidad. *Dipsosaurus dorsalis* does not occur along the west coast of Baja California north of Laguna San Ignacio; nor is it known to occur in the Vizcaíno Desert, notwithstanding the distribution map of Hulse 1992. Specimens have been reported from the Sierra Santa Clara in the southern section of the Vizcaíno Desert, but lizards occur nowhere else on the Vizcaíno Peninsula (Grismer et al. 1994). With the exception of the Sierra Santa Clara, their distribution only fringes the Vizcaíno Desert. It is likely that this cloudy, cool, windswept region is too cold for *D. dorsalis.* Its occurrence in the Sierra Santa Clara is restricted to low, flat, sandy arroyo bottoms between the isolated volcanic peaks (Grismer et al. 1994), which provide a barrier to the cool onshore weather patterns. *Dipsosaurus dorsalis* is also absent from the upper elevations of the Sierra la Asamblea. In the Gulf of California, it is known from Islas Ángel de la Guarda, Carmen, Cerralvo, Coronados, Espíritu Santo, Monserrate, Partida Sur, San José, San Luis, San Marcos, and Santiago (Grismer 1999b). *Dipsosaurus dorsalis* also occurs on the Pacific islands of Magdalena and Santa Margarita (map 4.4).

FIGURE 4.11 *Dipsosaurus dorsalis* (Desert Iguana) from 35 km south of Mexicali, BC

DESCRIPTION Moderately sized, with adults reaching 154 mm SVL; head relatively small, more triangular in adult males than in females and juveniles; 1 or 2 rows of scales separating rostral from nasals; scales of head small to slightly enlarged and juxtaposed; parietal scale not conspicuously enlarged, containing eye-spot; parietal scales grade abruptly into dorsal scales; dorsals small, slightly imbricate, and smooth to keeled, grading into smaller granular scales laterally that grade abruptly into larger flat ventral scales; enlarged middorsal scale row forming a low, poorly developed crest extending from parietals to posterior caudal region; 7 to 11 supralabials; 2 postmentals, usually not separated by small granular scale; gular scales beadlike, weakly imbricate posteriorly; gular fold extending up over forelimb insertion; 7 to 12 infralabials; forelimbs and hind limbs short and robust, but hind limbs longer than forelimbs; 31 to 48 femoral pores, most developed in adult males; 28 to 38 subdigital lamellae on fourth toe; tail slightly compressed laterally, 1.5 to 1.75 times SVL; caudal scales small, keeled, rectangular, occurring in whorls; enlarged postanal scales absent in males. **COLORATION** White to gray dorsal ground color overlaid with reddish-brown, semilineate body pattern, with paravertebral, reticulated ocelli; reticulations become more dense on sides of body; head unicolored; pattern on forelimbs faint to absent; reticulated reddish-brown pattern on hind limbs; thin, reddish-brown, caudal bands dorsally and laterally; faint to very dark lineate markings on gular region that coalesce centrally into a small, irregularly shaped blotch; ventrum pale, faded yellow; belly sometimes darker in adults.

GEOGRAPHIC VARIATION *Dipsosaurus dorsalis* from light-colored sandy areas north of Bahía de los Ángeles usually have a whitish ground color with a light, lineate lateral body pattern overlaying a reddish-brown paravertebral region, and the gular markings are faint to absent. The cismontane population in Valle la Trinidad, however, has a very bold color pattern despite its being in northern Baja California. Additionally, desert iguanas from darker substrates, such as that of the foothills west of Bahía San Luis Gonzaga, are darker laterally on the body and have relatively large white ocelli. In populations between Bahía de los Ángeles and San Ignacio, the dark, paravertebral coloration spreads ventrally on the body and surrounds the more laterally placed ocelli, encircling them with wide, dark rings and making the lizards appear darker over all. Gular streaking is faint between Bahía de los Ángeles and San Ignacio, but between San Ignacio and the Cape Region it becomes prominent.

The Cape Region populations are generally more boldly marked than other populations. The dorsal pattern is dark brown to nearly black, compared to the red-brown dorsal pattern of other populations. This darker pattern serves to highlight a wide, light-colored vertebral stripe with irregular scalloped borders. Many of the lizards of the Isla Cerralvo population have a distinct reddish tint on the rostrum that can be seen only on living specimens and first begins to appear in juveniles. Lizards of this population are also darker in overall ground color and have less distinct body markings and a smaller adult body size (mean SVL 94.3 vs. 109.1 mm; Scudder et al. 1983). Juveniles of the southernmost peninsular populations and those of Isla Cerralvo have distinctive orange markings on the sides of their heads and around their eyes. I have also noticed this coloration on juveniles from Isla Carmen.

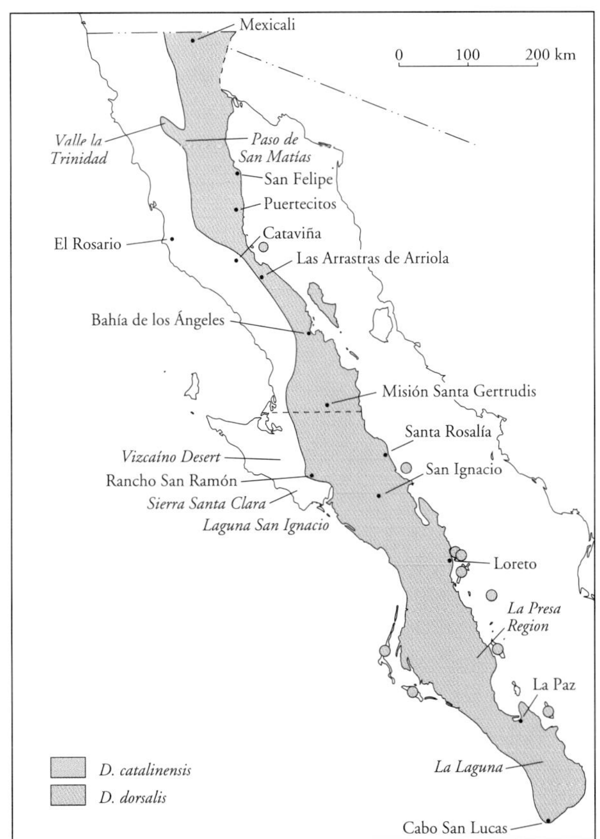

MAP 4.4 Distribution of *Dipsosaurus* (Desert Iguanas)

In populations north of Bahía de los Ángeles, the anterior dorsal body scales are smooth or only faintly keeled in some specimens. At Bahía de los Ángeles, keeling occurs in about 50 percent of the population, and from Misión Santa Gertrudis south into the Cape Region keeling is distinctive. In desert iguanas from Isla Coronados, keeling is generally weaker than that observed in populations from the adjacent peninsula. On Isla Carmen, however, keeling is generally more distinctive than in specimens from the adjacent peninsula.

Dipsosaurus dorsalis from Isla San José tend to be washed-out in overall coloration compared to those populations of the adjacent peninsula. Also, juveniles tend to lack the orange coloration around the eyes and along the side of the head. Juveniles from Isla Espíritu Santo also lack orange around the eyes but have slightly orangish tails. The population from Isla Santa Margarita, off the southern Pacific coast of Baja California, is distinctive. Most individuals are very dark in dorsal coloration, and in some the body is solid reddish-brown anteriorly. Most have a solid reddish-brown pattern on the posterior region of the body, overlaid with large white circular spots from the remaining ocelli. Others resemble the peninsular populations of that latitude in coloration and pattern. The gular region in the Isla Santa Margarita population is darkly streaked and approaches the condition observed in *D. catalinensis* (see below).

The peninsular populations of *D. dorsalis* from San Ignacio south have been reported to differ from the northern populations by having the rostral scale separated from the nasal ring by one rather than two rows of scales (Van Denburgh 1920; Smith and Holland 1971; Hulse 1992). There is no consistency in this character state, however. In all populations along the length of the peninsula, the rostral and nasal ring are separated by one or two scale rows.

NATURAL HISTORY *Dipsosaurus dorsalis* occurs in a wide variety of habitats. It is very common in the extremely hot and arid creosote bush scrub community of the Lower Colorado Valley Region of northeastern Baja California and is continuously distributed south into the (more humid) Arid Tropical Region and lower elevations of the Sierra la Laguna Region of the Cape Region (Alvarez et al. 1988). *Dipsosaurus dorsalis* does not occur along the cool Pacific coast north of Laguna San Ignacio (Grismer et al. 1994). One population does extend west, however, from the Lower Colorado Desert of northeastern Baja California through Paso de San Matías into the cooler, cismontane chaparral–creosote bush ecotone of the eastern end of Valle la Trinidad. Along very rocky regions of the peninsula, such as that near San Ignacio, or heavily vegetated areas such as parts of the Cape Region, *D. dorsalis* is generally confined to sandy arroyo bottoms or to other open places with loose soils (Alvarez et al. 1988). This is not always the case on Isla Cerralvo, where specimens have been found on the hillsides (Banks and Farmer 1963). Norris (1953) noted that in the rocky canyons near the town of San Bartolo, *D. dorsalis* occurred in the sandy washes but took refuge in rock cracks along the banks when threatened. I have also found individuals high on the boulder-covered hillsides of this same arroyo. At Las Arrastras de Arriola in northern Baja California, lizards commonly perch on the tops of large boulders. On the Magdalena Plain, *D. dorsalis* occurs almost exclusively in the open sandy flats (Norris 1953).

In northern Baja California, *D. dorsalis* is generally active from March through October, although I have observed adults active in early February near San Felipe during unseasonably warm weather. The seasonal activity period lengthens with decreasing latitude, and lizards are sometimes active year-round in the Cape Region, with minimal activity during December and January and most activity restricted to juveniles and subadults (Blázquez and Ortega-Rubio 1996). *Dipsosaurus dorsalis* is diurnal, emerging in the early morning to bask, often on small rocks. I have seen subadults abroad at night in the Cape Region, but I do not believe they were active. Lizards from Lower Colorado Valley Region populations are primarily herbivorous but shift toward eating arthropods during the drier months (Norris 1953). In the Cape Region, Asplund (1967) noted that during the wet season, when plant material is abundant, animal material still accounted for 41 percent of the dry weight of the stomach contents of adults and 68 percent of that of juveniles. The major components were isopterans and lepidopteran larvae. Galina (1994) reports that the diet of *D. dorsalis* populations on the well-vegetated flats of the Isthmus of La Paz is 100 percent lepidopteran larvae during the late summer rainy season. The ranchers at Rancho San Ramón in the Sierra Santa Clara told me that *D. dorsalis* climbs up into the Pitahaya Agria Cactus *(Stenocereus gummosus)* to eat the fruit.

In the northern populations south to approximately Santa Rosalía, breeding occurs during the spring and continues through midsummer, and hatchlings generally begin to appear in July. In the Cape Region, Asplund (1967) found no gravid females or females with enlarged ovarian eggs in August but noted that hatchlings were abundant, and I have noted that juveniles and subadults are abundant in September and October in the Cape Region and on Islas Cerralvo, Espíritu Santo, and San José. These observations suggest that both the northern and the southern populations have a spring-summer reproductive season. Asplund (1967) noted that the Cape Region populations showed little in the way of aggressive displays and that all size classes mixed freely. The Islas Cerralvo and Carmen populations are not nearly as wary as most peninsular populations, and lizards can be approached to within one to two meters before they flee. Asplund (1967) noted similar behavior in Cape Region populations.

E. Gergus and I discovered an interesting new insular population of *D. dorsalis* in July 1995 on Isla Santiago, a small, low-lying island west of Isla Coronados. This semicircular island measures only 80 by 300 m, with an elevation of 3 m. It is covered with a low-growing burrow bush *(Ambrosia)* and serves as a rookery for Yellow-Footed Gulls *(Larus livens)*. There are six to seven scattered patches of Pitahaya Agria in which the lizards reside. Four adults and two small juveniles were observed, a sign that this is a viable population.

FOLKLORE AND HUMAN USE The Seri refer to the Desert Iguana as *ziix tocázni heme imócoomjc,* which loosely translates to "deadly living thing that should not be carried back to camp" (Nabhan, forthcoming). Nabhan states that many Seri will not discuss this animal because it is so psychologically dangerous. Felger (1990) was told that it was toxic, and Nabhan was told that its bite would kill a person instantly.

FIG. 4.12

Dipsosaurus catalinensis

ISLA SANTA CATALINA DESERT IGUANA; CACHORA, CACHORÓN

Dipsosaurus catalinensis Van Denburgh 1922:83. Type locality: "Santa Catalina Island, Gulf of California [Baja California Sur], Mexico." Holotype: CAS 50505

IDENTIFICATION *Dipsosaurus catalinensis* can be distinguished from all other lizards in the region of study by a single, slightly enlarged, middorsal scale row forming a weak crest that continues from the neck along the length of the tail; a body rounded in cross section; small, rectangular caudal scales; and a solid brown central gular region in adults.

FIGURE 4.12 *Dipsosaurus catalinensis* (Isla Santa Catalina Desert Iguana) from Arroyo Mota, Isla Santa Catalina, BCS

RELATIONSHIPS AND TAXONOMY See *D. dorsalis.*

DISTRIBUTION *Dipsosaurus catalinensis* is endemic to Isla Santa Catalina, BCS, in the Gulf of California (see map 4.4). Hulse 1992 erroneously illustrates Isla San José as the locality for *D. catalinensis.*

DESCRIPTION Same as *D. dorsalis,* except for the following: adults reaching 129 mm SVL; head abruptly rounded and blunt in lateral profile; dorsal scales weakly keeled and more conical in shape anteriorly; 8 to 10 supralabials; postmentals usually separated by small granular scales; 8 to 10 infralabials; 35 to 41 femoral pores; 30 to 35 subdigital lamellae on fourth toe. **COLORATION** Same as *D. dorsalis,* except for the following: dark dorsal ground color pattern extensive in adults and juveniles; gular region solid dark brown in adults, faded or only darkly streaked in hatchlings and juveniles; large amounts of orange on sides of head in hatchlings; body generally light-colored, accentuating light shoulder patch; tail orange-yellow.

NATURAL HISTORY Nothing has been reported on the natural history of *D. catalinensis.* Isla Santa Catalina is a rocky, mountainous island with wide, sandy arroyos. It lies within the Central Gulf Coast Region and is dominated by a thorn scrub vegetation. *Dipsosaurus catalinensis* occurs in both the arroyo bottoms, where it takes refuge in the thorn scrub, and on the rocky hillsides. It is generally active year-round, with a minimum of activity, mainly in juveniles, occurring during the fall and winter. As with most diurnal lizards, *D. catalinensis* emerges in the early morning to bask and spends the remainder of the day going about its activities. It is omnivorous, and it is common to find groups of two to six individuals feeding on the fallen fruits of the Cardón Cactus *(Pachycereus pringlei).* This occurrence is most frequent where the cacti occur in stands. I found a carcass of an individual whose intestinal tracts were filled to capacity with the seeds of the Cardón Cactus. These seeds were mixed with several small rocks of the same approximate size, which appeared to be grinding away at the seeds' hard outer covering. In the large intestine, nearly all that was left was this outer covering, as the fleshy interior of the seeds had been digested. Sylber (1988) noted that gravel was purposely ingested by adult *Sauromalus hispidus* and *S. varius.* I have also seen iguanas as high as 1 m above the ground in Pitahaya Agria *(Stenocereus gummosus),* feeding on ripened fruits.

I observed a mating pair of *D. catalinensis* during early June and have seen hatchlings through late August. These observations suggest that hatching occurs from midsummer to early fall. I have seen what I suspect to be juveniles of the previous year's hatching from April through July.

This species is extremely wary and usually very difficult to approach. This wariness is something of an anomaly, because generally insular lizards in the Gulf of California have much shorter flight distances than their adjacent peninsular counterparts. During early April 1992, I noted that of approximately 90 lizards observed, 35 to 40 percent had freshly regenerated tails, which suggests that predation on this population may be high. This may be a reason for their uncharacteristically long flight distance. I have also observed a great deal of aggressive behavior (biting and chasing) among hatchlings and juveniles. The orange tail is the body part most often bitten, and many individuals of those age classes have broken or regenerated tails.

Sauromalus Duméril

CHUCKWALLAS

The genus *Sauromalus* is composed of a group of medium-sized to large, terrestrial to saxicolous, herbivorous lizards. It contains six living species that collectively range throughout the Sonoran and Mojave Deserts of the southwestern United States and northern Mexico (Hollingsworth 1998). *Sauromalus* are well-known for their peculiar escape behavior of wedging themselves into rock cracks and fissures and gulping air into their lungs to inflate their bodies. This action wedges the body against the opposing rock surfaces and makes it difficult for a predator to remove it. Some species of chuckwallas are commonly found in arroyo bottoms, cracks within Cardón Cactus *(Pachycereus pringlei),* and large thickets of thorny vegetation. One species, *S. obesus,* ranges widely throughout the arid, rocky regions of Baja California and occurs on several islands in the Gulf of California (Hollingsworth 1998; Grismer 1999b). Four other species are insular endemics known only from the Gulf of California (Hollingsworth 1998).

FIG. 4.13 **Sauromalus varius**

ISLA SAN ESTEBAN CHUCKWALLA; IGUANA

Sauromalus varius Dickerson 1919:464. Type locality: "San Esteban Island, Gulf of California [Sonora], Mexico." Holotype: USNM 64441

IDENTIFICATION *Sauromalus varius* can be distinguished from all other lizards in the region of study by its flat, wide, obese body, with loose folds of skin on the sides; smooth scales on the neck, dorsum, limbs, and tail; and a mottled color pattern of dark gray and yellow or orange.

RELATIONSHIPS AND TAXONOMY Hollingsworth (1998) placed *S. varius* as the sister species to all other species of *Sauromalus,* whereas Petron and Case (1997) place it as a terminal clade with *S. hispidus.* The most recent and comprehensive taxonomic and phylogenetic treatment of *Sauromalus* is by Hollingsworth (1998). Earlier works include Schmidt 1922, Shaw 1945, and Van Denburgh 1922.

DISTRIBUTION *Sauromalus varius* is known only from the Gulf of California islands of San Esteban, Sonora, and Roca Lobos, Baja California (Hollingsworth et al. 1997; Lawler et al. 1995; map 4.5).

DESCRIPTION Body large, obese, and dorsoventrally compressed; adults reaching 323 mm SVL; head flat, relatively small, and triangular in dorsal profile; no sexual dimorphic differences in head length, but head width greatest in adult males; head scales small and flat; no greatly enlarged interparietal; parietal eye-spot inconspicuous; temporals convex; parietals grade abruptly into smaller, convex, juxtaposed nuchals; no spinose scales above shoulders; dorsals weakly mucronate, grading abruptly into smaller flank scales; 28 to 41 middorsal scales in one head length; lateral body folds present, scales of fold not enlarged; ventrals equal in size to those of lateral body folds; 145 to 166 transverse rows of midventrals from vent to gular fold; 13 to 17 supralabials; 2 slightly enlarged postmentals; gulars granular; long parasagittal, weak midsagittal, antegular, and gular folds present; 15 to 19 infralabials; forelimbs short and robust, covered with smooth, juxtaposed scales; 53 to 61 scales encircling brachium; hind limbs short and robust, approximately 1.5 times length of forelimbs; scales of thigh smooth, smooth to faintly keeled on foreleg; 30 to 38 well-developed femoral pores in males; 23 to 26 subdigital lamellae on fourth toe; tail elliptical to oval in cross section, measuring 48 to 52 percent of SVL; caudals smooth and whorled, with 30 to 36 scales in single whorl two head lengths posterior to vent; postanal scales not enlarged in males. COLORATION ADULTS: Dorsal ground color of adults yellowish or orange, overlaid with a thick black mottling; no transverse body or caudal bands observable; gular and abdominal regions uniform yellow-orange; dark labial margins and occasional abdominal mottling present; side of head darker from ocular region to ear opening; pectoral region mottled. JUVENILES AND HATCHLINGS: Dorsal ground color of head and body brown; broad vertebral zone of light brown on body; five dark brown, middorsally constricted bands present when body is cold; after basking, dorsal pattern is a reticulum of small dark spots; six to eight dark caudal bands; ventrum uniform light brown except for weak speckling on forelimbs and thighs. Hatchling pattern transforms very early to adult pattern.

FIGURE 4.13 *Sauromalus varius* (Isla San Esteban Chuckwalla) from Arroyo Limantur, Isla San Esteban, Sonora

NATURAL HISTORY Various aspects of the ecology of *S. varius* have been studied in detail, and much of what is reported here comes from Case (1982) and conversations with Howard Lawlor of the Arizona-Sonora Desert Museum and Charles Sylber of the California State University at Fullerton.

Isla San Esteban occurs within the Central Gulf Coast Region, and its climate is hot and arid except during late summer and fall, when heavy rains are common. The flora is composed of both xerophilic and tropical deciduous species; during the rainy season, the foliage is vibrant, and grasses cover the arroyo bottoms. Chuckwallas are common in the arroyo bottoms, where they graze on grasses, and on the rocky hillsides. They take refuge in the large rock cracks and talus slopes on the lower foothills or in burrows dug into the banks along the arroyo bottoms. Juveniles and hatchlings are rarely seen in the arroyo bottoms or lower slopes of the hillsides. Instead, they are found at higher elevations on the ridges, where they reside beneath small rocks, within rock cracks, and beneath cap rocks.

Sauromalus varius is active year-round, with minimal activity during warmer days in the winter. Lizards usually pass the cold periods of winter in rock fissures and retreats along the lower slopes of the foothills. They emerge to bask one to two hours after sunrise and are frequently found beneath patches of cholla *(Opuntia).* Early in the day, individuals are wary; later, they are much more approachable. Some individuals may bask in front of their burrows the entire day, and lizards are often seen lying on top of one another. During the day, *S. varius* is common in the arroyo bottoms, often seen beneath the larger shrubs or in patches of *Opuntia,* where its mottled color pattern functions as camouflage in the dappled light (Case 1982). Hollingsworth (1998) also notes that this color pattern is cryptic on the worn and flaked surfaces of lava boulders on the hillsides. *Sauromalus varius* wanders much less than *S. hispidus* in search of food. Like *S. hispidus, S. varius* is a strict herbivore that favors flowers above all other parts of the plants and will seek them out before eating seeds and leaves (Sylber 1988). The Ironwood Tree *(Olneya tesota)* is the most important plant in its diet, although during the summer Rock Daisy *(Perityle emoryi)* blossoms are eaten more frequently. Cacti and grasses are also consumed but apparently not preferred (Sylber 1988). Many lizards are found with cholla spines in their lips and tongues.

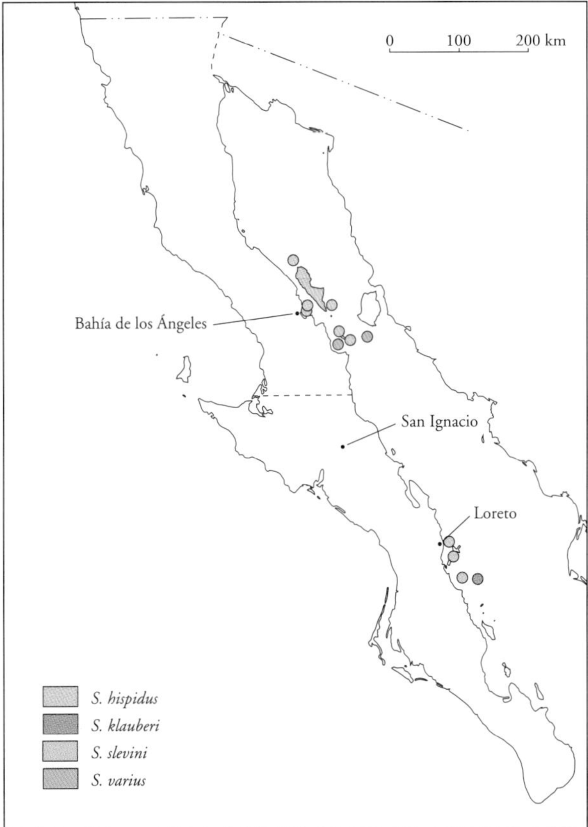

MAP 4.5 Distribution of the insular species of *Sauromalus* (Chuckwallas)

Sauromalus varius mates in the spring and lays its eggs during the spring and early summer, and hatchlings emerge in September and October (Case 1982; Sylber 1985a,b). Gravid females have been observed digging burrows in sandy, open areas in the arroyo bottoms. Sylber (1985b) reports clutch sizes of 21 and 22 eggs. I found a freshly killed female, dropped by a Red-Tailed Hawk *(Buteo jamaicensis),* that contained nine eggs. First- and second-year juveniles are noticeably absent from the arroyo bottoms. Howard Lawlor believes hatchlings migrate onto the rocky hillsides out of the arroyo bottoms to avoid predation by adult Spiny-Tailed Iguanas *(Ctenosaura conspicuosa).* I have found juveniles to be common on the upper hillsides in areas where *C. conspicuosa* were absent.

Successful recruitment is dependent on adequate rainfall. Often male-female pairs, trios, and quartets occupy the same overnight retreat for extended periods. Case (1982) observed one pair that remained together for six years. As in *S. hispidus,* aggressive interactions are very rare. Occasionally *C. conspicuosa* will occupy permanent burrows with *S. varius.* Case (1982) spent a week watching a pair (adult and juvenile males) of *C. conspicuosa* use the same den and basking sites as an adult female *S. varius.* At times, the juvenile *C. conspicuosa* would bask on the back of the *S. varius,* and the adult *C. conspicuosa* would chase away other adult *C. conspicuosa* that roamed into view.

Sauromalus varius also occur on Isla Roca Lobos. This tiny island (100 × 300 m) lies off the northwest end of Isla Salsipuedes, within the Central Gulf Coast Region. It is thus characterized by a hot and arid climate most of the year. *Sauromalus varius* feeds and takes shelter deep within the thick patches of cholla scattered through the eastern half of the island. There are usually runways leading from the periphery of the cholla patch to a central rocky burrow. Lizards may also be found in rock cracks some distance from the cholla patches. The *S. varius* of Isla Roca Lobos are much more wary and difficult to approach than those of Isla San Esteban. Hollingsworth et al. (1997) believe this population may be the result of a very recent human introduction not associated with aboriginal inhabitants. The presence of the carcasses of juveniles indicates that this is a viable population.

FOLKLORE AND HUMAN USE The original Seri of Isla San Esteban called their island *Coftécöl,* which was derived from the word *coof,* the Seri name for *S. varius.* The lizard figured prominently in the Seri diet (Felger and Moser 1985; Malkin 1962), and the Seri of Isla San Esteban were said to have hunted *S. varius* with the stems of the milkweed *Asclepias albicans* (Felger and Moser 1985). Long, slender branches were whipped through the air in rocky areas, and the sound scared the chuckwallas into moving to a different place. The sound of the lizard dragging its body across the rocks gave away its location, and the hunter then dug it out of the rock pile. *Sauromalus varius* is considered a delicacy among Seri and is prepared by frying (Malkin 1962). Nabhan (forthcoming) reports a recorded oral history that suggests supernatural punishment for those who senselessly kill or drown chuckwallas. It is stated that a "person who mistreats a chuckwalla by throwing it into the sea will be punished when he is at sea, perhaps by being subjected to strong winds." Having read this, I was reminded of a time on Isla Santa Catalina when I jokingly threw a chuckwalla into the water in front of Brad Hollingsworth while he was snorkeling. The *panga* ride back to Juncalito two days later was one of the windiest and most frightening trips I have had.

FIGURE 4.14 *Sauromalus hispidus* (Spiny Chuckwalla) from Isla Smith, BC

FIG. 4.14 **Sauromalus hispidus**

SPINY CHUCKWALLA; IGUANA

Sauromalus hispidus Stejneger 1891a:409. Type locality: "Angel de la Guarda Island, Gulf of California," Baja California, Mexico. Holotype: USNM 8563

IDENTIFICATION *Sauromalus hispidus* can be distinguished from all other lizards in the region of study by its flat, wide, obese body with loose folds of skin on its sides; sharp, conical, spinose scales on the neck, dorsum, and limbs (tuberculate in hatchlings and juveniles); a dark unicolored pattern in adults; and a weakly banded white punctate color pattern in hatchlings and juveniles.

RELATIONSHIPS AND TAXONOMY According to Hollingsworth (1998), *Sauromalus hispidus* is the sister species of a natural group containing *S. obesus, S. klauberi,* and *S. slevini.* Petron and Case (1997) place it as a terminal clade with *S. varius.* The most recent and comprehensive taxonomic and phylogenetic treatment of *Sauromalus* is by Hollingsworth (1998). Earlier works include Schmidt (1922), Shaw (1945), and Van Denburgh (1922).

DISTRIBUTION *Sauromalus hispidus* is known from the Gulf islands of Ángel de la Guarda, Cabeza de Caballo, Flecha, Granito, La Ventana, Mejía, Piojo, Pond, San Lorenzo Norte, San Lorenzo Sur, and Smith, BC (Hollingsworth 1998; Grismer 1999a; see map 4.5).

DESCRIPTION Body obese and dorsoventrally compressed; adults reaching 317 mm SVL; head flat, relatively small, and triangular in dorsal profile; head widest and longest in adult males; anterior head scales small and flat; temporals larger and conical; no greatly enlarged interparietal; parietal eyespot inconspicuous; parietals grade into larger, very spinose nuchals; nuchals larger than frontals; large, spinose scales present above shoulders; dorsal body scales strongly mucronate; dorsals grade laterally into slightly smaller flank scales; 17 to 25 middorsal scales in one head length; lateral body folds present; scales of body folds enlarged; flank scales grade into smaller ventrals; 103 to 139 transverse rows of midventrals from vent to gular fold; 12 to 16 supralabials; 2 slightly enlarged postmentals; gulars juxtaposed and weakly tuberculate; short parasagittal, weak midsagittal, antegular, and gular folds present; 11 to 18 infralabials; forelimbs short, robust, and covered with keeled, imbricate scales; 30 to 39 scales encircling brachium; hind limbs short and robust, approximately 1.5 times length of forelimbs, and covered with keeled, imbricate scales; 23 to 33 well-developed femoral pores in males; 19 to 26 subdigital lamellae on fourth toe; tail elliptical to oval in cross section and 47 to 54 percent of body length; caudals whorled, smooth ventrally and spinose laterally and dorsally; 20 to 31 scales per whorl two head lengths posterior to vent; postanal scales not enlarged in males. **COLORATION** Entire dorsum uniform dark olive green to dark brown or black in adults; no other dorsal pattern observable. Ground color of hatchlings and juveniles dark brown to black; side of head darker from ocular region to ear opening; four poorly developed, dark transverse bands on body, one on nape, three to five on tail; dorsal surface of head, body, limbs, and anterior portion of tail speckled with fine, faint, light-colored spots.

NATURAL HISTORY The ecology of *S. hispidus* is relatively well known, and much of what is reported here comes from the excellent study of Case (1982). The islands on which *S. hispidus* occurs, which are all within the Central Gulf Coast Region, are seasonally hot and arid and dominated by thorn scrub vegetation. *Sauromalus hispidus* is fairly ubiquitous on these islands, but hatchlings and juveniles have an overall preference for rocky areas, whereas adults are most common in, but not restricted to, deep burrows along hillsides, where they seek overnight and temporary refuge. I have observed adults out and feeding as early as mid-March on Isla Ventana, but *S. hispidus* is most active during late spring and early summer. During the hot periods of midsummer, activity diminishes during the day, and lizards bask in the early morning and at dusk. General activity is very low from late October to February, although if temperatures are high enough, adults emerge to bask. *Sauromalus hispidus* typically emerges from burrows one to two hours after sunrise during the spring and summer, and peak activity occurs during mid-morning. During this time, individuals may bask in groups before moving to additional basking areas closer to feeding sites. On Isla Mejía, there are well-worn pathways from burrows to feeding areas, and *S. hispidus* may travel as far as 900 m in one day to feed, returning to the same overnight retreat. These retreats are usually deep burrows or large cracks in rocks, which are often shared with as many as six other individuals. Rock piles in arroyo bottoms are also used, but on a more temporary basis. On Isla San Lorenzo Sur, adults are fairly ubiquitous, and I have observed individuals on some of the highest peaks of the island. Many adults dig burrows along the base of the hillsides, in patches of the Pitahaya Agria Cactus *(Stenocereus gummosus)*. Juveniles seem to be most common in the cobble along the beach, and they retreat into the spaces between the rocks when alarmed.

Intraspecific aggression is virtually nonexistent, with the exception of head-bobs before moving from one location to another. Juveniles climb as high as 3 m above the ground to feed on blossoms of Acacia (*Acacia* sp.), Desert Lavender *(Hyptis emoryi)*, Dye Weed (*Dalea* sp.), and Palo Verde *(Cercidium microphyllum)*. Sylber (1988) determined that *S. hispidus* is a strict herbivore that prefers flowers over all other parts of the plants and will seek them out before eating seeds and leaves. He reports that shrubs and forbs make up the most significant part of its diet. Cacti and grasses are also eaten but apparently not preferred. I even observed a juvenile eating moss off the rocks in the intertidal zone on Isla Smith and adults doing the same thing on Islas Mejía and Ángel de la Guarda. On Isla Smith I observed six adults feeding on grasses in a small valley. On my approach all ran 10 to 15 m to burrows along the bordering hillsides. Isla Pond is heavily vegetated with Cholla Cactus (*Opuntia* sp.), on which the chuckwallas feed. Shaw (1945) reported that many of the individuals on this island carry pieces of cactus stuck to their bodies, and some were seen with spines coming out of their eyes. On several occasions, I have removed spines from the temporal regions of adults. The spines had entered the roof of the mouth and were passing completely through the head.

Population densities of *S. hispidus* are low following and during drought years, and during such periods some individuals become emaciated. On 28 April 1989, I saw a female in an arroyo on Isla Mejía that had lost an estimated 70 to 80 percent of her body weight. On my return down the arroyo three hours later I found her dead. The breeding season of *S. hispidus* is correlated with rainfall, and recruitment following dry years is usually very low. Mating generally occurs in early spring, egg-laying in early summer, and hatching in the fall. I have seen gravid females during mid-June digging burrows in sandy arroyos at night on Isla San Lorenzo Sur and during the morning on Isla San Lorenzo Norte. It is likely that these females were excavating burrows for egg deposition. Sylber (1985a) reports a clutch size of 19 eggs and Carl and Jones (1979) a clutch size of 22 eggs for *S. hispidus*. I found an adult female, recently killed by Mexican fishermen on Isla San Lorenzo Norte during mid-June, that contained 28 well-developed eggs. The eviscerated carcasses of adults are commonly seen around the bases of Cardón Cactus *(Pachycereus pringlei)*. They are apparently the result of predation by Red-Tailed Hawks (*Buteo jamaicensis;* Case 1982). On Isla Mejía, I saw two adults run for cover when a raven flew overhead. I found the tail of a juvenile estimated to be 90 mm SVL impaled on the branch of a Desert Lavender *(Hyptis emoryi)* by a Loggerhead Shrike *(Lanius ludovicianus)*.

REMARKS Case (1982) and Petron and Case (1997) believe that the large size of *S. varius* and *S. hispidus* is a phenomenon of insular gigantism selected for by a lack of predators and a seasonally abundant food supply. Grismer et al. (1995) and Hollingsworth (1998) hypothesized that large body size is an ancestral characteristic of iguanid lizards in general and of *Sauromalus* in particular. Grismer et al. (1995) suggested that the large body size of *S. hispidus* and *S. varius* is maintained by the ecological parameters discussed by Case (1982), but that these were not selective forces causing its evolution. Based on a comparative analysis using other iguanids, Hollingsworth (1998) considered the large body size of *S. hispidus* and *S. varius* to be an inherited characteristic from an ancestral chuckwalla, and believes that what has evolved is dwarfism in *S. obesus, S. klauberi,* and *S. shawi.*

FOLKLORE AND HUMAN USE *Sauromalus hispidus* served as a food source for native Indians of Baja California (Aschmann 1959) as well as for the Seri from Sonora and Islas Tiburón and San Esteban (Malkin 1962). In fact, the Seri would go on shore on Baja California islands specifically to search for chuckwallas when they were in the vicinity. At Puerto Refugio, a Mexican fisherman from Guaymas told me that he and his friends eat this species while camped on the island. They prepare it by slitting the belly and removing the viscera, then placing it on the fire until cooked. Afterwards, they pull off the skin and eat the meat off the bone. He also stated that the oil from the fat bodies in the lizard's abdomen is good for the throat and ears.

FIGS. 4.15–16 **Sauromalus obesus**

NORTHERN CHUCKWALLA; IGUANA, IGUANA DE LA PIEDRA

Euphryne obesa Baird 1859d:253. Type locality: Fort Yuma, California, USA. Holotype: USNM 4172. (See Montanucci et al. 2001 for discussion of type locality and holotype.)

IDENTIFICATION *Sauromalus obesus* can be distinguished from all other lizards in the region of study by its flat, obese body, with granular to weakly imbricate dorsal scales and large lateral body folds; the lack of a mottled black and yellow dorsal pattern; the lack of very spinose neck and forelimb scales; and the presence of dark caudal bands in juveniles.

RELATIONSHIPS AND TAXONOMY According to Hollingsworth (1998), *Sauromalus obesus* is a basal member of a natural group containing *S. klauberi* of Isla Santa Catalina and *S. slevini* of Islas Monserrate, Carmen, and Coronados. Petron and Case (1997) place *S. obesus* as more basal in the evolution of *Sauromalus*. The latest and most comprehensive systematic work on *Sauromalus* is Hollingsworth 1998. Earlier works include Shaw 1945 and Robinson 1974.

Hollingsworth (1998) placed *Sauromalus obesus* (Baird 1859d) in the synonymy of *S. ater* (Duméril 1856) based on the absence of diagnostic characteristics discretely separating them from one another. Montanucci et al. (2001) petitioned the International Code of Zoological Nomenclature (ICZN) to conserve the name *obesus* over *ater* because of the much broader usage of the former. Under the conditions of Art. 23.9 of the 1999 edition of the Code, that usage is to be maintained until the ICZN renders its opinion.

DISTRIBUTION *Sauromalus obesus* is a common inhabitant of the Sonoran and Mojave deserts, ranging from eastern California, southern Nevada, and Utah south through western Arizona and southern California into northwestern Sonora and Baja California. In Baja California, it extends throughout the rocky areas of the arid Lower Colorado Valley Region of the northeast, south through the eastern portions of the peninsula to approximately 10 km south of La Paz

FIGURE 4.15 Ater pattern class of *Sauromalus obesus* (Northern Chuckwalla) from Isla Espíritu Santo, BCS

in the Cape Region (map 4.6). An isolated population occurs in the Sierra los Cucapás in northeastern Baja California. *Sauromalus obesus* tends to avoid the cool Pacific coast of Baja California, although it approaches the coastline in Arroyo San Javier north of Guerrero Negro (Bostic 1971) and occurs in isolated populations in the Sierra Vizcaíno and Sierra Santa Clara of the Vizcaíno Peninsula (Grismer et al. 1994). Additionally, I have observed scat in a large canyon just north of El Rosarito, 8 km from the west coast. *Sauromalus obesus* is absent from the relatively cool Magdalena Plain farther south. It is known from the Gulf islands of Ballena, Danzante, El Coyote (within Bahía Concepción), Espíritu Santo, Gallo, Partida Sur, San Cosme, San Diego, San Francisco, San José, San Marcos, Santa Cruz, Tiburón, and Willard (Grismer 1999a).

DESCRIPTION Moderately sized and obese; dorsoventrally compressed; adults reaching 224 mm SVL in northern populations; head flat, relatively small, and triangular in dorsal profile; head widest and longest in adult males; head scales small and flat to convex; no greatly enlarged interparietal; parietal eye inconspicuous; convex temporals and parietals grade abruptly into smaller, convex, juxtaposed nuchals; no spinose scales present above shoulders; dorsum weakly mucronate; wide vertebral row of scales slightly enlarged, grading abruptly into smaller scales laterally and extending onto flank; 22 to 41 middorsal scales in one head length; lateral body folds present, scales of fold smaller to slightly larger than adjacent flank scales; midventrals equal in size to those of lateral body folds; 155 to 195 transverse rows of midventrals; 9 to 18 supralabials; 2 slightly enlarged postmentals; gulars granular to convex; parasagittal and midsagittal folds weak to absent; antegular and gular folds present; 13 to 21 infralabials; forelimbs short and robust, covered with larger, smooth, juxtaposed scales; 32 to 55 scales encircling brachium; hind limbs short and robust, approximately 1.5 times length of forelimbs, covered with enlarged scales; anterior scales of hind limbs smooth, posterior scales keeled; 25 to 42 well-developed femoral pores in males; 18 to 29 subdigital lamellae on fourth toe; tail elliptical to oval in cross section and 45 to 56 percent of total length; caudals whorled, smooth to submucronate dorsally and laterally; 21 to 42 scales in single whorl two head lengths posterior to vent; postanal scales not enlarged in males. **COLORATION** Varies greatly with distribution, sex, and age. Generally, all hatchlings, juveniles, and adult females have a light gray-brown ground color and a darker banding pattern on the body and tail, with a lighter vertebral area in hatchlings and juveniles. Interspaces between the body bands contain dark speckling or blotching, but not the tail. This pattern is most distinct in juveniles and generally loses its vividness in adult females. Hatchlings and juveniles also tend to have dark snouts and a wide, dark postorbital stripe. Ventral ground color is generally lighter than the dorsal ground color, with varying degrees of dark streaking and speckling. See geographic variation (below) for coloration of adult males.

FIGURE 4.16 Obesus pattern class of *Sauromalus obesus* (Northern Chuckwalla) from Cataviña, BC

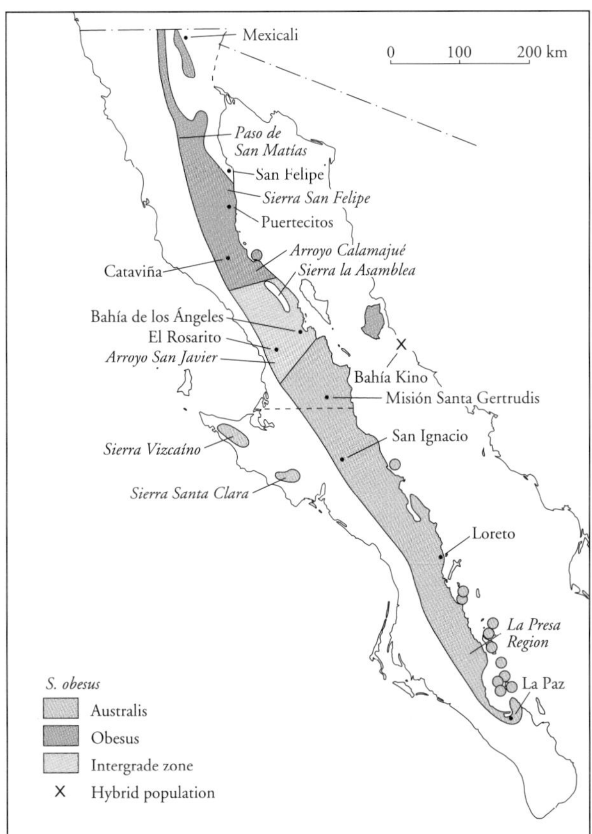

MAP 4.6 Distribution of the pattern classes of *Sauromalus obesus* (Northern Chuckwalla)

GEOGRAPHIC VARIATION The color pattern of adult male *S. obesus* shows considerable geographic variation. Information reported here comes mostly from Hollingsworth 1998. Populations from the eastern face of the Sierra Juárez are partially melanistic. The head, gular, pectoral, and pelvic regions are uniformly black, with a wide central light brown to brownish-yellow region on the back that contains varying numbers of small, black spots. The abdominal region has a dense black speckling, and the tail is light colored with faint banding. Farther south, this color pattern grades into a population characterized by a uniformly jet black head, body, and limbs and a cream-white tail lacking a banding pattern. This population first appears near the southern end of the Sierra San Pedro Mártir and extends south through the Sierra San Felipe to Arroyo Calamajué. These populations, previously considered *S. obesus obesus* (Shaw 1945), are referred to here as the pattern class Obesus.

Adult Obesus from an isolated population in the Sierra los Cucapás in northeastern Baja California have a dorsal color pattern similar to the partially melanistic populations in the adjacent Sierra Juárez but differ in that the central dorsal region is composed of dark transverse bands consisting of large blotches, rather than being a single wide, light-colored area lacking bands. Hatchlings from this population have a lemon yellow dorsal ground color.

Obesus intergrade with the Southern Banded Chuckwallas once considered *S. australis* (see Hollingsworth 1998) but referred to here as the pattern class Australis. Fifty kilometers south of Cataviña near the northern edge of the Sierra la Asamblea to just south of Bahía de los Ángeles, the solid black gular and pectoral regions of Obesus blend into the more darkly streaked pattern of Australis. Australis ranges from just south of Bahía de los Ángeles south into the northern areas of the Cape Region. In the northern portion of its range, from approximately Bahía de los Ángeles to Misión Santa Gertrudis, a transition in color pattern occurs (Hollingsworth 1998). Near Bahía de los Ángeles, adult males are similar in appearance to the partially melanistic populations of the Sierra la Asamblea and show only a slight ontogenetic change in color pattern, retaining much of the juvenile banding pattern into adulthood. The dorsal ground color of the head and body is dark brown to brownish-yellow with five irregular dark brown transverse body bands with lightened centers. South of here to Misión Santa Gertrudis, the interspaces between the body bands develop small dark spots, and the bands become obscured. Proceeding south, this pattern continues to develop to just north of Loreto. South of Loreto, the interspaces are marked with large dark blotches. In the Cape Region, the posterior bands often break up into a reticulated pattern. This pattern also occurs in lizards from the isolated populations in the Sierra Vizcaíno (Grismer et al. 1994) and many of the southern Gulf islands.

The pattern class Ater occurs on the Gulf islands of Ballena, Danzante, Espíritu Santo, Gallo, Pardo, Partida Sur, San Cosme, San Diego, San Francisco, San José, San Marcos, and Santa Cruz. Those of Isla San Diego are generally smaller than those of other populations (maximum 134 mm SVL) and usually lack the third dark dorsal body band. The first and second body bands are present, and the fourth is usually reduced to a large blotch between the hind limb insertions.

The Isla Tiburón population, considered *Sauromalus obesus townsendi* by Shaw (1945), is very similar in color pattern to the partially melanistic population from the Sierra los Cucapás of northern Baja California, and its members are referred to here as the pattern class Townsendi.

NATURAL HISTORY *Sauromalus obesus* is most commonly found in rocky habitats, ranging from large boulder-strewn outcrops to volcanic alluvia. It occurs in a broad range of environmental regions, from the extremely hot and arid Lower Colorado Valley Region of the northeast to the subtropical thorn forests of the Arid Tropical Region in the south. The presence of rocks and warm temperatures thus appear to be the environmental factors governing this species' distribution in Baja California.

In Baja California and mainland Mexico, *S. obesus* is generally saxicolous and has numerous morphological and behavioral adaptations for taking refuge in rock cracks at night or in the presence of predators (Hollingsworth 1998). In comparison, many insular populations seem to spend large amounts of time in burrows on the ground as well. On Isla Tiburón, individuals of *S. obesus* are as common on the ground, taking refuge in burrows beneath rocks, as they are on rocks in adjacent Sonora. The *S. obesus* of Isla Santa Cruz spend a great deal of time on the ground and take refuge in burrows beneath rocks on hillsides or beneath brush on the arroyo bottoms. Because there are no burrow-constructing mammals on this island (Lawlor 1983), it is logical to assume that the chuckwallas are constructing their own burrows. I found an adult *S. obesus* on Isla Espíritu Santo in December between two small rocks beneath a low-growing Elephant Tree *(Bursera microphylla).* It was on the top of a small, gravelly hill in an open plain on the southern end of the island, approximately 0.5 km from the nearest outcrop. Interestingly, *S. obesus* from the Sierra los Cucapás, another "insular" population, also spend a great deal of time on the ground, utilizing burrows beneath rocks as well as along earthen banks with no rocks. Chuckwallas are common on Isla San Francisco, occurring within piles of small rocks and beneath small rocks on the ground. I observed one lizard on numerous occasions basking on the ground outside the midden of a Desert Woodrat *(Neotoma lepida)* constructed on the edge of a large stand of Pitahaya Agria Cactus *(Stenocereus gummosus),* into which it always retreated on my approach. On the adjacent Isla San José, however, chuckwallas seem to be found only in areas with large rocks that are more typical of microhabitats in which the continental populations occur. This is not to say that continental populations do not use burrows beneath rocks or small rock piles in open areas, only that this behavior seems more prevalent in several insular populations. Chuckwallas of Isla San Diego are also found primarily in rock cracks along the crest of the island rather than in burrows.

Sauromalus obesus has been the focal point of several good ecological studies in the United States (Berry 1974; Beaman et al. 1997 and references therein) but no such studies have been published for Baja California populations. In Baja California, *S. obesus* is active from early spring to late fall, with peak adult activity occurring during the spring and early summer. In winter lizards hibernate in rock fissures or beneath cap rocks. I observed a specimen south of San Felipe in early February that had plant stains on its lower jaws, indicating that it recently had been feeding. This was much earlier than the usual seasonal activity period at this latitude, but the weather had been unseasonably warm. In northern Baja California, the majority of activity usually begins in mid-March to early April. By late summer, adult activity slows, and some have suggested that estivation occurs (e.g., Bostic 1971). In southern Baja California, activity is high in October, and I have seen adults active during early December on Isla Espíritu Santo, which suggests that the southern populations are active year-round as long as temperatures are sufficiently warm. Throughout southern Baja California south of San Ignacio, hatchlings are active at least into mid-October.

Chuckwallas come out to bask around mid-morning and are easily observed on the tops of large boulders. By late afternoon, or before the temperature gets too high, they retreat into large rock fissures. Occasionally, lizards reemerge after sundown to sit on the rocks. *Sauromalus obesus* is almost completely herbivorous and feeds on the flower buds and leaves of the vegetation near its basking site. In August on Isla San Diego, I saw an adult chuckwalla approximately 1.5 m above the ground in an *Opuntia* cactus, feeding on the fruit. I observed an adult male in the Sierra los Cucapás approximately 2 m above the ground in a Palo Verde tree *(Cercidium microphyllum)*, feeding on fresh blossoms. On my approach, the lizard dropped to the ground and took refuge beneath nearby rocks. Throughout Baja California, hatchlings generally begin to appear in late summer and continue through September, although I have observed hatchlings in the Sierra los Cucapás as early as late June. These observations suggest that breeding takes place in the spring and summer. I have observed small juveniles during mid-March near Juncalito in southern Baja California.

REMARKS On the small island of Alcatraz in Bahía Kino, along the Sonoran coast, Robinson (1972) reported on an introgressive hybrid population of *Sauromalus* involving *S. varius, S. hispidus,* and *S. obesus* (fig 4.17). The range of morphological variation in this population encompasses that of all three species. I have observed individuals within one arroyo whose color patterns match those of *S. varius, S. hispidus,* and *S. obesus* as well as everything in between. It seems that in every individual one of the three species tends to dominate overall appearance with respect to coloration and scale morphology, although characteristics of the other two are observable as well. Grismer (1994b) stated that *S.*

FIGURE 4.17 *Sauromalus obesus × varius × hispidus* hybrids from Isla Alcatraz, Sonora

obesus was most likely native to the island because it is common in the rocky foothills bordering Bahía Kino, and that the probable introduction of the larger, commonly eaten *S. hispidus* and *S. varius* onto Isla Alcatraz by Seri Indians resulted in this hybrid swarm.

FOLKLORE AND HUMAN USE This species composed part of the diet of the native inhabitants of Baja California and Sonora and is still eaten by the Seri Indians of coastal Sonora. These chuckwallas are considered a delicacy by the Seri and are prepared by skinning, gutting, and boiling. The Seri refer to this species as *ziix hast izj ano coom,* which loosely translates to "living thing that inflates between rocks" (Nabhan, forthcoming).

FIG. 4.18 *Sauromalus klauberi*

ISLA SANTA CATALINA CHUCKWALLA; IGUANA

Sauromalus klauberi Shaw 1941:285. Type locality: "Santa Catalina Island [Baja California Sur], Mexico." Holotype: SDSNH 6859

IDENTIFICATION *Sauromalus klauberi* can be distinguished from all other lizards in the region of study by its flat, obese body with granular to weakly imbricate dorsal scales and large lateral body folds; dorsal nuchal scales occurring in semilongitudinal rows; larger, lateral nuchal scales surrounded by smaller granular scales; the absence of a mottled black and yellow dorsal pattern; the lack of very spinose neck and forelimb scales; the presence of a faint reticulate dorsal pattern; and no dark caudal bands at any stage of its life.

RELATIONSHIPS AND TAXONOMY See *S. obesus.*

DISTRIBUTION *Sauromalus klauberi* is endemic to the Gulf island of Santa Catalina, BCS (see map 4.5).

FIGURE 4.18 *Sauromalus klauberi* (Isla Santa Catalina Chuckwalla) from Arroyo Blanco y Sol, Isla Santa Catalina, BCS

DESCRIPTION Same as *S. obesus,* except for the following: adults reaching 194 mm SVL; temporals and parietals convex and slightly larger than nuchals; dorsal nuchals occurring in somewhat longitudinal rows; lateral nuchals surrounded by much smaller granular scales; patch of large scales present above shoulders; 23 to 29 middorsal scales in one head length; scales of lateral body fold larger than adjacent flank scales; midventrals smaller than scales of lateral body folds; 126 to 146 transverse rows of midventrals; 10 to 13 supralabials; 11 to 15 infralabials; brachium with bluntly keeled dorsal scales; 35 to 40 scales encircling brachium; 27 to 33 femoral pores; 21 to 25 subdigital lamellae on fourth toe; tail 50 to 54 percent of total length; caudals whorled, smooth dorsally, and mucronate laterally; 23 to 27 scales in single whorl. **COLORATION** Ground color of dorsal surface dark gray, somewhat lighter vertebrally; dorsal pattern on body consisting of slightly darker, small, irregularly shaped blotches forming a dense reticulum; head, limbs, and tail unicolored; ventral ground color uniform light gray; gular region with distinct dark longitudinal streaks extending onto pectoral region. Hatchlings and juveniles lack a banding pattern and have a distinct postorbital stripe.

NATURAL HISTORY Isla Santa Catalina is a rugged, rocky island within the Central Gulf Coast Region that supports both a xerophilic and an arid tropical flora. *Sauromalus klauberi* is most common on the rocky hillsides and coastal bluffs, where it uses rock piles and fissures as retreats. Individuals occasionally can be found, however, in small piles of coral and beneath vegetation along the beach, and I have even found adults living on the open beach in the cobble only 10 m from the water's edge. I observed one individual along the edge of the beach beneath a cholla cactus *(Opuntia),* feeding on the fruit. When it was captured, I removed cactus spines from its mouth. *Sauromalus klauberi* may be active year-round, with a minimum of activity during the winter months, although I have never observed lizards later than October or observed adults abroad before mid-March. *Sauromalus klauberi* emerges approximately one hour after sunrise to bask, and it may remain abroad until just after sundown. I have seen adults foraging in the arroyo bottoms as well as a meter or so off the ground in Lomboy *(Jatropha cinerea).* Those observed in the arroyo bottom fled into burrows along the banks when approached. Shaw (1945) reported leaflets of Palo Verde *(Cercidium floridum),* fruits of Lomboy *(Jatropha cinerea),* Fescue (*Festuca* sp.), and the leaves of Catclaw Acacia *(Acacia gregii)* in the large intestines of three specimens. I have seen hatchlings in mid-March and early April that undoubtedly hatched in the late fall of the previous year, and I found a gravid female in mid-July. These observations suggest a spring breeding and a late summer or early fall hatching. Unlike other insular populations of *Sauromalus, S. klauberi* is uncommon and difficult to find. Individuals also have an unusually long flight distance. Interestingly, long flight distances also occur in the other large lizards on Isla Santa Catalina: the Isla Santa Catalina Spiny Lizard *(Sceloporus lineatulus)* and the Isla Santa Catalina Desert Iguana *(Dipsosaurus catalinensis).* This behavior may be an indication of predation by feral cats.

REMARKS This species is in great need of study.

FIG. 4.19 *Sauromalus slevini*

SLEVIN'S CHUCKWALLA; IGUANA

Sauromalus slevini Van Denburgh 1922:97. Type locality: "South end of Monserrate Island, [Baja California Sur] Gulf of California, Mexico." Holotype: CAS 50503

IDENTIFICATION *Sauromalus slevini* can be distinguished from all other lizards in the region of study by its flat, obese body with granular to weakly imbricate paravertebral dorsal scales; large lateral body folds; the absence of a mottled black and yellow dorsal pattern; the absence of spinose nuchal and posterior humeral scales; and a poorly defined dorsal band between the forelimb insertions.

RELATIONSHIPS AND TAXONOMY *Sauromalus slevini* is composed of three allopatric insular populations; it is most closely related to *S. klauberi* (Hollingsworth 1998).

FIGURE 4.19 *Sauromalus slevini* (Slevin's Chuckwalla) from Isla Carmen, BCS

DISTRIBUTION *Sauromalus slevini* is endemic to the Gulf islands of Coronados, Monserrate, and Carmen, BCS (see map 4.5).

DESCRIPTION Same as *S. obesus,* except for the following: adults reaching 209 mm SVL; nuchals conical to spinose; 21 to 31 middorsal scales in one head length; scales of lateral body folds larger than adjacent flank scales; midventrals smaller than those of lateral body folds; 105 to 133 transverse rows of midventrals; 10 to 15 supralabials; gulars granular; 11 to 17 infralabials; brachium covered with smooth, juxtaposed scales anteriorly and bluntly keeled scales posteriorly; 30 to 38 scales encircling brachium; 24 to 38 femoral pores; 19 to 24 subdigital lamellae on fourth toe; tail 50 to 56 percent of body length; dorsal and lateral caudals keeled and spinose; ventral caudals smooth; 19 to 29 scales in single whorl two head lengths posterior to vent. **COLORATION** Ground color of dorsal surfaces gray to tan; adults with darker, unicolored head and faint banding pattern anteriorly on body; rest of body covered with large dark, round spots, coalescing somewhat to form a reticulum, more so anteriorly than posteriorly; forelimbs inconspicuously speckled; hind limbs and tail unicolored; ground color of gular and pectoral regions dark brown; prominent dark gular streaks in subadults and some adults; ground color of remaining ventral surfaces tan. **JUVENILES**: wide, dark postorbital stripe present; five transverse body bands; caudal bands extremely faint to absent.

NATURAL HISTORY The islands on which *S. slevini* occurs lie within the Central Gulf Coast Region. They are hot, seasonally arid, and dominated by both xerophilic and tropical deciduous vegetation. *Sauromalus slevini* is most abundant in rocky areas. On Isla Monserrate, a low, volcanic, oceanic island, *S. slevini* is commonly found along ridges and escarpments bordering the arroyos. On Isla Carmen, a large, mountainous, landbridge island, *S. slevini* is abundant along the rocky foothills and along the cliff faces. On Isla Coronados, an emergent volcano, *S. slevini* is found primarily within the rocky talus of the lava flows. *Sauromalus slevini* is the least studied of all the chuckwallas. Nothing has been reported concerning its natural history, with the exception of SVL and tail break frequencies (Case 1982). This species is diurnal and, like most other *Sauromalus,* emerges to bask after sunrise. Its scats are commonly found in high, prominent areas along rocky ridges. I have observed juveniles in early April on Isla Carmen that undoubtedly hatched the year before, which suggests a spring breeding, summer egg-laying, and late summer or fall hatching cycle, as in other *Sauromalus* of this latitude.

REMARKS This species is in great need of study.

Phrynosomatidae

SAND, ROCK, HORNED, SPINY, FRINGE-TOED, BRUSH, AND SIDE-BLOTCHED LIZARDS

The Phrynosomatidae (*sensu* Frost and Etheridge 1989) are a New World group of lizards containing nine genera (Reeder and Wiens 1996) and over 100 species. Collectively, phrynosomatids range from southern Canada south through the United States, Mexico, and Central America to northern Panama. Phrynosomatids are the lizards commonly referred to as side-blotched lizards *(Uta),* brush lizards *(Urosaurus),* spiny lizards *(Sceloporus),* fringe-toed lizards *(Uma),* sand lizards *(Callisaurus, Cophosaurus,* and *Holbrookia),* rock lizards *(Petrosaurus),* and horned lizards *(Phrynosoma);* all but *Cophosaurus* and *Holbrookia* occur in Baja California. Where they occur, phrynosomatids form a conspicuous part of the ecosystem. They are diurnal insectivores that rely on keen vision and ambush foraging to obtain food. The Phrynosomatidae contain a wide variety of adaptive types, ranging from obligate sand dwellers to rock-climbing and arboreal forms, each of which has a number of highly modified characteristics adapted to its specialized lifestyle. Phrynosomatids are unique among lizards in that they have a sink-trap nasal chamber and an enlarged posterior lobe on the hemipenes (Frost and Etheridge 1989).

Callisaurus Blainville

ZEBRA-TAILED LIZARD

The genus *Callisaurus* contains the single small to medium-sized terrestrial species *C. draconoides.* It ranges throughout much of the combined distribution of the Sonoran and Mojave deserts of the southwestern United States and northern Mexico and occurs on several Gulf islands. *Callisaurus* is well-known for its extreme speed and the curious habit of curling its tail up over its back, exposing the black and white bands, and waving its tail nervously from side to side before running off. Wherever it is found, it is often the most abundant and conspicuous reptile species. *Callisaurus* is part of a natural group that includes the horned lizards *(Phrynosoma),* earless lizards *(Holbrookia* and *Cophosaurus),* and the fringe-toed lizards *(Uma)* (de Queiroz 1992; Reeder and Wiens 1996; Wilgenbusch and de Queiroz 2000).

FIGS. 4.20–22 **_Callisaurus draconoides_**
ZEBRA-TAILED LIZARD; AREÑERA, CACHORITA BLANCA, CACHIMBA

Callisaurus draconoides Blainville 1835a:286. Type locality: "Californie"; restricted to "Cape San Lucas, Baja California," Mexico (Smith and Taylor 1950:322). Holotype: MNHN 812

IDENTIFICATION *Callisaurus draconoides* can be distinguished from all other lizards in the region of study by strongly keeled supralabial scales; granular, homogeneous scales on the back of the thigh; and two to three black ventral body bars in adult males.

RELATIONSHIPS AND TAXONOMY *Callisaurus draconoides* is the sister species of a natural group that includes the earless lizards *(Holbrookia* and *Cophosaurus).* Wilgenbusch and de Queiroz (2000) discuss phylogeography.

DISTRIBUTION *Callisaurus draconoides* ranges throughout the Sonoran and Mojave deserts of the southwestern United States and northern Mexico south to southern Sinaloa and all the arid regions of Baja California. In northeastern Baja California, it extends west through Paso de San Matías into Valle la Trinidad. From here, it forms a broad, cismontane distribution edging the western foothills of the Sierra Juárez and Sierra San Pedro Mártir and extends to nearly the U.S.-Mexican border. *Callisaurus draconoides* first contacts the Pacific Coast near El Rosario and continues south into the Cape Region. It is known from Islas Ángel de la Guarda, Carmen, Cerralvo, Coronados, Espíritu Santo, Partida Sur, Patos, San Francisco, San José, San Luis, San Marcos, Santa Inez, Smith, and Tiburón in the Gulf of California and Islas Magdalena and Santa Margarita in the Pacific (map 4.7).

DESCRIPTION Dorsoventrally compressed; small to medium-sized, with adults reaching 109 mm SVL; frontal region weakly concave to weakly convex; snout wedge-shaped in lateral profile; dorsal head scales moderately sized and numerous; frontonasals undifferentiated; 4 to 6 superciliaries; large interparietal; parietal eye-spot conspicuous; interparietal grades abruptly into granular dorsals; 6 to 10 supralabials, all obliquely keeled and projected laterally; fringe scales of eyelids elongate and triangular; single postmental; gulars granular; antegular and gular folds present; 9 to 15 infralabials; dorsal body scales smooth, small, slightly enlarged medially, and weakly imbricate, grading into smaller granular scales laterally and into larger imbricate ventrals; lateral skin fold extends from axilla to groin; forelimbs long; scales of shoulder region enlarged and imbricate; fringes formed from lateral scales on toes present or absent; hind limbs long and muscular, approximately equal in length to SVL; 29 to 50 subdigital lamellae on fourth toe; digits with or without elongate, triangular, lateral fringe scales; 20 to 44 well-developed femoral pores in males; tail longer than body, flattened, conspicuously wider anteriorly; dorsal caudal scales weakly imbricate; two (rarely three or four) enlarged postanal scales in males. **COLORATION** Highly variable (see below); dorsal ground color whitish to dark gray; ground color sometimes overlaid with network of lighter spots; spots similarly sized on neck and body, and absent in middorsal region; usually a series of paired dark paravertebral blotches extending from neck to base of tail; adults have yellow inguinal region and orange axillary region; gular region gray with red center in adult males; tail with dark bands, bands solid black below; limbs variably banded; dark postfemoral line extending from knee to hind limb insertion, bordered below by white; ventrum immaculate to heavily mottled; two to three pairs of laterally directed black ventral body bars in adult males; ventral body bars set within turquoise to deep blue body patch in adult males; bars lighter in females.

FIGURE 4.20 Draconoides pattern class of *Callisaurus draconoides* (Zebra-Tailed Lizard) from Santiago, BCS

GEOGRAPHIC VARIATION Geographic variation in *Callisaurus draconoides* is extensive throughout its range: six different subspecies have been recognized in the region of study (de Queiroz 1989), which are treated here as pattern classes. There is considerable geographic variation among individuals within the pattern classes. Where individuals of different pattern classes come into contact, they freely interbreed. The Draconoides pattern class occurs in the Cape Region of Baja California (see map 4.7). Individuals of this pattern class are generally defined by the lack of well-developed fringes on the third and fourth toes; the posterior margins of the dark, paravertebral body blotches are pointed; the posterior body blotches are connected medially across the midline, forming dark, sinuous bands; the dark postfemoral line is bordered below by white; the dorsal caudal bands have pointed posterior margins; the indistinct posterior dorsal caudal bands are approximately half the width of the dorsal caudal interspaces; the distinct divergent black ventral body bars are equal in width and usually followed by one or two large black spots; and the medial ends of the ventral body bars are rounded. Males and females reach a maximum SVL of 72 mm and 67 mm respectively. Much of the variation in coloration and pattern between populations of Draconoides is due to substrate matching. Draconoides from La Burrera, at 500 m on the western slopes of the Sierra la Laguna, have bold dorsal markings with only faint evidence of dorsal caudal banding. A similar pattern is observed in specimens along the upper elevations of Ramal a los Naranjos just south of La Burrera. In La Paz, lizards vary in the distinctiveness of spotting posterior to the ventral body bars, which ranges from absent to distinct. In lizards at El Coyote, however, just 15 km east of La Paz, posterior spotting is always distinct, but here pectoral coloration ranges from immaculate to heavily mottled. The most colorful peninsular population occurs at the base of Sierra la Trinidad on the southeastern fringe of the Cape Region. This range is composed primarily of reddish sandstone, and Draconoides from here have an overall dorsal ground color of rusty orange with a darker, boldly marked dorsal pattern.

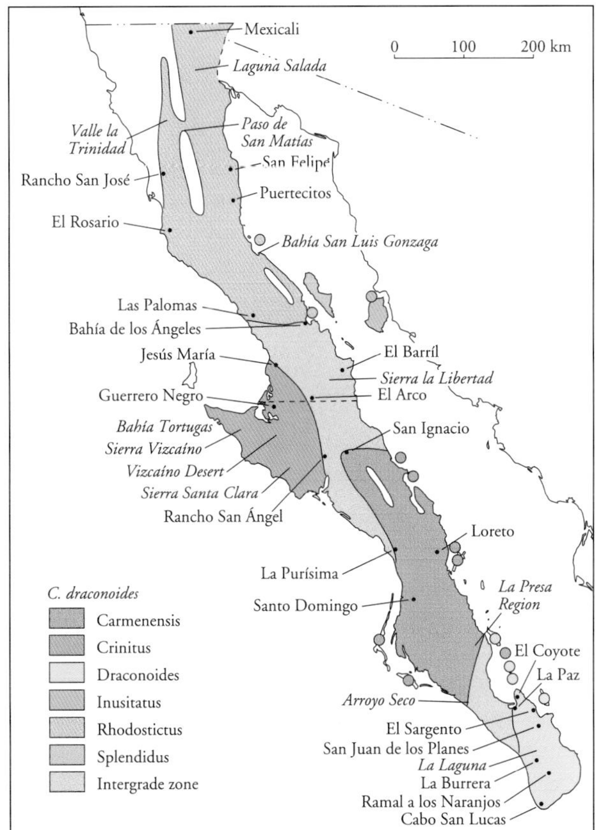

MAP 4.7 Distribution of the pattern classes of *Callisaurus draconoides* (Zebra-Tailed Lizard)

Most of the Draconoides from Isla Cerralvo have a well-defined dorsal pattern but lack dark blotching posterior to the ventral body bars. Additionally, the gular and pectoral regions of adult males are nearly solid black, and the belly patches are so dark that they do not highlight the ventral body bars. Juvenile males also have black gular regions. Both males and females usually have two or three undeveloped femoral pores distally on each leg, a characteristic not usually seen in other populations.

Draconoides from Isla Espíritu Santo are also strikingly colored. The lateral body coloration is nearly black, which serves to highlight a dorsal network of small, sometimes lemon-yellow spots. The pectoral regions are usually heavily mottled with gray, and there is dark blotching posterior to the ventral body bars. Because juveniles do not show such a contrast either dorsally or ventrally, this color pattern may develop ontogenetically. Females from Islas Espíritu Santo and Partida Sur lack the bold ventral coloration observed in males and are nearly immaculate. They are also lighter dorsally and do not exhibit the highlighted lateral network of yellow spots on the sides. All juveniles and females have an obliquely banded gular region, as opposed to the dark and unicolored gular region of adult males.

Carmenensis occur in central Baja California from south of San Ignacio in the north to the La Presa region in the south (see map 4.7). Carmenensis are similar to Draconoides, with which they intergrade widely (see below). They differ in that the dorsal paravertebral spots are not always pointed posteriorly and are never connected medially on the body, but are connected on the tail. Additionally, the equally sized ventral body bars are parallel rather than divergent, posteromedially directed, and usually not followed by a dark spot. Males and females reach a maximum SVL of 86 mm and 68 mm respectively. The major aspects of the dorsal color pattern of Carmenensis are reasonably consistent throughout their range, but ground color tends to vary locally in response to substrate matching. Near San Pedro de la Presa in the southeastern portion of their range, intergradation with Draconoides

FIGURE 4.21 Crinitus pattern class of *Callisaurus draconoides* (Zebra-Tailed Lizard) from 11 km south of Bahía Asunción, BCS

first becomes evident. Although the overall ventral coloration of this population is lighter than that of Draconoides, the dorsal pattern is more boldly marked and the dorsal blotches more elongate posteriorly and occasionally connected medially, as with Draconoides. Some males show a slight divergence in the first and second ventral body bars as well as spotting posterior to the bars, as with Draconoides. These trends continue to develop in the more southerly populations in the Isthmus of La Paz. In the west, the first signs of intergradation with Draconoides appear in the vicinity of Arroyo Seco, approximately 60 km west of La Paz. In some specimens, the ventral body bars become less parallel and more divergent, as in Draconoides, although both the Carmenensis and Draconoides pattern types occur. In some cases, both pattern types may occur in the same individual, one on each side. Some specimens also have faint spotting posterior to the ventral body bars, as in Draconoides. The La Paz lizards are essentially Draconoides.

Most Carmenensis males from Isla San Marcos are dark ventrally, and in some cases the ventral body bars are poorly defined, resembling the pattern in Inusitatus (see below). The pectoral region is heavily marked laterally with dark gray areas that are usually well-defined medially and are an extension of the blue lateral body patches that eventually grade into the gular region. The gular region is usually solid gray and rarely lineate laterally. Females and juveniles generally have an immaculate pectoral region. Males from Isla Coronados are similar in that their ventral body bars may be very faded. The Isla Carmen population is typical of that on the adjacent peninsula, except that the anteriormost ventral caudal bars are poorly defined and outlined in gray. The Isla San José population appears to be intermediate between Carmenensis and Draconoides, as is the adjacent peninsular population at that latitude at San Pedro de la Presa. Unlike the San Pedro de la Presa population, however, Isla San José specimens are intermediate principally in the high frequency of dark spotting posterior to the ventral body bars. This occurs in approximately 50 percent of the adult males, ranges from very faint to prominent, and may consist of one or two spots. The ventral caudal bands of Isla San José specimens are bold. In the adjacent Isla San Francisco population, the anterior ventral caudal bars are very well-defined and lack the fuzzy gray outline usually present in the Isla Carmen population. The ventral caudal bars are usually narrower than in most other populations, and the posterior bars have pointed posterior margins. The dorsal caudal bars are light. The ventral body bars are typical of Carmenensis, with some evidence of posterior spotting. Females and juveniles usually have a prominent and obliquely lineate lateral gular region.

The Crinitus pattern class, restricted to the Vizcaíno Desert of west central Baja California, contains some of the most distinctive individuals (see map 4.7). They have well-developed fringes on the third and fourth toes; a ground color overlaid with a dense network of lighter spots; paravertebral blotches that are faint to absent; a postfemoral line that is not bordered below in white; and three distinct, equally sized, parallel ventral body bars that are occasionally followed by a black spot. They are similar to Draconoides and Carmenensis in their caudal banding pattern. Males and females reach a maximum SVL of 81 mm and 69 mm respectively. Being relatively circumscribed in distribution, the geographic variation of Crinitus is generally restricted to differences in overall ground color related to substrate matching. Populations from coastal areas north of Guerrero Negro tend to have a pale or grayish dorsal ground color similar to that of the dunes they inhabit. Further inland, toward the central portion of the Vizcaíno Desert, the dorsal ground color takes on a golden hue, matching the changing color of the substrate. Populations from coastal areas west of the Sierra Vizcaíno, especially in the vicinity of Bahía Tortugas, are much darker dorsally and ventrally. This coloration is especially evident in the ventra of adult males, where the gular and pectoral regions are clouded with gray pigment. In the northwestern portion of their range, Crinitus populations intergrade with Rhodostictus (see below). In the northeastern portion of their range, east of El Arco, Crinitus intergrade with the intergrades of Rhodostictus and Carmenensis. Populations from this locality have a mosaic of characteristics common to Rhodostictus, Carmenensis, and Crinitus, although farther east in the Sierra la Libertad only Rhodostictus-

Carmenensis intergrades are found. In the eastern portions of their range, roughly 65 km west of San Ignacio, Crinitus begin to increase in SVL and have a smaller third ventral body bar. Additionally, the first two body bars take on a more medial orientation, similar to that found in Carmenensis, and the lamellar fringes on the third and fourth toes are not nearly as well-developed as in the more westerly populations of Crinitus. At Rancho San Ángel, approximately 32 km west of San Ignacio, specimens are clearly intermediate between Crinitus and Carmenensis. Here, SVL increases and begins to approach that of Carmenensis; some adult males may have three ventral body bars on one side and only two on the other; specimens with two or three ventral body bars on each side occur in the same population; and the lamellar fringes are only slightly larger than those of Crinitus. Mosauer (1936) noted a similar intermediacy in the specimens from this area. Zebra-Tailed Lizards from San Ignacio are generally Carmenensis, although some specimens show weak indications of intergradation with Crinitus. In the southern part of their range, north of La Purísima along the fringes of the western edge of the volcanic badlands, Crinitus and Carmenensis also intergrade. Schmidt (1922) reported two specimens of Crinitus from Santo Domingo. Additional material I examined from Santo Domingo reveals that the lizards occurring in this area are Carmenensis and show no signs of intergradation with Crinitus. Examination of Schmidt's specimens reveals that they are intergrades between Crinitus and Carmenensis, and thus his locality is suspect.

Rhodostictus occur in the northern third of Baja California, from Bahía de los Ángeles north into southern California (see map 4.7). They lack well-developed lamellar fringes; the posterior margins of the paravertebral blotches are rounded rather than pointed and are not connected medially; the postfemoral dark line is not bordered below in white; the dorsal caudal bands are distinct, with straight posterior margins, and equal in width to the white interspaces. Rhodostictus have two posteromedially directed parallel ventral body bars, the second of which is wider than the first and is not followed by a dark spot. Males and females reach a maximum SVL of 94 mm and 83 mm respectively. There is considerable geographic variation among Rhodostictus. Most of it is due to substrate matching, although minor variations in scalation also exist. Populations from the Lower Colorado Valley Region, from Laguna Salada south to the vicinity of Puertecitos, are light in overall coloration. In the vicinity of San Felipe, some specimens may have moderately developed fringes on the third and fourth toes. Others, however, have poorly developed fringes on the fourth toe and none on the third toe. Lizards from this area also tend to have larger, more elongate supraorbitals than other populations. Rhodostictus from the cismontane areas of Valle la Trinidad and Rancho San José are considerably darker in overall ground color both dorsally and ventrally. Lizards from here have a well-defined dorsal pattern and show no evidence of lamellar fringes. Lizards from between Puertecitos and Bahía San Luis Gonzaga have a dark gray ground color that matches the dark volcanic terrain dominating this area. Their anterior ventral caudal bars are larger, more prominent, and much closer to the vent than those of more northerly populations, a characteristic extending south all the way to El Barríl and also observed in Pacific coastal populations from El Rosario south to Las Palomas.

Populations at Bahía de los Ángeles begin to intergrade with Carmenensis. Here, the dorsal caudal bands are not black and well-defined, as in Rhodostictus farther north, but appear lighter, wavy posteriorly, and faded anteriorly as in Carmenensis. This patterning is most obvious in the anterior caudal bands. Most specimens from Bahía de los Ángeles lack the normal medial expansion and posterior extension of the second ventral body bar in adult males of Rhodostictus, and thus anterior and posterior bars are nearly equal in thickness, as are those of Carmenensis. Lizards of the Bahía de los Ángeles population also have a darker pectoral region than typical Rhodostictus. Further south, at El Barríl, all lizards appear completely intermediate between those of Rhodostictus and Carmenensis. Lizards have the typical Carmenensis dorsal pattern of paired, well-defined paravertebral blotches edged in dark and light lines. The second ventral body bar is more typical of Carmenensis in that the posterior extension is obscure and the dorsal caudal banding resembles that of Carmenensis. At San Ignacio, the transition to the Carmenensis pattern appears to be complete. On the west coast, Rhodostictus range at least as far south as Las Palomas before intergrading with Crinitus between here and Jesús María. Some intergrades have three ventral body bars on both sides, as in Crinitus, whereas others have three on one side and two on the other. They are also intermediate in SVL, the distance of the first ventral caudal bar to the vent, lamellar fringe development, and number of femoral pores. The posterior ventral body bar is expanded, like that of Rhodostictus, in all lizards. This intergrade zone is narrow (10 to 15 km) compared to that between Rhodostictus and Carmenensis (115 km), and it occurs where the northwestern edge of the Vizcaíno Desert contacts the Sierra Columbia.

FIGURE 4.22 Inusitatus pattern class of *Callisaurus draconoides* (Zebra-Tailed Lizard) from Ensenada de los Perros, Isla Tiburón, Sonora

The Rhodostictus from Isla San Luis have almost no lamellar fringe development, and the typical, paired, paravertebral blotches are obscure to absent. The dorsal ground color is a dark, ashy gray with many moderately distinct, small, cream-colored spots. This ground color matches the volcanic ash substrate of the island. This population also tends to have a higher number of ventral caudal bars.

The Isla Ángel de la Guarda population has been considered by some to be the subspecies *Callisaurus draconoides splendidus* (see Grismer 1999b for a review) and is considered here as the pattern class Splendidus. These lizards are smaller than adjacent peninsular populations of Rhodostictus, reaching a maximum SVL of 73 mm, compared with 94 mm in Rhodostictus. The dorsal ground color of this population is grayish to reddish-brown and overlaid with a dense network of dark and light spots. They have two distinct, posteromedially directed, black ventral body bars that are very thin dorsally and much wider medially than laterally. The second bar is always wider than the first. What little geographic variation is present in this population is due primarily to substrate matching. In areas of reddish volcanic substrate at the northern end of the island, near Puerto Refugio, lizards have a red-brown hue with a well-defined darker blotching pattern. This is most evident in adult males. In areas with long, wide sandy beaches, this population is similar in overall appearance to Rhodostictus from the adjacent peninsula.

Inusitatus occur in Sonora, Mexico, and on Islas Patos and Tiburón and do not contact any of the Baja California pattern classes (see map 4.7). Adults attain a maximum SVL of 109 mm and 89 mm in males and females, respectively. Adult males have orangish sides with a longitudinal row of poorly defined, dark blotches and a yellowish vertebral region. Females have yellowish belly patches, and juveniles have a lime-green tint to the body. The two parallel and posteromedially directed dark ventral body bars are very faint and often difficult to discern. The ventral body bars have lightened centers, are poorly set off from a turquoise belly patch, and blend imperceptibly into the mottled axillary body pattern.

NATURAL HISTORY Pianka and Parker (1972) reported extensively on the natural history of *C. draconoides* from desert regions in the United States, and much of this information may be applicable to populations in extreme northeastern Baja California. However, many aspects of its natural history appear to vary greatly among pattern classes. *Callisaurus draconoides* ranges through nearly all of Baja California and is found in a diverse array of habitats. Many of the distinctive habitat types in which this species occurs are geographically concordant with the distribution of the four peninsular pattern classes.

Draconoides range throughout the Arid Tropical Region of the Cape Region, reaching nearly 1,200 m elevation, and approach the oak woodland of the upper elevations of the Sierra la Laguna (Alvarez et al. 1988). They are also common on open flood plains, on beaches, and in sandy arroyos. At higher elevations, Draconoides are commonly seen on rocky and shaded hillsides covered with leaf-litter microhabitats not normally utilized by other pattern classes. Draconoides are active year-round (Blázquez and Ortega-Rubio 1996; Leviton and Banta 1964), and I have observed numerous individuals in late December and early January on the western slopes of the Sierra la Laguna. They are diurnal and can be observed basking in the mornings and scurrying about after temperatures have risen sufficiently. Asplund (1967) reported that nearly 50 percent of the diet of Draconoides consisted of lepidopteran larvae. Also eaten were ants, beetles, flies, and termites. Asplund reported females with oviductal eggs and males with enlarged testes during August. He also noted that during August no hatchlings were observed, which suggests that this was the beginning of the breeding season. I have observed hatchlings from October through December.

Carmenensis range throughout the Magdalena and Central Gulf Coast regions, their adjacent islands, and the Arid Tropical Region of the Sierra la Giganta. Although the environmental diversity in these three regions is great, Carmenensis are primarily found in relatively flat, open areas, where they can use their speed to escape predators. Although most common in open sandy washes, they are by no means restricted to such habitats and are often observed on hardpan and areas of desert pavement so long as it is not cluttered with large rocks or vegetation. At San Pedro de la Presa, Carmenensis are commonly found on the large, flat boulders in the arroyo bottoms. They do not usually forage or bask on steep rocky hillsides, although they can easily and swiftly ascend such terrain in escape. Carmenensis are active from March through November, depending on the weather. From San Ignacio, they are generally active year-round. Carmenensis are diurnal and easily observed in the morning basking on the tops of small rocks, which later in the day serve as lookouts. Tevis (1944) noted that Carmenensis differed from Rhodostictus in that they did not forage on the beach despite being a common inhabitant of the shores; I have also noted this to some extent on Isla San Marcos. Nothing has been reported on the reproductive biology of Carmenensis; however, breeding must take place during the spring, because I have observed numerous hatchlings from mid-July through mid-December. Tevis (1944) stated that Carmenensis do not elevate and wave their tails above their bodies, but I have observed this behavior in several populations.

Crinitus are restricted to the Vizcaíno Desert of west central Baja California. Here, they are strictly associated with sand dunes and open areas composed of fine windblown sand and widely spaced vegetation, which offers open areas for running. Crinitus seem to avoid areas with hard-packed substrate. In coastal areas, they prefer the leeward sides of dunes, where temperatures are much higher. In the western portion of their range, their distribution fringes the rocky foothills of the Sierra Vizcaíno and Sierra Santa Clara. Here, lizards remain on the flats and do not venture into the rocky canyons or arroyos as would the other pattern classes (Grismer et al. 1994). Crinitus are active during the warmer months, generally from April to September, depending on the weather. They bask in the morning and forage after they have reached their preferred body temperature. Early in the day, Crinitus are usually observed sunning on east-facing sand hummocks or dunes, with their bodies perpendicular to the sun's rays and pressed flat against the substrate. Crinitus are common lizards within the Vizcaíno Desert, but they are usually observed only after they have begun to run. Their cryptic coloration camouflages them well, and lizards usually go undetected if they remain motionless. Early authors (Mosauer 1936; Tevis 1944) stated that Crinitus do not elevate their tails over their backs to display the black and white caudal bands, as do other *C. draconoides.* However, I have found this behavior to be common in both coastal and inland populations. Crinitus usually rely on their speed and camouflage to escape predators. I observed a specimen in sand dunes north of Guerrero Negro far from vegetative cover. On my approach, it repeatedly ran over the crest of the nearest dune. When just out of view, it turned back 180 degrees and buried itself in the opposite side of the crest. When I dug the lizard up and released it, it continued to flee and bury itself in the same manner. Mosauer (1936) reported small beetles along with "Hymenoptera, Diptera, Neuroptera, small grasshoppers, insect larvae, a small cicada," and flower buds from the stomachs of 16 specimens. Nothing has been reported on the reproductive biology of Crinitus.

Rhodostictus are common inhabitants of flat, open areas. In the Lower Colorado Valley Region of northeastern Baja California, they are one of the most conspicuous lizards of the creosote bush scrub community. Along the beaches of the Gulf coast, they are commonly found in association with the local vegetation. Rhodostictus are not restricted to desert regions, however, and range well into the cismontane areas of the California Region, following sandy corridors through Paso de San Matías. Here they are most prevalent in open, gravelly areas in association with chaparral communities. On Isla San Luis, I have observed individuals high on the narrow rim of the ashen crater at the southeast end of the island. In the northern portion of the Vizcaíno Region, Rhodostictus are nearly ubiquitous, absent only from steep rocky hillsides.

Rhodostictus are generally active from February to November, provided the weather is sufficiently warm. During the day, lizards are commonly seen basking on the ground or on surface objects. Along the Gulf coast, these include shells and animal carcasses. Some lizards are occasionally seen as high as 1 m above the ground on the tops of shrubs. As with most phrynosomatid lizards, Rhodostictus are generally ambush feeders whose diets comprise small arthropods and an occasional small lizard. Tevis (1944) observed a specimen on the beach at Bahía de los Ángeles using its front feet to dig beneath a ray carcass to get at beetles. On the west coast, where cool onshore breezes are generally continuous and fogs are common, Rhodostictus are not common inhabitants of the beaches but are more active immediately inland, in many of the large, open arroyos leading down to the coast. Individuals that venture onto the beaches usually confine themselves to the warmer leeward side of coastal dunes.

Rhodostictus lay eggs from June to August and may have multiple clutches in the region of study, as they do in other portions of their range. Each clutch usually consists of 2 to 15 eggs (Stebbins 1985). I observed a breeding pair at Bahía San Luis Gonzaga during mid-July. Hatchlings are first seen in September and remain active through November. The previous year's offspring are numerous the following spring. When disturbed or about to flee, Rhodostictus commonly arch their tails up over their backs and wave them slowly back and forth, displaying the banded underside. This may be one of the best ways to identify *C. draconoides* in the field. Bayard Brattstrom (personal communication, 1990) has observed Rhodostictus in southern California arching their tails over their backs and moving the tips back and forth very slowly to lure insects closer for capture. On Isla San Luis, Rhodostictus exist in very high densities. In June 1993, over three hundred individual lizards per person-hour of searching were counted. Lizards occur on rocks, in bushes, and within the Brown Pelican *(Pelicanus occidentalis)* breeding colony. *Uta stansburiana* is noticeably rare on Isla San Luis, and I have only observed one individual in 15 visits. It appears that Rhodostictus have taken over the island and saturated all the available diurnal lizard niches.

Splendidus of Ángel de la Guarda are most common along the beaches but also inhabit other flat, open areas, including narrow arroyos, as long as they are not choked with vegetation. Presumably their patterns of annual and daily activity, diet, and reproductive biology are not significantly different from those of the adjacent peninsular populations. Lizards forage in the intertidal zone for marine isopods, a behavior reported for Inusitatus from Sonora (Quijada 1992). In the morning, individuals come down from the surrounding vegetation inland from the beach toward the shore to bask on rocks, drift objects, or marine mammal carcasses. When approached, most individuals will head-bob and inflate their throats before curling their tails over their backs and running off. During their escape, lizards gain sufficient momentum to carry them over the tops of sheer-faced dunes that can be as high as 3 m. They often become airborne when passing over the crest.

On Isla Tiburón, Inusitatus are common inhabitants of flat, open areas of the Central Gulf Coast Region. They are most common in washes and along the beaches but also occur in rocky inland areas where vegetation is not too dense. Inusitatus are active from March through late November, with juveniles emerging before the adults at the beginning of spring. Inusitatus are commonly observed basking on small rocks during the morning and remain active throughout the day, during very warm temperatures, when most other lizards have retreated to the shade. They are also commonly abroad at dusk. Inusitatus of Isla Tiburón are much easier to approach than those of adjacent Sonora and exhibit the typical caudal wag of all other *Callisaurus*. Reproduction begins in late March and probably lasts through June. I have observed many different size classes from late March through April, and the smallest probably represented hatchlings of the previous August and September. Juveniles are abundant in October, which indicates a previous hatching period. Nothing has been published on the diet of Inusitatus of Isla Tiburón, but it is probably not significantly different from that of Inusitatus of adjacent Sonora. Quijada (1992) observed a coastal population in northwestern Sonora feeding in the intertidal zone and preying predominantly on amphipods. He also noted that spiders and orthopterans were consumed in relatively large numbers.

REMARKS Adest (1987) presented allozyme data suggesting that the El Sargento population of the eastern Cape Region is a recent colonizer of Isla Cerralvo. El Sargento lies on the eastern edge of San Juan de los Planes, a large flood plain draining the Sierra las Canoas and the Sierra el Carrizalito. Despite the large size of Isla Cerralvo (roughly 95 km^2), *C. draconoides* is known from only a small sandy beach on the southeast corner, just opposite El Sargento. Adest also stated that *C. draconoides* of the Cape Region shows no evidence of divergence from more northerly populations.

Many authors have noted the *Uma*-like morphology of Crinitus, which have flatter heads, bodies, and tails; nearly unicolored patterns; and fringed digits. This similarity led to the specific designation of this pattern class (Cope 1896a; Norris 1958). Although Crinitus are distinct among Zebra-Tailed Lizards in these respects, they intergrade with Rhodostictus and Carmenensis and thus are not independent lineages. Moreover, Adest (1987) could find no allozyme differences between Crinitus and other populations and stated that the genetic differentiation within *C. draconoides* as a whole was very low, indicating a high level of gene flow.

Inusitatus are considerably larger than any other *C. draconoides* of Baja California or the Gulf of California. This increase in SVL can be traced as a clinal trend, beginning with small Draconoides in the Cape Region. Proceeding north, SVL increases in the Carmenensis of central Baja California and again in the Rhodostictus of northern Baja California. It continues to increase in areas extending around the head of the Gulf of California and then south into the Inusitatus populations of western Sonora.

FOLKLORE AND HUMAN USE The Seri Indians refer to the Zebra-Tailed Lizard as *ctamófi,* and in one Seri song, the lizard sings to the people and tells them they should bury themselves in the winter to escape the cold (Nabhan, forthcoming).

Petrosaurus Boulenger

BANDED ROCK LIZARDS

The genus *Petrosaurus* is composed of a group of medium-sized to large, dorsoventrally flattened, diurnal, insectivorous lizards. It contains four living species that collectively range through the rocky areas of southernmost California and south throughout Baja California. *Petrosaurus* species are renowned for their rock-climbing abilities and are common inhabitants of deep canyons with steep, rocky cliffs. One species, *P. slevini,* is endemic to Islas Mejía and Ángel de la Guarda in the Gulf of California. The other three, *P. mearnsi, P. repens,* and *P. thalassinus,* are peninsular endemics with insular representatives. Wiens (1993a) and Reeder and Wiens (1996) demonstrated that *Petrosaurus* may be the sister group to the brush lizards *(Urosaurus)* and spiny lizards *(Sceloporus).* Reeder (1995) placed *Petrosaurus* as basal within the Phrynosomatidae.

FIGURE 4.23 *Petrosaurus mearnsi* (Banded Rock Lizard) from Cañón Santa Isabel, BC

FIG. 4.23 ***Petrosaurus mearnsi***

BANDED ROCK LIZARD; LAGARTIJA DE LA PIEDRA

Uta mearnsi Stejneger 1894:589. Type locality: "Summit of Coast Range, United States and Mexican boundary line [San Diego County], California." Holotype: USNM 21882

IDENTIFICATION *Petrosaurus mearnsi* can be distinguished from all other lizards in the region of study by having a flat head and body; enlarged, flat head scales; a distinct black collar not bordered in white; large keeled, spinose caudal scales; and a dark throat overlaid with widely separated white spots.

RELATIONSHIPS AND TAXONOMY *Petrosaurus mearnsi* is most closely related to the insular endemic *P. slevini* of Islas Mejía and Ángel de la Guarda. Cliff (1954b) and Grismer (1999b) discuss taxonomy.

DISTRIBUTION *Petrosaurus mearnsi* occurs along the arid portions of the Peninsular Ranges from San Gorgonio Pass, Riverside County, California, south to at least Bahía de los Ángeles, Baja California. In northeastern Baja California, it is also known from the Sierra las Pintas, but there it is restricted to mountaintops. An isolated population occurs on El Pedregoso, a lone, gigantic rock pile in Valle de Gato south of Cataviña in north-central Baja California. *Petrosaurus mearnsi* also occurs on the island of El Muerto in the Gulf of California (map 4.8).

DESCRIPTION Moderately sized, with adults reaching 86 mm SVL; body and head dorsoventrally compressed; head triangular; snout pointed in dorsal profile; head scales weakly convex and moderately enlarged; medial frontonasal undifferentiated; 7 to 9 superciliaries; temporal scales smaller and more granular; enlarged interparietal scale with conspicuous parietal eye; abrupt transition between occipitals and dorsals; 1 or sometimes 2 scales between nasal and first supralabial; auricular scales large; dorsals granular, grading smoothly into larger flat, hexagonal ventrals; 5 or 6 supralabials; 2 postmentals; 2 rows of sublabials beginning at third to fifth (usually fourth) postmental; gulars flat, circular, and juxtaposed; 2 or 3 (usually 3) prominent folds in gular region; 7 or 8 infralabials; limbs covered with small keeled scales; forelimbs long and thin; hind limbs long and muscular, longer than forelimbs; 19 to 26 well-developed femoral pores in males; 21 to 26 subdigital lamellae on fourth toe; tail 1.50 to 1.75 times length of body, oval in cross section with enlarged keeled, spinose, and whorled scales larger than dorsal body scales; postanal scales enlarged in males. **COLORATION** Ground color of dorsum dull bluish-gray to dark brown; posterior portion of head unicolored to moderately banded; distinct straight, black collar terminating at shoulders; dorsum of body weakly punctate, with small white dots; dots most obvious in darker specimens; five weak to moderately distinct bands between forelimbs and base of tail; central body bands expanded medially and often with lightened centers; tail banded; forelimbs and fingers weakly banded; hind limbs weakly banded to reticulated; gular region dark gray posteriorly, with large, widely spaced round spots; gular pattern faint anteriorly; central region of belly, thighs, and tail light-colored and edged laterally by invading bluish-gray ground color, often most extensive in adult males; central portion of caudal scales usually heavily suffused with turquoise; sexually active females develop orange on head and anterior of body.

GEOGRAPHIC VARIATION Geographic variation in *Petrosaurus mearnsi* is most evident in overall dorsal ground color, which corresponds to the color of the local rock type. Populations living on light-colored granitic rocks usually have a bluish-gray cast to their dorsal ground color, with a prominent banding pattern overlaid with inconspicuous white punctations. Lizards occurring on darker volcanic rocks usually have an overall dark brown dorsal ground color with inconspicuous dark body bands. The white punctations, however, are prominent, giving the lizards a more spotted and less banded appearance. *Petrosaurus mearnsi* from Isla el Muerto occur on light brown volcanic rock and have a brown dorsal ground color with distinctive white punctations and dark body bands. Lizards of this population also lack the turquoise caudal scales common to all other peninsular populations, regardless of the rock type they inhabit, and the anterior gular region is more boldly patterned.

NATURAL HISTORY *Petrosaurus mearnsi* is saxicolous and generally confined to the hot and arid Lower Colorado Valley Region of northeastern Baja California. However, some populations extend west through Paso de San Matías and reach the chaparral and piñon-juniper woodlands along the western face of the Sierra San Pedro Mártir. They have been reported from similar habitats along the eastern escarpment at elevations of 1,200 m (Welsh 1988). *Petrosaurus mearnsi* is a common inhabitant of areas with large boulders and is less frequently seen on volcanic flows or alluvial bajadas composed of smaller rocks. This species is most abundant in canyons with high rocky cliffs, where its flat body and head are adaptive characteristics that allow it to use rock fissures and cracks for retreats. Welsh (1988) stated that *P. mearnsi* was most abundant in the Sierra San Pedro Mártir bordering riparian areas in canyons. I have noted a similar abundance in the Sierra Juárez, Arroyo San Luis, Misión San Fernando Velicatá, and Misión Santa María.

Petrosaurus mearnsi is generally active from at least mid-March through October in the northern portion of its range, in the Lower Colorado Valley Region, although it may emerge earlier if the weather is sufficiently warm. From at least Cataviña south, lizards are also active on warm days from November through February. *Petrosaurus mearnsi* hibernates during the winter in rock fissures and under exfoliations. I observed two *P. mearnsi* and nine Peninsular Leaf-Toed Geckos *(Phyllodactylus xanti)* under the same exfoliation in late December in Arroyo San Luis. During the summer, individuals emerge early in the morning, often before 0600, and position themselves on a rock in anticipation of sunrise. Groups of lizards can be observed basking together, with males defending their territories by head-bobbing (De Lisle 1991). During mid-May in the Sierra Juárez, I saw one male grab another by the throat, shake it, and throw it off a 1.5 m–high rock. When the attacked male hit the ground, it lay motionless for several minutes before running away. During late June, in a canyon just west of Bahía de los Ángeles, I observed a specimen at midday on the sandy arroyo bottom in the shade. On seeing me it immediately retreated to the vertical rocky cliff face. At dusk, *P. mearnsi* is abundant in riparian areas where large rocks or cliffs approach the water's edge. Here lizards can be seen on the ground chasing flying insects and can be approached quite closely. This species feeds primarily on arthropods but has been known to eat juvenile Side-Blotched Lizards (*Uta stansburiana;* Cozens 1974). During the breeding season, females develop an orange coloration on the head, throat, and forebody. During May, I have seen gravid females at Cataviña whose heads and bodies were almost completely orange. I have seen hatchlings in August, September, and October, which indicates that breeding takes place in the spring, egg-laying in the summer, and hatching in the late summer and fall.

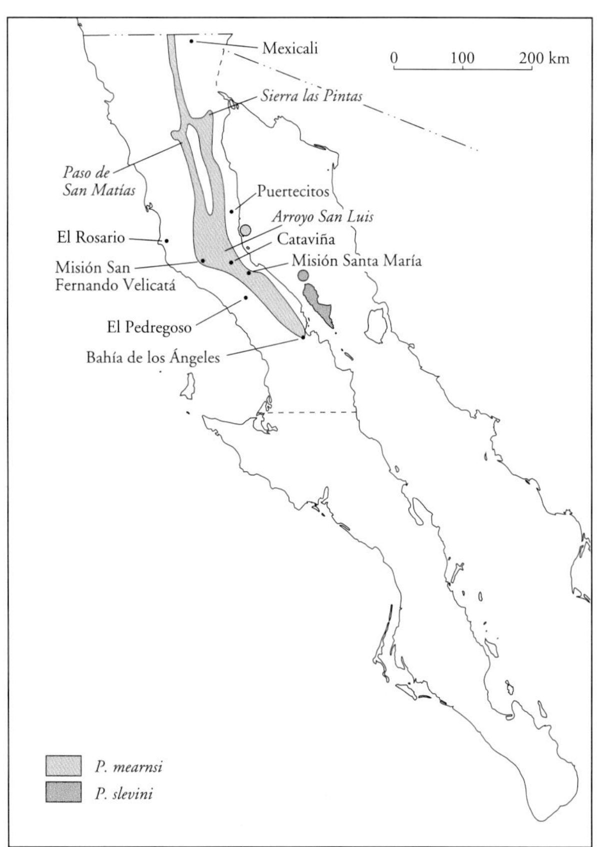

MAP 4.8 Distribution of *Petrosaurus mearnsi* (Banded Rock Lizard) and *P. slevini* (Slevin's Banded Rock Lizard)

FIG. 4.24 **Petrosaurus slevini**

SLEVIN'S BANDED ROCK LIZARD; LAGARTIJA DE LA PIEDRA

Uta slevini Van Denburgh 1922:194. Type locality: "Mejía Island, Gulf of California [Baja California], Mexico." Holotype: CAS 50506

IDENTIFICATION *Petrosaurus slevini* can be distinguished from all other lizards in the region of study by having a flat head and body; enlarged, convex head scales; a distinct black collar not bordered in white; large keeled, spinose caudal scales; and a dark gular region overlaid with a dense light-colored reticulum.

RELATIONSHIPS AND TAXONOMY See *P. mearnsi.*

DISTRIBUTION *Petrosaurus slevini* is known only from the Gulf islands of Ángel de la Guarda and adjacent Isla Mejía, Baja California (see map 4.8).

DESCRIPTION Same as *P. mearnsi,* except for the following: adults reaching 101 mm SVL; head scales moderately enlarged and convex; 5 or 6 supralabials; 2 rows of sublabials beginning at third or fourth postmental; 7 to 9 infralabials; 18 to 23 femoral pores; 11 to 15 subdigital lamellae on fourth toe. COLORATION Collar slightly arched anteriorly at midline; dorsum of body strongly punctate with small white dots; 5 to 6 poorly defined bands between forelimbs and base of tail; tail faintly banded; limbs and digits not banded; hind limbs weakly banded to reticulated; gular region dark with a dense pattern of small dark spots.

NATURAL HISTORY Islas Ángel de la Guarda and Mejía lie within the Central Gulf Coast Region off the coast of Bahía de los Ángeles. Both are rocky, mountainous, and topographically complex. Isla Ángel de la Guarda is the second largest island in the Gulf of California, and Mejía is a small satellite island that caps its northern tip. On Isla Ángel de la Guarda, *P. slevini* is commonly found on the scattered large boulders that cover the hillsides and extensive bajadas. Lizards are not restricted to these areas, however, and can often be found on the rocky cliff faces at higher elevations in the island's interior. In fact, this is the preferred habitat within the deep arroyos on Isla Mejía, which lacks the extensive boulder-covered hillsides and bajadas common on Isla Ángel de la Guarda.

Petrosaurus slevini is active from mid-March through October, as is *P. mearnsi* on the adjacent peninsula. *Petrosaurus slevini* is diurnal and can be observed basking on rocks and cliff faces during the morning. I have seen territorial behavior in males in late April, and gravid females have been collected in early June. These observations suggest a spring mating, which is likely followed by a late summer and early fall hatching. I found grasshoppers and beetles in the stomachs of four specimens collected during mid-March.

FIGURE 4.24 *Petrosaurus slevini* (Slevin's Banded Rock Lizard) from Isla Mejía, BC

REMARKS This species is in great need of study.

FIGS. 4.25–27 **Petrosaurus repens**

CENTRAL BAJA CALIFORNIA BANDED ROCK LIZARD; LAGARTIJA DE LA PIEDRA

Uta repens Van Denburgh 1895:102. Type locality: "Comondu, Lower [Baja] California [Sur, Mexico]." Holotype: CAS 633

IDENTIFICATION *Petrosaurus repens* can be distinguished from all other lizards in the region of study by a flat head and body; enlarged, flat head scales; one scale between the nasal and supralabials; a distinct black collar not bordered in white; small, weakly keeled caudal scales; and, usually, a distinct thin, dark postorbital stripe.

RELATIONSHIPS AND TAXONOMY *Petrosaurus repens* is most closely related to *P. thalassinus.* Grismer (1999b) discusses taxonomy.

DISTRIBUTION Jennings (1990b) noted that the northernmost locality of *P. repens* is at the north end of the Sierra Calamajué y San José, although his distribution map shows it occurring just south of El Rosario, much farther northwest. Bostic (1971) reported on a population from 17.4 km northeast of Rancho Santa Catarina on the northwest central coast, approximately 80 km farther north, which is considered its northernmost locality, except for a population on the west coast at Mesa San Carlos. *Petrosaurus repens* extends south to Arroyo Seco, 13 km north of La Paz in the Sierra la Giganta. It occurs on the tops of the Sierra la Asamblea, just north of Bahía de los Ángeles, and the tops of the Sierra la Libertad. It is absent from the Vizcaíno Pen-

FIGURE 4.25 *Petrosaurus repens* (Central Baja California Banded Rock Lizard) from San José Comondú, BCS

insula west of San Ignacio (Grismer et al. 1994) as well as from rocky areas within the Magdalena Plain. Its range borders the eastern edge of the Vizcaíno Desert, as evidenced by a specimen I observed in a narrow canyon at La Junta near the west coast, south of Laguna San Ignacio. South of Bahía de los Ángeles, *P. repens* is generally restricted to the Peninsular Ranges and its offshoots. It is also known from the Gulf island of Danzante (map 4.9).

DESCRIPTION Large, with adults reaching 129 mm SVL; body and head dorsoventrally compressed; head triangular; snout pointed in dorsal profile; head scales flat, moderately enlarged; medial frontonasals undifferentiated; lateral frontonasal scales square to rectangular; 7 to 9 superciliaries; temporal scales smaller; enlarged interparietal with conspicuous parietal eye; abrupt transition between occipitals and dorsals; 1 or 2 scales between nasal and first supralabial; 1 or sometimes 2 scale rows between subocular and labials; auricular scales small; dorsals granular, grading smoothly into larger flat, hexagonal ventrals; 5 to 7 supralabials; 2 postmentals; 1 or 2 rows of sublabials beginning at second postmental; gulars flat, circular, and juxtaposed; antegular and gular folds prominent; 6 to 8 infralabials; forelimbs long, moderately stout, and keeled dorsally; hind limbs muscular, longer than forelimbs, and keeled dorsally; 23 to 28 well-developed femoral pores in males; 23 to 29 subdigital lamellae on fourth toe; tail 1.50 to 1.75 times length of body, elliptical in cross section; caudals small, same size as dorsals, keeled, and whorled; postanal scales enlarged in males. **COLORATION** Dorsal ground color ashy gray to dark olive brown; dorsum overlaid with fine peppering of whitish to turquoise spots; large, sky-blue to turquoise spots usually on head and neck, largest posteriorly; postorbital stripe usually present; 6 to 11 dark bands on body between neck and base of tail; first band is a well-defined collar, central band well-defined, and remaining bands faded to varying degrees and some incomplete dorsally; dark, transverse nuchal blotch; interspaces between anterior bands have varying degrees of orange or yellow; bands on tail incomplete ventrally; anterior one-quarter of tail usually rust-colored dorsally and bluish to turquoise posteriorly; limbs gray with distinctive dark flecking; throat sooty gray to black, with color often spreading onto chest and proximal portions of forelimbs; grayish, chevron-shaped bands usually overlie a yellowish chest; sides of belly dark; black inguinal patches extend onto femora; postfemoral and anterior one-third of subcaudal region salmon-colored in adult males; anterior gular region and chest orange in gravid females. Hatchlings have large yellow spots on the dorsum and distinct white caudal bands.

GEOGRAPHIC VARIATION *Petrosaurus repens* vary greatly from population to population. This variation has largely to do with substrate matching of the parent rock. Lizards from the northernmost population, near Rancho Santa Catarina, have a very dark dorsal ground color with dense yellowish peppering that often obscures the banding pattern to some degree. Such a pattern makes this population resemble *P.*

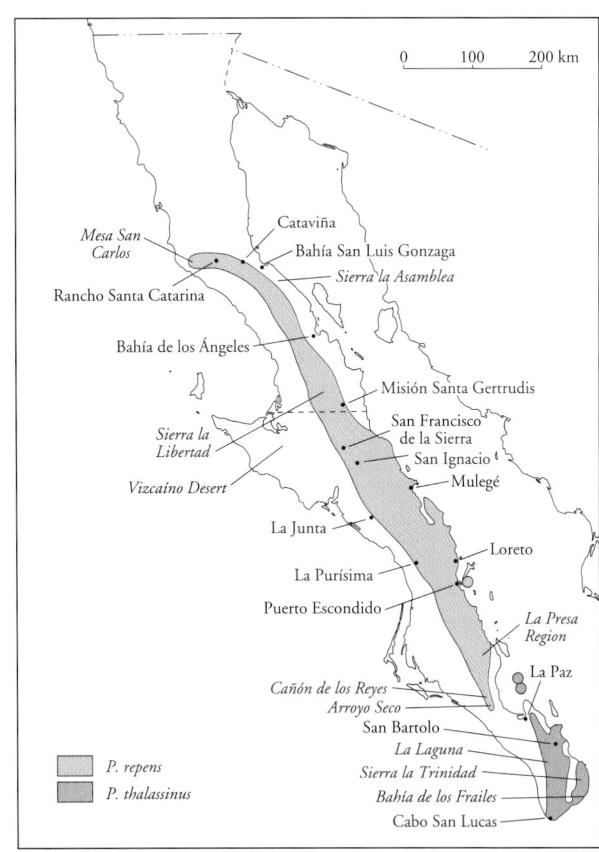

MAP 4.9 Distribution of *Petrosaurus repens* (Central Baja California Banded Rock Lizard) and *P. thalassinus* (San Lucan Banded Rock Lizard)

mearnsi. The thoracic region is burnt orange and the gular region reddish-orange. Lizards from this population also have distinctive dark spots on the head, a prominent postorbital stripe, and a distinctive nuchal blotch. Ventrally, the dark chevron bands of the chest are overlaid with large, round, open areas of ground color. Juveniles have thin dark bands on the neck and chest. This pattern extends to the western edge of the Sierra la Asamblea, approximately 103 km to the southeast.

Lizards from the volcanic highlands in the upper elevations of the Sierra la Libertad are dark overall. Those from the lower elevations in the granitic arroyos of the same mountains are lighter and have a distinctive yellowish lime-green hue to the anterior portion of the body, with conspicuous dark blotches on the head and a lyre-shaped marking on the neck. In the vicinity of San Francisco de la Sierra, *P. repens* lacks the large turquoise nuchal spots and has a faded, grayish overall ground color. This condition is accentuated in a population living on the walls of Misión Santa Gertrudis just north of the 28th parallel. The mission walls were constructed from a very light, ash-colored volcanic rock dug out of a nearby hillside. The lizards here have an extremely light, ashy ground color and virtually no turquoise or blue on any part of their bodies, a coloration that renders them cryptic on the mission walls. It is likely that these lizards have been living on the walls since the time of the mission's construction in the mid-1750s and are substrate-matching. Apparently this is the only native population of vertebrates in Baja California that benefited from the Jesuit invasion.

Lizards from the lower elevations of the volcanic sections of the Magdalena Region are generally more colorful, with a fine speckling of turquoise covering their bodies and well-developed turquoise nuchal spots. Those of the upper elevations of the Sierra Guadalupe have a deep brown ground color and appear very dark overall. The turquoise nuchal blotches stand out, although the yellow countershading of the dark body bands is only weakly evident. The dark markings on the top of the head, the dark nuchal blotch, and the overall coloration tend to fade in lizards from lower elevations along the Central Gulf Coast Region to the east and south. This condition reaches an extreme in the southernmost population from Cañón de los Reyes, 19 km north of La Paz (Grismer and Mahrdt 1996a), and Arroyo Seco, 13 km north of La Paz, at the southern end of the Sierra la Giganta. The canyon walls are composed of a light-colored conglomerate rock, and the yellow coloration in the lizards becomes so accentuated that some adults appear sulfur in color. The large turquoise nuchal blotches are absent, and the only turquoise coloring that remains is in small punctations within the dark dorsal bands. The base of the tail and hind limbs are pinkish. The belly is suffused with dark pigment invading from the sides of the body. Sexual dichromatism is also weak.

FIGURE 4.26 *Petrosaurus repens* (Central Baja California Banded Rock Lizard) from Cañón de los Reyes, BCS

Lizards from Isla Danzante look much the same as those from other populations in Baja California Sur, except for the number of body bands. All peninsular populations have six or seven dark dorsal body bands. Only two of these bands, however, are distinct, and one of these is the collar. The remaining bands are observable but faded to varying degrees. Lizards of the Isla Danzante population have ten or eleven bands that are generally thinner and much more clearly defined, making this population distinctive in overall appearance.

NATURAL HISTORY *Petrosaurus repens* ranges throughout much of the Vizcaíno Region in the north and the Central Gulf Coast and Magdalena regions in the south. Within these regions, it occurs in a variety of habitats, as long as large rocks are present. Its flat body and head are adaptive characteristics that allow it to enter the rock cracks and fissures it uses for shelter and overnight retreats. I have observed individuals in man-made wells, on large granitic boulders and outcroppings, on the escarpments of lava plateaus, on sheer cliffsides in steep canyons, and on the walls of several stone missions. They are most abundant in rocky canyons and on the lava escarpments. In an arroyo along the east face of the Sierra la Giganta opposite Puerto Escondido, I observed several specimens on a vertical cliff face approximately 70 m tall. Lizards were effortlessly running up and down the face at maximum velocity. At San Francisco de la Sierra I found an adult male 1.5 m above the ground, resting head-up on the trunk of a dead tree. In the bottoms of the canyons at San Francisco de la Sierra, lizards are common on the canyon walls and in caves and are of-

FIGURE 4.27 *Petrosaurus repens* (Central Baja California Banded Rock Lizard) from Cumbre de San Pedro in the Sierra Guadalupe, BCS

ten seen running along the cave paintings. I have seen lizards foraging in arroyo bottoms in the Sierra la Libertad and the southern portion of the Sierra la Giganta. On my approach they fled to the nearby canyon wall.

Petrosaurus repens is active from mid-March to October in the northern portion of its range, generally north of San Ignacio. It is active year-round in southern areas, provided the weather is warm, but is only minimally active during the winter. I have observed both adults and juveniles abroad in late December and early January between Loreto and La Purísima. *Petrosaurus repens* is diurnal and often emerges from overnight retreats as the sun is rising, despite the relatively cool ambient temperatures. During the day, it can be observed basking on rocks or on the walls of cliffs. Tevis (1944) noted *P. repens*'s ability to cling upside down to overhanging rock faces south of Mulegé, and I have seen lizards run at top speeds on such formations. Individuals from populations that occur near water are often seen foraging for insects along the water's edge. I observed six specimens at one time foraging for insects along a small *tinaja* (a rock pool along a stream bed), approximately 2 m in diameter, in the Sierra la Giganta near Loreto. In the Sierra la Libertad, I have seen lizards run into holes beneath rocks to escape and have found specimens beneath debris in trash piles. I have often observed this species out at night, long after the sun has set, where water is present. I suspect these lizards were feeding at the water's edge at dusk, as does *P. mearnsi,* and became "trapped" in the darkness before returning to overnight refuges. However, I have also observed lizards out at night in canyons along the western base of the Sierra la Asamblea in the absence of water. In late August, I observed several specimens abroad during a light rainstorm along the east face of the Sierra la Giganta, near Puerto Escondido. In Cañón de los Reyes, lizards are commonly seen foraging in the arroyo bottom and retreating to the cliffs when alarmed. Here they will ascend rapidly to heights of over 20 m to escape.

Bostic (1971) noted the omnivorous diet of *P. repens* from coastal north-central Baja California, stating that among stomach contents he found small black seeds resembling those of the nearby Barrel Cactus (*Ferrocactus* sp.) and Fish Hook Cactus (*Mammilaria* sp.), as well as other unidentifiable vegetation and beetle carapaces. I observed individuals in Cañon de los Reyes eating composite heads, cactus fruit (*Opuntia* sp.), and leaves. I also saw lizards feeding on beetles and one eating the yellow flowers of a Palo Verde tree *(Cercidium microphyllum)* near the rocky cliff face on which it was basking. In the Sierra la Libertad, groups of three to four lizards are commonly seen inhabiting a single outcrop. The groups are usually composed of a single adult male and various juveniles and adult females.

During the breeding season, females develop an orange coloration on the head, neck, and chest. I have observed adult females beginning to develop breeding coloration in early April in the Sierra la Asamblea and displaying breeding colors in mid-May at San Ignacio. I have observed hatchlings at San Francisco de la Sierra in late July, when no gravid females were found. Hatchlings less than 30 mm SVL have been observed in late August south of Loreto. These observations suggest that a spring breeding, summer laying, and late summer to early fall hatching takes place.

REMARKS Jennings (1990b) indicates this species' presence along the east coast of the peninsula at Bahía San Luis Gonzaga. It probably does not reach the east coast, however, until somewhere south of Bahía de los Ángeles. All *Petrosaurus* I have observed on the east coast in the vicinity of Bahía San Luis Gonzaga have been *P. mearnsi.* Ottley and Murphy (1981) state that the distributions of *P. repens* and *P. mearnsi* indicate that they are sympatric. This, however, is usually not the case. *Petrosaurus mearnsi* is more xerophilic and confined primarily to the lower, more arid Lower Colorado Valley Region and eastern margin of the Vizcaíno Region, whereas *P. repens* is more mesophilic and, in northern Baja California, is common in cooler, wetter areas at higher elevations. Although their distributions overlap considerably in latitude, they are not known actually to be sympatric except for populations at Jaraguay (Ottley and Murphy 1981).

Both species have been reported along the road to Bahía de los Ángeles. Close examination, however, reveals that *P.*

repens occurs at the upper west end of the road in the Vizcaíno Region, where the maritime influence is greater and morning fogs during spring and early summer are common. In fact, the vegetation in this area is covered with the bromeliad *Tillandsia recurvata. Petrosaurus mearnsi,* on the other hand, is common along the lower, eastern portion of the road in the Lower Colorado Valley and Central Gulf Coast regions, where the climate is warmer and considerably more arid. The farthest east I have found *P. repens* on this road is at km marker 50 (i.e., 50 km west of the town of Bahía de los Ángeles), and the farthest west I have found *P. mearnsi* is at km marker 39. Therefore, at least an 11 km gap separates these two species. This gap coincides with an ecotone between the hot, arid Lower Colorado Valley Region and the cooler, moister Vizcaíno Region. Further observations are necessary to assess parapatry between these species.

FIG. 4.28 ***Petrosaurus thalassinus***

SAN LUCAN BANDED ROCK LIZARD;

COCODRILO, LAGARTIJA AZUL

Uta thalassina Cope 1863:104. Type locality: "Cape St. Lucas [Cabo San Lucas]," Baja California Sur, Mexico. Holotype: USNM 5302

IDENTIFICATION *Petrosaurus thalassinus* can be distinguished from all other lizards in the region of study by having a flat head and body; enlarged, flat head scales; a distinct black collar complete across the back; small, weakly keeled caudal scales; no postorbital stripe; and two scales between the nasal and supralabials.

RELATIONSHIPS AND TAXONOMY See *P. repens.*

DISTRIBUTION *Petrosaurus thalassinus* is restricted to the Cape Region of Baja California, where it occurs in at least four disjunct populations: one in the Sierra la Laguna and contiguous ranges, another in the Sierra la Trinidad, and one each on the Gulf islands of Espíritu Santo and Partida Sur (Jennings 1990b; see map 4.9).

DESCRIPTION Same as *P. repens,* except for the following: adults reaching 149 mm SVL; lateral frontonasal scales thin and elongate; 2 scale rows between subocular and labials; 6 or 7 supralabials; 2 rows of sublabials beginning at second postmental; 8 or 9 infralabials; 15 to 21 well-developed femoral pores in males; 21 to 26 subdigital lamellae on fourth toe. **COLORATION** Dorsal ground color bluish; top and sides of head turquoise; eyes surrounded by sulfur yellow blotch; widely spaced, large, immaculate turquoise blue blotches on neck and anterior of body; ground color of

FIGURE 4.28 *Petrosaurus thalassinus* (San Lucan Banded Rock Lizard) from Arroyo el Aguaje, BCS

body gray, with a dense reticulum of thin black lines; turquoise color filling in spaces of reticulum; ground color light brown to reddish in sacral region; tail turquoise, with faint banding pattern; 6 dark bands on body; anterior 1 to 3 bands solid black, usually bordered posteriorly with yellow; first band a collar; second band incomplete laterally; posterior 3 to 6 bands very faint, more distinct in juveniles; transverse row of large yellow spots between bands 2 and 3, sometimes bordering them anteriorly; large round, dark spots on sides of body; limbs gray to tan dorsally, faintly banded; gular region yellow to orange with sooty black center; pectoral region yellow to orange with darker blotches that may or may not coalesce into chevron-shaped bands; remaining ventral surfaces dull white. **REPRODUCTIVE FEMALES:** Sides of neck suffused with reddish-orange, usually extending ventrally; dorsal body bands bordered posteriorly in red to bright orange.

GEOGRAPHIC VARIATION Geographic variation in *P. thalassinus* is generally restricted to an overall darkening or lightening of the dorsal ground color due to elevation of habitat or substrate matching and degrees of intensity in the yellow body bands. Lizards from La Laguna, at high elevations in the Sierra la Laguna, are generally darker than lizards from lower elevations, perhaps because the large granitic boulders in this valley are covered with thick layers of dark gray and green lichens. However, they have immaculate bright yellow body bands countershading the black body bands. The most brilliantly colored lizards come from coastal populations along the cliffs at Cabo San Lucas and Bahía de los Frailes. The bright color pattern of lizards from these areas contrasts enough with the light-colored rock to be seen from a distance of nearly 40 m. Lizards from San Bartolo are also relatively light in color, and the majority of the granitic boulders in this region are light gray. Lizards from the population at Rancho Remegio, just south of La Paz, have much more blue on their bodies than lizards from other populations and tend to lack the yellow posterior borders edging the dark body bands. The extremely attractive *P. thalassinus* from Islas Espíritu Santo and Partida Sur are light blue in overall coloration. They tend to lack the yellow posterior borders of the dark body bands, which are replaced with a dull white, and the interspaces between the anterior three bands are generally more heavily suffused with black.

NATURAL HISTORY Like other species of *Petrosaurus, P. thalassinus* is restricted to rocky areas and is common in deep canyons with granitic cliffsides and on boulder-strewn hillsides. In such areas, lizards are most often found on rocks near or under large trees or other vegetation. Lizards are rarely seen on boulders that are not in close proximity to vegetation. *Petrosaurus thalassinus* ranges throughout the appropriate habitats in the Arid Tropical Region of the Cape Region and to elevations of 2,020 m in the pine-oak woodlands of the Sierra la Laguna Region (Alvarez et al. 1988). It takes refuge within large rock cracks and fissures and is exceedingly wary and difficult to approach. Individuals usually retreat into a crevice on the far side of the rock when one approaches within 5 or 10 m. The climbing abilities of *P. thalassinus* are remarkable. I have seen adults run up vertical cliff faces on the beach at Cabo San Lucas seemingly as fast as the Zebra-Tailed Lizards *(Callisaurus draconoides)* run across the sand. Van Denburgh (1922, *in litt.* from J. Slevin) states that specimens have been seen leaping as far as "four feet" from boulder to boulder.

Petrosaurus thalassinus is active year-round. I have observed lizards basking on the cliff faces at San Bartolo, La Laguna, and Cabo San Lucas in December, and Leviton and Banta (1964) reported specimens out basking in mid-January at San Bartolo. Its activity, however, probably decreases at higher elevations in the Sierra la Laguna during the winter. I observed only one adult male at La Laguna in late December. *Petrosaurus thalassinus* basks in the morning, and it is common to see many individuals on the same boulder or cliff face. I have observed individuals out at night in April and June at San Bartolo and during August in Boca de la Sierra. I have even seen active adults in the Sierra la Victoria during light rain with fog.

The food habits of *P. thalassinus* have not previously been studied. It is omnivorous, as is its closest relative, *P. repens.* I examined several scats at San Bartolo and found them to contain approximately 50 percent grasshopper and beetle remains and 50 percent plant material, including large seeds. I have found leaves, flowers, seeds, and various arthropods in the stomachs of museum specimens. At Rancho Ancón, I observed an individual eating the leaves of Coralvine *(Antigonon leptopus)* and cockroaches. I observed an adult male (SVL 130 mm) on Isla Espíritu Santo chase down, capture, kill, and eat a subadult male (SVL 63 mm) that climbed onto the boulder on which he was basking. (The lizard was captured, forced to regurgitate the subadult so that measurements could be made, and subsequently released.) Asplund (1967) speculated that this species mates before midsummer. I have examined gravid females collected from mid-April to early June and seen hatchlings from mid-July to mid-August at San Bartolo. From early December through January, I have found juveniles but no hatchlings, supporting Asplund's speculations that *P. thalassinus* has a spring to early summer breeding season.

REMARKS Despite the popularity of this handsome lizard in the illegal pet trade (Mellink 1995), due primarily to one notorious commercial collector in Baja California, it remains in great need of ecological study.

Phrynosoma Wiegmann

HORNED LIZARDS

The genus *Phrynosoma* consists of a group of small to medium-sized, morphologically specialized lizards. It contains about 13 species that range collectively from the western and central United States south to Guatemala. Commonly referred to as horned lizards, *Phrynosoma* are generally slow runners with wide, flat, spiny bodies, and they rely on cryptic coloration and behavior to escape detection. Most horned lizards feed almost exclusively on ants, and many have the peculiar habit of squirting blood from the corners of their eyes when threatened. Despite the extreme morphological adaptations of this group, *Phrynosoma* occupies a wide variety of terrestrial habitats, ranging from barren sand dunes to dry tropical forests. *Phrynosoma* is the basal lineage of a larger group of sand lizards that includes fringe-toed lizards *(Uma),* zebra-tailed lizards *(Callisaurus),* and earless lizards (*Holbrookia* and *Cophosaurus)* (de Queiroz 1992; Reeder and Wiens 1996). Four species of *Phrynosoma* occur in the region of study, one of which *(P. coronatum)* is transpeninsular and is also known from the Pacific island of Cedros. Two others *(P. platyrhinos* and *P. mcallii)* occur in northeastern Baja California and another *(P. solare)* on Isla Tiburón in the Gulf of California. *Phrynosoma* have a number of unique adaptations for their specialized lifestyle, the most obvious of which are the spines covering the head, body, and tail.

FIGS. 4.29–30

Phrynosoma coronatum

COAST HORNED LIZARD; CAMALEÓN

Agama (Phrynosoma) coronata Blainville 1835a:284. Type locality: "Californiae"; restricted to "Cape San Lucas, Baja California, Mexico" by Smith and Taylor (1950:322). Holotype: MNHP 1921

IDENTIFICATION *Phrynosoma coronatum* can be distinguished from all other lizards in the region of study by a wide, flat body covered with spines; occipital spines not contacting at their bases; and six to eight longitudinal rows of greatly enlarged gular scales.

RELATIONSHIPS AND TAXONOMY *Phrynosoma coronatum* is the sister species of a natural group containing *P. cornutum, P. solare, P. mcallii, P. platyrhinos,* and *P. modestum* (Montanucci 1987; Reeder and Montanucci 2001) that collectively ranges throughout much of the American southwest and northern Mexico. Reeve 1952, Grismer and Mellink 1994, and Brattstrom 1997 discuss taxonomy.

DISTRIBUTION *Phrynosoma coronatum* ranges west of the Sierra Nevada crest from Shasta County, California, south through all of southern California west of the Mojave and Sonoran deserts and all of Baja California with the exception of the Lower Colorado Valley Region of the northeast (Jennings 1988a; map 4.10). It is also known from Isla Cedros off the west coast of central Baja California (Grismer and Mellink 1994; Jennings 1988b).

DESCRIPTION Relatively small, with adults reaching 107 mm SVL; head small and truncate in lateral profile; body and tail dorsoventrally compressed; scales on top of head flat to convex and smooth to weakly rugose; frontonasals undifferentiated; large, laterally protuberant superciliary ridge with a posterior spine; 5 superciliaries; 2 long, ridged, and occasionally curved occipital spines separated by much shorter interoccipital spine; 3 to 5 temporal spines, increasing in size posteriorly; bases of temporal spines not in contact; tympanum visible and free of scaly integument; 7 to 9 supralabials; postmental usually single; 4 or 5 large, spinose chinshields, increasing in size posteriorly; 3 to 5 longitudinal rows of enlarged gulars on each side of gular region, lateral rows largest; postrictal spine present, reduced, or absent; subrictal spine present; dorsal body scales and caudals keeled, subimbricate to juxtaposed, irregularly sized and with larger keeled, spinose scales dispersed among them in more or less distinct longitudinal rows; 2 rows of small spines in lateral body fringe, ventral row reduced; forelimbs stout, moderate in length; brachium covered with enlarged keeled, spinose scales, smaller spinose scales on forearm; hind limbs moderate in girth and approximately

FIGURE 4.29 Coronatum pattern class of *Phrynosoma coronatum* (Coast Horned Lizard) from Santiago, BCS

same length as forelimbs; large keeled, widely spaced spinose scales on thigh and foreleg; single row of widely spaced scales along edge of tail; ventral scales smooth to slightly keeled anteriorly and imbricate throughout; 28 to 48 femoral pores, most developed in males; femoral pore series separated medially; enlarged postanal scales in males.

COLORATION Extremely variable (see geographic variation section). Dorsal ground color dull white, light gray, dark gray, yellowish, light brown or brick red; occipital spines usually dark brown or black; large pair of gray to black nuchal blotches followed by 3 or 4 smaller, sometimes less distinctive dark blotches on sides that are often bordered posteriorly by lighter colors; blotches unite in pelvic region and continue down tail as 5 to 7 poorly defined bands; limbs weakly banded to mottled, banding more distinctive on hind limbs; keels on dorsal body spines usually black; ventrum white to cream-colored with small dark punctations; orange or yellowish coloration often present midventrally. Juveniles tend to be more colorfully marked.

GEOGRAPHIC VARIATION The geographic variation in *P. coronatum* is extensive and, for the most part, correlates well with the distribution of the phytogeographic regions. Four different subspecies in the region of study have been recognized (Reeve 1952) and are used here as pattern classes. Coronatum are restricted to the Cape Region of Baja California (see map 4.10). They are generally the most vividly marked and least variable in color pattern of all the pattern classes. Coronatum have five laterally projecting temporal spines, and the postrictal scales are absent to moderately developed. The subrictal scales vary in position from directly in line with the chinshields to slightly above them. The interoccipital spine is relatively long, and the parietal region is elongate. There are five enlarged longitudinal rows of gular scales on each side and 30 to 34 femoral pores. The dorsal ground color varies from light gray to medium brown. The nuchal blotches are often large enough to cover the neck, or may be reduced in size to appear as arcuate bands reaching the brachium. The frontal scales are rugose, black (less vivid in juveniles), and outlined with a thin white line, making the top of the head look like a tiled floor. The nape and body blotches are edged posteriorly in white, setting them off from the rest of the body. There is also a tendency for the lower sides to be heavily pigmented, and in some individuals they are solid black.

Coronatum intergrade with Jamesi across the Isthmus of La Paz (Grismer 1994c), and the latter extend north to the vicinity of El Rosario (see map 4.10). The postrictal scales in Jamesi are moderately developed and increase in size from south to north. There are usually four (sometimes five) laterally or anteriorly projecting temporal spines. The fourth from the rear is often reduced, leaving a gap in the continuity. The interoccipital spines vary in size, with Gulf coast populations having somewhat longer spines than Coronatum. The elongate parietal region in the Gulf coast population is similar to that seen in Coronatum. In Jamesi, the subrictal scales are situated slightly above the row of chinshields. There are four enlarged longitudinal rows of gular scales on each side and 32 to 35 femoral pores. The overall dorsal pattern of Jamesi is generally less contrasting than that of Coronatum, mostly because of a reduction in, or absence of, the white countershade markings posterior to the dark dorsal body blotches. The frontal scales, which vary in size, are slightly rugose to weakly conical, and black (less vivid in juveniles). In most of the southern populations and those along the Gulf coast, they are not edged by thin white lines. This coloration makes lizards appear to have a large black blotch on the top of the head. In populations from the Sierra la Libertad along the eastern edge of the Vizcaíno Desert, many individuals have outlined frontal scales, as in Coronatum, and a generally more contrasted dorsal pattern than adjacent populations from the Vizcaíno Desert. There is a considerable degree of interpopulational color variation among lizards in Jamesi that is correlated to substrate matching. Horned lizards from light-colored sandy areas in the Magdalena Plain have a nearly white ground color, whereas those from the Vizcaíno Desert farther north, where the sand is light brown, are generally straw-colored or in some cases yellow. Horned lizards from within the granitic areas of the Sierra la Libertad are relatively dark and heavily pigmented with dark punctations. One population from the top of a reddish-colored volcanic mesa at Rancho San Juan de Dios has a brick-red dorsal ground color.

Cerroense from Isla Cedros were originally considered a separate species, *P. cerroense* (Stejneger 1893b). Grismer and Mellink 1994 placed them in the synonymy of *P. coronatum* because of the lack of discrete diagnostic characteristics. Cerroense are very similar to Jamesi of the adjacent peninsula (Reeve 1952), differing only in having four rather than four to five chinshields, a generally larger space between the last chinshield and the subrictal, and a reduced fourth temporal spine.

Jamesi intergrade with Schmidti in the vicinity of El Rosario (Reeve 1952), and the latter continues north to the vicinity of Ensenada (see map 4.10). Schmidti have well-developed postrictal scales, and the subrictals are above the chinshields. There are usually five laterally projecting temporal spines. The fourth spine is often reduced, and the bases of the third and fifth may contact one another. The interoccipital spine and parietal regions are relatively short. There are three enlarged longitudinal rows of gular scales and 32 to 34 femoral pores. The overall dorsal pattern of Schmidti is generally darker than that of Jamesi because of the darker substrate on which they occur. There is very little white countershading of the dark dorsal body blotches. The frontal scales are small, convex, and rugose in popu-

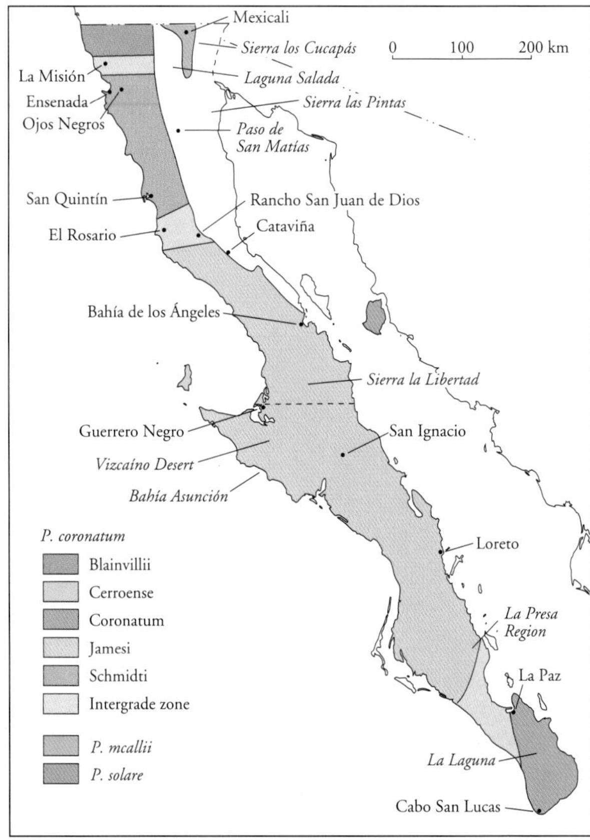

MAP 4.10 Distribution of the pattern classes of *Phrynosoma coronatum* (Coast Horned Lizard), *P. mcallii* (Flat-Tailed Horned Lizard), and *P. solare* (Regal Horned Lizard)

lations from El Rosario to San Quintín and grade into larger, smooth scales in the north, similar to those found in Blainvillii. The frontal scales also lack black pigmentation. There is a reasonable degree of interpopulational color pattern variation within this pattern class which, as in the others, is correlated with substrate matching. Horned lizards that extend down to the eastern base of the Sierra Juárez have a much lighter dorsal ground color that matches the lighter granitic substrate of that area. In contrast, some horned lizards from the mountainous areas east of Ojos Negros in Valle la Trinidad have an unmottled, dark slate gray ground color and a heavily pigmented ventrum.

Schmidti intergrade with Blainvillii in the vicinity of La Misión, between Ensenada and Tijuana (Reeve 1952), and lizards of the two pattern classes are nearly indistinguishable. Blainvillii range north into the Los Angeles Basin of southern California (see map 4.10). The frontal scales are convex and devoid of black pigmentation. The postrictal scale is well-developed, and the subrictal is above the chinshields. There are usually five posteriorly projecting temporal spines, and the interoccipital spine and parietal regions are relatively short. The overall ground color of Blainvillii from Baja California is generally straw-colored except for some darker individuals in the vicinity of Neji. There is little to no white countershading of the dark dorsal body blotches, but in some specimens the posterior edges of the dorsal blotches are very well-defined, much like those of Coronatum but unlike those of most Jamesi and southerly populations of Schmidti.

FIGURE 4.30 Schmidti pattern class of *Phrynosoma coronatum* (Coast Horned Lizard) from Arroyo Santo Domingo, BC

NATURAL HISTORY *Phrynosoma coronatum* is distributed through a wide range of climates and habitats from the California Region in the north, south through the Vizcaíno, Central Gulf Coast, Magdalena, and Arid Tropical regions. Within these varied phytogeographic regions, *P. coronatum* is generally restricted to locally open areas with loose soil, in which it burrows. It is common along the edges of sandy arroyo bottoms and is often seen perched on small rocks along roadsides in sandy areas. Because of its somewhat restrictive habitat requirements, its abundance in the various phytogeographic regions is localized. It is more common in the sandy, open Vizcaíno Desert and Magdalena Plain on the west coast than in the more thickly vegetated California and Arid Tropical regions or the rocky Central Gulf Coast Region. In these latter areas, it is most commonly observed along dirt roads and arroyos. Alvarez et al. (1988) noted that it is absent from the Sierra la Laguna Region.

Information on the natural history of California populations of *P. coronatum* has accumulated over the years (cf. Heath 1965; Hager 1992; Hager and Brattstrom 1997; Jennings 1988a), but very little has been published on its natural history in Baja California. In the north, Blainvillii are active relatively early in the year, and I have observed adults out in late February at Misión Santo Domingo and as late in the year as early October east of Ensenada. These observations correlate with the activity patterns observed by Hager and Brattstrom (1997) in southern California. Hager and Brattstrom also indicated that the activity peak for adults was from April through July and that hatchlings first appeared during July and remained active until November. Much the same is probably true for the Baja California populations. By early April through May, Jamesi are abundant in the vicinity of Bahía de los Ángeles and other northern areas south through the Vizcaíno Desert. During this period, individuals are commonly seen on Highway 1 in the vicinity of Guerrero Negro as well as on the dirt roads in the surrounding area. Horned lizards are seen less frequently as the year progresses, and activity diminishes sharply from summer through fall. In southern Baja California, Coronatum are active year-round, with minimal activity during the winter months, and are abundant until early November. Van Denburgh (1895) reported individuals active November through January in the Cape Region, Leviton and Banta (1964) reported on active juveniles in the Cape Region in mid-January, and Yarrow (1882b) reported Cape Region activity in February.

Phrynosoma coronatum emerges in the morning to bask and is often found in the open, away from vegetation. Horned lizards are generally approachable and usually prefer to lie flat and remain still, relying on their cryptic coloration to avoid detection. Occasionally, however, individuals run for the cover of nearby vegetation. Tevis (1944) reported on a specimen near El Rosario that ran into a hole beneath an agave. *Phrynosoma coronatum* feeds primarily on ants, but other arthropods are taken less frequently. I have observed gravid females during mid-March from Bahía de los Ángeles, during early May near Laguna Hanson, and during mid-June at Bahía Asunción. I observed a wild-caught female from La Paz lay 12 eggs during mid-August and have found hatchlings in the Cape Region during late August. This may suggest that populations from Bahía de los Ángeles and north have a breeding season that extends from March through May and that the breeding season of the more southerly populations is somewhat later. There are unconfirmed reports (V. Velazquez, personal communication, 1988) that some Cape Region populations have live birth and that fully formed hatchlings emerge from the female in a transparent egg sac.

REMARKS As with other species of *Phrynosoma, P. coronatum* may squirt blood from the membranes surrounding the eyes when threatened. Middendorf and Sherbrooke (1992) demonstrated that this behavior is exhibited mostly in response to canid predators. Curiously, 5 out of 11 *P. coronatum* I have collected in the foothills of the Sierra la Libertad have squirted blood, but I have been squirted by only one other horned lizard in Baja California in over 20 years of observations.

FOLKLORE AND HUMAN USE Most residents of Baja California believe that *Phrynosoma* is dangerous and that its spines will burn human skin upon contact.

FIG. 4.31 ***Phrynosoma mcallii***

FLAT-TAILED HORNED LIZARD; CAMALEÓN

Anota M'callii Hallowell 1852:182. Type locality: "Great Desert of the Colorado, between Vallicita [Vallecita] and Camp Yuma, about 160 miles east of San Diego"; restricted to "close to the present town of Caléxico," Imperial County, California, USA, by Klauber (1932:100). Holotype: ANSP 8680

IDENTIFICATION *Phrynosoma mcallii* can be distinguished from all other lizards in the region of study by a wide, flat body covered with spines; a tympanum usually concealed by scales; and a prominent, narrow black vertebral stripe.

RELATIONSHIPS AND TAXONOMY Montanucci (1987) placed *Phrynosoma mcallii* as the sister species of a natural group composed of *P. platyrhinos* of western United States and northwestern Mexico and *P. modestum* of the south-central United States and northern Mexico. Reeder and Montanucci (2001) place it as the sister species of *P. platyrhinos.*

DISTRIBUTION *Phrynosoma mcallii* ranges throughout southeastern California, southwestern Arizona, northeastern Baja California, and northwestern Sonora, Mexico (Funk 1981). In Baja California, it extends at least 80 km south of

FIGURE 4.31 *Phrynosoma mcallii* (Flat-Tailed Horned Lizard) from Laguna Salada, BC

the U.S.-Mexican border along the Río Hardy. It is likely, however, that it extends farther south through the Laguna Salada basin into a sandy valley formed between the Sierra las Pintas to the west and Sierra las Tinajas to the east (see map 4.10).

DESCRIPTION Relatively small, with adults reaching 73 mm SVL; head small and truncate in lateral profile; body and tail dorsoventrally compressed; scales on top of head slightly enlarged and weakly conical; frontonasals undifferentiated; laterally protuberant, spinose superciliary ridge; 4 to 6 superciliaries; 2 relatively long occipital spines and 3 long temporal spines on each side, forming a continuous ring of spines around the back of head; tympanum usually concealed by scaly integument; 8 to 10 supralabials; enlarged, spinose lateral series of scales in gular region; dorsal scales small, imbricate, and varying in size; 7 or 8 longitudinal rows of larger keeled, spinose scales dispersed among them, of which 4 extend onto tail; largest spinose body scales occur in 3 paravertebral pairs; ventral scales smooth; 2 (sometimes 3) lateral abdominal spinose scale rows; forelimbs relatively long and slender; brachium covered with enlarged keeled, spinose scales, smaller spinose scales on forearm; hind limbs moderate in length and relatively slender; large keeled, spinose scales on thigh, and smaller spinose scales on foreleg; 2 distinct rows of enlarged spinose scales on dorsum of tail; single, conical row of closely spaced scales along edge of tail; 34 to 48 femoral pores, most developed in males; pores continue medially onto body; enlarged postanal scales in males. **COLORATION** Ground color white to ashy gray; distinct thin, dark vertebral line from head to base of tail; dark elongate patches on side of neck, often continuous with generally darker ground color on sides of body; weak dark speckling on top of head and occipital and temporal spines; spinose chinshields often white and conspicuous; several pairs of dark spots on back usually bordered posteriorly in white, especially on tail; each dark spot covering a single enlarged keeled, spinose scale; several pairs of faint spots on tail uniting into transverse bands posteriorly; ventral surfaces immaculate white to cream.

NATURAL HISTORY *Phrynosoma mcallii* is most abundant in the sandy, flat washes within creosote bush scrub communities of the Lower Colorado Valley Region, but it is by no means restricted to such areas and can be observed less frequently in similar habitats outside washes. It is relatively common along the northeastern edge of Laguna Salada near the western base of the Sierra los Cucapás. Here it occurs within the sandy, open areas between Salt Cedar trees (*Tamarisk* sp.). It can also be found perched on low tree stumps along the edge of the dry lake.

Much is known about the natural history of this species in the United States (Mayhew 1965; Turner and Medica 1982), and, given its circumscribed distribution in the vicinity of the head of the Gulf of California, much of what has been reported for U.S. populations undoubtedly applies to those of northeastern Baja California. *Phrynosoma mcallii* is active from early spring through fall and hibernates during the winter months (Mayhew 1965). It is diurnal, emerging very early in the morning to bask, and may remain active into the early evening. Much of the time it occurs on the ground, where its cryptic coloration makes it very difficult to detect. Less frequently, however, it can be found by scanning the tops of small rocks or dead stumps of vegetation, on which it occasionally basks. In contrast to other horned lizards, *P. mcallii* has long, slender limbs that enable it to run quite fast. It spends much of the day in the vicinity of the anthills where it feeds. Pianka and Parker (1975) noted that ants made up 78 percent (by volume) of this lizard's diet. Reproduction occurs in the spring, and one to two clutches of 7 to 10 eggs are laid in May and June (Stebbins 1985). Hatching occurs in late summer.

REMARKS Several populations of *P. mcallii* have become extirpated in southern California because of habitat destruction as a result of housing construction and agriculture (Turner and Medica 1982). It is conceivable that the remote, undeveloped areas of northeastern Baja California and northwestern Sonora may become strongholds for this species.

FIG. 4.32 *Phrynosoma platyrhinos*

DESERT HORNED LIZARD; CAMALEÓN

Phrynosoma platyrhinos Girard 1852 in Baird and Girard 1852b:361. Type locality: "Great Salt Lake," Utah, USA. Syntypes: USNM 189 (three adults; one male, two unsexed) and MCZ 5948

FIGURE 4.32 *Phrynosoma platyrhinos* (Desert Horned Lizard) from Bahía San Luis Gonzaga, BC

IDENTIFICATION *Phrynosoma platyrhinos* can be separated from all other lizards in the region of study by a flat body covered with spines; occipital spines that are not in contact at their bases; one slightly enlarged row of longitudinal gular scales; and no narrow black vertebral line.

RELATIONSHIPS AND TAXONOMY See *P. mcallii.* Reeve (1952) and Pianka (1991) discuss taxonomy.

DISTRIBUTION *Phrynosoma platyrhinos* ranges from southeastern Oregon and southern Idaho south through Nevada, western Utah, southeastern California, western Arizona, northwestern Sonora, and northeastern Baja California (Pianka 1991). In Baja California, it ranges east of the Sierra Juárez and Sierra San Pedro Mártir, extending as far south as Bahía San Luis Gonzaga (map 4.11). It may be syntopic with *P. coronatum* in Paso de San Matías and northwest of the Sierra la Asamblea, and extends as far south as Bahía de los Ángeles.

DESCRIPTION Relatively small, with adults reaching 85 mm SVL; head small and truncate in lateral profile; body and tail dorsoventrally compressed; scales on top of head slightly enlarged and weakly convex; frontonasals undifferentiated; large, laterally protuberant, spinose superciliary ridge; 4 to 6 superciliaries; 2 long, stout, parallel occipital spines; 5 or 6 temporal spines, the posteriormost largest; tympanum free of scaly integument, sometimes hidden by fold of skin; 9 supralabials; postmental single; 7 to 9 large spinose chinshields; gulars flat and small, with slightly enlarged spinose series of lateral scales; 2 lateral neck patches with enlarged scales; dorsal scales of body and tail flat, subimbricate to juxtaposed, and irregularly sized, with larger keeled, spinose scales dispersed among them in more or less distinct longitudinal rows; largest spinose scales generally occurring in single paravertebral rows; 1 row of small spines in lateral fringe; forelimbs stout, moderate in length; brachium covered with enlarged keeled, spinose scales, smaller spinose scales on forearm; hind limbs moderate in girth and approximately same length as forelimbs; large keeled, spinose scales on thighs, smaller spinose scales on forelegs; single conical row of widely spaced scales along edge of tail; ventral scales smooth and imbricate throughout; 14 to 16 femoral pores, most developed in males; pores continue medially onto body; enlarged postanal scales present in males. **COLORATION** Color pattern extremely variable. Ground color light gray to tan; large pair of gray to black nuchal blotches; usually three wide, gray to black dorsal blotches forming indistinct transverse bands; six dark, irregular bands on tail; limbs weakly banded; ventrum cream-colored; large, dark precloacal spots usually present; similar spots often on posteroventral region of femora.

GEOGRAPHIC VARIATION Geographic variation in *P. platyrhinos* centers on its remarkable ability to match the substrate. Lizards from the light-colored, sandy areas in Laguna Salada are nearly white. On the creosote flats between El Huerfanito and Bahía San Luis Gonzaga, the banding pattern is indistinct, and most individuals have a dark, reticulated dorsal pattern. Along the dark volcanic rocks between Puertecitos and El Huerfanito, individuals have a bold, dark, mottled pattern tending to form transverse bands, overlying a reddish ground color. I have observed light purple lizards completely lacking banding along the western foothills of the Sierra las Tinajas in areas where the substrate has a maroon tint.

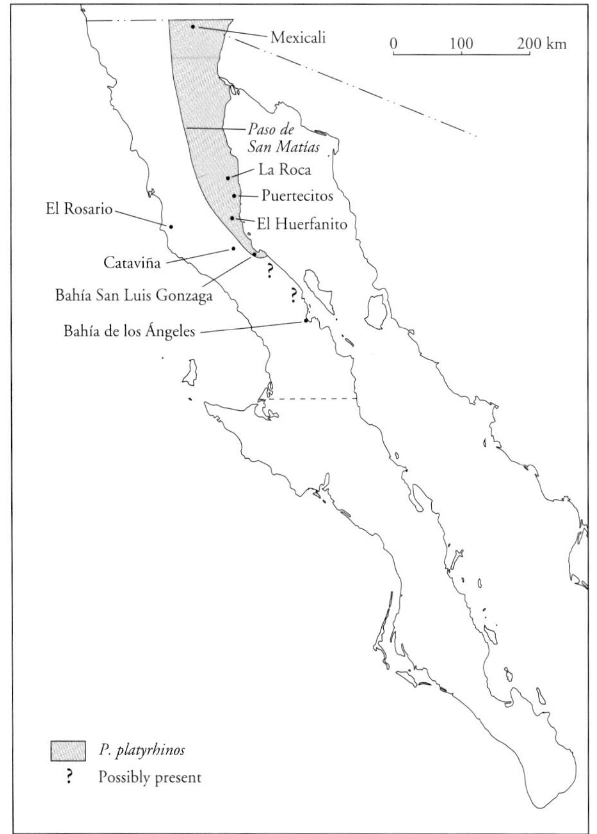

MAP 4.11 Distribution of *Phrynosoma platyrhinos* (Desert Horned Lizard)

NATURAL HISTORY In Baja California, *P. platyrhinos* is found only within the Lower Colorado Valley Region. Its preferred habitat seems to be sandy areas with small scattered rocks. It is most commonly observed in sandy washes coursing through rocky arroyos. It can also be found in bajadas where small rocks are present. *Phrynosoma platyrhinos* is active from early spring through early fall. I observed 16 individuals on two mornings in early April between Campo Cinco Islas and Bahía San Luis Gonzaga along the Gulf coast. Lizards emerge in the morning to bask and are commonly observed sitting on the tops of small rocks or on roadside berms. During the early part of its activity season, *P. platyrhinos* remains active all day, and I have observed individuals at dusk in late May west of La Roca. This species feeds primarily on ants but will also eat small arthropods. I have found gravid females in May, which suggests a spring breeding, spring and summer egg-laying, and summer hatching cycle.

FIG. 4.33 ***Phrynosoma solare***

REGAL HORNED LIZARD; CAMALEÓN

Phrynosoma solaris Gray 1845:229. Type locality: "California"; restricted to Tucson [Pima County], Arizona [USA] by Schmidt (1953:136). Holotype: BM 55111.125.d

IDENTIFICATION *Phrynosoma solare* can be distinguished from all other lizards in the region of study by a wide, flat body covered with spines and four occipital spines that connect at their bases to form a crown.

RELATIONSHIPS AND TAXONOMY Montanucci (1987) placed *Phrynosoma solare* as most closely related to a natural group containing *P. mcallii, P. platyrhinos,* and *P. modestum.* Reeder and Montanucci (2001) place it as the sister species to a natural group containing *P. platyrhinos* and *P. mcallii.* Reeve (1952) discusses taxonomy.

DISTRIBUTION *Phrynosoma solare* ranges from central Arizona south through central Sonora to northern Sinaloa and portions of western Chihuahua, Mexico. It also occurs on Isla Tiburón off the coast of Sonora in the Gulf of California (Parker 1974; see map 4.10).

DESCRIPTION Same as *P. platyrhinos,* except for the following: adults reach 117 mm SVL; spines on abdominal fringe short, long, interrupted, or absent, and a second, shorter

FIGURE 4.33 *Phrynosoma solare* (Regal Horned Lizard) from Ensenada de los Perros, Isla Tiburón, Sonora

row of much smaller spines below; hind limbs relatively small, moderate in girth, and slightly shorter than forelimbs; abdominal scales smooth; scales of anterior pectoral region weakly keeled and imbricate; 34 to 38 femoral pores.

COLORATION Color pattern variable. Dorsal ground color grayish-tan to reddish; points of semiconical head scales often darkly colored; large pair of elongate, gray to black nuchal blotches usually extending posteriorly along sides; dark paravertebral spots variable, often appearing as three undulate, transverse bands; six dark, irregular bands on tail; limbs weakly banded; ventrum cream-colored throughout, with widely spaced small, dark abdominal spots.

NATURAL HISTORY On Isla Tiburón, *P. solare* is most often found in gravelly areas along washes and foothills in association with the scrub vegetation of the Central Gulf Coast Region. It emerges relatively early from hibernation. I have observed adults and juveniles active from mid-March through early July in the vicinity of Bahía Kino and Punta Chueca, just opposite the coast of Isla Tiburón. However, activity extends at least into late October. Baharav (1975) showed that in southern Arizona, adults are most active from April to June, and hatchlings and juveniles are most active from October to November. Lizards emerge in the morning to bask and are commonly seen in open areas. *Phrynosoma solare* feeds primarily on Harvester Ants *(Pogonomyrmex rugosus)* and generally does not travel far from either the ant trails or the openings of their mounds (Baharav 1975). Stebbins (1985) reports that 7 to 28 eggs are laid in July through August, which suggests a late spring to early summer breeding season.

FOLKLORE AND HUMAN USE Malkin (1962) noted that the lizard classification of the Seri Indians is remarkably consistent with that of contemporary lizard systematists, except with regard to the horned lizard. The Seri believe the horned lizard is not a lizard but the undeveloped juvenile of an entirely different kind of animal. Malkin was unable to determine what that animal was. He was told that this species was viviparous and had a scrotum. The majority of the Seri believe that *P. solare* has a venomous bite that causes swelling but not death, although Seri children often keep them as pets (Rosenberg 1997). Nabhan (forthcoming) was told that, when aggravated, the horned lizard shoots a fluid from the spines crowning the head into the joints of the assailant. If a Seri individual unintentionally disturbs a horned lizard, his or her companions must immediately chant "Haxáx coquépe!" (Ay, let us comfort you!), so that the animal does not become offended or frightened (Nabhan, forthcoming). The local Mexican inhabitants believe it is deadly.

Sceloporus Wiegmann

SPINY LIZARDS

The genus *Sceloporus* is composed of a group of small to medium-sized, insectivorous, scansorial lizards occurring in a wide variety of habitats and representing a broad array of adaptive types. *Sceloporus* contains at least 84 living species (Wiens and Reeder 1997), which collectively range from southern Canada south to central Panama. *Sceloporus* are a common component of many ecosystems and for the most part are conspicuous, diurnal ambush feeders that are commonly seen basking and displaying in plain sight on elevated perches. Sexual dichromatism is marked in many species, and adult males of most species have brightly colored abdominal patches used in sexual discrimination. All species are easily recognized by their keeled spinose scales. There are 10 species of *Sceloporus* in the region of study, and some of these form endemic species complexes (Wiens and Reeder 1997). A basal complex comprises the insular endemics, *S. angustus* of Islas San Diego and Santa Cruz and *S. grandaevus* of Isla Cerralvo. There are also two peninsular complexes: the *S. orcutti* complex, containing the transpeninsular *S. orcutti* and the Cape Region endemics *S. licki* and *S. hunsakeri;* and the *S. magister* complex, containing *S. magister,* the endemic *S. lineatulus* of Isla Santa Catalina, and the transpeninsular *S. zosteromus.*

FIGS. 4.34–35

Sceloporus angustus

ISLA SANTA CRUZ SPINY LIZARD; CACHORA

Sator angustus Dickerson 1919:469. Type locality: "Santa Cruz Island, Gulf of California, Mexico." Holotype: AMNH 5712

IDENTIFICATION *Sceloporus angustus* can be distinguished from all other lizards in the region of study by its lack of an axillary spot; keeled, spinose body scales; and small flank

FIGURE 4.34 Adult male *Sceloporus angustus* (Isla Santa Cruz Spiny Lizard) from near the lighthouse on Isla Santa Cruz, BCS

scales that blend smoothly into much larger dorsal body scales.

RELATIONSHIPS AND TAXONOMY *Sceloporus angustus* is the sister species of *S. grandaevus,* and together they form a basal lineage within *Sceloporus* (Wiens and Reeder 1997).

DISTRIBUTION *Sceloporus angustus* is known only from Islas San Diego and Santa Cruz in the Gulf of California (map 4.12).

DESCRIPTION Moderately sized, with adults reaching 87 mm SVL; body stout; head triangular, snout pointed in dorsal profile; head scales large and platelike; frontals usually fused; temporals small, weakly keeled to rounded; enlarged interparietal scale with conspicuous eye-spot; abrupt transition between occipitals and dorsals; body somewhat laterally compressed and with slight vertebral ridge; dorsals imbricate, keeled, and occurring in parallel rows; dorsals grade smoothly into much smaller, keeled flank scales, which grade medially into flat, imbricate ventrals; lateral body fold between limbs; 5 to 7 elongate supralabials; 2 postmentals; gulars smooth and imbricate; distinct antegular fold; gular fold incomplete medially but present laterally; 6 or 7 elongate infralabials; forelimbs thin and relatively long; humerals and forelimb scales keeled dorsally and imbricate ventrally as well as keeled distally; hind limbs thicker than forelimbs and slightly less than twice as long; femorals and tibials keeled dorsally and imbricate ventrally; postfemorals small, and granular to convex; 11 to 17 femoral pores; 29 to 33 subdigital lamellae on fourth toe; tail approximately 1.6 times SVL and laterally compressed; caudal scales keeled, spinose, and whorled; no enlarged postanal scales in males. **COLORATION** ADULT MALES: Dorsal ground color of head, body, and limbs dark brown to nearly black; dorsal ground color of tail light brown to brown; network of irregularly sized cream-colored spots covering dorsal surfaces on head and body, occasionally forming broken, poorly defined crossbars; no ocelli on sides; distinct black oval-shaped blotch on shoulder bordered posteriorly by thin, cream-white to yellow lines not extending onto brachium or across back; dark oblique lines on gular region; thick dark crossbars on sides of abdomen; light bands on limbs. FEMALES AND JUVENILES: Dorsal ground color of body dark brown with central row of light-colored spots or a wide, light-colored line; a similar line is present immediately dorsal to lateral body fold; light bands on tail and limbs; top and sides of head pink to orange in adult females, most intense during reproductive season.

GEOGRAPHIC VARIATION There is no geographic variation between the Isla Santa Cruz and Isla San Diego populations.

NATURAL HISTORY Islas Santa Cruz and San Diego are within the Central Gulf Coast Region and, as such, are dominated by desert scrub and arid climates. On both islands, *S. angustus* is ubiquitous and can be found on rocks, within rocky caves along the beach, in beach cobble, on the branches of small bushes, on the limbs of large Cardón Cacti *(Pachyce-*

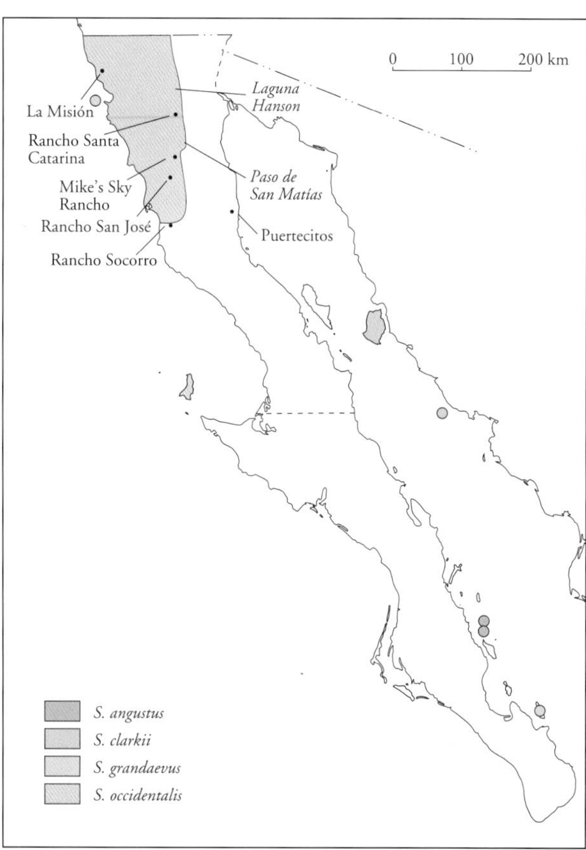

MAP 4.12 Distribution of *Sceloporus angustus* (Isla Santa Cruz Spiny Lizard), *S. grandaevus* (Isla Cerralvo Spiny Lizard), *S. clarkii* (Clark's Spiny Lizard), and *S. occidentalis* (Western Fence Lizard)

reus pringlei), and even in sandy areas along the beach. It is as abundant inland, climbing several meters up the vertical rock faces in some of the larger and steeper arroyos, as it is along the coast. Lizards are probably active year-round, with peak activity from March through September or October. During the day, *S. angustus* can be observed basking on several substrates, including turtle shells and the carcasses of sea lions. I observed an adult male 2 m above the ground on top of a Cardón Cactus, basking like a spiny-tailed iguana *(Ctenosaura).* During June and July, it appears that larger rocks and cardón trunks are preferred as basking and perching sites by only the largest males. Later in the year, after breeding has taken place, females are also found in such areas. I observed individuals abroad at night during early July and September in the center of an arroyo bottom a great distance from cover.

Joseph R. Slevin (in Van Denburgh 1922:258) stated that he saw several specimens along the beach feeding on the flies attracted to dead turtles and sharks left by Mexican fishermen. In a rocky canyon, he observed one individual about 5 m above the ground on top of a Cardón Cactus. It was feeding on flies or bees that were attracted to the blossoms. When the lizard sighted Slevin, it immediately turned head downward and pressed itself into one of the deep longitudinal grooves on the trunk of the Cardón, steadied itself by pressing its long tail against the sides of the groove, and rushed down behind the spines. Case (1983) noted that 60 percent of the contents of nine stomachs he examined was plant material. I observed several individuals feeding on large diurnal moths and other arthropods that are common on the islands during early September.

During the reproductive season in late summer and early fall, the crimson-orange to pink coloration of the head and lateral gular region in adult females intensifies. I have seen some gravid females whose entire bodies were bright orange, with the labial and gular regions nearly blood red. Their stomachs were immaculate white, and the two colors met on the ventrolateral area of the body, forming a conspicuous, abrupt union. Generally speaking, gravid females from Isla Santa Cruz are brighter orange than those from Isla San Diego. Observations of this color in the field suggest that the breeding season extends through spring and summer. Museum specimens collected in early September contained five to seven eggs.

It is common to find lizards in pairs or trios (one male, two females), with the females doing as much displaying and head-bobbing as the males. I noted a similar behavior in *S. grandaevus,* the closest relative of *S. angustus* from Isla Cerralvo.

FIGURE 4.35 Gravid female *Sceloporus angustus* (Isla Santa Cruz Spiny Lizard) from near the lighthouse on Isla Santa Cruz, BCS

REMARKS The reason for the ubiquitous habitat preference of *S. angustus* is likely the lack of competition from other *Sceloporus*-group lizards such as *Petrosaurus, Uta, Urosaurus,* and other species of *Sceloporus.* This lack of competition may have allowed *S. angustus* to exploit all the available niches of its insular habitat that on other islands are filled by these other species (Case 1983). This species is in great need of study.

FIGS. 4.36–37

Sceloporus grandaevus

ISLA CERRALVO SPINY LIZARD; CACHORA

Sator grandaevus Dickerson 1919:665. Type locality: "Cerralvo Island, Gulf of California, Mexico." Holotype: AMNH 5491

IDENTIFICATION *Sceloporus grandaevus* can be distinguished from all other lizards in the region of study by its lack of an axillary spot; keeled, spinose, body scales; and small flank scales that are abruptly differentiated from the much larger dorsal body scales.

RELATIONSHIPS AND TAXONOMY See *S. angustus.*

DISTRIBUTION *Sceloporus grandaevus* is endemic to Isla Cerralvo in the Gulf of California (see map 4.12).

DESCRIPTION Same as *S. angustus,* except for the following: adults reaching 81 mm SVL; temporal scales small; upper temporals keeled; body more strongly compressed laterally; vertebral ridge much more distinct; dorsals occurring in distinct parallel rows; dorsals grade abruptly into flank scales; 6 (rarely 7) elongate supralabials; distinct V-shaped antegular fold present; 6 to 8 elongate infralabials; postfemorals small and sharply conical; 14 to 22 femoral pores; 28 to 31 subdigital lamellae on fourth toe. **COLORATION** ADULT MALES: Dorsal ground color of head and body light brown; ground color of sides and limbs dark gray; tail cream-colored dorsally and ventrally; network of light blue to turquoise spots

FIGURE 4.36 Adult male *Sceloporus grandaevus* (Isla Cerralvo Spiny Lizard) from Arroyo Viejos, Isla Cerralvo, BCS

FIGURE 4.37 Gravid female *Sceloporus grandaevus* (Isla Cerralvo Spiny Lizard) from Arroyo Viejos, Isla Cerralvo, BCS

present on vertebral region of neck and body; 3 to 5 poorly defined dark ocelli on flank; distinct black, hourglass-shaped blotch on shoulder, edged posteriorly by thin cream-white to yellow line that extends onto brachium and occasionally makes medial contact with the corresponding line from opposite side; gular region yellow to tan; ventral surface of body and limbs dark gray ADULT FEMALES: Dorsal ground color of body light brown to charcoal gray with broad dorsolateral yellowish stripe; line of light-colored vertebral spots on body sometimes fuses into a complete or broken line; head and tail beige; yellowish bands on tail and hind limbs; forelimbs lightly spotted; ventral surfaces light-colored; abdomen edged laterally in gray; top and sides of head pink to orange, most intense during reproductive season. JUVENILES: Very similar to females and capable of considerable color change; at high body temperatures, the ground color is black with a wide yellowish dorsolateral stripe on each side of the body and a series of yellowish vertebral spots appearing as an incomplete vertebral stripe. At lower body temperatures, the ground color is nearly unicolored tan, and the striping is not as conspicuous.

NATURAL HISTORY Isla Cerralvo, lying within the Central Gulf Coast Region, is dominated by xerophilic scrub and arid climates even though a significant component of its flora is derived from the dry tropic forests of the adjacent peninsular Arid Tropical Region. *Sceloporus grandaevus* is most commonly found in rocky arroyos. Lizards are also abundant on rocks, vegetation, vegetated coastal dunes, and open flats so long as vegetation is nearby. They avoid wide, sandy areas lacking vegetation. When threatened, *S. grandaevus* usually seeks cover deep within nearby vegetation. I watched a fleeing juvenile jump from branch to branch with remarkable agility. *Sceloporus grandaevus* is active year-round, with peak activity occurring from March through October. During the winter, hatchlings and juveniles are the most abundant age class. Although *S. grandaevus* basks and forages during the day, it is commonly observed at night on arroyo bottoms or sleeping on the branches of vegetation. Case (1983) reported that 41 percent of its diet consisted of plant material. It is common to find what are presumably breeding pairs of adults together in the same bush. I have observed gravid females with corresponding coloration from June through October and hatchlings from June through mid-December, which indicates that the reproductive season runs from spring to late fall, with perhaps more than one clutch of eggs being laid in a single season.

REMARKS *Sceloporus angustus* and *S. grandaevus* were originally described as the sole members of the genus *Sator* (Dickerson 1919). As proposed by Dickerson and demonstrated by Wyles and Gorman (1978) and Wiens and Reeder (1997), *Sator* is a lineage that branched out within the genus *Sceloporus*. Therefore, to be consistent with the evolutionary history of *Sceloporus* and its recognition as a natural evolutionary group whose classification logically indicates that its members are each other's closest relatives, Wiens and Reeder (1997) proposed that *Sator* be placed in the synonymy of *Sceloporus*.

FIG. 4.38

Sceloporus clarkii

CLARK'S SPINY LIZARD; BEJORI

Sceloporus clarkii Baird and Girard 1852d:127. Type locality: "province of Sonora," Mexico; restricted to "Santa Rita Mountain" by Smith and Taylor (1950:354). Lectotype: USNM 2940 (Smith 1940).

IDENTIFICATION *Sceloporus clarkii* can be distinguished from all other lizards in the region of study by keeled, spinose scales on the body that are nearly all the same size, and dark bands on the wrists and tops of the hands.

RELATIONSHIPS AND TAXONOMY Larson and Tanner (1975) and Wiens and Reeder (1997) indicate that *Sceloporus clarkii* is the sister species to *S. melanorhinus* of southwestern Mexico. Smith (1939) discusses taxonomy.

DISTRIBUTION *Sceloporus clarkii* ranges from east central Arizona and southwestern New Mexico south to northern Jalisco, Mexico (Sites et al. 1992). In the Gulf of California, it is known from Islas Tiburón and San Pedro Nolasco off the coast of central Sonora (Grismer 1999a; see map 4.12).

DESCRIPTION Moderately sized, with adults reaching 130 mm SVL; head and body stout, not dorsally or laterally compressed; head triangular, somewhat pointed in dorsal profile; scales of top of head large and platelike; 5 large supraoculars, fourth and fifth contacting medial head scales; temporals wide, flat, spinose, imbricate, and weakly keeled; interparietal large, with conspicuous eye-spot; interparietals and parietals abruptly contact dorsal scales; dorsals wide, imbricate, and strongly keeled, with denticulate posterior margins; dorsals grade ventrally into flat, imbricate, denticulate ventrals; 4 to 5 supralabials; 2 postmentals; first sublabial not contacting mental; gulars wide, flat, imbricate, and denticulate; antegular fold absent; gular fold interrupted medially but present laterally; 5 or 6 infralabials; forelimbs stout, moderate in length, covered with keeled spinose, imbricate scales; hind limbs stout, thicker than forelimbs, all but infrafemoral region covered with keeled, imbricate, spinose scales; infrafemorals smooth; 10 to 16 well-developed femoral pores in males; 18 to 21 subdigital lamellae beneath fourth toe; tail approximately 1.25 times length of body, round in cross section; dorsal caudals keeled, spinose, imbricate, and approximately same size as dorsals; subcaudals smooth anteriorly, keeled posteriorly; enlarged postanal scales in males. **COLORATION** ADULT MALES: Dorsal ground color dark brown to gray-brown; indistinct dark collar anterior to forelimbs, narrowly complete dorsally; collar usually bordered dorsoposteriorly by thin, light-colored line; faint undulate banding pattern dorsally; broad reticulum of thin dark lines on sides; light turquoise flecks prominent on sides but generally absent in the vertebral region; turquoise scales on lateral surface of tail; distinct, well-defined black bands on wrists and tops of hands; posteromedial gular region with large blue-green spot bordered by black; gular region with gray lines anteriorly and laterally; abdominal patches turquoise-green laterally, bordered medially by blue and medially again by black; abdominal patches contacting medially; pectoral and pelvic regions and ventral surface of limbs and tail light-colored, often immaculate. FEMALES AND JUVENILES: Very similar to adult males except collar less distinct; dorsal banding pattern more prominent and edged posteriorly in white; light turquoise flecks on body generally absent; dark bands on tail distinct; ventral surfaces white, except in adult females, which have faint bluish abdominal patches; adult females have pale blue gular patch with distinct lines anteriorly; large, light-colored spots on hatchlings.

FIGURE 4.38 Adult male *Sceloporus clarkii* (Clark's Spiny Lizard) from Isla San Pedro Nolasco, Sonora

NATURAL HISTORY Islas Tiburón and San Pedro Nolasco lie within the Central Gulf Coast Region and are dominated by thorn scrub vegetation. Tinkle and Dunham (1986) studied the natural history of *S. clarkii* from central Arizona, but nothing has been reported on the populations from Islas Tiburón and San Pedro Nolasco. On these islands, *S. clarkii* is distinctly scansorial and found more often on rocks and in trees than on the ground. On Isla Tiburón, *S. clarkii* seems to be more prevalent in trees along the arroyo bottoms than on boulders, perhaps because of the presence of *S. magister*, which is commonly seen on boulders. Smith (1939) and Tinkle and Dunham (1986) report similar observations of these two species partitioning the habitat in mainland areas.

Isla San Pedro Nolasco has a large component of thorn scrub vegetation and lacks *S. magister*. Here, *S. clarkii* is very common on rocks as well as in crevices and small caves along the cliffs. Adults are most common in rocky canyons, whereas hatchlings and juveniles are more abundant on talus slopes near vegetation. I have observed only juveniles and subadults during late March on Isla San Pedro Nolasco, which suggests that this population may emerge from hibernation earlier than central Arizona populations (Tinkle and Dunham 1986). Adults on Isla San Pedro Nolasco generally emerge a month or so later. During the summer, *S. clarkii* is seen early in the day basking on rocks. As temperatures begin to rise, lizards usually seek shelter and by midday are found only in shaded areas. At sunset, activity increases sharply once again, and lizards remain active until dark.

Sceloporus clarkii is known to feed on a variety of small arthropods and flower buds (Stebbins 1985), and I suspect that on Isla San Pedro Nolasco it may also feed on the Isla San Pedro Nolasco Side-Blotched Lizard *(Uta nolascensis).* I found a specimen of the latter with bite scars on its body that were too big to have been made by another *U. nolascensis* and too wide to be from the Isla San Pedro Nolasco Whiptail *(Cnemidophorus bacatus).* I observed a juvenile *S. clarkii* jumping from a bush to catch a grasshopper, and I have seen adults eating the fruit of *Mammillaria* cactus. I have observed gravid females during early July and hatchlings during early October on Isla San Pedro Nolasco. These observations would suggest a late spring–early summer breeding season and a late summer–early fall hatching period.

REMARKS Fishermen I have interviewed on Isla San Pedro Nolasco believe that *S. clarkii* is venomous and that its bite or the prick of its spines causes severe itching.

FOLKLORE AND HUMAN USE See the Desert Spiny Lizard *(S. magister).*

FIG. 4.39 *Sceloporus occidentalis*

WESTERN FENCE LIZARD; BEJORI

Sceloporus occidentalis Baird and Girard 1852:175. Type locality: "California," USA; restricted to "Benicia," California, by Grinnell and Camp (1917:159). Neotype (Bell 1954): MVZ 59874

IDENTIFICATION *Sceloporus occidentalis* can be distinguished from all other lizards in the region of study by moderately keeled dorsal body and caudal scales of nearly equal size, with a posteriorly projecting spine; no distinct bands on the forearms; and no distinct black shoulder bar.

RELATIONSHIPS AND TAXONOMY Wiens and Reeder (1997) demonstrated that *Sceloporus occidentalis* has evolved out of what is currently recognized as *S. undulatus,* a broadly distributed species that ranges throughout most of the southern and central United States and northern Mexico. Wiens and Reeder (1997) discuss taxonomy of the problematic *S. undulatus* group.

DISTRIBUTION *Sceloporus occidentalis* ranges throughout the western United States from central Washington and southwestern Idaho south to northwestern Baja California, Mexico (Bell and Price 1996). In Baja California, it ranges throughout the cismontane areas as far south as 15 km east of Rancho Socorro. It is also known from the Pacific islands of Todos Santos, off the coast of Ensenada, and Isla Cedros, off the coast of central Baja California (Grismer 1989a; Grismer and Mellink 1994; see map 4.12).

FIGURE 4.39 Adult male *Sceloporus occidentalis* (Western Fence Lizard) from Rancho Nopalito, BC

DESCRIPTION Small, with adults reaching 83 mm SVL; head and body not dorsally or laterally compressed; head triangular, rounded in dorsal profile; scales on top of head large and platelike; 5 or 6 large supraoculars, fourth though sixth not contacting medial head scales; temporals moderate in size, imbricate, and weakly keeled; interparietal large, with conspicuous parietal eye; interparietal abruptly borders dorsal scales of body; dorsals imbricate, keeled, and spinose; dorsals grade ventrally into flat imbricate, denticulate ventrals; 5 to 7 supralabials; 2 postmentals; first sublabial not contacting mental; gulars flat, imbricate, and denticulate; antegular fold interrupted medially; gular fold absent; 6 or 7 infralabials; forelimbs stout, moderate in length, covered with keeled imbricate, spinose scales dorsally; infrahumerals small, smooth, and imbricate; ventral forearm scales imbricate and keeled distally; hind limbs stout, thicker than forelimbs, covered with keeled imbricate, spinose scales dorsally and smooth, imbricate scales ventrally; postfemorals small, keeled, imbricate, and spinose; 13 to 19 well-developed femoral pores in males; 9 or more scales between femoral pore series; 23 to 27 subdigital lamellae on fourth toe; tail approximately 1.25 times length of body, round in cross section; dorsal caudals keeled, spinose, imbricate, and slightly larger than dorsals; subcaudals weakly keeled; enlarged postanal scales in males.

COLORATION Individuals of this species are capable of considerable color change and sexual dichromatism. When basking, all individuals are nearly solid black. ADULT MALES

(NOT BASKING): Dorsal ground color dark gray to nearly black; head darkly mottled, blue flecks occasionally present; blue to turquoise area in center of vertebral and paravertebral dorsal scales; yellowish spots on flank scales; indistinct row of black paravertebral markings; markings sometimes fuse to form dorsolateral and lateral stripes bordering a central light area; center of dorsal caudals may be turquoise, or only lateral caudals turquoise; dorsal surfaces of limbs dark and indistinctly banded; turquoise occasionally present on femora; postfemoral region yellow; gular region deep blue, may or may not be bordered anteriorly by black; pectoral region may or may not be suffused with black; short black shoulder bar present; abdominal patches blue, medially grading to turquoise laterally, usually bordered medially in black; ventral ground color of limbs, central abdomen, and tail gray; infrahumeral regions yellow. FEMALES AND JUVENILES (NOT BASKING): Dorsal ground color light gray to dark gray; head mottled; dark, distinct postorbital stripe extending irregularly onto flank; paravertebral blotches bordered posteriorly with white to pale blue; blotches occasionally fuse to form stripes; limbs faintly to moderately banded; ground color of ventral surfaces dull white; gular and abdominal patches greatly reduced.

GEOGRAPHIC VARIATION The geographic variation in *Sceloporus occidentalis* is limited primarily to differences in the coloration of adult males; only a general characterization is possible because of the considerable intrapopulational variation. Generally, populations from the higher elevations of the Sierra Juárez, in the vicinity of Laguna Hanson, are darker overall because they lack the extensive blue and turquoise coloring on the dorsum seen in other populations. However, adult males from Río las Flores, just west of Laguna Hanson, are nearly all blue dorsally. Some of the large females from just south of La Misión have tan-colored bands. Adult males from Rancho San José are brilliantly colored, with striking metallic blue abdominal patches and gular regions. Often the ventrum of adult males from Islas Todos Santos is nearly all black. Males from Isla Cedros tend to have more green in their abdominal patches and more yellow on the ventral surfaces of their limbs and chests.

A distinctive color pattern polymorphism exists in *S. occidentalis* whereby the paravertebral blotches are divided and become elongate, fused in an anterior-to-posterior orientation, forming a pair of distinct longitudinal stripes on each side of the body. In males, the vertebral region between the stripes is often blue, and the field between the stripes on the sides of the body is cream-colored. In females, the ground color generally remains the same. Intermediate conditions can be found wherein the blotches have become elongate but not yet fused, forming only dashed lines. This polymorphism is most prevalent in populations from the interior foothill chaparral regions. It has been observed in populations from Japa, just west of Laguna Hanson in the north, south through Rancho Santa Catarina to Mike's Sky Rancho and Rancho San José in the foothills of the Sierra San Pedro Mártir. This lineate pattern has not been observed in the Sierra Juárez or any of the insular populations.

NATURAL HISTORY *Sceloporus occidentalis* is a habitat generalist found within the cismontane coastal sage scrub and chaparral communities of the California Region as well as the piñon woodlands and mixed conifer forests in the Sierra Juárez of the Baja California Coniferous Forest Region. It is very common in riparian habitats in the western foothills of the Sierra San Pedro Mártir and far less so in the California Region–Baja California Coniferous Forest Region ecotone (Welsh 1988). On Isla Cedros, it is known from a narrow belt of chaparral isolated along the upper elevations (ca. 900 m) of the southwestern side of the island (Grismer and Mellink 1994). On Islas Todos Santos, *S. occidentalis* is most abundant among the rocks. At Laguna Hanson in the Sierra Juárez, I have observed *S. occidentalis* to be more saxicolous than in other areas. Lizards are very adept at jumping from rock to rock and from rocks to trees. While running across large boulders, they flatten out their bodies and splay their limbs as do banded rock lizards *(Petrosaurus).* Linsdale (1932) reports collecting *S. occidentalis* from Laguna Hanson, one from a mouse trap set among the rocks and another from a log. Welsh (1988) and Linsdale noted that *S. occidentalis* was "highly arboreal" in the riparian woodland on the lower western foothills of the Sierra San Pedro Mártir.

Cismontane populations of *S. occidentalis* are active year-round, with a minimum of activity during the winter; however, montane populations hibernate. In the spring and summer, lizards can be seen basking on rocks and logs as soon as the sun rises. During this period, lizards often turn jet black and show no signs of a dorsal pattern. On Isla Cedros, I have noticed that lizards perch in a basking location with the proper body orientation prior to the clearing of the daily morning fog. *Sceloporus occidentalis* generally remains active all day and is usually not seen after sunset. Its food consists mainly of small arthropods. Welsh (1988) reported seeing lizards descend to the ground from their basking locations to catch food and then return. I have observed *S. occidentalis* mating in May and June and have found gravid females in late July on Isla Cedros. The earliest I have observed hatchlings is 4 August, at Mike's Sky Rancho. This observation would suggest a spring and early summer reproductive season for the Baja California populations.

FIG. 4.40 **_Sceloporus magister_**
DESERT SPINY LIZARD; BEJORI

Sceloporus magister Hallowell 1854:93. Type locality: "near Fort Yuma, at junction of Colorado and Gila, also near Tucson in Sonora"; restricted to "Fort Yuma, Yuma County, Arizona," USA, by Smith and Taylor (1950:355). Montanucci et al. (2001) indicated that Fort Yuma was in California. Holotype: USNM 2967

IDENTIFICATION *Sceloporus magister* can be distinguished from all other lizards in the region of study by having strongly keeled dorsal scales of nearly all the same size with a prominent, posteriorly projecting spine; the first sublabial usually contacting the mental; and the sides being heavily suffused with black in adult males.

RELATIONSHIPS AND TAXONOMY *Sceloporus magister* is the sister species of a natural group composed of *S. zosteromus* of Baja California and *S. lineatulus* of Isla Santa Catalina in the Gulf of California (Wiens and Reeder 1997). Grismer and McGuire (1996) discuss taxonomy.

DISTRIBUTION *Sceloporus magister* ranges from central Nevada, southern Utah, and northwestern New Mexico, south through southeastern California and into northeastern Baja California. It also occurs on Isla Tiburón in the Gulf of California (Parker 1982; map 4.13). In northeastern Baja California, it is known as far south as La Roca (Grismer 1994a) and probably ranges south to Puertecitos. It also extends west into cismontane habitats in the eastern end of Valle la Trinidad through Paso de San Matías (Grismer and McGuire 1996).

DESCRIPTION (Based on material from northeastern Baja California. See geographic variation section for the Isla Tiburón population.) Moderately sized, with adults reaching 116 mm SVL; head and body robust, not dorsally or laterally compressed; head triangular, sharply rounded in dorsal profile; scales on top of head large and platelike; 4 to 6 large supraoculars, fourth and fifth supraoculars contacting medial head scales; temporals wide, flat, spinose, imbricate, and keeled; interparietal large, with conspicuous parietal eye; interparietals and parietals abruptly border dorsal scales; dorsals wide, imbricate, spinose, and strongly keeled, with denticulate posterior margins; dorsals grade ventrally into flat, imbricate, denticulate ventrals; 4 to 6 supralabials; 2 postmentals; first sublabial usually contacting mental; gulars wide, flat, imbricate, and denticulate; antegular fold interrupted medially; gular fold ab-

FIGURE 4.40 Subadult male Uniformis pattern class of *Sceloporus magister* (Desert Spiny Lizard) from Paso de San Matías, BC

sent; 5 to 8 infralabials; forelimbs stout, moderate in length, covered with keeled, spinose, imbricate scales, except for smooth infrahumerals; hind limbs stout, thicker than forelimbs, dorsal surfaces keeled, imbricate, and spinose; ventral surfaces smooth; postfemorals small; 10 to 16 well-developed femoral pores in males; 21 to 24 subdigital lamellae on fourth toe; tail approximately 1.5 times length of body, round in cross section; dorsal caudals keeled, spinose, imbricate, and approximately same size as dorsals; subcaudals smooth anteriorly, strongly keeled posteriorly; enlarged postanal scales in males. **COLORATION** ADULT MALES: Dorsal ground color light brown; head essentially unicolored; back light brown, paravertebral blotches faint to absent; sides heavily suffused with black, in some cases solid with isolated yellow scales; tail turquoise-green; black shoulder patch not bordered posteriorly in white; limbs sooty gray; ventral ground color dull white; gular region suffused with black; gular patch turquoise-green; throat white, occasionally with black markings; anterior surface of humeral region black, rarely extending onto chest and not continuous with abdominal patches; interior scales of abdominal patches turquoise centrally and outlined in black; abdominal patches outlined in black medially and yellow laterally; black extends onto inguinal region and anteroventral surface of femora; tail dull white ventrally. FEMALES AND JUVENILES: Similar in dorsal ground color to adult males; paravertebral blotches distinct; lateral blotches on sides larger and often coalescing; shoulder patch small and faint; ventral surfaces dull white. Females take on a faint orangish suffusion of the head during the reproductive season. Some juveniles also have orangish heads.

GEOGRAPHIC VARIATION The various subspecies of *S. magister* all intergrade widely where their ranges come into contact (Phelan and Brattstrom 1955; Tanner 1955). In consequence, recent accounts have recognized these forms not as formal taxonomic units (i.e., lineages) but as pattern classes (Grismer and McGuire 1996). Therefore the subspecific epithets are used here as pattern class designations. These names are useful in that they delimit the geographic distribution of certain character states (i.e., individuals in a given area generally look alike and can be distinguished from other individuals from different areas).

There is no significant geographic variation within the Baja California population of *S. magister* (pattern class Uniformis). However, the differences between adult males of Uniformis and those of the Isla Tiburón pattern class (Magister) are substantial (Phelan and Brattstrom 1955). Magister are characterized by having a wide, deep purple to black middorsal stripe bordered by light stripes. Their sides are generally light in color and lack a black suffusion. The ventral surfaces of the forelimbs are heavily suffused in black. These characteristics do not occur in Uniformis.

NATURAL HISTORY *Sceloporus magister* ranges throughout the Lower Colorado Valley Desert Region of northeastern Baja California and enters the eastern edge of the California Region through Paso de San Matías (Grismer and McGuire 1996). It is a scansorial species, commonly occurring on vegetation or large rocks, and is generally not found in areas lacking trees or shrubs. On the desert floor of northern Baja California, it is most common in arroyos that are lined with Palo Verde *(Cercidium microphyllum)* and Ironwood *(Olneya tesota)* trees. I observed an adult male approximately 5 m above the ground in a Desert Agave *(Agave deserti)* in the Sierra San Felipe. Along the eastern foothills of the Peninsular Ranges, *S. magister* is most common in vegetated rocky areas. Linsdale (1932) reports specimens from the Río Colorado Delta area among the leaves of Willow thickets (*Salix* sp.), head down on the trunks of willow trees, or sleeping on horizontal branches of Mesquite trees (*Prosopis* sp.). Welsh (1988) reports finding this species in association with scrub vegetation, and Welsh and Bury (1984) observed specimens on the ground near sandy hummocks that had formed around the bases of palo verde, mesquite, and other bushes. *Sceloporus magister* can also be found beneath boards (Linsdale 1932) and other surface litter, provided there is sufficient vegetation nearby.

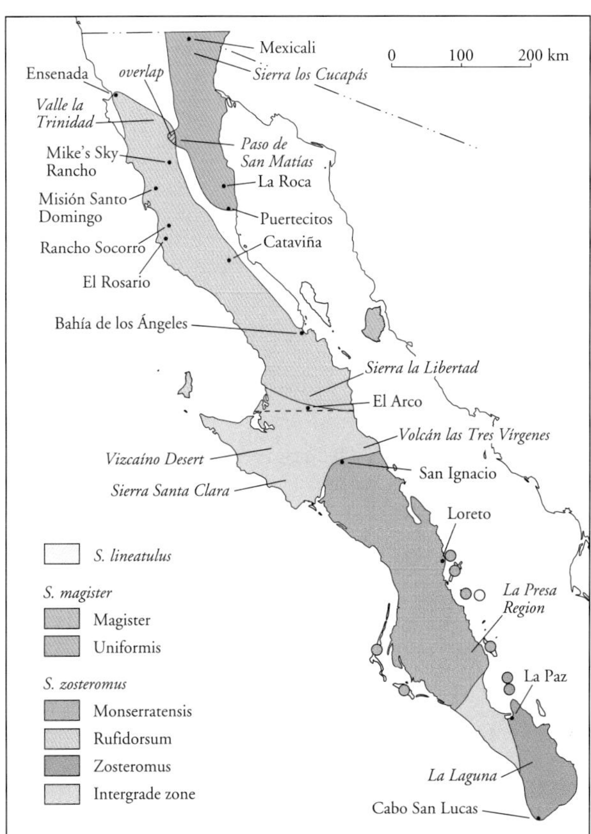

MAP 4.13 Distribution of the pattern classes of *Sceloporus magister* (Desert Spiny Lizard), *S. zosteromus* (Baja California Spiny Lizard), and *S. lineatulus* (Isla Santa Catalina Spiny Lizard)

A great deal has been published concerning the natural history of *S. magister* in the United States (see Dunham et al. 1994; Greene 1994; Parker 1982; Vitt et al. 1974 and references therein). Comparatively little has been reported for this species in Baja California, although it is not expected that the Baja California populations would vary significantly. I have observed adults and juveniles active from early March to early November. *Sceloporus magister* emerges in the mornings to bask and is conspicuous when perched on rocks, fence posts, or the tops of shrubs. It is far more cryptic, however, on the limbs or trunks of large trees. It remains active throughout the day, although during the midday hours in the summer it is more commonly observed in the shade near the bases of trees or among the branches.

Sceloporus magister eats a wide variety of foods, including arthropods, other lizards, leaves, and flower buds. During an uncharacteristically lush spring bloom in April 1992, I saw several juvenile and subadult *S. magister* sitting on the tops of blooming Rabbit Bush *(Chrysothamnus nauseosus)* in Valle San Felipe feeding on insects attracted by the flowers as well as on the flowers themselves. *Sceloporus magister* has a typical spring to early summer breeding season. I have observed gravid females in May in Paso de San Matías and hatchlings in July in the Sierra los Cucapás.

FOLKLORE AND HUMAN USE Malkin (1962) noted that Seri Indian boys would occasionally keep *S. magister* on a leash. However, they were kept for food. Nabhan (forthcoming) reports that the push-up displays of Desert and Clark's Spiny Lizards are the subject of many lewd jokes. He states that O'odham women will not urinate behind woodpiles known to contain spiny lizards for fear they might run up their skirts and painfully lodge themselves in the women's vaginas.

FIG. 4.41

Sceloporus zosteromus

BAJA CALIFORNIA SPINY LIZARD; BEJORI, CANARRO

Sceloporus zosteromus Cope 1863:105. Type locality: "Cape St. Lucas [Cabo San Lucas]," Baja California Sur, Mexico. Syntypes: USNM 5298 (23 specimens) and 69472 to 88

FIGURE 4.41 Adult male of Monserratensis pattern class of *Sceloporus zosteromus* (Baja California Spiny Lizard) from Rancho San Ángel, BCS

IDENTIFICATION Adult males can be distinguished from all other lizards in the region of study by having strongly keeled dorsal scales of nearly all the same size with a prominent, posteriorly projecting spine; the first sublabial usually contacting the mental; and large white scales on the sides of the head and neck.

RELATIONSHIPS AND TAXONOMY *Sceloporus zosteromus* is the sister species of *S. lineatulus* (Wiens and Reeder 1997). Grismer and McGuire (1996) discuss taxonomy.

DISTRIBUTION *Sceloporus zosteromus* is endemic to Baja California and ranges from at least Ensenada in the north, southward to Cabo San Lucas (Grismer 1994d; map 4.13). It does not reach the Gulf coast or occur east of the Peninsular Ranges north of Bahía de los Ángeles. It extends only a short distance east through Paso de San Matías towards the desert floor (Grismer and McGuire 1996). *Sceloporus zosteromus* occurs on Islas Cedros, Magdalena, and Santa Margarita along the Pacific coast and Islas Coronados, Carmen, Monserrate, San José, Partida Sur, and Espíritu Santo in the Gulf of California.

DESCRIPTION Moderately sized, with adults reaching 131 mm SVL; head and body stout, not dorsally or laterally compressed; head triangular, sharply rounded in dorsal profile; scales on top of head large and platelike; 5 or 6 large supraoculars, fourth and fifth contacting medial head scales; temporals wide, flat, spinose, imbricate, and weakly keeled; interparietal large, with conspicuous parietal eye; interparietals and parietals abruptly border dorsal scales of body; dorsals wide, imbricate, spinose, and strongly keeled, with denticulate posterior margins; dorsals grade ventrally into flat, imbricate, denticulate ventrals; 5 or 6 supralabials; 2 postmentals; first sublabial usually contacting mental; gulars wide, flat, imbricate, and denticulate; antegular fold absent to weak medially and present laterally; gular fold absent; 6 to 8 infralabials; forelimbs stout, moderate in length, covered with keeled, spinose, imbricate scales, except for smooth infrahumerals; hind limbs stout, thicker than forelimbs, all but ventral surface covered with keeled, imbricate, spinose scales; postfemorals small; 15 to 22 well-developed femoral pores in males; 21 to 27 subdigital lamellae on fourth toe; tail approximately 1.25 times length of body, round in cross section; dorsal caudals keeled, spinose, imbricate, and approximately same size as dorsals; subcaudals smooth to weakly keeled anteriorly, strongly keeled posteriorly; enlarged postanal scales in males. **COLORATION** Varies geographically, sexually, and ontogenetically. ADULT MALES: Dorsal ground color light to dark brown; head unicolored with white markings; wide or narrow vertebral stripe, light brown to reddish; faint paravertebral blotches present or absent; scales on sides white or orange to straw-colored centrally and outlined with thick black lines; tail same color as body; black shoulder patch bordered posteriorly with white or yellow; forelimb ground color gray to yellow and overlaid with distinct black lines; ground color of hind limbs yellow to orange, lineation less distinct; ventral ground color white; gular patch composed of alternating black and metallic green longitudinal lines; throat white, occasionally with black markings; antehumeral region black, extending onto forelimbs and chest, continuous with black outlined abdominal patches; interior scales of patches turquoise to metallic green centrally; patches outlined in black medially and sometimes yellow laterally; black extends onto inguinal region, anteroventral surface of femora, and preanal region as isolated blotches; tail dull white to tan ventrally. This pattern develops ontogenetically in males. FEMALES AND JUVENILES: Similar in dorsal coloration to adult males; paravertebral blotches distinct; sides not brightly colored; shoulder patch small and faint; ventral surfaces dull white, with faint abdominal and gular patches in adult females.

GEOGRAPHIC VARIATION When *S. zosteromus* was considered to be part of *S. magister,* it was recognized as containing three subspecies, based on variation in adult male coloration (Phelan and Brattstrom 1955; Parker 1982). These subspecies are used here as pattern classes. Rufidorsum range from Ensenada in the California Region south into the Vizcaíno Region to the northern edge of the Vizcaíno Desert (see map 4.13). They are characterized by distinct, transversely oriented white markings on the rear of the head; a narrow yellowish vertebral stripe bordered by gray to faint reddish paravertebral stripes; gray forelimbs with a bold, dark lineate pattern; and yellowish-orange hind limbs. Generally, Rufidorsum are much more heavily suffused with black on most ventral surfaces than are Monserratensis and Zosteromus (see below). Specifically, Rufidorsum are differentiated from the latter two by having a chest that is often solid black, and abdominal patches that are in wide medial contact but are not outlined laterally in yellow. Rufidorsum on Isla Cedros generally have lighter-colored belly patches with a deeper blue medial border and more orange on the sides and hind limbs. Also, the adult females more closely resemble adult males in coloration than adult females from other populations do. Hatchlings from Isla Cedros have a moderate amount of orange on the head.

The Monserratensis intergrade widely with Rufidorsum through the Vizcaíno Desert, although the intergrade zone is difficult to pinpoint because the differences between these pattern classes are subtle. Monserratensis extend south through the Magdalena Region to the Isthmus of La Paz. They are characterized by distinct, transversely oriented white markings on the rear of the head; a narrow yellowish-red vertebral stripe bordered by paravertebral stripes that are gray to faint reddish or orangish, with occasionally slightly darker paravertebral blotches; and yellow limbs with a bold, dark lineate pattern on the forelimbs. Ventrally, Monserratensis are distinguished by more white on the chest that is bordered laterally by thick black lines arcing around the shoulders and extending posteriorly to form the medial border of the abdominal patches. The abdominal patches do not contact or are only narrowly in contact medially, and they are outlined laterally in yellow. Intergrades from the Sierra Santa Clara have an overall lime-green hue. Subadult males from the Magdalena Plain and Islas Magdalena and Santa Margarita tend to have a more broken or scattered dark ventral blotching pattern than other populations. In the dune area of Isla Magdalena, adults are very light in overall coloration and have small black punctations in many of their dorsal scales. The adult females have lemon-yellow flank scales with reddish keels. Adult males from Isla San José are suffused with bright yellow dorsally and orange laterally.

Zosteromus range through the Arid Tropical Region of the Cape Region (see map 4.13) and are seen less frequently in the higher elevations of the Sierra la Laguna Region (Alvarez et al. 1988). Zosteromus intergrades widely with Monserratensis in southern Baja California, most notably through the Isthmus of La Paz. In the Cape Region, Zosteromus are characterized by generally unicolored heads; wide, straw-colored backs lacking paravertebral stripes but with occasional bluish paravertebral blotches; and yellow limbs with only a faint lineate pattern on the forelimbs. Ventrally, Zosteromus resemble Monserratensis.

NATURAL HISTORY *Sceloporus zosteromus* is a habitat generalist, ranging through the low-lying scrub of the Arid Tropical Region of southern Baja California (Alvarez et al. 1988) to coastal sand dunes of north-central Baja California (Bostic 1971; Tevis 1944) and chaparral habitats of the California Region in northern Baja California (Grismer and McGuire 1996). It is most common in rocky, vegetated terrain such as that within the Sierra la Libertad and the deep arroyos on Isla Cedros. I have observed *S. zosteromus* in dune habitats at Rancho Socorro and on Isla Magdalena. Bostic (1971) found *S. zosteromus* in the vicinity of sand hummocks that had surrounded the bases of Tree Sunflowers *(Encelia ventorum),* and Tevis (1944) noted that lizards would retreat into Wood Rat (*Neotoma* sp.) middens when alarmed. Bostic also noted that lizards from coastal areas of north-central Baja California were most common in the vicinity of impenetrable thickets of Thorn Bush (*Lycium* sp.) and Pitahaya Agria *(Stenocereus gummosus).* Additionally, *S. zosteromus* occurs in well-vegetated arroyos and grassy and plowed fields, where it is often observed on fence posts. It is equally abundant on the flat, sandy expanses of the Vizcaíno Desert, where lizards are usually seen along the edge of the highway. Linsdale (1932) reports catching several specimens from such areas in mammal traps. Grismer and McGuire (1996) report *S. zosteromus* and *S. magister* as syntopic in foothill chaparral areas northeast of Mike's Sky Rancho along the northwestern base of the Sierra San Pedro Mártir. Here lizards are commonly observed sitting on the tops of rocks and shrubs. I have also observed *S. zosteromus* throughout the cismontane areas of Valle la Trinidad from Ensenada to Paso de San Matías. On Isla Cedros, *S. zosteromus* is ubiquitous, although Grismer and Mellink (1994) note that it is less abundant in the presence of *S. occidentalis.*

Sceloporus zosteromus is occasionally active during the winter from October through February in reduced numbers in the southern portion of its range, south of Guerrero Negro (Blázquez and Ortega-Rubio 1996; Leviton and Banta 1964). Winter activity from these southern areas is generally restricted to juveniles. Peak activity occurs throughout its range from March through September. In more northerly areas, where winters are not as mild, *S. zosteromus* apparently hibernates, and I have observed lizards deep within rock fissures during December and January at Cataviña. Linsdale (1932) reports six "hibernating" lizards being dug out of the ground from a common burrow during late December at El Rosario. The earliest I have observed this species active in the north is mid-February at Misión Santo Domingo, and Linsdale (1932) reports a specimen active on 25 February at Rancho Socorro. General activity in the north begins in mid-March, even if the weather is cool. Adults remain active until at least late September.

Sceloporus zosteromus is principally diurnal, emerging in the morning to bask. During this period, lizards are commonly seen on small rocks and boulders, although barrel cactus *(Ferrocactus)* and yucca *(Yucca)* stumps seem to be favored where they are present. Lizards are often seen 1 to 1.5 m above ground, basking on rocks or within vegetation. On the top of the Sierra la Asamblea, *S. zosteromus* is much more saxicolous than elsewhere. Despite the presence of large shrubs, it is more commonly seen on rocks, and its agility matches that of rock lizards *(Petrosaurus)* as it runs swiftly for cover in sheer rocky crevices. During the hot afternoons of summer, *S. zosteromus* is commonly observed in the shade of vegetation or in the low branches of brush and trees. Tevis (1944) reports seeing a specimen east of El Rosario, foraging out in the open at night during early June. I have observed lizards out at night as well, but they did not seem to be foraging. *Sceloporus zosteromus* utilize fallen Cardón Cactus *(Pachycereus pringlei),* the axils of agaves, and surface debris for overnight retreats. I have often put Cardón logs on campfires during cold spring nights only to have *S. zosteromus* come scurrying out. I have seen individuals abroad during periods of light summer rain at the base of Volcán las Tres Vírgenes. Nothing has been reported on the diet of *S. zosteromus,* but it is probably very similar to that of *S. magister* (see above). I have seen adults in the Sierra la Libertad eating beetles.

Sceloporus zosteromus appears to have a spring breeding season and a summer egg-laying season. Bostic (1971) reported females from north-central Baja California with oviducal eggs in early July, and I have observed gravid females on Isla Cedros during mid-July. Shaw (1952) reports gravid females from just south of El Rosario laying 7 to 11 eggs in June. Another female from near El Arco laid 18 eggs on 12 May (Shaw 1952). I have seen hatchlings on Islas Cedros and Magdalena during late July and early August, and at Rancho Socorro and along the western sides of the Sierra la Asamblea in early November, which would be in accord with a spring and summer reproductive season.

FIG. 4.42 *Sceloporus lineatulus*

ISLA SANTA CATALINA SPINY LIZARD; BEJORI

Sceloporus lineatulus Dickerson 1919:467. Type locality: "Santa Catalina Island, Gulf of California, Mexico." Holotype: USNM 64263

IDENTIFICATION *Sceloporus lineatulus* can be distinguished from all other lizards in the region of study by having strongly keeled dorsal scales of nearly all the same size with a prominent, posteriorly projecting spine; the first sublabial usually contacting the mental; a nearly unicolored golden dorsal pattern with no yellow on the sides; and a black shoulder patch lacking a distinct white posterior border.

RELATIONSHIPS AND TAXONOMY See *S. zosteromus.*

DISTRIBUTION *Sceloporus lineatulus* is endemic to Isla Santa Catalina in the Gulf of California, off the coast of southern Baja California, Mexico (see map 4.13).

DESCRIPTION Same as *S. zosteromus,* except for the following: adults reaching 115 mm SVL; 6 or 7 supralabials; 7 or 8 infralabials; 17 to 20 well-developed femoral pores in males; 23 or 24 subdigital lamellae on fourth toe; tail approximately 1.25 times length of body. **COLORATION** ADULT MALES: Dorsal ground color gray to golden tan; head unicolored; body nearly unicolored, vertebral region light; sides darker, with scales marked with varying degrees of turquoise, and broken parallel dark lines running parallel to keels; tail uniform tan, with scales at base occasionally edged in yellow; black shoulder patch not or only indistinctly bordered posteriorly in white; limbs sooty gray; ventral ground color dark gray; gular region suffused with black; gular patch black, with a few turquoise-centered scales anteriorly; throat black; anterior surface of forelimbs and chest black, continuous with black outline of abdominal patches; interior scales of patches turquoise centrally and outlined in black; black extends onto inguinal region and anteroventral surface of thighs; tail gray ventrally. FEMALES AND JUVENILES: Very similar in dorsal coloration to adult males; much lighter ventrally. Hatchlings have a light brown ground color with a pattern of light, thin, chevron-shaped crossbars; black shoulder bars outlined posteriorly in white; and a dark throat.

NATURAL HISTORY Isla Santa Catalina lies within the Central Gulf Coast Region, and its climate is generally hot and

FIGURE 4.42 Adult female *Sceloporus lineatulus* (Isla Santa Catalina Spiny Lizard) from Arroyo Mota, Isla Santa Catalina, BCS

FIGURE 4.43 Adult male *Sceloporus orcutti* (Granite Spiny Lizard) from Cañón Guadalupe, BC

arid. Virtually nothing is known about the natural history of *S. lineatulus.* It is most commonly found on the flats along the inland edges of the beaches and in the arroyo bottoms. I have never observed this species on the rocky hillsides, although it is likely to occur there in limited numbers. *Sceloporus lineatulus* has an unusually long flight distance relative to the peninsular spiny lizards (e.g., *S. zosteromus*) and is difficult to observe in the open. Lizards generally remain near large shrubs and cacti, into which they retreat when threatened. Most are observed crawling on the lower branches deep within the centers of the plants. I have even seen hatchlings deep within patches of *Opuntia* cacti, crawling on the branches. *Sceloporus lineatulus* is active from at least early April to early July. I have never observed any lizards abroad later than September. They may, however, be active year-round in limited numbers (juveniles more so than adults). I have seen hatchlings and small juveniles from early July to early August but never later in the year. Subadults are common in April, which suggests a spring to early summer breeding period with a summer to early fall hatching period.

FIG. 4.43 **Sceloporus orcutti**

GRANITE SPINY LIZARD; BEJORI, CANARRO

Sceloporus orcutti Stejneger 1893a:181. Type locality: "Milquatay Valley, San Diego County, Calif. [USA], . . . just bordering the Mexican boundary, 50 miles east of San Diego by wagon road," correctly rendered "Campo Valley, San Diego County, California," by Higgins (1959); restricted to "the flat just east of Campo, San Diego Co., Calif.," by Klauber in Hall and Smith (1979:24). Holotype: USNM 16330

IDENTIFICATION *Sceloporus orcutti* can be distinguished from all other lizards in the region of study by wide, flat, weakly keeled dorsal scales with a short, posteriorly projecting spine and the absence of light markings immediately posterior to the black shoulder patch.

RELATIONSHIPS AND TAXONOMY *Sceloporus orcutti* is most closely related to the sister species *S. hunsakeri* and *S. licki* (Wiens and Reeder 1997), which are endemic to the Cape Region of Baja California (Hall and Smith 1979).

DISTRIBUTION *Sceloporus orcutti* ranges throughout montane and cismontane southern California, from southern San Bernardino County (Weintraub 1980) south throughout Baja California to Cañón de los Reyes, 19 km west of La Paz (map 4.14). In the north, it reaches the Pacific coast of Baja California 46 km north of Ensenada and the Gulf coast at Puertecitos. It is absent from the upper elevations of the Sierra San Pedro Mártir but is common in the upper elevations of the Sierra la Asamblea, as well as in mountains farther south. Grismer et al. (1994) noted that it was one of the saxicolous species absent from the Vizcaíno Peninsula. *Sceloporus orcutti* occurs on the islands of Carmen, Coronados, San Francisco, San Ildefonso, San José, San Marcos, and Tortuga in the Gulf of California.

DESCRIPTION Moderately sized, with adults reaching 115 mm SVL; head and body stout, not dorsally or laterally compressed; head triangular, somewhat pointed in dorsal profile; scales on top of head large and platelike; 5 large supraoculars, fourth and fifth contacting medial head scales; fifth supraocular usually contacting superciliaries; temporals wide, flat, spinose, imbricate, and weakly keeled; interparietal large, with conspicuous parietal eye; interparietals and parietals abruptly border dorsal scales of body, with little to no gradation; frontoparietals usually separated; dorsals wide, imbricate, spinose, and strongly keeled, with denticulate posterior margins; dorsals grade ventrally into flat, imbricate, denticulate ventrals; 5 or 6 supralabials;

2 postmentals; first sublabial not contacting mental; gulars wide, flat, imbricate, and denticulate; antegular fold faint in adults medially but present laterally; gular fold reduced to absent; 6 or 7 infralabials; subnasal usually present; forelimbs stout, moderate in length, and covered with keeled, spinose, imbricate scales, except for posthumerals, which are smooth; hind limbs stout, thicker than forelimbs, covered with keeled, imbricate, spinose scales dorsally and smooth scales ventrally; postfemorals small; 20 to 48 well-developed femoral pores in males; 18 to 25 subdigital lamellae on fourth toe; 16 or 17 escutcheon scales anterior to vent; tail approximately 1.25 times length of body, round in cross section; dorsal caudal scales keeled, spinose, imbricate, and approximately same size as dorsals; subcaudals smooth anteriorly, keeled posteriorly; enlarged postanal scales in males. **COLORATION** ADULT MALES: Dorsal ground color turquoise or gray; light turquoise to straw-colored spots on top of head; usually a wide blue to purple vertebral stripe extending from shoulders to base of tail; flank scales turquoise or gray and outlined in black; limbs and tail gray, black, or turquoise; black triangular shoulder patch narrowly extending onto brachium; shoulder patch lacks light central spot and light posterior border; ground color of ventral surfaces dull white; gular patch usually iridescent turquoise anteriorly and blue posteriorly, with faint dark, oblique lines; sides of throat black; abdominal patches iridescent turquoise anteriorly and laterally, and may or may not be in contact medially or posteromedially; when not contacting, midventral surface suffused with gray pigment; abdominal patches extend posteriorly onto femoral region, forming black inguinal patches; abdominal patches not continuous with gular patch; ventral surfaces of limbs and tail light-colored. FEMALES AND JUVENILES: Dorsal ground color light gray to brownish and fading to dark gray posteriorly; body, limbs, and tail with distinctive thin, brownish oblique bands; middorsal scales purple in adult females; gular region often light blue with dark oblique lines; oblique lines most prominent in juveniles; in adult females, weak, faint abdominal patches that may or may not extend onto the femoral region.

GEOGRAPHIC VARIATION Geographic variation in the color pattern of *S. orcutti* is limited and localized. It is discussed here from north to south. Lizards from the U.S.-Mexican border south to Valle la Trinidad tend to have a tan ground color anteriorly. Lizards from populations south of Valle la Trinidad have a slightly darker ground color. Lizards from the northern portion of the Sierra San Felipe are melanistic. *Sceloporus orcutti* from the upper elevations of the San Francisco de la Sierra and the Sierra Guadalupe have a dark brown ground color, and those from adjacent low-lying regions are lighter. In the vicinity of the Isthmus of La Paz, the middorsal scales of adult females are almost entirely orange, with the typical blue to purple coloration narrowly restricted to the centers of the scales. Females and juveniles south of San Ignacio tend to be less distinctly banded than those of northern areas.

Sceloporus orcutti from the U.S.-Mexican border south to El Rosario have 13 to 16 femoral pores, and lizards south of Bahía de los Ángeles have 10 to 13. Between El Rosario and Bahía de los Ángeles, lizards have 11 to 14 femoral pores. There is a weak cline in SVL of *S. orcutti:* lizards at the northernmost extent of its distribution have a maximum SVL of 115 mm, and those from the southernmost localities have a maximum SVL of 109 mm.

Various insular trends are as follows: adult males from Isla San Marcos have a relatively narrow purple or blue dorsal stripe. Adult males from Isla San Ildefonso have extensive blue pigment on the ventrum, and the belly patches are in broad medial contact. Adult males from Isla Carmen have a grayish ground color on the sides, limbs, and tail, in contrast to the turquoise ground color common in most

MAP 4.14 Distribution of *Sceloporus orcutti* (Granite Spiny Lizard) and *S. hunsakeri* (Hunsaker's Spiny Lizard)

other populations. Adult males from Isla Tortuga tend to be brownish in ground color with green throats.

NATURAL HISTORY The natural history of *S. orcutti* from southern California, which is well-known (Mayhew 1963a,b,c; Weintraub 1968, 1969), is applicable to populations of northern Baja California, from approximately the U.S.-Mexican border south to the Sierra San Pedro Mártir. *Sceloporus orcutti* occurs in several phytogeographic regions throughout its distribution and is absent only from the Baja California Coniferous Forest Region of the northern mountains. It is nearly always found near large rocks and uses the cracks and large exfoliations for retreats and overnight refuge. From the Sierra la Libertad north, lizards usually inhabit large granitic boulders (Bostic 1971; Linsdale 1932; Welsh 1988). South of the Sierra la Libertad, however, *S. orcutti* is common on volcanic talus, lava flows, and cliff faces. Although this lizard has been described by some as being "entirely saxicolous" (e.g., Welsh 1988), this is actually not the case. Bostic (1971) reports a specimen running from a bush into the axils of an agave in a coastal area in north-central Baja California. I have found specimens in the axils of agaves in the Sierra San Felipe and from El Rosarito, 40 km south of Bahía de los Ángeles. Linsdale (1932) reports finding lizards on the limbs of willow trees, below *Opuntia* cactus, and on open ground. I have often found them on tree trunks. In Cañón Santa Isabel, along the eastern face of the Sierra Juárez, and at Misión Santa María, I observed several specimens on the trunks of palm trees that would take refuge beneath the palm fronds when threatened, and I commonly found lizards on fallen palm logs in the groves at San Ignacio. At Misión Santa Gertrudis I observed numerous individuals on small rocks in a dry streambed. At the southern end of Isla Carmen, *S. orcutti* is common on the rocky flats. On Islas Coronados and San Francisco, lizards are commonly found along the rocky beaches, and I have often seen them at the base of vegetation along the coastal dunes on Isla San Francisco. On Isla Tortuga, lizards are commonly found on the ground at the bases of bushes into which they retreat when threatened.

In northern Baja California, adults begin to emerge from hibernation in late January and February and are most active from March through June. Their activity generally ceases during October and November, when they enter deep fissures and cracks to hibernate. *Sceloporus orcutti* is one of a few species of lizards known to aggregate in hibernacula, and as many as 37 lizards have been found beneath a single exfoliation (Weintraub 1968). From approximately the Sierra la Libertad south, adults are commonly active into early November and during warm winter days later in the year; first-year hatchlings may be active year-round (Weintraub 1968). During spring and summer, lizards emerge from their overnight retreats to bask on the rocks in the morning, at which time they are jet black. They can be identified at a distance by the sheen on their scales. During the very hot summer months in southern Baja California, it is not uncommon to find lizards abroad during the evening. I have seen adults during the evening in mid-August at Canipolé and in late June just south of Loreto. I commonly see lizards abroad between the hours of 2100 and 0100 on the ground near the lava flows in the vicinity of San Ignacio during the summer and fall. I even observed one individual abroad at 2000 h in the Sierra Guadalupe that retreated into a rodent burrow. It is not known whether such individuals are active or simply did not retreat into overnight refuges with the onset of evening.

Sceloporus orcutti from northern populations become reproductively active during March and April and may continue breeding through July (Mayhew 1963a). Females lay 8 to 11 eggs per clutch and lay only one clutch per year (Shaw 1952; Mayhew 1963a). The breeding season ends in August, and hatchlings first appear in late July and continue to appear through mid-October. *Sceloporus orcutti* feeds predominantly on arthropods, although Murray (1955) captured a specimen at Alaska (BC) that was eating a Granite Night Lizard *(Xantusia henshawi).*

REMARKS A population of *S. orcutti* in the central portion of the Sierra San Felipe is nearly melanistic in coloration. Lizards here are also found more commonly among the agave than on the adjacent granite boulders. The absence of *S. orcutti* from the Cape Region is likely due to the presence of its close relatives *S. licki* and *S. hunsakeri.*

FIG. 4.44 ***Sceloporus hunsakeri***

HUNSAKER'S SPINY LIZARD; BEJORI, CANARRO

Sceloporus hunsakeri Hall and Smith 1979:452. Type locality: "3 mi. E of San Bartolo, 500 ft.," BCS, Mexico. Holotype: MVZ 73570

IDENTIFICATION *Sceloporus hunsakeri* can be distinguished from all other lizards in the region of study by having strongly keeled dorsal scales of nearly equal size, with a prominent, posteriorly projecting spine; the first sublabial not contacting the mental; the posterior supraoculars contacting the superciliaries; and usually lacking a white spot in the center of the black shoulder patch.

RELATIONSHIPS AND TAXONOMY *Sceloporus hunsakeri* is the sister species of *S. licki* of the Cape Region (Wiens and Reeder 1997).

FIGURE 4.44 Adult male *Sceloporus hunsakeri* (Hunsaker's Spiny Lizard) from Isla Espíritu Santo, BCS

DISTRIBUTION Ranges from La Paz south to El Triunfo and then farther south along the eastern foothills of the Sierra la Laguna to Cabo San Lucas (Hall and Smith 1979). At Cabo San Lucas, it extends northward, west of the Sierra la Laguna, to at least La Palma. *Sceloporus hunsakeri* is known from Islas Ballena, Espíritu Santo, Gallo, and Partida Sur (Grismer 1999a; Hall and Smith 1979; see map 4.14) in the southern portion of the Gulf of California.

DESCRIPTION Same as *S. orcutti,* except for the following: adults reach 86 mm SVL; fifth supraocular contacting superciliaries; 7 to 9 infralabials; 24 to 36 well-developed femoral pores in males; 2 or more rows of transverse escutcheon scales anterior to vent. **COLORATION** ADULT MALES: Dorsal ground color iridescent bronze-green to blue; light turquoise spots on top of head; side of head yellowish; wide purple stripe on back extending to base of tail; sides yellowish, lime green, or light orange and heavily mottled; limbs dark green and heavily mottled; tail green to turquoise; shoulder patch black, triangular, usually lacking light central spot, and bordered posteriorly by thin white line; ground color of ventral surfaces dull white; posterior gular patch iridescent green with faint dark chevrons anteriorly; throat and chest usually black; abdominal patches iridescent green to metallic dark blue, in broad medial contact anteriorly, and extending posteriorly onto femoral region; abdominal patches continuous with gular patch; ventral surfaces of limbs and tail light-colored. FEMALES AND JUVENILES: Dorsal ground color dark green to light brown; dark mottling pattern on back, purple markings faint to absent; light dorsal speckling on hatchlings, usually consisting of light, paired, paravertebral spots on dorsum; ventral ground color dull white; abdominal patches and dark chest faint to absent; dark oblique lines on gular region, parallel paramedial lines usually absent.

GEOGRAPHIC VARIATION Geographic variation is not extensive in *S. hunsakeri.* Adult males from Isla Espíritu Santo are occasionally slightly lighter in overall dorsal coloration than those of the peninsula and have a more lime-green to yellow hue to the dorsal pattern.

NATURAL HISTORY *Sceloporus hunsakeri* is a saxicolous species common in rocky areas throughout the Arid Tropical Region and lower elevations of the Sierra la Laguna Region (Hall and Smith 1979). In the Cape Region, *S. hunsakeri* is more commonly found in association with large rocks and boulders near trees than in small, open, alluvial rubble. On Isla Espíritu Santo, lizards are commonly found on sparsely vegetated hillsides and in small beach cobble. On Isla Ballena, *S. hunsakeri* is most common in the dense patches of Pitahaya Agria cactus *(Stenocereus gummosus)* that cover the island. Here, lizards are commonly seen on the ground and within the cactus, perched as high as 1.5 m, basking on the cactus limbs. *Sceloporus hunsakeri* is active year-round, with peak activity from March through October. Hatchlings and juveniles are more numerous than adults during the winter. After morning temperatures have risen sufficiently, *S. hunsakeri* is commonly seen basking and foraging on and among rocks. I have observed individuals active at night during June at San Bartolo. During midday in January, I found an adult male half-buried in the dirt, headfirst, at the base of a rock in an area that received continual shade. It may have been attempting to hibernate. Nothing on the diet of *S. hunsakeri* has been reported, but it presumably preys on many of the small arthropods within its habitat. I observed one individual eating small beetles.

Hall and Smith (1979) state that in captivity, an adult male *S. hunsakeri* repeatedly attempted to court an adult female and was sexually active from 3 August, when it was collected, to 24 September, when it was sacrificed. I have observed hatchlings in the Cape Region from late September through mid-December, with peak numbers usually occurring in October, and during late December on Isla Espíritu Santo. These observations suggest that *S. hunsakeri* has a late summer to fall reproductive season. The same has been reported for other Cape Region species and correlates with the onset of the summer rainy season (Asplund 1967).

REMARKS Although the distinctive dorsal color pattern differences between adult male *S. hunsakeri* and adult male *S. licki* (with which it is sympatric) make them easily distinguishable from one another, females and juveniles are more difficult to identify: sometimes individuals cannot be unequivocally identified. Some *S. licki,* like *S. hunsakeri,* may have contact between the last supraocular and the superciliaries and may lack paramedial lines in the gular region. Other characteristics listed by Hall and Smith (1979: 14–15) that generally separate these two species are also too variable for an unequivocal identification.

FIG. 4.45 ***Sceloporus licki***

CAPE SPINY LIZARD; BEJORI

Sceloporus licki Van Denburgh 1895:110. Type locality: "Sierra San Lazaro, Lower California," BCS, Mexico. Neotype: CAS 2987a (Smith 1939)

IDENTIFICATION *Sceloporus licki* can be distinguished from all other lizards in the region of study by having strongly keeled dorsal scales of nearly equal size, with a prominent, posteriorly projecting spine; the first sublabial not contacting the mental; the posterior supraocular usually not contacting the superciliaries; and usually a white spot in the center of the black shoulder patch.

RELATIONSHIPS AND TAXONOMY See *S. hunsakeri.*

DISTRIBUTION *Sceloporus licki* ranges along the mountainous foothill areas of the Cape Region from Rancho Ancón south to near La Soledad in the Sierra la Laguna (Grismer 1989b; Hall and Smith 1979; map 4.15).

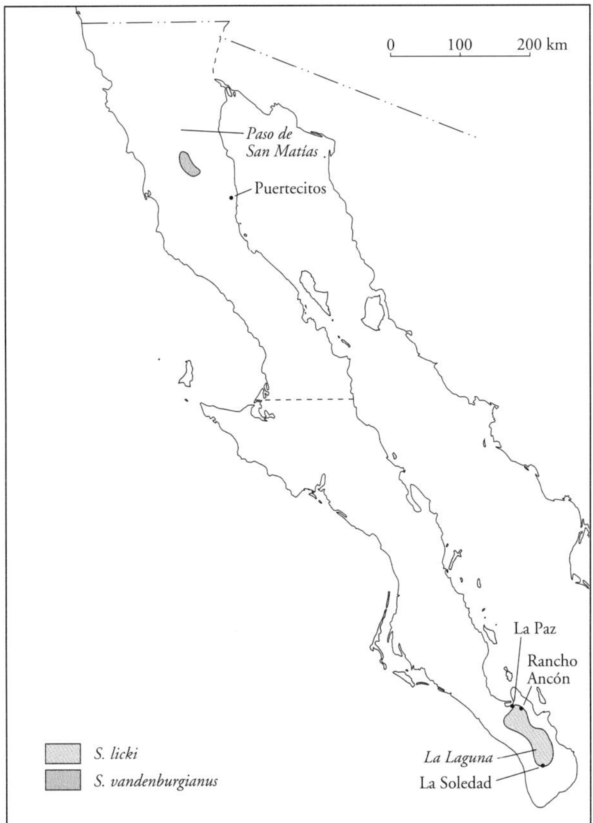

MAP 4.15 Distribution of *Sceloporus licki* (Cape Spiny Lizard) and *S. vandenburgianus* (Southern Sagebrush Lizard)

FIGURE 4.45 Adult male *Sceloporus licki* (Cape Spiny Lizard) from San Bartolo, BCS

DESCRIPTION Same as *S. orcutti,* except for the following: adults reach 78 mm SVL; 5 large supraoculars, fourth and fifth usually not contacting medial head scales; 5 to 7 supralabials; 6 or 7 infralabials; 14 to 18 well-developed femoral pores in males; 20 to 22 subdigital lamellae beneath fourth toe; single row of transverse escutcheon scales anterior to vent. **COLORATION** ADULT MALES: Dorsal ground color tan to dark gray; head generally unicolored; deep blue vertebral stripe extending from shoulders to base of tail; scales of sides dark gray with turquoise centers; turquoise centers generally absent in subadult males; wide, poorly defined, light-colored dorsolateral stripe beginning above shoulders and fading posteriorly at hind limb insertions; limbs black, heavily mottled in subadult males; tail turquoise in adults, brownish in subadult males; shoulder patch black, triangular, and usually with light central spot; shoulder patch bordered posteriorly by thin tan line; ground color of ventral surfaces dull white; gular region streaked with oblique dark lines and usually with parallel paramedial lines; throat often black; abdominal patches dark blue medially and posteriorly, turquoise centrally, and bordered by black medially and posteriorly; black extends onto femoral region; abdominal patches not continuous with gular patch; ventral surfaces of limbs and tail light-colored, suffused with black. FEMALES AND JUVENILES: Dorsal ground color gray to brown; vertebral stripe absent; body and limbs mottled; abdominal patches faint; tail grayish brown; dark oblique lines on the gular region, and paramedial parallel lines usually present.

NATURAL HISTORY *Sceloporus licki* inhabits brushy areas of the Arid Tropical and Sierra la Laguna regions, and its distribution is generally associated with the central portion of the Sierra la Laguna. *Sceloporus licki* is scansorial and occurs on rocks and logs near to low trees (Alvarez et al. 1988). According to Hall and Smith 1979, *S. licki* is most often found on logs. I believe this assertion is somewhat overstated. Some populations, such as those in the vicinity of San Bartolo, are most often seen on logs, not rocks. However, those in Arroyo Aguaje, at the base of La Victoria, along the eastern side of the Sierra la Laguna, are commonly found on rocks. On the west face of the Sierra la Laguna I have observed *S. licki* only on rocks.

Sceloporus licki is active year-round, with peak activity occurring from spring through fall. This species is not often observed basking in the open like *S. hunsakeri* but instead tends to remain under the cover of trees. Nothing has been reported on its diet, which presumably consists of small arthropods. I found a centipede in the stomach of a dead specimen on the road near San Bartolo. Hall and Smith (1979) state that the testes of an adult male in early August were completely regressed, which suggests that this species has a spring-summer breeding period, unlike *S. hunsakeri,* which has a late summer breeding season.

REMARKS See *S. hunsakeri.*

FIG. 4.46 ***Sceloporus vandenburgianus***

SOUTHERN SAGEBRUSH LIZARD; BEJORI, LAGARTIJA

Sceloporus vandenburgianus Cope 1896b:834. Type locality: "Summit of Coast Range, San Diego County, California," USA. The holotype has additional data "Campbell's Ranch, Laguna," San Diego County, California, USA (Cochran 1961:144). Holotype: USNM 21931

IDENTIFICATION *Sceloporus vandenburgianus* can be distinguished from all other lizards in the region of study by having keeled dorsal body scales nearly all the same size, with a weak, posteriorly projecting spine, and dorsal caudal scales nearly twice the size of the dorsal body scales.

RELATIONSHIPS AND TAXONOMY *Sceloporus vandenburgianus* forms a natural group with *S. arenicolus* of western Texas and southeastern New Mexico and *S. graciosus* of the western United States (Wiens and Reeder 1997). Smith et al. (1992), Sites et al. (1992), and Wiens and Reeder (1997) discuss taxonomy.

DISTRIBUTION *Sceloporus vandenburgianus* ranges from the Coast Ranges in Los Angeles County, California, south to southern San Diego County. A disjunct population occurs in the Sierra San Pedro Mártir of northern Baja California, Mexico (Censky 1986; Welsh 1988; see map 4.15).

DESCRIPTION Small, with adults reaching 63 mm SVL; head and body not dorsally or laterally compressed; head triangular, rounded in dorsal profile; scales on top of head large and platelike; 4 or 5 large supraoculars, fourth and fifth not contacting medial head scales; temporals small, imbricate,

FIGURE 4.46 Adult male *Sceloporus vandenburgianus* (Southern Sagebrush Lizard) from Campamento Forestal, Sierra San Pedro Mártir, BC

and keeled; interparietal large, with conspicuous parietal eye; interparietal abruptly borders dorsal scales of body; dorsals imbricate with distinctive low keel; dorsals not spinose and grading ventrally into flat, imbricate ventrals; 5 to 8 supralabials; 2 postmentals; first sublabial rarely contacting mental; gulars flat, imbricate, and denticulate; antegular fold absent medially, present laterally; gular fold absent; 7 to 9 infralabials; forelimbs stout, moderate in length, and covered with keeled, imbricate scales dorsally; infrahumerals smooth and imbricate; ventral forearm scales imbricate and keeled distally; hind limbs stout, thicker than forelimbs, covered with keeled, imbricate spinose scales dorsally and smooth imbricate scales ventrally; postfemorals granular to convex; 15 to 18 well-developed femoral pores in males; 23 to 28 subdigital lamellae beneath fourth toe; tail approximately 1.25 times length of body, round in cross section; dorsal caudals keeled, spinose, imbricate, and nearly twice the size of dorsal body scales; subcaudals weakly keeled; enlarged postanal scales in males. **COLORATION** Individuals are capable of considerable color change, and all lizards are nearly solid black when basking. ADULT MALES (NOT BASKING): Dorsal ground color dark brown to dark gray; head unicolored; blue to turquoise small spot in center of dorsal scales; double row of darker paravertebral blotches not always apparent; greenish yellow spots in center of flank scales; tail dark dorsally, blue laterally; dorsal surfaces of limbs dark, nearly unicolored; no yellow postfemoral region; gular region deep blue; pectoral region suffused with black, extending dorsally to form short black shoulder bar; abdominal patches pale blue, usually bordered medially in black, sometimes contacting medially; ventral surface of limbs and proximal portion of tail suffused with varying degrees of black. FEMALES AND JUVENILES (NOT BASKING): Dorsal ground color light gray to dark gray; head with or without small dark markings; faint to distinct wide, dark postauricular stripe; faint to distinct wide, light-colored dorsolateral stripe; double row of dark paravertebral blotches apparent; limbs faintly to strongly blotched; ventral surfaces dull white, occasionally with slight blue tint; belly patches present in juvenile males; gular region covered with dark, oblique lines; sides of head orange in gravid females.

GEOGRAPHIC VARIATION The intensity of color pattern in adult male *Sceloporus vandenburgianus* appears to correlate with altitude. In living specimens, populations along the crest of the Sierra San Pedro Mártir, from the observatory to near Picacho del Diablo, between approximately 3,000 and 3,100 m elevation, are lighter in overall coloration and tend to have a bolder, more contrasted color pattern. Populations from lower elevations, such as Las Tasajeras, Vallecitos, and Campamento Forestal, at approximately 2,100 to 2,450 m, are generally darker, and their color pattern is not as contrasted.

NATURAL HISTORY *Sceloporus vandenburgianus* is confined to the Baja California Coniferous Forest Region of the Sierra San Pedro Mártir and the chaparral ecotone above 1,980 m (Welsh 1988). Depending on the temperature, it generally emerges from hibernation in late April. It is diurnal and can be active at surprisingly low temperatures. Welsh (1988) reports *S. vandenburgianus* to be active from 0500 to 1800 hours (time of year not specified). Activity level is likely dependent on overnight low temperatures. During an unseasonably cold summer in 1988, when June temperatures commonly dropped below 15°C, lizards generally did not become active until mid-morning. *Sceloporus vandenburgianus* is the most conspicuous reptile in the Sierra San Pedro Mártir (Welsh 1988) and is commonly seen basking on rocks and logs, periodically dashing down to catch one of the many small arthropods on which it feeds. It has not been observed in open meadows, although Welsh (1988) reports that juveniles are found more often on the ground. I have observed gravid females in early June and females with gravid coloration in late June. These observations suggest that breeding begins sometime in May and lasts through July. I have measured gravid females as small as 48 mm SVL.

Welsh (1988) believes that the distribution of *S. vandenburgianus* in the upper elevations of the Sierra San Pedro Mártir may be influenced by competition with *S. occidentalis,* which occurs at lower elevations. He states that where these two species are narrowly sympatric, they may be separated ecologically in their selection of perching and foraging sites. According to Welsh, *S. vandenburgianus* favors perching on rocks and descending onto the ground to feed, whereas *S. occidentalis* remains on the ground. Interestingly, *S. vandenburgianus* displays this same behavior at higher elevations in the absence of *S. occidentalis* and has even been observed foraging along the edges of streams (J. Grismer 1994).

REMARKS The Sierra Juárez separates the Sierra San Pedro Mártir population of *S. vandenburgianus* from that of the Laguna Mountains of San Diego County, California, just over 200 km north (Censky 1986). *Sceloporus occidentalis* occurs within the montane habitat of the Sierra Juárez and probably displaced *S. vandenburgianus,* causing the observed disjunction in distribution of the latter.

Uma Baird

FRINGE-TOED LIZARDS

The genus *Uma* is composed of a group of small to medium-sized, terrestrial, arenicolous (sand-dwelling) lizards containing five allopatric species. All *Uma* have a suite of morphological and behavioral adaptations for living within, and moving about on, a substrate of fine, windblown sand (Stebbins 1944). The most notable adaptations are the enlarged lateral digital scales on the toes, which form a fringe-like structure that increases the surface area of the foot; the strongly keeled supralabial scales, which break the surface tension of the sand at the onset of burrowing; and a countersunk lower jaw, which prevents sand from entering the mouth while burrowing. Three closely related species are restricted to sand dune areas in the deserts of southern California, western Arizona, and northern Mexico. Two other species are distributed in similar habitats in the Chihuahuan Desert of southeastern Chihuahua, northeastern Durango, and western Coahuila, Mexico. One species, *U. notata,* ranges into northeastern Baja California.

FIG. 4.47 ### *Uma notata*

COLORADO DESERT FRINGE-TOED LIZARD; CACHORA

Uma notata Baird 1859d:253. Type locality: "Mojave Desert"; restricted to "vicinity of Yuma, Arizona," USA, by Smith and Taylor (1950:355). Holotype: USNM 4124

IDENTIFICATION *Uma notata* can be distinguished from all other lizards in the region of study by having obliquely keeled supralabial scales; a wide, flat body with granular dorsal scales; and a patch of enlarged postfemoral scales.

RELATIONSHIPS AND TAXONOMY *Uma notata* is part of a natural group containing the two other southern California species, *U. scoparia* and *U. inornata* (de Queiroz 1992). Wilgenbusch and de Queiroz (2000) demonstrated that some populations of *U. notata* are more closely related to *U. inornata* than to other populations of *U. notata.* De

FIGURE 4.47 *Uma notata* (Colorado Desert Fringe-Toed Lizard) from Sierra las Pintas, BC

Queiroz (1989) and Wilgenbusch and de Queiroz (2000) discuss taxonomy.

DISTRIBUTION *Uma notata* occurs in southeastern California, southwestern Arizona, and northwestern Sonora. In Baja California, it ranges no farther east than the western sides of the Sierra los Cucapás south to at least the northern end of the Sierra las Pintas (Pough 1977; map 4.16). Lizards have been reported only from the northernmost part of the dune system in the Sierra las Pintas because this is the only portion that is accessible by Highway 5. *Uma notata* almost certainly extends farther south along the western edge of this mountain system where dunes have formed, but gaining access to this area is difficult. There is a hiatus between the Sierra los Cucapás population and the Sierra las Pintas population across the barren and occasionally flooded southeastern edge of Laguna Salada, although these populations are probably continuous around the eastern sides of the Sierra Juárez.

DESCRIPTION Small to medium-sized, dorsoventrally compressed; adults reaching 121 mm SVL; frontal region flat; snout sharply pointed in lateral profile; external nares narrowly elliptical, situated posteriorly on snout, and directed dorsally; auricular lobes large; lower jaw deeply countersunk within upper jaw; dorsal head scales small and numerous; 4 to 6 superciliaries; slightly enlarged interparietal with conspicuous parietal eye; postparietals grade smoothly into granular dorsals; 8 or 9 obliquely keeled and laterally projecting supralabials; eyelid fringe scales large; single postmental; gulars subimbricate and flat, with midventral rows slightly enlarged anteriorly; pregular and gular folds present; 12 to 15 infralabials; dorsal scales smooth and granular, grading into larger, flat, weakly imbricate ventrals; lateral skin fold from axilla to groin; forelimbs moderate in length, with small lateral digital scales forming fringes; scales of shoulder and anterohumeral region enlarged and imbricate; hind limbs muscular, approximately 1.25 times length of forelimbs; patch of enlarged, pointed, postfemoral scales present; 26 to 34 subdigital lamellae on fourth toe; fourth toe with greatly enlarged triangular, fringes; enlarged fringes less well developed on third toe; 46 to 55 well-developed femoral pores in males; tail approximately same length as body, flat, and very wide anteriorly; dorsal caudal scales weakly imbricate; vent bordered posteriorly by small granular scales; 2 enlarged postanal scales in males; slightly enlarged postanal scales present in females. **COLORATION** Dorsal ground color whitish, overlaid with network of dark ocelli forming lineate pattern on body; limbs and tail spotted; large black ventrolateral body bar equidistant between limbs; thin dark chevron patterns on gular region; ventral surfaces of body, limbs, and anterior tail immaculate; ventral caudal bands extending approximately from midpoint of tail to tip; faint to bold broad orange stripe on side of belly posterior to body bar in adults; smaller orange spot occasionally present anterior to body bar; orange most vivid during breeding season; orange occasionally on eyelids.

NATURAL HISTORY *Uma notata* has many specialized morphological adaptations for its highly arenicolous lifestyle

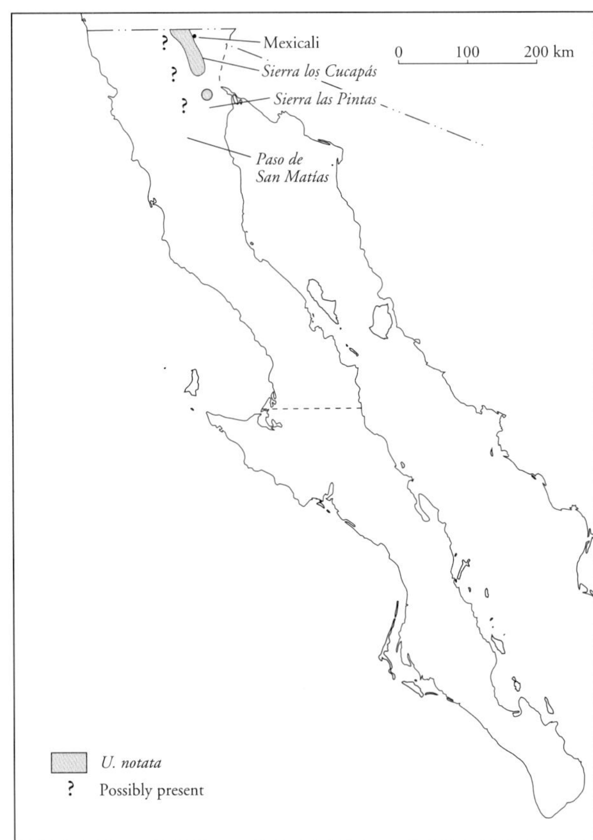

MAP 4.16 Distribution of *Uma notata* (Colorado Desert Fringe-Toed Lizard)

(Carothers 1986; Norris 1958, 1967; Pough 1969; Stebbins 1944) and is restricted to areas of fine, windblown sand. These habitats are disjunct in the Lower Colorado Valley Region of northeastern Baja California. Windblown sand accumulates at the northeastern edge of Laguna Salada at the western base of the Sierra los Cucapás. Although sand does not accumulate in large enough quantities to form dunes, it forms extensive hummocks around the bases of vegetation, and these are sufficient to support *U. notata.* This lizard also occurs in sand dunes along the irregular and deeply serrated western edge of the Sierra las Pintas, approximately 25 km southeast of the Sierra el Mayor.

Much has been published on the natural history and ecology of *U. notata* in the United States (e.g., Carothers 1986; Dunham et al. 1994; Green 1994; Pough 1970, 1977, and references therein). Although there are no reports on the Baja California populations, their geographic proximity to the United States populations and the continuum of habitat between them suggest that they should not differ significantly. *Uma notata* is generally active from spring through fall, and I have observed lizards in the Sierra las Pintas from early March through late September. Lizards emerge in the early morning to elevate their body temperatures by exposing only the tops of their heads above the surface of the sand into which they burrowed the previous late afternoon. Blood heated in the head circulates through the body, and when sufficiently warm, lizards emerge and may continue to bask on the dunes. *Uma notata* is cryptically colored and often goes unobserved unless frightened into running. It feeds on a wide variety of arthropods and is known to eat the juveniles of other lizards. Its diet also includes plant material. Females become gravid in the spring and lay clutches of one to five eggs every four to six weeks from May through August (Stebbins 1985). I have observed hatchlings in the Sierra las Pintas during mid-September.

Urosaurus Hallowell

BRUSH LIZARDS

The genus *Urosaurus* is composed of eight small terrestrial to scansorial species that collectively range from southwestern United States and northern Mexico south along the west coast of Mexico to Chiapas and throughout Baja California. There are two additional endemic species in the Revillagigedo Archipelago, approximately 520 km off the coast of Colima, Mexico. Brush lizards are usually conspicuous elements of the environments they inhabit. Adult males are commonly seen displaying from elevated perches, bobbing up and down and exposing their brightly colored abdominal patches. *Urosaurus* is most closely related to *Sceloporus* (Reeder and Wiens 1996). Four species of *Urosaurus* occur in Baja California, of which two, *U. nigricaudus* and *U. lahtelai,* are endemic, and two others, *U. graciosus* and *U. ornatus,* are more widespread but are found only in northeastern Baja California.

FIG. 4.48 ***Urosaurus graciosus***

LONG-TAILED BRUSH LIZARD; LAGARTIJA

Uro-saurus graciosus Hallowell 1854:92. Type locality: "Lower California"; restricted to "Winterhaven [Fort Yuma], Calif.," USA, by Smith and Taylor (1950:355). Syntypes: ANSP 8550, 8551

IDENTIFICATION *Urosaurus graciosus* can be distinguished from all other lizards in the region of study by the presence of eyelids; gular folds; enlarged, platelike head scales; nonspinose dorsal scales; enlarged and keeled middorsal scale rows; abdominal scales that are neither enlarged nor rectangular; supraocular scales that are flat to weakly convex; and a tail that is at least twice the length of the body.

RELATIONSHIPS AND TAXONOMY Wiens (1993b) placed *Urosaurus graciosus* as the sister species to all the other *Urosaurus.* Reeder and Wiens (1996), however, place *U. graciosus* as the basal member of a monophyletic group that contains *U. ornatus,* from the southwestern United States and northern Mexico, and the Islas Revillagigedo endemics, *U. auriculatus* and *U. clarionensis.*

DISTRIBUTION *Urosaurus graciosus* ranges from southern Nevada south through southeastern California and western Arizona to northwestern Sonora and northern Baja California, Mexico (Stebbins 1985). In Baja California, *U. graciosus* ranges throughout the northeastern portion of the peninsula east of the Peninsular Ranges south to at least Bahía San Luis Gonzaga (Grismer 1989c; map 4.17). No specimens of *U. graciosus* have been observed between San Felipe and Bahía San Luis Gonzaga, a distance of approximately 145 km. I believe this species ranges continuously to at least Puertecitos, 45 km south of San Felipe, but perhaps no farther. The region from Puertecitos south to El Huerfanito is composed of a rocky volcanic terrace, 25 km wide, on which I have never found *U. graciosus* and on which I do not believe it occurs. I know of no continuum of lowland desert habitat around the west side of this volcanic terrace, which forms the eastern slopes of the Sierra Santa Rosa. Therefore, *U. graciosus* either occurred south of the volcanic area before its formation or dispersed down the coast from the north, over water, subsequent to the formation of the volcanic area.

DESCRIPTION Small, with adults reaching 58 mm SVL; body somewhat slender; head and body not flattened; head tri-

FIGURE 4.48 *Urosaurus graciosus* (Long-Tailed Brush Lizard) from Bahía San Luis Gonzaga, BC

angular; snout smoothly rounded in dorsal profile; head scales enlarged and platelike; 6 postrostrals; internasal not contacting rostral; frontal usually divided; 6 flat to weakly convex supraoculars; interparietal enlarged, with conspicuous parietal eye; postparietals grading abruptly into granular nuchal scales; dorsal scales granular with wide vertebral series of enlarged, keeled, uniformly sized scales; granular dorsals grade smoothly into larger flat, imbricate ventrals; single paravertebral fold extending from nuchal region to base of tail containing enlarged scales; 6 to 9 supralabials; 2 postmentals; gulars flat and juxtaposed; incomplete antegular and complete gular fold; 9 or 10 infralabials; forelimbs thin, moderate in length, covered with large keeled, spinose scales dorsally and granular scales ventrally; hind limbs thin, approximately 1.25 times length of forelimbs; anterior of thigh and posterior of foreleg covered with large keeled, imbricate scales; granular scales covering posterior region of thigh; large keeled, imbricate scales covering dorsal and anterior portion of foreleg; 21 to 23 well-developed femoral pores in males; 20 to 28 subdigital lamellae on fourth toe; tail 1.8 to 2.4 times SVL, conical in cross section; caudals whorled, strongly keeled, and spinose; pair of enlarged postanal scales in males. **COLORATION** Dorsal ground color gray; thin dark line extends from external nares to eye; wide dark line extends from eye along flank to hind limb insertion, bordered ventrally by pale stripe; dorsal body pattern occasionally consists of paired, oblique, dark paravertebral lines; thin, sinuous, incomplete vertebral lines occasionally present; ventral surfaces light gray to white and faintly spotted to immaculate; abdominal patches in males pale blue to green, with light-colored spots; yellow to orange gular region in males and females; thin dark bands on forelimbs; hind limbs variably mottled; tail immaculate.

NATURAL HISTORY *Urosaurus graciosus* is primarily an arboreal species, ranging throughout the Lower Colorado Valley Region of northeastern Baja California. It is usually found on vegetation in flat, open sandy areas or in wide arroyo bottoms. Lizards are uncommonly seen in areas with rocky substrate. They are often observed on the stems of Creosote Bushes *(Larrea tridentata)* but frequent other vegetation as well. Welsh 1988 reports two specimens from the eastern base of the Sierra San Pedro Mártir on Ironwood Trees *(Olneya tesota),* Linsdale 1932 reports on a specimen beneath a "mesquite tree" at San Felipe, and Meek 1905 reports several individuals on "trees and bushes" at San Felipe. South of San Felipe, at Bahía San Luis Gonzaga, *U. graciosus* is found in association with the low-growing, halophytic vegetation along the beaches (Grismer 1989c). Lizards are abundant on Salt Bush (*Atriplex* sp.) in sand dunes along the western face of the Sierra las Pintas.

Other than the brief comments on microhabitat noted above, nothing has been reported on the natural history of *U. graciosus* in Baja California. Vitt and Omart (1978) studied its ecology and reproduction along the Lower Colorado River in southern Arizona, south to the U.S.-Mexican border. These populations should not differ significantly from

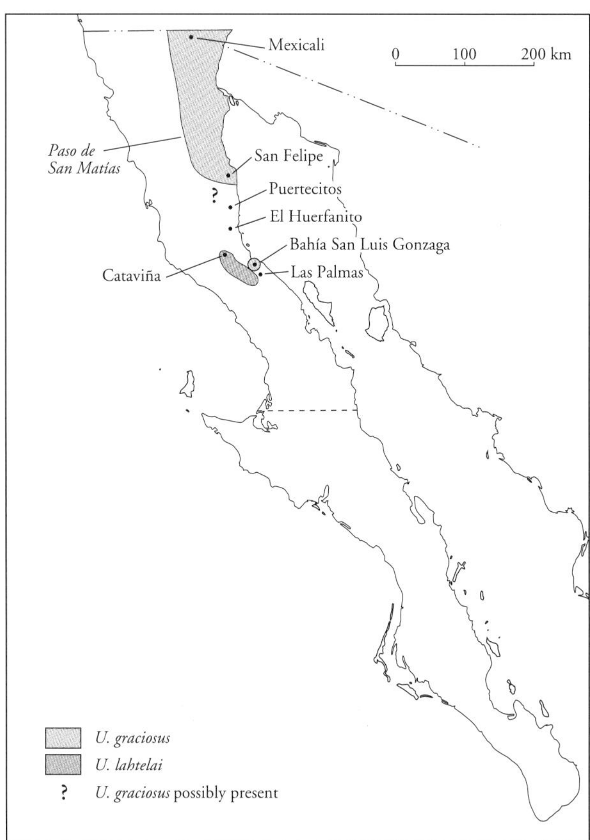

MAP 4.17 Distribution of *Urosaurus graciosus* (Long-Tailed Brush Lizard) and *U. lahtelai* (Baja California Brush Lizard)

those in northeastern Baja California. *Urosaurus graciosus* in Baja California is active as early as late winter. I observed several adults in the late afternoon in early March, basking on salt bushes (nearly always two per bush) in the northern portion of the Sierra las Pintas. Linsdale 1932 reports an active specimen at San Felipe on 20 March. *Urosaurus graciosus* emerges in the morning to bask on vegetation. This species is cryptic and capable of markedly varying its coloration. Lizards living on creosote bushes have a lineate dorsal pattern that is very similar to the thin, dark variegations of the stems on which they bask. Those on light green salt bushes have a distinctive greenish-golden hue. When frightened, lizards run down into the base of the bush to escape. *Urosaurus graciosus* feeds on a variety of small arthropods and may lay one or two clutches of two to ten eggs from May to August (Vitt and Omart 1978).

FIGURE 4.49 *Urosaurus lahtelai* (Baja California Brush Lizard) from Misión Santa María, BC

FIG. 4.49 **_Urosaurus lahtelai_**

BAJA CALIFORNIA BRUSH LIZARD; LAGARTIJA

Urosaurus lahtelai Rau and Loomis 1977:25. Type locality: "4 km N parador Cataviña (also known as Santa Inés) (114 50'W, 29 45'N), elevation 564m near Mexico Highway 1, State of Baja California, Mexico." Holotype: LACM 116541

IDENTIFICATION *Urosaurus lahtelai* can be distinguished from all other lizards in the region of study by the presence of eyelids; gular folds; enlarged, platelike head scales; nonspinose dorsal scales; enlarged and keeled middorsal scale rows; abdominal scales that are neither large nor rectangular; a paravertebral body fold lacking enlarged scales; no axillary spot; and long, thin, sinuous, dark paravertebral lines (in females) or shallow, U-shaped, thin blotches (in males) bordering a lighter vertebral region.

RELATIONSHIPS AND TAXONOMY Wiens (1993b) and Aguirre-L. et al. (1999) place *Urosaurus lahtelai* as the sister species of *U. nigricaudus.* Reeder and Wiens (1996) place *U. lahtelai* as the basal member of a group that contains *U. nigricaudus* and the southwestern Mexican forms *U. bicarinatus* and *U. gadovi.* Wiens (1993b) and Grismer (1994b,d, and 1999b) discuss taxonomy.

DISTRIBUTION *Urosaurus lahtelai* is known only from the granitic outcroppings in the vicinity of Cataviña, Las Arrastras de Arriola, and Las Palmas (see map 4.17).

DESCRIPTION Small, with adults reaching 58 mm SVL; body somewhat slender; head and body not flattened; head triangular; snout smoothly rounded in dorsal profile; head scales enlarged and platelike; 2 postrostrals; internasal contacting rostral; frontal single; 4 or 5 flat to weakly convex supraoculars; interparietal enlarged, with conspicuous parietal eye; postparietals grading abruptly into granular nuchal scales; dorsal scales granular, with median series of enlarged, keeled, uniformly sized scales; granular dorsals grade smoothly into larger, flat, imbricate ventrals; single paravertebral fold extending from nuchal region to base of tail and lacking enlarged scales; 4 to 7 supralabials; 2 postmentals; gulars flat and juxtaposed; incomplete antegular and complete gular fold; 7 to 10 infralabials; forelimbs covered with large keeled, spinose scales dorsally and granular and larger flat, imbricate scales ventrally in humeral and forearm regions, respectively; hind limbs approximately 1.00 to 1.25 times length of forelimbs; suprahumeral and supratibial surfaces of foreleg covered with large keeled, imbricate scales; granular scales covering postfemoral region; large imbricate scales covering infratibial region; 22 to 28 well-developed femoral pores in males; 23 to 25 subdigital lamellae on fourth toe; tail 1.6 to 2.2 times SVL, conical in cross section; caudals indistinctly whorled, strongly keeled, and spinose; pair of enlarged postanal scales in males. **COLORATION** ADULT MALES: Dorsal ground color dark gray; dorsal body pattern of adult males consisting of a wide, dark, sometimes inconspicuous paravertebral line beginning as a shoulder bar and terminating at the hind limb insertion; nuchal blotch bordered anteriorly and posteriorly in yellow; short yellow transversely diagonal lines occurring at regular intervals along the dark paravertebral line; vertebral region light and bordered in black; patternless, ground-colored patch occurring immediately posterior to forelimb insertion; yellowish spots on sides and upper portions of limbs; thin, light-colored bands on forearms; posterior two-thirds of tail black; ventral ground color light gray; gular region greenish yellow to orange; lateral belly patches metallic green and enclosing lighter spots. FEMALES AND JUVENILES: Similar to adult males except for general lack of the yellow coloration; tail light-colored; females have long, thin, sinuous, dark paravertebral lines and lack gravid coloration.

NATURAL HISTORY *Urosaurus lahtelai* is scansorial and reported to occur on both rocks and trees (Linsdale 1932; Rau and Loomis 1977). In the vicinity of Cataviña and Rancho Santa Inés, *U. lahtelai* is most commonly seen on large boulders or the branches of adjacent Mesquite *(Prosopis glandulosa)* trees. During mid-April, I observed 15 individuals at Misión Santa María, approximately 9 km northeast of Cataviña in fan palms *(Washingtonia filifera* and *Brahea armata)* associated with an oasis (Grismer and McGuire 1993). Thirteen individuals were observed on palm trunks from 0.5 to 1.5 m in height, with their heads facing down. Another was seen on a large rock within the grove and another on the ground. *Urosaurus lahtelai* was the only lizard seen in the palm grove and occurred only where the density of the trees was greatest. Where the palms began to thin out, downstream to the east, lizards were not found. Linsdale (1932:362) reports seeing a specimen (reported as *Uta microscutata*) in a "semi-dormant condition on wet sand of river bed" at Cataviña, and I have observed lizards on the ground at Las Palmas, east of Bahía San Luis Gonzaga. In a palm grove just north of Cataviña, I observed several *U. lahtelai* as high as 6 m above the ground on the trunks of the palms, just below the edges of the dead palm fronds. When frightened, the lizards retreated beneath the fronds. It was common to see one pair of adults per tree.

Urosaurus lahtelai has been observed abroad as early as mid-March (Linsdale 1932). I have seen individuals abroad during mid-March and early April at Cataviña and as late as mid-September. These observations suggest that *U. lahtelai* has a typical spring to early fall activity period and late fall to winter dormancy. During mid-December at Las Palmas, approximately 12 km southeast of Bahía San Luis Gonzaga, I found a hibernating adult male under a granitic exfoliation. Beneath the same exfoliation during late December, I found three adult *U. lahtelai* hibernating, along with a juvenile *Petrosaurus mearnsi. Urosaurus lahtelai* emerges early in the morning to bask on rocks and large boulders. During the hotter summer days its activity decreases, and it is most often seen only in the mornings and afternoons. During the spring, it is observable throughout the day.

I have found gravid females from mid-April to late May and observed hatchlings during mid-September. During mid-April, I observed four pairs of adults in close proximity on the trunks of palm trees. This suggests a typical spring and early summer breeding period followed by a summer to early fall egg-laying and hatching season. Nothing is known of the diet of *U. lahtelai,* but it presumably eats a variety of small arthropods.

REMARKS The small, circumscribed distribution of *U. lahtelai* is completely surrounded by that of its closest relative, *U. nigricaudus* (Grismer 1994a; Rau and Loomis 1977). It is not yet known whether contact occurs between these two species or where the contact zone might be. I have observed *U. nigricaudus* on the volcanic rocks in Arroyo de Rumbaugh, 4 km west of Bahía San Luis Gonzaga, and *U. lahtelai* on the granitic rocks at Las Palmas, approximately 16 km south of Bahía San Luis Gonzaga. This brings these two species to within approximately 14 km. The habitat between them is continuous along the rocky eastern sides of the Sierra Santa Isabel.

FIG. 4.50 ***Urosaurus nigricaudus***

BLACK-TAILED BRUSH LIZARD; LAGARTIJA

Uta nigricauda Cope 1864:176. Type locality: "Cape St. Lucas, Lower California," Baja California Sur, Mexico. Syntypes: USNM 5307

IDENTIFICATION *Urosaurus nigricaudus* can be distinguished from all other lizards in the region of study by the presence of eyelids; gular folds; enlarged, platelike head scales; nonspinose dorsal scales; lack of supranasals; abdominal scales that are not large and rectangular; a dorsolateral body fold lacking enlarged scales; no axillary spot; no long, thin, sinuous, dark paravertebral lines or shallow, U-shaped, thin blotches; and a ground color in the vertebral region that is not conspicuously lighter than the rest of the dorsum.

RELATIONSHIPS AND TAXONOMY See *U. lahtelai.*

DISTRIBUTION *Urosaurus nigricaudus* ranges along the eastern side of the Peninsular Ranges from San Diego County, California, south to the Cape Region of Baja California. In northern Baja California, *U. nigricaudus* occurs on both sides of the Peninsular Ranges. In the northwest, it extends from Tijuana to San Quintín, reaching 1,525 m in elevation in the Sierra San Pedro Mártir (Welsh 1988). In the northeast, *U. nigricaudus* extends along the rocky eastern faces of the Sierra Juárez and Sierra San Pedro Mártir in the Lower Colorado Valley Region, reaching 2,120 m along the Sierra San Pedro Mártir (Welsh 1988). South of the Sierra San Pedro Mártir, its distribution is continuous into the Cape Region (map 4.18). *Urosaurus nigricaudus* does not occur in the Vizcaíno Desert of central Baja California but is present on the Pacific islands of Magdalena and Santa Margarita and the Gulf islands of Ballena, Carmen, Cayo, Coronados, Danzante, El Coyote, El Requesón, Espíritu Santo, Gallina, Gallo, Gaviota, Islitas, Las Ánimas, Pardo, Partida Sur, San Cosme, San Damian, San Francisco, San José, San Marcos, and Tijeras (Grismer 1993, 1999a,b). Aguirre-L. et al. (1999) erroneously claim this species to be present on Isla Cedros (Grismer 2001a).

DESCRIPTION Same as *U. lahtelai,* except for the following: adults reaching 50 mm SVL; postparietals grading abruptly into granular to convex nuchal scales; scales of paravertebral fold undifferentiated or slightly larger than adjacent flank scales; 7 to 10 infralabials; forearm covered with flat, imbricate scales ventrally; 21 to 26 well-developed femoral pores in males; 20 to 23 subdigital lamellae on fourth toe; tail 1.2 to 2.0 times SVL. **COLORATION** ADULT MALES: Dorsal ground color gray to dark gray; faint to conspicuous paired, paravertebral, obliquely oriented black blotches beginning as a shoulder bar; blotches usually bordered posteriorly by pale blue to turquoise spots or bars; shoulder bar bordered posteriorly and anteriorly by pale blue to turquoise spots or bars; dark flecks on top of head; limbs indistinctly banded; posterior three-quarters of tail black; anterior of sides with occasional brownish-orange tint; sides usually covered with yellowish spots; ground color of ventral surfaces light gray; vivid metallic green to bluish-green abdominal patches with or without light speckling; abdominal patches may or may not contact medially; gular region variable, often uniform green, sulfur yellow, lemon yellow, orange, or light green centrally, surrounded by yellow or burnt orange, but occasionally yellow or orange centrally surrounded by blue. FEMALES AND JUVENILES: Dorsal color pattern similar to that of males, but dark blotching reduced and blue spotting generally absent; gular region usually with pale yellowish-orange throat patch.

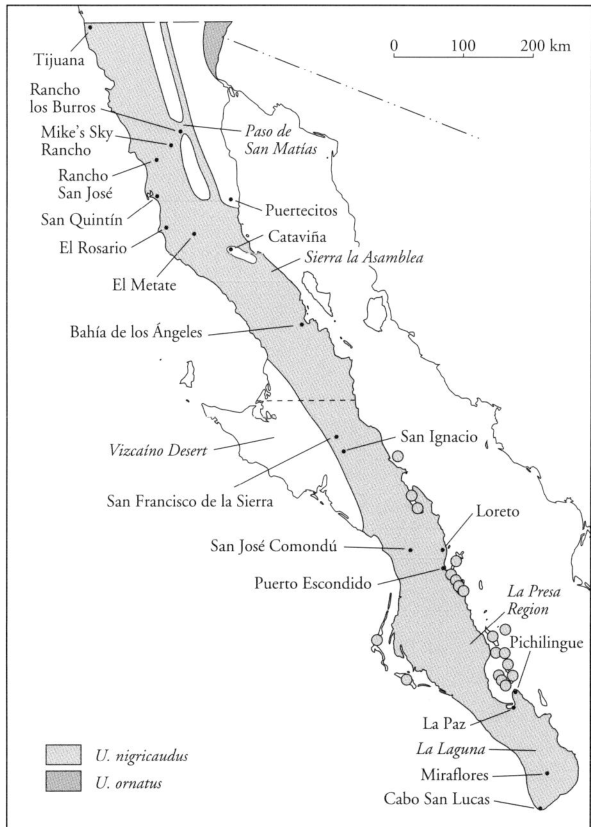

MAP 4.18 Distribution of *Urosaurus nigricaudus* (Black-Tailed Brush Lizard) and *U. ornatus* (Tree Lizard)

FIGURE 4.50 Adult male *Urosaurus nigricaudus* (Black-Tailed Brush Lizard) from Jaraguay, BC

GEOGRAPHIC VARIATION *Urosaurus nigricaudus* exhibits a modest degree of geographic variation throughout its distribution in Baja California as well as unique color pattern anomalies on some Gulf islands. The most notable variation in squamation is the size and degree of keeling of the median dorsal scales. The median dorsal scales of lizards from the Comondú area and to the north (once considered *U. microscutatus;* Mittleman 1942) are only slightly larger than the adjacent dorsal scales (28 to 36 dorsals in one head length, measured from the tip of the snout to the posterior margin of the interparietal) and faintly keeled. South, near La Presa, the median dorsals are slightly larger (25 to 36 per head length), and the keeling is more distinct. In populations from Islas Espíritu Santo and Gallo, scales are larger still (22 to 33), and the keeling is much stronger. Median dorsals are largest (18 to 24) and very strongly keeled in Cape Region populations, which also tend to have slightly enlarged scales in the paravertebral folds. This condition is barely visible in populations from the vicinity of La Presa and does not occur in populations farther north.

The dorsal blotches in lizards from the Comondú area and north tend to be smaller than the shoulder bar and less well-defined, and generally do not extend far enough ventrally to contact the dorsolateral fold. From the vicinity of

La Presa south, the dorsal blotches are the same size as the shoulder bar and well-defined, and they usually make contact with or extend below the dorsolateral fold. Intermediate pattern types are observable in lizards between La Presa and the Comondú area, 150 km to the north.

Gular region coloration in adult males is highly variable and noted here from north to south (descriptions are taken from lizards captured while active during the day): uniform metallic green at Rancho San José; uniform sulfur yellow at Bahía de los Ángeles; light green centrally surrounded by lemon yellow from San Francisco de la Sierra to Puerto Escondido; light green centrally surrounded by burnt orange 69 km south of Loreto; cream-colored centrally surrounded by yellow, or yellow to light green centrally surrounded by burnt orange, in the vicinity of La Presa; and uniform lemon yellow at Miraflores in the Cape Region. North to south insular variation in gular coloration is as follows: light green centrally surrounded by yellow on Isla San Marcos; light green centrally surrounded by yellow on Isla Bocana; light yellow centrally surrounded by burnt orange on Islas Danzante and San Damian; uniform light orange on Isla San Cosme; sulfur-yellow gular spot on Isla Ánimas; uniform light yellow to lemon yellow on Isla Cayo; light green centrally surrounded by burnt orange on Isla Ballena; tricolored gular marking with an orange border surrounding a light greenish interior with a central blue spot on Isla Gallina; solid metallic reddish-orange on Isla Gallo; and either a solid burnt orange or sulfur yellow on Isla Espíritu Santo. Additionally, adult males from the adjacent islands of Cayo, San José, and San Francisco have extensive orange coloration on the flanks.

Lizards from the montane areas in the northern portion of the species' range, near Rancho los Burros in the Sierra San Pedro Mártir, are less boldly marked and generally lack the vivid turquoise blue spots characteristic of other populations. In many cismontane populations in northwestern Baja California lizards are boldly marked, but the blue spotting is generally pale and much less intense. Farther south, lizards from Isla Ánimas, a very small (0.5 km × 0.3 km), rocky island covered with white guano, appear faded and washed-out in dorsal coloration. They also tend to lack turquoise spots on the dorsum, and their blue abdominal patches are separated medially by yellow. *Urosaurus nigricaudus* from Isla Danzante tend to lack the turquoise borders surrounding the dark shoulder bar. Adult males from Islas Espíritu Santo and Partida Sur are very colorful, with immaculate yellow-white oblique bars extending the length of the body.

NATURAL HISTORY *Urosaurus nigricaudus* ranges throughout a wide variety of habitats in Baja California, from the California Region in the north to the Arid Tropical Region in the south. This species is scansorial, and many anecdotal comments have been made concerning microhabitat variation (Linsdale 1932; Murray 1955; Tevis 1944; Van Denburgh 1922). These have led to the general misconception that populations north of the Isthmus of La Paz (once considered *U. microscutatus;* see Aguirre-L. et al. 1999 and Grismer 1999b) primarily prefer (or are restricted to) rocks, whereas populations from the Cape Region prefer vegetation (Linsdale 1932; Mittleman 1942). I have not found this distinction to be valid and have observed this species, throughout the peninsula, to be equally abundant on both. Asplund (1967), Alvarez et al. (1988), and, to some extent, Leviton and Banta (1964) indicate that *U. nigricaudus* is a habitat generalist in the Cape Region, as well as on Isla San José (Asplund 1967). I have observed lizards on the branches of shrubs and on rocks on Islas Espíritu Santo and San José; Welsh (1988) reports *U. nigricaudus* on vegetation along the base of the Sierra San Pedro Mártir. On many occasions I have observed specimens on rocks and logs along the stream bank of Arroyo San Rafael in the foothill chaparral east of Mike's Sky Rancho. On Isla San Marcos, *U. nigricaudus* is ubiquitous and is as common on vegetation as on rocks. I have observed adult males displaying from 2 m above the ground on the trunks of Cardón Cacti *(Pachycereus pringlei)* just south of Bahía Concepción and on Isla Espíritu Santo. I observed an adult male at San Ignacio jump from a rock wall to a thin Mesquite *(Prosopis glandulosa)* 20 to 25 cm above its position to avoid my approach. I have often observed individuals at Bahía de los Ángeles nimbly running and jumping throughout the thin branches of brushy vegetation. On Isla Ánimas, I watched an adult male climbing on a narrow twig of a dead Burrow Bush (*Ambrosia* sp.) as though it were searching for something. It jumped approximately 15 cm to the twig of another bush behind it, requiring a mid-air back flip. Its agility was distinctly anole-like. On Isla Bocana, lizards were found on the beach cobble running from bush to bush.

On the small island of Gallo (0.4 km × 0.5 km), *Urosaurus nigricaudus* is sympatric with *Uta stansburiana.* Here, *Uta* remain on the ground, and adult *U. nigricaudus* are principally confined to the Guayacán Bush *(Vizcainoa geniculata).* On this island *Vizcainoa geniculata* are often used as nesting sites by Brown Pelicans *(Pelecanus occidentalis),* whose guano whitens most of the shrub. The *U. nigricaudus* that use these shrubs are faded and whitish in coloration and cryptic within this microhabitat. Hatchlings and juveniles are often seen on the ground. It appears that *Uta* and *Urosaurus* are avoiding competition by selecting different microhabitats.

From San Ignacio south, *U. nigricaudus* is active year-round. This is especially true in the Cape Region (Asplund 1967; Blázquez and Ortega-Rubio 1996; Leviton and Banta 1964). I have even seen lizards basking at 1,700 m elevation in the Sierra la Laguna in late December. In northern Baja California, lizards generally emerge from hibernation during mid-February to early March, and hatchlings remain active into November. *Urosaurus nigricaudus* basks in the morning and remains abroad throughout the day during the spring. During the hotter summer months, activity is restricted to early morning, late afternoon, and dusk, except when it is overcast or has recently rained. On Isla San José, I have observed *U. nigricaudus* active during moderately heavy rain showers during late August. I have occasionally observed lizards out at night as well.

Urosaurus nigricaudus eats a wide variety of small arthropods. Asplund (1967) reports that in the Cape Region, termites are the most commonly eaten arthropods, and Galina-Tessaro (1994) and Galina-Tessaro et al. (2000) report ants, moth and butterfly larvae, beetles, and bees as common food items. Although giving no locality, Case (1983) lists ants as the most commonly eaten food, followed by termites and beetles.

The reproductive season of *U. nigricaudus* varies with latitude. In northern Baja California, I have observed gravid females from early April at El Metate to mid-June at Rancho San José. Hatchlings begin to appear in mid-June in northern Baja California and have been observed as late as early November in the western foothills of the Sierra la Asamblea. Throughout central Baja California, I have seen gravid females during mid-July. Murray (1955) reports hatchlings from Mesquital in central Baja California during early July. In the Cape Region, Asplund (1967) reports gravid females from mid-July through late August and notes that males are sexually active during this period as well, stating that *U. nigricaudus* synchronizes its reproductive season with late summer rains. Romero-Schmidt et al. (1999) indicate that synchronized reproduction results from a May–June and August reproductive peak in Cape Region populations. In early August I observed a mating pair approximately 0.5 m above the ground on the trunk of a Plumeria tree *(Plumeria acutifolia)*. The male approached the female from behind and grabbed her tail in his mouth. He then moved forward, grabbed her nape, and twisted his body beneath hers so as to place their vents in apposition. I have seen hatchlings in the Cape Region from late August to mid-October; in the Sierra Guadalupe during late August; and at San Ignacio during mid-October. I have observed juveniles during late December and early January in the Sierra la Laguna and Pichilingue in the Cape Region. Hatchlings have been reported from mid-September through November from San José in the Cape Region (Asplund 1967), and I have seen them on Isla Ánimas in early October and on Isla Espíritu Santo throughout December.

Adult male *U. nigricaudus* have a push-up display wherein the entire body is elevated by extending both forelimbs and hind limbs (Rau 1980). While displaying, lizards lower the gular region, laterally compress the abdomen, and expose their bright metallic blue abdominal patches. In direct sunlight, these colors glisten most remarkably, as though they were electric. I have noticed that some males from Bahía de los Ángeles finish their display sequence with a violent lateral vibration of the tail.

REMARKS On Islas Ánimas and Cayo in the southern portion of the Gulf of California, *U. nigricaudus* reaches tremendous densities. Both of these very small islands serve as seabird rookeries, and *U. nigricaudus* is the only diurnal lizard present on either. Densities of 750 lizards per person-hour have been estimated in October on Isla Ánimas. During this period, all lizards were emaciated, perhaps because they were no longer obtaining nourishment from insects attracted to birds' nests as they did during the birds' reproductive season.

FIG. 4.51 *Urosaurus ornatus*

TREE LIZARD; LAGARTIJA

Uta ornata Baird and Girard 1852:126 (part). Type locality: "On the Rio San Pedro (Texas) and the province of Sonora"; restricted to "Rio San Pedro, (Devil's River), Val Verde County, Texas," USA, by Mittleman (1942:133). Syntypes: USNM 2750 (2 specimens)

IDENTIFICATION *Urosaurus ornatus* can be distinguished from all other lizards in the region of study by the presence of eyelids; gular folds; enlarged, platelike head scales; strongly convex supraocular scales; nonspinose dorsal body scales; enlarged and keeled middorsal scale rows; abdominal scales that are neither large nor rectangular; and paravertebral body folds encrusted with enlarged scales.

RELATIONSHIPS AND TAXONOMY See *U. graciosus*. Wiens (1993b) discusses taxonomy.

DISTRIBUTION *Urosaurus ornatus* ranges from southwestern Wyoming south through western Colorado, central and western New Mexico, and southwest Texas to northern Chihuahua, Coahuila, Nuevo León, and Tamaulipas, Mexico, in the east. In the west, it ranges south through southeastern Nevada, all of Arizona, and southeastern California to northeastern Baja California and southern Sinaloa, Mexico. In Baja California, it is strictly associated with ri-

FIGURE 4.51 Schottii pattern class of *Urosaurus ornatus* (Tree Lizard) from Isla Tiburón, Sonora

parian vegetation along the Río Colorado and its tributaries (see map 4.18). *Urosaurus ornatus* also occurs on Isla Tiburón in the Gulf of California.

DESCRIPTION Same as *U. graciosus,* except for the following: adults reaching 60 mm SVL; 2 postrostrals; internasals usually contacting rostral; 4 to 5 strongly convex supraoculars; dorsal scales granular, with wide median series of enlarged, keeled scales invaded by a vertebral series of smaller scales; series of 4 to 7 obliquely arranged rows of enlarged conical scales on sides of body present or absent; 5 to 9 supralabials; 6 to 11 infralabials; forelimbs covered with large keeled, spinose scales dorsally and granular and larger flat, imbricate scales ventrally in humeral and forearm regions, respectively; hind limbs approximately 1.0 to 1.25 times length of forelimbs; suprahumeral and supratibial surface of foreleg covered with large keeled, imbricate scales; granular scales covering postfemoral region; large imbricate scales covering infratibial region; 25 to 29 femoral pores; 22 to 29 subdigital lamellae on fourth toe; tail 1.2 to 1.9 times SVL. **COLORATION** Dorsal ground color light to dark gray, dull yellow, or tan; top of head unicolored or with thin, dark transverse lines; series of dark paravertebral blotches between limb insertions; blotches usually bordered posteriorly by lighter color; blotches occasionally in contact, forming dark, sinuous lines of varying thickness; conspicuous wide, light-colored area extending from external nares posteriorly below eye to hind limb insertion; tail and limbs faintly banded; adult males with vivid blue to bluish-green abdominal patches, with or without light speckling; abdominal patches may or may not contact medially; gular region yellow, green, or pale blue to green, and may or may not contact abdominal patches; adult females with whitish, yellow, or orange gular region; abdominal region gray, with or without faint speckling; juveniles uniform gray throughout, with or without faint speckling.

GEOGRAPHIC VARIATION There is a considerable difference in overall squamation and color pattern between *U. ornatus* from northeastern Baja California and Isla Tiburón. Previous authors considered these populations as different subspecies (Mittleman 1942), although Wiens (1993) found no discrete diagnostic differences between them or other subspecies and considered *U. ornatus* to be monotypic. The previously described subspecies from the region of study are used here as pattern classes. Schottii ranges throughout the state of Sonora, including Isla Tiburón. The dorsal body and caudal scales of Schottii from Isla Tiburón are strongly keeled, especially the caudal scales and the median series of dorsal scales, where the keels often form a nearly continuous ridge down the back. These same scales in Symmetricus from northeastern Baja California are less strongly keeled and do not form ridges. Schottii has a series of obliquely arranged rows of enlarged conical scales along the side of the body, whereas the flank scales in Symmetricus are nearly uniform in size, with only some scales slightly enlarged. In Schottii, the scales of the paravertebral fold are much larger than those in Symmetricus.

Schottii has a much darker and more vivid dorsal color pattern, being well marked with thin black lines on the head and paravertebral blotches. Symmetricus is much lighter in overall coloration, and in comparison the dorsal pattern appears washed-out. While basking, Schottii may appear jet black; Symmetricus from northeastern Baja California are never jet black.

NATURAL HISTORY In Baja California, *U. ornatus* is associated with riparian vegetation along the Río Colorado and its tributaries. Lizards are often seen basking on trunks of Salt Cedar (*Tamarisk* sp.) and other vegetation near the water's edge, as much as 3 m above the ground. *Urosaurus ornatus* is also common on rocks, so long as the rocks are not too far from vegetation. On Isla Tiburón, *U. ornatus* is commonly seen on Ironwood trees *(Olneya tesota)* on the flats and bajadas, as well as in the arroyo bottoms. This population seems particularly adept at jumping from branch to branch to escape.

Several authors have reported on aspects of the natural history of *U. ornatus* from the United States (Ballinger 1983; Dunham 1982, 1983; Dunham et al. 1994; Smith 1977, and references therein), however, nothing has been reported on its natural history in Baja California or on Isla Tiburón. To this I can only add that I have observed *U. ornatus* on Isla Tiburón abroad during late March and never past September. I have observed gravid females in Baja California in June and July. Based on the above U.S. studies, *Urosaurus ornatus* emerges in early spring, lays one to six clutches of eggs from May to September, and feeds on small arthropods.

Uta Baird and Girard

SIDE-BLOTCHED LIZARDS

The genus *Uta* contains seven species of small terrestrial lizards. One species, *Uta stansburiana,* is a habitat generalist that ranges widely throughout the western United States and northern Mexico and occurs on several offshore islands. It is usually the most conspicuous and abundant species of lizard where it is found, and its densities on islands can be very high. The remaining six species of *Uta* are endemic to different islands in the Gulf of California. *Uta* is distinguished from other phrynosomatid genera by having a dark axillary spot and sublabials that contact the second infralabial scales (Wiens 1993a). *Uta* is the basal member of a natural group that contains the rock lizards *(Petrosaurus),* the brush lizards *(Urosaurus),* and the spiny lizards *(Sceloporus)* (Reeder and Wiens 1996).

FIG. 4.52 **_Uta stansburiana_**

SIDE-BLOTCHED LIZARD; LAGARTIJA

Uta stansburiana Baird and Girard 1852:69. Type locality: "Valley of Great Salt Lake," Utah, USA. Syntypes: USNM 2753 (four specimens)

IDENTIFICATION *Uta stansburiana* differs from all other lizards in the region of study by having a dark axillary spot, unicolored head, and flat prefrontal region; lacking a uniform green dorsal color pattern; and having fewer than 75 transverse rows of ventral scales.

RELATIONSHIPS AND TAXONOMY Grismer (1994a,b) suggests that *U. stansburiana* has independently given rise to the insular endemics *U. nolascensis, U. palmeri, U. squamata,* and the ancestor of the *U. encantadae* complex.

DISTRIBUTION *Uta stansburiana* ranges widely across the western United States and northern Mexico from central Washington southeast to western Texas and south to northern Sinaloa, Durango, Nuevo León, and Tamaulipas, Mexico (Ballinger and Tinkle 1972). *Uta stansburiana* ranges throughout Baja California and occurs on the Pacific islands of Asunción, Cedros, Coronado (Norte, Media, and Sur), Magdalena, Natividad, San Benito (Oeste, Centro, and Este), San Gerónimo, San Martín, San Roque, Santa Margarita, and Todos Santos, and the Gulf islands of Alcatraz, Ángel de la Guarda, Ballena, Bota, Cabeza de Caballo, Cardonosa Este, Carmen, Cerraja, Coronados, Danzante, Dátil, El Pardito, Espíritu Santo, Flecha, Gallo, Granito, Lagartija, La Ventana, La Rasa, Las Galeras,

FIGURE 4.52 Adult male *Uta stansburiana* (Side-Blotched Lizard) from Isla Tortuga, BCS

Mejía, Mitlán, Monserrate, Partida Norte, Partida Sur, Patos, Piojo, Pond, Roca Lobos, Salsipuedes, San Lorenzo Norte, San Francisco, San Ildefonso, San Lorenzo Sur, San Luis, San José, San Marcos, Smith, Tiburón, Tortuga, and Willard (Grismer 1993, 1999a,b, 2001a; map 4.19).

DESCRIPTION Small, with adults reaching 69 mm SVL; body stout; head triangular, snout rounded in dorsal profile, pointed in lateral profile; prefrontal region slightly convex; head scales large and platelike; single prefrontal on each side usually in contact medially; 2 to 4 scales between supraoculars and fourth superciliary; frontoparietals elongate to square, contacting medially or not; parietal sulcus absent; temporal scales smaller and more granular; enlarged interparietal scale, with conspicuous parietal eye; abrupt transition between interparietal and dorsals; 79 to 132 transverse rows of dorsals; dorsals juxtaposed and convex anteriorly grading into flat, weakly imbricate scales posteriorly, with vertebral series weakly keeled; 2 to 132 transverse rows of nonkeeled dorsals; dorsals grade ventrally into granular flank scales, which grade medially into flat, imbricate ventrals; 55 to 78 transverse rows of ventrals; 5 to 7 enlarged supralabials; 2 postmentals; first sublabial contacting second infralabial; gulars flat, circular, and weakly imbricate; antegular fold present; gular fold extending dorsally to slightly above level of forelimb insertion; 8 to 11 infralabials; forelimbs stout and of moderate length; hind limbs stout, nearly twice as long as forelimbs; 24 to 33 well-defined femoral pores; 22 to 31 subdigital lamellae on fourth toe; tail approximately 1.6 times SVL, elliptical in cross section; dorsal and lateral caudal scales keeled and spinose, ventral caudals smooth and imbricate, caudals somewhat whorled; 2 enlarged postanal scales present in males. **COLORATION** Color pattern is extremely variable because of substrate matching and differs by sex and age. **ADULT MALES:** Dark ground color on dorsum; dorsolateral light lines present on neck, extending onto body and occasionally forming two bold dorsolateral stripes extending length of body and uniting at base of tail; paired dark paravertebral chevron markings on back, bordered posteriorly with lighter markings; dorsum generally covered with turquoise spots centrally and yellowish spots on sides; dark axillary spot usually distinct; lighter spotting on limbs usually present; tail with dense green flecking; gular region dark with orange or sometimes yellow spots laterally; ventrum usually light gray. FEMALES AND JUVENILES: Similar to males except dorsal spotting pattern absent, making paravertebral blotches more distinctive; chevron pattern usually coalesces on tail and continues to tip as darker bands. Females develop orange gravid coloration on sides of head, throat, and gular region, which fades after oviposition.

GEOGRAPHIC VARIATION The geographic variation of adult male *U. stansburiana* in Baja California and the Pacific and Gulf islands is extensive. Variation in females and juveniles is minimal and is not discussed. Lizards from cismontane and montane areas generally have a dark and boldly marked dorsal color pattern, with golden sides and a dense network of turquoise speckling. Lizards from arid, low-lying desert areas have a light overall ground color and lack the dense turquoise speckling and golden sides.

Notable localized variations are discussed from north to south. Lizards along the Pacific coast of northwestern Baja California usually have four dark blotches arranged transversely across the neck. Lizards from cismontane northwestern Baja California, near Mike's Sky Rancho at the northwestern base of the Sierra San Pedro Mártir and Laguna Hanson in the Sierra Juárez, are generally dark and vividly marked. Lizards from the upper peaks of the Sierra la Asamblea, near 1,000 m elevation, have prominent, light-colored dorsolateral stripes. *Uta stansburiana* from the Sierra Santa Clara have dark gular regions and bellies. Lizards from the sandy areas surrounding Estero Salinas, on the Pacific coast of the Magdalena Plain, are faded in color and nearly lack an axillary spot. Thirty-two kilometers to the east, however, the habitat is more rocky, and lizards have a distinct axillary spot.

The most significant departures from the "typical" color pattern occur in insular populations. Variation on islands along the Pacific coast from north to south is as follows. *Uta stansburiana* from Isla San Martín are very dark overall, matching the dark substrate of the volcanic island. The three insular populations of Islas San Benito, once considered to be a distinct species *(U. stellata),* show a modest degree of interisland variation given their proximity. Lizards of Isla San Benito del Oeste are dark overall, with a peppering of very fine turquoise spots centrally and yellowish spots laterally. Those of Isla San Benito del Centro are lighter in overall dorsal coloration and have a more typical *U. stansburiana* dorsal color pattern, lacking the fine peppering. Lizards from Isla San Benito del Este have both pattern types, and some specimens show varying degrees of intermediacy. *Uta stansburiana* from Isla San Roque have blood-red gular regions. On Isla Magdalena, adult males tend to lack a side blotch and have a distinct salmon-colored ventrum.

Variation on islands in the Gulf of California from north to south is as follows. Many individuals from Isla Willard have a uniform dull red dorsum, whereas others have a more typical color pattern. On Isla Cabeza de Caballo in Bahía de los Ángeles, lizards are very dark, matching the dark volcanic rock of this island. Lizards from Isla Partida Norte have a reddish hue with a weak banding pattern. Lizards from the guano-covered flats of Isla Cardonosa are much more faded than the populations on the adjacent hillsides. *Uta stansburiana* from Isla la Rasa is most commonly found on the black volcanic rock within patches of

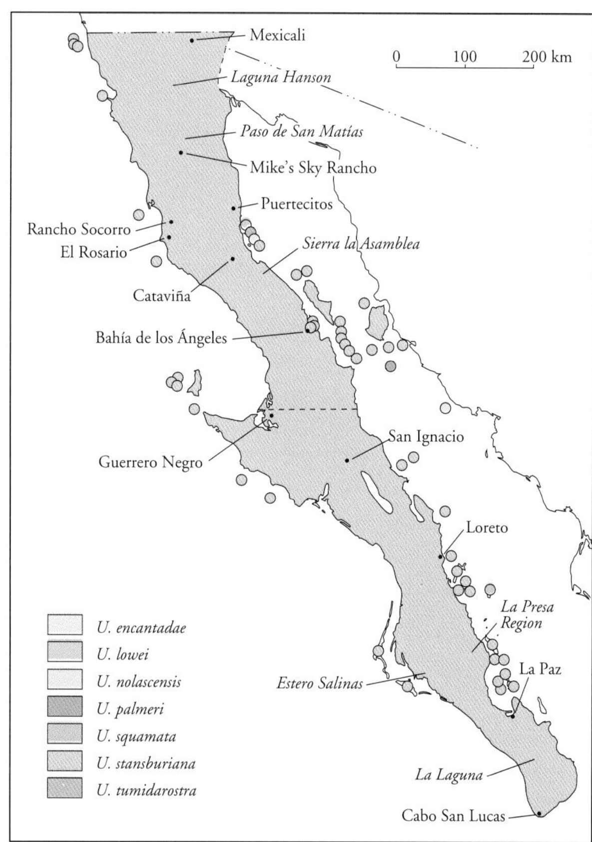

MAP 4.19 Distribution of *Uta* (Side-Blotched Lizards)

Cholla (*Opuntia* sp.), where it is dark and quite cryptic. *Uta stansburiana* from Islas Salsipuedes, San Lorenzo Norte, and San Lorenzo Sur (once recognized as *U. antiqua*) have very dark gular regions and bellies. Adult males from Isla Alcatraz in Bahía Kino, Sonora, are boldly marked and at a distance resemble juvenile Clark's Spiny Lizards *(Sceloporus clarkii)*. Lizards from Isla Tortuga have a dark brown ground color that matches the volcanic rock of the island. The throat and chest are nearly black, and the turquoise spots on the dorsum stand out in marked contrast to the dark ground color. This makes lizards of this population strikingly handsome. The dark axillary spot is often weak to absent in lizards from the adjacent islands of Carmen, Danzante, Coronados, Monserrate, and Las Galeras. The Isla Carmen population tends to have prominent dark shoulder markings and dark gray ventra. *Uta stansburiana* from Isla Pardito, off the south end of Isla San José, are brilliantly colored with gold and yellow. Adult males on Isla San Francisco tend to have a double axillary spot.

NATURAL HISTORY *Uta stansburiana* is a terrestrial generalist. In northern Baja California, it occurs in habitats ranging from the coastal sage scrub and chaparral communities of the California Region to the coniferous forest of the Sierra Juárez and creosote bush scrub communities of the Lower Colorado Valley Region. Lizards are common in dune areas along the Pacific coast at Rancho Socorro and just north of Guerrero Negro in the Vizcaíno Region; the flat, open Vizcaíno Desert; the volcanic badlands of the Magdalena Region; the Magdalena Plain; the Sierra la Giganta; and the Arid Tropical Region of the Cape Region (Alvarez et al. 1988). The only portions of Baja California that *U. stansburiana* does not inhabit are the Baja California Coniferous Forest Region of the Sierra San Pedro Mártir (Welsh 1988) and the upper elevations of the Sierra Guadalupe. In the Cape Region, it is relatively rare in regions of dense vegetation along the eastern face of the Sierra la Laguna but common in man-made forest-edge habitats along road cuts and agricultural fields.

Uta stansburiana does not seem to prefer any particular substrate or microhabitat, although it is generally more abundant near rocks. It is ubiquitous on every island on which it occurs, including the intertidal areas of Islas Willard, Ángel de la Guarda, Salsipuedes, San Lorenzo Norte, Roca Lobos, and San Lorenzo Sur. I have observed individuals on the crests of some of the highest peaks of Islas Cedros, Willard, and San Lorenzo Sur. On some islands, lizards are less common inland than in coastal areas, especially on Islas San Marcos and Carmen. On Isla Coronados, *U. stansburiana* is common inland on the volcanic rock as well as near shrubs on the sandy beaches. In the latter habitat, *U. stansburiana* behave like, and have very similar patterning to, Zebra-Tailed Lizards *(Callisaurus draconoides)*. They are light-colored like the sand and nearly lack an axillary spot. When threatened, they run to the bases of bushes and burrow in the sand to escape, or flee into burrows. On the extremely small Isla Las Galeras Oeste, which resembles a flat mesa, lizards are common along the rocky, clifflike periphery of the island but rare in the thick grass of the interior. On Isla Espíritu Santo, its distribution seems to be largely restricted to the sandy flats of the beaches because the rocky hillsides are occupied by Black-Tailed Brush Lizards *(Urosaurus nigricaudus)*.

On the Pacific islands, *U. stansburiana* is abundant on the outlying Islas San Benito archipelago along the Pacific coast as well as on the guano-covered rocks of Isla San Roque and Isla Asunción off the west coast of the Vizcaíno Peninsula. On Islas San Benito, *U. stansburiana* is common in the burrows of the Black Storm Petrel *(Oceanodroma melania)*, which it uses as retreats (Murray 1955). Lizards are also abundant in the garbage on Isla San Benito del Oeste but when approached usually run to a nearby low shrub for cover. On this island, I have observed lizards on rocks of all sizes, garbage, and sea lion carcasses, up to 1 m above the ground in agaves, and on the tops of low-growing shrubs.

In Baja California and the Gulf of California, *U. stansburiana* is active year-round. Activity generally peaks from March through October, but lizards are commonly seen basking and foraging from November through February (Blázquez and Ortega-Rubio 1996; Leviton and Banta 1964). During the cooler months of late fall and winter, *U. stansburiana* is strictly diurnal. During spring, and more so in summer and early fall, lizards are often observed abroad at night. It cannot be said exactly what they are doing, only that they are not using the overnight retreats frequented during the colder months. I have observed lizards abroad at night on all the Gulf islands on which they occur and in the desert regions of the peninsula from April through September.

During the spring, *U. stansburiana* emerges from beneath rocks, vegetation, or loose soil at sunrise to bask and begin its daily activities; it is most active during mid-morning. During the summer, lizards are often seen abroad long before sunrise. Some of these may be lizards that remained out all night or individuals beginning their daily activities before their thermoregulatory capacities are exceeded; they are also frequently found abroad at dusk and on overcast days. I have seen lizards out during light to moderate rainstorms in early August just south of Loreto and during late August near San Ignacio.

Uta stansburiana feeds on a variety of small arthropods (Asplund 1967; Case 1983). Whereas *U. palmeri, U. nolascensis,* and *U. squamata* feed extensively on plant material (see below), Case (1983) reported that plant material accounted for no more than 5 percent of the stomach contents of *U. stansburiana.* On Isla Ángel de la Guarda I observed one individual chase a moth down a hillside and another hanging by its forelimbs in a Brittle Bush (*Encelia* sp.) whose flowers were attracting flying insects. On Islas Roca Lobos, Salsipuedes, San Lorenzo Norte, and San Lorenzo Sur, lizards are commonly seen in the intertidal zone feeding on Isopods (*Ligia* sp.). On Isla Las Galeras Oeste, a juvenile was taken from the stomach of a large adult male.

Uta stansburiana mates year-round. I have observed copulating pairs from mid-February to mid-December in northern Baja California and from late March to mid-October in southern Baja California. Gravid females have been observed from mid-March through mid-July throughout the peninsula and Gulf islands. Gravid females ranged from 47 to 77 mm SVL and contained two to five eggs. Hatchlings have been observed from March through November throughout the peninsula and on various Gulf and Pacific islands. Interestingly, females with breeding coloration have been observed only from early April through June.

REMARKS *Uta stansburiana* has a circumgulf distribution and is a good example of an ecological generalist, ranging throughout a great diversity of habitats from sea level to 2,400 m elevation. These qualities make it especially capable of colonizing islands in the Gulf of California and along the Pacific coast, and at least six endemic insular species have independently evolved from various segments of *U. stansburiana* (Upton and Murphy 1997; Hollingsworth 1999).

FIG. 4.53 ***Uta nolascensis***

ISLA SAN PEDRO NOLASCO SIDE-BLOTCHED LIZARD; LAGARTIJA

Uta nolascensis Van Denburgh and Slevin 1921c:395. Type locality: "San Pedro Nolasco Island, Gulf of California [Sonora], Mexico." Holotype: CAS 50508

IDENTIFICATION *Uta nolascensis* differs from all other lizards in the region of study by having a dark axillary spot; fewer than 74 rows of ventral scales; conspicuous, dark inguinal patches; an aquamarine-colored ventrum; and a nearly green or greenish-brown, uniformly colored dorsum.

RELATIONSHIPS AND TAXONOMY See *U. stansburiana.*

FIGURE 4.53 Adult male *Uta nolascensis* (Isla San Pedro Nolasco Side-Blotched Lizard) from Isla San Pedro Nolasco, Sonora

DISTRIBUTION *Uta nolascensis* is endemic to Isla San Pedro Nolasco, Sonora, Mexico (see map 4.19).

DESCRIPTION Same as *U. stansburiana,* except for the following: adults reach 55 mm SVL; frontoparietals irregularly shaped, usually squarish; 1 or 2 scales between supraoculars and fourth superciliary; 95 to 108 transverse rows of dorsals; 64 to 72 transverse rows of ventrals; 17 to 41 transverse rows of nonkeeled dorsals; 6 or 7 supralabials; 8 or 9 infralabials; hind limbs stout, 1.5 times as long as forelimbs; 28 to 36 well-defined femoral pores in adult males; 25 to 30 subdigital lamellae on fourth toe; caudals whorled. **COLORATION** **ADULT MALES:** Ground color of head golden tan; ground color of body and tail green, usually overlaid with turquoise flecks; flecking occasionally on hind limbs; ground color of sides more golden, merging with that of head; dark antebrachial blotches usually present; dark inguinal patches present; gular region green centrally and yellowish-gold laterally, blending with aquamarine-colored abdomen. **FEMALES AND JUVENILES:** Similar to males, except dorsal coloration much more brown, with less turquoise flecking; antebrachial and inguinal patches not as distinctive.

NATURAL HISTORY Isla San Pedro Nolasco is a small, steep-sided, rocky island covered with fairly dense tropical deciduous thorn scrub (Felger and Lowe 1965). *Uta nolascensis* is common on and among rocks of all sizes but does not frequent the rocky beaches or the steep cliffs. It remains primarily in the arroyos and the upper, more level areas of the island. Here, lizards are more common on rocks in the vicinity of shrubs than on open rock faces or on rocks in grassy areas.

Uta nolascensis apparently has a typical seasonal activity period, with peak activity occurring in the late spring and summer. In late March, lizards are relatively uncommon, and no gravid females have been observed. In early July, lizards are very common, and many females are gravid. I have observed small juveniles in October, which indicates that hatching occurs in August and probably September as well. Gravid females with four eggs have been examined in museum collections. *Uta nolascensis* is unique among all other species of *Uta,* and most other species of phrynosomatids, in that females lack an orange breeding coloration. Eggs are presumably laid in late July and August.

During the summer, *U. nolascensis* emerges from its overnight retreats to bask on rocks in the morning sun. During this period, lizards are wary and difficult to approach. Quite often they are observed basking on rocks under the cover of vegetation. At dusk they are more easily approached. One night I observed an adult male approximately 1 m above the ground, sleeping while clinging to a thin vertical branch, as anole lizards do.

Uta nolascensis feeds on a variety of small arthropods, including scorpions, and according to Case (1983), 34 percent of its diet is plant material. I have observed individuals eating ants and found seeds in their scat.

REMARKS The green color pattern of *U. nolascensis* clearly has nothing to do with camouflage. In fact, individuals are conspicuous from great distances. I know of no predators on Isla San Pedro Nolasco to select against such coloration, and I suspect that it may have something to do with sexual selection, because adult males are a much brighter green than females and juveniles.

FIG. 4.54 ***Uta palmeri***

ISLA SAN PEDRO MÁRTIR SIDE-BLOTCHED LIZARD; LAGARTIJA

Uta palmeri Stejneger 1890a:106. Type locality: "San Pedro Martir Island, Gulf of California," Sonora, Mexico. Holotype: USNM 16002

IDENTIFICATION *Uta palmeri* differs from all other lizards in the region of study by having a dark axillary spot; more than 75 ventral scales; and more than 103 dorsal scales.

RELATIONSHIPS AND TAXONOMY See *U. stansburiana.*

DISTRIBUTION *Uta palmeri* is endemic to Isla San Pedro Mártir, Sonora, Mexico (see map 4.19).

DESCRIPTION Same as *U. stansburiana,* except for the following: adults reach 83 mm SVL; 2 or 3 scales between supraoculars and fourth superciliary; 103 to 121 transverse rows of dorsals; 38 to 60 nonkeeled transverse rows of dorsals; 75 to 86 transverse rows of ventrals; 5 to 7 supralabials; 9 or 10 infralabials; forelimbs slender and relatively long;

FIGURE 4.54 Adult male *Uta palmeri* (Isla San Pedro Mártir Side-Blotched Lizard) from Isla San Pedro Mártir, Sonora

hind limbs approximately 1.25 times as long as forelimbs; 30 to 37 well-developed femoral pores in adult males; 25 to 30 subdigital lamellae on fourth toe; tail approximately 1.5 times length of body; caudals whorled. **COLORATION ADULT MALES:** Capable of marked degree of color change depending on their thermoregulatory, reproductive, and behavioral status. Dorsal coloration ranges from jet black to dark gray with entire dorsum overlaid with dense network of turquoise spots; head and forelimbs lack turquoise spots; ground color of sides dark brown when dorsum spotted; lateral gular region gray and darkened centrally; rest of ventrum light-colored; breeding adult males develop orange hue similar to that of gravid females (see below), but less intense. **FEMALES AND JUVENILES:** Dorsal ground color light brown; dorsum with light-colored spots and lacking asymmetrically arranged blotches; ventrum light except for darkened posterior gular region; gravid females develop a bright pink to red stripe from axillary spot to hind limb insertion, which continues posteriorly to encompass tail.

NATURAL HISTORY Lying in the center of the Gulf of California, Isla San Pedro Mártir is the most isolated island in the region of study. It is part of the Central Gulf Coast Region, but it lacks extensive plant growth, primarily because it is covered with guano. This island is little more than an emergent rocky peak edged with precipitous bluffs. *Uta palmeri* is ubiquitous on Isla San Pedro Mártir, occurring on the rocky hillsides as well as in the intertidal zone. I observed one individual basking on the top of a Cardón Cactus *(Pachycereus pringlei)* branch approximately 2 m above the ground. Lizards on this island are unusually numerous: standing in one location in the intertidal zone, I counted 20 individuals within a few meters.

Uta palmeri has a most remarkable lifestyle, much of

which is tied to the seasonal nesting of the Blue-Footed Booby *(Sula nebouxi).* Much of what is reported here comes from the studies of Wilcox (1980) and Hews (1990). The reproductive cycle of *Uta palmeri* is synchronized to the seasonal nesting of boobies, and its local density is positively correlated with that of seabird nesting sites. *Uta palmeri* is active year-round, with peak activity extending from February to September. During the spring and summer, individuals emerge to bask shortly after sunlight reaches their basking sites. Emergence is slightly delayed on cloudy or overcast days. Activity usually peaks in the morning and drops sharply by noon, when temperatures begin to exceed the lizard's thermoregulatory capabilities. When temperatures drop during the late afternoon, activity resumes. Territorial males may remain active throughout the day but usually confine their activities to the shade. They show much less bimodality in their activity pattern and are among the first individuals seen in the morning and the last before sunset.

Uta palmeri eats a wide range of prey, and its local densities are correlated with prey abundance. During the breeding season of the Blue-Footed Booby, from late February through mid-May, lizards concentrate around the nests. A species of small black fly, referred to as *bobito* by local fishermen, breeds abundantly in the associated guano. So dense are the *bobitos* that they have driven visitors off the island; fishing boats that venture too close to the shore retreat out to sea. Lizards are often seen catching *bobitos* off the white breasts of nesting boobies. During a visit in late March, my field party became soaked with sea water on a particularly rough boat ride to the island. On landing, we were forced to lay our clothes out along the shore to dry. *Bobitos* landing on the clothes were very conspicuous, and lizards immediately began feeding on them. One adult male set up a feeding territory, which he defended by chasing off all potentially competing lizards. *Uta palmeri* also feeds on fresh and rotting fish scraps regurgitated by female boobies after they return to the nest. Lizards congregate around these "leftovers" in large numbers and feed frantically. Some individuals have also been observed eating seabird feces and feeding on dead hatchlings. In the intertidal areas, *U. palmeri* feeds heavily on marine isopods *(Ligia occidentalis).* Lizards have been observed to climb into vegetation and feed on flower heads, petals, and insects, a behavior uncommon in other species of *Uta.* Hews and Dickhaut (1989) reported cannibalism by an adult male (75 mm SVL) on a juvenile (40 to 50 mm SVL): the adult male was observed carrying the limp body of the juvenile. Whether this was the result of active predation or a case of carrion ingestion is not clear. Wilcox (1980) reported a similar observation.

As a result of the El Niño weather pattern of 1992, productivity in the Gulf of California was low, and breeding of the Blue-Footed Booby on Isla San Pedro Mártir was virtually nonexistent. I visited the island during late June of that year and saw no breeding birds but high densities of *U. palmeri;* all lizards appeared to be in very good health. Lizards were attracted to a red backpack my field partner had set down near some rocks. A large adult male entered the backpack and aggressively displayed at other lizards that approached. Later we found that lizards had been feeding on the fallen, ripened fruits of the Cardón Cactus *(Pachycereus pringlei),* whose fleshy interior was bright red like the backpack. The fruits of the Cardón Cactus are likely responsible for the attraction of this species to the color red, as reported by Wilcox (1980), and may have little to do with sexual behavior (see below). During this period, lizards would take pieces of kiwi fruit from our hands, and one individual actually crawled into my palm to drink fresh water. Apparently, in the absence of breeding birds, *U. palmeri* rely heavily on plant material.

Females reach sexual maturity at 54 to 58 mm SVL, within their first year of life. Males do not mature until reaching 67 mm SVL in their second year of life. This sexual bimaturism, which is not known to occur in *U. stansburiana,* may be unique to *U. palmeri,* although other insular species have not been studied. Females generally become reproductively active in May, although some gravid females have been found in April. By June, almost all adult females are gravid. Gravid female coloration peaks in development during the oviducal/corpora-luteal stage of the reproductive cycle and serves to inhibit aggressive behavior in territorial males. During this period, the coloration is most intense on the tail and is accompanied by a submissive tail-raising posture. Thus this color serves as a rejection rather than an attraction cue, although Wilcox (1980) hypothesized that in the early stages of its seasonal development, it may also serve to attract males. This hypothesis, however, was never tested. Clutch size ranges from one to five eggs, with peak egg deposition in July. Very rarely, *U. palmeri* lays two clutches in the same season. Hatching begins in late August and peaks in mid-September. After hatching, juveniles grow at the tremendous rate of 0.5 mm per day.

FIG. 4.55 ***Uta squamata***

ISLA SANTA CATALINA SIDE-BLOTCHED LIZARD; LAGARTIJA VERDE

Uta squamata Dickerson 1919:471. Type locality: "Santa Catalina Island, Gulf of California [Baja California Sur], Mexico." Holotype: AMNH 5424

FIGURE 4.55 Adult male *Uta squamata* (Isla Santa Catalina Side-Blotched Lizard) from Arroyo Mota, Isla Santa Catalina, BCS

IDENTIFICATION *Uta squamata* differs from all other lizards in the region of study by having a faint axillary spot, bright malachite-green tail, and turquoise spots on the top of the head and rostrum.

RELATIONSHIPS AND TAXONOMY See *U. stansburiana.*

DISTRIBUTION *Uta squamata* is endemic to Isla Santa Catalina in the Gulf of California, Baja California Sur, Mexico (see map 4.19).

DESCRIPTION Same as *U. stansburiana,* except for the following: adults reach 57 mm SVL; snout weakly pointed in dorsal profile; prefrontal region slightly concave; 4 or 5 scales between supraoculars and fourth superciliary; dorsals keeled and imbricate anteriorly, grading into larger, more strongly keeled, imbricate scales posteriorly; 70 to 84 transverse rows of dorsals; 2 to 12 transverse rows of non-keeled dorsals; dorsals grade abruptly into granular flank scales; 58 to 67 transverse rows of ventrals; 5 or 6 supralabials; 8 or 9 infralabials; hind limbs approximately 1.75 times as long as forelimbs; 25 to 33 well-developed femoral pores in males; 25 to 30 subdigital lamellae on fourth toe; tail approximately 1.5 times SVL; dorsal and lateral caudal scales strongly keeled, spinose, and somewhat whorled.

COLORATION Adult males and females lack sexual dichromatism, except for the ephemeral reproductive female coloration (see below). Dorsal ground color dark gray, overlaid with whitish to turquoise flecks centrally and yellow laterally; dorsal spotting extends anteriorly onto head and rostrum; conspicuous yellow to cream-colored stripe extends from nasal region through eye onto anterior portion of body; turquoise flecks on limbs; tail solid malachite green with a thin, dark vertebral stripe anteriorly; antebrachial blotches absent; inguinal patches absent; axillary spot very reduced; gular region dark with yellow spots laterally; remainder of ventral surfaces generally dark gray. Reproductively active females develop orange spots on the sides of their heads, gular regions, and anterior portions of brachium, which fade after oviposition.

NATURAL HISTORY Isla Santa Catalina lies with the Central Gulf Coast Region and is dominated by xerophilic thorn scrub. With the exception of water-edge microhabitats, *U. squamata* is ubiquitous. Lizards appear to be equally abundant on rocky hillsides, in sandy arroyo bottoms, and in the thick thorn scrub on the small bajadas. *Uta squamata* is far more common within 500 m of the coast than in the island's interior, although I have observed lizards on the crests of some of the highest peaks. Inland, lizards are almost always found in the vicinity of larger rocks and never on open ground, as they are near the coast. This species often occurs in pairs, especially during the spring, when reproductive behavior is also observed. During this period, males are often seen chasing females and following them around.

Uta squamata is probably active year-round, with peak activity extending from March through September. Activity decreases noticeably by October. During the summer, *U. squamata* emerges before sunrise, and I have observed individuals active at 0530, even in the presence of overcast skies and light rain. By midday during the summer, lizards usually retreat to the shade of rocks or up to 1 m off the ground in shaded vegetation. *Uta squamata* generally retreats to overnight refuges just prior to sundown. I have seen individuals out at night in April, but I could not determine whether they were active.

Uta squamata no doubt feeds on a variety of arthropods, and Case (1983) reported stomach contents to be 31 percent plant material. I observed an adult with the ripened fruit of a Fish Hook Cactus (*Mammillaria* sp.) in its mouth and have found the seeds of the Barrel Cactus *(Ferrocactus diguetii)* in the guts of others.

I have observed several females with gravid female coloration in April, and three eggs were found in a specimen collected in early July. By late July to mid-August, gravid females are rare, and the orange coloration has faded. Hatchlings are abundant from mid-July but become increasingly uncommon by early October. By early September, many females are gravid again and suffused with orange. One female collected during this period had three eggs, and another had four. This strongly indicates that *U. squamata* lays more than one clutch of eggs per year and would account for the very small individuals observed during the spring of the following year.

FIGURE 4.56 Adult male *Uta encantadae* (Enchanted Side-Blotched Lizard) from Isla Encantada, BC

FIG. 4.56 **Uta encantadae**

ENCHANTED SIDE-BLOTCHED LIZARD; LAGARTIJA

Uta encantadae Grismer 1994e:459. Type locality: "Isla Encantada, Gulf of California, Baja California, Mexico." Holotype: UA 49549

IDENTIFICATION *Uta encantadae* can be distinguished from all other lizards in the region of study by a dark axillary spot; a greatly inflated nasal region; and narrow, elongate frontoparietal scales.

RELATIONSHIPS AND TAXONOMY *Uta encantadae* is a member of the *U. encantadae* group, which also contains *U. lowei* and *U. tumidarostra* from the nearby Islas El Muerto and Coloradito, respectively (Grismer 1994d). Grismer (1994e) discusses the taxonomy of the *U. encantadae* group.

DISTRIBUTION *Uta encantadae* is known only from Isla Encantada and the adjacent rocks, Islotes Blancos, in the northern Gulf of California, BC, Mexico (see map 4.19).

DESCRIPTION Small, with adults reaching 69 mm SVL; body stout; head triangular, snout rounded in dorsal profile, truncate in lateral profile; prefrontal region greatly enlarged and protuberant; head scales large and platelike; single prefrontal on each side not contacting medially; 1 or 2 scales between supraoculars and fourth superciliary; frontoparietals elongate, contacting medially; deep parietal sulcus present at union of frontoparietals and interparietal; temporal scales smaller and more granular; enlarged interparietal scale, with conspicuous parietal eye; abrupt transition between interparietal and dorsals; 102 to 124 transverse rows of dorsals; dorsals juxtaposed and convex anteriorly, grading into flat, weakly imbricate scales posteriorly; vertebral series weakly keeled; 28 to 67 transverse rows of nonkeeled dorsals; dorsals grade ventrally into granular flank scales, which grade medially into flat, imbricate ventrals; 63 to 71 transverse rows of ventrals; 5 or 6 supralabials; 2 postmentals; first sublabial contacting second infralabial; gulars flat, circular, and weakly imbricate; gular and antegular fold present; 9 to 11 infralabials; forelimbs stout and of moderate length; hind limbs stout, twice as long as forelimbs; 23 to 30 femoral pores; 22 to 34 subdigital lamellae on fourth toe; tail approximately 1.6 times SVL, elliptical in cross section; dorsal and lateral caudal scales keeled and spinose; subcaudals smooth and imbricate, caudals not whorled; 2 enlarged postanal scales present in males and occasionally in females. **COLORATION** ADULT MALES: Dorsal ground color dark gray; faint pattern of darker offset paravertebral blotches visible; ground color of sides light brown to dull orange; vertebral region overlaid with network of small turquoise spots, grading laterally into yellowish-white spots on sides; large black axillary spot present; dorsal and lateral regions of tail nearly solid turquoise; gular and pectoral regions dark gray; lateral gular region usually with orange spots; remaining ventral surfaces blue-gray to very dark gray. FEMALES AND JUVENILES: Dorsal ground color gray to dark brown; dorsal pattern consisting of asymmetrically arranged, small, dark dorsal blotches bordered posteriorly by light markings; ventrum as in adult males. When reproductively active, females take on an increased orange suffusion in the gular region and the sides of the head, which apparently fades after oviposition.

NATURAL HISTORY *Uta encantadae* is almost completely confined to the intertidal areas of Isla Encantada and Islotes Blancos. Here, lizards are most abundant along the base of the cliffs in areas below seabird nests. *Uta encantadae* basks, forages, and takes refuge among the large dark boulders at the water's edge. It is less frequently observed within the island's guano-covered, rocky interior. At high tides, lizards do not simply move inland to escape the rising waters; they become scarce, but where they go is not known.

Uta encantadae is active year-round but with only minimal activity from November through February. During the spring, lizards are concentrated below the cliffside nesting sites of the Brown Boobies *(Sula leucogaster)* and Blue-Footed Boobies *(Sula nebouxi)*. Here there is a seasonal abundance of fish scraps and other carrion that fall from the nest and attract flies on which this lizard feeds. Other items found in the guts of this species are isopods, spiders, beetles, scorpions, ants, sea mites, shed skin, and plant material (Grismer 1994e). From June through mid-September, lizards emerge from their overnight retreats just before sunrise and are abundant. After basking for a short while, lizards begin foraging for food, which consists mostly of the intertidal isopod *Ligia occidentalis* (Grismer 1994e).

This species is wary and difficult to approach during the day. It runs toward the cliffs and takes refuge beneath large rocks. Lizards can be more easily approached at dusk and in early evening, when they are commonly found lying with their bodies pressed flat against the rocks as though trying to absorb their heat. Often large, dark males are observed lying on rocks covered with white guano, making them very conspicuous. Before dark, lizards retreat beneath the boulders or into cracks on the cliff.

I have observed gravid females during April. Hatchlings are generally not seen until late June and are abundant in late July and August. These observations seem to indicate that *U. encantadae* has a typical spring and early summer reproductive season with a summer hatching period. Hatchlings are not abundant in areas of the intertidal zone with large boulders as are adults and subadults, but they are quite common on stretches of open beach cobble away from the cliffs.

REMARKS *Uta encantadae* and its close relatives, *U. lowei* and *U. tumidarostra,* are unique among lizards, with the exception of the Galápagos Marine Iguana *(Amblyrhynchus cristatus),* in that they have hypertrophied (overdeveloped) nasal glands. In reptiles, this gland eliminates excessive salts from the blood and helps to maintain electrolyte balance (Dunson 1976). It seems likely that this hypertrophy evolved in response to these species' consumption of the isopod *Ligia occidentalis,* which has a salt concentration in its body higher than that of the surrounding sea water (Grismer 1994e; Hazard et al. 1998). These species are generally seen with copious salt crystals surrounding their external nares. Many individuals of *U. tumidarostra* have been observed with dried salt solution covering their entire bodies, giving them a white appearance. Grismer (1994d) suggested that the intertidal lifestyle of these species may have evolved in response to nesting seabirds covering the interior of the island with guano and rendering it nearly sterile. Thus, in the absence of primary production, the only food sources were in the intertidal zone.

FIG. 4.57

Uta lowei

DEAD SIDE-BLOTCHED LIZARD; LAGARTIJA

Uta lowei Grismer 1994e:464. Type locality: "Isla El Muerto, Gulf of California, Baja California, Mexico." Holotype: UA 49564

IDENTIFICATION *Uta lowei* can be distinguished from all other lizards in the region of study by a dark axillary spot; a greatly inflated nasal region; and diamond-shaped frontoparietal scales that do not contact one another at their corners on the midline or contact only narrowly.

FIGURE 4.57 Adult male *Uta lowei* (Dead Side-Blotched Lizard) from Isla El Muerto, BC

RELATIONSHIPS AND TAXONOMY See *U. encantadae.*

DISTRIBUTION *Uta lowei* is endemic to Isla El Muerto in the northern Gulf of California, BC, Mexico (see map 4.19).

DESCRIPTION Same as *U. encantadae,* except for the following: adults reach 66 mm SVL; frontoparietal scales not, or only narrowly, contacting at their corners on the midline; parietal sulcus absent; 91 to 113 transverse rows of dorsals; 14 to 36 transverse rows of nonkeeled dorsals; 59 to 70 transverse rows of ventrals; 8 to 11 infralabials; 24 to 31 well-developed femoral pores in males; 24 to 29 subdigital lamellae on fourth toe. **COLORATION** Same as *U. encantadae,* except for the following. ADULT MALES: Dense network of turquoise spots covering dorsum, not restricted to vertebral region; lateral gular region with yellow spots; remaining ventral surfaces off-white to light gray. FEMALES AND JUVENILES: Same as *U. encantadae.*

NATURAL HISTORY *Uta lowei* is abundant in the intertidal zone, although individuals can be found less frequently in inland rocky areas. In the intertidal zone, lizards are most common where the steep sides of the island meet the water. Adult *U. lowei* are generally absent from open intertidal areas that lack large boulders, such as the small beaches on the southwest side of the island. Adults are also uncommon in intertidal areas where rocks are generally smaller than 18 cm; they are usually found only in the vicinity of much larger rocks.

Uta lowei is active year-round but with only minimal activity from November through February. From June through mid-September, juveniles appear to emerge at sunrise, before adults, in contrast to *U. encantadae* and *U. tumidarostra* populations, in which adults are also active before sunrise. Adult *U. lowei* are abundant later in the morning as the temperature rises. *Uta lowei* forage for food in the intertidal zone and prey mostly on the intertidal isopod *Ligia occidentalis.* Other food sources include flies, ants, and pseudoscorpions (Grismer 1994e). Lizards forage at the water's edge and are generally difficult to approach during the day. When approached, they quickly take refuge beneath large rocks as they run toward the interior of the island. To escape my approach, one individual actually jumped into the water to reach a rock approximately 2 m away. From midday to late afternoon, lizards are more commonly seen in the shade of large rocks than in direct sunlight. Unlike *U. encantadae* and *U. tumidarostra, U. lowei* is not active at dusk and generally takes refuge beneath rocks before sundown.

Gravid females have been observed during April. Hatchlings are generally not seen until late June and are extremely abundant in late July and August. These observations suggest that *U. lowei* has a typical spring and early summer breeding season with a summer hatching period.

REMARKS See *U. encantadae.*

FIG. 4.58

Uta tumidarostra

SWOLLEN-NOSED SIDE-BLOTCHED LIZARD; LAGARTIJA

Uta tumidarostra Grismer 1994e:461. Type locality: "Isla Coloradito, Gulf of California, Baja California, México." Holotype: UA 49587

IDENTIFICATION *Uta tumidarostra* can be distinguished from all other lizards in the region of study by a dark axillary spot, a greatly inflated nasal region, and transversely divided prefrontal scales.

RELATIONSHIPS AND TAXONOMY See *U. encantadae.*

DISTRIBUTION *Uta tumidarostra* is known only from Isla Coloradito in the northern Gulf of California, BC, Mexico (see map 4.19).

DESCRIPTION Same as *U. encantadae,* except for the following: adults reach 74 mm SVL; prefrontals nearly always divided transversely on each side; frontoparietals square and in broad contact at midline; 110 to 134 transverse rows of dorsals; 31 to 74 transverse rows of nonkeeled dorsals; 61 to 78 transverse rows of ventrals; 8 to 10 enlarged infralabials; 22 to 30 femoral pores; 23 to 32 subdigital lamellae on fourth toe. **COLORATION** ADULT MALES: Color polymorphism exists, with some males having dark bodies and bluish spots in the lateral gular region and others having a color pattern more typical of *U. lowei,* with orangish sides and orange spots in the gular region. Faint pattern of darker, offset paravertebral blotches not visible as in other *Uta;* dorsum of body overlaid with network of small turquoise spots, grading laterally into yellow-white spots on sides. FEMALES AND JUVENILES: Same as *U. encantadae.*

FIGURE 4.58 Adult male *Uta tumidarostra* (Swollen-Nosed Side-Blotched Lizard) from Isla Coloradito, BC

NATURAL HISTORY *Uta tumidarostra* is almost entirely confined to the intertidal regions of Isla Coloradito, although individuals can occasionally be found inland in the rocky areas all the way to the top of the island. Here, they are lighter in coloration, more nearly matching the guano-covered substrate. Lizards are most abundant where the island's steep cliffs meet the water and large boulders are present. Generally, only juveniles and hatchlings occur in open areas lacking large boulders unless there are carcasses of sea lions or dolphins on which to establish feeding stations. *Uta tumidarostra* basks, forages, and takes refuge on the cliffs and among the large dark boulders in the intertidal zone.

Uta tumidarostra is active year-round, with minimal activity from November through February. From June through mid-September, lizards emerge from overnight retreats just before sunrise, although many return to cover if the day is overcast or if there is a cool breeze. They are abundant during the early morning and are most commonly observed along the bases of the cliffs. After basking a short while, lizards begin foraging for food, which during the summer consists almost entirely of the intertidal isopod *Ligia occidentalis.* During the spring, lizards are concentrated on, within, and around the carcasses of decaying California Sea Lions *(Zalophus californicus)* and dolphins that wash up along the beach. Here, lizards catch flies that lay eggs on the carrion. They also eat the fly larvae, which they retrieve by crawling inside the decaying carcasses. Lizards are often found with an accumulation of sticky, rotting tissues on the bottoms of their feet, on their bellies, and around their mouths. I have observed as many as 12 lizards on a single sea lion carcass less than 2 m long. On 10 April 1993, I counted 33 carcasses on the small beach (100 m long), all of which had at least five lizards on them. Lizards appear to set up feeding and lookout territories on the carcasses from which they watch and pursue prey. Once the prey is captured and consumed, the lizards return to their original lookout. From these lookouts, they can be seen head-bobbing at neighboring lizards. Often, the head-bob display is followed by a conspicuous vibration of the entire body. For some reason, carcasses are not used later in the year. When feeding on isopods in the intertidal zone, lizards dash from lookout perches on nearby rocks and chase the isopods over the boulders and cobble, often down to the water's edge. I have even seen lizards running through some of the small pools trapped in the rocks in pursuit of prey. On capture, lizards often carry the isopods back to their perching spot before eating them. Other items found in the guts of *U. tumidarostra* include bugs, beetles, spiders, shed skin, rocks, feathers, and turbinid shells (Grismer 1994e). Additionally, a large portion of this species' diet (16 percent by volume) is composed of sea mites *(Acari)* that are probably associated with the large sea lion colony on the island.

Unlike *U. encantadae* and *U. lowei, U. tumidarostra* is generally easy to approach. With a sizable colony of California Sea Lions occupying the same habitat, *U. tumidarostra* may be somewhat accustomed to large organisms moving around in its environment. Sea lion colonies do not exist on Islas Encantada and El Muerto, where the lizards have an inordinately long flight distance. At dusk, lizards are commonly found lying with their bodies flat against the rocks as though trying to absorb their heat. Before dark, they retreat beneath the boulders or into crevices on the cliff. Occasionally, exposed individuals are observed at night sleeping on the walls of the cliffs.

Gravid females have been observed during April, and I witnessed an attempted mating on 12 April 1992. Hatchlings are generally not seen until late June and are abundant in late July and early August. By September, hatchlings are no longer found. These observations suggest that *U. tumidarostra* has a typical spring and early summer breeding season with a summer hatching period.

REMARKS Information concerning the intertidal lifestyle of this species is still greatly lacking. For example, it is not known where these lizards lay their eggs (presumably, above the high-tide mark) or whether the eggs have some special resistance to salinity. I visited Isla Coloradito on 17 September 1993, during an unusually high tide that covered all the optimal intertidal habitat for *U. tumidarostra.* Many lizards seemed to gather along the edges of the cliffs where the cliffs opened up onto the beach, but where the other lizards retreated to I could not determine.

Hazard et al. (1998) demonstrated that the hypertrophied nasal gland of *U. tumidarostra* was an effective organ for excreting the excessive salt load acquired by this species' diet of salty intertidal prey. They demonstrated that *U. tumidarostra* was able to take in quantities of salt that would likely kill *U. stansburiana.* In captivity, I have observed *U. tumidarostra* to drink sea water and avoid fresh water.

Eublepharidae

EYELIDDED GECKOS

The Eublepharidae are a peculiar group of ancestral geckos containing six genera and 24 to 26 species (Grismer 1988b, 1999b, 2001b; Grismer et al. 1999). Eublepharid geckos are terrestrial to scansorial species that actively forage for prey during warm evenings. They have the curious habit of elevating the tail and waving it back and forth just prior to grasping their prey. Despite their cosmopolitan representation, the Eublepharidae have a highly fragmented distribution, and many of the Old World species are endemic to localized continental regions or to islands. Eublepharids occur in sub-Saharan Africa, western Asia, southeast Asia, and the Far East. In the New World, they are represented by the banded geckos *(Coleonyx),* which range from the American Southwest to Panama. Eublepharid geckos are the only geckos with movable eyelids and a single parietal bone, and the only lizards whose prefrontal bones contact medially beneath the frontal bone (Grismer 1988b). In Baja California and the Gulf of California, eublepharid geckos are represented by three species of *Coleonyx.*

Coleonyx Gray

BANDED GECKOS

The genus *Coleonyx* is composed of eight species of terrestrial, nocturnal, insectivorous lizards (Grismer 1988b, 1999b). *Coleonyx* ranges through a variety of habitats from the Sonoran and Chihuahuan deserts of the United States and northern Mexico to the tropical and dry tropical forests of southern Mexico and Central America. Where they occur, *Coleonyx* are often the most common species of nocturnal lizards. Although they range throughout a wide variety of habitats, they are most common in the vicinity of rocks, where they forage for food and take refuge. Their large eyelids, with distinctive countershading, make them peculiar and handsome species. The three species in the region of study are *C. gypsicolus,* which is endemic to Isla San Marcos in the Gulf of California; *C. switaki,* which is endemic to the Baja California peninsula; and *C. variegatus,* which ranges widely throughout the Sonoran and Mojave deserts.

FIGURE 4.59 Peninsularis pattern class of *Coleonyx variegatus* (Western Banded Gecko) from 5 km south of San Bartolo, BCS

FIGS. 4.59–60

Coleonyx variegatus

WESTERN BANDED GECKO; SALAMANQUESA

Stenodactylus variegatus Baird 1859d:254. Type locality: "Colorado Desert"; restricted to "Winterhaven [Fort Yuma], Imperial County, California," USA (Klauber 1945; Smith and Taylor 1950). Holotype: USNM 3217

IDENTIFICATION *Coleonyx variegatus* can be distinguished from all other lizards in the region of study by having eyelids and elliptical pupils; lacking a gular fold; and having granular body scales without larger interspersed tubercles.

RELATIONSHIPS AND TAXONOMY *Coleonyx variegatus* is the sister species of a natural group containing *C. brevis* of the Chihuahuan Desert and *C. fasciatus* of the western foothills of the Sierra Madre Occidental, Mexico (Grismer 1988b). Klauber (1945), Grismer et al. (1994), and Grismer (1999b) discuss taxonomy.

DISTRIBUTION *Coleonyx variegatus* is distributed throughout the Sonoran and Mojave deserts of southwestern United States and northern Mexico from southern Nevada south through eastern Arizona, southwestern New Mexico, and western Sonora, west of the Sierra Madre Occidental (Dixon 1970). It ranges south from the desert and cismontane foothill regions of southern California throughout Baja California. The only regions in Baja California where it does not occur are the upper elevations of the Sierra Juárez, Sierra San Pedro Mártir, and Sierra la Laguna (Grismer 1994b; map 4.20). It occurs on the Pacific islands of Cedros and Santa Margarita and the Gulf islands of Ángel de la Guarda, Coronados, Danzante, Espíritu Santo, Partida Sur, San José, San Marcos, Santa Inez, and Tiburón (Grismer 1993, 1999a, 2001a).

DESCRIPTION Small, slightly elongate lizard of moderate stature; adults reach 71 mm SVL; head triangular, distinctly wider than neck, and pointed in dorsal profile; scales of head granular; eyelids present; 33 to 47 eyelid fringe scales; head scales grading smoothly into slightly larger, granular body scales; no enlarged tubercles interspersed among granular scales of nape and body; 5 to 8 supralabials; 4 to 9 postmentals; gulars granular, grading posteriorly into flat, semi-imbricate scales of ventrum; gular fold absent; 5 to 10 infralabials; forelimbs thin, moderate in length; hind limbs slightly thicker and longer than forelimbs; 16 to 27 subdigital lamellae on fourth toe; 4 to 8 precloacal pores in males; tail constricted at base, widest centrally; caudals flattened and whorled; subcaudals at tip not tuberculate; pair of large cloacal spurs at base of tail in males. **COLORATION** Color pattern extremely variable (see geographic variation section). Ground color of dorsum ranging from off-white to light yellow or light brown; head with or without darker spots; thin, light-colored nuchal loop, inserting behind each eye, may be present; distinct light bands on dorsum, or dorsum spotted to reticulated with no sign of banding; when banded, 4 to 6 rows present between limb insertions; limbs faintly spotted to unicolored; tail may be distinctly banded or not; males lack ephemeral breeding color.

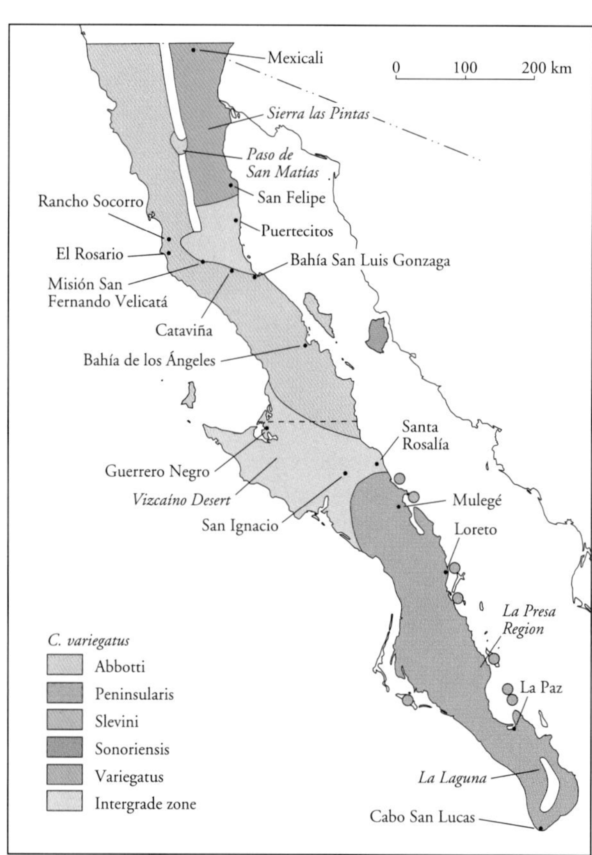

MAP 4.20 Distribution of the pattern classes of *Coleonyx variegatus* (Western Banded Gecko)

FIGURE 4.60 Sonoriensis pattern class of *Coleonyx variegatus* (Western Banded Gecko) from Ensenada de los Perros, Isla Tiburón, Sonora

GEOGRAPHIC VARIATION *Coleonyx variegatus* exhibits a great degree of geographic variation, and five subspecies in the region of study have been recognized (Klauber 1945), which are used here as pattern classes. Variegatus generally range throughout the Lower Colorado Valley Region of northeastern Baja California and are the most variable of all the pattern classes in dorsal pattern. Hatchlings and juveniles have a unicolored head and a well-defined nuchal loop and body bands. With increasing SVL, the head becomes mottled with darker spots, the nuchal loop generally becomes unrecognizable, and the body bands become clouded with ground color, giving the dorsum a variegated or reticulated pattern with only a faint indication of banding. Darkly patterned populations occur in the volcanic areas of the Sierra las Pintas and Cerro Prieto. Variegatus have the largest SVL of all the pattern classes in Baja California, reaching at least 71 mm.

Abbotti range throughout the California Region in northwestern Baja California, southward through much of the Vizcaíno and Central Gulf Coast regions, and are known from the Pacific island of Cedros (see map 4.20). Abbotti are smaller than Variegatus (maximum SVL 58 mm) and the distinct, light-colored bands of the adults are equal in width to or slightly wider than the darker interspaces. Adult Abbotti have a well-defined nuchal loop, and the head is usually unicolored. The bands of lizards from Isla Cedros are often irregularly shaped and transversely split.

Peninsularis range from the northern Magdalena Region in central Baja California south into the Arid Tropical Region of the Cape Region. Peninsularis closely resemble Abbotti, except that the light-colored body bands are thinner than the darker interspaces. From Loreto south, the bands are often irregular in shape or incomplete. Also, the bands become more yellow and the dark interspace edged with an even darker shade of brown. This coloration is most pronounced in the very handsome lizards of the Cape Region, from the eastern side of the Sierra la Laguna. Here, the bands and nuchal loop are lemon yellow, and the interspaces are mahogany brown and edged in a darker, deeper brown.

The peninsular pattern classes intergrade extensively where their members come into contact, and this intergradation is most evident in transitions of color pattern characteristics. Variegatus and Abbotti intergrade from approximately 18 km south of San Felipe to at least Bahía San Luis Gonzaga along the Gulf coast. Lizards from this area have well-defined bands, but the bands are mottled with ground color. The mottling is most pronounced in specimens from the northern portion of the zone. The nuchal loop is faded at Bahía San Luis Gonzaga and is absent in most adults from the vicinity of San Felipe. Head mottling shifts from a vividly mottled pattern in San Felipe to a nearly unicolored head at Bahía San Luis Gonzaga. Some intergradation is also seen in specimens from just east of El Rosario and Cataviña. Both these regions are connected to large arroyos, extending east onto the desert floor, through which gene flow occurs with Variegatus. Another zone of intergradation between populations of Abbotti and Variegatus occurs at Paso de San Matías (Welsh and Bury 1984).

Abbotti and Peninsularis intergrade widely throughout the Vizcaíno Desert (Grismer et al. 1994). Lizards from this area have an irregular, somewhat reticulated banding pattern. This pattern extends southeast to at least Santa Rosalía. Between Santa Rosalía and Mulegé, the bands become more defined and much thinner than the interspaces. This pattern, which is typical of Peninsularis, continues south into the Cape Region. Additionally, from San Ignacio to Santa Rosalía, lizards are generally dark, like the volcanic rock common to that region. At Mulegé, the interspaces are very dark brown and the thin bands very light, producing an extremely marked color pattern.

The population of *C. variegatus* from the flat, tiny Isla Santa Inez off the coast of Punta Chivato was once considered a separate subspecies, *C. v. slevini* (Klauber 1945). Grismer (1999b) demonstrated that Slevini were not discretely diagnosable from *C. variegatus* of the adjacent peninsula, although lizards do tend to have more bands (5 to 7 rather than 4 to 6) that are irregularly shaped and fewer postmental scales (3 to 5 rather than 4 to 8). Also, the body stature of Slevini is slightly more robust. These characteristics, however, occur within the range of variation observed in southern peninsular *C. variegatus.*

Sonoriensis range throughout western Sonora and occur on Isla Tiburón. They are characterized by wide dorsal bands split by a light-colored vertebral line. Hatchlings look very similar to hatchling Abbotti. Both have well-defined body bands, nuchal loops, and a unicolored head. In the slightly larger juvenile stage, the body bands divide longitudinally on the midline as the vertebral line develops. Populations of Sonoriensis intergrade with Variegatus and populations of Bogerti in northern Sonora.

The insular populations of *C. variegatus* closely resemble those populations from the nearest adjacent peninsular region (Grismer et al. 1994), with one exception. Lizards from Isla Ángel de la Guarda appear to be intergrades between Variegatus and Abbotti, although the adjacent peninsular populations at Bahía de los Ángeles are Abbotti. Interestingly, the geological origin of Ángel de la Guarda is near Bahía San Luis Gonzaga (Henyey and Bischoff 1973), where there is currently an intergrade zone between Variegatus and Abbotti.

NATURAL HISTORY In Baja California, *C. variegatus* ranges from sea level to approximately 2,000 m elevation. It occurs in a wide range of habitats, including chaparral communities of the California Region, sand dune areas and Creosote Bush *(Larrea tridentata)* scrub communities of the Lower Colorado Valley Region, and dunes along the Pacific coast from Rancho Socorro south to Guerrero Negro in the Vizcaíno Region. In central and southern Baja California, *C. variegatus* inhabits the flat, open Vizcaíno Desert; the volcanic badlands of the Magdalena Region; the flat, open Magdalena Plain; and the Arid Tropical regions of the Sierra la Giganta and the Cape Region (Alvarez et al. 1988). Additionally, I have found individuals beneath rocks at the water's edge just north of Bahía San Luis Gonzaga. Murphy and Ottley (1984) report finding specimens at Puerto Refugio on Isla Ángel de la Guarda, on desert hardpan among clumps of thick, low brush. The only habitats in Baja California where *C. variegatus* does not occur are the pine forests of Sierra San Pedro Mártir (Welsh 1988) and Sierra Juárez, in the Baja California Coniferous Forest Region, and the pine-oak woodlands of the Sierra la Laguna Region (Alvarez et al. 1988).

Although *C. variegatus* has been studied in the United States (Parker 1972; Parker and Pianka 1974; Kingsbury 1989), very little is known about its natural history in Baja California. It is a nocturnal, terrestrial habitat generalist. During the day, it seeks refuge within burrows and beneath rocks, large pieces of fallen vegetation, and trash. I have observed Variegatus abroad from mid-April through early November and Sonoriensis on Isla Tiburón from mid-May through early November. North of San Ignacio, Abbotti and Variegatus hibernate during the winter, and I have found inactive specimens beneath rocks from November through February. In the Cape Region, I have found Peninsularis beneath small rocks and cardboard during mid-December. Given the shallowness of the shelter, I do not believe these lizards were hibernating. Peninsularis from the Cape Region may be active during the winter for short periods if weather conditions are appropriate; however, none of the lizards I observed had food in their stomachs.

Coleonyx variegatus feeds on a variety of small arthropods, for which it actively forages. In Baja California, *C. variegatus* probably mates during late spring and early summer. I have observed mating pairs during June near San Ignacio and gravid females throughout the peninsula during June, July, and August. I have found hatchlings from mid-July to mid-October throughout the peninsula. On Isla Tiburón, I have observed a number of hatchlings during late March.

Slevini from Isla Santa Inez occur in very high densities in rocky areas where vegetation is scant. Nearly every other rock or large bush has a gecko beneath it. During one afternoon in mid-June, I found 13 specimens in half an hour of turning rocks. That night, I found 36 additional specimens abroad in less than two hours of walking with a lantern.

FOLKLORE AND HUMAN USE The local inhabitants of Baja California refer to all *Coleonyx* and other geckos as *salamanquesa.* They are very frightened of these species, believing that their feet are not only venomous but also necrophagous. Many people have told me that if a gecko walks across your skin or clothing, the areas contacted by the feet will slough away. A rancher at Misión San Fernando told me that when a *C. variegatus* bit his cow on the nose, the cow died immediately.

The Seri Indians believe that the flesh and blood of *C. variegatus* are highly toxic, fatal to both man and Seri (Malkin 1962). The Seri state that it will not bite but will cause a fatal lung illness if touched. In addition, touching a gecko will cause the flesh to fall off one's hands and body (Malkin 1962). Its saliva is believed to cause rashes, and if a person lies down where a gecko is buried, its silhouette will be burned into the victim's flesh (Nabhan, forthcoming).

FIGS. 4.61–63

Coleonyx switaki

BAREFOOT BANDED GECKO; SALAMANQUESA

Anarbylus switaki Murphy 1974:87. Type locality: "5.5 miles west of San Ignacio (27 'N, 112 51'W) along Mexican Highway 1, Baja California Sur, Mexico, 500 feet elevation." Holotype: CAS 139472

IDENTIFICATION *Coleonyx switaki* can be distinguished from all other lizards in the region of study by having eyelids; fewer than 41 eyelid fringe scales; no gular fold; and tubercles interspersed among granular body scales.

RELATIONSHIPS AND TAXONOMY *Coleonyx switaki* is the sister species of *C. gypsicolus,* which is endemic to Isla San Marcos in the Gulf of California. Taxonomy is discussed by Grismer and Ottley (1988) and Grismer (1999b).

DISTRIBUTION *Coleonyx switaki* ranges along the desert foothills of the Peninsular Ranges from at least northern San Diego County, California, south to just north of Santa Rosalía (Grismer 1990a; map 4.21).

DESCRIPTION Small, with adults reaching 86 mm SVL; head triangular, distinctly wider than neck, and pointed in dorsal profile; scales of head granular; eyelids present; 32 to 40 eyelid fringe scales; head scales grading smoothly into slightly larger, granular body scales; enlarged tubercles interspersed among granular scales of nape and back; 9 to 13 supralabials; 5 to 10 postmentals; gulars granular, grading posteriorly into flat, semi-imbricate scales of ventrum; gular fold absent; 10 to 14 infralabials; forelimbs thin, moderate in length; hind limbs slightly thicker and longer than forelimbs; 17 to 19 subdigital lamellae on fourth digit; 2 to 8 precloacal pores in males; tail constricted at base, widest centrally; caudals flattened and whorled, terminal subcaudals semituberculate; pair of cloacal spurs at base of tail in males. **COLORATION** Color pattern extremely variable. Ground color of dorsum ranging from light gray to dark brown; head with grayish to yellow or white spots; dorsum banded or spotted or with coalescing spots, tending to form bands; 4 to 9 bands or rows of transverse spots between limb insertions; limbs faintly spotted with light and dark markings; tail consisting of black and white rings; black rings often with lightened centers; males develop an ephemeral lemon-yellow wash on body during breeding season.

FIGURE 4.61 Adult male *Coleonyx switaki* (Barefoot Banded Gecko) from San Ignacio, BCS

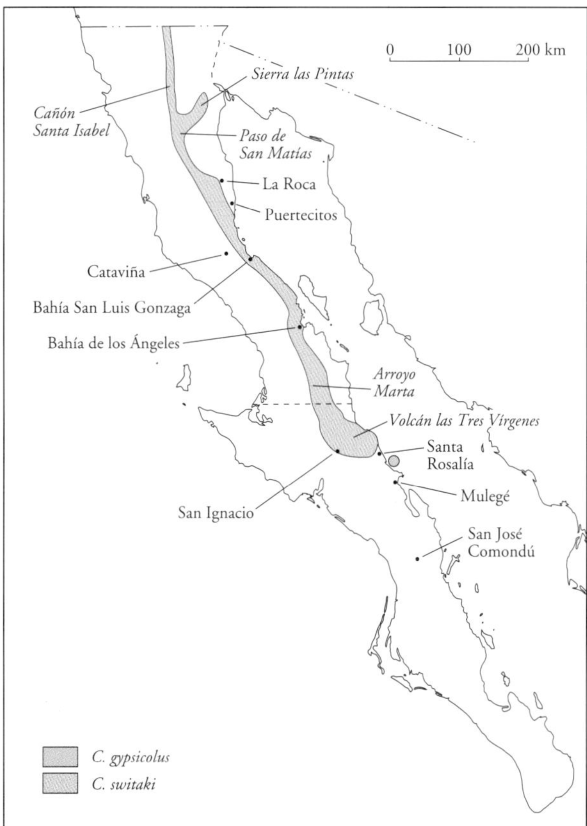

MAP 4.21 Distribution of *Coleonyx switaki* (Barefoot Banded Gecko) and *C. gypsicolus* (Isla San Marcos Barefoot Banded Gecko)

GEOGRAPHIC VARIATION *Coleonyx switaki* exhibits clinal variation in many characteristics along its latitudinal extent in Baja California. Grismer and Ottley (1988) demonstrated that the northern populations from southern California have smaller scales than the southern populations of the Tres Vírgenes region near San Ignacio, resulting in significantly higher mean scale counts. Additional material from several geographically intermediate localities indicates that these differences are clinal. Also, southern populations have larger and more pronounced tubercles that are more numerous on the nape of the neck and body. In the northern populations, the anterior scales of the rostrum and gular region are two to three times larger than adjacent posterior scales, whereas those scales in the southern populations are approximately the same size.

The ground color of *C. switaki* varies with the color of the substrate and appears to exhibit little clinal variation (Grismer 2001b). Lizards from granitic areas have a grayish ground color, whereas those from andesitic volcanic regions have a deep, dark brown ground color. Interestingly, within the andesitic Tres Vírgenes region, there is a deep (ca. 100 m), narrow (ca. 5 m), light-colored sandstone gorge just northwest of Santa Rosalía. Here I found a specimen whose

FIGURE 4.62 *Coleonyx switaki* (Barefoot Banded Gecko) from La Roca, BC. Adult male above, in breeding coloration; gravid female below.

FIGURE 4.63 Juvenile *Coleonyx switaki* (Barefoot Banded Gecko) from San Ignacio, BCS

ground color was very light, quite unlike lizards of the surrounding populations but similar to the sandstone substrate.

The banding pattern of the northern Baja California populations is vivid and well-defined. This condition extends from at least Mortero Wash in San Diego County south to the vicinity of Cañón Santa Isabel in the central portion of the Sierra Juárez (Grismer 2001b). South of Cañón Santa Isabel, the bands begin to break up into a series of transversely arranged spots. This trend continues southward and is pronounced in lizards from the Sierra las Pintas and Paso de San Matías, where the spots are distinct and transversely arranged. Farther south at La Roca, just south of San Felipe, a banding pattern is difficult to discern, and the overall dorsal pattern is generally reticulated. Populations farther to the south at Bahía San Luis Gonzaga, Bahía de los Ángeles, and Arroyo Marta show varying degrees of banding and spotting (Grismer 2001b). In the Tres Vírgenes region, however, the dorsal pattern is strictly spotted.

The color pattern of the caudal rings may be clinal (Grismer 2001b). Southern populations at Arroyo Marta and the Tres Vírgenes region have wide, light-colored, scallop-edged black rings with lightened centers that are wider than the white rings. Geographically intermediate populations from Bahía de los Ángeles to La Roca have black and white caudal rings of roughly the same width. The black rings are variable in outline and central coloration. The remaining populations to the north have dark, evenly edged black rings that are the same size as or thinner than the white caudal rings.

NATURAL HISTORY *Coleonyx switaki* is found along the rocky desert foothills of the Peninsular Ranges in the Lower Colorado Valley and Central Gulf Coast regions of northern Baja California as well as in some of the smaller, adjacent rocky ranges to the east. It also inhabits the large andesitic volcanic flows of the northern portion of the Magdalena Region in central Baja California. This species is most common on the talus slopes of volcanic hillsides and terraces where the mean rock size is slightly less than 1 m and the vegetation is not too dense. The sharp, angular nature of such rocks provides adequate overnight and overwinter retreats. *Coleonyx switaki* is occasionally found under cap rocks and in deep cracks. It is less common in areas with large granitic outcroppings.

Fritts, Snell, and Martin (1982) and Grismer and Edwards (1988) reported on some aspects of the natural history of *C. switaki*. These findings are summarized and supplemented here. *Coleonyx switaki* is active from at least early May to mid-October. It is a strictly crepuscular to nocturnal species. I have found specimens abroad just west of Bahía de los Ángeles in early May, when temperatures were as low as 20°C. I have also found individuals active on bright, moonlit nights and during very windy periods. On such occasions, *C. switaki* is usually the only species observed. No field data exist on its diet. Presumably it is a typical insectivore.

Coleonyx switaki is unique among eublepharid geckos in that the males develop a bright yellow breeding color during the reproductive season (Grismer 1988b). Males smaller than 65 mm SVL do not develop this coloration and are presumably not reproductively mature. Grismer and Edwards (1988) stated, on the basis of behavior of captive specimens, that females mate once per season and display aggression toward males that attempt subsequent matings. In captivity, one female actually killed a male that attempted a second mating. After a single mating, up to three fertile clutches of two eggs each may be laid within 20 days of oviposition of the first clutch. Based on field observations, the reproductive season lasts from mid-May through late July. I found a gravid female and a male in breeding coloration during mid-May at La Roca. I found a male in breeding coloration during early June in Cañón Santa Isabel. I also observed a gravid female and a male in breeding coloration during mid-June in Arroyo Marta. I found a male whose breeding coloration was beginning to fade in Arroyo de Rumbaugh during late July, and an adult male during early October near San Ignacio whose breeding color had completely faded. Near San Ignacio, I have seen juveniles during early August and hatchlings and post-partum females from mid-August to early October. These observations suggest that *C. switaki* has a typical spring to early summer breeding season and a late summer to early fall hatching period.

REMARKS The various populations of *C. switaki* are generally distinct from one another. Often discretely diagnosable in some characteristics, they express clinal variation in others, which suggests that gene flow between them is unrestricted (Grismer 1999b, 2001b). Therefore it would be pointless to consider any as a distinct taxon.

Although the appropriate habitat for *C. switaki* extends much farther south, I am reasonably certain that this species' distribution extends no farther south than the vicinity of Santa Rosalía. I have searched the extensive volcanic flows and talus slopes along the terraces between Mulegé and San José Comondú but have found no specimens. This region receives enough summer rain to support a lush, arid tropical thorn scrub, and *C. switaki* seems to prefer areas with less vegetative cover. In northern Baja California, I have noticed that *C. switaki* is not found in the vicinity of brushy hillsides, although it can be found less than 100 m away where the vegetation is generally sparse.

FIG. 4.64 *Coleonyx gypsicolus*

ISLA SAN MARCOS BAREFOOT BANDED GECKO; SALAMANQUESA

Coleonyx switaki gypsicolus Grismer and Ottley 1988:150. Type locality: "Arroyo de la Taneria, Isla San Marcos, Baja California Sur, Mexico." Holotype: BYU 37643

IDENTIFICATION *Coleonyx gypsicolus* can be distinguished from all other lizards in the region of study by having eyelids; more than 41 eyelid fringe scales; no gular fold; tubercles interspersed among granular body scales; and a light-colored vertebral stripe.

RELATIONSHIPS AND TAXONOMY See *C. switaki.*

DISTRIBUTION *Coleonyx gypsicolus* is endemic to Isla San Marcos in the Gulf of California, BCS (see map 4.21).

FIGURE 4.64 *Coleonyx gypsicolus* (Isla San Marcos Barefoot Banded Gecko) from Arroyo la Taneria, Isla San Marcos, BCS

DESCRIPTION Same as *C. switaki,* except for the following: adults reach 89 mm SVL; 42 to 48 eyelid fringe scales; 9 to 14 supralabials; 6 to 8 postmentals; 11 to 15 infralabials; 19 to 22 subdigital lamellae on fourth toe; 6 to 8 precloacal pores in males. **COLORATION** Ground color of dorsum light gray to light brown, covered with dark, irregularly shaped spots and circular, white to yellow, transversely arranged spots; 8 or 9 transverse rows of spots between limb insertions; limbs faintly spotted with dark markings; tail marked with black and white rings in juveniles; black rings of adults break up into a series of oblong blotches; male breeding color extremely bright yellow.

NATURAL HISTORY Isla San Marcos lies within the Central Gulf Coast Region and is dominated by xerophilic scrub. The island is composed of sedimentary sandstone with a southerly situated gypsum base. Very little is known about the natural history of *C. gypsicolus,* but presumably it is similar to that of the adjacent southern populations of *C. switaki. Coleonyx gypsicolus* has been collected only from within Arroyo la Taneria, where it is principally associated with large outcroppings of gypsum rocks on the hillsides, in steep canyons, and in smaller tributary arroyos (Murphy and Ottley 1984). Grismer and Ottley (1988) reported this species to be common around the garbage dump near the arroyo bottom. The north end of the island is more volcanic, with a much darker substrate, and lizards from this area may be darker in overall ground color.

Coleonyx gypsicolus is nocturnal and likely feeds on a variety of arthropods. It may also feed on the much smaller *C. variegatus.* When I placed a large adult *C. gypsicolus* in a 20 l plastic container with several *C. variegatus,* a number of them began squealing and running around. Some of them even dropped their tails. I found a gravid female with two eggs during early July, and on the same night I found an adult male in breeding color. These observations suggest that the reproductive season is similar to that of adjacent peninsular populations of *C. switaki.*

FOLKLORE AND HUMAN USE See *C. switaki.*

Gekkonidae

GECKOS

The Gekkonidae are a large group of lizards containing approximately 106 genera and 950 species (Kluge 2001). It is worldwide in distribution, with the exception of Arctic and Antarctic regions, and is represented by a tremendous diversity of adaptive types that is paralleled by an equally diverse range of morphological specializations. Most geckos are nocturnal and are equipped with expanded subdigital lamellae that aid their scansorial lifestyles. There are, however, large radiations of diurnal, terrestrial groups with unmodified digits, as well as limbless, burrowing, and grass-swimming species. The Gekkonidae are the only group of lizards with brittle eggshells and eggs that are nearly round instead of oblong in shape (Kluge 1987). There are approximately 15 New World genera, of which three, the stump-toed geckos *(Gehyra),* the house geckos *(Hemidactylus),* and the leaf-toed geckos *(Phyllodactylus),* occur in the region of study. In the New World, *Gehyra* and *Hemidactylus* are the result of human introductions.

Gehyra Gray

STUMP-TOED GECKOS

The genus *Gehyra* is composed of approximately 33 species of small to moderately sized, scansorial, nocturnal lizards (Kluge 2001). *Gehyra* is a remarkable island colonizer and is known from Madagascar, islands throughout the Indian Ocean, south and southeast Asia, Japan, the Philippines, the Indo-Australian Archipelago, Australia, many Pacific islands, and western Mexico (Bauer 1994). This worldwide insular distribution is due in large part to its commensal relationships with humans, who have facilitated its colonization in many regions. One species, *G. mutilata,* occurs in the Cape Region of Baja California.

Gehyra mutilata

STUMP-TOED GECKO; SALAMANQUESA

Hemidactylus (P. [Peropus]) mutilatus Wiegmann 1835:238. Type locality: "Manilla, [Luzon, Philippines]." Lectotype: ZMB 370 (Bauer and Günther 1991)

IDENTIFICATION *Gehyra mutilata* can be distinguished from all other lizards in the region of study by its lack of movable eyelids; terminal series of subdigital lamellae greatly expanded and occurring in paired rows; and granular dorsal body scales not interspersed with larger tubercles.

RELATIONSHIPS AND TAXONOMY Ineich (1987) discusses systematics.

DISTRIBUTION *Gehyra mutilata* is one of the world's most widely distributed geckos, occurring in New Guinea, Melanesia, Micronesia, Polynesia, India, Sri Lanka, Indochina, Japan, the Philippines, Indonesia, Madagascar, and the Mascarene Islands. It has been introduced into the United States (Smith and Kohler 1978), Sinaloa, Nayarit (Hardy and McDiarmid 1969; Taylor 1922), and La Paz, BCS, Mexico (Reynoso 1990a; map 4.22).

DESCRIPTION Based on specimens from La Paz, BCS. Small, with adults reaching 50 mm SVL; head and body flat; head narrowly triangular in dorsal profile, slightly wider than neck; eyes large, eyelids absent; head and body scales granular, with no enlarged tubercles; dorsal scales grading ventrally into larger flat, imbricate ventral scales; 8 or 9 supralabials from rostral to point below center of eye; 2 postmentals bordered laterally by enlarged, rectangular sublabials; gulars flat; 7 infralabials; limbs covered with flat, imbricate scales; forelimbs moderately sized, hind limbs larger than forelimbs; terminal series of subdigital lamellae greatly expanded and occurring in paired rows; first digit

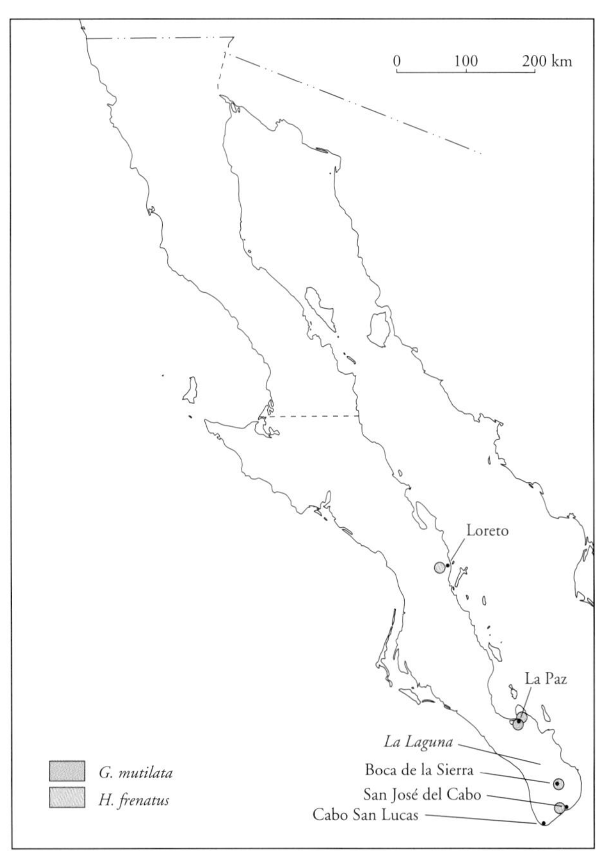

MAP 4.22 Distribution of *Gehyra mutilata* (Stump-Toed Gecko) and *Hemidactylus frenatus* (Common House Gecko)

on hands and feet well-developed but lacking a claw; tail subelliptical in cross section; caudal scales granular and juxtaposed; vertebral series of subcaudals enlarged and imbricate; enlarged, rectangular, submucronate scales form weak ventrolateral caudal fringe; tail constricted at base; 30 to 34 femoral and preanal pores, series separated medially by at least one scale. **COLORATION** Dorsal ground color beige; scales of entire dorsum with 1 to 5 minute dark brown spots; indistinct, thin, light-colored transverse caudal bands; paravertebral series of light-colored spots occasionally present; ventrum immaculate; ground color dull white except for darker subcaudal region.

NATURAL HISTORY Nothing on the natural history of *G. mutilata* from La Paz has been reported. In Baja California, *G. mutilata* is known only from the walls of houses within the city of La Paz. It is commonly seen in some areas on warm evenings crawling on walls beneath lights, feeding on insects.

REMARKS Taylor 1922 speculates that *G. mutilata* was introduced into southwestern mainland Mexico no earlier than the sixteenth or seventeenth century, when shipping lanes between the Philippines and Mexico became well-established. Since that time, *G. mutilata* has become a common species in several seaports of western mainland Mexico (Hardy and McDiarmid 1969). Its introduction into the port of La Paz (Reynoso 1990a) is probably correlated with the establishment of shipping lanes between La Paz and ports of southwestern Mexico.

Hemidactylus Gray

HOUSE GECKOS

The genus *Hemidactylus* is composed of approximately 78 species of small to moderately sized scansorial, nocturnal lizards (Kluge 2001), many of which live in close association with humans. *Hemidactylus* is a wide-ranging genus known from much of Africa, Asia, Japan, the Philippines, the Indo-Australian Archipelago, many Pacific Basin islands, northeastern South America, and the West Indies (Bauer 1994); it has been introduced into Australia, the United States, and Mexico. In these areas, *Hemidactylus* is usually a common household occupant. One species, *H. frenatus,* has been introduced into Baja California.

FIG. 4.65 *Hemidactylus frenatus*

COMMON HOUSE GECKO; SALAMANQUESA

Hemidactylus frenatus Schlegel, in Duméril and Bibron 1836:366. Type locality: "l'Afrique australe, et . . . tout l'archipel des grandes Indes" (restricted to "Java" by Loveridge 1947). Lectotype: MNHM 5135, designated by Wells and Wellington 1985

FIGURE 4.65 *Hemidactylus frenatus* (Common House Gecko) from La Paz, BCS

IDENTIFICATION *Hemidactylus frenatus* can be distinguished from all other lizards in the region of study by its lack of movable eyelids; greatly expanded subdigital lamellae; and vestigial first digit on the hands and feet.

RELATIONSHIPS AND TAXONOMY Moritz et al. (1993) discuss taxonomy.

DISTRIBUTION *Hemidactylus frenatus* is one of the world's most widely distributed geckos, ranging throughout all tropical and subtropical regions (Bauer 1994). It has been introduced into Australia, East Africa, islands in the Indian Ocean, the United States, and Mexico (Bauer 1994; Hardy and McDiarmid 1969; Reynoso 1990b). In Baja California, it is common in the city of La Paz in the Cape Region, and there have been unconfirmed reports of its occurrence farther south in the city of San José del Cabo (see map 4.22). I have also seen individuals on walls at night beneath lights in the city of Loreto.

DESCRIPTION Based on specimens from La Paz, BCS. Same as *G. mutilata* except for the following: adults reach 59 mm SVL; slightly enlarged, flat tubercles on body; 8 to 10 supralabials; postmentals bordered by one enlarged sublabial; 7 to 9 infralabials; first digit on hands and feet vestigial and usually lacking a claw; tail subelliptical in cross section, lacking ventrolateral fringe; dorsal caudal scales flat and imbricate; caudal tubercles whorled; tail not constricted at base; 27 to 31 femoral and preanal pores, series usually not separated medially; pores transversely elongate.

COLORATION Dorsum dull tan; two faint, brownish postorbital stripes; rostrum somewhat darker than head; numerous small brown flecks on dorsum; ventrum beige and immaculate.

NATURAL HISTORY Nothing on the natural history of *H. frenatus* from Baja California has been reported. In La Paz, *H. frenatus* is known primarily from the walls of houses, where the males can be heard making crisp staccato chirps from about April through October. I observed three individuals beneath a bridge at Boca de la Sierra, 100 km south of La Paz, which indicates that this species is no longer restricted to the larger shipping ports. Residents state that this gecko is active at night and that individuals congregate in lighted areas to feed on insects. Specimens have been collected from early February through early October, which suggests that its activity is minimal during the winter. Gravid females usually contain two eggs and have been observed from mid-April through mid-July; and hatchlings have been collected from mid-May through mid-October. Eggs are usually laid behind hanging structures such as pictures or within the cracks of walls. Several clutches of eggs were found in an open light socket in one household (P. Galina, personal communication, 1996).

REMARKS The introduction of *H. frenatus* into La Paz (Reynoso 1990b), presumably from mainland Mexico, probably resulted from circumstances similar to those surrounding the introduction of *G. mutilata.*

Phyllodactylus Gray

LEAF-TOED GECKOS

The genus *Phyllodactylus* is composed of 42 species of small, scansorial, nocturnal lizards (Grismer 1999b; Kluge 2001). In the New World, *Phyllodactylus* ranges throughout a variety of habitats, from the Sonoran Desert of North America south through the semiarid tropics of Mexico, Central America, and coastal South America to northern Chile. *Phyllodactylus* are often cryptically colored microhabitat specialists that occur on or within vegetation or rocks. Here they are usually the most abundant species of lizards and a conspicuous component of the environment.

The systematics of *Phyllodactylus* has always been controversial, and the species within the region of study are no exception. The fine works of Dixon (1964, 1966) for the *Phyllodactylus* in the region of study served as a solid foundation for an early taxonomy of the group. Dixon recognized 9 subspecies and 10 species, most of which were insular endemics. More recent evaluations of this group (e.g., Murphy 1983a), based on alternative concepts of species (e.g., Grismer 1999b), now consider the region of study to contain at least five species, of which two *(P. bugastrolepis* and *P. partidus)* are Gulf island endemics, two others *(P. xanti* and *P. unctus)* are endemic to the Baja California peninsula, and one *(P. homolepidurus)* occurs on Isla San Pedro Nolasco and adjacent Sonora (see Grismer 1994a, 1999b).

FIG. 4.66 *Phyllodactylus homolepidurus*

SONORAN LEAF-TOED GECKO; SALAMANQUESA

Phyllodactylus homolepidurus Smith 1935:121. Type locality: "five miles southeast of Hermosillo," Sonora, Mexico. Holotype: CNHM 10011

IDENTIFICATION *Phyllodactylus homolepidurus* from Isla San Pedro Nolasco can be distinguished from all other lizards in the region of study by its greatly expanded terminal subdigital lamellae; 41 to 48 paravertebral tubercles from the rear of the head to the base of the tail; caudal tubercles whorled; relatively small abdominal scales; and 21 to 24 scales across the snout between the third supralabials.

RELATIONSHIPS AND TAXONOMY The relationships of *P. homolepidurus* have not been addressed in a phylogenetic systematic context, but it is likely this species is most closely related to *P. partidus,* given that they were once considered to be conspecific (Dixon 1964, 1966). Dixon (1966) and Grismer (1999b) discuss the taxonomy of *P. homolepidurus* and *P. partidus.*

DISTRIBUTION *Phyllodactylus homolepidurus* ranges throughout much of western Sonora and occurs on Isla San Pedro Nolasco, which lies approximately 15 km southwest of Punta San Pedro, north of Guaymas, Sonora (Dixon 1964; map 4.23)

DESCRIPTION Small, with adults reaching 62 mm SVL; head and body flat; head triangular in dorsal profile, slightly wider than neck; eyes large, movable eyelids absent; anterior head scales granular; scales bordering supralabials and external nares enlarged; 21 to 24 scales across snout between third supralabials; scales on posterior of head and body granular and interspersed with larger tubercles; 14 to 16 longitudinal rows of tubercles on body; 41 to 48 paravertebral tubercles between posterior margin of ear opening and base of tail; granular dorsals abruptly differentiated from larger flat, imbricate ventral scales; 6 or 7 supralabials from rostral to point below center of eye; 2 or 3 postmentals; gulars flat; 5 or 6 infralabials; forelimbs moderately sized and covered with flat, imbricate scales dorsally and granular scales ventrally; tubercles often present on forearm; hind limbs larger than forelimbs; suprafemorals circular, juxtaposed, and interspersed with tubercles; supratibials conical and interspersed with tubercles; terminal pair of subdigital lamellae of hands and feet square and greatly expanded; tail round in cross section; caudal scales subconical and interspersed with longitudinal rows of tubercles; caudals whorled; 2 to 4 longitudinal rows of tubercles at second whorl of tubercles; 3 or 4 enlarged postanal tubercles in males. **COLORATION** Postnasal stripe extending to anterior margin of eye and from posterior margin of eye to forelimb insertion; ground color of dorsum brown to dark gray with a darker, weak crossbanding pattern; limbs weakly mottled; ventrum whitish.

NATURAL HISTORY Isla San Pedro Nolasco lies within the Central Gulf Coast Region, and although it is dominated by arid-adapted vegetation, it retains a significant component of tropical scrub as well. I have found *P. homolepidurus* beneath rocks in shallow earthen depressions. It is not strictly saxicolous on Isla San Pedro Nolasco but is most common beneath granitic exfoliations and within rock fissures and shallow cracks. The density of *P. homolepidurus* seems to decrease with distance from the shoreline. Most of the specimens I have observed have been within 100 m of the coast.

FIGURE 4.66 *Phyllodactylus homolepidurus* (Sonoran Leaf-Toed Gecko) from Isla San Pedro Nolasco, Sonora

Phyllodactylus homolepidurus is probably not active year-round, and it is difficult to find even during early spring. During the warmer months, however, it is easy to see 10 or so individuals in one to two hours of lantern walking near the shore. Nothing has been reported on the reproductive or food habits of this species. Presumably it is oviparous, with a spring to summer breeding season. I have seen hatchlings in early October. *Phyllodactylus homolepidurus* probably eats a variety of small arthropods; I have observed individuals feeding on ants at dusk.

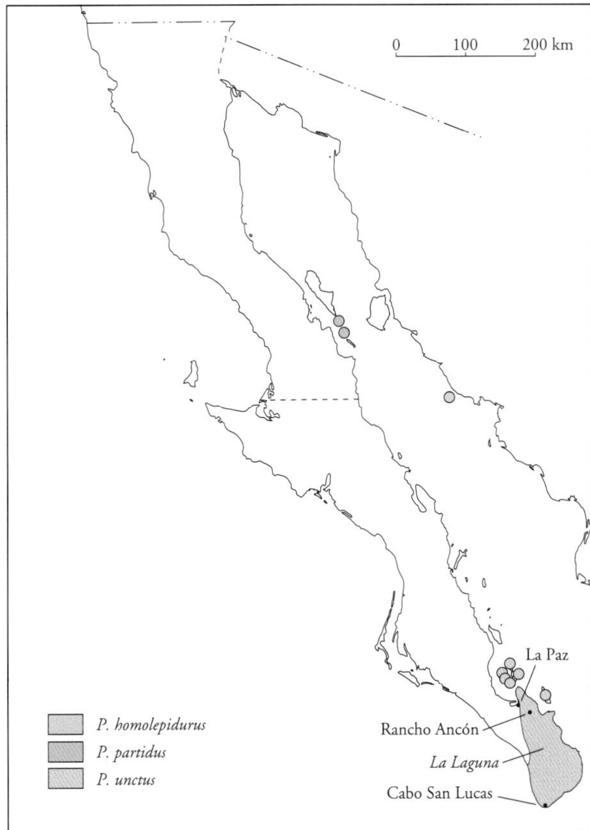

MAP 4.23 Distribution of *Phyllodactylus homolepidurus* (Sonoran Leaf-Toed Gecko), *P. partidus* (Isla Partida Norte Leaf-Toed Gecko), and *P. unctus* (San Lucan Leaf-Toed Gecko)

FOLKLORE AND HUMAN USE Local inhabitants generally display terror in a most animated way at the sight of any *Phyllodactylus.* They emphatically insist that their touch is venomous. In particular, they believe that when the expanded toe pads come into contact with human skin (some believe clothing as well), the skin will decay and slough off.

FIG. 4.67 *Phyllodactylus partidus*

ISLA PARTIDA NORTE LEAF-TOED GECKO; SALAMANQUESA

Phyllodactylus partidus Dixon 1966:445. Type locality: "Isla Partida (N.), Baja California," Mexico. Holotype: CAS 98429

IDENTIFICATION *Phyllodactylus partidus* can be distinguished from all other lizards in the region of study by its greatly expanded terminal subdigital lamellae; 31 to 41 paravertebral tubercles from the rear of the head to the base of tail; relatively small abdominal scales; 23 to 28 scales across the snout between the third supralabials; and dark brown coloration.

RELATIONSHIPS AND TAXONOMY See *P. homolepidurus.*

FIGURE 4.67 *Phyllodactylus partidus* (Isla Partida Norte Leaf-Toed Gecko) from Isla Cardonosa Este, BC

DISTRIBUTION *Phyllodactylus partidus* is known only from Islas Partida Norte and Cardonosa Este, BC (Dixon 1964; Grismer 1999a; see map 4.23).

DESCRIPTION Same as *P. homolepidurus,* except for the following: adults reach 67 mm SVL; 23 to 28 scales across snout; 12 to 16 longitudinal rows of tubercles; 32 to 39 paravertebral tubercles; 2 postmentals. **COLORATION** Ground color capable of changing from transparent gray to dark gray or chocolate brown; brown postnasal stripe extending to anterior margin of eye and from posterior margin of eye to rear of head; head and body mottled with large, squarish, brown to gray paravertebral blotches; blotches forming more obvious paravertebral rows in juveniles; thin, light-colored, poorly defined vertebral stripe with light transverse bars in the form of poorly defined bands; ventrum cream-yellow to tan.

GEOGRAPHIC VARIATION *Phyllodactylus partidus* from Isla Cardonosa Este, a small island less than 1 km southeast of Partida Norte, are darker in dorsal and ventral ground color than individuals from Isla Partida Norte.

NATURAL HISTORY Partida Norte and Cardonosa Este are adjacent rocky islands within the Central Gulf Coast Region in the northern Gulf of California. They are generally hot and dry for most of the year and dominated by xerophilic scrub. On Isla Partida Norte, *P. partidus* is principally saxicolous but occasionally observed on the ground beneath decaying Cardón Cactus *(Pachycereus pringlei).* At night, geckos are commonly seen on rocks or in rock cracks. Dixon (1966) reported finding a specimen in rocky rubble on the beach, and it is clear that its distribution is islandwide. While abroad during the evening, *P. homolepidurus* is light gray in coloration. When uncovered during the day in their refugia, lizards are dark. I observed two small juveniles in early October and no gravid females out of approximately 30 individuals sighted. This suggests that the breeding season had passed.

REMARKS The presence of *P. partidus* on Isla Partida Norte seems anomalous. Isla Partida Norte is centrally located within the western midriff islands, lying between Isla Ángel de la Guarda to the north and Isla Salsipuedes to the south. All *Phyllodactylus* populations from the surrounding islands are members of the peninsular *P. xanti,* whereas *P. partidus* appears to be derived from *P. homolepidurus,* which occurs on adjacent mainland Sonora (Dixon 1966). Grismer (1994c) discussed how the midriff islands may serve as stepping-stones for dispersing continental populations from both sides of the Gulf.

FOLKLORE AND HUMAN USE See *P. homolepidurus.*

FIG. 4.68 ***Phyllodactylus xanti***

PENINSULAR LEAF-TOED GECKO; SALAMANQUESA

Phyllodactylus xanti Cope 1863:102. Type locality: "Cape St. Lucas [Cabo San Lucas], Lower California," BCS, Mexico. Holotype: none designated

IDENTIFICATION Can be distinguished from all other lizards in the region of study by its greatly expanded terminal subdigital lamellae; 32 to 40 paravertebral tubercles from the rear of the head to the base of the tail; relatively small abdominal scales; and 14 to 19 scales across the snout between the third supralabials.

RELATIONSHIPS AND TAXONOMY *Phyllodactylus xanti* is probably closely related to the *P. tuberculosus* group (*sensu* Dixon 1964) of southwestern Mexico (see Grismer 1994b). Dixon (1964), Murphy (1983a), and Grismer (1999b) discuss taxonomy. Dixon (1966) considered the populations from Islas Ángel de la Guarda, Mejía, and Pond as *P. angelensis.* The populations from Islas Ánimas, Santa Cruz, and La Rasa were considered the species *P. apricus, P. santacruzensis,* and *P. tinklei,* respectively (Dixon 1966). Grismer (1999b) placed these populations in the synonymy of *P. xanti* because they lacked characteristics unequivocally distinguishing them from *P. xanti.*

DISTRIBUTION *Phyllodactylus xanti* extends along the Peninsular Ranges from Riverside County, California, south throughout nearly all of Baja California. In Baja California, *P. xanti* usually occurs in all rocky desert areas, but it is conspicuously absent from the Sierra los Cucapás and Sierra el Mayor in the northeast (Grismer 1994b; McGuire 1994) and the Vizcaíno Peninsula in central Baja California (Grismer et al. 1994). In northern Baja California, *P. xanti* ranges west of the Peninsular Ranges from at least Arroyo Santo Domingo south through Valle San José to El Rosario. South of El Rosario, *P. xanti* extends farther west, toward the Pacific coast (Bostic 1971; Welsh 1988). In the Cape Region, *P. xanti* is absent from the upper elevations of the Sierra la Laguna but does inhabit the fringes of the oak woodland (Alvarez et al. 1988). In the Gulf of California, *P. xanti* has been reported from 32 islands (Grismer 1999a) and from Islas Magdalena and Santa Margarita along the Pacific coast (Grismer 2001a; map 4.24).

DESCRIPTION Small, with adults reaching 76 mm SVL; head and body flat; head triangular in dorsal profile, slightly wider than neck; eyes large, lacking movable eyelids; anterior head scales granular; scales bordering supralabials and external nares enlarged; 14 to 19 scales across snout between third supralabials; scales on posterior of head and body granular and interspersed with larger tubercles; 12 to 14 longitudinal rows of tubercles on body; 32 to 40 paravertebral tubercles between posterior margin of ear opening and base of tail; granular dorsal scales abruptly differentiated from larger flat, imbricate ventral scales; 6 supralabials from rostral to point below center of eye; 2 or 3 postmentals; gulars flat; 5 infralabials; forelimbs moderately sized; suprahumerals large and circular; infrahumerals granular; tubercles often present on forearm; hind limbs larger than forelimbs; dorsal surfaces granular and tuberculate; ventral surfaces granular; terminal pair of subdigital lamellae on hands and feet square and greatly expanded; tail round in cross section; caudal scales subconical and interspersed with longitudinal rows of tubercles; caudals whorled; 2 to 4 longitudinal rows of tubercles at second whorl of tubercles; 2 or 3 enlarged postanal tubercles in males. **COLORATION** Dorsal ground color gray to dark gray; head and limbs mottled with elongate, irregularly shaped dark blotches; some populations have immaculate white tubercles scattered on limbs and bodies, contributing to a punctate appearance; 5 to 7 dark, irregularly shaped cross bands on body between nape and caudal constriction, often separated by narrow, cream-colored interspaces or a double row of paravertebral spots; original tail distinctly banded; ventrum light gray.

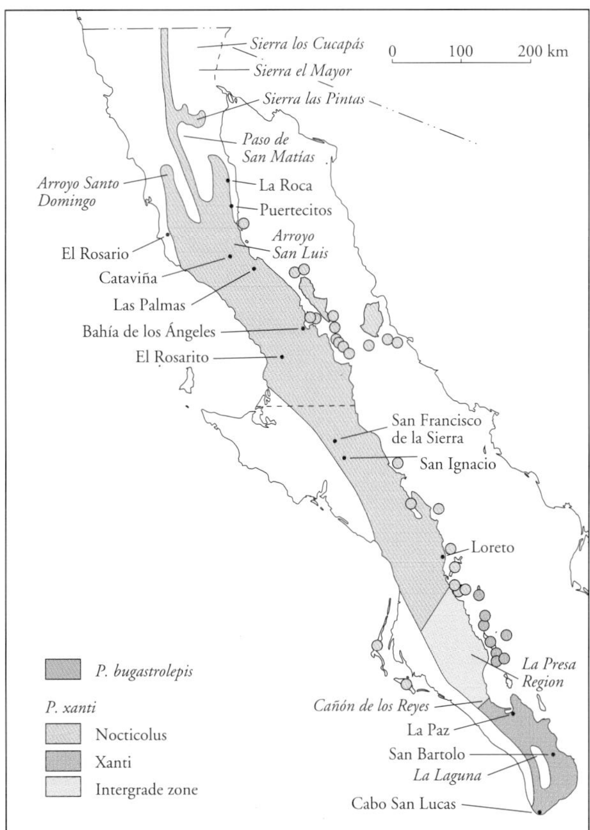

MAP 4.24 Distribution of the pattern classes of *Phyllodactylus xanti* (Peninsular Leaf-Toed Gecko) and *P. bugastrolepis* (Isla Santa Catalina Leaf-Toed Gecko)

FIGURE 4.68 Xanti pattern class of *Phyllodactylus xanti* (Peninsular Leaf-Toed Gecko) from Arroyo Aguaje, BCS

GEOGRAPHIC VARIATION Dixon (1966) considered as distinct subspecies *Phyllodactylus xanti acorius* from Isla San Diego; *P. x. angulus* from Islas Salsipuedes, San Lorenzo Norte, and San Lorenzo Sur; *P. x. circus* from Isla San Ildefonso; *P. x. coronatus* from Isla Coronados, and *P. x. estebanensis* from Islas San Esteban, Tiburón, and Alcatraz. However, the differences between these forms and adjacent peninsular *P. xanti* are so subtle and variable that they are not used here as pattern classes.

Dixon (1964) noted differences between Cape Region populations of Leaf-Toed Geckos, which he referred to as *Phyllodactylus xanti xanti,* and those north of Loreto, which he referred to as *P. x. nocticolus.* These subspecies intergrade from at least Loreto (Dixon 1964) to Cañón de los Reyes, 170 km to the south, and are referred to here as pattern classes.

Xanti attain an SVL of at least 76 mm, whereas the maximum SVL of Nocticolus is 62 mm. Adult Xanti have a fairly distinctive banding pattern, whereas Nocticolus generally have a dorsal pattern consisting of double rows of paravertebral blotches that appear as broken bands. This same color pattern occurs in intergrades of juvenile Xanti and Xanti × Nocticolus from north of the Cape Region. Nocticolus also lack the large white body and femoral tubercles present in Xanti.

Geographic variation in dorsal body pattern within Nocticolus is extensive, and every arroyo and insular population shows some subtle variation of the typical paravertebral rows of blotches (see Bostic 1971; Dixon 1964, 1966). Blotches can vary in size, degree of middorsal separation, and coloration, ranging from brown to gray. The color variation appears to be related to substrate matching. Lizards with gray blotches are usually associated with granitic rock, whereas those with brown blotches are more prevalent on dark volcanic rock. Variation can also be very localized. I have seen specimens from the sedimentary cliffs of San Francisco de la Sierra that are beige with a nearly unicolored dorsal pattern, and specimens just a few meters away on the dark volcanic rock that are much darker and have a prominent blotching pattern. Some lizards from the white, guano-covered Isla las Ánimas tend to have gold flecking in their dorsal pattern during their dark phase.

NATURAL HISTORY Although *P. xanti* ranges throughout several phytogeographic regions, it is always restricted to rocky areas. It is common in areas with large granitic boulders, on bajadas containing volcanic rubble or talus, and within the cobble of beaches. During the day, lizards usually reside beneath exfoliating sheets of granite and in the fractures of other rocks. They can often be seen clinging upside down to the bottoms of cap rocks and basking in the morning light at the edges of cracks. Additionally, I have found Nocticolus beneath surface rocks at La Roca, within the leaves of a dead Coastal Agave *(Agave shawii)* at El Rosarito, and beneath trash at Santa Rosalía. Xanti may be less restricted to rocky habitats than Nocticolus (Dixon 1964). Leviton and Banta (1964) commonly found Xanti beneath fallen Cardón Cactus *(Pachycereus pringlei)* as well as rocks. When active at night, *P. xanti* forages throughout a wide range of habitats. I have observed lizards on open beaches on Isla Smith and in sandy arroyo bottoms on Isla Ángel de la Guarda. On Isla El Muerto, I found an individual at the top of a lighthouse 60 m above the ground. On Isla San José, I have commonly found lizards within and beneath the bark of decaying Cardón Cactus and Leatherplant *(Jatropha cuneata),* as well as abroad at night on the coastal dunes. On Isla la Rasa, it is commonly seen foraging for insects on the walls beneath the lights of the field station. I have seen individuals on the trunks of Elephant Tree *(Bursera microphylla)* east of San Ignacio.

Phyllodactylus xanti is nocturnal and often abroad in temperatures as low as 20°C. It is usually the most abundant reptile observed and may be active during moonlit periods while other species remain hidden. I even observed a specimen abroad at 0830 on an overcast morning at La Presa during early July. During the evening, lizards are usually observed running across the tops of small rocks or on rocky ground. In an attempt to escape the light, they leap aimlessly into the air and are often quite difficult to catch. Xanti are probably active year-round, with a peak activity period from March through October. I have seen gravid females during July and August. During mid-December, I observed an individual at San Bartolo basking beneath a granitic exfoliation at midday. Nocticolus, however, from at least Bahía de los Ángeles north, hibernate during the winter. I have seen Nocticolus active at night no earlier than mid-March and no later than October. During late December in Arroyo San Luis, I found nine Nocticolus and two juvenile Banded Rock Lizards *(Petrosaurus mearnsi)* beneath a single exfoliation. I have also observed hibernating lizards in rock cracks at Las Palmas and in the Sierra las Pintas during December and January. In contrast, Linsdale (1932) reported finding a specimen from Bahía de Calamajué that may have been active during late January. Nocticolus from the southern portion of its range are probably active year-round, depending on the weather.

Nocticolus generally emerge from hibernation during March. They breed during the spring and summer, and gravid females are commonly seen from May through July. Hatchlings are abundant on the peninsula north of the Isthmus of La Paz and the adjacent islands from late July through September. They may be found as late as mid-October but are not nearly as common. During mid-October on Isla San José, I found a clutch of eggs that had already hatched.

Phyllodactylus xanti can lay several clutches of one to two eggs, which are usually stuck to a rocky surface beneath an exfoliation or within a crack. On Isla San José, I have found egg clutches beneath the exfoliating bark of decaying Cardón Cactus. *Phyllodactylus xanti* feeds on small arthropods.

FOLKLORE AND HUMAN USE See *P. homolepidurus.*

FIG. 4.69 **_Phyllodactylus bugastrolepis_**

ISLA SANTA CATALINA LEAF-TOED GECKO; SALAMANQUESA

Phyllodactylus bugastrolepis Dixon 1966:447. Type locality: "Isla Santa Catalina, Baja California [Sur]," Mexico. Holotype: CAS 98485

IDENTIFICATION *Phyllodactylus bugastrolepis* can be distinguished from all other lizards in the region of study by its greatly expanded terminal subdigital lamellae; 38 to 46 paravertebral tubercles from the rear of the head to the base of the tail; very large abdominal scales; and 16 to 22 scales across the snout between the third supralabials.

RELATIONSHIPS AND TAXONOMY *Phyllodactylus bugastrolepis* is probably the sister species of *P. xanti* of Baja California (see Grismer 1994c).

DISTRIBUTION *Phyllodactylus bugastrolepis* is endemic to Isla Santa Catalina in the Gulf of California, BCS (map 4.24).

DESCRIPTION Same as *P. xanti,* except for the following: adults reach 63 mm SVL; 16 to 22 scales across snout between third supralabials; 12 to 15 longitudinal rows of tubercles on body; 38 to 46 paravertebral tubercles; ventral scales very large; 6 or 7 supralabials; 2 or 3 postmentals; 5 or 6 infralabials. **COLORATION** Dorsal ground color dull white to light gray; irregularly shaped dark stripe extending from nostril to ear opening; head, back, and limbs mottled with irregularly shaped dark blotches; tail somewhat banded; ventrum cream-colored.

NATURAL HISTORY Isla Santa Catalina is a rocky island lying within the Central Gulf Coast Region. Its climate is hot and arid for much of the year, and the island is dominated by xerophilic scrub. *Phyllodactylus bugastrolepis* is not as restricted to rocky areas as is *P. xanti.* Although quite common beneath cap rocks and exfoliations and within fracture cracks, it is equally abundant within dead Cardón Cactus *(Pachycereus pringlei)* and Giant Barrel Cactus *(Ferrocactus diguetii).* While abroad at night, *P. bugastrolepis* can be found on a number of substrates. I have observed lizards on the ground in sandy washes, on the trunks of living Cardón Cactus, on rocks of various sizes and shapes, within small beach cobble along the shore, and jumping from branch to branch within dead Leatherplant *(Jatropha cuneata).*

The seasonal activity period of *P. bugastrolepis* is not well-documented. It probably extends from March through at least October. I have observed active specimens from early April through early September. *Phyllodactylus bugastrolepis* is nocturnal, emerging during the evenings to forage and mate, although I observed one individual

FIGURE 4.69 *Phyllodactylus bugastrolepis* (Isla Santa Catalina Leaf-Toed Gecko) from Arroyo Blanco y Sol, Isla Santa Catalina, BCS

sunning itself at the edge of a rock crack at 0500 during mid-July. I have found gravid females carrying one or two eggs from late May through early July, hatchlings as early as late July, and juveniles as late as mid-October. These observations suggest that *P. bugastrolepis* has a summer reproductive season. *Phyllodactylus bugastrolepis* presumably feeds on small arthropods. During one evening in mid-June, I found many places along the bank of an arroyo where several winged termites were above ground. At each site, I observed several fat *P. bugastrolepis* that had been feeding.

FIG. 4.70 **_Phyllodactylus unctus_**

SAN LUCAN LEAF-TOED GECKO; SALAMANQUESA

Diplodactylus unctus Cope 1863:102. Type locality: "Cape St. Lucas [Cabo San Lucas], Lower California," BCS, Mexico Holotype: USNM 5304

IDENTIFICATION *Phyllodactylus unctus* can be distinguished from all other lizards in the region of study by its greatly expanded terminal subdigital lamellae and absence of body tubercles.

RELATIONSHIPS AND TAXONOMY *Phyllodactylus unctus* is probably the sister species of the *P. delcampi* group (*sensu* Dixon 1964) of southwestern Mexico. Grismer (1994a) discusses the taxonomy.

DISTRIBUTION On the peninsula, *P. unctus* is restricted to the Cape Region. In the Gulf of California, it has been reported from Islas Partida Sur, Espíritu Santo, Ballena, Gallo, Gallina, and Cerralvo (Grismer 1999a; see map 4.23).

DESCRIPTION Small, with adults reaching 52 mm SVL; head and body flat; head triangular in dorsal profile, slightly wider than neck; eyes large, movable eyelids absent; ante-

FIGURE 4.70 *Phyllodactylus unctus* (San Lucan Leaf-Toed Gecko) from Cañón San Bernardo, BCS

rior head scales granular; scales bordering supralabials and external nares slightly enlarged; 18 scales across snout between third supralabials; dorsal scales granular and lacking tubercles; dorsals abruptly differentiated from larger flat, imbricate ventral scales; 7 supralabials from rostral to point below center of eye; 2 postmentals; gulars flat; 6 infralabials; forelimbs moderately sized, hind limbs slightly larger; limbs covered with smooth, flat, rounded scales; terminal pair of subdigital lamellae on hands and feet square and greatly expanded; tail oval in cross section; caudals flat and rectangular; subcaudals larger and platelike; supracaudals whorled; 3 or 4 enlarged postanal tubercles in males. **COLORATION** Ground color capable of changing from transparent gray to dark gray or brown; postnasal stripe extending to anterior margin of eye; postorbital stripe extending from posterior margin of eye to forelimb insertion; lineate blotches on top of head; semi-irregular to irregularly shaped bands on body; usually 6 body bands; scattered, indistinct spots on limbs; banding more distinctive in juveniles; ventrum dull white.

GEOGRAPHIC VARIATION Specimens from the Rancho Ancón area, east of La Paz, are nearly 1.5 times the size of lizards from other populations.

NATURAL HISTORY *Phyllodactylus unctus* is a common lizard in the Arid Tropical Region, ranging up to the edge of the oak woodland of the Sierra la Laguna Region. It is scansorial, and in the Cape Region and Isla Cerralvo has been found beneath exfoliating slabs of granite, under rocks, in rock cracks, in caves, beneath the bark of dead vegetation, beneath and within rotting Cardón Cactus *(Pachycereus pringlei)* and other large cacti, and in thatched roofs made from palm leaves (Banks and Farmer 1963; Dixon 1964; Etheridge 1961; Leviton and Banta 1964; Murray 1955). I have observed specimens beneath rocks on Isla Gallo and within rotting Cardón Cactus on Islas Espíritu Santo and Gallina. At night, *P. unctus* is commonly seen on various substrates and structures. On Isla Cerralvo, it seems to be most abundant in the steep-sided arroyos, where it forages in the open on the rocky arroyo floor, vegetation, and rocks, and along the rocky cliff faces. Lizards are not found on the open, sandy flats.

Leviton and Banta (1964) reported that *P. unctus* is one of the most common species of lizards found in the Cape Region during winter. I have observed active specimens during December and January on Isla Espíritu Santo and during late March on Isla Cerralvo. Asplund (1967) observed hatchlings throughout August, and on Isla Cerralvo I have found gravid females during late August, and hatchlings and juveniles from August through mid-October. On Isla Espíritu Santo, I have found juveniles during late December. These observations suggest that *P. unctus* has an early summer breeding season and late summer to fall hatching period. Asplund (1967) noted that the incidence of many larval insects on which Cape Region lizards feed is correlated with the onset of summer rains, and many species of lizards in the Cape Region hatch during this period and feed on the larvae.

REMARKS In the Cape Region, *P. unctus* and the much larger *P. xanti* are often found beneath the same granitic exfoliations. These species probably exist in syntopy because of their difference in body size. Case (1983) demonstrated that among other sympatric, closely related lizards with different body sizes, different-sized prey are eaten, thus reducing competition. *Phyllodactylus unctus* does not range north of the Cape Region, where *P. xanti* is roughly the same size as *P. unctus.* Interestingly, in areas of the Cape Region such as Rancho Ancón, where *P. xanti* does not occur, *P. unctus* is much larger than in localities where it is sympatric with *P. xanti.*

FOLKLORE AND HUMAN USE See *P. homolepidurus.*

Teiidae

WHIPTAILS AND RELATIVES

The Teiidae are a large radiation of New World lizards containing nine genera and approximately 105 species (Pough et al. 2001) that range collectively from the northern United States south through Mexico, Central America, and the West Indies, to the southern regions of South America. The Teiidae are an ecologically diverse group that occupies a number of habitats, ranging from desert scrub to tropical rain forests. Two genera, *Dracaena* and *Crocodylurus,* are semiaquatic, and the former is specialized to feed on mollusks. All teiids are active foragers that rely heavily on olfaction to locate prey. Teiids are well differentiated from all other lizards by a host of morphological characteristics (see Estes et al. 1988).

Cnemidophorus Wagler

WHIPTAILS

The genus *Cnemidophorus* is composed of over 50 species of small to medium-sized terrestrial lizards (Maslin and Secoy 1986) that collectively range from northern United States to southern South America. *Cnemidophorus* are conspicuous diurnal components of many habitats and are commonly seen darting from bush to bush, their tongues flicking in search of prey. *Cnemidophorus* forms a significant portion of the lizard fauna of Baja California and the Gulf of California. It is represented by 14 species (13 of which are endemic; Grismer 1999b) in three distinct species groups (Lowe et al. 1970): the *deppei* group, with *C. ceralbensis, C. hyperythrus, C. espiritensis, C. carmenensis, C. pictus, C. danheimae,* and *C. franciscensis;* the *sexlineatus* group, with *C. labialis;* and the *tigris* group, with *C. tigris, C. bacatus, C. canus, C. catalinensis, C. celeripes,* and *C. martyris.*

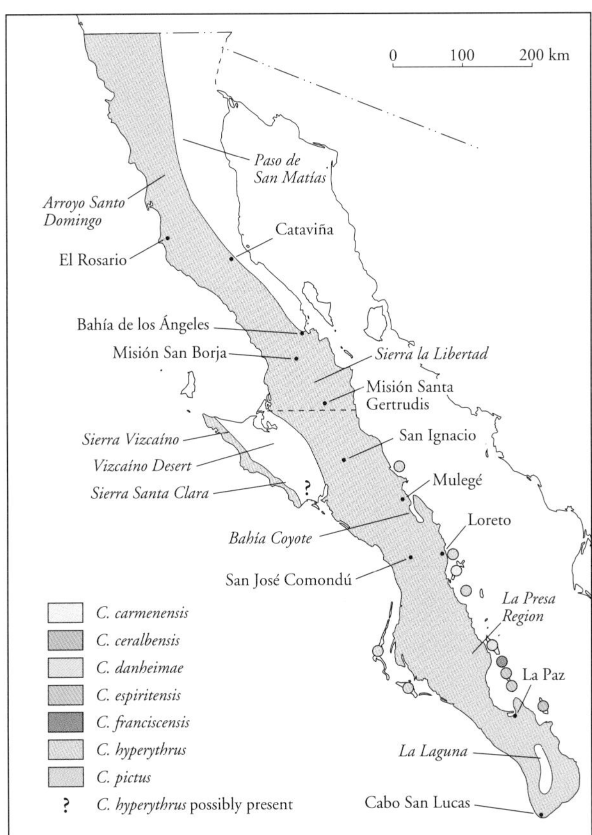

MAP 4.25 Distribution of *Cnemidophorus hyperythrus* (Orange-Throated Whiptail), *C. ceralbensis* (Isla Cerralvo Whiptail), *C. carmenensis* (Isla Carmen Whiptail), *C. danheimae* (Isla San José Whiptail), *C. espiritensis* (Isla Espíritu Santo Whiptail), *C. franciscensis* (Isla San Francisco Whiptail), and *C. pictus* (Isla Monserrate Whiptail)

FIGURE 4.71 *Cnemidophorus ceralbensis* (Isla Cerralvo Whiptail) from Arroyo Viejos, Isla Cerralvo, BCS

FIG. 4.71

Cnemidophorus ceralbensis

ISLA CERRALVO WHIPTAIL; GÜICO

Verticaria ceralbensis Van Denburgh and Slevin 1921c:396. Type locality: "Ceralbo Island, Gulf of California [BCS], Mexico." Holotype: CAS 50510

IDENTIFICATION *Cnemidophorus ceralbensis* can be distinguished from all other lizards in the region of study by having movable eyelids; granular dorsal scales; large rectangular, imbricate ventral scales; a single frontoparietal scale; and light crossbars in the dorsolateral fields (spaces between the stripes).

RELATIONSHIPS AND TAXONOMY *Cnemidophorus ceralbensis* is the sister species of the *C. hyperythrus* complex of Baja California and adjacent islands (Lowe et al. 1970; Radtkey et al. 1997; Robinson 1973; see Grismer 1999c for discussion). Grismer (1999c), Walker et al. (1966b), and Walker and Taylor (1968) discuss taxonomy.

DISTRIBUTION *Cnemidophorus ceralbensis* is endemic to Isla Cerralvo in the Gulf of California, BCS (map 4.25).

DESCRIPTION Small, with adults reaching 93 mm SVL; body rounded in cross section, somewhat elongate and fusiform; head not compressed, sharply pointed in lateral and dorsal profile; dorsal head scales enlarged and plate-like; nasal scales not contacting second supralabials; frontoparietal single; 3 supraocular scales; granular scales absent between second supraocular and frontal; 1 large parietal on either side of large interparietal, which lacks an eye spot; smaller row of occipitals abruptly contacting granular dorsal body scales; dorsal scales abruptly contacting very large, rectangular, imbricate ventral scales; 85 to 107 longitudinal rows of dorsal scales; 5 to 12 granules separating paravertebral stripes; ventrolateral body fold extending from axilla to groin; 5 supralabials; single large postmental; sublabials very large; gulars granular anteriorly, larger and rounded centrally, and much smaller and granular posteriorly; mesoptychial scales moderately sized; gular fold complete; antegular fold interrupted medially; 6 or 7 infralabials; forelimbs stout and relatively short, with large, imbricate rectangular scales on anterior surface; hind limbs approximately twice the size of forelimbs; antefemorals and infratibials greatly enlarged and imbricate; 13 to 20 well-developed femoral pores in males; 28 to 32 subdigital lamellae on fourth toe; tail cylindrical, pointed, and 2.5 to 3.0 times length of body; caudal scales keeled, whorled, and submucronate; caudal keels forming distinct longitudinal ridges running length of tail; subcaudals large, flat, and imbricate; postanals not enlarged. **COLORATION** Dorsal ground color of head and forelimbs gray-brown, body and hind limbs black, tail brown to blue; 6 well-defined stripes present on body; one pair of tan paravertebral stripes running from parietals to base of tail; a pair of white dorsolateral stripes running from superciliaries to base of tail; a pair of white lateral stripes running from orbits to groin; usually a very poorly defined tan vertebral stripe running from interparietal to base of tail, most often appearing as a lightened area between paravertebral stripes; striping absent on tail; fields black with light spots or transverse bars, less marked in juveniles; forelimbs with light irregular markings; hind limbs with large white spots; gular region and chest dark sooty gray; abdomen heavily mottled with black; tail less mottled; hatchlings vividly striped; tail turquoise.

NATURAL HISTORY The mountainous Isla Cerralvo lies within the Central Gulf Coast Region and is dominated by xerophilic vegetation. But because it is situated off the coast of the Cape Region, a large component of its flora is also derived from the Arid Tropical Region, and it has a marked wet and dry season. *Cnemidophorus ceralbensis* is a common species that prefers open, sandy areas with scattered clumps of thick vegetation, into which it retreats when pursued. Activity and foraging are often confined to the edges of vegetation where the sunlight is dappled. Lizards are common on dunes along the beach at the southern end of the island, as well as in the rocky washes that extend deep into the island's interior, so long as vegetative cover is present. *Cnemidophorus ceralbensis* is not restricted to low-lying areas, and lizards are not uncommon on the rocky hillsides. I even observed one specimen climbing up a vertical cliff face from the bottom of the arroyo.

Cnemidophorus ceralbensis is active year-round, with minimal activity occurring during the winter. From mid-December through March, all age classes are active, but juveniles are by far the most abundant. Like other *Cnemidophorus, C. ceralbensis* is diurnal and forages for small prey. Lizards are often seen digging beneath bushes and placing their heads inside the holes to extract various arthropods and their larvae. I have found cockroaches, walking sticks, and pieces of the fruit of Coyote Melon *(Ibervillea sonorae)* in the stomachs of several individuals in mid-October. During this period, the ripened fruits drop to the ground and are soft enough for the lizards to eat. I noticed that nearly every fruiting *I. sonorae* had a number of whiptails nearby.

Banks and Farmer (1963) observed a copulating pair during mid-May and noted that males had enlarged testes during late May and early June. I observed a courting pair in late August. Taylor and Walker (1987) noted gravid females in June and August, and Walker et al. (1966b) noted that no hatchlings were present during early August. I have observed hatchlings from late August to mid-December. These observations suggest that reproduction takes place from late spring to at least early summer, and hatching coincides with the summer rainy season. Taylor and Walker (1987) noted that females were probably reclusive during the reproductive season and reported clutch sizes of one to four eggs. Case (1983) reported finding only a single egg in a gravid female he examined.

FIG. 4.72 *Cnemidophorus hyperythrus*
ORANGE-THROATED WHIPTAIL; GÜICO

Cnemidophorus hyperythrus Cope 1863:103. Type locality: "Cape St. Lucas [Cabo San Lucas]," BCS, Mexico. Holotype: USNM 5290

IDENTIFICATION *Cnemidophorus hyperythrus* can be distinguished from all other lizards in the region of study by having movable eyelids; granular dorsal scales; large, rectangular, imbricate ventral scales; a single frontoparietal scale; no light crossbars in the body fields (spaces between the stripes); large mesoptychial scales; and orange on the ventrum of adult males.

RELATIONSHIPS AND TAXONOMY *Cnemidophorus hyperythrus* forms a natural group with the Gulf island endemics *C. carmenensis, C. danheimae, C. espiritensis, C. franciscensis,* and *C. pictus* (Grismer 1999c; Radtkey et al. 1997; Walker and Taylor 1968). Grismer (1999c) and Walker and Taylor (1968) discuss taxonomy.

DISTRIBUTION *Cnemidophorus hyperythrus* ranges from Orange and San Bernardino counties of southern California south to Cabo San Lucas (Grismer 1999c; Thompson et al. 1998). In Baja California, it ranges throughout the cismon-

FIGURE 4.72 *Cnemidophorus hyperythrus* (Orange-Throated Whiptail) from Tijuana, BC

tane northwestern portion of the peninsula, arcing east to contact the Gulf coast south of the Sierra la Asamblea in the vicinity of Bahía de los Ángeles. It then continues south around the eastern border of the northern portion of the Vizcaíno Desert and into the Cape Region (see map 4.25). Although *C. hyperythrus* has not been reported or observed in the northern Vizcaíno Desert (Grismer 1994c), it does occur in the rocky mountainous portion of the Vizcaíno Peninsula to the west, and Grismer et al. (1994) believe that this may be a disjunct population. *Cnemidophorus hyperythrus* does not occur in the upper elevations of the Sierra la Laguna (Alvarez et al. 1988) or the Sierra la Asamblea. *Cnemidophorus hyperythrus* is known from the Gulf islands of San Marcos and Coronados and the Pacific islands of Magdalena and Santa Margarita (Grismer 1993, 1999a).

DESCRIPTION Small, with adults reaching 72 mm SVL; body rounded in cross section, somewhat elongate and fusiform; head not compressed, sharply pointed in lateral and dorsal profile; dorsal head scales enlarged and platelike; nasal scales not contacting second supralabials; frontoparietal single; 3 to 5 (usually 4) supraocular scales; granular scales may or may not be present between second supraocular and frontal; 1 large parietal on either side of large medial interparietal, which lacks an eye spot; smaller row of occipitals abruptly contacting granular dorsal body scales; dorsal scales abruptly contacting large, rectangular, imbricate ventral scales; 66 to 90 longitudinal rows of dorsal scales; 0 to 13 granules separating paravertebral stripes; ventrolateral fold on body extending from axilla to groin; 5 or 6 supralabials; single large postmental; sublabials very large; gulars granular anteriorly, larger and rounded centrally, and much smaller and granular posteriorly; mesoptychial scales large; gular fold complete; antegular fold interrupted medially; 5 to 7 infralabials; forelimbs stout, relatively short, and with large imbricate, rectangular scales on anterior surface; hind limbs approximately twice the size of forelimbs; antefemorals and infratibials greatly enlarged and imbricate; 12 to 22 well-developed femoral pores in males; 25 to 33 subdigital lamellae on fourth toe; tail cylindrical, pointed, and 2.5 to 3.0 times length of body; caudal scales keeled, whorled, and submucronate; caudal keels forming distinct longitudinal ridges running length of tail; subcaudals large, flat, and imbricate; postanals not enlarged.

COLORATION Dorsal ground color of head and forelimbs gray-brown, body and hind limbs dark brown to black, tail brown to blue; body fields generally immaculate; 5 to 7 well-defined stripes present at midbody (the seventh stripe is an occasional, poorly defined vertebral stripe running from interparietal to base of tail, most often appearing as a lightened area between paravertebral stripes); a pair of tan paravertebral stripes running from parietals to base of tail when 6 or 7 stripes present; a single cream-white vertebral stripe when 5 stripes present; a pair of yellowish, dorsolateral stripes running from superciliaries to base of tail; a pair of white lateral stripes running from orbits to groin; striping present on tail; paravertebral fields tan, suffused with black, solid black in juveniles; dorsolateral fields black to brown; forelimbs with or without light irregular markings; hind limbs with sinuous longitudinal stripes; gular region and chest orange in adult males, less so in females; most intense in males during breeding season; orange may extend onto abdomen and tail (see below); ground color of abdomen and tail uniform dull white to gray. HATCHLINGS: Hind limbs and tail bright blue; striping vivid; some with orangish forelimbs.

GEOGRAPHIC VARIATION The geographic variation reported within *C. hyperythrus* focuses on four characteristics: number of dorsal body granules, the degree of separation of the second supraocular from the frontal by granular scales, the number of body stripes, and the position of the vertebral fork of the middorsal stripe (see Bostic 1971; Burt 1931; Linsdale 1932; Murray 1955; Van Denburgh 1922; Welsh 1988; Walker and Taylor 1968). Based on these characteristics, various authors have designated three subspecies of peninsular *C. hyperythrus: C. h. beldingi* in the north, *C. h. schmidti* in central Baja California, and *C. h. hyperythrus* in southern Baja California. The distributional limits of these subspecies, however, have always been a source of contention because of the poor discriminating value of these "diagnostic" characteristics (see Welsh 1988). Therefore it is uninformative to consider these subspecies even as pattern classes. The variation of these characteristics is most efficiently treated by distinguishing populations from the regions of northern (from the U.S.-Mexican border to El Rosario), central (El Rosario to Loreto), and southern (Loreto to Cabo San Lucas) Baja California, as all populations are morphologically variable and blend with adjacent populations.

There is almost no variation in the number of dorsal body granules between the peninsular populations (see Bostic 1971). The separation of the second supraocular from the frontal by granular scales is most prevalent in the north, occurring in at least 50 percent of the specimens examined (Van Denburgh 1922). In central Baja California, few specimens show complete separation, and there is no separation in southern populations.

The number of stripes at midbody ranges from five to seven. All *C. hyperythrus* have a pair of lateral and a pair of dorsolateral stripes (see above). Lizards with five stripes have a single well-defined vertebral stripe with an anteriorly located fork. Lizards with six stripes lack a vertebral stripe because it has bifurcated to form a pair of paravertebral stripes. Here the fork of the vertebral stripe occurs posteriorly on the body, usually in the sacral region, or fades beyond recognition on the base of the tail. Occasionally, a faded seventh vertebral stripe appears between the paravertebral stripes. Populations from the north have five to seven stripes, depending on the location of the vertebral fork and the presence or absence of a faded vertebral stripe. The tendency in the north is for the fork to occur anteriorly, and thus five stripes is the most common pattern. In central Baja California, the fork tends to occur between the forelimb and hind limb insertions, so the number of stripes depends on where the count is taken. Tevis (1944) reported that in a series collected between Bahía Coyote (along the western edge of Bahía Concepción) and "Comondú," five specimens had a single vertebral stripe, two had a pair of paravertebral stripes, and one had a pair of paravertebral stripes along with a faded vertebral stripe. In seven specimens I observed from the La Presa region in southern Baja California, the fork occurred anterior to the forelimb insertions in two specimens, just posterior to the forelimbs insertions in three specimens, at midbody in one specimen, and not at all in another. In the Cape Region, the faded vertebral stripe very rarely forks, and the paravertebral stripes extend the length of the body. Thus, lizards from the Cape Region usually have seven stripes.

The degree of ventral orange coloration in adult males also varies geographically. In northern populations, the orange is generally confined to the anterior half of the body. From approximately San Ignacio south, the orange coloration is much more extensive on the abdomen, hind limbs, and tail, although some adult males from the Cape Region have orange only in the gular region. Some adult males from the Magdalena Plain lack orange ventrally, and some adult females from the Cape Region and Isla Espíritu Santo have small amounts of orange in the gular region (Grismer 1999c).

Lizards from the white beaches of the northwestern shore of the Islas Coronado have a faded dorsal color pattern, and the sides and the ventral surfaces of their limbs are bluish. This bluish coloration also occurs in many adults from the Magdalena Plain and Islas Magdalena and Santa Margarita.

NATURAL HISTORY *Cnemidophorus hyperythrus* ranges throughout most of the phytogeographic regions of Baja California. It is absent, however, from the cooler high elevations of the Baja California Coniferous Forest Region in the north; the hot, arid Lower Colorado Valley Region in the northeast; the sandy Vizcaíno Desert in central Baja California; and the densely vegetated Sierra la Laguna Region in the south. The species is most commonly found in areas with open spaces between clumps of thick vegetation on fine-grained, loose soils, such as rocky hillsides bordering arroyos or the lower slopes of foothills (Tevis 1944; Welsh 1988). Alvarez et al. (1988) reported that *C. hyperythrus* is ubiquitous in the Arid Tropical Region of the Cape Region, and Van Denburgh (1922) states that in the Cape Region it is commonly seen among fallen cactus and clumps of vegetation.

The natural history of *C. hyperythrus* from the northern portion of its range has been well studied (Bostic 1965, 1966a–d; Case 1983). In the north, *C. hyperythrus* enters hibernation burrows from late July through September (Bostic 1966a) and emerges from hibernation during March. Yearlings, however, may remain active as late as December, depending on the weather. *Cnemidophorus hyperythrus* is active year-round in southern Baja California from the vicinity of Mulegé south. Blázquez and Ortega-Rubio (1996) and Leviton and Banta (1964) reported that at lower elevations it is one of the most commonly seen lizards in the Cape Region during the winter. I have observed all size classes during late December and early January in the lower elevations of the Cape Region and at higher elevations in the western foothills of the Sierra la Laguna on the edge of the oak woodland belt. The majority of winter activity in the Cape Region, however, is that of hatchlings and juveniles. Adult females are occasionally observed, and adult males are rarely seen. I have also observed adult females abroad in the Sierra Santa Clara in central Baja California during mid-December.

Bostic (1966a) noted that the daily activity of *C. hyperythrus* occurs within a narrow range of substrate temperatures and that it clearly avoids high midday temperatures. Lizards emerge from overnight retreats to bask before foraging. Tevis (1944) reported seeing lizards pushing dirt out of their burrows with their forelimbs in the early morning. He went on to report that they are most active during the morning and early afternoon and were seen foraging at the edge of clumps of vegetation. Leviton and Banta (1964) stated that *C. hyperythrus* was most active in the early afternoon in the Cape Region during the winter. In the La Presa region during early July, when the sun is out, *C. hyperythrus* is active only during the early morning. On overcast days, however, lizards remain active throughout the day. During late June in the Sierra Santa Clara, lizards are active during the early morning and retreat into their burrows at approximately 1000 hours, when temperatures become too high. On one occasion, clouds developed at around 1200 hours, and the temperature dropped significantly. Shortly afterward, *C. hyperythrus* resumed activity and remained active until the skies cleared at approximately 1500 hours, when lizards once again became inactive. During midday in late June at Misión Santa Gertrudis, I observed an adult male *C. hyperythrus* deliberately moving along the edge of some vegetation to remain in the shade. It would not even cross sunlight areas that were less than half a meter wide, but instead followed the shade line around the open areas, even though this required traversing a much greater distance.

Cnemidophorus hyperythrus is most commonly seen moving through stands of vegetation using short, rapid, jerky movements, tongue-flicking as it goes. It regularly stops moving momentarily to probe for food within leaf litter, shallow burrows, and loose soil. *Cnemidophorus hyperythrus* stays primarily within the cover of vegetation when moving about, and when confronted with open spaces darts rapidly across them. This behavior may be an attempt to avoid predators or heat. In the Comondú region, I have often seen whiptails foraging in the leaf litter beneath Mesquite trees (*Prosopis* spp.) and in the Sierra Santa Clara at the base of Datilillo *(Yucca valida).* At San Ignacio, I watched an individual repeatedly climb into the lower branches of a small bush in search of food or to cool off. In a rocky arroyo in the Sierra la Libertad, I saw an individual drinking water from the edge of a pool. Cape Region populations are commonly seen on steep rocky hillsides with an abundance of brush.

Throughout its distribution, *C. hyperythrus* feeds primarily on termites. Bostic (1966b) noted that in southern California, termites made up nearly 100 percent of its diet. Case (1983) noted that termites composed nearly 70 percent of the diet of Baja California populations. Beetles, orthopterans, silverfish, and spiders made up most of the remainder. Asplund (1967) reported that approximately 70 percent of the diet of Cape Region populations was termites, and I have found termites and cockroaches in the guts of Cape Region specimens. At Misión San Borja, I watched lizards moving awkwardly through tall green grass next to a well, attempting to catch the abundant grasshoppers, and I found grasshoppers in the gut of a museum specimen from Arroyo Santo Domingo. Galina (1994) reported that *C. hyperythrus* from the Isthmus of La Paz eats primarily termites and the larvae of beetles, moths, and butterflies during the spring and summer. During the fall, she noted, the primary diet was the larvae of moths and butterflies.

The reproductive season of *C. hyperythrus* seems to vary slightly with latitude. In southern California, Bostic (1966c) noted that the breeding season lasts from May through mid-July. Two to three eggs are laid (possibly two clutches per year in older females), and hatchlings appear from mid-August through early September. In the Cape Region, Asplund (1967) noted that no hatchlings were seen in August, but that females had ovarian and oviducal eggs, and copulations were frequently observed. Romero-Schmidt and Ortega-Rubio (2000) note females with oviducal eggs in July and males having the greatest testicular size in June. Van Denburgh (1922) reported that in the Cape Region, copulations had been observed during the first week of July. I have noticed that during mid-October, most of the *C. hyperythrus* in the Cape Region are hatchlings and very small juveniles. This would suggest that the Cape Region populations have a later reproductive season than those of southern California. Between these two regions, I have observed hatchlings and gravid females from late June to mid-July in the Sierra Santa Clara and Sierra Vizcaíno; gravid females on Islas Coronado during mid-July; hatchlings during early August in the vicinity of Loreto; and hatchlings during late September at San José Comondú.

FOLKLORE AND HUMAN USE North of the Cape Region, indigenous people's reactions to *Cnemidophorus* are somewhat indifferent. In the Cape Region, however, many are frightened by whiptail lizards and believe them to be venomous. In fact, Alvarez et al. (1988) state that in general people in the Cape Region fear *güicos* more than rattlesnakes.

FIG. 4.73 *Cnemidophorus carmenensis*

ISLA CARMEN WHIPTAIL; GÜICO

Verticaria caerulea Dickerson 1919:472. Type locality: "Carmen Island, Gulf of California, [BCS], Mexico." Holotype: AMNH 5517

Cnemidophorus hyperythrus carmenensis Maslin and Secoy 1986:21. New name for *C. h. caeruleus* because of preoccupation by *C. caeruleus* (Laurenti 1768).

FIGURE 4.73 *Cnemidophorus carmenensis* (Isla Carmen Whiptail) from Isla Carmen, BCS

IDENTIFICATION *Cnemidophorus carmenensis* can be distinguished from all other lizards in the region of study by having movable eyelids; granular dorsal scales; large rectangular, imbricate ventral scales; a single frontoparietal scale; no light crossbars in the body fields (spaces between the stripes); small mesoptychial scales; faded stripes on the body; and no markings on the hind limbs and tail.

RELATIONSHIPS AND TAXONOMY See *C. hyperythrus*. *Cnemidophorus carmenensis* is the sister species of *C. pictus* (Grismer 1999c; Radtkey et al. 1997).

DISTRIBUTION *Cnemidophorus carmenensis* is endemic to Isla Carmen in the Gulf of California, BCS (see map 4.25).

DESCRIPTION Same as *C. hyperythrus,* except for the following: adults reach 70 mm SVL; 2 to 4 (usually 3) supraocular scales; 72 to 88 longitudinal rows of dorsal scales; mesoptychial scales small; 14 to 21 femoral pores; 25 to 32 subdigital lamellae on fourth toe. **COLORATION** ADULTS: Generally faded in appearance; dorsal ground color of head and body dark brown, that of limbs and tail dark, sooty gray; 5 (rarely 6) faded stripes present at midbody; a poorly defined vertebral stripe forked anterior to forelimb insertions; one pair of tan dorsolateral stripes and one pair of white lateral stripes; striping absent on tail; paravertebral fields uniformly gray-brown; dorsolateral fields uniformly light brown; limbs unicolored; ventrum bluish-gray to sky blue with occasional orange pigment in femoral region. HATCHLINGS AND JUVENILES: Overall coloration and pattern brighter and more vibrant; tail bright turquoise.

NATURAL HISTORY Isla Carmen is dominated by Central Gulf Coast Region vegetation, and *C. carmenensis* is most commonly found on the low-lying, flat areas where stands of floristic dominants such as Creosote Bush *(Larrea tridentata),* Leatherplant *(Jatropha cuneata),* Pitahaya Agria *(Stenocereus gummosus),* and Cholla *(Opuntia cholla)* are separated by open spaces. The bases of these shrubs provide ample habitat for foraging, and whiptails are commonly seen darting from bush to bush. Lizards also occur along the rocky foothills of the island's steep, mountainous axis, although they are not as common in such areas.

Cnemidophorus carmenensis is active year-round, with activity peaking from March through early November. From mid-November through February, hatchlings and juveniles are the most abundant age class. Adults resume activity during March. *Cnemidophorus carmenensis* is diurnal and emerges in the morning to bask at the edge of vegetation, beneath which it forages for small arthropods. I observed two females in early April that appeared to be gravid.

REMARKS *Cnemidophorus carmenensis* and *C. pictus* are sister species occurring on the adjacent islands of Carmen and Monserrate, respectively (Grismer 1999c; see map 4.25). Isla Carmen is a shallow-water landbridge island, whereas Isla Monserrate is an oceanic island (Gastil et al. 1983; Gastil and Krummenacher 1977); the two islands have never been connected. The fact that *C. carmenensis* is primitive relative to *C. pictus* (Grismer 1999c) suggests that overwater dispersal may have been from Isla Carmen to Monserrate. The physiography of Isla Carmen would certainly facilitate such an event. Isla Carmen is a large island dominated by a steep central mountain range reaching 480 m. This range runs nearly the entire length of the island, and runoff from its slopes has formed a large, deltaic fan that faces east toward Isla Monserrate. It would serve as a logical staging point for overwater dispersal following large storms.

FOLKLORE AND HUMAN USE See *C. hyperythrus.*

FIG. 4.74

Cnemidophorus danheimae

ISLA SAN JOSÉ WHIPTAIL; GÜICO

Verticaria sericea Van Denburgh 1895:132. Type locality: "San José Island, Gulf of California," BCS, Mexico. Holotype: CAS 435

Cnemidophorus hyperythrus danheimae Burt 1929:153 to 156. The name *sericea* was preoccupied by *C. gularis sericea* (Cope 1892a) and Burt (1929) submitted *danheimae* as the replacement name.

IDENTIFICATION *Cnemidophorus danheimae* can be distinguished from all other lizards in the region of study by having movable eyelids; granular dorsal scales; large, rectangular, imbricate ventral scales; a single frontoparietal scale; no light crossbars in the body fields (spaces between the stripes); small mesoptychial scales; stripes on the body; light markings on the hind limbs; and reddish-brown dorsolateral field.

FIGURE 4.74 *Cnemidophorus danheimae* (Isla San José Whiptail) from Isla San José, BCS

RELATIONSHIPS AND TAXONOMY See *C. hyperythrus. Cnemidophorus danheimae* is the sister species of *C. franciscensis* (Radtkey et al. 1997).

DISTRIBUTION *Cnemidophorus danheimae* is endemic to Isla San José in the Gulf of California, BCS (see map 4.25).

DESCRIPTION Same as *C. hyperythrus,* except for the following: adults reach 66 mm SVL; 3 or 4 (usually 3) supraocular scales; 73 to 87 longitudinal rows of dorsal scales; mesoptychial scales small; 14 to 19 femoral pores; 26 to 33 subdigital lamellae on fourth toe. **COLORATION** Dorsal ground color of head and body dark brown, limbs and tail gray; 5 strikingly vivid, light-colored wide stripes present at midbody; one tan vertebral stripe forked anterior to forelimb insertions; one pair of yellowish dorsolateral stripes; one pair of white lateral stripes; striping present on tail; paravertebral fields nearly black; dorsolateral fields deep reddish-brown; forelimbs generally uniform gray; hind limbs gray with light markings dorsally; distinct dark and light postfemoral lines; ventrum dark blue to bluish-gray with occasional orange pigment in femoral region of adults. Hatchlings and juveniles have an equally brilliant color pattern but differ from adults in that the tail is bright blue.

GEOGRAPHIC VARIATION *Cnemidophorus danheimae* from the flat, sandy areas of the southern end of Isla San José are more brightly colored than those of the rocky areas to the north. The blue ventrum tends to be brighter and lighter in overall hue, and the dorsolateral fields are noticeably more red-orange.

NATURAL HISTORY Isla San José is dominated by Central Gulf Coast Region vegetation. *Cnemidophorus danheimae* is commonly found on the low-lying, flat, sandy areas or rocky flats of the island and less frequently along the rocky foothills. It is most common where there are stands of vegetation separated by open space.

The species is active year-round, with a peak activity period from April through October. During winter and early spring, hatchlings and juveniles are the most active age class. I have observed courting and copulating pairs and hatchlings during late August and early September. This suggests that *C. danheimae* has a summer reproductive season.

Cnemidophorus danheimae is diurnal and forages for arthropods beneath stands of vegetation. I have found ant lions and termites in the stomachs of adults and watched an adult climb within low vegetation to capture bees.

REMARKS The holotype of *C. danheimae* was destroyed by the 1906 fire following the San Francisco earthquake, and the name is currently without a type specimen.

FOLKLORE AND HUMAN USE See *C. hyperythrus.*

FIG. 4.75 **Cnemidophorus espiritensis**

ISLA ESPÍRITU SANTO WHIPTAIL; GÜICO

Verticaria espiritensis Van Denburgh and Slevin 1921c:397. Type locality: "Espiritu Santo Island, Gulf of California, [BCS], Mexico." Holotype: CAS 50511

IDENTIFICATION *Cnemidophorus espiritensis* can be distinguished from all other lizards in the region of study by having movable eyelids; granular dorsal scales; large rectangular, imbricate ventral scales; a single frontoparietal scale; no light crossbars in the body fields (spaces between the stripes); large mesoptychial scales; and a gray-white abdomen in adult males.

RELATIONSHIPS AND TAXONOMY See *C. hyperythrus.*

DISTRIBUTION *Cnemidophorus espiritensis* is endemic to Islas Espíritu Santo and Partida Sur in the Gulf of California, BCS (see map 4.25).

DESCRIPTION Same as *C. hyperythrus,* except for the following: Adults reach 60 mm SVL; usually 3 or 4 supraocular scales; 71 to 89 longitudinal rows of dorsal scales; 14 to 19 femoral pores; 26 to 31 subdigital lamellae on fourth toe. **COLORATION** Five (rarely 7) stripes present at midbody; vertebral stripe tan, wide, and diffuse, often appearing to touch dorsolateral stripes; dorsolateral and lateral stripes cream-colored and sharply delimited from adjacent fields; dorsolateral fields light brown; forelimbs with or without light irregular markings; hind limbs with sinuous longitudinal stripes; gular region and throat orange in adult males, less vivid in females; abdomen and ventral surface of tail grayish-white. Hatchlings have black fields and bright blue tails and hind limbs.

FIGURE 4.75 *Cnemidophorus espiritensis* (Isla Espíritu Santo Whiptail) from Isla Espíritu Santo, BCS

FIGURE 4.76 *Cnemidophorus franciscensis* (Isla San Francisco Whiptail) from Isla San Francisco, BCS

NATURAL HISTORY Islas Espíritu Santo and Partida Sur are dominated by Central Gulf Coast Region vegetation. *Cnemidophorus espiritensis* is commonly found on the low-lying flat, brushy areas of the island where rocks are sparse. It also occurs in the coastal dune systems built up along the back edge of the beaches and in the arroyo bottoms. Lizards are less commonly found on the rocky hillsides. The majority of specimens I have observed were found in the arroyo bottoms, foraging among clumps of vegetation and fallen Cardón Cactus *(Pachycereus pringlei).*

Adults are uncommon during March, although juveniles are abundant. Adult activity usually begins sometime in late April and lasts through October. Hatchlings and juveniles remain active through the winter, provided the days are sufficiently warm. *Cnemidophorus espiritensis* is diurnal and emerges in the morning to bask in the sun at the edge of vegetation, beneath which it forages for arthropods. I have observed very small juveniles from mid- to late December, which suggests that this species has a summer reproductive season.

FOLKLORE AND HUMAN USE See *C. hyperythrus.*

FIG. 4.76 *Cnemidophorus franciscensis*

ISLA SAN FRANCISCO WHIPTAIL; GÜICO

Verticaria franciscensis Van Denburgh and Slevin 1921c:397. Type locality: "San Francisco Island, Gulf of California [BCS], Mexico." Holotype: CAS 50513

IDENTIFICATION *Cnemidophorus franciscensis* can be distinguished from all other lizards in the region of study by having movable eyelids; granular dorsal scales; large rectangular, imbricate ventral scales; a single frontoparietal scale; temporal region of adult males with an orangish-yellow wash; no light crossbars in the body fields (spaces between the stripes); small mesoptychial scales; stripes on the body; light markings on the hind limbs; and a faded dorsal pattern.

RELATIONSHIPS AND TAXONOMY See *C. danheimae.*

DISTRIBUTION *Cnemidophorus franciscensis* is endemic to Isla San Francisco in the Gulf of California, BCS (see map 4.25).

DESCRIPTION Same as *C. hyperythrus,* except for the following: adults reach 62 mm SVL; usually 3 supraocular scales; 72 to 84 longitudinal rows of dorsal scales; mesoptychial scales small; 14 to 18 femoral pores; 25 to 31 subdigital lamellae on fourth toe. **COLORATION** Dorsal ground color of head brown, with body, limbs, and tail gray; temporal region of adult males with slight orangish-yellow wash; 5 slightly faded stripes present at midbody; one tan vertebral stripe forked anterior to forelimb insertions and often possessing a sacral loop; one pair of whitish dorsolateral stripes; one pair of whitish lateral stripes; striping present on tail; color of body fields not contrasting greatly with color of stripes; paravertebral fields brown, with slightly lighter lateral fields; forelimbs generally uniform gray; hind limbs gray with very light markings dorsally; faded dark and light postfemoral lines; chin and throat yellowish; other ventral surfaces bluish in adult males and grayish-white in females and subadults; tail bluish in adult males; vivid dorsal pattern in hatchlings and juveniles; tail bright turquoise-blue.

NATURAL HISTORY Isla San Francisco is a small island dominated by Central Gulf Coast Region vegetation. *Cnemidophorus franciscensis* is most commonly found where sand hummocks have accumulated at the foot of the rocky hills along the beaches or in dune areas (provided that the vegetative cover is sufficient for protection yet sparse enough to afford running room). Lizards also occur on the rocky foothills and flats but are not nearly as common there. No lizards have been found on the tops of the low, rocky hills that dominate the island.

Like *C. espiritensis*, adult *C. franciscensis* are uncommon during March, although juveniles are abundant. Adult activity usually begins during April and lasts through October. Hatchlings and juveniles remain active throughout the winter, provided the days are sufficiently warm. This species is diurnal, and I have observed specimens foraging beneath the vegetation along the edge of the beach. *Cnemidophorus franciscensis* presumably feeds on a variety of small arthropods, and I have found small beetles in the stomachs of adults. I have observed hatchlings during mid-July, which suggests that this species has a spring breeding season.

FOLKLORE AND HUMAN USE See *C. hyperythrus.*

FIG. 4.77 *Cnemidophorus pictus*

ISLA MONSERRATE WHIPTAIL; GÜICO

Verticaria picta Van Denburgh and Slevin 1921b:98. Type locality: "Monserrate Island, Gulf of California [BCS], Mexico." Holotype: CAS 49155

IDENTIFICATION *Cnemidophorus pictus* can be distinguished from all other lizards in the region of study by having movable eyelids; granular dorsal scales; large rectangular, imbricate ventral scales; a single frontoparietal scale; generally unicolored dorsal pattern; and sky-blue ventrum.

RELATIONSHIPS AND TAXONOMY See *C. carmenensis.*

DISTRIBUTION *Cnemidophorus pictus* is endemic to Isla Monserrate in the Gulf of California, BCS (see map 4.25).

DESCRIPTION Same as *C. hyperythrus,* except for the following: adults reach 68 mm SVL; 3 supraocular scales; 72 to 92 longitudinal rows of dorsal scales; mesoptychial scales small; 16 to 20 femoral pores; 26 to 32 subdigital lamellae on fourth toe. **COLORATION** Dorsal ground color of head gray, with body light brown anteriorly, blending smoothly to light sooty gray on sacrum, hind limbs, and tail; light stripes absent from dorsum and tail; vestiges of dorsolateral stripes on side of head; one pair of wide, poorly defined reddish-brown lateral stripes, corresponding in position and color to the dorsolateral fields of other *C. hyperythrus* group species; extremely faint dorsolateral stripes occasionally present anteriorly; forelimbs uniform blue-gray; no light markings on hind limbs; all ventral surfaces sky blue. Hatchlings and juveniles lack reddish-brown lateral stripes; tail bright blue.

FIGURE 4.77 *Cnemidophorus pictus* (Isla Monserrate Whiptail) from Isla Monserrate, BCS

NATURAL HISTORY Isla Monserrate is a small, volcanic, oceanic island dominated by Central Gulf Coast Region vegetation. *Cnemidophorus pictus* is common in the arroyo bottoms among dense vegetation in the island's interior. Lizards are also common along the sand dune–like areas behind the beach that are dominated by clumps of Leatherplant *(Jatropha cuneata).*

Juvenile *C. pictus* are far more abundant than adults during March. By the end of April, adults are common, and they remain active through September. Hatchlings and juveniles remain active through the winter, provided the weather is sufficiently warm. *Cnemidophorus pictus* forages slowly through the arroyos under cover of the dense vegetation. Those occurring in the more open dune areas forage among the Leatherplant and move rapidly across the open spaces. I have seen termites and ants in the stomachs of adults.

FOLKLORE AND HUMAN USE See *C. hyperythrus.*

FIG. 4.78 *Cnemidophorus labialis*

BAJA CALIFORNIA WHIPTAIL; GÜICO

Cnemidophorus labialis Stejneger 1890b:643. Type locality: "Cerros Island [Isla Cedros], Lower California," BC, Mexico. Holotype: USNM 15596

IDENTIFICATION Can be distinguished from all other lizards in the region of study by having movable eyelids; granular dorsal scales; large rectangular, imbricate ventral scales; a divided frontoparietal scale; contact of nasal scales and second supralabials; and no spotting or reticulations between the dorsal stripes.

RELATIONSHIPS AND TAXONOMY *Cnemidophorus labialis* may be most closely related to *C. inornatus* (see Grismer 1994a) and is considered a basal member of the *C. sexlineatus* group (Lowe et al. 1970; Duellman and Zweifel 1962).

FIGURE 4.78 *Cnemidophorus labialis* (Baja California Whiptail) from 16 km south of Guerrero Negro, BCS

DISTRIBUTION *Cnemidophorus labialis* ranges along a narrow strip of the Pacific coast, from Punta San José just south of Ensenada south to at least 6 km southeast of Guerrero Negro. It generally extends no more than 16 km inland (Bostic 1968; map 4.26) in the northern Vizcaíno Desert and even less so farther north. Stebbins (1985) presents *C. labialis* as ranging along the entire coastal margin of Bahía Sebastián Vizcaíno; however, this assertion is not based on specimens and remains unconfirmed.

DESCRIPTION Small, with adults reaching 63 mm SVL; body rounded in cross section, somewhat elongate and fusiform; head not compressed, sharply pointed in lateral and dorsal profile; dorsal head scales enlarged and platelike; nasal scales contact second supralabials; frontoparietal divided; usually 4 supraocular scales; second supraocular contacts frontal; one large parietal on either side of large medial interparietal, which lacks an eye spot; smaller row of occipitals abruptly contacting granular dorsal body scales; dorsals abruptly contacting large rectangular, imbricate ventrals; 53 to 65 longitudinal rows of dorsal scales; 7 to 13 granules separating paravertebral stripes; ventrolateral body fold extending from axilla to groin; 5 or 6 supralabials; single large postmental; sublabials large; gulars convex, elongate, and juxtaposed, slightly larger centrally and smaller posteriorly; mesoptychial scales large; gular fold complete; antegular fold interrupted medially; 6 or 7 infralabials; forelimbs stout, relatively short, and with large imbricate, rectangular scales on anterior surface; hind limbs approximately twice the size of forelimbs; antefemorals and infratibials greatly enlarged and imbricate; 9 to 15 well-developed femoral pores in males; 26 to 32 subdigital lamellae on fourth toe; tail cylindrical, pointed, and 2.0 to 2.5 times length of body; caudal scales keeled, whorled, and submucronate; caudal keels forming distinct longitudinal ridges running length of tail; subcaudals large, flat, and imbricate; postanals not enlarged. **COLORATION** Dorsal ground color gray to bluish-gray on head (adult males have blue heads), limbs, and tail, brownish-gray on body; 7 stripes present at midbody; vertebral stripe sinuous, usually forked immediately behind head, and extending from interparietal to base of tail; one pair of yellowish paravertebral stripes running from parietals to base of tail when stripes present; one pair of dorsolateral stripes running from superciliaries onto basal third of tail; one pair of cream-white to cream-yellow lateral stripes running from orbits to groin; striping on tail; paravertebral and dorsolateral fields gray-brown to dark brown or nearly black; lateral fields and sides deep reddish-brown; forelimbs blue anteriorly and ventrally and gray posteriorly, with light yellow stripes; hind limbs grayish with 1 or 2 thin, well-defined, sinuous longitudinal stripes; ventral surfaces immaculate, powder blue in males and white in females.

GEOGRAPHIC VARIATION Geographic variation in *C. labialis* appears to consist principally of an overall lightening and darkening of the dorsal color pattern in relation to substrate coloration. Individuals from the southernmost portion of its distribution, in the light-colored, sandy Vizcaíno Desert, are generally much lighter and more faded in overall col-

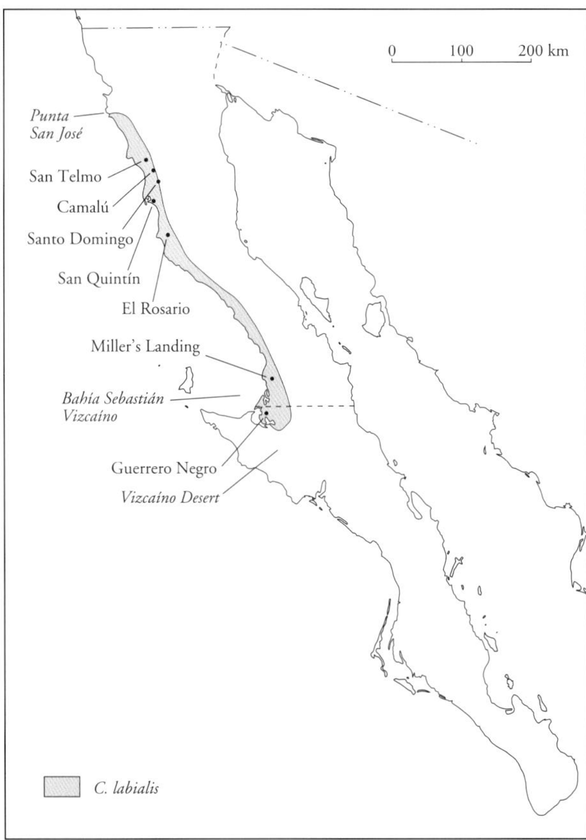

MAP 4.26 Distribution of *Cnemidophorus labialis* (Baja California Whiptail)

oration than those farther north, which also have a more vivid striping pattern.

NATURAL HISTORY *Cnemidophorus labialis* occupies a unique strip of habitat along the Pacific coast of the California and Vizcaíno regions. This is a cool, windblown, and often fog-drenched region that may go for extended periods without direct sunlight. In the northern portion of its distribution, near Camalú, San Telmo, and Santo Domingo, Bostic (1968) found *C. labialis* to be locally abundant in areas with fine, compacted, well-drained soil lacking either dense or very sparse vegetation. Noting that *C. labialis* rarely occurred on gravelly or sandy soils, Bostic believed that the "clumped pattern of local distribution" was correlated with the availability of suitable egg-laying sites. Farther south, near San Quintín, Walker (1966) found *C. labialis* to be abundant in flat, sandy areas (as opposed to nearby rocky terrain) dominated by stands of Bursage *(Ambrosia chenopodifolia)* and Desert Thorn Bush *(Lycium californicum),* none exceeding 1.5 m in height, which formed a sparse ground cover. Walker also noticed the clumped distribution pattern of *C. labialis* around certain stands of *A. chenopodifolia* and *L. californicum* and observed that on the east side of Bahía San Quintín, *C. labialis* occurred all the way to the water's edge. Bostic (1971) collected a series of specimens along the coastal strip between El Rosario and Miller's Landing and reported that the habitat requirements of *C. labialis* from this locality were similar to those of the populations he studied farther north (Bostic 1968). I have observed lizards in the Vizcaíno Desert approximately 6 km south of the 28th parallel, which extends the distribution south from Miller's Landing (Bostic 1968b, 1971) approximately 62 km. Here, lizards are common on the fine, windblown, sandy hummocks surrounding the Thorn Bush.

Most of what is known about the natural history of *C. labialis* comes from Bostic's work with the northern populations near San Telmo (1966b, 1968). Lizards emerge from hibernation in mid-March and enter into dormancy in late July; by late September, adult activity has ceased. All immature *C. labialis* enter hibernation by late December. *Cnemidophorus labialis* overnights in self-constructed burrows whose entrances are blocked by soil pushed up from within by the forelimbs (Tevis 1944). Lizards commence daily activities between 0800 and 1000 hours, a time that generally coincides with the dissipation of early morning fog. Daily activity peaks between 1000 and 1300 hours, the warmest period of the day between the early morning fogs and the late afternoon cloud cover and cool onshore breezes. Farther south and inland in the Vizcaíno Desert, away from the coastal breezes, I have noticed that activity is low in the early afternoon, perhaps because of high substrate temperatures.

Cnemidophorus labialis actively forages for food among the vegetation and is often observed basking for short periods. Nothing has been reported on its diet, but presumably it consists of small arthropods.

I have seen gravid *C. labialis* from mid-June through late July. The earliest I have seen hatchlings is early August, and very small juveniles are common in late August and early September. These observations suggest that *C. labialis* has a spring to summer breeding season.

REMARKS *Cnemidophorus labialis* is unique among whiptail lizards in its ability to be active at a low body temperature (Bostic 1968). Its voluntary maximum body temperature is approximately 37°C, which is roughly 1°C lower than that of other whiptail lizards. However, its voluntary minimum temperature of 30°C is 1 to 4°C lower than that of other *Cnemidophorus* (Bostic 1968). Bostic hypothesized that this lower temperature range may be an adaptation for living in an environment cooled by morning and afternoon cloud and onshore breezes.

Stejneger (1890a) gave Isla Cedros as the type locality for *C. labialis.* The specimens on which the description was based were collected by Lyman Belding in spring 1881. According to Savage (1954), most of Belding's material was labeled as coming from "Cerros Island [Isla Cedros] and San Quintín Bay" (Belding 1887:99). Because *C. labialis* has never been reported from Isla Cedros in over 100 years, I agree with Savage that specimens from Isla Cedros and Bahía San Quintín were mixed.

FIGS. 4.79–80

Cnemidophorus tigris

WESTERN WHIPTAIL; GÜICO

Cnemidophorus tigris Baird and Girard 1852a:69. Type locality: "Valley of the Great Salt Lake"; restricted to "Salt Lake City, Utah," USA (Smith and Taylor 1950). Holotype: USNM 5299

IDENTIFICATION *Cnemidophorus tigris* can be distinguished from all other lizards in the region of study by having movable eyelids; granular dorsal scales; large rectangular, imbricate ventral scales; a divided frontoparietal scale; no contact between the nasals and second supralabials; usually a distinct lineate dorsal pattern; dorsal ground color not black; adult males lacking an apricot-colored gular region; and thin, sinuous dorsal stripes.

RELATIONSHIPS AND TAXONOMY *Cnemidophorus tigris* forms a natural group with the insular endemic whiptails *C. bacatus, C. canus, C. catalinensis, C. celeripes,* and *C. martyris*

FIGURE 4.79 Maximus pattern class of *Cnemidophorus tigris* (Western Whiptail) from San Bartolo, BCS

(Wright 1993; Radtkey et al. 1997). Wright (1993), Grismer et al. (1994), and Grismer (1999b) review and discuss taxonomy.

DISTRIBUTION *Cnemidophorus tigris* ranges widely throughout the Sonoran and Mojave deserts of the southwestern United States and northern Mexico. It is transpeninsular and ubiquitous in Baja California (map 4.27), and it is known from 14 islands in the Gulf of California (Grismer 1999c) and 6 Pacific islands (Grismer 1993, 2001a; Grismer and Hollingsworth 1996).

DESCRIPTION Moderate to large, with adults reaching 137 mm SVL; body rounded in cross section, somewhat elongate and fusiform; head not compressed, pointed in lateral and dorsal profile; dorsal head scales enlarged and platelike; nasal scales not contacting second supralabials; frontoparietal divided; 3 to 5 supraocular scales; granular scales absent between second supraocular and frontal; 1 large parietal on either side of large interparietal, which lacks an eye spot; smaller row of occipitals abruptly contacting granular dorsal body scales; dorsals abruptly contacting large rectangular, imbricate ventrals; 74 to 134 granules around midbody; 148 to 269 dorsal granules between occipitals and base of tail; ventrolateral body folds extending from axilla to groin; 5 to 7 supralabials; single large postmental; sublabials large; gulars small anteriorly, larger and rounded centrally, and abruptly smaller posteriorly; mesoptychial scales small; gular and antegular fold complete; 5 to 8 infralabials; forelimbs stout, relatively short, and with large imbricate, rectangular scales on anterior surface; hind limbs approximately twice the size of forelimbs; antefemorals and infratibials greatly enlarged and imbricate; 16 to 28 well-developed femoral pores in males; 26 to 37 subdigital lamellae on fourth toe; tail cylindrical, pointed, and 2.5 to 3.0 times length of body; caudal scales keeled, whorled, and submucronate; caudal keels generally form distinct longitudinal ridges running length of tail; anteriormost subcaudals large, flat, and imbricate; postanals not enlarged. **COLORATION** *C. tigris* varies greatly in color pattern (see geographic variation section). The following is only a general description. Ground color of head, forelimbs, and tail brown; temporal region may have dark markings; black reticulate pattern on forelimbs; dorsal ground color of body and hind limbs black to dark brown; black reticulum on hind limbs; gular and pectoral regions grayish to pink, with or without black mottling; abdominal scales grayish and edged in black; subcaudal region dull yellow to tan and generally immaculate. ADULTS: Dorsal pattern usually broken and irregular; some degree of gray striping often present, but usually obscured by blotches in body fields merging with stripes; striping becomes less vivid with age; stripes often so broken that a faint to distinct banding pattern emerges; tail mottled with dark pigment, or may become solid dark gray distally. HATCHLINGS AND JUVENILES: Five to 11 distinct cream to bright yellow stripes on dorsum; spots or bars in fields faint to absent; posterior half of tail orange, blue, blue-green, or turquoise; dark reticulum on limbs persists through adulthood.

GEOGRAPHIC VARIATION Burt (1931) provides an extensive discussion on the concordance of color pattern variation in *C. tigris* with habitat. Geographic variation in *C. tigris* in Baja California is extensive, and four peninsular and five insular subspecies have been recognized (Burt 1931; Walker 1981a; Walker and Maslin 1981; Walker et al. 1966a; Taylor and Walker 1996). Used here as pattern classes, these correspond well with the phytogeographic regions and physiography of the peninsula. Maximus of the Cape Region attain the greatest SVL, reaching 137 mm. According to Walker 1966, Maximus have the most numerous dorsal granules around midbody (112 to 134) and between the occiput and the base of the tail (220 to 269). Additionally, they have 18 to 28 femoral pores and an arched, downward-sloping rostrum. The adult dorsal color pattern consists of five irregular light gray to light brown stripes. The paravertebral and dorsolateral fields are chocolate brown with evenly spaced, large, light-colored spots. The ventrolateral fields are black and contain a network of large, irregular grayish bars. The tail is reddish, with brown pigment generally occurring at the base of every other caudal anulus, giving it a somewhat banded appearance. The bands fade and the overall caudal coloration becomes lighter distally. The ventrum is generally light-colored, with varying degrees of gular and abdominal mottling. Hatchlings have five distinct, bright yellow stripes with black fields. The vertebral stripe is usually sinuous and often appears as two stripes intertwined with one another but separated anteriorly. Yellow spots are present in the dorsolateral and ventrolateral fields, and the tail is bright orange. Maximus from Islas Espíritu Santo and Partida Sur are similar to peninsular forms, except for having slightly wider and more undulate dorsal stripes and a faded, less colorful dorsal pattern.

Rubidus range from north of the Isthmus of La Paz to Punta San Francisquito (Burt 1931) and are confined primarily to the Magdalena and Central Gulf Coast regions. Rubidus attain an SVL of 105 mm. The rostrum also arches forward, but not usually as much as that of Maximus, especially those from the more northerly populations of its distribution. Rubidus have 91 to 121 midbody granules, 181 to 220 granules between the occiput and the base of the tail, and 16 to 23 femoral pores. The adult dorsal color pattern usually consists of distinct black and gray-brown banding. This results from the light bars in the body fields becoming more transversely arranged during ontogeny, to the point where some individuals lack all traces of striping, having only distinctive body bands. This patterning is most pronounced in lizards from the Magdalena Plain and the foothills and western arroyos of the Sierra la Giganta. Anteriorly, the gray-brown bands may widen, producing a nearly unicolored pattern in some extreme cases, as in many individuals from the Pacific islands of Magdalena and Santa Margarita. More often, however, wide, transversely aligned light-colored spots occur in the body fields, producing an almost checkerboard appearance. The lower temporal and lateral gular regions often develop a pinkish-purple cast. The central gular region is usually mottled with orangish-purple blotches. The tails of Rubidus are somewhat banded and generally darker than those of Maximus, although they are suffused with pink that extends onto the preanal and femoral regions. The undersides of the tail and hind limbs are dull orange. The ventrum is generally dark because the abdominal scales are edged in black.

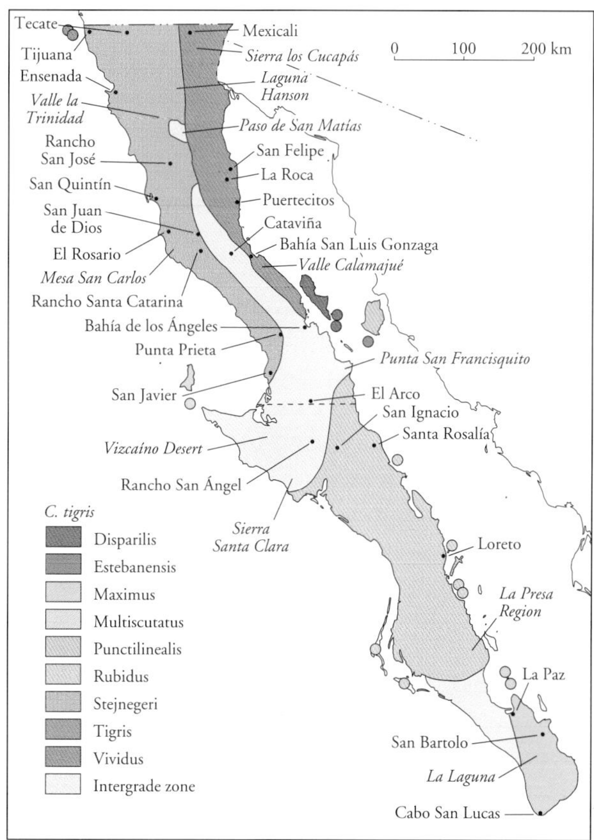

MAP 4.27 Distribution of the pattern classes of *Cnemidophorus tigris* (Western Whiptail)

Populations along the Gulf coast from Loreto north to Santa Rosalía do not show the same degree of banding as lizards from the Magdalena Plain populations. Here, lizards retain an observable striping pattern to varying degrees. Lizards from Islas San Cosme, Carmen, Coronados, Danzante, and San Marcos also have this color pattern and tend to be smaller in SVL. Additionally, lizards from Isla San Marcos are reddish overall and lack lateral banding, and the juveniles have orangish tails. Where the Magdalena Plain grades into the southern end of the Vizcaíno Desert, individuals have a dorsal pattern with dark reticulations. This pattern continues northwest through the Vizcaíno Peninsula (Grismer et al. 1994), where intergradation with Stejnegeri of the California and Vizcaíno regions begins. Hatchling Rubidus have six irregular light-colored stripes and black fields. This pattern results from a complete bifurcation of the vertebral stripe seen in Maximus. In place of the vertebral stripe is a vertebral field with a line of yellow spots. Spotting is generally absent from the paravertebral fields, but the dorsolateral and ventrolateral fields contain prominent light-colored bars. The tail is bright orange.

Populations of Rubidus and Maximus intergrade through the Isthmus of La Paz, where the Arid Tropical Region of the Cape Region transforms into the drier Magdalena Region. Here, the light-colored stripes of Maximus begin to break up because of the expansion and fusion of the light-colored spots in the body fields. This leaves fragmented areas of the dark fields appearing as large black spots. Whiptails from this area lack the purplish-pink hue on the sides of the head found in lizards from the La Presa region 35 km to the north. This intergrade pattern also occurs in populations from the Pacific islands of Magdalena and Santa Margarita.

Rubidus intergrade in the north with both Stejnegeri of northwestern Baja California and Tigris of northeastern Baja California. The zone of intergradation with Stej-

FIGURE 4.80 Tigris pattern class of *Cnemidophorus tigris* (Western Whiptail) from San Felipe, BC

negeri begins near the Sierra Santa Clara (Grismer et al. 1994) and San Ignacio. From San Ignacio north through El Arco and at least as far as Punta San Francisquito, all three pattern classes intergrade. Here, adult lizards begin to show distinct striping in their dorsal color pattern, and the spots within the body fields become irregularly shaped and more randomly arranged, giving rise to a more reticulated dorsal pattern overall. Also, the tail becomes less banded and more unicolored. At San Javier, on the Pacific coast, lizards are predominantly Stejnegeri. This pattern class continues north, west of the Peninsular Ranges, into southern California. At Bahía de los Ángeles, lizards are predominantly Tigris, and this pattern class continues north throughout the Lower Colorado Valley Region.

Stejnegeri range throughout the California Region and south into the cismontane portion of the northern Vizcaíno Region. They attain an SVL of 100 mm, and their rostra are flat to weakly arched. Stejnegeri have 74 to 113 midbody granules, 161 to 242 granules between the occiput and the base of the tail, and 19 to 25 femoral pores. The adult dorsal pattern generally consists of six gray-brown, faint to moderately distinct stripes. Considerable mottling in the fields with large light-colored blotches produces a thick reticulum on a black background. The distinctiveness of the striping tends to increase from south to north and is most pronounced between Ensenada and Tijuana. More easterly populations, from Rancho Santa Catarina north through Rancho San José, Valle la Trinidad, Laguna Hanson, and Tecate, tend to show a bolder, denser dorsal reticulum. This patterning results from a general expanse of the black ground color, giving the lizards a darker overall cast. Stejnegeri generally lack the pinkish coloration found in the subcaudal and lower temporal region of many Rubidus. Instead, the ventrum is generally gray to dark gray, with the abdominal scales edged in black in the darker individuals. The tails of Stejnegeri are weakly banded in southern populations from the vicinity of San Javier. Farther north, from Mesa de San Carlos to El Rosario, the banded pattern gives way to weakly mottled to unicolored tails. From San Quintín north, a faint striping pattern occurs on tails from a darkening of the caudal keels. Hatchling Stejnegeri have vivid stripes, composed of cream-colored, wavy lines and spotted body fields, and a turquoise tail. This pattern transforms into the adult pattern at SVLs around 40 mm.

Stejnegeri intergrade with Tigris of the Lower Colorado Valley Region of northeastern Baja California through Paso de San Matías, which cuts through the northern Peninsular Ranges and through Santa Ana and Calamajué valleys at the southern end of the Sierra San Pedro Mártir. Between these areas, the Peninsular Ranges apparently provide a barrier to gene flow, although intergradation probably occurs through some of the deeper canyons that drain the eastern face. South of Valle Calamajué, the California and Lower Colorado Valley regions give way to the Vizcaíno Region, where lower elevations in the southern regions of the Peninsular Ranges permit gene flow between Tigris from the east and Stejnegeri from the west, as inferred through color pattern variation. Stejnegeri × Tigris intergrades have wavy dorsal stripes and evidence of tiger striping in the lateral fields. Intergrades from the vicinities of El Arco (which are also intergrading with Rubidus) and Punta Prieta north to San Juan de Dios show only faint lateral striping and the beginnings of the thin striping characteristic of Tigris. Lizards from Cataviña tend to be dark anteriorly and have wide, dark, chevron-shaped bands in the gular region. Murray (1955) stated that intergradation was also occurring in the Sierra los Cucapás, but I could find no evidence to support this claim. All the specimens I have seen have been Tigris.

Tigris attain an SVL of 99 mm, and their rostra are flat to weakly arched. They have 74 to 100 midbody granules, 148 to 188 granules between the occiput and the base of the tail, and 17 to 25 femoral pores. The adult dorsal color pattern generally consists of four to five weakly defined, light-colored stripes on a black background. The sides are covered with numerous thin, light bands that give the lizards a tiger-striped appearance. Striping on the sides is absent. The extensive light-colored mottling in the fields produces a thin, dense reticulum on a black background. The light overlying color pattern is gray anteriorly and blends to light brown posteriorly. The density of the reticulum and indistinctness of the lateral stripes increases geographically from south to north. Lizards from Punta San Francisquito in the south have a thick reticulate pattern that appears as large dark spots on a light background, and faint lateral stripes resulting from intergradation with Rubidus. Lizards from Bahía de los Ángeles north have a narrow, dense reticulum and maintain the faint lateral stripes. Very large specimens lose their striping, develop undulate dorsal markings posteriorly, and become nearly unicolored anteriorly. This trend is carried over into lizards from the adjacent Isla Smith population, which develop these characteristics at smaller SVLs and are even more unicolored posteriorly (see below). At Bahía San Luis Gonzaga and Puertecitos, the lateral stripes have faded, and the sides are beginning to develop the tiger-striped effect. The reticulum of the dorsal pattern also becomes finer. From La Roca to San Felipe north, lizards are typical Tigris, although their coloration appears somewhat washed out. The tails of Tigris are moderately mottled and light brown anteriorly, transforming to a unicolored dark gray posteriorly. Tigris lack the pinkish coloration found in the subcaudal and lower temporal regions of many Rubidus. Instead, the ventrum is generally gray to dark gray, with the abdominal scales edged in black in the darker individuals. Hatchling Tigris have a more distinctive striping and bright blue tails. This pattern transforms into the adult pattern relatively early.

Cnemidophorus tigris from Isla Smith within Bahía de los Ángeles were at one time allocated to *C. dickersonae* (Van Denburgh 1922; *C. t. disparilis* of Taylor and Walker 1996), a population once considered endemic to Islas Ángel de la Guarda, Pond, and Partida Norte (Walker 1981b). Murphy and Ottley (1984) placed this population in the synonymy of *C. t. tigris,* a decision that was subsequently supported with data (Grismer 1999b). Adults from Isla Smith have 78 to 91 midbody granules and 143 to 155 granules between the occiput and the base of the tail. The adult dorsal color pattern consists of fine gray to light brown reticulations on a dark background, which appears almost unicolored in some specimens. Striping is evident only in smaller individuals. The sides are covered with a light and dark barring. The ventrum is light gray, with black edging usually confined to the lateral abdominal scales. Black coloration may or may not occur beneath the tail. Striping is more vivid in juveniles, although a fine reticulum is the dominant pattern.

Whiptails from the Islas Coronado, off the Pacific coast from Tijuana, described as *C. tigris vividus* by Walker (1981a), are recognized here as a pattern class. The diagnostic characteristics of this population are also found in adjacent peninsular populations of Stejnegeri, although the striping of Vividus tends to be more distinct than that of Stejnegeri. Lizards from the southern island of this archipelago have a black dorsal ground color; a gular region lacking apricot coloring; no dark markings in the temporal region; and bold, wide, light-colored stripes on the body. Hatchlings and juveniles have six distinct stripes and no spots or bars in the fields. Transverse bars appear in the fields with increasing SVL, giving all the stripes an undulate appearance and often completely disrupting the lineate margins of the lateral stripes. The posterior half of the tail is bright blue and becomes grayish with age.

Whiptails from the Pacific islands of Natividad and Cedros have been considered the subspecies *C. tigris multiscutatus* (see Walker 1981a). I was unable to find consistent, discrete differences separating these populations from Stejnegeri of the adjacent peninsula. I therefore follow Burt 1931 in not according them formal taxonomic recognition; rather, I consider them a pattern class. Adult Multiscutatus have scattered black blotches on the head, with distinct white markings in the temporal region. There are 5 to 7 (occasionally 9) tan to cream-white, sinuous, incomplete stripes anteriorly on the body. Spots and vertical bars occur in the body fields, making the stripes extremely irregular and somewhat reticulated posteriorly. There are bold, irregularly shaped white markings on the limbs, and apricot coloration suffuses the gular and pectoral regions in adult males and the underside of the tail in males and females. The gular region and the sides of the head are boldly marked with dark blotches that are most prominent in whiptails from Isla Natividad. The dorsal pattern in juveniles is slightly more vivid, and the distal portion of the tail is blue-green. Multiscutatus from the low, flat, sandy southern end of Isla Cedros are lighter in overall coloration than those from the rocky arroyos of the center of the island. Multiscutatus from Isla Natividad generally have wider, more prominent, lighter dorsal markings than whiptails from Isla Cedros. The narrower, lighter markings of the latter give the dorsal pattern a more finely reticulated overall appearance. Also, lizards of the Isla Natividad population generally have larger and more numerous black, irregularly shaped markings on their ventral surfaces and the sides of their heads.

Cnemidophorus tigris from Islas Ángel de la Guarda, Partida Norte, Pond, San Esteban, and Tiburón are all more similar to one another than to peninsular populations. Their similarity is likely due to their common origin from the Sonoran populations of *C. tigris,* which are darker and smaller than the peninsular populations (see Grismer 1994c). The populations from Islas Ángel de la Guarda, Partida Norte, and Pond have been considered the subspecies *C. t. disparilis* (Taylor and Walker 1996), which is treated here as a pattern class. The whiptails of Islas San Esteban and Tiburón have been considered the subspecies *C. t. esteba-*

nensis and *C. t. punctilineatus,* respectively (Walker et al. 1966a; Walker and Maslin 1981). These too are treated here as pattern classes.

Disparilis differ from peninsular Tigris (Walker 1981b) in having 64 to 90 midbody granules and 138 to 186 granules between the occiput and the base of the tail. The adult dorsal color pattern consists of a black ground color, with highly irregular yellowish-brown stripes that fade posteriorly, and profusely spotted sides. The ventrum has extensive amounts of black, especially in the gular, pectoral, and caudal regions. This gives lizards from these populations a much darker overall appearance. Hatchlings have 7 to 11 light-colored, irregularly shaped stripes and turquoise-blue tails. Whiptails from Islas Ángel de la Guarda and Pond show no significant differences from one another, although those from Partida Norte tend to have a somewhat faded dorsal pattern. Whiptails from Partida Norte also have slightly less black on the ventrum, and prominent paravertebral stripes.

Estebanensis, from Isla San Esteban, differ from peninsular *C. tigris* (Walker et al. 1966a) in that they reach a maximum SVL of 84 mm, have 71 to 83 midbody granules, and have 168 to 205 granules between the occiput and the base of the tail. The adult dorsal color pattern consists of four irregularly shaped brownish stripes on a black ground color. In many individuals striping is not evident, and only a light reticulum is present. The black sides are overlaid by large whitish spots. The abdomen is usually light posteriorly and dark anteriorly and blends into nearly solid black pectoral and gular regions. Extensive black coloration is present beneath the tail. Striping is more persistent in juveniles but not really vivid. Juveniles also have a bluish cast to their lower jaws and gular regions and turquoise-green tails.

Punctilinealis from Isla Tiburón differ from peninsular *C. tigris* (Walker et al. 1966a) in having 71 to 91 midbody granules and 168 to 200 granules between the occiput and the base of the tail. The adult dorsal color pattern consists of six irregular to relatively distinct, brownish-gray stripes on a black ground color. In many adults, striping is moderately distinct, and the vertebral and paravertebral fields contain regularly spaced light-colored spots. In others, only a light reticulum is present. The sides are overlaid by large expanses of light-colored markings that almost completely obscure the black ground color. The gular, pectoral, and ventral surfaces of the forelimbs, as well as the anterior portion of the abdomen, are usually solid black. Extensive black coloration is present beneath the tail. Vivid lemon-yellow stripes occur in hatchlings and juveniles, and paravertebral markings tend to be absent. Hatchlings and small juveniles have turquoise to greenish tails.

NATURAL HISTORY *Cnemidophorus tigris* is fairly ubiquitous in Baja California. It prefers brushy areas with open spaces on loose, gravelly soil but is by no means restricted to such conditions. Stejnegeri range throughout the California, Baja California Coniferous Forest, and western Vizcaíno regions and adjacent islands off northwestern Baja California. Bostic (1971) reports that Stejnegeri prefer the soft soil of washes to the rocky substrate of the adjacent flats. Linsdale (1932) noted that Stejnegeri are common on the sandy flats among sagebrush at Rancho San José, and both Linsdale and Welsh (1988) reported that lizards were commonly seen in the washes and brushy areas along the eastern base of the Sierra San Pedro Mártir. In Cañón El Cajón, I have observed Stejnegeri foraging among the rocks along the streams and beneath the Willow trees (*Salix* sp.). Welsh (1988) indicated that Stejnegeri range throughout all habitats in the Sierra San Pedro Mártir, occurring on both sides of the peninsular divide and extending down to the desert floor. Stejnegeri from Islas Natividad and Cedros are also ubiquitous in distribution. Isla Cedros is the largest and most environmentally diverse island on the Pacific coast of Baja California (Oberbauer 1993). Its lower elevations are dominated by Vizcaíno Region flora and chaparral, and some of its higher peaks host conifer woodlands. I have observed Stejnegeri in both. On Isla Natividad, Burt (1931) reported Stejnegeri as being abundant around seabird nests. I found lizards also to be common in the narrow, sandy washes that transect the center of the island.

Stejnegeri emerge from hibernation in spring, and I have observed adults active as early as mid-March. Adults cease activity in early October, but juveniles remain active into early November. Whiptails emerge during the late morning, after temperatures have risen sufficiently, and begin actively foraging from bush to bush (Bostic 1971) in search of food, which consists mostly of termites and the larvae of other arthropods. I have seen adults basking in the sand on overcast days on Isla Natividad.

Isla Coronado Sur is dominated by the low-growing maritime sage scrub of the California Region (Thorne 1976). Zweifel (1958) reports that Stejnegeri appear to be uniformly distributed over the island and that whiptails are abundant during April when the sun is out. On overcast days, lizards are scarce to absent. Zweifel observed one lizard taking refuge beneath a rock.

During mid-July on Islas Cedros and Natividad, I observed gravid females and saw several courting pairs, with the males closely following the females around and occasionally biting them on the neck. Throughout the range of Stejnegeri, hatchlings begin to appear in late July and peak through August. This indicates that they have a spring and summer reproductive season.

Tigris are usually confined to the low-lying, hot, arid Lower Colorado Valley Region of northeastern Baja California and midriff islands, where they are generally ubiquitous. Seasonal activity begins in late winter to early spring and continues into fall. Juveniles often remain active in low numbers throughout the winter, depending on the weather. I have observed juveniles foraging during midday in early November in the Sierra los Cucapás and in late December in Arroyo San Luis, just north of Bahía San Luis Gonzaga. Tigris are commonly seen on the desert flats, darting from bush to bush as they forage for food. Whiptails are often found beneath surface litter (Linsdale 1932) and use rodent burrows for shelter. On the midriff islands, where flat, open space is not as common, whiptails are found on the rocky hillsides and take refuge in thickets of thorny plants and beneath rocks. Tigris emerge from overnight retreats during mid-morning to bask and forage. I have seen juveniles basking on fallen Cardón Cactus *(Pachycereus pringlei)* on Isla Partida Norte. I have often found Tigris beneath desert shrubs that contained termites, their principal food (Case 1983). I watched an adult on Isla Tiburón foraging beneath a bush by pushing aside all the leaf litter, moving out of the way to allow the termites to emerge, and then rushing back in to eat them. Many other arthropods are also eaten, and the larvae of larger species seem to be preferred. I saw one large adult Tigris at the base of a rocky hillside near Bahía de los Ángeles with what appeared to be a freshly killed juvenile Night Snake *(Hypsiglena torquata)* hanging from its mouth.

Tigris have a spring and summer reproductive season. I have observed matings during mid-March on Isla Tiburón, and hatchlings first begin to appear in July in all areas. Hatchlings continue to appear through early September on Isla Tiburón, and juveniles are common in the Sierra los Cucapás and Isla Partida Norte during early October. I have observed all size classes on Isla San Esteban during mid-October. The previous year's hatchlings are common as juveniles in March and April.

Rubidus range throughout the Magdalena and Central Gulf Coast regions of south central Baja California. They are commonly found in rocky canyons, sandy arroyos, and on gravelly soil where reasonably thick stands of vegetation occur nearby. Along the Pacific coast, Rubidus are a conspicuous component of the brushy flats. Whiptails are generally active year-round in the southern portion of their range, with a minimum of activity occurring during the winter. North of San Ignacio, I have not observed active lizards earlier than late March. Lizards are often observed during the day foraging for food in the vicinity of brush, poking their heads into holes and looking beneath surface objects. Mosauer (1936) remarked that Rubidus were common at Rancho San Ángel and were observed on several occasions to enter houses and search for food by digging beneath objects on the floor. I have often observed whiptails in the Sierra la Giganta foraging beneath rocks and at the bases of bushes. I found two squamate eggs and a grub in the stomach of a road-killed specimen east of San Ignacio. Rubidus have a spring breeding season, and hatchlings can be found from July through early October.

Maximus range throughout the Cape Region and are confined to the Arid Tropical Region and lower foothills of the Sierra la Laguna (Alvarez et al. 1988). These very timid lizards are predominantly found on firm soils in the vicinity of rocks and are quite adept at ascending steep, rocky hillsides. Asplund (1967) observed whiptails at San Bartolo on rock ledges and reported that they would retreat into vertical crevices when alarmed. Maximus are common in the thorn scrub woodlands of the low-lying flats where vegetation is not too scarce and an overstory is present (Asplund 1967; Linsdale 1932). In such areas, lizards remain in the vicinity of thorny thickets. Where the thorn scrub woodland grades into the low, desert scrub of the Magdalena Plain on the western side of the Sierra la Laguna, Maximus are rare (Asplund 1974). Juveniles, however, generally occupy areas where the vegetation is more open.

Maximus are active year-round, with a peak activity period from March through October. I have observed adults foraging in late December on Isla Espíritu Santo and in early January at the base of the Sierra la Laguna. Juveniles, however, are more commonly seen than adults during the winter. Maximus emerge in the morning from beneath vegetation or from within burrows beneath rocks when temperatures are warm enough to allow them to forage. Asplund (1967) reports that hatchlings and juveniles are often observed basking. During August and September, activity is restricted to the shade, and lizards often go to great lengths to avoid moving into the sunlight.

Maximus spend a great deal of time foraging for their

primary food sources, termites and arthropod larvae (Asplund 1967), beneath and at the base of vegetation (Murray 1955), and lizards are commonly seen in pairs. Burt (1931) reports an observation of an adult Maximus carrying a dead Orange-Throated Whiptail *(Cnemidophorus hyperythrus),* and I observed an adult during early September on Isla Espíritu Santo carrying a large worm. Maximus have been observed copulating in late May (Murray 1955), and Asplund (1967) reported hatchlings being common in late August. I have observed several hatchlings from early August through September, when the onset of summer rains results in increased arthropod activity (Asplund 1967).

FOLKLORE AND HUMAN USE Many of the local inhabitants of the Cape Region believe that Maximus are aggressive and venomous. Some report that the lizard forages around chicken coops to feed on chicks and kills the mother hens. *Cnemidophorus tigris* is eaten by the Seri Indians of Sonora (Malkin 1962).

FIG. 4.81 *Cnemidophorus bacatus*

ISLA SAN PEDRO NOLASCO WHIPTAIL; GÜICO

Cnemidophorus bacatus Van Denburgh and Slevin 1921b:97. Type locality: "San Pedro Nolasco Island, Gulf of California, [Sonora,] Mexico." Holotype: CAS 49152

IDENTIFICATION *Cnemidophorus bacatus* can be distinguished from all other lizards in the region of study by having movable eyelids; granular dorsal scales; large rectangular, imbricate ventral scales; a divided frontoparietal scale; no contact between the nasals and second supralabials; dark brown to black ground color with white to light blue, well-defined spots on the hind limbs and sides; and generally no spotting on the body anterior to the forelimbs or in the vertebral region.

RELATIONSHIPS AND TAXONOMY See *C. tigris.* Based on overall similarity, Cliff (1954b) and Walker and Maslin (1981) considered *C. bacatus* to be most closely related to *C. catalinensis.* Considering, however, the geological history of the islands on which these species occur, their similarities are most likely due to common ancestry from a continental predecessor. Taxonomy is discussed by Burt (1931), Walker and Maslin (1969 and 1981), and Grismer (1999b).

DISTRIBUTION *Cnemidophorus bacatus* is endemic to Isla San Pedro Nolasco, Sonora (map 4.28).

DESCRIPTION Same as *C. tigris,* except for the following: adults reach 81 mm SVL; 4 supraocular scales present, with the fourth greatly reduced; 66 to 88 granules around midbody; 153 to 180 dorsal granules between occipitals and base

FIGURE 4.81 *Cnemidophorus bacatus* (Isla San Pedro Nolasco Whiptail) from Isla San Pedro Nolasco, Sonora

of tail; 7 or 8 supralabials; central gulars slightly enlarged and rounded; 8 or 9 infralabials; 15 to 22 femoral pores; 26 to 34 subdigital lamellae on fourth toe. **COLORATION** Head unicolored brown to gray; dorsal ground color of body and tail brown with black sides; limbs dark brown to black; a network of well-defined, distinctive white to light blue spots on body posterior to forelimb insertions; spots also on hind limbs and anterior portion of tail; spots largest on sides, lateral regions of tail, and hind limbs; all ventral surfaces solid, shiny black except for dark brown posterior portion of tail and occasionally grayish chinshields; lateral stripes absent in all size classes. **HATCHLINGS AND JUVENILES:** One pair of vivid cream to cream-yellow dorsolateral and paravertebral stripes present in lizards less than 52 mm SVL, stripes absent in specimens larger than 51 mm SVL; tail bright turquoise; spots present in fields between the stripes; spots more boldly defined in juveniles than in adults, occasionally occurring on sides of neck; middorsal spotting becomes less vivid with increasing SVL.

NATURAL HISTORY Isla San Pedro Nolasco is a steep, rocky, continental island covered with Central Gulf Coast vegetation (Felger and Lowe 1976). *Cnemidophorus bacatus* is generally ubiquitous on the island, avoiding only the areas with excessively steep cliff faces. It is commonly seen among rocks, in grassy areas, and crawling among branches. On one occasion, I approached a lizard approximately 0.75 m off the ground among the branches of a low shrub. Instead of dropping to the ground to flee, it climbed higher into the bush.

Cnemidophorus bacatus is probably active year-round, with a minimum of activity during the winter. During early spring, juveniles are more commonly seen than adults. *Cnemidophorus bacatus* is diurnal and emerges only when temperatures are sufficiently warm: in early spring, lizards are generally not observed until late into the morning. During the summer, however, lizards can be observed emerging at sunrise.

Van Denburgh (1921) reports that *C. bacatus* was commonly seen among the colonies of nesting Brown Boobies *(Sula leucogaster)*, feeding on the flies and other insects attracted by decaying fish and other refuse. Similar feeding habits are known for *C. martyris* (see below), but this food source is available only during the spring and early summer. The rest of the year, *C. bacatus* can be observed foraging among the rocks and fallen vegetation, as well as within the low branches of shrubs, in search of other arthropods. I observed some unique bite marks on a small Isla San Pedro Nolasco Side-Blotched Lizard *(Uta nolascensis)* that suggested that it might have been preyed on by a *C. bacatus.*

Walker and Maslin (1969) reported the presence of approximately one-month-old hatchlings and gravid females during August and suggested that *C. bacatus* may produce two clutches per year of one to three eggs (Walker 1980). I have observed hatchlings into early October. Furthermore, Walker and Maslin believed, on the basis of the simultaneous occurrence of one-month-old hatchlings and females with yoked follicles, corpora lutea, and oviducal eggs, that *C. bacatus* may lack a distinct reproductive season. Instead, they suggested, these lizards' reproductive efforts may be correlated with their activity cycles.

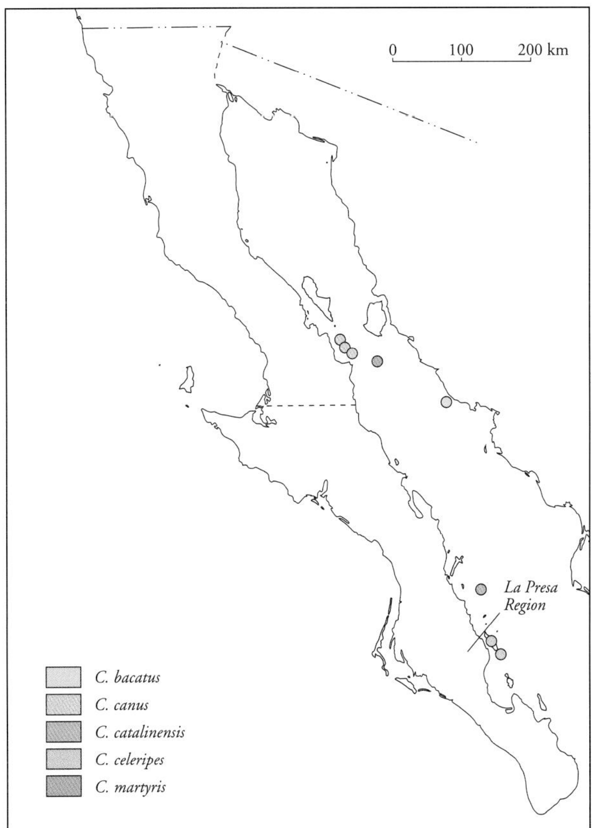

MAP 4.28 Distribution of the endemic insular species of the *Cnemidophorus tigris* group (Western Whiptail)

FIG. 4.82 **Cnemidophorus canus**

ISLA SALSIPUEDES WHIPTAIL; GÜICO

Cnemidophorus canus Van Denburgh and Slevin 1921b:97. Type locality: "Sal Si Puedes [Salsipuedes] Island, Gulf of California, [BC,] Mexico." Holotype: CAS 49152

IDENTIFICATION *Cnemidophorus canus* can be distinguished from all other lizards in the region of study by having movable eyelids; granular dorsal scales; large rectangular, imbricate ventral scales; a divided frontoparietal scale; no contact between the nasals and second supralabials; and a faded to nearly unicolored dorsal pattern.

RELATIONSHIPS AND TAXONOMY See *Cnemidophorus tigris.* Based on overall similarity, Burt (1931) considered *C. canus* to be most closely related to *C. martyris;* this was supported by Radtkey et al. (1997). Taxonomy is discussed in Burt 1931, Walker and Maslin 1981, and Grismer 1999b.

DISTRIBUTION *Cnemidophorus canus* is endemic to the Islas Salsipuedes, San Lorenzo Norte, and San Lorenzo Sur, BC (see map 4.28).

DESCRIPTION Same as *C. tigris,* except for the following: adults reach 78 mm SVL; usually 4 supraoculars; 65 to 88 granules around midbody; 144 to 187 dorsal granules between occipitals and base of tail; 5 or 6 supralabials; central gulars generally same size as lateral gulars; 6 to 8 infralabials; 16 to 20 femoral pores; 25 to 33 subdigital lamellae on fourth toe. **COLORATION** Dorsal ground color of head, body, limbs, and tail tan to light brown; body and limbs overlaid with an extremely fine, light-colored dorsal reticulation; very small, widely scattered black flecks on dorsum; dorsal pattern often approaches a unicolored condition; reticulations most notable in small individuals; tail unicolored and darker posteriorly; ventral surfaces of body and limbs slate gray to sooty black; subcaudals sooty black, often appearing as a wide medial stripe; abdominal and ventral hind limb scales gray and edged in black.

GEOGRAPHIC VARIATION *Cnemidophorus canus* exists as three geographically proximate, allopatric insular populations on the San Lorenzo block. There is an obvious cline in overall coloration, from the lighter, more unicolored Isla Salsipuedes population in the north to a darker, more reticulated Isla San Lorenzo Sur population in the south, passing through an intermediate, less reticulated condition in the

FIGURE 4.82 *Cnemidophorus canus* (Isla Salsipuedes Whiptail) from Isla San Lorenzo Sur, BC

Isla San Lorenzo Norte population (Burt 1931). This coloration is most noticeable in the ventral pattern as a darkening of the gular and pectoral regions.

NATURAL HISTORY Islas Salsipuedes, San Lorenzo Norte, and San Lorenzo Sur are north-to-south-tending continental islands formed from the same continental block (Gastil et al. 1983). They share the same mountainous spine and lie in close proximity. All three are steep and rocky and dominated by a depauperate Central Gulf Coast Region flora. On Islas Salsipuedes and San Lorenzo Norte, *C. canus* is common in arroyo bottoms and on rocky hillsides. It is a wary species that spends most of its time foraging near brush. On Isla San Lorenzo Sur, *C. canus* seems to be most common on the beaches and bottoms of the larger arroyos, where it forages beneath the vegetation. I have seen adults from San Lorenzo Sur climb into the branches of low-growing shrubs in search of food. Farther inland, in the arroyo bottoms, it is less common. *Cnemidophorus canus* is diurnal and most active during the spring, summer, and early fall. I have observed juveniles in early October, which suggests that this species has a spring to summer reproductive season.

REMARKS Burt (1931) was the first to point out the clinal nature of the variation in the three insular populations of *C. canus.* He extended this cline south to include *C. martyris* of Isla San Pedro Mártir, approximately 51 km south of Isla San Lorenzo Sur, stating that the two species differed only in intensity of the color pattern and that the Isla San Lorenzo Sur population represented intergrades between the two extremes. Both species have a finely reticulated color pattern, although that of *C. martyris* is bolder. Additionally, *C. martyris* has a much more contrasted color pattern than *C. canus* and is much darker overall. However, the two species are superficially very similar and most likely derived from the same mainland Mexican ancestor. Burt also believed that *C. canus* intergraded in the north with *C. tigris* through the population on Isla Smith in Bahía de los Ángeles, approximately 62 km to the north. His hypotheses of intergradation, however, assumed that Isla Smith, the San Lorenzo block islands, and Isla San Pedro Mártir once existed as a single land mass. This, however, is not the case (Gastil et al. 1983), and the similarities of these populations are most likely attributable to parallel evolution.

FIG. 4.83

Cnemidophorus catalinensis

ISLA SANTA CATALINA WHIPTAIL; GÜICO

Cnemidophorus catalinensis Van Denburgh and Slevin 1921c:396. Type locality: "Santa Catalina Island, Gulf of California [Sonora], Mexico." Holotype: CAS 50507

IDENTIFICATION *Cnemidophorus catalinensis* can be distinguished from all other lizards in the region of study by having movable eyelids; granular dorsal scales; large rectangular, imbricate ventral scales; a divided frontoparietal scale; no contact between the nasals and second supralabials; a dorsal pattern of diffuse, poorly defined, yellowish to cream-colored spots; and a general lack of an ontogenetic change in color pattern.

RELATIONSHIPS AND TAXONOMY See *C. bacatus.* Radtkey et al. (1997) placed *C. catalinensis* basal within the *tigris* group.

DISTRIBUTION *Cnemidophorus catalinensis* is endemic to Isla Santa Catalina, BCS (see map 4.28).

DESCRIPTION Same as *C. tigris,* except for the following: adults reach 84 mm SVL; usually 4 supraocular scales; 69 to 83 granules around midbody; 148 to 184 dorsal granules between occipitals and base of tail; 5 to 7 supralabials; central gulars not noticeably enlarged and rounded; 6 or 7 infralabials; 15 to 20 femoral pores; 27 to 33 subdigital lamellae on fourth toe. **COLORATION** Head unicolored brown to dark gray or black; dorsal ground color of body, hind limbs, and tail brown; sides black; forelimbs dark brown to black; a network of diffuse, poorly defined, yellowish to cream-colored spots on body, hind limbs, and faintly on forelimbs; dorsal spotting more vivid in juveniles; keels of proximal caudal scales tipped with gray; ventral surfaces of head, body, and limbs solid, shiny black except for light suffusion of gray on chinshields, posterior abdominal region, and hind limbs; subcaudals brownish, often appearing as a wide medial stripe. HATCHLINGS AND JUVENILES: Color pattern same as adults, but brighter and better-defined.

NATURAL HISTORY Isla Santa Catalina, a rocky island in the Central Gulf Coast Region, is dominated by arid-adapted thorn scrub. *Cnemidophorus catalinensis* is most commonly

FIGURE 4.83 *Cnemidophorus catalinensis* (Isla Santa Catalina Whiptail) from Arroyo Mota, Isla Santa Catalina, BCS

FIGURE 4.84 *Cnemidophorus celeripes* (Isla San José Western Whiptail) from Isla San José, BCS

found in scrub vegetation within arroyos and along the edges of beaches, in the cobble. It is less frequently found on the rocky, open hillsides. Density of whiptails becomes lower toward the interior of the island, although I have seen lizards on some of the island's highest peaks.

Cnemidophorus catalinensis may be active year-round, with a minimum of activity during the winter. Lizard density is highest from April through July and tapers off during fall. This lizard is diurnal and, like other whiptails, requires warm ambient temperatures before emerging from overnight retreats; it will not emerge on overcast days. It is an active forager, searching for food beneath the scrub vegetation. It rarely emerges from the cover of vegetation to enter open, unprotected areas but will forage right up to the edge of the brushy overhang, where it can be relatively easily approached. I observed an adult foraging in an Elephant Tree *(Bursera microphylla)* nearly 1 m above the ground. *Cnemidophorus catalinensis* presumably feeds on small arthropods. I observed a juvenile unsuccessfully attempt to eat a June bug, and I watched an adult chasing and catching marine isopods in the beach cobble.

During March and April, adults and juveniles (of the previous year's hatching) are common, but hatchlings are very rare. I have observed mating pairs during early July and have seen numerous hatchlings from late June through September and some as late as mid-October. The simultaneous occurrence of hatchlings and matings may indicate that *C. catalinensis* has a lengthy, poorly defined reproductive season similar to that proposed for *C. bacatus* (Walker and Maslin 1969; see above).

FIG. 4.84 *Cnemidophorus celeripes*

ISLA SAN JOSÉ WESTERN WHIPTAIL; GÜICO

Cnemidophorus celeripes Dickerson 1919:472. Type locality: "San Jose Island [BCS], Mexico." Holotype: USNM 64444

IDENTIFICATION *Cnemidophorus celeripes* can be distinguished from all other lizards in the region of study by having movable eyelids; granular dorsal scales; large rectangular, imbricate ventral scales; a divided frontoparietal scale; no contact between the nasals and second supralabials; black dorsal ground color on the body and hind limbs; more than 109 granules at midbody; and conspicuous dark blotches in the temporal region.

RELATIONSHIPS AND TAXONOMY *Cnemidophorus celeripes* is purported to be derived from *C. tigris* of southern Baja California, based on similarities in color pattern, scale morphology, and mitochondrial DNA (Radtkey et al. 1997; Walker 1966). Grismer (1999b) and Walker (1966) discuss taxonomy.

DISTRIBUTION *Cnemidophorus celeripes* is endemic to Isla San José and the adjacent Isla San Francisco (see map 4.28).

DESCRIPTION Same as *C. tigris,* except for the following: adults reach 127 mm SVL; 3 or 4 (usually 4) supraocular scales; granular scales usually present between second supraocular and frontal, at least posteriorly; 109 to 126 granules around midbody; 214 to 236 granules between occipitals and base of tail; 6 or 7 supralabials; gulars slightly larger and rounded centrally; 6 or 7 infralabials; 19 to 23 femoral pores; 27 to 33 subdigital lamellae on fourth toe. **COLORATION** Ground color of head, forelimbs, and tail brown; dark markings in temporal region and often in preorbital region; thin black reticulate pattern on forelimbs; dorsal ground color of body and hind limbs black; wide cream-colored reticulum on hind limbs; gular and pectoral regions grayish with slight reddish suffusion; abdominal scales grayish and edged in black; subcaudal region dull yellow to tan and generally immaculate. **ADULTS:** Four to 6 wide, irregular, cream-yellow stripes; one pair of paravertebral stripes originating posterior to parietals, and one pair of dorsolateral stripes originating in temporal region, both pairs terminating at base of tail; often a detectable pair of lateral stripes originating in temporal region and terminating in groin; spots and vertical bars usually present in body fields contacting stripes, giving stripes an irregular, undulate appearance and often completely disrupting lateral stripes; a vertebral row of spots usually present; posterior third of tail pinkish. HATCHLINGS AND JUVENILES: Six distinct yellowish stripes in hatchlings; no spots or bars in fields; posterior half of tail orange; spots and bars appear in fields with increasing SVL, giving stripes an undulate appearance; posterior of tail turns reddish; pinkish pigments on ventrum.

NATURAL HISTORY Islas San José and San Francisco lie within the Central Gulf Coast Region and are dominated by xerophilic thorn scrub. Isla San José is a very large landbridge island, with a rocky, mountainous keel running along its eastern flank and an extensive bajada to the west. *Cnemidophorus celeripes* is most commonly observed in rocky areas of the island's interior (Asplund 1967). I have never observed lizards on the beach, where the Isla San José Whiptail *(C. danheimae)* is relatively common, or on the flats of the island's interior. I have found it to be most common on the rocky hillsides and in the arroyo bottoms. Lizards are far less abundant in open, sandy areas. They are commonly observed moving in the cover of large thickets and shrubs among the rocks and are quite adept at scaling rocky hillsides in escape. Isla San Francisco is a much smaller island that lacks the wide bajadas of Isla San José. Here, *C. celeripes* is most common in the many east-to-west-tending rocky arroyos along the northern part of the island and, as on Isla San José, is not usually observed on the flats or along the beaches.

Cnemidophorus celeripes is probably active year-round, with a minimum of activity occurring during the winter. It is a diurnal, active forager, leaving its burrow in the morning, after temperatures have become sufficiently warm, to search for arthropods. I have observed very small juveniles and gravid females during early September. This may indicate that *C. celeripes* has an extended reproductive season that coincides with the summer rainy season.

FIG. 4.85
Cnemidophorus martyris
ISLA SAN PEDRO MÁRTIR WHIPTAIL; GÜICO

Cnemidophorus martyris Stejneger 1891b:407. Type locality: "San Pedro Martir Island, Gulf of California, [Sonora], Mexico." Holotype: USNM 15620

IDENTIFICATION *Cnemidophorus martyris* can be distinguished from all other lizards in the region of study by having movable eyelids; granular dorsal scales; large rectangular, imbricate ventral scales; a divided frontoparietal scale; no contact between the nasals and second supralabials; lacking an ontogenetic change in color pattern; and having a very dark, finely reticulate dorsal pattern.

FIGURE 4.85 *Cnemidophorus martyris* (Isla San Pedro Mártir Whiptail) from Isla San Pedro Mártir, Sonora

RELATIONSHIPS AND TAXONOMY See *C. canus.*

DISTRIBUTION *Cnemidophorus martyris* is endemic to Isla San Pedro Mártir, Sonora (see map 4.28).

DESCRIPTION Same as *C. tigris,* except for the following: adults reach 80 mm SVL; 4 or 5 supraocular scales; 71 to 85 granules around midbody; 149 to 183 dorsal granules between occipitals and base of tail; 6 supralabials; 4 to 8 infralabials; 33 to 41 femoral pores; 30 to 35 subdigital lamellae on fourth toe. COLORATION Dorsal ground color generally black; forelimbs black; fine, dense, dark brown reticulum on body and hind limbs; anterior portion of tail dark brown; ventral surfaces heavily suffused with black; gular region smoke-colored, with small, light-colored spots often extending onto pectoral region, which is usually black; posterior abdominal scales and underside of hind limbs dark gray and edged with black; subcaudal region dark brown; no ontogenetic change in color pattern occurs. Hatchlings and juveniles are very similar to adults in coloration and pattern.

GEOGRAPHIC VARIATION See *C. canus.*

NATURAL HISTORY Isla San Pedro Mártir lies in the center of the Gulf of California and is the most isolated island in the region of study. Technically it lies within the Central Gulf Coast Region, but it lacks extensive plant growth, primarily because it is covered with guano. Isla San Pedro Mártir is little more than a steep, emergent, rocky peak edged with precipitous bluffs. *Cnemidophorus martyris* is found throughout the island in shallow arroyos and on steep rocky hillsides. It is uncommon on the narrow, rocky beaches.

Cnemidophorus martyris is generally active from late April to early November, provided that temperatures are sufficiently high. Its activity peaks during the summer, and individuals are most abundant during the hottest part of the day. *Cnemidophorus martyris* is commonly observed foraging along the rocky hillsides, sifting through the substrate between and below surface objects in search of food, moving quickly and sporadically through its habitat. One of the food items it may be searching for is the eggs of the Isla San Pedro Mártir Side-Blotched Lizard *(Uta palmeri)*. In fact, Wilcox (1980) believes that the low survivorship of eggs or hatchlings of *U. palmeri* may be due to predation by *C. martyris*. During the reproductive season of the Blue-Footed Boobies *(Sula nebouxi)*, whiptails are commonly found near nests, eating fish scraps, and will aggressively defend this resource from *U. palmeri* (Wilcox 1980). Lizards also have been observed "picking off" the small black flies *(bobitos)* that land on the birds and nearby objects.

Walker (1980) reported that *C. martyris* has the smallest clutch size (1 or 2 eggs) and largest eggs of all *Cnemidophorus*. He noted the absence of hatchlings and subadults during August but the presence of gravid females, indicating that reproductive maturity is reached in less than one year. He also speculated that this species lays two clutches per year.

Xantusiidae

NIGHT LIZARDS

The Xantusiidae are a peculiar group of lizards comprising three extant genera and approximately 18 species (Pough et al. 2001). Xantusiids range disjunctly throughout the southwestern United States south through Mexico and Central America to Panama. An endemic, monotypic genus, *Cricosaurua,* occurs in southern Cuba. Xantusiids are renowned for their secretive, sedentary lifestyle in restrictive microhabitats. Most species are adapted for living beneath exfoliating objects such as decaying bark, leaves, and rock. In spite of their common name, xantusiids are not exclusively nocturnal. They possess a host of derived character states separating them from all other lizard groups, including unique circular, depressed microscopic scale organs (Peterson and Bezy 1985; Estes et al. 1988).

Xantusia Baird

NIGHT LIZARDS

The genus *Xantusia* is a small group of 7 to 9 secretive, terrestrial North American lizards. *Xantusia* ranges disjunctly throughout much of the southwestern United States and northern Mexico. Two species, *X. henshawi* and *X. vigilis,* occur in Baja California.

FIG. 4.86 ***Xantusia henshawi***

GRANITE NIGHT LIZARD; SALAMANQUESA

Xantusia henshawi Stejneger 1893b:467. Type locality: "Witch Creek, San Diego County, California." Holotype: USNM 20339

IDENTIFICATION *Xantusia henshawi* can be distinguished from all other lizards in the region of study by lacking eyelids; having normal digits that lack transversely expanded lamellae; and having a very flat body covered dorsally with large black blotches and a yellow reticulum.

RELATIONSHIPS AND TAXONOMY *Xantusia henshawi* is most closely related to a large complex of species currently referred to as *X. vigilis* (Bezy and Sites 1987).

DISTRIBUTION *Xantusia henshawi* ranges from western Riverside County, California, south into northwestern Baja California, to at least Cañón el Cajón of the Sierra San Pedro Mártir in the east and Valle la Trinidad in the west (Lee 1976; Welsh 1988; map 4.29). Pappenfuss et al. (2001) suggest alternative relationships.

DESCRIPTION Small, with adults reaching 70 mm SVL; body dorsoventrally compressed; head triangular in dorsal profile, slightly wider than neck, more so in adult males; eyelids absent; dorsal head scales large and platelike; upper temporals bordering parietal, and postparietal enlarged; interparietal large, with conspicuous eye spot; one pair of greatly enlarged postparietals abruptly bordering granular, juxtaposed dorsal scales; dorsals abruptly grade into flat, square juxtaposed ventrals; 14 longitudinal rows of ventral scales; lateral body fold extends from axilla to groin; vertical folds often present in axilla and groin; 5 to 8 supralabials; 2 large postmentals followed by 3 or 4 large sublabials; gulars round, flat, and juxtaposed; antegular and gular fold present; mesoptychial scales large and square; 4 to 6 infralabials; forelimbs moderate in length, somewhat robust, covered with flat, subimbricate scales; hind limbs robust, longer than forelimbs, covered with small juxtaposed to subimbricate scales dorsally and larger flat, imbricate scales ventrally; 12 to 30 femoral pores; pores most developed in adult males; 19 to 29 subdigital lamellae on fourth toe; tail slightly longer than body, flattened in cross section; caudal scales elongate, imbricate, and whorled; no enlarged postanal scales in males.

COLORATION Capable of considerable lightening and darkening. DARK PHASE: Dorsal ground color of head and back black, covered with a yellow reticulum; ground color of limbs dark gray, yellow reticulum generally fainter on

FIGURE 4.86 *Xantusia henshawi* (Granite Night Lizard) from Valle la Trinidad, BC

limbs; tail with irregular dull white bands; ventral surfaces cream-colored, with minute black spots; spots most prevalent on sides of abdomen and hind limbs; subcaudal region mottled. LIGHT PHASE: Dorsal ground color gray; large, irregular dark blotches on neck, back, and tail; yellow reticulum present; ventrum same as in dark phase.

GEOGRAPHIC VARIATION Geographic variation in *Xantusia henshawi* is limited to a slight decrease in the size of the dark dorsal blotches from north to south. This is most obvious when populations from southern California are compared with those from Valle la Trinidad and the western populations of the Sierra San Pedro Mártir. Such variation was also observed by Murray (1955) in specimens from La Grulla.

NATURAL HISTORY *Xantusia henshawi* is a common inhabitant of areas containing extensive granitic outcroppings. It is a microhabitat specialist that resides beneath the thin exfoliating layers of rock. *Xantusia henshawi* is most common in the foothill chaparral habitat of the California Region, extending east into the Baja California Coniferous Forest Region of the Sierra Juárez. In the Sierra San Pedro Mártir, *X. henshawi* has been found as high as 2,100 m in the conifer belt along the west face at Arroyo Encantado (Welsh 1988). Lizards also occur along the rocky eastern desert escarpments of the Sierra San Pedro Mártir and Sierra Juárez in association with piñon-juniper woodland.

Interestingly, this species has not been reported farther south than the Cañón el Cajón divide in the Sierra San Pedro Mártir (Welsh 1988). Welsh attributes this limit to an abrupt change in geomorphology. *Xantusia henshawi* is known from disjunct localities throughout its range (Lee 1975; Grismer and Galvan 1986), and it is surprising that it has not been found in the southern Sierra San Pedro Mártir, Cerro el Matomí, Cataviña, the upper elevations of the Sierra la Asamblea, or anywhere in the Sierra la Libertad, especially because other disjunct chaparral inhabitants (e.g., *Elgaria multicarinata*) have been discovered in some of these southern areas (Grismer and Mellink 1994).

The natural history and ecology of *X. henshawi* in southern California has been well studied (Brattstrom 1965; Mautz and Case 1974; Lee 1974, 1975), and it is assumed that little variation occurs in the Baja California populations. Much of its activity is diurnal, especially during the winter (Mautz and Case 1974). Lizards are seldom seen, however, because activity is restricted to rock crevices. Here, lizards are able to thermoregulate by moving to areas of the overlying exfoliation that are heated by the sun. I have also seen lizards basking in the sunlight at the edge of exfoliations on cold winter mornings in December in Valle la Trinidad. Mautz and Case (1974) noted that lizards feed on crevice-dwelling arthropods during February. During the summer, Lee (1975) observed several lizards out foraging from dusk to midnight.

Xantusia henshawi bears one or two live young, and both males and females become reproductively active in their third year of life. Mating takes place in late spring and early summer, and birth occurs from mid-September through mid-October.

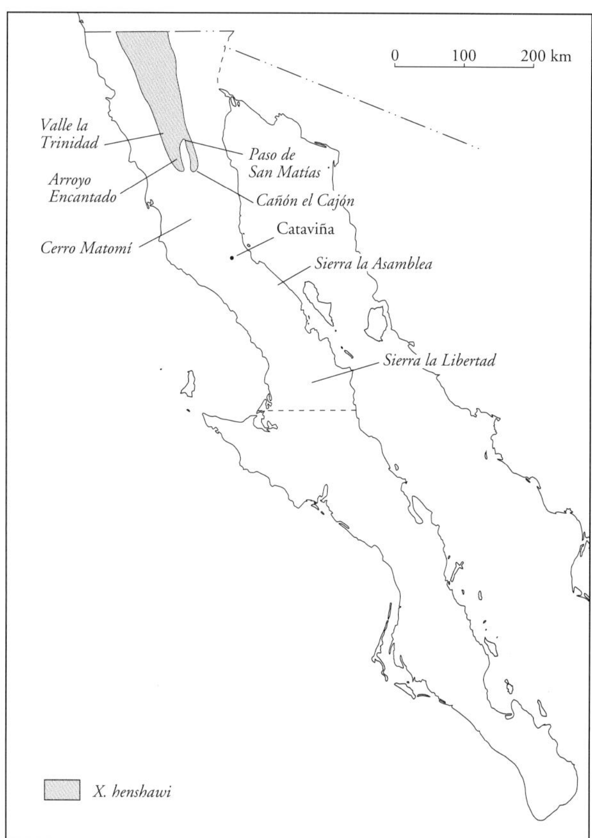

MAP 4.29 Distribution of *Xantusia henshawi* (Granite Night Lizard)

REMARKS *Xantusia henshawi* and the Leaf-Toed Gecko *Phyllodactylus xanti* are rarely sympatric and almost never syntopic, even though they exist in the same microhabitat. I suspect this may have to do with predation on *P. xanti* eggs by *X. henshawi* (Grismer and Galvan 1986).

FIG. 4.87

Xantusia vigilis

DESERT NIGHT LIZARD; SALAMANQUESA

Xantusia vigilis Baird 1859d:254. Type locality: "Fort Tejon [Kern County] Cal. [California]," USA. Holotype: USNM 3060 (3 specimens)

IDENTIFICATION *Xantusia vigilis* can be distinguished from all other lizards in the region of study by having a cylindrical body and short limbs; lacking eyelids; having normal digits that lack expanded lamellae; and having a brown ground color with or without small black markings.

RELATIONSHIPS AND TAXONOMY As currently constituted, *X. vigilis* comprises a complex that appears to be paraphyletic with respect to the other members of the genus (Bezy and Sites 1987).

DISTRIBUTION *Xantusia vigilis* ranges through the desert areas of southern California, southern Nevada, southern Utah, western Arizona, and western Sonora, Mexico (Bezy 1982). In Baja California, it occurs throughout the desert areas in the northeast and south to at least Rancho San Ángel in central Baja California. From San Felipe south, its distribution extends west and arches around the southern end of the Sierra San Pedro Mártir, making contact with the west coast in the vicinity of Punta Baja (Bostic 1971). From here, it extends north along the western foothills of the Sierra San Pedro Mártir, through an arid cismontane valley created by the Sierra San Miguel and Sierra Peralta (Grismer 1997), north as far as San Telmo (Welsh 1988). This population may even contact the more northwesterly populations that extend west through Paso de San Matías (Meek 1905; Welsh and Bury 1984) into Valle Picacho (Welsh 1988) along the northwestern flank of the Sierra San Pedro Mártir, to just east of the town of Valle la Trinidad. *Xantusia vigilis* ranges along the west coast of the peninsula, from Punta Baja south to just north of Santa Rosalillita (Bostic 1971) and east to Ejido Vizcaíno. Curiously, however, *X. vigilis* appears to be absent from the Vizcaíno Peninsula, despite the presence of what appears to be ideal habitat (Grismer et al. 1994). It does occur farther south, in coastal localities at Bahía San Juanico and La Poza Grande. *Xantusia vigilis* is absent from the upper elevations of the Sierra la Asamblea just north of Bahía de los Ángeles, the Sierra San Borja to the south of Bahía de los Ángeles, San Francisco de la Sierra, and the Sierra Guadalupe. It does occur in the upper elevations of the Sierra la Laguna in the Cape Region (map 4.30).

FIGURE 4.87 Wigginsi pattern class of *Xantusia vigilis* (Desert Night Lizard) from western foothills of the Sierra la Asamblea, BC

DESCRIPTION Small, with adults reaching 44 mm SVL; body cylindrical; head triangular in dorsal profile, slightly wider than neck, more so in adult males; eyelids absent; dorsal head scales large and platelike; upper temporals bordering parietal, and postparietal not greatly enlarged; interparietal large, with conspicuous eye spot; one pair of greatly enlarged postparietals abruptly bordering granular, juxtaposed dorsal scales; dorsals abruptly grade into flat, square, juxtaposed ventrals; usually 12 longitudinal rows of ventrals; lateral body, axillary, and inguinal folds weak to absent; 7 to 9 supralabials; seventh supralabial equal in height to, taller than, or shorter than sixth; eye relatively small; 2 large postmentals followed by 3 or 4 large sublabials; gulars round, flat, and juxtaposed; antegular and gular fold present; mesoptychial scales large and square; 4 to 7 infralabials; forelimbs moderate in length, somewhat robust, covered with flat, subimbricate scales; hind limbs robust, longer than forelimbs, and covered with small, juxtaposed to subimbricate scales dorsally and larger flat, imbricate scales ventrally; 4 to 11 femoral pores; pores most developed in adult males; 15 to 23 subdigital lamellae on fourth toe; tail approximately equal to SVL in length, fat and round in cross section; caudal scales elongate, imbricate, and whorled; no enlarged postanal scales in males. **COLORATION** Extremely variable. Dorsal ground color of head, body, and tail gray-brown to rusty brown; light dorsolateral stripes bordered with black spots extending from eyes onto anterior portion of body; stripes extend further in juveniles; caudal striping usually very faint; body variably covered in small black spots, often forming distinct thin black lines or elongate narrow blotches; dots on tail larger; some specimens lack dots and are unicolored; ventrum of body dull yellow with few punctations; occasionally heavy punctations in gular region.

GEOGRAPHIC VARIATION Geographic variation of *X. vigilis* in Baja California has been characterized by recognition of three subspecies (Bezy 1982), which are used here as pattern classes. The differences are mostly limited to color and color pattern, which are highly variable (differing shades of brown in the ground color, and the density and lineate tendencies of the black dorsal markings), and to the height of the seventh supralabial in relation to the sixth.

Lizards from the U.S.-Mexican border south to the vicinity of Puertecitos correspond to the subspecies *X. vigilis vigilis* (Bezy 1972). Vigilis have a dark brown dorsal ground color and tend to have larger, elongate markings on the body. Lizards south of El Rosario eventually correspond to the subspecies *X. v. wigginsi,* which is referred to here as the pattern class Wigginsi. Wigginsi are generally tan to light gray-brown in dorsal ground color, and those from between El Rosario and Paso de San Matías intergrade with Vigilis (Bezy 1982; Welsh 1988; Welsh and Bury 1984). South of El Rosario, there is a general tendency in Wigginsi for the blotches to be broken into smaller black spots, although lizards from the Vizcaíno Region, west of Bahía de los Ángeles, may be either dark with a well-defined lineate pattern and postorbital stripes or lighter overall and more yellowish, lacking a lineate pattern and having only faint postorbital stripes.

Lizards from the Sierra la Laguna are referred to as the pattern class Gilberti. In Gilberti the seventh supralabial is usually as high as or higher than the sixth supralabial, whereas in Vigilis and Wigginsi it is equal in height or shorter; the body and tail are occasionally more orangish; the body is variably covered with small black spots, more so on sides than dorsally; the anterior portion of the tail is striped, and the caudal spots are larger; the midventral caudal region is light and the lateral region heavily marked with black punctations.

NATURAL HISTORY *Xantusia vigilis* occurs throughout much of the Lower Colorado Valley, Vizcaíno, and extreme northern Magdalena regions. It also occurs in the ecotone between the Lower Colorado Valley and the California regions in the vicinity of Paso de San Matías. Welsh (1988) reported a population on the edge of the piñon-juniper woodland just east of Mike's Sky Rancho. *Xantusia vigilis* is a microhabitat specialist, living within and beneath decaying vegetation of the locally dominant plant species. In the Lower Colorado Valley Region, these include various species of agave *(Agave).* Near Paso de San Matías, *X. vigilis* is common beneath and within Datilillo *(Yucca schidigera)* (Meek 1905; Murray 1955; Welsh and Bury 1984) and Desert Agave *(Agave deserti).* In the Vizcaíno and Magdalena regions, night lizards are most common in the basal leaves of dead agave and Datilillo *(Yucca valida),* and less common beneath fallen Boojum Trees *(Fouquieria columnaris)* and other cacti (Bostic 1971). I found one specimen under a piece of cardboard in the dump at Bahía San Juanico. The surrounding habitat was atypical for *X. vigilis* in that it lacked large plants.

The ecology and natural history of *X. vigilis* from the Mojave Desert of southern California has been well studied (e.g., Miller 1951, 1954; Zweifel and Lowe 1966), but very little has been reported on its natural history in Baja California. It is probably active year-round, with activity peaks in spring and fall and diminished activity during midsummer. Although it is diurnal (Lapointe 1967), lizards are rarely observed abroad during the day because their activity is confined to the shelter of the debris within which they reside (Mautz and Case 1974). Klauber (1939) remarks on never seeing lizards abroad at night; neither were any such observations made by Zweifel and Lowe (1966) in the Mojave Desert. *Xantusia vigilis* is a remarkably sedentary

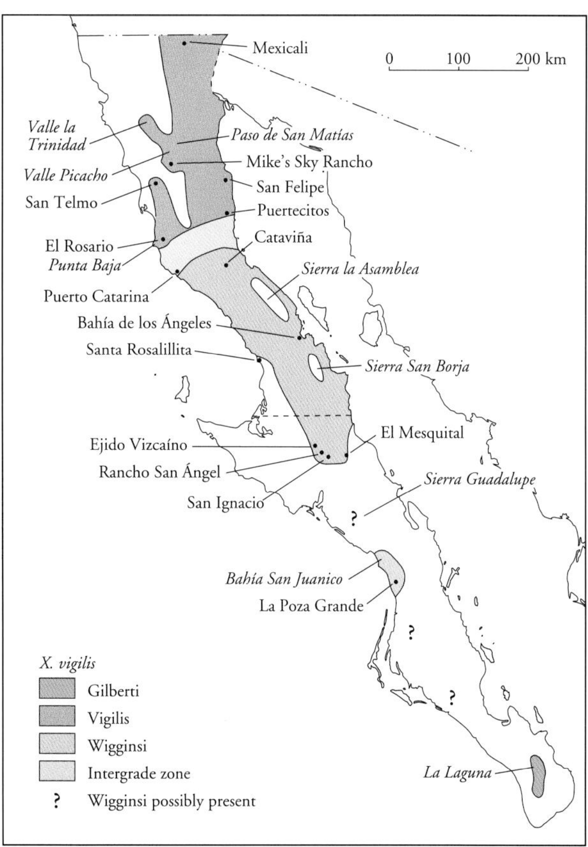

MAP 4.30 Distribution of the pattern classes of *Xantusia vigilis* (Desert Night Lizard)

species, and many individuals may spend their entire lives in a single stand of plant litter (Zweifel and Lowe 1966).

Xantusia vigilis feeds on small arthropods that inhabit the decaying vegetation in which it lives. It is a live-bearing species that usually gives birth to two young during the fall, which indicates that mating may occur throughout the summer.

Gilberti are restricted to the oak and pine-oak woodlands of the upper elevations of Sierra la Laguna, where they are common above 1,200 m (Alvarez et al. 1988). Lizards are often found beneath rocks at the base of Sotol *(Nolina beldingi).* They seem, however, to prefer residing beneath decaying pieces of vegetation. I have found them to be most common beneath oak and pine bark that is either on the ground or exfoliating from the trunk. Less often I have found them beneath decaying logs. Virtually nothing has been published on the natural history of Gilberti. Lizards are active year-round, with peak activity occurring during the spring, summer, and fall.

REMARKS Savage (1952) reports on three specimens from La Poza Grande, along the west coast in the Magdalena Plain of southern Baja California. He considered these to be intergrades between *X. vigilis gilberti* and *X. v. wigginsi.* Bezy (1982) examined additional specimens and came to the same conclusion. I found another coastal Magdalena Plain specimen at Bahía San Juanico, approximately 78 km north of La Poza Grande, that matched the description of the La Poza Grande population. I believe this is evidence that these populations are continuous around the eastern edge of the Vizcaíno Desert, with the southernmost reported population of Wigginsi at Rancho San Ángel, west of San Ignacio. I do not believe they are continuous with Gilberti, which are restricted to the pine-oak woodlands of the upper elevations of the Sierra la Laguna (Alvarez et al. 1988) and which, with additional molecular data, will probably be shown to be a separate species.

FOLKLORE AND HUMAN USE The Seri Indians believe that *X. vigilis* has toxic flesh and, if eaten, is fatal to both men and Seri (Malkin 1962).

Scincidae

SKINKS

The Scincidae are the most diverse family of lizards in the world; their cosmopolitan distribution includes the vast majority of all temperate and tropical regions. Although there are approximately 100 genera and over 1,100 species (Pough et al. 2001), skinks are poorly represented in the New World, and the vast majority of their radiation occurs in the western Pacific, Australia, Asia, and Africa. Skinks occupy a tremendous variety of habitats and have an equally remarkable number of adaptive types, even though the majority of skinks are terrestrial and many are brightly colored. Skinks are differentiated by the combination of a number of derived characteristics (Estes et al. 1988), the most obvious being the presence of smooth, shiny cycloid scales. In Baja California, skinks are represented by three species, all of which belong to the genus *Eumeces.*

Eumeces Wiegmann

SKINKS

The genus *Eumeces* is an ancestral radiation of the Northern Hemisphere composed of small to moderately sized terrestrial forms. It contains over 200 species (Greer 1970) and occurs in a wide range of habitats in both subtropical and temperate regions. *Eumeces* are generally secretive, fast-moving species that are often vividly striped and have brightly colored tails. Residing beneath surface litter such as rocks and logs, they go largely unnoticed. Three species of skinks occur in the region of study, one of which, *E. lagunensis,* is an endemic mesophilic relict of southern Baja California. The remaining two species, *E. gilberti* and *E. skiltonianus,* are restricted to the cooler, more mesic northwestern portion of the peninsula.

FIG. 4.88 ***Eumeces lagunensis***

SAN LUCAN SKINK; AJOLOTITO RAYADO

Eumeces lagunensis Van Denburgh 1895:134. Type locality: "San Francisquito, Sierra Laguna," BCS, Mexico. Holotype: CAS 400, destroyed in earthquake and fire of 1906. Neotype: USNM, Cat. No. 67398, collected "on trail between Loreto and Comondú" (Taylor 1935:433)

IDENTIFICATION *Eumeces lagunensis* can be separated from all other lizards in the region of study by having smooth, shiny cycloid scales and the last supralabial scale broadly contacting the upper secondary supratemporal.

RELATIONSHIPS AND TAXONOMY Taylor (1935) placed *Eumeces lagunensis* in the *E. skiltonianus* group, along with *E. gilberti* and *E. skiltonianus,* but stated it was "intermediate" to this group and the *E. brevirostris* group of southwestern Mexico. Grismer (1994a) suggested it may be more closely related to the latter group, based on biogeographical evidence. Grismer (1996a) and Rodgers and Fitch (1947) discuss taxonomy.

DISTRIBUTION *Eumeces lagunensis* has a disjunct distribution throughout southern Baja California (map 4.31). In the Cape Region, it is restricted to the Sierra la Laguna and associated eastern foothills. North of the Cape Region, *E. lagunensis* is known from four localities: the vicinity of the Comondús; Santa Águeda, approximately 150 km to the

FIGURE 4.88 *Eumeces lagunensis* (San Lucan Skink) from San Francisco de la Sierra, BCS

north (Tanner 1988); Los Coronados of the northern Sierra Guadalupe; and San Francisco de la Sierra (Grismer 1996a; Tanner 1988). *Eumeces lagunensis* undoubtedly occurs throughout the Sierra la Giganta.

DESCRIPTION Small, with adults reaching 60 mm SVL; body cylindrical in cross section and elongate; head pointed in dorsal profile; head scales large and platelike; large interparietal enclosed posteriorly by medial contact of large parietals; eye spot within interparietal; upper secondary temporal in broad contact ventrally with last supralabial; 7 or 8 supralabials, posteriormost largest; 2 postmentals; 6 infralabials; nuchals blending posteriorly with wide, cycloid, imbricate dorsal scales of body and tail; ventral scales of body similar to those of dorsum; ventrals blend smoothly with slightly smaller gular scales; single, central, longitudinal series of slightly wider subcaudal scales; forelimbs short, covered with cycloid, imbricate scales; 0 to 2 granular axillary scale rows; hind limbs very short, covered with cycloid imbricate scales; 12 to 14 subdigital lamellae on fourth toe; 2 large scales bordering vent anteriorly. **COLORATION** Ground color of dorsum very dark brown throughout life; limbs somewhat lighter, with scales darkly outlined; original tail bright pink (with or without a purple base) in hatchlings and juveniles, usually turning purple to orangish in adults; regenerated tail of adults bright pink with darker vertebral caudal stripe; one pair of cream-colored dorsolateral stripes originating on the rostrum, running length of body, and fading at base of tail; less distinct ventrolateral stripe running from rostrum to hind limb insertion; stripes not edged in black; gular region, undersides of limbs, vent, and tail cream-colored; belly and throat dark; pectoral region mottled.

NATURAL HISTORY *Eumeces lagunensis* occurs in the Arid Tropical Region of the foothills of the Sierra la Laguna as well as the oak and pine-oak woodlands of the Sierra la Laguna Region (Alvarez et al. 1988). Within these areas, it occurs in the relatively more humid and shady microhabitats and has been taken from beneath rocks and tree trunks. North of the Cape Region, *E. lagunensis* is confined to various isolated volcanic arroyos of the Magdalena and Vizcaíno regions that support permanent water and large palm groves. In these areas, such as Arroyo Comondú, from which the majority of specimens are known, *E. lagunensis* is found primarily beneath rocks in loose, dry soil within grassy or brushy areas and beneath fallen palm fronds. I have found hatchlings in small, loose rocks at the edges of streams in Arroyo Comondú, juveniles beneath fallen palm logs in the canyons of San Francisco de la Sierra, and adults within the hollows of fallen palm trees.

Eumeces lagunensis is active from at least February through November, with peak activity seeming to coincide with summer rainfall. It is secretive and seldom seen foraging abroad. I observed one specimen in early October in the town of San Miguel Comondú crossing the road at dusk, and a hatchling at Puente de Madera in early July moving among some small streamside rocks during mid-morning.

Eumeces lagunensis presumably feeds on a variety of

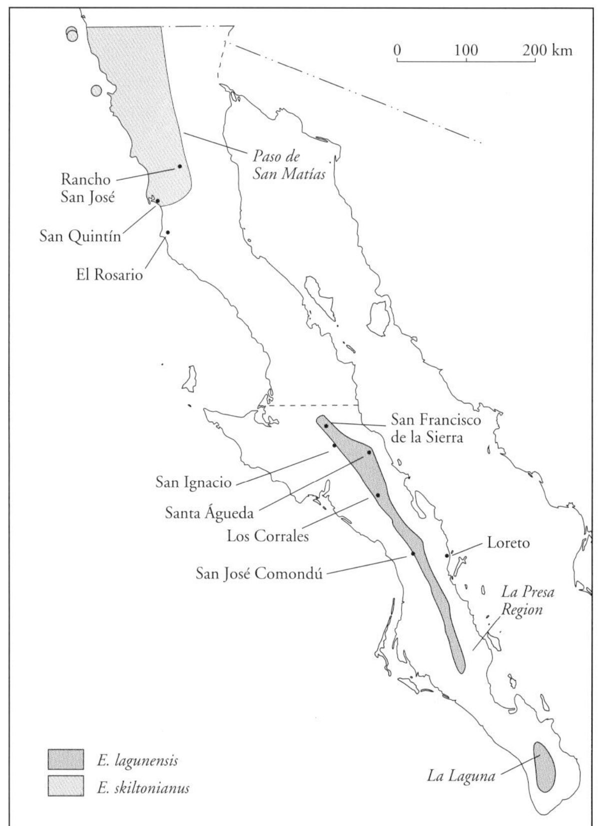

MAP 4.31 Distribution of *Eumeces lagunensis* (San Lucan Skink) and *E. skiltonianus* (Western Skink)

small arthropods. I observed a juvenile feeding on ants in San José Comondú and an adult with a small beetle in its mouth near Santiago in the Cape Region. I have observed hatchlings at San José Comondú during mid-July and juveniles from early August to mid-November in Cañón San Pablo at San Francisco de la Sierra. This suggests that *E. lagunensis* may have a spring to early summer reproductive season north of the Cape Region. The males do not develop the ephemeral bright orange throat and gular region during the breeding season that males of *E. skiltonianus* and *E. gilberti* do.

FOLKLORE AND HUMAN USE Many locals believe that *E. lagunensis* is venomous and are afraid of its brightly colored tail.

FIGURE 4.89 Juvenile *Eumeces skiltonianus* (Western Skink) from Isla Coronado Sur, BC

FIG. 4.89

Eumeces skiltonianus

WESTERN SKINK; AJOLOTE RAYADO

Plestiodon Skiltonianum Baird and Girard 1852a:69. Type locality: "Oregon," "[restricted to] the vicinity of the Dalles" on the Columbia River, Wasco Co., Oregon (Smith and Taylor 1950). Lectotype (Taylor 1935): USNM 3172b

IDENTIFICATION *Eumeces skiltonianus* can be distinguished from all other lizards in the region of study by having smooth, shiny cycloid scales; the last supralabial not or only barely contacting the upper secondary supratemporal; adults maintaining stripes; and hatchlings and juveniles having a bright blue tail.

RELATIONSHIPS AND TAXONOMY *Eumeces skiltonianus* may be most closely related to *E. lagunensis* (see Grismer 1994a, 1996a). Jones (1985) and Rodgers and Fitch (1947) discuss taxonomy.

DISTRIBUTION *Eumeces skiltonianus* ranges from southern British Columbia south through Washington and Oregon, into northwestern Baja California, and east to western Montana and Utah (Rodgers and Fitch 1947; Tanner 1988). In Baja California, *E. skiltonianus* is confined to the California and Baja California Coniferous Forest regions and ranges at least as far south as San Quintín, BC (Tanner 1988). Ted Papenfuss observed a specimen in the town of El Rosario. It also occurs on the adjacent Pacific islands of Islas Coronado (Norte and Sur) and Islas Todos Santos (see map 4.31).

DESCRIPTION Same as *E. lagunensis,* except for the following: adults reach 68 mm SVL; interparietal usually enclosed posteriorly by medial contact of large parietals; upper secondary temporal rarely contacting the last supralabial, but in broad contact with smaller, lower secondary temporal; posteriormost supralabial same size as preceding supralabial; 6 (rarely 5 or 7) infralabials; 3 to 6 (usually 4 or 5) rows of small, juxtaposed, axillary scales present; 14 to 19 subdigital lamellae on fourth toe. **COLORATION** Ground color of dorsum very dark brown in juveniles, becoming lighter and mottled with age; limbs same color as dorsum, with scales darkly outlined; tail bright blue in hatchlings and juveniles, usually turning reddish with adulthood; dark striping may or may not be evident on tail; dorsolateral stripes sometimes suffused with dark coloration; a less distinct ventrolateral stripe may or may not be present; stripes edged in dark coloration; pectoral region light-colored, mottled, or dark; gular and subcaudal orange coloration develops in adult males during breeding season.

GEOGRAPHIC VARIATION The majority of the geographic variation observed in *E. skiltonianus* occurs between populations of northern Baja California and those of the Pacific islands (Grismer 1996a). The upper secondary temporal scale in populations from Islas Todos Santos and Isla Coronado Norte does not contact the last supralabial scale, whereas in all other populations such contact is variable. Also, in the first two populations, the interparietal is enclosed posteriorly by the parietals, whereas in the other populations its enclosure is variable. More subtle differences in squamation are described by Grismer (1996a). Adults in the populations of the Islas Coronado reach a larger size than those of the adjacent peninsula and Isla Todos Santos. The largest known specimen from Isla Coronado Norte is a male (SVL 67 mm), and the largest specimen from Isla Coronado Sur is a female (SVL 68 mm). The largest peninsular specimen is a 61 mm female, and the largest specimen from Isla Todos Santos is a 61 mm male.

Lizards from the Sierra San Pedro Mártir tend to maintain the bright blue juvenile coloration of the tail into adulthood, whereas coastal and insular populations generally develop a reddish hue on the tail. *Eumeces skiltonianus* from

Rancho San José, at the base of the Sierra San Pedro Mártir, are intermediate in this respect. Also, *E. skiltonianus* from the Sierra San Pedro Mártir retain a more vivid color pattern into adulthood, whereas the color pattern of the adults from other localities becomes considerably lightened and mottled with increasing SVL. The change is especially evident in lizards from the Islas Coronado and to a lesser extent Islas Todos Santos. Here the dark, nearly unicolored ground color of the dorsal pattern between the dorsal stripes becomes so broken up and mottled that dark longitudinal lines edging the dorsolateral stripes are all that remain. Even the dorsolateral stripes become suffused with dark coloration, making them sometimes nearly indistinguishable. The development of orange breeding coloration in males is more extensive in the Islas Coronado population than in those of the mainland (Zweifel 1952a). Also, the skinks of the Islas Coronado population lose the blue juvenile caudal pattern at a much smaller SVL than mainland lizards, even though they reach a larger size (Zweifel 1952a).

NATURAL HISTORY *Eumeces skiltonianus* is abundant throughout the Sierra San Pedro Mártir, especially in the vicinity of rocks. In cismontane chaparral areas, this lizard is most common in rocky, riparian habitats, where it may reach its greatest density (Welsh 1988). Specimens are common beneath boards and rocks in grassy fields throughout Isla Coronado Sur (Zweifel 1952a).

Eumeces skiltonianus becomes active in early spring and remains active until late fall. In montane areas, the activity period may be somewhat shorter. This lizard is secretive; it has been observed in the Sierra San Pedro Mártir, moving about during the day in rocky washes in the forest (Welsh 1988). Its food consists of small arthropods. The reproductive season of *E. skiltonianus* in Baja California presumably lasts from spring through early summer, when adult males take on a crimson-orange coloration on their lower jaws and gular regions. One adult male with such coloration was observed in mid-June in Vallecitos Meadow of the Sierra San Pedro Mártir, and several with this coloration were seen in late March on Isla Coronado Sur.

FIG. 4.90 *Eumeces gilberti*

GILBERT'S SKINK; AJOLOTE

Eumeces gilberti Van Denburgh 1896:350. Type locality: "Yosemite Valley, Mariposa County, California." Holotype: CAS 4139

IDENTIFICATION *Eumeces gilberti* can be distinguished from all other lizards in the region of study by having the combination of smooth, shiny cycloid scales; the last supralabial scale not broadly contacting the upper secondary supratemporal; and an adult SVL greater than 95 mm. If the SVL is less than 96 mm, the tail is pink.

FIGURE 4.90 Adult male *Eumeces gilberti* (Gilbert's Skink) from Paso de San Matías, BC

RELATIONSHIPS AND TAXONOMY May be most closely related to *E. lagunensis* and *E. skiltonianus* (see Grismer 1994a). Rodgers and Fitch (1947) and Jones (1985) discuss taxonomy.

DISTRIBUTION *Eumeces gilberti* ranges from central northern California south into montane and cismontane northern Baja California. It also occurs in disjunct populations in the Mojave and Sonoran deserts of southern Nevada, eastern California, and western Arizona (Jones 1985). In Baja California, *E. gilberti* is known from La Grulla Meadow in the Sierra San Pedro Mártir and as far south as San Antonio del Mar along the Pacific coast (map 4.32). In both areas, it probably ranges farther south.

DESCRIPTION Same as *E. lagunensis,* except for the following: moderately sized, with adults reaching 113 mm SVL; body cylindrical and elongate; head triangular, much wider in adult males than in females and juveniles; 6 or 7 infralabials; 4 to 7 rows of small, juxtaposed axillary scales; 15 to 17 subdigital lamellae on fourth toe. **COLORATION** Color pattern varies greatly with SVL. AT 33 TO 37 MM SVL: Color pattern in hatchlings is sharp and contrasted; ground color of dorsum dark brown, with pair of cream-colored, dorsolateral stripes running length of head and body, coalescing and turning bright pink on tail; ground color extends short distance onto base of tail; scales on dorsal surfaces of limbs weakly edged in darker color; ventrum uniformly cream-colored. AT 38 TO 51 MM SVL: Striping less vivid in small juveniles; dark coloration begins to invade edges of dorsal scales; ground color at base of tail extends half the length of tail; ventrum begins to darken. AT 52 TO 59 MM SVL: Striping on body in larger juveniles becomes obscured because of further invasion of dark coloration at scale edges; overall body color becomes dark green to olive drab; striping on head more observable; dorsal surface of tail dark green, beginning to extend ventrally; dorsal limb surfaces uniformly dark green, with color beginning to extend into the postgular region. AT 60 TO 70 MM SVL: Striping on body faint in subadults but slightly more vivid on head; body color uniformly dark green, with stripes present only as shaded areas; lateral areas of tail and gular region darker. AT 71 TO 81 MM SVL: Striping extremely faint in larger subadults and lacking defined edges; postgular region completely dark. AT 82 MM SVL: Adult coloration usually attained; dorsum uniform dark green and lacking stripes; ventrum dark green to dull gray except for gular, pectoral, pelvic, and ventromedial portions of tail, which may remain cream-colored; adult males develop crimson-orange gular and infralabial regions during breeding season.

NATURAL HISTORY *Eumeces gilberti* ranges throughout the California and Baja California Coniferous Forest regions. It appears to be most common within riparian corridors (Welsh 1988), although it has been found in grassy areas with relatively soft soils. This species also extends into desert areas along riparian corridors, as evidenced by its occurrence in Cañón de Guadalupe (Welsh 1988).

Eumeces gilberti is active from mid-February through mid-October, with activity peaking from May through July. It is diurnal and secretive, occasionally seen moving among rocks and brush. The reproductive season lasts from mid-May through mid-July, with hatchlings appearing from late July through late September. *Eumeces gilberti* eats a wide variety of arthropods, including butterflies, beetles, flies, bees, bugs, spiders, scorpions, solpugids, isopods, and occasionally its own young.

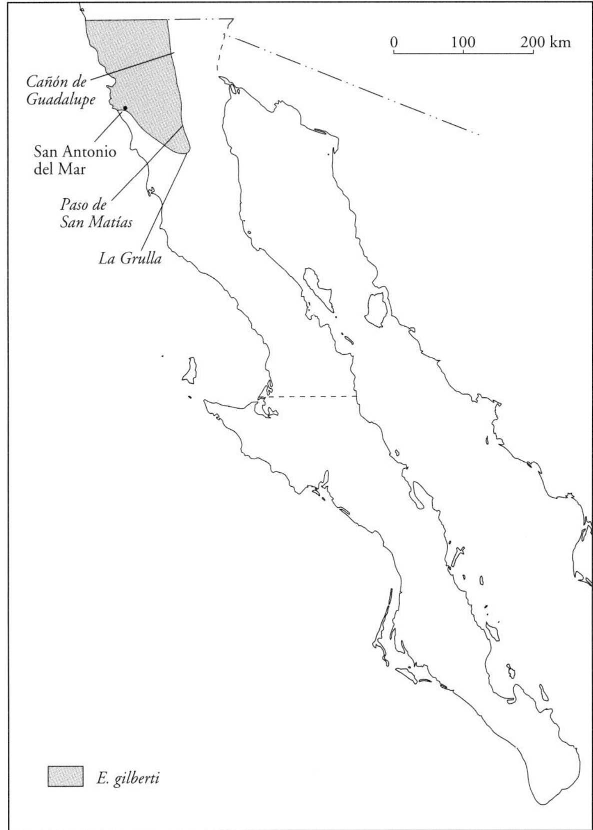

MAP 4.32 Distribution of *Eumeces gilberti* (Gilbert's Skink)

Anguidae

LEGLESS AND ALLIGATOR LIZARDS

The Anguidae are a New World and Eurasian group of lizards containing approximately 15 genera and 102 species (Pough et al. 2001). New World anguids range throughout western North America and the eastern United States south to northern South America. Anguids are absent from the Amazon Basin but occur again in southern Brazil, Argentina, Paraguay, Uruguay, Bolivia, and Peru. Eurasian anguids range across much of Europe to central Asia and from India to east Asia. The Anguidae are ecologically diverse, composed of small to moderately sized limbless fossorial, terrestrial, and prehensile-tailed arboreal species. All are carnivorous and prey on a variety of arthropods and small vertebrates. Anguidae are a lineage of anguinimorphan lizards defined by a number of derived characteristics and closely related to varanoids and xenosaurids (Estes et al. 1988). In Baja California, anguids are represented by two species of legless lizards *(Anniella)* and five species of alligator lizards *(Elgaria)*.

Anniella Gray

LEGLESS LIZARDS

The genus *Anniella* is composed of two limbless fossorial, insectivorous species. *Anniella* ranges from the San Francisco Bay region of California south through western California into cismontane northwestern Baja California. *Anniella* are locally common where they are found but are seldom seen because of their fossorial lifestyle. Both species, *A. pulchra* and *A. geronimensis*, occur in the region of study.

FIG. 4.91 ***Anniella geronimensis***

BAJA CALIFORNIA LEGLESS LIZARD; CULEBRA

Anniella geronimensis Shaw 1940:225. Type locality: "Isla San Geronimo Island, Lower California [BC], Mexico." Holotype: SDSNH 7543

IDENTIFICATION *Anniella geronimensis* can be distinguished from all other lizards in the region of study by its lack of legs; the fourth supralabial being the largest; and a gray, mottled ventrum.

RELATIONSHIPS AND TAXONOMY *Anniella geronimensis* is the sister species of *A. pulchra*. Bezy et al. (1977) discuss rela-

FIGURE 4.91 *Anniella geronimensis* (Baja California Legless Lizard) from Rancho Socorro, BC

tionships and taxonomy. The nomenclatural history of *A. geronimensis* and *A. pulchra* and the designation of appropriate holotypes have been rather convoluted. For a complete discussion of this interesting exercise on nomenclatorial rules and protocol, see Hunt (1983) and Murphy and Smith (1985 and 1991).

DISTRIBUTION *Anniella geronimensis* ranges along the coastal aeolian dune regions of northwestern Baja California, from approximately 6 km north of Colonia Guerrero south to just south of Punta Baja at the northern edge of Bahía Rosario, BC (Hunt 1983; Shaw 1949). It is also known from the Pacific islands of San Gerónimo within Bahía Rosario and San Martín (Sánchez-P. and Mellink 2001; map 4.33).

DESCRIPTION Small, limbless lizard with an elongate body and pointed head, reaching 142 mm SVL; head small, indistinct from body, frontal region steeply sloped downward in lateral profile, giving rise to sharply pointed rostrum; rostrum shovel-shaped in dorsal profile; head scales large and platelike, frontal and frontoparietal largest; rostral scale wide; 6 or 7 supralabials, fourth supralabial largest; external nares completely enclosed within first supralabial; eye deeply sunk within head; eye opening reduced to a thin, horizontal aperture concealed beneath 3 scales, forming a movable lower lid; lower jaw deeply countersunk; 5 or 6 infralabials; mental large; 2 postmentals; sublabials and chinshields decrease in size posteriorly, grading into ventral body scales; body subcylindrical and slightly concave ventrally; dorsal body scales smooth, cycloid, and imbricate, grading ventrally into similarly shaped and sized ventral scales; 24 to 26 scale rows around midbody, 4 to 6 rows between upper lateral stripes; preanal scales slightly enlarged; tail cylindrical, tapering to blunt tip, and 29 to 37 percent of total length; 72 to 86 dorsal caudal scales. **COLORATION** Dorsal ground color light copper to silver-gray; head grayish and darkly mottled; 7 or 8 brown to black lateral stripes running length of body along scale margins, often reduced to 6 or 7 stripes on tail; dorsum generally immaculate except for dark vertebral stripe extending length of body and tail; lighter paravertebral lines occasionally present; ventral scales grayish with darkened edges; preanal scales immaculate.

NATURAL HISTORY *Anniella geronimensis* ranges along the western edge of the ecotone between the California and the Vizcaíno regions of northwestern Baja California. This lizard is extremely arenifossorial, occurring in a narrow strip of fine-grained, aeolian coastal dunes 87 km in length. *Anniella geronimensis* ranges only 4 km inland (Shaw 1953) in Arroyo Socorro where the onshore breezes have carried sand inland. Otherwise, it is restricted to the coastal margins. Shaw states that *A. geronimensis* should occur farther inland and that its coastal localities simply reflect the ease with which it is found in dune areas. I have searched farther inland for suitable habitat and lizards but have found neither. This absence and the extreme morphological specializations of the species seem to indicate that it is highly and exclusively adapted to the fine-grained coastal dunes of this region. Legless lizards are commonly found at the base of

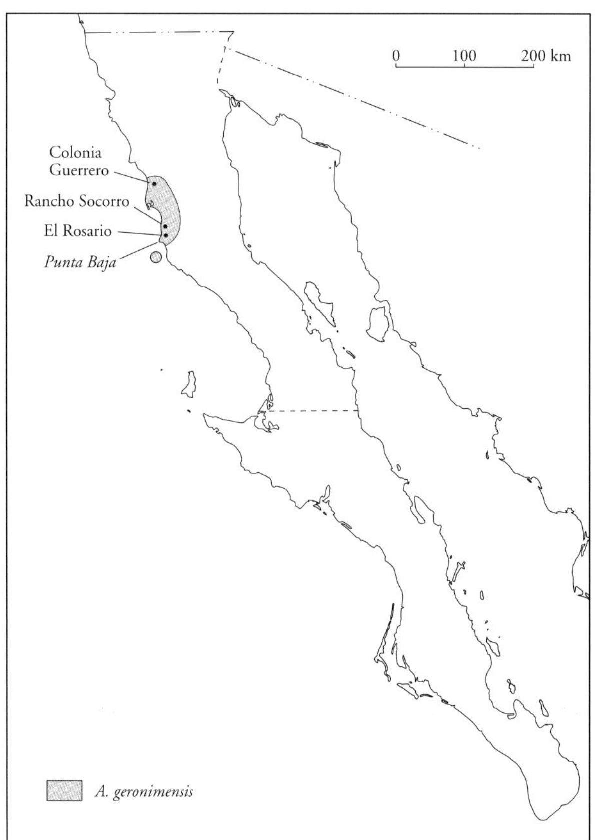

MAP 4.33 Distribution of *Anniella geronimensis* (Baja California Legless Lizard)

Brittlebush (*Encelia* sp.), Bursage (*Ambrosia* sp.), and Locoweed (*Astragulus* sp.), which grow on the dunes. I have also found specimens beneath the trunks of fallen Coastal Agave *(Agave shawii)*. Shaw 1949 reports finding specimens beneath "desert shrubs" on the edges of the dunes, and the tracks of *A. geronimensis* are commonly seen all over the dunes which they inhabit.

Anniella geronimensis is generally active from February through October, depending on the weather. It is encountered only rarely during other months. No specimens were found at Rancho Socorro during March 1991. The winter had been unseasonably dry, resulting in scant dune vegetation and associated arthropods, and this may explain the lizards' scarcity. During other years, *A. geronimensis* was commonly found in March. Shaw (1953) noted that during the morning, before the coastal fogs burn off, *A. geronimensis* remains buried relatively deep in the sand, avoiding the cooler surface temperatures. As soon as the sun strikes the dunes, lizards move to just below the surface, taking advantage of the thin layer of warming sand. At Rancho Socorro I have noticed that late in the day, lizards are difficult to find on the windward side of the dunes, where continuous cool, onshore breezes keep the surface layers cool. They are also difficult to find on the leeward sides of the dunes, which receive no onshore breezes and are exposed to direct sunlight; here surface temperatures are probably too high. During late afternoon, however, lizards can easily be found on the tops of the dunes, where the temperature falls between the two extremes. Shaw (1953) reports seeing a specimen he believed to be basking above the sand in the stems of an *Encelia.*

Nothing has been reported on the food habits of *A. geronimensis,* but presumably it feeds on the abundant arthropod fauna beneath the dune vegetation. Shaw (1953) reports finding two gravid females on 12 August, each with a single embryo lacking pigment. This indicated that the embryos were not near term. Hunt (1983) reports on a gravid female collected between 2 and 22 October, with a single near-term embryo, and I have observed juveniles from mid-May through August. These observations suggest that mating takes place during the spring and summer and that birth occurs from summer to early fall.

REMARKS *Anniella geronimensis* seems to be more specialized for living in fine, windblown sand (Bezy et al. 1977) than *A. pulchra,* with which it is narrowly syntopic in the dunes west of Colonia Guerrero. Where both species are found together, *A. geronimensis* is far more abundant (Shaw 1953). *Anniella pulchra* ranges much farther inland and occurs in various types of loose or loamy soil (see below).

FIGURE 4.92 *Anniella pulchra* (California Legless Lizard) from Isla Coronado Sur, BC

FIG. 4.92 *Anniella pulchra*

CALIFORNIA LEGLESS LIZARD; CULEBRA

Anniella pulchra Gray 1852:440. Type locality: "0.8 km southeast of Pinnacles National Monument, San Benito County, California" (Murphy and Smith 1991). Neotype: MVZ 64656 (Murphy and Smith 1991)

IDENTIFICATION *Anniella pulchra* can be distinguished from all other lizards in the region of study by lacking legs; the second supralabial being the largest supralabial; and having yellow sides and belly.

RELATIONSHIPS AND TAXONOMY See *A. geronimensis.*

DISTRIBUTION *Anniella pulchra* ranges throughout western California, from the San Francisco Bay region south into northwestern Baja California. In Baja California, it ranges widely through the northwestern portion of the peninsula (Bury 1983; Hunt 1983) and may extend along sandy arroyos into the lower foothills of the Sierra San Pedro Mártir (Welsh 1988). It is also known to extend east along Mexican Highway 2 to at least La Rumarosa, north of the Sierra Juárez (Hunt 1983). *Anniella pulchra* ranges as far south as Arroyo Pabellon, approximately 17 km south of San Quintín (Hunt 1983; map 4.34). This species is also known from the Pacific archipelagos of the Islas Coronado (Norte and Sur) and Todos Santos (Isla Sur).

DESCRIPTION Same as *A. geronimensis,* except for the following: Adults reach 165 mm SVL; frontal region shallowly sloped downward in lateral profile, giving rise to an acutely rounded rostrum; second supralabial largest; 28 to 30 scale rows around midbody; 10 to 12 rows between upper lateral stripes; tail length 34 to 42 percent of total length; 80 to 94 dorsal caudal scales. **COLORATION** Dorsal ground color light to dark gray; one brown to black lateral stripe running length of body and tail; ventrolateral portion of body and abdomen yellow; dark middorsal stripe reduced or absent; preanal scales edged and centrally punctated with black.

NATURAL HISTORY Miller (1944) provides an excellent treatment of the natural history of *A. pulchra* from the northern portion of its distribution in California. In Baja California, *A. pulchra* is a common inhabitant of the loose, fine-textured, sandy, and loamy soils of the California Region. In such habitats, it is commonly found beneath stones and other surface debris (Klauber 1932; Linsdale 1932; Zweifel 1952a). Shaw (1953) reports finding lizards beneath dune vegetation in the southern, coastal portion of its distribution, and I have found them beneath Coastal Agave *(Agave shawii)* near Colonet and beneath trash and bursage *(Ambrosia)* on the beach at Colonia Guerrero. Zweifel (1952a) reports seeing tracks of *A. pulchra* in landlocked sea caves hundreds of feet above the water on Isla Coronado Sur, and I have found specimens beneath stones in loose soil on the top of the island.

Anniella pulchra can be found year-round. During the winter, lizards are most commonly found beneath stones and are probably inactive. I have found active individuals in mid-February near Colonet. Miller (1944) noted that lizards approach the surface during the morning to bask in the warm subsurface layers and then emerge to forage for small arthropods. Pregnant females usually carry two embryos. Shaw (1953) reports a pregnant female from near Colonia Guerrero on 12 August that contained a single embryo that was not near term. Klauber (1932) reported on a female from San Diego County with two near-term embryos on 31 August. These observations suggest that the reproductive season for the Baja California populations begins in late spring to early summer, as suggested for the more northerly California populations (Miller 1944).

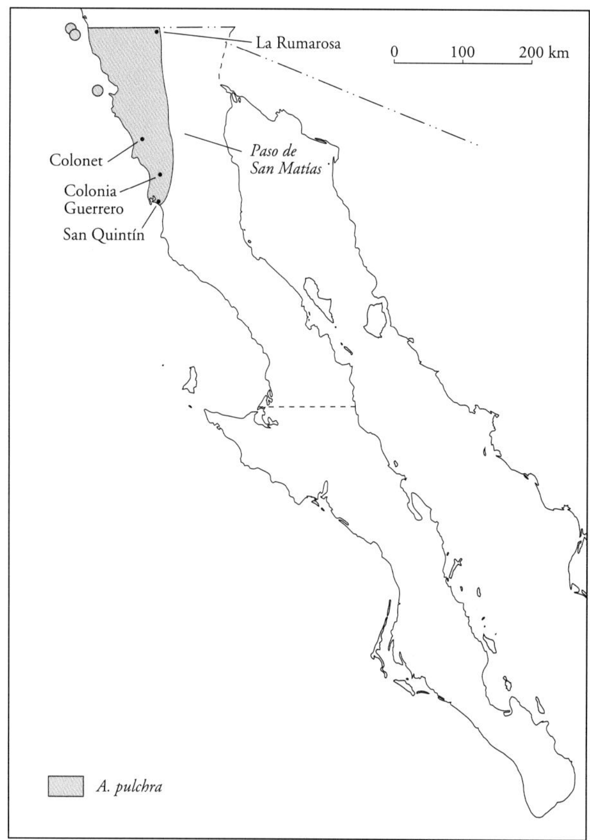

MAP 4.34 Distribution of *Anniella pulchra* (California Legless Lizard)

REMARKS See *A. geronimensis.*

Elgaria Gray

ALLIGATOR LIZARDS

The genus *Elgaria* is composed of eight species of moderately sized lizards ranging through a variety of habitats, from the Pacific northwestern United States south to the tip of Baja California. *Elgaria* occur again in the mountainous portion of central Arizona and range south along the Sierra Madre Occidental to Jalisco, Mexico. *Elgaria* are secretive, terrestrial to semiarboreal species that are well-known for their generally fierce appearance and aggressive nature. Five species occur in the region of study, of which four *(E. cedrosensis, E. nana, E. paucicarinata,* and *E. velazquezi)* are endemic and the fifth *(E. multicarinata)* ranges into the United States.

FIG. 4.93 *Elgaria cedrosensis*

ISLA CEDROS ALLIGATOR LIZARD; AJOLOTE

Gerrhonotus cedrosensis Fitch 1934a:6. Type locality: "Cañon on southeast side of Cedros Island, Lower California, Mexico." Holotype: CAS 56187

IDENTIFICATION *Elgaria cedrosensis* can be distinguished from all other lizards in the region of study by having a ventrolateral body fold; large keeled, imbricate dorsal body scales; 14 to 16 (usually 14) longitudinal dorsal scale rows; black and white labial markings; and no light-colored transverse body bands.

RELATIONSHIPS AND TAXONOMY *Elgaria cedrosensis* is probably the sister species of *E. velazquezi* of Central Baja California (Grismer and Hollingsworth 2001) and part of a natural group containing *E. paucicarinata, E. cedrosensis, E. panamintina, E. kingii,* and *E. parva* (Good 1988a,b), from the Cape Region; west central Baja California; the Panamint and Inyo Mountains of California; southern Arizona and the Sierra Madre Occidental of northwestern Mexico; and central Nuevo León, respectively. Macey et al. (1999) provide an alternative hypothesis that includes *E. multicarinata* as part of Good's natural assemblage. Grismer (1988a) and Grismer and Hollingsworth (2001) discuss taxonomy.

DISTRIBUTION *Elgaria cedrosensis* ranges continuously along the Pacific coast of north central Baja California, from approximately 30 km south of El Rosario south to at least 0.8 km north of San Javier (Bostic 1971), and inland as far as Nuevo Rosarito. *Elgaria cedrosensis* has also been taken from various localities throughout Isla Cedros, off the west coast of central Baja California (Lais 1976a; map 4.35).

DESCRIPTION Moderately sized and elongate, with adults reaching 96 mm SVL; head triangular, wider than neck, widest in males, and sharply rounded in lateral profile; head scales enlarged and platelike; interparietal large and containing eye spot; parietals abruptly contacting much smaller dorsal body scales; 24 to 31 eyelid fringe scales; 10 to 12 supralabials; 2 asymmetrical postmentals; 10 to 11 infralabials; squarish gular scales grading into rectangular ventrals; dorsal scales square, imbricate, and weakly keeled; 14 to 16 (usually 14) rows of longitudinal dorsals; 48 to 54 rows of transverse dorsals; 5 to 13 longitudinal rows of keeled dorsals; dorsals abruptly contacting granular scales of ventrolateral body fold; forelimbs small, covered with relatively large, smooth, imbricate scales; axillary pockets present; hind limbs slightly larger than forelimbs; antefemoral and antetibial regions covered by large, weakly keeled scales; postfemoral and posttibial regions covered by smaller, smooth scales; femoral pores absent; postfemoral pockets present; 16 to 18 subdigital lamellae on fourth toe; plantar scales relatively small; tail long and pointed, 1.75 to 2.25 times body length, round in cross section, and with 1 to 6 weakly keeled dorsal rows; caudal scales elongate and imbricate; no enlarged postanal scales. **COLORATION** Ground color of dorsum dark olive brown to dull brick-red or tangold; adult females tend to be more grayish overall, and males are often brick-red; ground color of sides dark brown to gray; top of head usually with distinct dark markings; distinct dark postorbital stripe usually present; distinct black and white markings on supralabials and to some extent infralabials; transverse banding always on sides but may be present or absent on back and tail; bands often consist of thin, zigzagged dark lines countershaded posteriorly by white scales; dorsal banding complete or incomplete; when incomplete, dorsum is a unicolored tan to rusty red; limbs usually immaculate; small amount of black pigmentation in lateral fold encircles round, whitish spots; ventrum dull gray with faint to distinct dark longitudinal stripes; hatchlings and juveniles have a nearly unicolored dorsum.

FIGURE 4.93 *Elgaria cedrosensis* (Isla Cedros Alligator Lizard) from Arroyo Vargas, Isla Cedros, BC

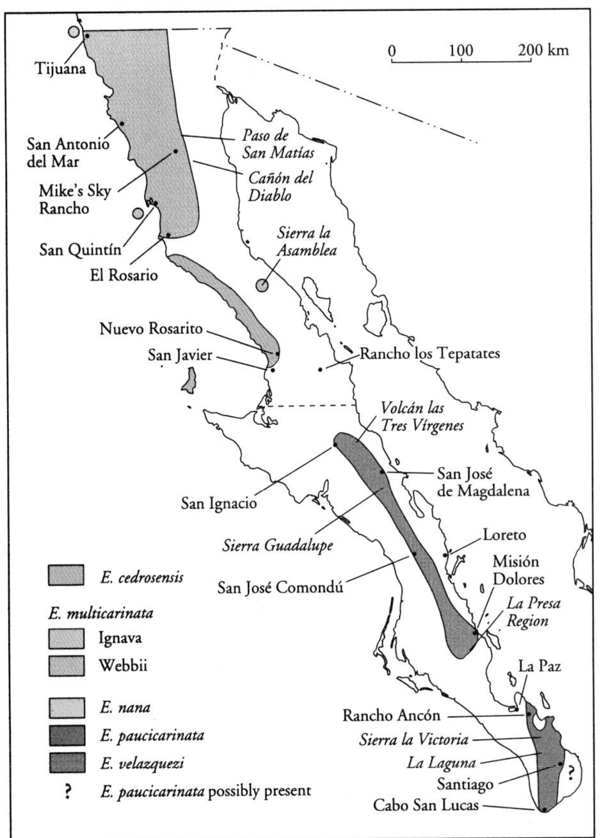

MAP 4.35 Distribution of the pattern classes of *Elgaria multicarinata* (Southern Alligator Lizard) and the remaining species of *Elgaria*

GEOGRAPHIC VARIATION *Elgaria cedrosensis* from Isla Cedros tend to be smaller than lizards from the peninsular populations. This population also has a less distinctive banding pattern because the dark dorsal bands are thinner and poorly defined and generally lack the white countershading; these lizards have a lighter ground color and are occasionally unicolored dorsally as adults (Grismer 1988a). I collected one adult male that had a nearly unicolored brick-red dorsum. Lizards from peninsular populations have a prominent ventral postorbital stripe that only rarely occurs in lizards from Isla Cedros. Also, keeling in the peninsular lizards is generally more distinctive than keeling in lizards from Isla Cedros.

NATURAL HISTORY *Elgaria cedrosensis* commonly occurs in areas where localized moisture is relatively high. Along the Pacific coast of Baja California, opposite Isla Cedros, Bostic (1971) reported finding specimens beneath dead or partially dead Coastal Agave *(Agave shawii)* and surface debris. This hilly to mountainous area is frequently interrupted by marine terraces. Continuous cool onshore breezes keep air temperatures relatively low, and morning fogs add significantly to the area's precipitation (Bostic 1971; Grismer 1994b; Meigs 1966). The weather is much the same farther inland at Nuevo Rosarito, where the hilly terrain is dominated by *Agave shawii.* Alligator lizards live within the leaf axils and beneath the agaves. On Isla Cedros I have found specimens beneath rocks and debris in rocky canyons both with and without water. Lizards have also been collected from Arroyo Vargas, a well-watered canyon on the west side of the island. In all cases, lizards are found close to brush. Residents report finding alligator lizards in town in some of the watered gardens around the houses.

Elgaria cedrosensis has been observed abroad from late May through late July on Isla Cedros and from mid-March through mid-October on the adjacent peninsula. It is a diurnal lizard, occasionally observed foraging among vegetation or rocks. At Nuevo Rosarito I found an individual abroad at 0630 on a cool, overcast morning during mid-March. Later, during mid-morning after the cloud had cleared, I observed alligator lizards basking on the leaves of *Agave shawii.* At my approach, they retreated into the bases of the agaves. I also observed lizards abroad on the ground, moving from one stand of agaves to another. I observed a specimen foraging among the rocks in an arroyo on Isla Cedros.

The diet of *E. cedrosensis* is primarily arthropods. I observed an individual with a grub in its mouth at Nuevo Rosarito. Bostic (1971) reported females collected in July to have convoluted oviducts but no enlarged ova, which indicates that these were postpartum females.

FIGS. 4.94–95

Elgaria velazquezi

CENTRAL BAJA CALIFORNIA ALLIGATOR LIZARD; AJOLOTE

Elgaria velazquezi Grismer and Hollingsworth 2001: 493. Type locality: "41.5 km west by Mexican Highway 1 of Santa Rosalía, Baja California Sur, Mexico." Holotype: SDSNH 68769

FIGURE 4.94 Adult male *Elgaria velazquezi* (Central Baja California Alligator Lizard) from 41.5 km west of Santa Rosalía, BCS

IDENTIFICATION *Elgaria velazquezi* can be distinguished from all other lizards in the region of study by having a ventrolateral body fold; large keeled, imbricate dorsal body scales; light-colored transverse body bands in juveniles and almost always in adults; and no dark postorbital stripe.

RELATIONSHIPS AND TAXONOMY See *E. cedrosensis.*

DISTRIBUTION Although *E. velazquezi* probably ranges continuously from at least San Ignacio south to the Isthmus of La Paz, throughout the central desert region of Baja California, lizards are known only from isolated populations occurring at San Ignacio, the base of Volcán las Tres Vírgenes, the Sierra Guadalupe, San José de Magdalena, San José Comondú, and Misión Dolores (Grismer 1988a; Grismer and McGuire 1993; Grismer and Hollingsworth 2001; see map 4.35). It is not known how far north this species extends, although it probably reaches to at least San Francisco de la Sierra, 30 km northwest of San Ignacio. A rancher in the Sierra la Libertad, at Rancho los Tepatates, described a lizard to me that was likely an *Elgaria.* It is not known whether it was *E. cedrosensis* or *E. velazquezi.*

DESCRIPTION Same as *E. cedrosensis,* except for the following: adults reach 124 mm SVL; 28 to 35 eyelid fringe scales; dorsal scales moderately to strongly keeled; 46 to 59 rows of transverse dorsals; 14 longitudinal rows of keeled dorsals; tail with 2 to 8 weakly keeled dorsal rows. **COLORATION** Ground color of dorsum dark olive brown to dull brick red or tan-gold; ground color of sides usually lighter and more gray; top of head with distinct dark spots; no distinct postorbital stripe; distinct black and white markings on supralabials and to some extent infralabials; light-colored transverse bands on body and anterior of tail bold and immaculate in hatchlings and juveniles, usually faded (rarely absent) in adults, bands exist as transversely aligned white tips of black scales on sides of body in most adults; light countershading of dark transverse body bands only weakly present; limbs may or may not have small black flecks dorsally; white bars in lateral body fold bordered by black, some random dark flecking occasionally present; ventrum dull gray with no striping, white bars of ventrolateral fold extend onto lateral ventral scales.

GEOGRAPHIC VARIATION *Elgaria velazquezi* from San Ignacio to Misión Dolores, approximately 100 km north of the Isthmus of La Paz, almost always retain the prominent juvenile banding pattern into adulthood. One large adult male from Cumbre de San Pedro, in the Sierra Guadalupe, had no trace of transverse banding left save for the white tips on the scales along the sides of the body. This may be a case of matching the greenish, lichen-covered rocks among which these lizards are often found. Banding is distinct in all populations but becomes less so with age. Bands can be most pronounced in lizards from the dark lava flows at the base of Volcán las Tres Vírgenes. Here, the white countershading may be transformed into thin, immaculate white bands on a very dark brown ground color. A specimen found in the Sierra Guadalupe had no color pattern and was a uniform dull gray. This coloration is considered to be anomalous; the same anomaly occasionally occurs in the *E. paucicarinata* from La Victoria in the Sierra la Laguna.

NATURAL HISTORY *Elgaria velazquezi* is a semiscansorial saxicolous species. Ottley and Murphy (1983) reported on a specimen from San Ignacio that was collected beneath palm fronds near the water's edge, and Karl Switak (personal communication, 1995) collected a specimen at Misión Dolores beneath a rock in a moist, well-vegetated arroyo bottom. He also found the shed skin of an adult nearby. From June through October, *E. velazquezi* is nocturnal and abroad from dusk to at least 0200. If overcast is present, however, and temperatures are low enough, lizards will emerge during the day. From March through May, they are probably diurnal, although I have never made such observations. At the base of Volcán las Tres Vírgenes, I found an alligator lizard abroad at 0120 hours during mid-August, foraging through and weaving among the small volcanic rocks in a lava flow. During evenings in early to mid-October, I observed additional individuals at the same locality. All were foraging in the lava flow in the vicinity of vegetation. During one evening in mid-October at the same locale, several individuals were periodically observed over five hours crawling among the same rocks. Alligator lizards collected from the Sierra Guadalupe were found in the cracks of the lichen-covered boulders, beneath logs in a

FIGURE 4.95 Juvenile *Elgaria velazquezi* (Central Baja California Alligator Lizard) from 41.5 km west of Santa Rosalía, BCS

stream bed under the canopy of an oak forest, and near rocks surrounded by brush along a road cut. At San José de Magdalena and Misión Dolores, specimens were found beneath rocks in a dry stream bed.

Brad Hollingsworth collected a gravid female during mid-May at the base of the Sierra Guadalupe. Karl Switak collected a small juvenile during mid-August at Misión Dolores. I have observed hatchlings from late August through early September at Volcán las Tres Vírgenes. These findings suggest that *E. velazquezi* may have a breeding season lasting from at least April to June.

REMARKS Grismer (1988a) and Grismer and McGuire (1993) hypothesized that *Elgaria velazquezi* was restricted in distribution to well-watered areas throughout its range in central Baja California. The collection of this species in the lava flows at the base of Volcán las Tres Vírgenes and the upper elevations of the Sierra Guadalupe contradicts this hypothesis. Both these localities are several kilometers from water, and it is now believed that this species is well-adapted to seeking moist microhabitats deep within rock cracks and lava flows and is continuously distributed throughout the central Peninsular Ranges to at least the Isthmus of La Paz (Grismer and Hollingsworth 2001).

FIG. 4.96

Elgaria paucicarinata

SAN LUCAN ALLIGATOR LIZARD; AJOLOTE

Gerrhonotus paucicarinatus Fitch 1934b:173. Type locality: "Todos Santos, Lower California, Mexico." Holotype: MVZ 11768

IDENTIFICATION *Elgaria paucicarinata* can be distinguished from all other lizards in the region of study by having a ventrolateral body fold; large keeled, imbricate, dorsal body scales; 16 longitudinal dorsal scale rows; and black and white labial markings.

FIGURE 4.96 *Elgaria paucicarinata* (San Lucan Alligator Lizard) from La Victoria, BCS

RELATIONSHIPS AND TAXONOMY See *E. cedrosensis.*

DISTRIBUTION *Elgaria paucicarinata* is restricted to the mountains and foothill regions of the Cape Region of Baja California, BCS (Lais 1976c; see map 4.35).

DESCRIPTION Same as *E. cedrosensis,* except for the following: Adults reach 110 mm SVL; 26 to 31 eyelid fringe scales; 16 (rarely 14) rows of longitudinal dorsal scales; 50 to 59 rows of transverse dorsals; 8 to 14 longitudinal rows of keeled dorsals; antefemoral and antetibial scales moderately keeled; 2 to 6 moderately keeled dorsal caudal rows. **COLORATION** Faint to distinct dark postorbital stripe; extensive black pigmentation in ventrolateral body fold, bordering irregularly shaped yellowish or orange markings; unicolored brown dorsum in hatchlings and juveniles.

GEOGRAPHIC VARIATION Adult male *E. paucicarinata* from the western slopes of the Sierra la Laguna tend to be darker in overall dorsal ground color than lizards from other areas. They also have orangish markings in the ventrolateral body fold and yellowish anterior supralabials. Grismer (1988a) noted that adult alligator lizards from the vicinity of Rancho Ancón, approximately 30 km southeast of La Paz, had a tendency to retain the juvenile pattern of a broad, tan-colored longitudinal stripe on the dorsum. This adult pattern is not unique to the Rancho Ancón area, but it is more prevalent there than elsewhere.

NATURAL HISTORY *Elgaria paucicarinata* is endemic to the Cape Region of Baja California and ranges throughout the Arid Tropical and Sierra la Laguna regions (Lais 1976c; Richmond 1965). Within the Sierra la Laguna Region, *E. paucicarinata* is fairly ubiquitous and most common in areas with sufficient leaf litter (Alvarez et al. 1988). Lizards are found beneath rocks near streams (Murray 1955) and beneath fallen Sotol *(Nolina beldingi),* pine logs, and palm fronds (Richmond 1965). I have found *E. paucicarinata* to be very common in these microhabitats as well as within rock cracks. At Rancho Ancón, I have found lizards as high as 2 m above the ground in the hollows of oak trees (*Quercus* sp.) and within rotting logs at La Victoria in the Sierra la Laguna. Alvarez et al. (1988) reported *E. paucicarinata* from lowland thorn scrub areas in the Arid Tropical Region, west of the Sierra la Laguna. Here, however, lizards are not as common as in the Sierra la Laguna and seem to be restricted to localized mesic areas.

Elgaria paucicarinata is active year-round, with an activity peak from March to late October. Linsdale (1932) reported seeing lizards abroad during November. Richmond (1965) reports seeing lizards out basking during the day at a time when nighttime temperatures were below freezing; in the Cape Region, this would most likely be in December or January. I have found lizards abroad during the day in shaded areas of Arroyo El Aguaje during late August. Lizards bask in open sunlight during the morning. John Wright (personal communication, 1983) informed me that he saw several basking on Cholla pads (*Opuntia* sp.) in the Sierra la Laguna. After reaching an appropriate body temperature, lizards forage for food. Because they are secretive, they are not often seen, although Richmond (1965) reports seeing specimens moving about in a grassy meadow, and I have observed lizards in the afternoon on sidewalks in the town of Santiago. I have observed adults feeding on various species of beetles. Hatchlings have been collected during August, which indicates that this species probably has a late spring mating and summer hatching season.

FOLKLORE AND HUMAN USE *Elgaria paucicarinata* is intensely feared by most of the inhabitants of the Cape Region. It is believed to seek out and enter the rectal cavities of small children, causing various types of sickness.

FIG. 4.97

Elgaria multicarinata

SOUTHERN ALLIGATOR LIZARD; AJOLOTE

Cordylus (*Gerrhonotus*) *multi-carinatus* Blainville 1835b:37. Type locality: "Californie," restricted to "Monterey, [Monterey Co., California]" by Fitch (1934b:173). Holotype: MNHN 2002

IDENTIFICATION *Elgaria multicarinata* can be distinguished from all other lizards in the region of study by having a ventrolateral body fold; large keeled, imbricate dorsal body scales; no black and white labial markings; and an adult SVL greater than 115 mm.

RELATIONSHIPS AND TAXONOMY *Elgaria multicarinata* and its sister species *E. nana* (Grismer 2001a) are most closely related to a group containing *E. cedrosensis* of Isla Cedros and central Baja California, *E. velazquezi* of southern Baja California, *E. paucicarinata* of the Cape Region, and *E. panamintina, E. kingii,* and *E. parva* (Good 1988a,b; Grismer and Hollingsworth 2001), from the Panamint and Inyo Mountains of California; southern Arizona and the Sierra Madre Occidental of northwestern Mexico; and central Nuevo León, respectively. Grismer (1988a, 2001a) discusses taxonomy.

DISTRIBUTION *Elgaria multicarinata* ranges along the Pacific coast of North America, from southern Washington south through western Oregon and California into northwestern Baja California (Lais 1976b). In Baja California, it ranges continuously south along the Sierra Juárez, Sierra San Pedro Mártir, and western foothills and coastal plains to at least 13.5 km south of El Rosario. *Elgaria multicarinata* is also known from some of the well-watered canyons on the desert side of the Sierra Juárez and Sierra San Pedro Mártir (Grismer 1988a; Linsdale 1932; Welsh 1988). Grismer and Mellink (1994) reported on an isolated population at the top of the Sierra la Asamblea, approximately 175 km south of El Rosario. *Elgaria multicarinata* is also known from the Pacific island of San Martín, off the coast of northwestern Baja California (see map 4.35).

DESCRIPTION Same as *E. cedrosensis,* except for the following: adults reach 166 mm SVL; 29 to 38 eyelid fringe scales; temporal scales keeled; 10 to 12 supralabials; 9 to 11 infralabials; dorsal scales strongly keeled; 38 to 52 rows of transverse dorsals; 12 to 14 longitudinal rows of keeled dorsals; dorsal scales of forelimbs weakly keeled, those of hind limbs moderately keeled; 18 to 21 subdigital lamellae on fourth toe; tail 2.0 to 2.5 times length of body with 6 to 15 strongly keeled, dorsal rows. **COLORATION** Ground color of dorsum variable, being dark olive-brown, light gray, tan-gold, or dull brick red; ground color of sides dark brown to gray; top of head immaculate or boldly marked with dark blotches; postorbital stripes usually absent; supralabials and infralabials edged posteriorly with dark markings but lacking white markings; transverse banding on back, sides, and tail, consisting of thin, zigzagged dark lines often bordered posteriorly by white scales; limbs usually immaculate; white irregularly shaped markings present in lateral fold but black pigmentation absent; ventrum dull gray, rarely with faint longitudinal or sinuous stripes; hatchlings and juveniles have a unicolored brown, gold, or copper-colored dorsum and very dark sides.

GEOGRAPHIC VARIATION Dorsal color pattern variation within peninsular populations of *E. multicarinata* (pattern class Webbii) can be extensive, and geographic variation between

FIGURE 4.97 Webbii pattern class of *Elgaria multicarinata* (Southern Alligator Lizard) from Tecate, BC

populations is generally limited to dorsal ground color. Grismer (1988a) noted that near the southern end of its distribution, from the vicinity of San Quintín to El Rosario, and at its eastern limits, in the Sierra San Pedro Mártir, *E. multicarinata* has a grayish ground color with a faded banding pattern. From El Rosario south, specimens maintain the light grayish color laterally but tend to have yellowish-green dorsums.

The Isla San Martín population has been considered the subspecies *E. multicarinata ignava* (Van Denburgh 1905) and is considered here as the pattern class Ignava. Adults reach 126 mm SVL and have 14 longitudinal rows of keeled dorsals. The ground color of the head and body is olive-gray, and that of the sides is occasionally darker. The top of the head usually has dark to light brown markings, and a postorbital stripe is usually present. There are few or no white markings middorsally, but a brownish, interband mottling is often present. The limbs are covered dorsally with a dark reticulum. The ventrum is dull gray and occasionally has faint longitudinal or sinuous stripes.

NATURAL HISTORY A great deal is known about the natural history of *E. multicarinata* in southern California (Kingsbury 1993, 1994, 1995; Lais 1976b and references therein), and it would be surprising if the populations from northwestern Baja California showed significant departures from what has already been reported. *Elgaria multicarinata* ranges widely throughout the California and Baja California Coniferous Forest regions, where alligator lizards appear to be most prevalent in riparian habitats near rocks (Linsdale 1932; Murray 1955; Welsh 1988), especially in the western foothills of the Peninsular Ranges (Welsh 1988). Linsdale (1932) reported specimens from the desert slopes of the Sierra San Pedro Mártir in Cañón del Diablo and Cañón el Cajón. Here, lizards are most frequently found in and around Willow thickets (*Salix* sp.). On the coastal plain, Welsh (1988) reported seeing lizards in the "desert scrub" near San Antonio del Mar. I have observed lizards in grassy fields just north of San Quintín and in Vizcaíno Region scrub south of El Rosario. I observed one juvenile and several shed skins of adults beneath dead Sotol *(Nolina beldingi)* in a relict chaparral community on the top of the Sierra la Asamblea, at approximately 1,350 m elevation (Grismer and Mellink 1994).

Elgaria multicarinata is active from February through November, with a peak activity period of March through October. Alligator lizards are generally abroad during the day and are able to continue their activities at remarkably low ambient temperatures brought on by overcast skies (Kingsbury 1994) or the onset of evening. I observed two individuals on a cool evening (19°C) in early May at Mike's Sky Rancho, at the base of the Sierra San Pedro Mártir. Lizards are usually seen foraging on the ground but can often be found in the branches of low vegetation.

Cunningham (1956) reported that *E. multicarinata* primarily consumed carabid beetles and orthopterans. Also eaten were isopods, spiders, ants, and various larvae. He also noted that *Eumeces skiltonianus* (often just the tails) and other alligator lizards were occasionally consumed, as well as small mammals and birds. Based on captive observations, Cunningham (1956) noted that *E. multicarinata* probably eats carrion, Pacific Treefrogs *(Hyla regilla),* and their tadpoles.

Gravid *E. multicarinata* have been observed from April through August (Atsatt 1952; Fitch 1970; Shaw 1943), and Burrage (1965) and Shine (1994) indicate that two, and occasionally three, clutches of eggs per year are laid. Breeding occurs during early spring (Fitch 1970). First clutches are usually laid in June (Burrage 1965; Shaw 1952), with the first hatchlings appearing in August (Shaw 1952). Burrage (1965) also reports finding hatchlings in November. I observed a hatchling that was probably hibernating beneath a deeply buried pine log in the Sierra Juárez during late December. Shine (1994) noted that female *E. multicarinata* share egg-laying sites and that large groups of eggs have been found with several females nearby.

Ignava is known from Isla San Martín, a small island 5 km off the coast of San Quintín in an ecotone between the California and the Vizcaíno regions. The northernmost peak of a chain of six volcanic cones bordering the western and northern edges of Bahía San Quintín, the island is covered with lava flows, volcanic ash, lava tubes, and pits. Alligator lizards seem to be restricted to the volcanic areas of the island. Murray (1955) stated that they have been found only on the lava, not on the flats or sand dunes. Lizards are most active from March through September. Murray reported specimens as being abroad during both the day and the night, and Zweifel (1958) reports lizards active with body temperatures as low as 18.7°C.

FIG. 4.98 **Elgaria nana**

ISLAS CORONADO ALLIGATOR LIZARD; AJOLOTE

Gerrhonotus scincicauda nanus Fitch 1934a:7. Type locality: "South Island, Los Coronados Islands, Lower California, Mexico." Holotype: MVZ 5402

IDENTIFICATION *Elgaria nana* can be distinguished from all other lizards in the region of study by having a ventrolateral body fold; large, keeled, imbricate dorsal body scales; no black and white labial markings; and an adult SVL less than 115 mm.

RELATIONSHIPS AND TAXONOMY See *E. multicarinata.*

DISTRIBUTION *Elgaria nana* is known only from the Islas Coronado Norte and Sur, off the coast of northwestern Baja California (Grismer forthcoming; see map 4.35).

DESCRIPTION Same as *E. cedrosensis,* except for the following: adults reach 115 mm SVL; 24 to 35 eyelid fringe scales; temporal scales keeled; 10 to 12 supralabials; 9 to 11 infralabials; dorsal scales strongly keeled; 33 to 49 rows of transverse dorsals; 10 or 11 longitudinal rows of keeled dorsals; dorsal scales of forelimbs weakly keeled, those of hind limbs moderately keeled; 17 to 20 subdigital lamellae on fourth toe; tail 2.0 to 2.5 times length of body, with 6 to 15 strongly keeled dorsal rows. **COLORATION** Ground color of dorsum generally darker overall than *E. multicarinata,* being dark olive-gray or dull brick red; ground color of sides usually darker; top of head usually with dark blotches; postorbital stripes usually distinct; supralabials and infralabials heavily edged posteriorly with brown or black but lacking white markings; transverse banding on back, sides, and tail consisting of thin, zigzagged dark lines, rarely bordered posteriorly by white scales; brownish interband mottling often present; limbs usually covered dorsally with a dark reticulum; white, irregularly shaped markings in lateral folds, but black pigmentation absent; ventrum dull gray with dark longitudinal lines; hatchlings have a color pattern similar to that of *E. multicarinata,* but grayer and with less bold lateral markings. The adult color pattern emerges at around 42 mm SVL.

NATURAL HISTORY The Islas Coronado are a narrow archipelago composed of three small, steep, rocky islands lying 13 km off the coast of Tijuana (the middle island is actually made up of two closely spaced islets). These are the erosional remnants of a once-larger island (Lamb 1992). They are dominated by coastal scrub and are part of the California Region. *Elgaria nana* is common and seems to be ubiquitous (Zweifel 1952a). It is usually abroad from March through October, but its yearly activity varies with weather conditions. *Elgaria nana* is most active during overcast periods (Zweifel 1952a), when it forages for food and mates. Most individuals are seen near rocks and beneath stands of *Opuntia* cactus as well as climbing on the branches. I have seen *E. nana* pairs copulating during March, and Zweifel (1952a) observed copulation in April. Clutches of 1 to 18 eggs have been laid as early as May and June (Burrage 1964, 1965). Burrage (1965) noted that marked field specimens, which had laid two clutches of eggs earlier in the season, were gravid again in September.

FIGURE 4.98 *Elgaria nana* (Islas Coronado Alligator Lizard) from Isla Coronado Sur, BC

REMARKS Fitch (1934a) described *E. nana* as a subspecies of *E. multicarinata,* which he separated from *E. m. webbii* of San Diego County, California, by its being smaller, more weakly keeled, and having fewer transverse ventral scale rows, and by subtle differences in the head and body color pattern. However, only adult body size discretely differentiates these two species, and, according to Fitch, "no overlapping size could be shown among the 49 adult specimens from Los Coronados Islands and 35 from San Diego County." Average SVL for *E. nana* was 93.3 mm with a maximum SVL of 114 mm; for *E. multicarinata* from San Diego County, average SVL was 135 mm, with a maximum of 166 mm (Fitch 1938). Fitch was aware that body size differences could be the result of the ecological effect of an insular environment rather than a genetically based difference, as has been proposed by more recent authors for other insular lizard groups (e.g., Petren and Case 1997 and citations therein). However, Fitch opted for a genetically based difference, noting ontogenetic changes and sexual dimorphism in *E. multicarinata:* specifically, that the largest individuals have thicker, more heavily keeled scales, and males have much wider temporal regions. Fitch pointed out that the largest individuals of *E. nana* showed these characters, indicative of advanced age, even though they were smaller than the small adults of *E. multicarinata.* Additionally, Zweifel (1952a) observed a 79 mm SVL female *E. nana* copulating, and Burrage (1965) indicated that the hatchling size range of the Islas Coronado population is 32 to 35 mm SVL, essentially the same as that of hatchling *E. multicarinata* (30 to 36 mm) from San Diego County. These observations strongly suggest that adulthood, with its accompanying morphological changes, is reached at a much smaller SVL in *E. nana* than in *E. multicarinata* from the adjacent peninsula.

5

WORM LIZARDS

WORM LIZARDS MAKE UP A BIZARRE GROUP OF FOSSORIAL REPTILES KNOWN AS THE AMPHISBAENIA. THIS GROUP CONTAINS 24 GENERA AND OVER 143 SPECIES (POUGH ET AL. 2001), WHICH ARE HIGHLY ADAPTED FOR BURROWING WITH SMALL, HEAVILY OSSIFIED SKULLS, CYLINDRICAL BODIES, AND SHORT TAILS. ADDITIONALLY, BECAUSE THEIR BODY SCALES OCCUR IN ANNULI, THEY CAN MOVE EASILY FORWARD AND BACKWARD WITHIN THEIR SELF-CONSTRUCTED BURROW SYSTEMS. WORM LIZARDS ARE USUALLY LIMBLESS, ALTHOUGH ONE GROUP, THE BIPEDIDAE, RETAIN THEIR FORELIMBS. THE GROUP IS REPRESENTED IN THE REGION OF STUDY BY A SINGLE ENDEMIC SPECIES OF BIPEDID, WHICH HAS NO INSULAR DERIVATIVES.

Bipedidae

Bipes Latreille

TWO-FOOTED WORM LIZARDS

The Bipedidae are a New World group of lizards containing one genus *(Bipes)* and three species. Bipedids have a fragmented distribution, with one species, *B. biporus,* occurring in southern Baja California and the remaining two, *B. canaliculatus* and *B. tridactylus,* being allopatrically distributed in the states of Guerrero and Michoacán in southwestern Mexico (Papenfuss 1982). Bipedids are unique among amphisbaenians in that they have forelimbs with five *(B. biporus),* four *(B. canaliculatus),* or three *(B. tridactylus)* claws. All other amphisbaenians are limbless. Bipedids are burrowing lizards, found most frequently in loose, well-aerated soils; they rarely surface above ground. All bipedids are arthropodivorous and prey on a variety of species. Ironically, bipedids are probably the most numerous lizards in the areas where they are found, but because of their fossorial lifestyle they are seldom seen.

FIG. 5.1 ***Bipes biporus***

FIVE-TOED WORM LIZARD; AJOLOTE

Euchirotes biporus Cope, 1894:436. Type locality: "Cape San Lucas, Lower California," Mexico. Holotype: USNM 8568

IDENTIFICATION *Bipes biporus* can be distinguished from all other lizards in the region of study by having forelimbs and lacking hind limbs.

RELATIONSHIPS AND TAXONOMY Cole and Gans (1987) suggest that *B. biporus* is the sister species of the Mexican mainland lineage composed of *B. canaliculatus* and *B. tridactylus.* Kim et al. (1976) discuss genetic variation.

DISTRIBUTION *Bipes biporus* ranges throughout the western portion of the southern half of Baja California, west of the Peninsular Ranges, from approximately 17 km north of Jesús María, where the Sierra Columbia contacts the Pacific coast, south to Cabo San Lucas (Papenfuss 1982). At the Isthmus of La Paz, its distribution extends east across the low, sandy flats and contacts the Gulf coast at Bahía de La Paz (map 5.1).

DESCRIPTION Body cylindrical in cross section, reaching 240 mm in total length; head bluntly rounded, not distinct from neck; eyes small, pupils rounded; tail short, thick, and blunt, approximately 12 percent of SVL; anterior head scales large and platelike, posterior head scales small and square to rectangular; forelimbs short and stout, covered with square to rectangular scales; 5 long claws on forelimbs; 242 to 261 dorsal annuli; 143 to 167 ventral annuli; 3 to 6 lateral annuli; 0 to 9 (usually 4 to 6) intercalated quarter annuli; 24 to 31 caudal annuli; 27 to 32 dorsal segments at midbody; 24 to 30 ventral segments at midbody; 2 preanal pores. **COLORATION** Head, body, and tail whitish-pink in adults and pinkish in juveniles.

NATURAL HISTORY Papenfuss (1982) presented a superb study on the ecology and systematics of *Bipes;* the majority of the information presented below comes from that study. The distribution of *B. biporus* is highly concordant with the Vizcaíno Desert and Magdalena Plain. Both areas are relatively cool and dominated by loose, sandy soils and scrub vegetation. *B. biporus* constructs an elaborate system of burrows just below the surface of the ground, and this system is usually centered on stands of vegetation. The systems generally run horizontally in various directions, occasionally opening to the surface beneath rocks and logs; *B. biporus* can be found when these objects are overturned. I even found an individual beneath a sheet of tin in the Sierra Santa Clara. In the Vizcaíno Desert and Magdalena Plain, *B. biporus* is most commonly found within the sand hummocks that build up around the bases of vegetation. In the Sierra Santa Clara, *B. biporus* follows sandy arroyo bottoms along the bases of the volcanoes, into the interior of the mountain range (Grismer et al. 1994). In the Cape Region,

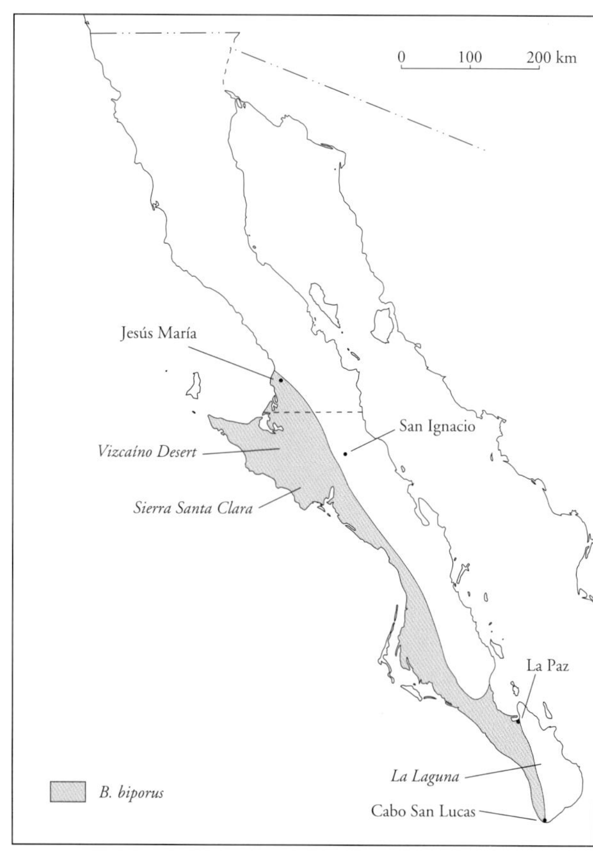

MAP 5.1 Distribution of *Bipes biporus* (Five-Toed Worm Lizard)

B. biporus is common in the loose soil beneath mesquite trees (*Prosopis* sp.), especially if there is an abundance of leaf litter.

Bipes biporus is active year-round. In the morning, worm lizards lie just below the surface in sections of their burrows that are exposed to the sun. As the day proceeds and temperatures increase, worm lizards travel deeper into the soil or move into a portion of the burrow that is shaded. Papenfuss (1982) indicated that *B. biporus* rarely surfaces and that during the evening it retreats deeper into its burrow system. He noted that of the 2,719 specimens he collected, only three were found above ground. I observed five freshly killed specimens on the road one evening in mid-June in the Vizcaíno Desert just east of San Ignacio. I believe, however, that their emergence was the result of a rather strange and rapidly developing weather pattern. Earlier that day, temperatures had exceeded 38°C, but in the late afternoon the weather began to cool rapidly and became very windy as a fast-moving bank of advectional fog moved in from the west coast. Interestingly, several Western Blind Snakes (*Leptotyphlops humilis*) and many rarely seen fossorial arthropods were also observed that evening, which suggests that this climatic anomaly stimulated the surface activity of several fossorial species. Dial et al. (1987) reported that the tracks of *B. biporus* were occasionally found beneath leaf litter in the Cape Region.

Bipes biporus feeds on a variety of arthropods, many of which are also fossorial. Females reach maturity at a total length of approximately 185 mm and become gravid only during June and July. Clutch size ranges from one to four eggs, but usually two are laid. Hatchlings first begin to appear in late September.

FOLKLORE AND HUMAN USE In my experience, no reptile strikes more fear in the hearts of the local inhabitants than this *ajolote*. Almost anyone with knowledge of this animal will tell you that its suppository-shaped head facilitates its penetration into the human anus (other animals are sometimes believed to be attacked as well, but usually it's humans). Its long claws grip the walls of the rectum and serve to anchor it within the intestinal tract. The short limbs are used to locomote deeper into the body. To my knowledge, Papenfuss has not yet tested this hypothesis.

6

SNAKES

SNAKES FORM A DIVERSE COMPONENT OF THE HERPETOFAUNA OF BAJA CALIFORNIA AND THE ISLANDS OF NORTHWESTERN MEXICO. THEY ARE REPRESENTED BY SIX MAJOR GROUPS. THE COLUBRIDAE ARE THE MOST SPECIOSE GROUP IN THE REGION OF STUDY, COMPRISING 17 GENERA AND 36 SPECIES. THERE ARE ALSO 15 SPECIES OF RATTLESNAKES (VIPERIDAE), ONE SPECIES OF BLIND SNAKE (LEPTOTYPHLOPIDAE), ONE BOA (BOIDAE), AND ONE CORAL SNAKE AND ONE SEA SNAKE (ELAPIDAE). THE SNAKES WITHIN THE REGION OF STUDY SHOW A BROAD RANGE OF ADAPTIVE TYPES, INCLUDING THE FOSSORIAL BLIND SNAKE *LEPTOTYPHLOPS HUMILIS,* THE ARENICOLOUS SHOVEL-NOSED SNAKE *CHIONACTIS OCCIPITALIS,* THE SIDEWINDING RATTLESNAKE *CROTALUS CERASTES,* SEMIARBOREAL TERRESTRIAL WHIPSNAKES *(MASTICOPHIS),* THE SAXICOLOUS LYRE SNAKE *TRIMORPHODON BISCUTATUS,* AND BOTH MARINE AND FRESHWATER AQUATIC FORMS SUCH AS THE SEA SNAKE *PELAMIS PLATURUS* AND GARTER SNAKES (*THAMNOPHIS).* AT PRESENT COUNT, THE SNAKE FAUNA CONTAINS 22 GENERA, ONE OF WHICH IS ENDEMIC, AND 55 SPECIES, OF WHICH 23 (45 PERCENT) ARE ENDEMIC. OF THESE, 11 SPECIES (48 PERCENT) ARE KNOWN ONLY FROM ISLANDS. THUS, A SIGNIFICANT NUMBER OF THE SNAKES OF BAJA CALIFORNIA AND THE GULF OF CALIFORNIA OCCUR NOWHERE ELSE IN THE WORLD. SNAKES ARE WELL REPRESENTED ON BOTH PACIFIC AND GULF ISLANDS (GRISMER 1993, 1999A, 2001A) AND THE MAJORITY OF THESE ARE MOST CLOSELY RELATED TO POPULATIONS FROM THE GEOGRAPHICALLY MOST PROXIMATE PENINSULAR OR MAINLAND SOURCE. HOWEVER, SOME TRANSGULF COLONIZATIONS MAY HAVE OCCURRED (GRISMER 1994C).

Leptotyphlopidae

BLIND SNAKES

The Leptotyphlopidae are a peculiar group of snakes containing two genera and approximately 87 species (Pough et al. 2001). In the New World, leptotyphlopids range widely throughout the southwestern United States, south through Mexico, Central America, and the West Indies to southern South America. In the Old World, they range through much of southwest Asia and nearly all of Africa. Leptotyphlopids are slender, fossorial species with immovable, edentate maxillae; four to five teeth on each dentary; and vestigial eyes lying beneath an ocular scale (List 1966).

Leptotyphlops Fitzinger

BLIND SNAKES

The genus *Leptotyphlops* contains over 75 species that collectively range widely throughout the American southwest, Mexico, Central America and much of South America (Hahn 1979a). In the Old World, *Leptotyphlops* ranges throughout all of Africa and extends east throughout the Arabian Peninsula into southwest Asia. Because of the ameliorating effects of subterranean microhabitats on ambient conditions, *Leptotyphlops* is able to exist in a variety of habitats, ranging from extremely arid deserts to humid tropical forests. In fact, blind snakes may be one of the most common species of snakes in any particular area, but because of their fossorial lifestyle, they are the least often observed. In Baja California and the Gulf of California, the Leptotyphlopidae are represented by *Leptotyphlops humilis.*

FIG 6.2

Leptotyphlops humilis

WESTERN BLIND SNAKE; CULEBRITA CIEGA

Rena humilis Baird and Girard 1853:143. Type locality: "Valliecitas, Cal." Restricted to the "vicinity of Vallecito, eastern San Diego County" by Klauber (1931a) and to the "Upper Sonoran Life Zone of the Vallecito area" by Brattstrom (1953a). Holotype: USNM 2101

IDENTIFICATION *Leptotyphlops humilis* can be distinguished from all other snakes in the region of study by its long, wormlike, cylindrical body; degenerate eyes; and lack of transverse ventral scales.

RELATIONSHIPS AND TAXONOMY Klauber (1931a, 1940a), Hahn (1979b), Murphy (1975), and Grismer (1999b) discuss taxonomy.

DISTRIBUTION *Leptotyphlops humilis* ranges throughout the southwestern United States south into the Chihuahuan Desert of north-central Mexico, along the west coast of Mexico to Colima, west of the Sierra Madre Occidental (Hahn 1979b), and into Baja California in the west. In Baja

FIGURE 6.2 Humilis pattern class of *Leptotyphlops humilis* (Western Blind Snake) from Arroyo Blanco y Sol, Isla Santa Catalina, BCS

California, it is ubiquitous (Grismer 1994d) but has not been found in the upper elevations of the Sierra Juárez and the Sierra San Pedro Mártir (Welsh 1988; map 6.1). *Leptotyphlops humilis* is also known from the Pacific island of Cedros and the Gulf islands of Carmen, Cerralvo, Danzante, San Marcos, Santa Catalina, and Santa Cruz (Grismer 1999a)

DESCRIPTION Body cylindrical, long and thin, reaching 389 mm total length; head bluntly rounded, indistinct from body; eyes small, degenerate, and covered by large ocular scales; tail short, blunt, and terminating in prominent recurved, keratinous spine; rostral scale large and curving around front of head; prefrontal, frontal, interparietal, interoccipital, temporals, and postoccipitals undifferentiated from dorsal body scales; nasals, parietals, and occipitals enlarged; anterior supralabial and posterior supralabial separated by ocular; mental small and indented; 4 infralabials; chinshields smaller, grading into ventrals; dorsal scales cycloid, imbricate, and undifferentiated from ventrals; 237 to 305 longitudinal dorsals; anal plate enlarged and single; 12 to 21 undifferentiated subcaudals. **COLORATION** Dorsal ground color ranges from pink to dark brown, with pigments concentrated on 5 to 7 dorsal scale rows; ventrum immaculate, ground color pink to cream; labia and rostrum generally lack pigmentation.

GEOGRAPHIC VARIATION The variation of dorsal coloration and squamation in *L. humilis* of Baja California has resulted in the designation of three subspecies (Klauber 1940a), which are treated here as pattern classes. Humilis range throughout cismontane northwestern Baja California south to the Isthmus of La Paz. They have a dark gray-brown ground color with seven pigmented, transverse dorsal scale rows, a maximum length of 315 mm, and 253 to 285 longitudinal dorsal scale rows (Grismer 1999b). They tend to show a general lightening and increase in dorsal scale counts from north to south (Klauber 1940a). Blind snakes from Isla Santa Catalina were described as the subspecies *L. h. levitoni* (Murphy 1975) but conform to the pattern class Humilis (Grismer 1999b), along with those from Isla Danzante. The taxonomic status of the single specimen from Isla Carmen described as *L. h. lindsayi* (Murphy 1975) remains questionable (Grismer 1999b).

Populations of Humilis intergrade with Cahuilae of northeastern Baja California through Valle la Trinidad (Klauber 1940a; Welsh 1988) and along the east side of the Peninsular Ranges south of the Sierra San Pedro Mártir to at least Bahía de los Ángeles. They intergrade with Boettgeri of the Cape Region through the Isthmus of La Paz. Boettgeri have a dark brown dorsum; five pigmented, transverse dorsal scale rows; 244 to 269 longitudinal dorsal scale rows; and a maximum size of 253 mm. Cahuilae range widely through northeastern Baja California in the Lower Colorado Valley Region south to at least Puertecitos. Cahuilae have a pink dorsal ground color with 5 or 7 pigmented, transverse dorsal scale rows (Klauber 1940a), 280 to 305 longitudinal dorsal scale rows, and a maximum length of 389 mm. Cahuilae from the Vizcaíno Desert tend to be light

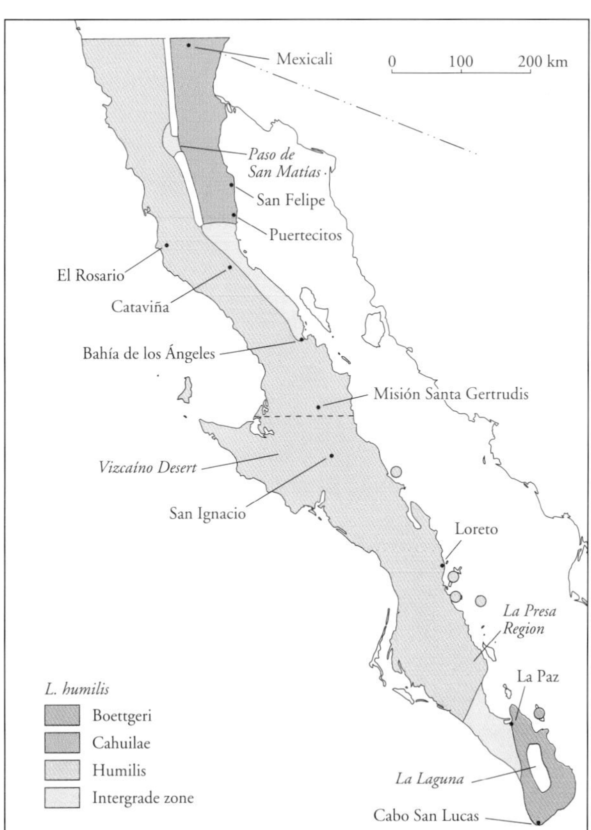

MAP 6.1 Distribution of the pattern classes of *Leptotyphlops humilis* (Western Blind Snake)

brown in dorsal coloration (Mosauer 1936). Klauber (1940a) stated that perhaps this population resembled Humilis because of parallel evolution, but it is more likely the result of intergradation with Humilis.

NATURAL HISTORY *Leptotyphlops humilis* is fossorial, ranging throughout all phytogeographic regions except the Sierra la Laguna (Alvarez et al. 1988) and the Baja California Coniferous Forest (Welsh 1988) regions. It is restricted locally in distribution to areas of loose soil, usually in the vicinity of vegetation and rocks. Welsh (1988) reported specimens from open gravel areas in coastal scrub in northwestern Baja California. Murray (1955) observed specimens on a sparsely vegetated sandy bluff near the water's edge at San Felipe. I have observed several specimens crawling on loose soil among the boulders at Cataviña and in arroyo bottoms within stands of thick vegetation at Misión Santa Gertrudis and on Isla Santa Catalina. Mosauer (1936) reported blind snakes being found in dune areas in the Vizcaíno Desert. Banks and Farmer (1962) reported pit fall trapping specimens on Isla Cerralvo in coarse sand and loose gravel near dense vegetation. Murphy and Ottley (1984) report finding a specimen on Isla Danzante in an open area near a rock outcrop.

Leptotyphlops humilis takes refuge within cracks in the ground and beneath rocks. B. D. Hollingsworth and I observed a specimen on Isla Santa Cruz in an arroyo bottom attempting to escape into a rock crevice in a cliff. We found another beneath the surface of the sand below a Leatherplant *(Jatropha cuneata)* along the beach. I also found a specimen beneath an andesitic cap rock in a lava flow at San Ignacio. Mosauer (1936) reported capturing a specimen as it was entering a crack in the woody trunk of a desert shrub.

North of San Ignacio, *L. humilis* is generally active from March through September. In the more southerly regions of the peninsula, it may be active year-round, with peak activity occurring from April through October. Blind snakes are nocturnal and emerge from their subterranean refugia, or from beneath rocks or logs, shortly after sunset. During this period, they forage for food, which consists primarily of ants, termites, and their larvae. I found a specimen at San Ignacio beneath a cap rock that also harbored ants and their larvae. On Isla Santa Catalina, I observed two specimens abroad in an arroyo bottom in the middle of a swarm of winged termites, and blind snakes at the base of Cardón Cactus *(Pachycereus pringlei)* in the midst of a swarm of winged ants. Banks and Farmer (1963) report finding ants and their larvae in the stomach of an individual collected on Isla Cerralvo.

The reproductive biology of *L. humilis* from Baja California is unknown. I have observed hatchlings on Isla Santa Catalina from early May through late June and at the base of Volcán las Tres Vírgenes during early September. This suggests that breeding and egg-laying occur during the spring and summer.

Boidae

BOAS AND PYTHONS

The Boidae are a diverse group of snakes containing nearly 20 genera and approximately 61 species (Kluge 1989, 1991, 1993a,b). In the New World, boids range throughout the western United States, Mexico, the West Indies, Central America, and much of South America. In the Old World, they range across north Africa and sub-Saharan Africa west throughout most of Asia and into Australia and parts of Oceania. Boids have a varied lifestyle, encompassing fossorial, aquatic, terrestrial, and arboreal species, and inhabit regions ranging from cool northern coniferous forests to hot, arid deserts and the warm, wet tropics. Wherever they are found, boids are specialized for ambushing, constricting, and swallowing large prey. In Baja California and the Gulf of California, boids are represented by the monotypic genus *Lichanura*.

Lichanura Cope

ROSY BOA

The genus *Lichanura* contains the species *L. trivirgata,* which ranges through most of the Sonoran and Mojave deserts as well as coastal scrub regions of southern California and northwestern Baja California. *Lichanura trivirgata* is a small, slow-moving, docile, terrestrial snake with a highly variable color pattern. It ranges through nearly all of Baja California and occurs on at least one Pacific island and five islands in the Gulf of California (Grismer 1993, 1999a, 2001a).

Lichanura trivirgata is the sister species of the Rubber Boa, *Charina bottae,* and together they form the sister group to the Calabar Burrowing Python, *Calabaria reinhardtii* (Kluge 1993a). Kluge placed all of these species in the genus *Charina* on the basis that it was taxonomically the most efficient and emphasized the historical connection between the Old and New worlds. However, recognizing these species as congeneric or not does not affect their relationship to one another, nor does it affect the evolutionary soundness of the classification within eurycine boas, as proposed by Kluge (1993a). All have a long-established history within the original classification. Therefore, in the interest of taxonomic clarity and information retrieval (taxonomic efficiency aside), I maintain these species as separate genera.

FIGS. 6.3–5

Lichanura trivirgata

ROSY BOA; SOLQUATE, DOS CABEZAS

Lichanura trivirgata Cope 1861:304. Type locality: "The southern region of Lower California." Smith and Taylor (1945) indicate the type locality as "Cape San Lucas, Baja California," although this was not a formal restriction. Given that the specimen was collected by John Xantus, who was stationed at Cabo San Lucas from April 1859 to August 1861, the holotype is undoubtedly a Cape Region specimen. Thus the type locality is restricted here to Cabo San Lucas, BCS, Mexico. Holotype: USNM 5023

IDENTIFICATION *Lichanura trivirgata* can be distinguished from all other snakes in the region of study by having small scales on the top of the head; narrow, transverse ventral scales; and a blunt tail.

RELATIONSHIPS AND TAXONOMY See *Lichanura* above. Various subspecies have been described within *L. trivirgata* and are summarized in Spiteri 1986.

DISTRIBUTION *Lichanura trivirgata* ranges widely throughout the Mojave and Sonoran deserts of the southwestern United States and northern Mexico, as well as coastal regions of southern California (Yingling 1982). In Baja California, *L. trivirgata* occurs in all areas (map 6.2) except for the upper elevations of the northern Peninsular Ranges (Welsh 1988) and the Sierra la Laguna (Alvarez et al. 1988). Although it has not been reported from the Vizcaíno Peninsula, Grismer et al. (1994) suspect its occurrence. It is known from Islas Santa Margarita (Reynoso 1990c) and Cedros along the Pacific coast and, from an unconfirmed report (Bostic 1975), from the Pacific island of Natividad. In the Gulf of California, *L. trivirgata* occurs on Islas Ángel de la Guarda, Mejía, Tiburón, San Marcos, and Cerralvo (Grismer 1999a).

DESCRIPTION Body somewhat cylindrical in cross section, stout and relatively short, reaching 1,100 mm SVL; head and tail relatively blunt; head small, slightly distinct from neck, elongate triangular; eyes small with elliptical pupils, surrounded by 7 to 12 ocular scales; preocular often enlarged; 0 to 2 suboculars; tail very short, often incomplete; rostral large and triangular; head scales small; 12 to 15 supralabials; postmentals absent; 13 to 17 infralabials; dorsal body scales smooth and imbricate, 36 to 44 rows at midbody; 210 to 235 narrow ventral scales in males and females; anal plate undivided; 39 to 48 undivided, subcaudal scales in males, 41 to 50 in females; pelvic spur on each side of the vent in males. **COLORATION** Dorsal color pattern variable (see geographic variation section). Dorsal ground color dark gray, brown, dull yellow, or tan; one dorsal and two lateral, serrate, or well-defined stripes; stripes may be nearly black, dark brown, brick red, orange, or tan, and may or may not be distinct from dorsal ground color; iris silver, gray, or orangish.

GEOGRAPHIC VARIATION *Lichanura trivirgata* shows marked geographic variation in Baja California, which corresponds well with the different phytogeographic regions throughout which *L. trivirgata* ranges. Three subspecies have been recognized in the region of study to account for this variation (Spiteri 1988) and are considered here as pattern classes. Roseofusca occur in the California Region and range from the U.S.-Mexican border south to the vicinity of San Quintín. Roseofusca are generally uniform in color, with very poorly defined, obscure, deeply serrated stripes. Between Santo Tomás and San Vicente, the ground color is dull yellow to straw-colored. Elsewhere, it is dark gray to varying shades of brown. Roseofusca have a gray iris and an enlarged preocular scale, and may or may not have suboculars.

Populations of Roseofusca intergrade with Saslowi throughout Valle la Trinidad in the east and between San Quintín and El Rosario in the south. Saslowi occupy much of the north-central portion of Baja California and extend from Cañón Santa Isabel along the lower eastern sides of the Sierra Juárez south to just north of Arroyo San Regis, north of Guerrero Negro. They also occur on the Gulf islands of Ángel de la Guarda and Mejía. Saslowi have a gray to dull white ground color with well-defined orangish stripes, 17 to 23 oculars, 215 to 235 ventrals (males and females), 41 to 48 subcaudals (males and females), 39 to 44 midbody scale rows, and an orangish iris. Boas ranging along the east coast of the peninsula tend to have brighter orange stripes than boas from the midpeninsular or west coast populations. Stripes are brightest in boas ranging from the vicinity of San Felipe north to Cañón Santa Isabel. Some individuals from near Bahía San Luis Gonzaga have darker stripes than boas from further north and from Bahía de los Ángeles to the south. Boas from Islas Ángel de la Guarda and Mejía have a much less contrasting color pattern, and the stripes are not greatly differentiated from the grayish ground color.

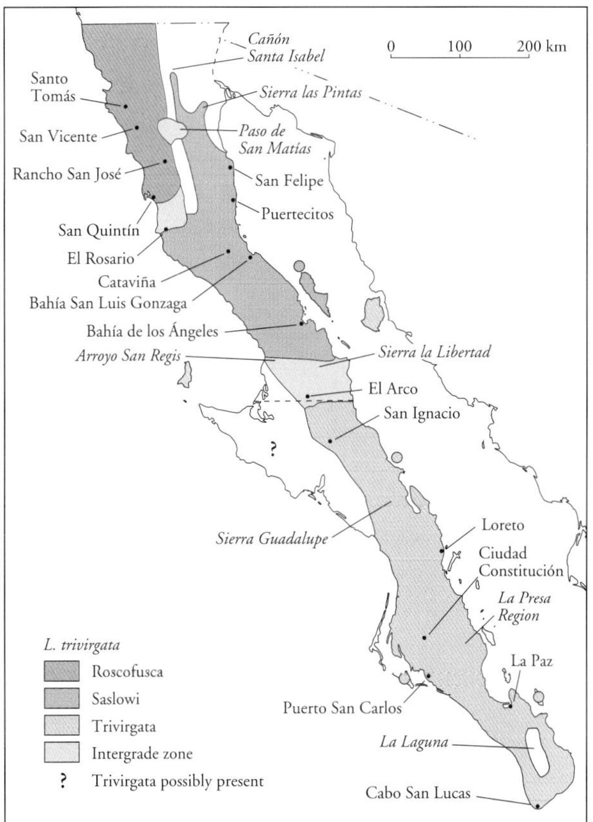

MAP 6.2 Distribution of the pattern classes of *Lichanura trivirgata* (Rosy Boa)

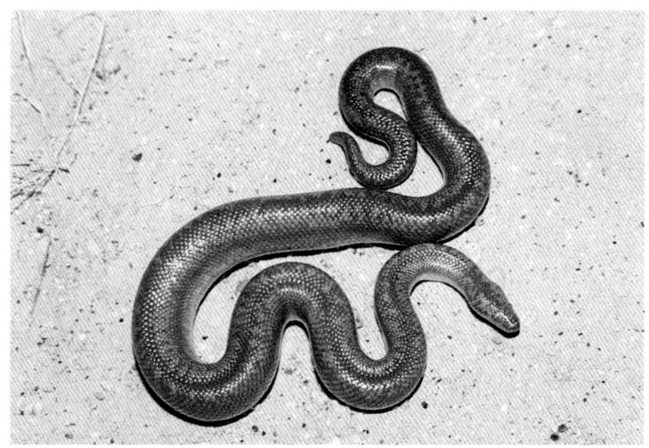

FIGURE 6.3 Roseofusca pattern class of *Lichanura trivirgata* (Rosy Boa) from Valle la Trinidad, BC

Populations of Saslowi intergrade with Trivirgata through the Sierra la Libertad from at least the vicinity of Arroyo San Regis to El Arco (Ottley et al. 1980). Those from Arroyo San Regis tend to have burnt-orange stripes. Trivirgata range south through the rest of the peninsula and probably occur in the Vizcaíno Peninsula (Grismer et al. 1994). They have a tan to dull white ground color with well-defined dark brown to lighter chocolate brown stripes and a light brown iris. Boas from Isla San Marcos have silver irises. Some individuals from San Ignacio have noticeably wider stripes. Trivirgata from the Magdalena Plain from at least Ciudad Constitución to Puerto San Carlos have a somewhat blue-gray cast to the ground color and thinner stripes.

The Isla Cedros population of Trivirgata was originally described as the subspecies *Lichanura trivirgata bostici* (Ottley 1978). Spiteri (1992, 1994) demonstrated with additional specimens, however, that the characteristics of narrow stripes, yellow ventrum, and high lateral stripes were

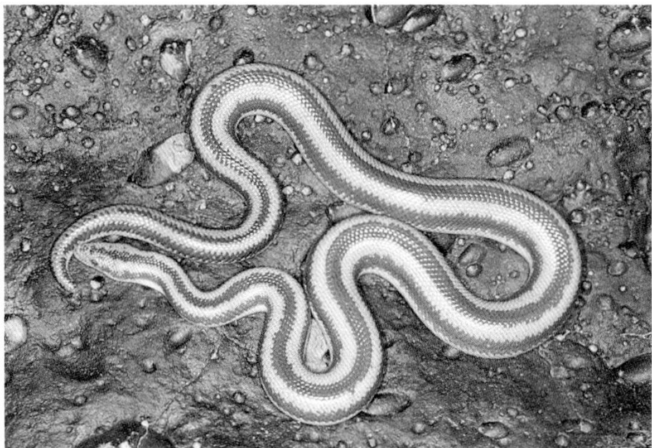

FIGURE 6.4 Saslowi pattern class of *Lichanura trivirgata* (Rosy Boa) from Cañón Santa Isabel, BC

FIGURE 6.5 Trivirgata pattern class of *Lichanura trivirgata* (Rosy Boa) from Cumbre de San Pedro, Sierra Guadalupe, BCS

too variable to be of taxonomic significance. Young Isla Cedros boas do have a considerable amount of yellow in the interspaces between the stripes, but so do some peninsular and mainland populations. Additionally, Isla Cedros boas have noticeably narrower stripes, but the range of variation overlaps the variation observed in boas from adjacent peninsular populations.

NATURAL HISTORY *Lichanura trivirgata* ranges through all the phytogeographic regions of Baja California except for the Baja Coniferous Forest and Sierra la Laguna regions. It is most commonly found near rocks, and I have observed several specimens at night on rocky volcanic slopes along the east face of the Sierra Juárez and in the Sierra las Pintas, Arroyo de Rumbaugh west of Bahía San Luis Gonzaga, 20 km west of Bahía de los Ángeles, Sierra la Libertad, and San Ignacio. Bostic (1971) reports finding a specimen in a grain field 30 km southeast of El Rosario at Rancho San Vicentito, and I found a specimen on the road between agricultural fields 21 km east of Ciudad Constitución. I have found others west of Ciudad Constitución on the open, sandy Magdalena Plain along the road to Bahía Magdalena. Welsh (1988) reports *L. trivirgata* as being common along the western slopes of the Sierra San Pedro Mártir, and I have seen boas on the edge of the pine belt east of Rancho San José and all through Valle la Trinidad to the desert floor east of San Matías.

Lichanura trivirgata is active from approximately mid-March through October, and into November in the Cape Region. Boas are by far most commonly observed during June and early July. During late January and February, specimens are often found during the day beneath small rocks. During April and May, it is not uncommon to see boas out during the late morning and late afternoon, crawling among rocks. During the much warmer summer, *L. trivirgata* delays its activity until the evening, spending the daylight hours beneath rocks or within rodent burrows. At this time of the year, boas are most commonly seen foraging at dusk and early evening. If daytime temperatures are cool as a result of overcast, boas can be found abroad during summer. I even found a specimen at 0930 hours above 1500 m near the top of the Sierra Guadalupe that was abroad during periods of wind and light rain. Unlike other nocturnal desert reptiles, *L. trivirgata* is relatively common during half-moon to full-moon phases.

Lichanura trivirgata may feed almost exclusively on small mammals (Klauber 1933; Merker and Merker 1995; Rodríguez-Robles, Bell, and Greene 1999a), and Klauber reports dual constriction (constricting more than one prey item at a time) in a captive specimen. *Lichanura trivirgata* breeds early in the year, from March through early May. Two to eight live young are born from August through early October.

FOLKLORE AND HUMAN USE Although not well understood by the Seri Indians, the Rosy Boa is referred to as *haas ano cocázni,* which loosely translates to "thing living among mesquite like a rattlesnake" (Nabhan, forthcoming).

Colubridae

GLOSSY SNAKES, RAT SNAKES, SAND SNAKES, SHOVEL-NOSED SNAKES, RING-NECKED SNAKES, NIGHT SNAKES, KINGSNAKES, WHIPSNAKES, LEAF-NOSED SNAKES, GOPHER SNAKES, LONG-NOSED SNAKES, PATCH-NOSED SNAKES, GROUND SNAKES, BLACK-HEADED SNAKES, GARTER SNAKES, LYRE SNAKES, AND THEIR OLD WORLD RELATIVES

The Colubridae are a very large, diverse group of advanced snakes containing approximately 320 genera and over 1,700 species, or nearly two-thirds of the world's snakes (Pough et al. 2001). This group cannot be characterized by any unique characteristics, and it is certainly not a natural assemblage but rather an amalgam of species whose inclusions are the result of convenience rather than closeness of relationship. Although colubrids are cosmopolitan in distribution, they are noticeably absent from central sub-Saharan Africa and most of Australia. As would be expected of such a large assemblage of species, colubrids have extremely diverse lifestyles and are found in virtually every kind of habitat. In Baja California and the Gulf of California, colubrids are generally terrestrial, arid-adapted species that are often specialized for surviving in a particular microhabitat or consuming a particular type of prey, or both. In the region of study, 16 genera are represented by 33 species.

Arizona Kennicott

GLOSSY SNAKES

The genus *Arizona* contains two species that collectively range throughout most of southwestern United States and northern Mexico (Dixon and Fleet 1976). *Arizona* are moderately sized terrestrial snakes with a generally faded color pattern and smooth, shiny scales. They are handsome, docile species that may be active during cool evenings, long after other snakes have retreated for the night. Glossy snakes are common in arid regions and abundant in areas of loose soil. Both species of glossy snake *(A. elegans* and *A. pacata)* occur in the region of study; *A. pacata* is endemic to southern Baja California.

FIG 6.6

Arizona elegans

GLOSSY SNAKE; CULEBRA

Arizona elegans Kennicott (in Baird) 1859b:18. Type locality: "Rio Grande," restricted to "lower Rio Grande" by Yarrow 1883. Lectotype: USNM 1722 (Blanchard 1924:4)

IDENTIFICATION *Arizona elegans* can be distinguished from all other snakes in the region of study by having large head plates; an undivided anal plate; smooth dorsal scales; a rostral scale that is not greatly enlarged; round pupils; and 51 to 83 short, transverse body blotches.

RELATIONSHIPS AND TAXONOMY *Arizona elegans* is presumably the sister species of *Arizona pacata* of southern Baja California. Together they are most closely related to a large natural group that contains kingsnakes *(Lampropeltis),* long-nosed snakes *(Rhinocheilus),* scarlet snakes *(Cnemophora),* short-tailed snakes *(Stilosoma)* (Keogh 1996), and perhaps rat snakes *(Elaphe* and *Bogertophis)* (Rodríguez-Robles and Jesús-Escobar 1999). Klauber (1946a) discusses taxonomy.

Some researchers (e.g., Liner [1994]) consider the western populations of *Arizona* to be *A. occidentalis* rather than *A. elegans.* The rationale for this supposition, however, has yet to be expressed. Therefore, *A. elegans* is recognized here.

FIGURE 6.6 Eburnata pattern class of *Arizona elegans* (Glossy Snake) from 10 km south of San Felipe, BC

DISTRIBUTION *Arizona elegans* ranges throughout much of the southwestern United States and northern Mexico south into northern Baja California (Dixon and Fleet 1976). In northern Baja California, it is absent from the upper elevations of the Sierra Juárez and Sierra San Pedro Mártir but extends south at least to 8 km north of the turnoff to Bahía de los Ángeles (Seifert 1980) on the west side of the Peninsular Ranges, and at least to 23 km south of San Felipe (Grismer 1994d) on the desert side. Undoubtedly it extends along the east coast at least as far south as Puertecitos in the northeast (map 6.3).

DESCRIPTION Body elongate, rounded above, and truncate at its base in cross section, reaching 1,000 mm SVL; head slightly distinct from neck; pupils round; tail relatively short; rostral triangular; head scales large and platelike; 6 to 8 (usually 8) supralabials; genials contact medially; 11 to 15 (usually 13) infralabials; dorsal body scales smooth and imbricate, 27 scale rows at midbody; 207 to 233 ventral scales in males, 215 to 241 in females; anal undivided; 47 to 54 undivided subcaudals in males, 43 to 50 in females. **COLORATION** Dorsal ground color light green to dull white; dark interorbital bar on head; dark postorbital stripe present; often a dark line on medial edges of occipitals; one pair of dark, elongate paravertebral blotches on nape; 51 to 83 short, dark transverse blotches on body; blotches often divided medially and offset; dark lateral blotches may or may not extend ventrally to contact edges of ventral scales; scales on sides of body dark centrally, making scale edges appear to be outlined in a lighter color; vertebral scales lack darkened centers, giving appearance of a wide, poorly defined vertebral stripe; ventrum usually immaculate.

GEOGRAPHIC VARIATION Notable geographic variation of color pattern in *A. elegans* from the southwest led Klauber (1946a) to designate two subspecies *(A. e. eburnata* and *A.*

e. occidentalis) that extend into northern Baja California and are considered here as pattern classes. Occidentalis are cismontane in distribution, ranging ubiquitously throughout the California Region south into the Vizcaíno Region, to at least 8 km north of the turnoff to Bahía de los Ángeles (Seifert 1980). Occidentalis are generally much darker overall than Eburnata, and the lateral body blotches extend ventrally to make contact with the edges of the ventral scales. Occidentalis from the vicinity of San Quintín north tend to have a light, olive drab overall cast to their ground color and dorsal blotches. Farther south, the ground color lightens, and the blotches tend to be more grayish, although snakes from Cerro Prieto, a large volcanic region, have dark brown dorsal blotches. Populations of Occidentalis intergrade with Eburnata through Paso de San Matías and south of Puertecitos, where snakes are intermediate in coloration and pattern between the two pattern classes.

Eburnata have a light ground color ranging from beige to dull white. The lateral body blotches are narrow and do not extend ventrally to make contact with the edges of the ventral scales. Eburnata occur in the Lower Colorado Valley Region, extending from the U.S.-Mexican border south to at least 23 km south of San Felipe along the east coast,

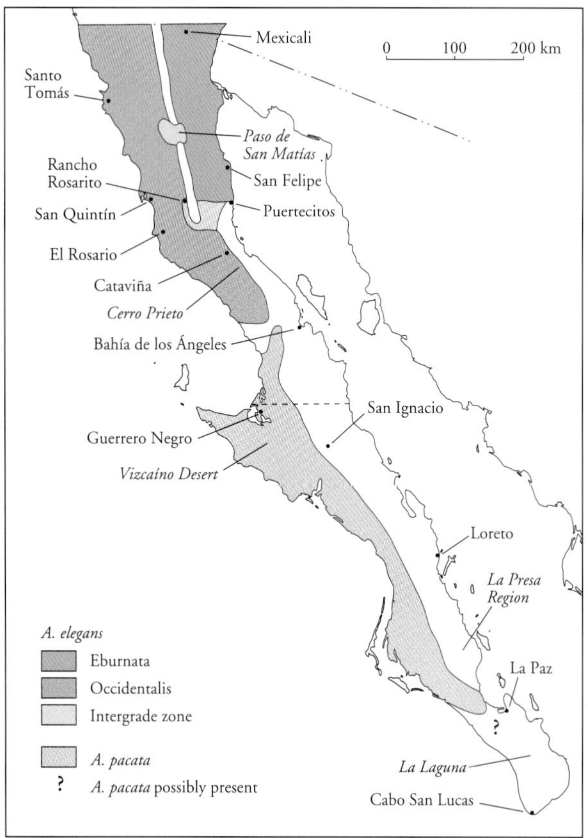

MAP 6.3 Distribution of *Arizona pacata* (Baja California Glossy Snake) and the pattern classes of *A. elegans* (Glossy Snake)

but probably farther south to at least Puertecitos. Welsh (1988) reported on a specimen from Rancho Rosarito on the Pacific side of the Peninsular Ranges. This area lies at the southern end of a narrow valley formed between the Sierra San Miguel and Sierra San Pedro Mártir, whose locally hot, arid climate has promoted the northerly advance of southern Vizcaíno Region species and the invasion of desert species around the southern end of the Sierra San Pedro Mártir (Grismer 1997).

NATURAL HISTORY *Arizona elegans* is commonly found in coastal scrub and foothill chaparral areas of the California Region of northwestern Baja California, extending south into the more arid scrublands of the Vizcaíno Desert. *Arizona elegans* is also a common inhabitant of the Creosote Bush *(Larrea tridentata)* and Bursage *(Ambrosia dumosa)* flats of the extremely hot and arid Lower Colorado Valley Region of northeastern Baja California. Glossy Snakes, however, are absent from the intervening mountainous Baja California Coniferous Forest Region (Welsh 1988). They are most common in open areas where the soil is loose, and they are rarely found on hardpan or rocky slopes. Glossy Snakes are active from late February through November, depending on the weather. This species is generally nocturnal and can be found on very cool nights (air temperature 20°C; Klauber 1946a) long after other nocturnal reptiles have retreated. Several authors have reported observing *A. elegans* abroad during daylight hours (see Brattstrom 1952). Linsdale (1932) notes finding a specimen at Santo Tomás at 0600, just as the sun was rising. However, dietary studies by Rodríguez-Robles, Bell, and Greene (1999b) indicate that *A. elegans* is primarily nocturnal and feeds mostly on sleeping phrynosomatid lizards and nocturnal rodents. Larger specimens include birds and occasionally snakes in their diet (Rodríguez-Robles, Bell, and Greene 1999b). Linsdale (1932) reports finding a specimen consuming a *"Sceloporus."* Glossy Snakes subdue their prey by constriction. Reynolds (1943) reports an Occidentalis laying three eggs in late July, and Klauber (1946a) reports a female as having laid seven eggs. Cowles and Bogert (1944) observed an Eburnata that laid 23 eggs in early July. These dates indicate that *A. elegans* has a spring breeding season.

FIG 6.7

Arizona pacata

BAJA CALIFORNIA GLOSSY SNAKE; CULEBRA

Arizona pacata Klauber 1946a:379. Type locality: "Santo Domingo (lat. 25° 30' N.), Baja California [Sur], Mexico." Holotype: SDSNH 17652

IDENTIFICATION *Arizona pacata* can be distinguished from all other snakes in the region of study by having large head

FIGURE 6.7 *Arizona pacata* (Baja California Glossy Snake) from 20 km south of Guerrero Negro, BCS

plates; an undivided anal plate; smooth dorsal scales; a rostral scale that is not greatly enlarged; round pupils; and 36 to 41 elongate, subcircular body blotches.

RELATIONSHIPS AND TAXONOMY Seifert (1980) reported *Arizona elegans occidentalis* from 20.9 km and 8 km north of the turnoff to Bahía de los Ángeles. Seifert (1980) indicated that their dorsal blotch counts showed no signs of intergradation with *A. e. pacata,* which he reported from "only a few kilometers south and west" of these records. I have found *A. e. pacata* 11 km west of Mexican Highway 1 on the road to Bahía de los Ángeles (LSUPCS-0233); this specimen also showed no signs of intergradation with *A. e. occidentalis.* This closes the gap between these two subspecies to a straight-line distance of less than 10 km through continuous habitat. I therefore hypothesize that these are not subspecies but different species, as did Klauber (1946a). I believe this zone represents the southern extent of *A. elegans* and the northern extent of *A. pacata.* This area is relatively cool and lies within a north-to-south-tending valley that opens onto the Pacific coast through the Vizcaíno Desert. As such, it is cloaked in advectional morning fogs for much of the spring, summer, and fall; these burn off and give way to relatively hot, arid days. This climatic regime apparently provides sufficient moisture and cool enough temperatures to support *A. pacata,* which is adapted to the cool Pacific coast of southern Baja California (see below), as well as enough aridity to marginally support *A. elegans.*

DISTRIBUTION *Arizona pacata* ranges along the Pacific coast of the southern two-thirds of Baja California from at least the turnoff to Bahía de los Ángeles south to 20 km north of La Paz (Reynoso 1990d; see map 6.3).

DESCRIPTION Same as *A. elegans,* except for the following: adults reach 1,100 mm SVL; 6 to 8 supralabials; 10 to 15 infralabials; 199 to 229 ventral scales in males, 217 to 245 in females; 41 to 49 undivided subcaudals in males, 43 to 50 in females. **COLORATION** Dorsal ground color beige to dull white; 36 elongate, dark subcircular blotches on body; dark lateral blotches do not extend ventrally to contact edges of ventral scales; vertebral scales and scales on sides of body have darkened centers; vertebral scales only slightly lighter than lateral scales; wide, light-colored vertebral stripe absent.

GEOGRAPHIC VARIATION Geographic variation in *A. pacata* is minimal. Individuals from north of the Vizcaíno Desert tend to have a lighter, more whitish ground color than snakes from the south. Glossy Snakes from the Vizcaíno Desert have a beige ground color matching the substrate, whereas snakes from the Magdalena Plain tend to have a dull white ground color.

NATURAL HISTORY *Arizona pacata* inhabits the cool Pacific coastal desert of western Baja California, which encompasses the Vizcaíno Desert in the north and the Magdalena Plain in the south. It does extend north of the Vizcaíno Desert to at least the turnoff to Bahía de los Ángeles. Here the vegetation is cloaked with epiphytes, which indicates a cool, moist climate. Like other glossy snakes, *A. pacata* is most commonly found in areas with open vegetation and loose soil. It is particularly abundant in the Vizcaíno Desert north of Guerrero Negro.

Virtually nothing is known about the natural history of *A. pacata.* Banta and Leviton (1963) reported collecting a specimen in early April from the Vizcaíno Desert, and I have observed the most specimens throughout its distribution during April. The earliest I have seen specimens abroad is mid-March, just south of Guerrero Negro. Reynoso (1990d) collected a specimen during late October from the Isthmus of La Paz. Nothing has been reported concerning this snake's food or reproductive habits.

Bogertophis Dowling and Price

DESERT RAT SNAKES

Snakes of the genus *Bogertophis* are moderately sized terrestrial species occupying arid regions of the southwestern United States and northern Mexico. *Bogertophis* are modestly elongate and beautifully patterned, with handsomely shaped, distinctive squarish heads and large protuberant eyes. Their large size places them as one of the ecosystem's top predators, and therefore they are not nearly as abundant as many of the other smaller colubrid snakes with which they are sympatric. Currently, *Bogertophis* contains two disjunct species: *B. subocularis* of the Chihuahuan Desert of the south-central United States and northern

Mexico, and *B. rosaliae* of the Peninsular Ranges of southern California and Baja California (Dowling and Price 1988).

FIG 6.8 **_Bogertophis rosaliae_**

BAJA CALIFORNIA RAT SNAKE; RATONERA

Coluber rosaliae Mocquard 1899:321. Type locality: "Santa Rosalia, Distrito Sur, Baja California," BCS, Mexico. Holotype: MHNP 92-438

IDENTIFICATION *Bogertophis rosaliae* can be distinguished from all other snakes in the region of study by having large protuberant eyes; a row of suboculars (loriolabials); and a nearly unicolored brownish adult ground color (faintly banded in juveniles).

RELATIONSHIPS AND TAXONOMY *Bogertophis rosaliae* is the sister species of *B. subocularis* (Dowling and Price 1988) of the Chihuahuan Desert.

DISTRIBUTION *Bogertophis rosaliae* ranges throughout the rocky slopes of the Peninsular Ranges from Mountain Springs, San Diego County, California, south to Cabo San Lucas (map 6.4). North of Cataviña, *B. rosaliae* is restricted to the eastern desert foothills of the Peninsular Ranges, but in the vicinity of Cataviña it spreads west, through suitable habitat, across the low, mountainous areas on the Pacific side of the Peninsular Ranges. *Bogertophis rosaliae* is not ubiquitous in southern Baja California, as is claimed by Dowling and Price (1988) and Price (1990). It occurs on the Gulf island of Danzante.

DESCRIPTION Body elongate, rounded above, and truncate in cross section, reaching over 1,500 mm SVL; head moderately distinct from neck; eyes protuberant, pupils round; 3 to 6 suboculars (loriolabials); tail relatively short, 19 percent of SVL; rostral triangular; head scales large and platelike; 10 or 11 supralabials; genials contact medially; 12 to 15 infralabials; dorsal body scales relatively small and imbricate; 0 to 9 vertebral rows, keeled; 31 to 35 scale rows at midbody; 271 to 287 ventral scales; anal divided; 79 to 94 divided subcaudals. **COLORATION** ADULTS: Dorsal ground color brownish; dark skin between scales often visible, giving scales a dark-edged appearance; ventrum usually immaculate. HATCHLINGS AND JUVENILES: Transversely arranged banding pattern composed of thin, light-colored lines.

GEOGRAPHIC VARIATION *Bogertophis rosaliae* from the Cape Region and areas with dark volcanic rock, such as San Ignacio, tend to be darker in overall coloration, with the dorsal scales more boldly outlined. Within the Cape Region, snakes from some populations, such as those from Pichilingue and San Antonio, tend to have an overall greenish hue. Specimens along the Gulf coast from Loreto to Mulegé are generally straw-colored overall, with less outlining of the scales. Rat snakes to the west, and above Loreto in the Sierra la Giganta, are almost mustard yellow.

FIGURE 6.8 *Bogertophis rosaliae* (Baja California Rat Snake) from Santa Rosalía, BCS

NATURAL HISTORY *Bogertophis rosaliae* ranges throughout a number of phytogeographic regions along the Peninsular Ranges, extending from the hot, arid Lower Colorado Valley Region in the north to the Arid Tropical Region at the southern tip of Baja California. Within these areas, Rat Snakes are associated with rocky habitats. Price (1990) and others have generally indicated that this species is restricted to, or closely associated with, mesic areas such as oases and palm groves. This appears to be the case only because of the association of humans with such areas; it is from these areas that the majority of specimens have been collected. I have found *B. rosaliae* to be equally common throughout its range in regions of open thorn scrub where granitic or volcanic rocks are abundant. Additionally, Murphy and Ottley (1984) reported two specimens taken from the arid northern end of Isla Danzante. Therefore I do not believe this species has any special affinity for palm groves.

Bogertophis rosaliae is active from at least late February through October in the northerly portions of its range. I have observed adults abroad in the Cape Region during December, which indicates that it is active year-round at this latitude. *Bogertophis rosaliae* is both diurnal and nocturnal, depending on daytime temperatures. Hunsaker (1965) observed two individuals crawling in direct sunlight in Cañón Guadalupe, approximately 60 km south of the U.S.-Mexican border along the eastern face of the Sierra Juárez, during early March, and Dowling (1957) observed a specimen abroad during the day in early July at San

Bartolo, which Ottley and Jacobsen (1983) suggested was crawling in the shade. Ottley and Jacobsen reported on a specimen abroad during the day in the shade of a palm grove at San José Comondú. I have observed specimens during the day in the shade of palm groves at San Ignacio and Mulegé during April, June, and August. This species is also common at night, and the majority of specimens collected have been taken during evenings from April through October.

Bogertophis rosaliae is a large constricting snake that feeds primarily on small rodents and birds. I found scales of Granite Spiny Lizards *(Sceloporus orcutti)* in the scat of a wild-caught juvenile. Little is known about the reproductive biology of *B. rosaliae* other than information obtained from illegally acquired captive populations. Ottley and Jacobsen (1983) reported on a gravid female collected in late May and on hatchlings and juveniles being found during June through late October. I observed a small juvenile during late June at Cataviña. These observations may indicate that wild *B. rosaliae* breeds from early spring through early summer, with hatching from at least early June through late October.

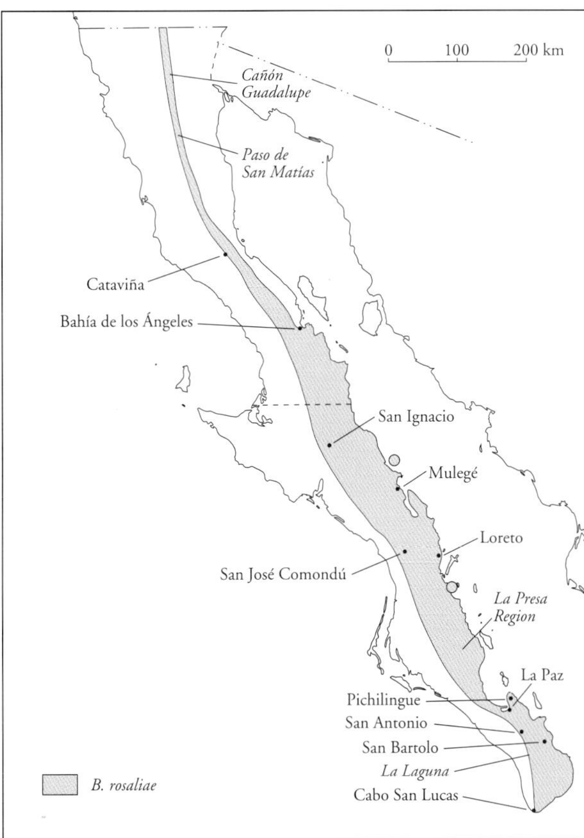

MAP 6.4 Distribution of *Bogertophis rosaliae* (Baja California Rat Snake)

Chilomeniscus Cope

SAND SNAKES

The genus *Chilomeniscus* contains two species of semifossorial, arenicolous snakes, *C. stramineus* and *C. savagei,* both of which occur in the region of study (Grismer et al. 2002). *Chilomeniscus* is closely related to the shovel-nosed snakes *(Chionactis)* and the ground snakes *(Sonora),* and like them exhibits extensive color pattern variation. Sand snakes are common in areas of loose soil and an important component of many dune ecosystems. Their specialized burrowing morphology of a countersunk lower jaw, valved nostrils, and smooth scales aids them in swimming both above and below the surface of the sand. Sand snakes are nocturnal and are usually not observed unless actively sought out. However, their presence is obvious during the early morning hours, before midday winds blow away the numerous tracks left on the dunes the night before.

FIGS. 6.9–11

Chilomeniscus stramineus

SAND SNAKE; CULEBRITA DE LA ARENA, CORALILLO

Chilomeniscus stramineus Cope 1860:339. Type locality: "Cape St. Lucas, Lower California [BCS]," Mexico. Holotype: USNM 4674

IDENTIFICATION Can be distinguished from all other snakes in the region of study by having a sharply pointed rostral scale; deeply countersunk lower jaw; no loreal scales; and prefrontal scales in contact with one another.

RELATIONSHIPS AND TAXONOMY Much of the previous taxonomy of *Chilomeniscus* was based on color pattern variation (Banta and Leviton 1963), and most of this centered on the banded *(C. cinctus, C. punctatissimus,* and *C. savagei)* and unbanded *(C. stramineus)* patterns. Grismer et al. (2002) demonstrated that nearly all color pattern types, as well as intermediate forms, could actually be found at many localities throughout the range of the genus, which indicates that these variations were in part a color pattern polymorphism of just a single species, *C. stramineus.* Additionally, Grismer et al. reported on an ontogenetic series from a single beach on Isla Espíritu Santo where the hatchlings were banded, and the bands faded and broke up with increasing SVL into an unbanded blotched pattern.

DISTRIBUTION *Chilomeniscus stramineus* occurs in two disjunct populations. The eastern population ranges through most of southwestern Arizona southward, west of the Sierra Madre Occidental of Sonora, Mexico, to northern Sinaloa. The western population is restricted to Baja Cali-

FIGURE 6.9 *Chilomeniscus stramineus* (Sand Snake) from 40 km west of Santa Rosalía, BCS

fornia and occurs from Valle la Trinidad and Arroyo San Antonio west of the Peninsular Ranges south to Cabo San Lucas. It is absent from the Lower Colorado Valley Region, where it is presumably replaced by *Chionactis occipitalis* (Grismer 1997) and from the Sierra la Laguna Region (Alvarez et al. 1988). Its northernmost record along the Gulf coast is at Bahía de los Ángeles. *Chilomeniscus stramineus* occurs on the Pacific islands of Cedros and Magdalena (Wong 1997) and the Gulf islands of Danzante, Espíritu Santo, Monserrate, Partida Sur, San José, San Marcos, and Tiburón (map 6.5). Powers and Banta (1974) report a specimen being collected from Isla Cerralvo, which Murphy and Ottley (1984) and Grismer et al. (2002) considered erroneous.

DESCRIPTION Body relatively stout and subcircular in cross section, reaching 229 mm SVL; head small, flat, sharply pointed in lateral view, and not distinct from neck; eyes small, pupils round; lower jaw deeply countersunk; tail moderate, 13 to 20.2 percent of SVL in males, 11 to 15 percent in females; head scales large and platelike; rostral greatly enlarged and pointed in lateral profile; internasals separated by rostral; prefrontals large, contacting medially; 5 to 8 (usually 7) supralabials; genials contact medially; 6 to 10 (usually 8) infralabials; dorsal body scales smooth and imbricate; 12 to 14 (usually 13) scale rows at midbody; 104 to 118 ventral scales in males, 114 to 127 in females; anal plate divided; 23 to 31 divided subcaudals in males, 22 to 28 in females. **COLORATION** Extremely variable (see geographic variation section). Ground color ranges from cream to dark brown; dark bands range from 1 to 4 dorsal scales in width or may not be present; number of bands on body and tail ranges from 17 to 46, with first band encompassing back of head; interspaces between bands may be immaculate or blotched and reddish to cream-colored; unbanded forms may be blotched or have a single black apical spot on the dorsal scales in rows 3 to 11.

GEOGRAPHIC VARIATION Variation in color pattern of *C. stramineus* is extensive (Grismer et al. 2002) and is discussed here from north to south. From the California Region, Sand Snakes have unbroken bands 3 or 4 scale rows wide, with no blotching in the interspaces between the bands. In the Vizcaíno Region, between El Rosario and San Ignacio, many *C. stramineus* have incomplete, broken bands, and near San Ignacio, brownish blotches appear in the interspaces. This last color pattern predominates to the south in the interior of the Magdalena Region, east of the Magdalena Plain, and may be a response to living on a darker volcanic substrate. On the lighter-colored, sandy Magdalena Plain, the color pattern becomes variable once again, and many snakes, especially in coastal areas, have a blotched color pattern or a combination of blotches and broken bands. Although no unbanded forms occur in the inland portions of the Magdalena Region, they do occur on the adjacent Gulf coast. Along the Gulf coast from Santa Rosalía to Misión Dolores in the La Presa region, the color pattern is highly variable, ranging from blotched, unbanded forms to solid, banded forms with immaculate in-

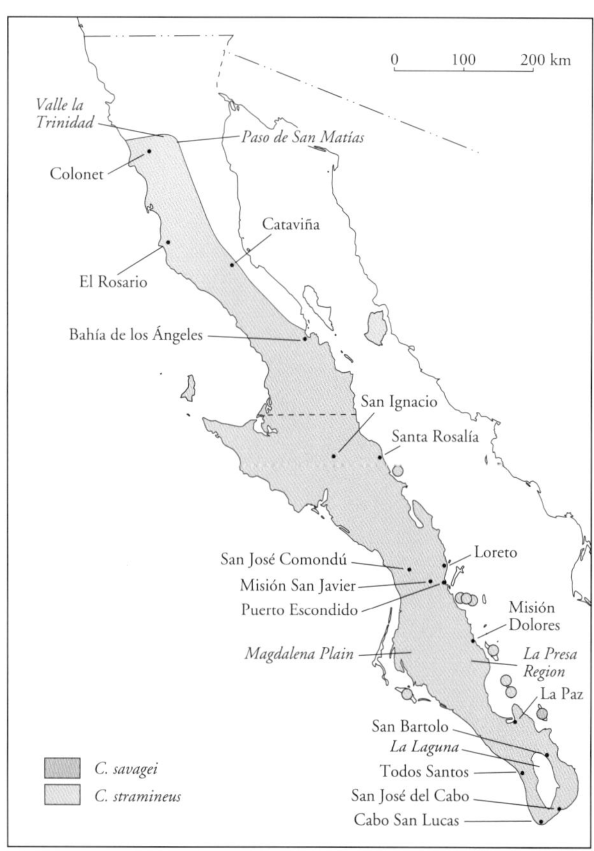

MAP 6.5 Distribution of *Chilomeniscus* (Sand Snakes)

FIGURE 6.10 *Chilomeniscus stramineus* (Sand Snake) from Isla Espíritu Santo, BCS

FIGURE 6.11 *Chilomeniscus stramineus* (Sand Snake) from Isla Tiburón, Sonora

terspaces. North of the Isthmus of La Paz, there is a correlation of unbanded forms with coastal dune systems. However, in the northern and western portion of the Cape Region, generally west of the Sierra la Laguna south to Todos Santos, the color pattern remains variable but is dominated by unbanded forms with varying degrees of blotching. Southeast of here, the color pattern stabilizes and is usually nearly unicolored, with one small, dark apical spot on the scales of rows 3 to 11 (although I did find a specimen with orangish bands at San José del Cabo). Sand snakes here usually have darkened postorbital regions, and some may have a dark brown rather than cream ground color. *Chilomeniscus stramineus* from insular populations are generally similar in color pattern to snakes of the adjacent peninsula, and often both banded and unbanded forms occur on the same island (e.g., Islas Espíritu Santo and Tiburón; Grismer et al. 2002).

NATURAL HISTORY *Chilomeniscus stramineus* is widely distributed throughout Baja California and occurs in every major phytogeographic region except the Lower Colorado Valley Region and Baja California Coniferous Forest Region of the north and the Sierra la Laguna Region of the Cape (Alvarez et al. 1988; Grismer 1997; Welsh 1988). Within these varied habitats, *C. stramineus* is commonly associated with loose soils even in rocky terrain (see Bostic 1971; Leviton and Banta 1964; Mosauer 1936). As such, *C. stramineus* is often found in arroyo bottoms, open sandy desert areas, and coastal dune systems. The dune systems are the easiest place to observe Sand Snakes. Their tracks crisscross the open areas between shrubs, and in some places tracks literally cover the dunes. Here Sand Snakes take refuge at the base of shrubs, presumably within the root systems. *Chilomeniscus stramineus* can also be found beneath surface objects, such as rocks, wood, and other debris (Leviton and Banta 1964; Van Denburgh and Slevin 1921a). However, *C. stramineus* is not restricted to habitats with loose soil. I have found specimens high up on the sides of the boulder-covered hills surrounding San Bartolo in the Cape Region, and I even found one individual approximately 1 m above the ground, deep in a rock crack, in a lava flow near San José Comondú.

North of San Ignacio, *C. stramineus* is active from at least March through early October. In the southerly portion of its range, it may be active year-round, especially in the Cape Region (Leviton and Banta 1964), with peak activity from mid-March to late August. I have seen active specimens or fresh tracks on Isla Tiburón from mid-March through early October. *Chilomeniscus stramineus* is generally nocturnal, but I have seen individuals abroad in shaded areas during the late afternoon in early April near Puerto Escondido, just south of Loreto. At night, *C. stramineus* can be active when temperatures are very low and most other reptiles are inactive. I have seen specimens abroad in central Baja California at ambient temperatures of 17°C.

Nothing has been reported on the reproductive biology of *C. stramineus* from the region of study, and there is little to add here. Four hatchlings were found one night during late August on Isla Espíritu Santo, and I have seen hatchlings beneath rocks at San Ignacio and Misión San Javier during early to mid-January. These observations may indicate that *C. stramineus* has a mid- to late summer breeding season.

Nothing has been reported on the diet of *C. stramineus* from the region of study. I have commonly found sand snakes at the base of bushes on Isla Espíritu Santo that contained ant and termite nests. Given that ants compose part of the diet of its close relative *C. savagei* from Isla Cerralvo

(Banks and Farmer 1963), I strongly suspect that these individuals were using these nests as a food source.

FOLKLORE AND HUMAN USE The Seri Indians show great affection for this snake. According to Nabhan (forthcoming) its name, *hapéquet camízj,* loosely translates to "sleek baby doll" or "smooth-skinned darling." The Seri believe they must protect this species from the predatory Western Coral Snakes *(Micruroides euryxanthus)* and uncaring humans. Nabhan was told that to ensure a baby will be born with lovely skin, "a Sand Snake is captured, and passed across the belly or the small of the back of the expectant mother. This ritual, performed less and less frequently, is intended to give pregnant women hopeful feelings and good luck."

FIG 6.12 ***Chilomeniscus savagei***

ISLA CERRALVO SAND SNAKE; CULEBRITA DE LA ARENA, CORALILLO

Chilomeniscus savagei Cliff 1954a:71. Type locality: "Southwest coast of Cerralvo Island," BCS, Mexico. Holotype: CAS-SU 14031

IDENTIFICATION *Chilomeniscus savagei* can be distinguished from all other snakes in the region of study by having a sharply pointed rostral scale; deeply countersunk lower jaw; no loreal scales; and prefrontal scales that do not contact one another.

RELATIONSHIPS AND TAXONOMY See *C. stramineus.*

DISTRIBUTION *Chilomeniscus savagei* is endemic to the Gulf island of Cerralvo (see map 6.5).

DESCRIPTION Same as *C. stramineus,* except for the following: tail 13 to 14 percent of SVL in males, 12 to 15 percent in females; internasals widely contacting enlarged frontal scale; prefrontals small, separated by frontal; 6 or 7 (usually 7) supralabials; 4 to 8 (usually 8) infralabials; 13 scale rows at midbody; 127 to 129 ventral scales in males, 130 to 135 in females; 27 to 28 divided subcaudals in males, 24 to 27 in females. **COLORATION** Dorsal ground color gray; 29 to 36 wide black bands on body and tail, first band on back of head; dorsal portion of interspaces between bands orangish; juveniles tend to have a more vibrant color pattern.

NATURAL HISTORY Although Isla Cerralvo lies within the Central Gulf Coast Region, a large portion of its flora consists of Arid Tropical Region species. As such, Isla Cerralvo is generally more vegetated than the other Gulf islands. The distribution of *C. savagei* is not well known, and it has actually been reported from only three localities at the south end of the island. Here, there is a significant alluvial outflow from a large arroyo, which, in conjunction with onshore winds and tides, has created a small series of dunes and sandy areas. The majority of the specimens collected have been from this southern beach (Banks and Farmer 1963; Etheridge 1961), and their tracks cover the dunes in some places. Banks and Farmer report another small series of dunes, approximately 1 km up the coast, where the tracks of *C. savagei* have also been found. They state that there are no sandy corridors between these dune systems and that they are separated by precipitous, rocky bluffs. Banks and Farmer did find specimens inland in the arroyo bottom in leaf litter along the side of a cliff, and I have found individuals in the same sort of microhabitat in the same arroyo. In the sandy areas, I have found *C. savagei* to be most common at the base of vegetation. However, given that *C. stramineus* occur on the rocky hillsides at San Bartolo in the adjacent Cape Region, *C. savagei* may be more widely distributed on Isla Cerralvo than originally believed.

FIGURE 6.12 *Chilomeniscus savagei* (Isla Cerralvo Sand Snake) from Arroyo Viejos, Isla Cerralvo, BCS

I have observed *C. savagei* to be active from at least late March to late August, but it is likely that this species is active year-round. Banks and Farmer (1963) report finding ants, scorpions, and other chitinous material in the stomachs of the specimens they collected. They also found grains of sand, which they believe were incidentally ingested.

Chionactis Hallowell

SHOVEL-NOSED SNAKES

The genus *Chionactis* contains two arenicolous species whose entire range is contained within the Mojave Desert and the northern portion of the Sonoran Desert (Stebbins 1985). *Chionactis* specializes in eating arthropods and is generally restricted to sandy desert basins. It possesses a number of remarkable morphological and behavioral adaptations that suit its burrowing lifestyle (Cross 1979). Only one species, *C. occipitalis,* occurs in the region of study.

FIG 6.13 **Chionactis occipitalis**

WESTERN SHOVEL-NOSED SNAKE; CULEBRA DE LA ARENA

Rhinostoma occipitale Hallowell 1854:95. Type locality: "Mojave Desert," California, USA. Holotype: USNM 2105.

IDENTIFICATION *Chionactis occipitalis* can be distinguished from all other snakes in the region of study by having a sharply pointed rostral scale; deeply countersunk lower jaw; loreal scales; and internasals that contact medially.

RELATIONSHIPS AND TAXONOMY *Chionactis occipitalis* is the sister species of *C. palarostris* of southern Arizona and northern Mexico (Klauber 1951). Klauber (1951) discusses taxonomy.

DISTRIBUTION *Chionactis occipitalis* ranges throughout the Mojave Desert and extends south through most of the northern portion of the Sonoran Desert of the southwestern United States and northern Mexico (Mahrdt et al. 2001). In Baja California, *C. occipitalis* is restricted to the Lower Colorado Valley Region, extending 34 km south of San Felipe (Grismer 1989d; map 6.6). It is likely that *C. occipitalis* ranges south to at least Puertecitos. Welsh and Bury (1984) indicated that no specimens had been found more than 14.2 km inland from the Gulf of California. I have seen several specimens in Paso de San Matías, and one was within a desert-chaparral ecotone, extending the distribution of this species approximately 58 km to the west.

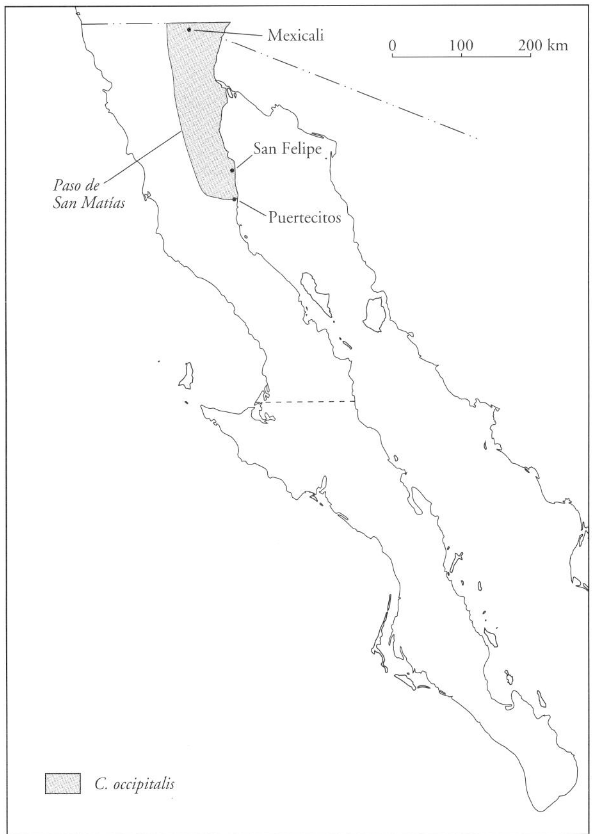

MAP 6.6 Distribution of *Chionactis occipitalis* (Western Shovel-Nosed Snake)

FIGURE 6.13 *Chionactis occipitalis* (Western Shovel-Nosed Snake) from 33 km south of Mexicali, BC

DESCRIPTION Body relatively stout and subcircular in cross section, reaching 369 mm SVL; head small, flat, sharply pointed in lateral view, and not distinct from neck; eyes small, pupils round; lower jaw countersunk; tail moderate, 17.9 to 20.2 percent of SVL in males, 15.7 to 18.4 percent in females; head scales large and platelike; rostral greatly enlarged and pointed in lateral profile; internasals contact medially; 6 or 7 (usually 7) supralabials; genials contact medially; 6 to 8 (usually 7) infralabials; dorsal body scales smooth and imbricate; 15 scale rows at midbody; 143 to 161 ventral scales in males, 155 to 166 in females; anal plate divided; 42 to 48 divided subcaudals in males, 35 to 44 in females. **COLORATION** Ground color ranges from cream to white; black, arc-shaped occipital band extends from posterior corner of one eye to posterior corner of other eye; 20 to 32 black dorsal body bands, usually not as wide as interspaces; posterior bands may completely encircle body, or a black spot may occur on ventrum between ends of bands; interspace mottling occasionally present; orange bands often occupy interspaces.

NATURAL HISTORY Nothing has been reported on the natural history of *C. occipitalis* from Baja California, but presumably it is no different from that of United States populations described by Stebbins (1985). *Chionactis occipitalis* is most common in low-lying, sandy areas dominated by Creosote Bush *(Larrea tridentata)* and Bursage *(Ambrosia dumosa)*. It is not uncommon to find *C. occipitalis* on the open sand dunes at the northern end of Sierra las Pintas or climbing in low vegetation on the Creosote Bush–Bursage flats. *Chionactis occipitalis* is generally nocturnal but is often sighted abroad during the early morning or just before sunset (Mahrdt and Banta 1996). It can occasionally be found abroad on overcast spring days, if the temperature is not too hot. Shovel-nosed snakes eat arthropods and prey mainly on spiders, scorpions, and centipedes. I found a specimen (301 mm SVL) dead on the road near Mexicali that had eaten a solpugid (abdomen first) that was nearly 1.5 times as wide as its own body. Breeding generally occurs during the spring, and 2 to 4 eggs are laid during the summer (Goldberg 1997; Goldberg and Rosen 1999).

Diadophis Baird and Girard

RING-NECKED SNAKES

The genus *Diadophis* contains a single, widely distributed species *(D. punctatus)* that ranges over much of North America. Additional taxonomic work will likely reveal that the western races constitute a separate species *(D. amabilis),* as originally proposed by Blanchard (1942). *Diadophis punctatus* is a semifossorial snake, and although it is common in backyards and gardens in some parts of its range, its lifestyle makes it inconspicuous and seldom seen. Ring-Necked Snakes have the curious behavior of coiling their tails and displaying their bright orange undersides when threatened (Gehlbach 1970). It has been suggested that such behavior is aposematic—that it signals to a potential predator that the snake may possess a noxious quality.

FIG 6.14 **_Diadophis punctatus_**

RING-NECKED SNAKE; CULEBRITA

Coluber punctatus Linnaeus 1766:376. Type locality: California, USA. Holotype: None designated

IDENTIFICATION *Diadophis punctatus* can be distinguished from all other snakes in the region of study by having a long, slender body; a brightly colored ventrum; and a yellow to orange neck ring.

RELATIONSHIPS AND TAXONOMY The relationships between *D. punctatus* and other snakes remain enigmatic. Cadle (1984a) hypothesizes that *D. punctatus* may be representative of one of the more ancestral lineages of xenodontines. Blanchard (1942) discusses taxonomy.

DISTRIBUTION *Diadophis punctatus* ranges widely throughout the eastern United States and south through mainland Mexico (Conant and Collins 1998; Stebbins 1985). There are several disjunct populations of *D. punctatus* in the western United States and one large population in the Pacific northwest that ranges from southern Washington south into northwestern Baja California to at least Rancho San José, BC. The species is also known from the Islas Todos Santos (south island) and San Martín in the Pacific (map 6.7).

FIGURE 6.14 *Diadophis punctatus* (Ring-Necked Snake) from Arroyo Santo Domingo, BC

DESCRIPTION Body elongate and subcircular in cross section, reaching 524 mm SVL; head flat, not distinct from neck; eyes moderately sized, pupils round; tail moderate, 17 to 21 percent of SVL in males, 12 to 19 percent in females; head scales large and platelike; usually 8 supralabials; genials contact medially; usually 7 infralabials; dorsal body scales smooth and imbricate; 15 scale rows at midbody; 182 to 207 ventral scales in males, 191 to 209 in females; anal plate divided; 54 to 68 divided subcaudals in males, 48 to 60 in females. COLORATION Dorsal ground color of body olive drab; head darker than body; labials pale yellow and darkly mottled; bright yellow to red-orange neck ring bordered posteriorly by black dots; ventral surface of body bright orange, with several well-defined, moderately sized black spots on the posterior edges of the ventrals; subcaudal region bright red and not nearly as spotted. Hatchlings usually darker with a cream to white rather than orange neck ring.

GEOGRAPHIC VARIATION The three specimens from Islas Todos Santos were originally described as *D. anthonyi* and differ from snakes of the adjacent peninsula in that the body tends to be more stout or robust (Blanchard 1942) and the neck ring obscure with poorly defined edges (Van Denburgh and Slevin 1923). Subsequent authors have considered it to be *D. punctatus.*

NATURAL HISTORY *Diadophis punctatus* is very common in coastal areas of cismontane northwestern Baja California but, because of its secretive habits, it is seldom seen. Schmidt (1922) recorded a specimen from Paso de San Matías that Welsh (1988) believed came from a spring in the piñon-juniper woodland farther south. Welsh reported Ring-Necked Snakes as being present in riparian areas at Rancho San José and irrigated slopes in Arroyo San Telmo and concluded that this species is probably restricted to riparian corridors in cismontane areas. This conclusion is in accord with Klauber's (1928) observations of this species in San Diego County, California, just to the north. Klauber concluded that *D. punctatus* was very common in coastal areas but rare in inland valleys and montane habitats.

Ring-Necked Snakes are semifossorial. Some authors have considered them burrowers, but they probably do little burrowing themselves. Rather, they occur in areas of expansive soils and take refuge in small cracks in the ground or in the burrows of other animals. One specimen has been reported from Isla San Martín off the coast of San Quintín (Blanchard 1942). Isla San Martín is one of six volcanoes in the San Quintín area, and its terrain is extremely rugged, rocky, and covered with cactus. This habitat seems uncharacteristic of *D. punctatus,* and it would be interesting to know exactly where on the island the specimen was collected. Ring-Necked Snakes are diurnal and most commonly found beneath surface objects such as rocks, logs, and boards. Unlike most other snakes, Ring-Necked Snakes are communal and are often found in groups beneath the same object.

Ring-Necked Snakes are quite active during cold temperatures (Blanchard 1942), and Klauber (1928) noted that in San Diego County they could be found year-round.

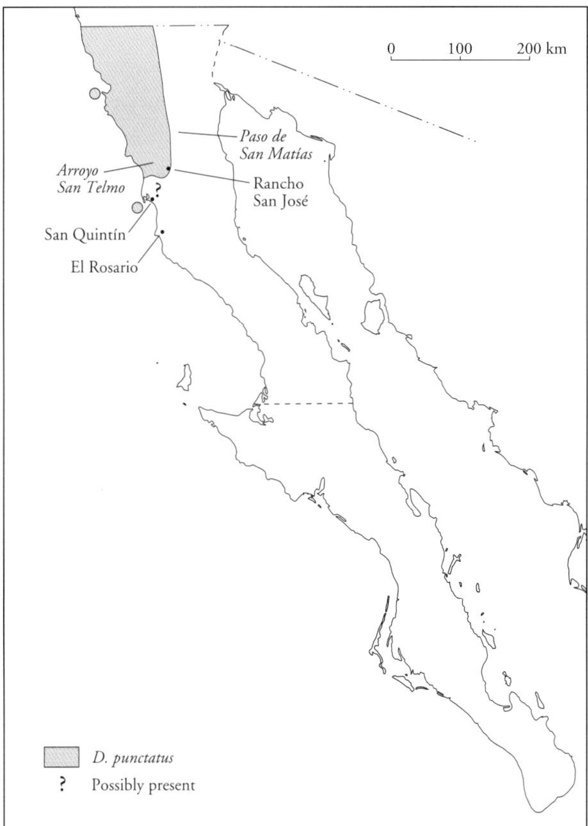

MAP 6.7 Distribution of *Diadophis punctatus* (Ring-Necked Snake)

However, their peak activity occurs during the cooler months, from February through April. Mating presumably occurs during this period, and hatchlings emerge during the summer. *Diadophis punctatus* preys on earthworms, termite and ant larvae, Slender Salamanders *(Batrachoseps major),* and small lizards. Prey is envenomated with toxic saliva by way of enlarged rear teeth.

Eridiphas Leviton and Tanner

BAJA CALIFORNIA NIGHT SNAKES

The genus *Eridiphas* is composed of two species of small, rear-fanged terrestrial snakes endemic to the region of study. *Eridiphas* was originally described as a night snake (*Hypsiglena;* Tanner 1943), but was later recognized as morphologically intermediate between *Hypsiglena* and the Cat-Eyed Snakes *(Leptodeira)* of southern Mexico, and subsequently given generic status (Leviton and Tanner 1960). In his monograph on *Leptodeira* (1958), Duellman not only hypothesized that *Hypsiglena* and *Leptodeira* shared a common ancestor but also gave a prophetic morphological description of that ancestor. Amazingly, his description very closely resembled *Eridiphas.* More recently, Cadle (1984b) has suggested that *Eridiphas* and *Leptodeira* are more closely related to each other than to *Hypsiglena.* Dowling and Jenner (1987) place *Eridiphas* as the sister genus to the neotropical banded night snake *(Pseudoleptodira),* and together they form the sister group to the night snakes *(Hypsiglena).*

FIG 6.15 *Eridiphas slevini*

SLEVIN'S NIGHT SNAKE; CULEBRA NOCTURNA

Hypsiglena slevini Tanner 1943:53. Type locality: "Puerto Escondido, Lower California [BCS]," Mexico. Holotype: CAS 5361

IDENTIFICATION *Eridiphas slevini* can be distinguished from all other snakes in the region of study by having large, protuberant eyes with elliptical pupils; a dorsal pattern composed of 51 to 63 dark blotches; and 184 to 195 ventral scales.

RELATIONSHIPS AND TAXONOMY *Eridiphas slevini* is the sister species of *E. marcosensis* of Isla San Marcos. Grismer (1999b) and Ottley and Tanner (1978) discuss taxonomy.

DISTRIBUTION *Eridiphas slevini* ranges continuously from at least Bahía de los Ángeles in the north to Cabo San Lucas in the south (Grismer 1996d; Mellink 1996; Ottley and Tanner 1978; map 6.8). It is also known from Isla Santa Margarita, off the west coast of Baja California (Reynoso 1990e), and from Islas Cerralvo and Danzante in the Gulf of California (Grismer 1999a). Ottley and Tanner (1978) indicate that the northernmost population of *E. slevini* may

FIGURE 6.15 *Eridiphas slevini* (Slevin's Night Snake) from Santiago, BCS

be disjunct. However, because of the continuum of habitat between Bahía de los Ángeles and San Ignacio through the Sierra la Libertad and their presence in the Sierra San Francisco, this is not the case.

DESCRIPTION Body elongate and subcircular in cross section, reaching 481 mm SVL; head flat, moderately distinct from neck; eyes large and protuberant, pupils elliptical; tail moderate, 17 to 19 percent of SVL; head scales large and platelike; rostral wider than high; parietals contacting both postoculars; 8 to 12 (usually 8) supralabials; genials contact medially; 10 to 12 (usually 10) infralabials; dorsal body scales smooth and imbricate; 23 scale rows at midbody; 184 to 195 ventral scales; anal plate divided; 55 to 68 divided subcaudals. **COLORATION** Dorsal ground color gray to light brown; bold, dark brown markings on top of head; usually a pair of dark brown, obliquely oriented nuchal blotches present; blotches occasionally contacting on midline; dark brown postoccipital blotch anterior to nuchal blotches; 51 to 63 dark brown rows of paired or offset dorsal blotches; 1 or 2 rows of lateral spots on sides of body; centers of interspace scales often have darkened centers; ventrum dull cream and immaculate.

GEOGRAPHIC VARIATION Individuals from mesic areas and areas with dark substrate tend to have darker body blotches and bolder markings. Such is the case in specimens from the Cape Region and along the west coast of the Magdalena Plain, where precipitation is relatively high and foliage relatively dense, and in those from the region between San Ignacio and Santa Rosalía, which is crisscrossed with dark, andesitic volcanic lava flows. *Eridiphas slevini* from other areas, such as Bahía de los Ángeles, the vicinity of Loreto, and the La Presa region, are generally lighter overall.

NATURAL HISTORY *Eridiphas slevini* ranges throughout the southern two-thirds of Baja California, occurring in the Central Gulf Coast Region south through the Magdalena Region, and up to the base of the pine-oak woodland of the Sierra la Laguna Region (Alvarez et al. 1988). Within these regions, *E. slevini* is primarily associated with rocky areas. I have observed specimens in lava flows; on boulder-strewn hillsides; on rocky hillsides lacking large boulders; and in rocky streambeds within shallow, wide arroyos that course through flat, featureless, sandy terrain as well as deep rocky canyons. At Aguas Calientes, in the Cape Region, I found a specimen 2 m above the ground beneath a granitic exfoliation. Ottley and Tanner (1978) reported finding a specimen coiled in a small basal fissure of a boulder in a dry, narrow, rock-strewn wash. Soulé (1961) collected a specimen beneath driftwood on a high beach at the south end of Isla Cerralvo. *Eridiphas slevini* has been observed abroad throughout its range from early March through October. During the winter, it is probably minimally active in the Cape Region and inactive in the northern portion of its distribution. Leviton and Banta (1964) reported finding a specimen beneath a rock during late December in the Cape Region that was not "especially active." *E. slevini* is nocturnal and has only been observed at night. During the day, snakes usually remain hidden beneath rocks or within rock cracks.

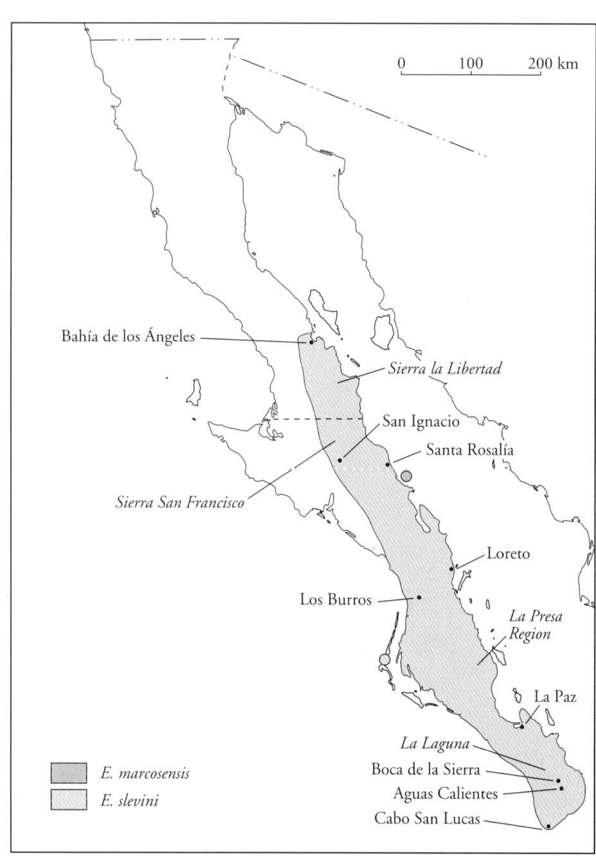

MAP 6.8 Distribution of *Eridiphas* (Baja California Night Snakes)

Lizards, which *E. slevini* subdue with their enlarged rear teeth and mildly toxic saliva, are a major prey item. I found a specimen at Boca de la Sierra, in the Cape Region, with a Black-Tailed Brush Lizard *(Urosaurus nigricaudus)* in its mouth that it had not yet started to eat. When I removed the lizard, I discovered that it had been freshly killed. One evening I found a specimen nearly 2 m above the ground on a granite boulder beneath an exfoliation. I believe it may have been searching for Leaf-Toed Geckos *(Phyllodactylus xanti),* which were seen on the same boulder. I coaxed an *E. slevini,* collected from beneath a rock in a streambed at Los Burros in the Magdalena Plain, to regurgitate a Pacific Treefrog *(Hyla regilla).* Ottley and Tanner (1978) reported finding a specimen foraging along a streambed near an aqueduct. Leviton and Banta (1964) reported finding a specimen on damp sand beneath a small rock by a trickle of water near a waterfall in the Cape Region. Brad Hollingsworth found two specimens in an arroyo bottom near San Francisco de la Sierra in the vicinity of ponds that harbored *H. regilla.* From these observations, I believe it is likely that amphibians compose a major portion of this species' diet in certain areas.

REMARKS Rodríguez-Robles, Mulcahy, and Greene (1999) believe that *Eridiphas* is part of a neotropical group of snakes that has retained the primitive behavior of feeding on frogs even though it is restricted to arid regions.

FIG 6.16 ***Eridiphas marcosensis***

ISLA SAN MARCOS NIGHT SNAKE; CULEBRA NOCTURNA

Eridiphas slevini marcosensis Ottley and Tanner 1978:407. Type locality: "South side of Arroyo de la Taneria," Isla San Marcos, BCS, Mexico. Holotype: BYU 34617

IDENTIFICATION *Eridiphas marcosensis* can be distinguished from all other snakes in the region of study by having large, protuberant eyes with elliptical pupils; a dorsal pattern composed of 65 dark blotches; and 198 to 200 ventral scales.

RELATIONSHIPS AND TAXONOMY See *E. slevini.*

DISTRIBUTION *Eridiphas marcosensis* is endemic to Isla San Marcos in the Gulf of California, BCS, Mexico (see map 6.8).

DESCRIPTION Same as *E. slevini,* except for the following: head 4.6 to 4.7 percent of SVL; 198 to 200 ventral scales; 65 dorsal body blotches.

NATURAL HISTORY Isla San Marcos lies within the Central Gulf Coast Region and is dominated by xerophilic scrub. The island is composed of sedimentary sandstone with a southerly-situated gypsum base. Nothing has been reported on the natural history of *E. marcosensis* other than the statement of Murphy and Ottley (1984) that they found a specimen among the gypsum boulders on a steep hillside.

FIGURE 6.16 *Eridiphas marcosensis* (Isla San Marcos Night Snake) from Arroyo de la Taneria, Isla San Marcos, BCS. (Photograph by Jesse L. Grismer)

REMARKS *Eridiphas marcosensis* was originally described as a subspecies of *E. slevini* on the basis of two specimens (Ottley and Tanner 1978). Its discrete diagnosability from *E. s. slevini* on the basis of ventral scale counts, number of body blotches, and head length prompted Grismer (1999b) to elevate it to a full species. Acquisition and examination of additional specimens may indicate that it does not merit species status.

Hypsiglena Cope

NIGHT SNAKES

The genus *Hypsiglena* contains two species: the widely distributed continental form, *H. torquata* (*sensu* Flores 1993), and the insular, endemic *H. gularis* (Grismer 1999b; Tanner 1954). *Hypsiglena* are small, nocturnal, terrestrial snakes occurring through much of Mexico and the United States, and they are well-represented on islands in the Gulf of California (Grismer 1999a). They are common, secretive species that usually emerge only at dusk and during the evening. *Hypsiglena* are often misidentified as very small rattlesnakes because they have triangularly shaped heads and vertical pupils. *Hypsiglena* pose no threat to humans; however, their enlarged rear teeth and mildly toxic saliva are used to subdue prey.

FIG 6.17 ***Hypsiglena torquata***

NIGHT SNAKE; CULEBRA NOCTURNA

Leptodeira torquata Günther 1860:170. Type locality: "Nicaragua, Island of Laguna." Holotype: BM 61–12–30, 97

IDENTIFICATION *Hypsiglena torquata* can be distinguished from all other snakes in the region of study by having elliptical pupils; a dark postorbital stripe; distinctive dark

FIGURE 6.17 Baueri pattern class of *Hypsiglena torquata* (Night Snake) from Isla Cedros, BC

nuchal blotches; and gular scales that are not enlarged and do not resemble the chinshields.

RELATIONSHIPS AND TAXONOMY *Hypsiglena torquata* is most closely related to *H. gularis* of Isla Partida Norte (Grismer 1999b). For generic relationships, see *Eridiphas slevini.* Dixon and Dean (1986), Tanner (1944, 1985), and Grismer (1999b) discuss taxonomy. The variability of this species has been a source of contention among systematists for several decades, and its taxonomy is not yet settled. Reviews of the continental forms can be found in Dixon and Dean 1986 and Tanner 1944 and 1985. Grismer (1999b) reviewed the Gulf of California populations.

DISTRIBUTION *Hypsiglena torquata* ranges widely throughout the western United States and Mexico. It is generally ubiquitous in Baja California, being absent only from the Vizcaíno Desert (Grismer et al. 1994). *Hypsiglena torquata* occurs on the Pacific islands of Coronado (Norte, Media, and Sur), San Martín, and Cedros, and the Gulf islands of Ángel de la Guarda, Carmen, Cerralvo, Coronados, Danzante, Mejía, Monserrate, Partida Sur, Salsipuedes, San Esteban, San Francisco, San José, San Lorenzo Sur, San Marcos, Santa Catalina, Smith, Tiburón, and Tortuga (Grismer 1993, 1999a, 2001a; McAfee and Gilardi 2000; map 6.9).

DESCRIPTION Body elongate and subcircular in cross section, reaching 642 mm SVL; head flat, moderately distinct from neck; eyes moderate, pupils elliptical; tail moderate, 15 to 19 percent of SVL; head scales large and platelike; rostral wider than high; parietals not contacting postoculars; 8 or 9 (usually 8) supralabials; gular scales not enlarged; genials contact medially; 10 or 11 (usually 10) infralabials; dorsal body scales smooth and imbricate; 21 scale rows at midbody; 161 to 184 ventral scales in males, 173 to 190 in females; anal plate divided; 42 to 63 divided subcaudals in males, 39 to 54 in females. **COLORATION** Dorsal ground color gray-brown; small dark flecks on top of head; tripartite nuchal blotch present, lateral portions often continuous with dark postorbital stripes; medial portion usually elongate, occasionally contacting parietal scales anteriorly; preorbital stripe often present and may or may not be highlighted below in white; 40 to 95 dark brown rows of paired to offset paravertebral blotches; 1 or 2 rows of offset lateral spots on sides of body; ventrum dull white and immaculate.

GEOGRAPHIC VARIATION *Hypsiglena torquata* is widespread throughout Baja California and ranges through several phytogeographic regions. Four intergrading peninsular subspecies have been recognized on the basis of combined ventral and caudal scale counts and the number of dorsal body blotches (Tanner 1985), which are used here as pattern classes. Additionally, *H. torquata* shows a great deal of localized variation in the overall contrast of the dorsal color pattern, which is usually a result of substrate matching. Generally, the darker, bolder dorsal patterns occur on Night Snakes from areas dominated by dark volcanic rocks. In areas with light-colored, granitic rocks or sandy substrates, Night Snakes have dorsal patterns that are faded and less distinct. There is also a general north-to-south trend toward fewer ventral and caudal scales and more dorsal spots on the body and tail.

Klauberi range throughout the cismontane areas of northwestern and central Baja California, extending through the California Region to the northern Vizcaíno Region. They have 207 to 236 (male) and 212 to 238 (female) ventral plus caudal scales and 40 to 61 dorsal spots. They also have a weak preorbital stripe, which is rarely highlighted below in white. Deserticola range throughout the Lower Colorado Valley Region of northeastern Baja California south to at least Bahía de los Ángeles. They have 226 to 261 (male) and 230 to 261 (female) ventral plus caudal scales and 49 to 75 dorsal spots. Deserticola have a well-defined preorbital stripe highlighted below in white. The central region of Baja California is occupied by Venusta, which range throughout the Magdalena and Central Gulf Coast regions from San Ignacio south to the Isthmus of La Paz. Venusta have 223 to 244 (male) and 232 to 251 (female) ventral plus caudal scales, 62 to 95 dorsal spots, and a well-defined preorbital stripe highlighted below in white. Venusta from the Sierra Santa Clara have a very faded color pattern, and often the nuchal blotch is reduced to a thin, dagger-shaped marking (Grismer et al. 1994:67). Some Venusta from San Francisco de la Sierra have very small, paired paravertebral spots. Ochrorhyncha, which range throughout the Cape Region, have 208 to 229 (male) and 217 to 236 (female) ventral plus caudal scales and 53 to 107 dorsal spots. Ochrorhyncha have a well-defined preorbital stripe highlighted below in white. All pattern classes intergrade where their populations come into contact.

There are at least 20 insular populations of *H. torquata* (Grismer 1999a; McAfee and Gilardi 2000). Some were originally described as distinct subspecies, and others represent insular populations of the adjacent peninsular pattern classes. For most populations, sample sizes are too small to show any meaningful statistical differences in scale counts; this discussion is concerned primarily with the overall coloration and pattern of insular populations that show trends differing from those of adjacent peninsular or mainland populations.

Night Snakes on Isla Ángel de la Guarda, from near Puerto Refugio at the north end of the island, tend to have a pinkish hue that is likely due to the reddish substrate on which they occur. These snakes have well-defined preorbital stripes highlighted below in white.

Hypsiglena torquata from the Pacific island of San Martín was described as *H. t. martinensis* (Tanner and

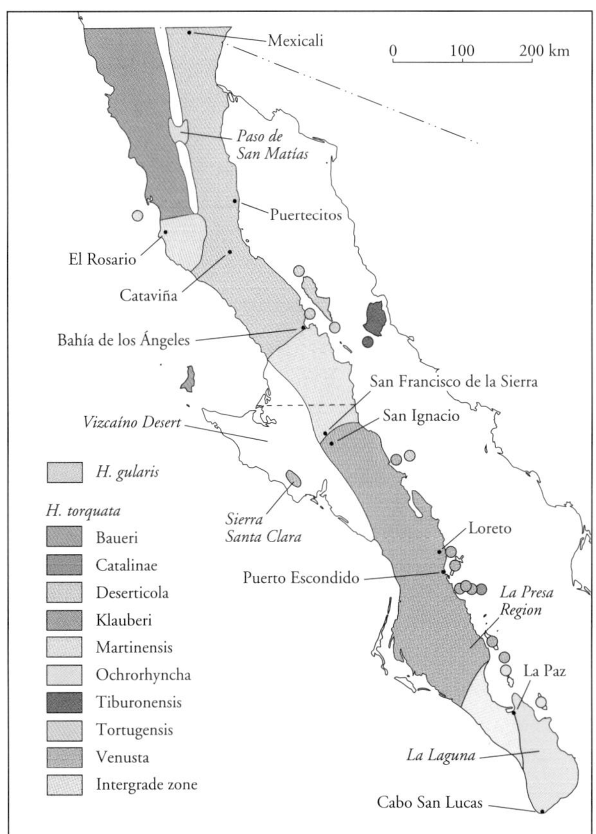

MAP 6.9 Distribution of *Hypsiglena gularis* (Isla Partida Norte Night Snake) and the pattern classes of *H. torquata* (Night Snake)

Banta 1962), a distinction that is used here as a pattern class. Martinensis tend to be darker overall than most other populations, undoubtedly because of their existence among the dark volcanic rocks of Isla San Martín.

Hypsiglena torquata from Isla Cedros was described as *H. t. baueri* (Zweifel 1958) and is considered here as a pattern class. Like other insular populations, Baueri tend to lack a dark preorbital stripe. Additionally, they have an overall reddish-brown tint to their dorsal coloration. In some specimens, the lateral body spots become so elongate that they connect to form a broken lateral stripe.

Hypsiglena torquata from Islas San Esteban and Tiburón was described as *H. t. tiburonensis* (Tanner 1981) and is considered here as the pattern class Tiburonensis. These snakes are similar to the Night Snakes from Isla San Lorenzo Sur and adjacent Sonora in that they have an overall light brown coloration with poorly defined dorsal spots and well-defined preorbital stripes highlighted below in white.

Hypsiglena torquata from Isla Tortuga was described as *H. t. tortugensis* (Tanner 1944) and is considered a pattern class. Tortugensis are similar to some Venusta in having a narrow nuchal blotch. They are generally very dark in overall dorsal coloration, matching the dark volcanic substrate on which they occur.

The population from Isla Santa Catalina was described as *H. t. catalinae* (Tanner 1966a) and is recognized here as a pattern class. Catalinae tend to have larger dorsal spots than Venusta from the adjacent peninsula, as illustrated by their lower spot counts (56 to 68 in the Isla Santa Catalina population compared with 67 to 95 in Venusta). Additionally, the dorsal spots tend to be less offset, often resulting in one large spot rather than two smaller spots. Snakes from this population have weak or absent preorbital stripes and tend to be larger than most other Night Snakes.

The remaining insular populations are only known from a few specimens, which show no detectable trends. All have the same generalities of squamation and color pattern of the adjacent peninsular pattern classes.

NATURAL HISTORY *Hypsiglena torquata* is nearly ubiquitous in Baja California, occurring in almost every habitat except perhaps the upper elevations of the Baja California Coniferous Forest Region of the Sierra San Pedro Mártir (Welsh 1988) and the central portion of the Vizcaíno Desert (Grismer et al. 1994). Where Night Snakes occur, they are most common in areas with rocks, thick brush, or surface litter, such as logs, boards, and garbage. In the California Region, Klauberi are commonly found beneath rocks and in trash piles. In other parts of the peninsula, I have found *H. torquata* to be abundant in rocky arroyos and on hillsides. I have found them beneath surface material such as dead Agave (*Agave* sp.; Bostic 1971), logs, and garbage, as have other authors (Banta and Leviton 1964; Zweifel 1958). On islands, *H. torquata* seem to be restricted to rocky areas and are most commonly seen in rocky washes and arroyos, but perhaps only because these are the places where most herpetologists work. High on a hillside on Ángel de la Guarda, I saw a Night Snake crawling down the face of a large boulder and found two other individuals foraging in the low branches of shrubs. On Isla Santa Catalina, Catalinae are common in the cobble of the beaches near the water's edge and can be found beneath rotten Cardón Cactus *(Pachycereus pringlei)* during the day. Zweifel (1958) reported finding a specimen of Baueri beneath a piece of cloth above the high-tide mark on Isla Cedros, and I have found Baueri in garbage piles and rocky arroyos.

Venusta are generally active year-round south of Loreto, although a minimum of activity occurs during the winter. Leviton and Banta (1964) reported active specimens of Ochrorhyncha in the Cape Region during January and December. North of Loreto, *H. torquata* is generally active from late February through early November, with an activity peak from April through September. Night snakes are often abroad during relatively cold nights when other species may be inactive. As the common name indicates, *H. torquata* is nocturnal, although individuals are occasionally observed during the day from February through April. I found one individual crossing the road in the center of town at Bahía de los Ángeles at 0745 hours when the sun was up and the air temperature was 28°C.

Rodríguez-Robles, Mulcahy, and Greene (1999) demonstrated that *H. torquata* eats mainly lizards and frequently squamate eggs. They also noted that it occasionally consumes anurans, snakes, and insects. Papenfuss (1982) reported a specimen from La Paz with a Two-Footed Worm Lizard *(Bipes biporus)* in its stomach. Lizard prey consists primarily of diurnal lizards that seek shelter beneath surface objects or within burrows or crevices at night (Rodríguez-Robles, Mulcahy, and Greene 1999). *Hypsiglena torquata* may locate sleeping lizards using chemosensory cues while foraging (Rodríguez-Robles, Mulcahy, and Greene 1999) and presumably locate buried squamate eggs in the same way (Diller and Wallace 1986). I observed a Night Snake on Isla Ángel de la Guarda on a coastal dune one evening in early April, crawling very slowly and tongue-flicking rapidly with its head cocked downward. It then thrust its head beneath the sand and withdrew a buried Zebra-Tailed Lizard *(Callisaurus draconoides)*, which it proceeded to eat. Rodríguez-Robles, Mulcahy, and Greene (1999) noted an incidence of diurnal ambush feeding in La Paz during mid-January, when an Orange-Throated Whiptail *(Cnemidophorus hyperythrus)* was found in the mouth of a Night Snake that was found beneath a board. I found a Baueri beneath a piece of wood along with a Western Banded Gecko *(Coleonyx variegatus)* and a Side-Blotched Lizard *(Uta stansburiana)*. Murphy and Ottley (1984) reported on a specimen collected during December with a Side-Blotched Lizard *(Uta stansburiana)* in its stomach. On Isla Santa Catalina, Catalinae are very common in the intertidal zone on the cobblestone beaches, where they forage for Isla Santa Catalina Leaf-Toed Geckos *(Phyllodactylus bugastrolepis)*. In areas where Treefrogs (*Hyla* sp.) are present, *Hypsiglena torquata* likely eats them as well. I found an engorged Baueri beneath a piece of tin at a well-watered area on Isla Cedros and saw several Pacific Treefrogs *(Hyla regilla)* nearby. Night Snakes have enlarged rear teeth with longitudinal grooves, and they use a chewing action to inject a mildly toxic saliva into their prey, causing paralysis.

Little has been reported on the reproductive biology of *H. torquata* in the region of study, but presumably it is not too different from that of U.S. populations, especially in the north. According to Stebbins (1985), *H. torquata* lays 2 to 9 eggs from April to August. Murphy and Ottley (1984) reported on a female collected during mid-April from Isla Mejía with three eggs. I have observed gravid females from near San Ignacio during mid-May, hatchlings at Mulegé during mid-August, and juveniles during mid-December, west of Puerto Escondido. These observations suggest that *H. torquata* has a spring breeding season and summer hatching period.

FIG 6.18 *Hypsiglena gularis*

ISLA PARTIDA NORTE NIGHT SNAKE; CULEBRA NOCTURNA

Hypsiglena torquata gularis Tanner 1954:54. Type locality: "Isla Partida, 28° 53' north latitude and 113° 3' west longitude, BC, Mexico." Holotype: CAS 51009

IDENTIFICATION *Hypsiglena gularis* can be distinguished from all other snakes in the region of study by having elliptical pupils; a dark postorbital stripe; distinctive dark nuchal blotches; and enlarged gular scales resembling chinshields.

RELATIONSHIPS AND TAXONOMY See *Hypsiglena torquata*.

DISTRIBUTION *Hypsiglena gularis* is endemic to Isla Partida Norte, BC (see map 6.9).

FIGURE 6.18 *Hypsiglena gularis* (Isla Partida Norte Night Snake) from Isla Partida Norte, BC

DESCRIPTION Same as *H. torquata,* except for the following: adults reach at least 342 mm SVL; at least one pair of enlarged gular scales resembles chinshields; 17 to 19 dorsal scale rows at midbody; 183 or 184 ventrals and 43 caudals in females. **COLORATION** Dorsal ground color reddish brown; small dark flecks on top of head; nuchal blotch reduced, not reaching parietal scales; preorbital stripe absent; postorbital stripe weak; 68 to 80 brownish, poorly defined dorsal blotches occasionally paired and offset, but more often appearing as one large spot; spots on tail not discernible; 1 or 2 rows of offset lateral spots on sides of body; ventrum dull white and immaculate.

NATURAL HISTORY Nothing has been reported on the natural history of *H. gularis.* I have observed two specimens, both adult females, during the day beneath small rocks at the top of the island. One was found during mid-April and the other during mid-October.

REMARKS The only unique characteristic of this population that separates it from *H. torquata* is the enlarged gular scales, which resemble chinshields (Tanner 1954). This population was described on the basis of two specimens, and Grismer (1999b) indicated that with the acquisition of additional material, this character may turn out to be variable. Both of the additional specimens I have observed in the field, however, had enlarged gular scales. Therefore, its status as a species is still followed here.

Lampropeltis Fitzinger

KINGSNAKES

The genus *Lampropeltis* is a morphologically and ecologically diverse group of very attractive snakes that contains at least 10 species (Garstka 1982; Grismer 1999b). *Lampropeltis* ranges through much of the United States, Mexico, Central America, and northern South America and is represented on both Pacific and Gulf islands in the region of study. The various species of *Lampropeltis* frequent a variety of habitats. One wide-ranging species, the Common Kingsnake *(L. getula),* is a ubiquitous habitat generalist, whereas others, such as the Mountain Kingsnake *(L. zonata),* are primarily restricted to coniferous forests. Kingsnakes are terrestrial and kill their prey by causing cardiac arrest through constriction. All species feed on small rodents and lizards, and others take in a wider variety of prey, including toads and other snakes.

The taxonomy of *Lampropeltis* has frustrated snake systematists for many decades and remains unsettled. The most recent and comprehensive reviews of various groups can be found in Blaney 1977, Garstka 1982, Williams 1988, and Zweifel 1952b. Keogh (1996) demonstrated that *Lampropeltis* formed a natural group with long-nosed snakes *(Rhinocheilus),* the scarlet snake *(Cnemophora),* and short-tailed snakes *(Stilosoma).* Rodríguez-Robles and Jesús-Escobar (1999) indicate that *Stilosoma* may have evolved out of *Lampropeltis.*

FIG 6.19 *Lampropeltis zonata*

CALIFORNIA MOUNTAIN KINGSNAKE; CORALILLO

[Coluber?] (Zacholus) zonatus Blainville 1835a:293. Type locality: "Californie." Holotype: Formerly in Paris museum but now not known to exist.

IDENTIFICATION *Lampropeltis zonata* can be distinguished from all other snakes in the region of study by having a prominent white band on the posterior portion of the head in contact with the parietal scales; several red bands on the body; and red on the snout.

RELATIONSHIPS AND TAXONOMY *Lampropeltis zonata* is most closely related to *L. herrerae* (see below) of Islas Todos Santos. Together these species form a natural group with the Gray-Banded Kingsnakes *(L. mexicana* and *L. alterna),* the Arizona Mountain Kingsnake *(L. pyromelana),* and the Milksnake *(L. triangulum)* (Keogh 1996; Rodríguez-Robles and Jesús-Escobar 1999). Taxonomy is discussed by Hayes (1975), Zweifel (1952b), and below (see *L. herrerae*).

DISTRIBUTION *Lampropeltis zonata* occurs in two disjunct populations: one from the upper elevations of the Sierra Juárez and the other from Sierra San Pedro Mártir (Zweifel 1975; map 6.10).

DESCRIPTION Body elongate and subcircular in cross section, reaching 638 mm SVL; head moderately flat, somewhat distinct from neck; eyes moderately sized, pupils

FIGURE 6.19 *Lampropeltis zonata* (California Mountain Kingsnake) from El Portrero, Sierra San Pedro Mártir, BC

round; tail moderate in length, 17 to 19 percent of SVL; head scales large and platelike; 7 or 8 supralabials; genials contact medially; 8 or 9 infralabials; dorsal body scales smooth and imbricate; 23 scale rows at midbody; 207 to 213 ventral scales; anal plate divided; 47 to 59 divided subcaudals. **COLORATION** Dorsal ground color black and overlaid with red and white rings in the sequence of white, black, red, black, white; 41 to 48 triads, with 63 to 100 percent split dorsally by red bands; red bands usually incompletely encircle belly; white and black bands completely encircle belly; snout heavily suffused with red; prominent white band on posterior portion of the head in contact with parietal scales and posteriormost supralabials.

GEOGRAPHIC VARIATION *Lampropeltis zonata* from the Sierra San Pedro Mártir tend to have more red on the body and snout than snakes from the Sierra Juárez. Both have been referred to as the subspecies *L. z. agalma* (Zweifel 1952b, 1975). Recent phylogeographic work by Rodríguez-Robles, DeNardo, and Staub (1999) using DNA sequence data, however, strongly suggests that the Sierra Juárez population is most closely related to populations from southern California (*L. z. pulchra*) and that the San Pedro Mártir population is most closely related to *L. herrerae*. If a phylogenetic analysis confirms this hypothesis, then the Sierra San Pedro Mártir population should also be accorded status as a species.

NATURAL HISTORY Very little has been reported on the natural history of *L. zonata* from Baja California. This species is most common in rocky areas within the Baja California Coniferous Forest Region. However, in the Sierra Juárez it is also common in rocky areas on the edge of the pine belt along the western slopes. Welsh (1988) has reported a similar occurrence of *L. zonata* below the pine belt on the western slopes of the Sierra San Pedro Mártir. Mountain Kingsnakes are most often observed beneath rocks on the ground near outcroppings or within rocky crevices. They are by no means restricted to such situations, however, and can also be found beneath logs and under the exfoliating bark of fallen trees.

Lampropeltis zonata is active from April through September, with peak activity occurring from late April through early July. During May snakes may be seen foraging during the day. Mountain Kingsnakes in Baja California feed primarily on lizards of the genera *Uta, Sceloporus, Eumeces,* and *Elgaria,* but also take small mice and baby birds. Mating usually occurs during the spring, and 3 to 13 eggs are laid during the summer.

FOLKLORE AND HUMAN USE Local residents believe this species is venomous. Skins of this snake are often tacked to the walls of the ranchos. In 1991, there was a skin on the wall at the ranger station in the Sierra San Pedro Mártir. The snake was unearthed by a bulldozer digging up large rocks. The park officials killed it, believing it to be venomous.

REMARKS The most serious threat to this species is its high

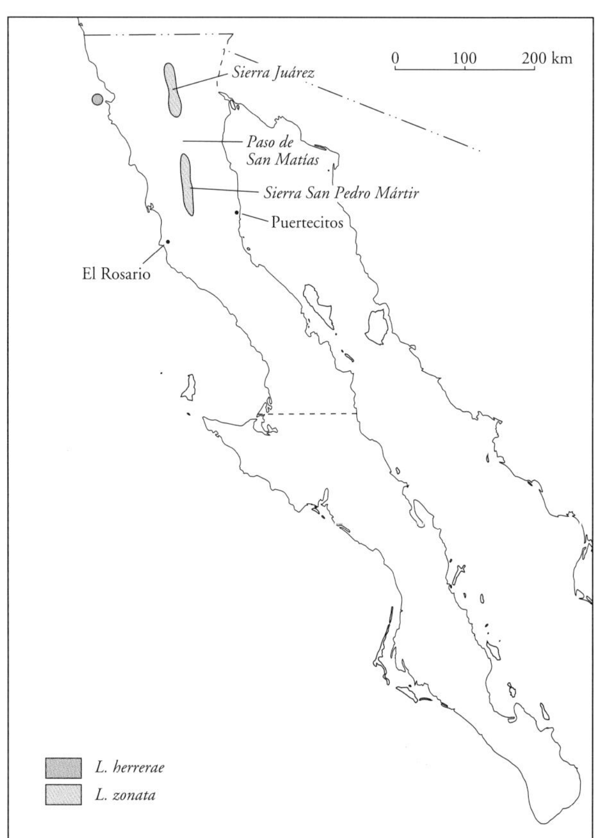

MAP 6.10 Distribution of *Lampropeltis zonata* (California Mountain Kingsnake) and *L. herrerae* (Islas Todos Santos Mountain Kingsnake)

commercial value. Much of its habitat in the Sierra Juárez has been destroyed by commercial snake hunters breaking open and turning over rocks. Kingsnakes are smuggled across the border and sold (Mellink 1995).

FIG 6.20 **Lampropeltis herrerae**

ISLAS TODOS SANTOS MOUNTAIN KINGSNAKE; CORALILLO

Lampropeltis herrerae Van Denburgh and Slevin 1923:2. Type locality: "South Todos Santos Island, Lower California [BC], Mexico." Holotype: CAS 56755

IDENTIFICATION *Lampropeltis herrerae* can be distinguished from all other snakes in the region of study by having a prominent white band on the posterior portion of the head in contact with the parietal scales; few or no red bands on the body; no red on the snout; and last supralabial touched with black.

RELATIONSHIPS AND TAXONOMY *Lampropeltis herrerae* is endemic to the Pacific island of Todos Santos Sur off the coast of Ensenada, BC, and was originally considered a subspecies of *L. zonata.* Rodríguez-Robles, DeNardo, and Staub (1999) noted on the basis of DNA sequence data that *L. herrerae* formed a unique lineage. They noted that Hayes (1975) recommended elevating *L. herrerae* to full species, but they felt that doing so on the basis of their DNA sequence data alone would be premature in the absence of a morphological analysis. In light of those findings, *L. herrerae* is distinguished from *L. zonata* by having the first white body band occurring relatively far forward on the head, making contact with the parietal scales, with its posterior margin extending posterior to the angle of the jaw. Additionally, it is the only member of the *L. zonata* complex that lacks red in its dorsal pattern (with the occasional exception of some specimens of *L. z. multicincta* from the Sierra Nevada and northern California) and has elongated vertebrae (Hayes 1975). Given this coloration and morphology and its insular isolation, Grismer (2001a) considered it a distinct species.

DISTRIBUTION *Lampropeltis herrerae* is endemic to the southern island of Isla Todos Santos, BC (Zweifel 1975; see map 6.10).

DESCRIPTION Same as *L. zonata,* except for the following: adults reach 851 mm SVL; 7 or 8 supralabials; 8 or 9 infralabials; supraoculars often fused to parietals; 217 to 220 ventral scales in males, 202 to 216 ventral scales in females; 59 to 77 subcaudals in males, 46 to 53 subcaudals in females.

COLORATION Dorsal ground color black and overlaid with white rings forming triads in the sequence of white, black, white; red bands reduced or absent, often present as 2 or 3

FIGURE 6.20 *Lampropeltis herrerae* (Islas Todos Santos Mountain Kingsnake) from Isla Sur of Islas Todos Santos, BC

white bands on the neck and only ventrolaterally placed, brownish patches in the anterior triads on the body; 36 to 41 triads, none of which are split dorsally by red markings; red areas usually present on belly; black bands completely encircle belly, and white bands are occasionally broken on the belly; snout black; prominent white band on the posterior portion of the head in contact with parietal scales and usually the posteriormost supralabials; last supralabial touched with black.

NATURAL HISTORY The Islas Todos Santos are two small islets totaling 21.2 km^2, lying 6 km off the coast of Ensenada and forming the northwestern emergent tip of Punta Banda. Individually these islets, known as Isla Norte and Isla Sur, are the erosional remnants of a larger island and are dominated by coastal scrub species of the California Region. Very little has been reported on the natural history of *L. herrerae.* It appears to be restricted to the rocky interior areas of Isla Sur. *Lampropeltis herrerae* is abroad from at least April through June and is active during periods of overcast. It is usually found near rocks. I observed a specimen crawling in the lower branches of vegetation in early May.

Grismer (1993) suggests that its occurrence on the island is relict and facilitated by the presence of rocky habitat and of the lizards *Sceloporus occidentalis* and *Eumeces skiltonianus,* which it eats, and the absence of the snakes *Pituophis catenifer* and *Lampropeltis getula,* with which it would compete.

REMARKS Unfortunately this species has a high commercial value, and illegal collection and smuggling into the United States continue. Its habitat has been considerably degraded by the breaking open of rock crevices and the overturning of rocks. Mellink (1995) reported finding snake traps on the Islas Todos Santos baited with live mice, and one notorious commercial collector took a gravid female off Isla Sur.

FIGS. 6.21–22

Lampropeltis getula

COMMON KINGSNAKE; BURRILA, SERPIENTE REAL

Coluber getulus Linnaeus 1766:382. Type locality: "Carolina," restricted by Klauber (1948:6) to "Charleston, South Carolina." Holotype: none designated

IDENTIFICATION *Lampropeltis getula* can be distinguished from all other snakes in the region of study by having round pupils; large head plates; a rostral scale that is not greatly enlarged; smooth dorsal scales; an entire anal plate; 50 to 100 percent of the subcaudals divided; the lack of a dark dorsal bar across the top of the head connecting the eyes; and the absence of small yellow dots on the sides of the body.

RELATIONSHIPS AND TAXONOMY *Lampropeltis getula* is most closely related to *L. catalinensis;* together they form the sister taxon to the Mole Kingsnake, *L. calligaster.* These taxa form the earliest branching lineage within *Lampropeltis* (Keogh 1996).

DISTRIBUTION *Lampropeltis getula* is nearly ubiquitous in Baja California and is absent only from the higher elevations of the Sierra San Pedro Mártir in northern Baja California (Welsh 1988; map 6.11). It is known from the Gulf islands of Ángel de la Guarda, Cerralvo, Monserrate, Salsipuedes, San Esteban, San Lorenzo Norte, San Lorenzo Sur, San Pedro Mártir, San Pedro Nolasco, Santa Cruz, and Tortuga (Grismer 1999a; Grismer and Hollingsworth 2000).

DESCRIPTION Body elongate and subcircular in cross section, reaching 1,813 mm SVL; head rounded in dorsal profile, not distinct from neck; eyes moderately sized, pupils round; tail relatively short, 11 to 16 percent of SVL in males, 10 to 14 percent in females; head scales large and platelike; 8 to 10 supralabials; genials contact medially; 8 to 10 (usually 9) infralabials; dorsal body scales smooth and imbricate; 23 or 25 scale rows at midbody; 213 to 250 ventral scales in males, 213 to 255 in females; anal plate single; 46 to 63 subcaudals in males, 44 to 57 in females, 50 to 100 percent of subcaudals divided. **COLORATION** This species has extremely wide variation in color pattern (see below). The populations that are not solid black (i.e., all except the insular populations of Islas San Pedro Mártir and San Pedro Nolasco) have a chocolate-brown to black dorsal ground color; head nearly all black or with light markings posteriorly; labials with white markings; 21 to 44 light cross bands on body or a vertebral stripe; belly banded or uniform chocolate brown or black.

FIGURE 6.21 *Lampropeltis getula* (Common Kingsnake) from Colonia Guerrero, BC

GEOGRAPHIC VARIATION From the California Region of northwestern Baja California south to the vicinity of San Quintín, *L. getula* has a banded pattern in which the scales of the bands are solid white or yellow and are usually not edged at their margins by the darker ground color. The ground color of kingsnakes from this region is either chocolate or very dark brown. The light bands usually continue across the ventrum, and there is usually a spot or some other light marking on the back of the head. Between El Rosario and Cataviña, Kingsnakes have a dark brown ground color with immaculate white bands, and these are some of the most handsome kingsnakes of all. South of Cataviña, adult Kingsnakes have a dark brown ground color, and the outer edges of the scales in the white bands are slightly edged with dark pigment. This condition is usually not present in juveniles but develops with age. This trend continues south through the Vizcaíno Desert and San Ignacio and is most pronounced from San José Comondú south to the Cape Region. The scales of the white bands of Kingsnakes from the Cape Region are profusely invaded by the ground color, which makes the snakes appear very dark overall.

In the Lower Colorado Valley Region of northeastern Baja California, near Mexicali, Kingsnakes are very dark. The bands are reduced to thin lines, and the scales of the bands are profusely clouded with dark coloration, often nearly obscuring them. Kingsnakes from this region have very dark heads with almost no light markings. Farther south at Bahía San Luis Gonzaga, Kingsnakes have a light brown ground color and very wide, white bands composed of dark-edged scales.

Lampropeltis getula occurs on several islands in the Gulf of California. Kingsnakes on most islands in the Midriff Region are derived from peninsular populations, as evidenced by their having bands rather than being nearly all black like the Kingsnakes of adjacent Sonora. Populations are known to occur on Islas Ángel de la Guarda, Salsipuedes, San Lorenzo Norte, San Lorenzo Sur, and San Esteban. Except for the Isla Ángel de la Guarda population, all look fairly similar, having a dark ground color, relatively narrow white bands composed of dark-edged scales, and generally black heads. Kingsnakes from Isla Ángel de la Guarda have very wide white bands with dark-edged white scales and commonly attain lengths of nearly 2 m. A single specimen is known from Isla Tortuga (Ruth 1974), and its color pattern is nearly melanistic. The ground color is very dark brown, and the head is nearly unicolored except for small, light-colored, well-separated spots on the labials. The dorsal bands are reduced to rows of small cream-yellow spots. The ventrum is dark with a faint, reduced network of irregular white markings. Kingsnakes from Islas Monserrate and Cerralvo are banded and conform with the banding pattern of snakes from the adjacent peninsula of southern Baja California, being dark and having heads with little light mottling, and bands composed of dark-edged scales. Common Kingsnakes inhabiting Islas San Pedro Mártir and San Pedro Nolasco are solid black and are derived from the black populations inhabiting coastal Sonora.

Interestingly, in Baja California, two independently derived striped color pattern phases occur (Zweifel 1982): one in the north and one in the south. In coastal northwestern Baja California as far south as Cabo Colonet, striped Common Kingsnakes have a light or dark brown ground color and a well-defined white or yellow vertebral stripe extending from the neck to the base of the tail. On the tail, the stripe is usually broken. The ventrum is usually solid yellow, but in some snakes it is solid brown. Striping also occurs in Kingsnakes from the Cape Region along the eastern foothills of the Sierra la Laguna from approximately Santiago to San José del Cabo. Here, striped Kingsnakes have a very clouded vertebral stripe with varying degrees of lateral striping, and the ventrum is always solid dark. In both the northern and southern populations, varying degrees of intermediacy can be found between the striped and banded phase. In fact, in the Cape Region, intermediacy is the most common condition; in the northern populations it is less common. Zweifel (1981, 1982) indicated that this polymorphism in the northern populations is inherited in simple Mendelian fashion and that although geographically restricted, the striped pattern is dominant.

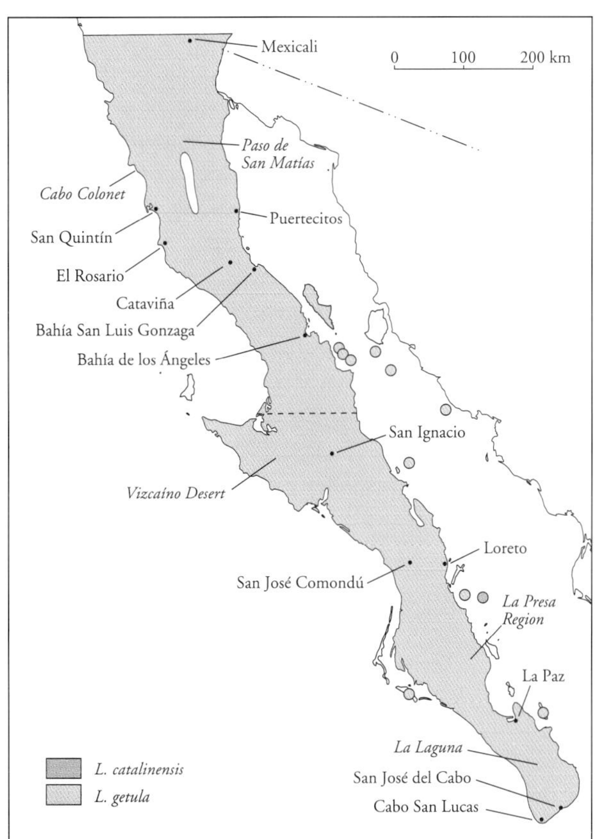

MAP 6.11 Distribution of *Lampropeltis getula* (Common Kingsnake) and *L. catalinensis* (Isla Santa Catalina Kingsnake)

FIGURE 6.22 *Lampropeltis getula* (Common Kingsnake) from El Chorro, BCS

NATURAL HISTORY *Lampropeltis getula* is found in virtually every conceivable habitat in Baja California, ranging from marshes and sand dunes to open desert flats and rocky arroyos. It may be absent only from the Baja Coniferous Forest Region in the upper elevations of the Sierra San Pedro Mártir (Welsh 1988). In northern Baja California Kingsnakes are inactive during the winter; however, south of San Ignacio they are active year-round, with a peak activity period from April through October. During climatically mild times of year, *L. getula* is commonly seen out foraging for food during the day. During the summer, however, daytime surface temperatures can exceed the thermal capacity for this species, and its activity becomes strictly nocturnal.

Kingsnakes eat a wide variety of prey but most commonly feed on small rodents, hatchling birds, and lizards. They feed less frequently on snakes (including rattlesnakes) and toads. I have found gravid females during the late spring and early to mid-summer in northern Baja California, which indicates a spring breeding season. In the Cape Region gravid females may be found during July and August.

Lampropeltis catalinensis

ISLA SANTA CATALINA KINGSNAKE

Lampropeltis catalinensis Van Denburgh and Slevin 1921c:397. Type locality: "Santa Catalina Isla [BCS], Gulf of California, Mexico." Holotype: CAS 50515

IDENTIFICATION *Lampropeltis catalinensis* can be distinguished from all other snakes in the region of study by having round pupils; large head plates; a rostral scale that is not greatly enlarged; smooth dorsal scales; an entire anal plate; no dark dorsal bar across the top of the head; and small yellow dots on the sides of the body.

RELATIONSHIPS AND TAXONOMY See *L. getula.* Blaney (1977) and Grismer (1999b) discuss taxonomy.

DISTRIBUTION *Lampropeltis catalinensis* is endemic to Isla Santa Catalina (see map 6.11).

DESCRIPTION This species is known only from a single adult male. Body elongate and subcircular in cross section, 984 mm SVL; head rounded in dorsal profile, not distinct from neck; eyes moderately sized, pupils round; tail relatively short, 16 percent of SVL; head scales large and platelike; 8 supralabials; genials contact medially; 9 infralabials; dorsal body scales smooth and imbricate; 23 scale rows at midbody; 228 ventral scales; anal plate single; 63 subcaudals. **COLORATION** Dorsal ground color purple; top of head immaculate; centers of paravertebral and lateral body scales yellow; edges widely bordered by ground color, making body appear as though it has small yellow spots on its sides; wide, nearly immaculate purple vertebral stripe; weak remnants of transverse banding present as small white spots on single scales, occurring at regular intervals; ventrum purple, edged laterally, to varying degrees, in yellow.

NATURAL HISTORY All that is known of this species' natural history is that one was dug out of the center of a cactus on 12 June. A shed skin of probable identity, reported by Cliff (1954b), was found near a small spring.

REMARKS Blaney (1977) demonstrated that this species is similar in pattern to *L. getula splendida,* clearly indicating that it was derived from a Sonoran progenitor. I have made many attempts during various times of year and climatic conditions to find additional specimens, but with no success. Some have questioned the validity of this record, since no other specimens have been found. The distinctiveness of the holotype from other *L. getula* clearly indicates, however, that this population is valid and has been on Isla Santa Catalina for some time.

Masticophis Baird and Girard

WHIPSNAKES AND RACERS

The genus *Masticophis* is composed of 10 to 12 moderately sized species of snakes (Camper and Dixon 1994; Grismer 1990b, 1999b; Johnson 1977; Wilson 1971) that collectively range throughout much of the United States, Mexico, Central America, and south into South America. *Masticophis* are elongate, semiarboreal to terrestrial snakes that are notoriously swift, agile, often aggressive diurnal predators. They chase down their prey with great speed, subdue them with their strong jaws, and swallow them alive. *Masticophis* commonly forage with the anterior quarter of their bodies elevated off the ground and their heads moving from side to side. Unlike many other snakes, which rely primarily on olfaction for initially locating prey, *Masticophis* detects movement as well. By elevating their heads above the ground, these snakes gain a visual advantage. The various species of *Masticophis* frequent a number of habitats, ranging from open deserts to tropical forests, and all are adept climbers and swimmers. Seven species of *Masticophis* occur in the region of study, representing three distinctive species groups: The *M. lateralis* group contains *M. lateralis, M. aurigulus,* and *M. barbouri;* the *M. flagellum* group contains *M. flagellum* and *M. fuliginosus,* and the *M. taeniatus* group contains *M. bilineatus* and *M. slevini.*

FIG 6.23 *Masticophis lateralis*

CALIFORNIA STRIPED RACER; CULEBRA RAYADA

Leptophis lateralis Hallowell 1853:237. Type locality: "California," restricted by Schmidt (1953) to "San Diego." Holotype: ANSP 5365

IDENTIFICATION *Masticophis lateralis* can be distinguished from all other snakes in the region of study by having a thin, elongate body; a dark olive drab ground color; no orange neck ring; conspicuous dorsolateral stripes on scale rows 3 and 4 continuing the length of the body; and mottling on the chin and throat.

RELATIONSHIPS AND TAXONOMY *Masticophis lateralis* forms a natural group with the sister species *M. aurigulus* of the Cape Region and *M. barbouri* of Islas Partida Sur and Espíritu Santo (Grismer 1990b). Grismer (1990b, 1999b) and Ortenburger (1928) discuss taxonomy.

DISTRIBUTION *Masticophis lateralis* ranges continuously from northern California south around both sides of the Central Valley to at least 25 km south of El Rosario in northwestern Baja California. South of El Rosario, it is known from disjunct localities throughout the peninsula south to Cañón de los Reyes, 37 km north of La Paz (Grismer 1990b 1994b; Grismer and Mahrdt 1996b; Grismer and McGuire 1993; Jennings 1983), but is probably continuous in distribution (map 6.12).

DESCRIPTION Body thin and elongate, subcircular in cross section, reaching 1,065 mm SVL; head large and elongate, moderately distinct from neck, somewhat pointed in dorsal profile; prominent orbital ridge; anterior loreal scale square; eyes large, pupils round; tail long, 34 percent of SVL in males and females; head scales large and platelike; 7 or 8 (usually 8) supralabials; genials contact medially; 9 to 11 (usually 10) infralabials; dorsal body scales smooth and imbricate; 17 scale rows at midbody; 182 to 216 ventral scales in males and females; 119 to 153 divided subcaudal scales in males and females; anal plate single. **COLORATION** Dorsal ground color dark olive drab, almost black, and extending onto lateral margins of ventral scales; ground color often fades to brown posteriorly; discontinuous cream-colored preorbital stripe running from anterior margin of one eye to anterior margin of other eye; top of head dark and nearly immaculate, except for light temporal spot; ground color of labials and gular region white; labials and gular region mottled; union of dark ground color of head and body with white color of labials and gular region forms an undulate border posterior to eye; light dorsolateral stripe on dorsal scale rows 3 and 4 continues to base of tail but fades rapidly on tail; dorsolateral stripe edged with slightly darker ground color; throat orangish, blending to yellowish on belly, then to pinkish on subcaudal region; paired dark spots on ventral scales of throat often present.

FIGURE 6.23 *Masticophis lateralis* (California Striped Racer) from Cañón de los Reyes, BCS

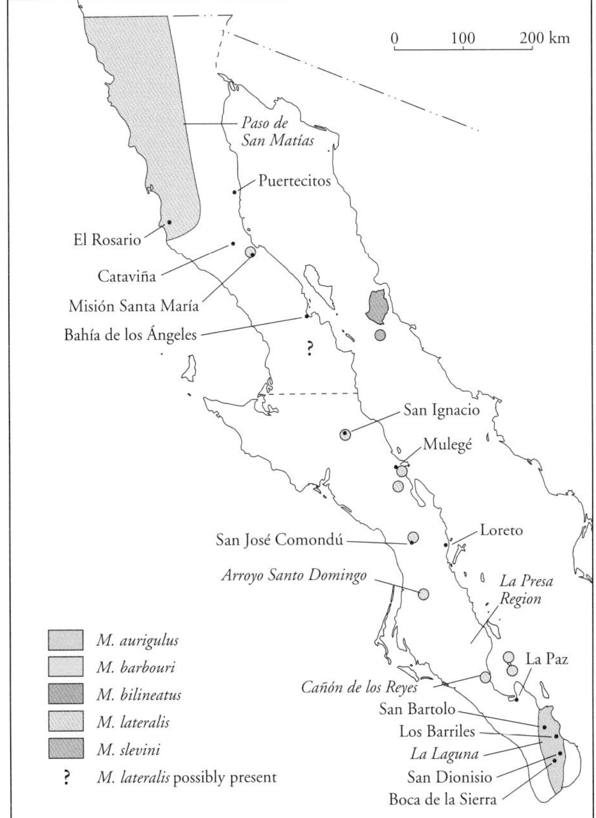

MAP 6.12 Distribution of *Masticophis lateralis* (California Striped Racer), *M. aurigulus* (Cape Striped Racer), *M. barbouri* (Isla Espíritu Santo Striped Racer), *M. slevini* (Isla San Esteban Whipsnake), and *M. bilineatus* (Sonoran Whipsnake)

GEOGRAPHIC VARIATION There is very little geographic variation in the color pattern of *M. lateralis* in Baja California. Specimens from the Magdalena Region tend to be slightly lighter in overall ground color, which highlights the darkened area of ground color bordering the dorsolateral stripes. Snakes from the Vizcaíno and California regions are generally darker and do not show the highlighting of the dorsolateral stripes.

NATURAL HISTORY In northern Baja California, *M. lateralis* is most common in the chaparral habitats of the California Region. It has also been found, however, in montane areas in the Baja California Coniferous Forest Region and in canyons along the eastern escarpment of the Sierra San Pedro Mártir (Linsdale 1932; Murray 1955; Welsh 1988). South of the California Region, striped whipsnakes seem to be restricted to the cooler or more mesic areas within the arid Vizcaíno and Magdalena regions. I observed a road kill in the Vizcaíno Region, 25 km south of El Rosario. At this point, Mexican Highway 1 courses through a narrow arroyo with a mixture of chaparral and xerophytic Vizcaíno flora. All the vegetation in this arroyo is cloaked with Spanish Moss *(Ramalina reticulata),* indicative of heavy precipitation from marine advectional fogs. Specimens have also been found in the oases of Misión Santa María, San Ignacio, Mulegé, San José Comondú, and Arroyo Santo Domingo (Grismer 1990a, 1994b). The southernmost specimen comes from just 37 km north of La Paz in Cañón de los Reyes (Grismer and Mahrdt 1996b). This is a canyon along the Gulf coast that appears to be relatively arid, and thus the occurrence of *M. lateralis* here was surprising.

Masticophis lateralis may be active year-round south of San Ignacio. In the north, its activity peaks from approximately late March through mid-October, depending on the weather. It is a quick, agile, diurnal hunter that emerges in the morning to bask and search for small frogs, lizards, birds, and rodents. Hatchlings may also eat arthropods. Welsh (1988) reported finding a specimen in a hollow oak log eating a deer mouse *(Peromyscus).* I observed a Striped Racer exiting a hollow palm log 2 m off the ground at Misión Santa María with a Baja California Brush Lizard *(Urosaurus lahtelai)* in its mouth.

In the California Region, *M. lateralis* is frequently observed moving through the chaparral in search of prey. It is alert, wary, and difficult to approach, and it flees from danger at surprising speeds. Often, when escaping predators, Striped Racers will ascend the branches of nearby vegetation. The breeding season of *M. lateralis* begins in late March, and I have found gravid females from May through July. Hatchlings are most commonly observed from late August through October.

FIG 6.24 *Masticophis aurigulus*

CAPE STRIPED RACER; CINTURIATA, CHIRRIONERA DEL CABO

Drymobius aurigulus Cope 1861:301. Type locality: "Cape San Lucas, Lower California," BCS, Mexico. Holotype: USNM 5793

IDENTIFICATION *Masticophis aurigulus* can be distinguished from all other snakes in the region of study by having a thin, elongate body; a dark olive drab ground color; no orange neck ring; conspicuous dorsolateral stripes on scale rows 3 and 4, which are expanded anteriorly at regular intervals and fade posteriorly two-thirds of the way down the body; and an immaculate chin and throat.

DISTRIBUTION *Masticophis aurigulus* is known only from the Cape Region of Baja California, along the eastern slopes of the Sierra la Laguna (Grismer 1990c; see map 6.12).

RELATIONSHIPS AND TAXONOMY See *M. lateralis.*

FIGURE 6.24 *Masticophis aurigulus* (Cape Striped Racer) from Boca de la Sierra, BCS

DESCRIPTION Same as *M. lateralis,* except for the following: adults reach 1,103 mm SVL; anterior loreal scale rectangular; 8 supralabials; 192 to 207 ventral scales; 120 to 165 divided subcaudals. **COLORATION** Continuous cream-colored preorbital stripe running from anterior margin of one eye to anterior margin of other eye; labials and gular region immaculate; union of dark ground color of head and body with white color of labials and gular region forms a straight border posterior to eye; top of head slightly mottled in large specimens; dorsolateral stripes either orange or lemon yellow; stripes fade posteriorly two-thirds of the way down the body; anterior portions of stripes expanded at regular intervals, with faint cross-banding between the expansions; 2 rows of faint secondary stripes present on ventrolateral portion of body; secondary stripes fade one-third to half the way down body; ventrum immaculate, orange to salmon-colored anteriorly, blending posteriorly to cream-yellow on belly and faint pink in caudal region.

NATURAL HISTORY *Masticophis aurigulus* appears to be primarily associated with the eastern foothills surrounding the Sierra la Laguna (Grismer 1990b,c) although it is occasionally found in the pine-oak woodland of the Sierra la Laguna Region (Alvarez et al. 1988). No specimens have yet been observed on the western side of the Sierra la Laguna. *Masticophis aurigulus* is most common in the heavily vegetated, rocky arroyos draining the eastern face of this range. These arroyos usually have permanent water and a mixture of Arid Tropical and Sierra la Laguna region flora. I did observe one specimen, however, on the beach at Los Barriles, crawling through the lower branches of sparse xerophytic vegetation. *Masticophis aurigulus* is probably active year-round, and I have observed specimens abroad from March through November. Like other racers, *M. aurigulus* is a swift and agile diurnal predator that forages primarily for lizards, small birds, and rodents. I have seen individuals hunting at the water's edge in Boca de la Sierra, San Bartolo, and Arroyo San Dionisio in areas where Pacific Treefrogs *(Hyla regilla)* were abundant. In Aguas Calientes, between San Dionisio and Boca de la Sierra, I observed the head of an individual protruding from a rock crack 3 m above the ground, with several Black-Tailed Brush Lizards *(Urosaurus nigricaudus)* nearby. Very little is known about the reproductive biology of this species.

Masticophis barbouri

ISLA ESPÍRITU SANTO STRIPED RACER; CINTURIATA

Coluber barbouri Van Denburgh and Slevin 1921b:98. Type locality: "Isla Partida, Espiritu Santo Island, Lower California [BCS], Mexico." Holotype: CAS 49157

IDENTIFICATION *Masticophis barbouri* can be distinguished from all other snakes in the region of study by having an elongate, thin body; a dark olive drab ground color; no orange neck ring; conspicuous dorsolateral stripes on scale rows 3 and 4, which are expanded anteriorly at regular intervals and continue the length of the body; and immaculate labial scales, gular region, and throat.

RELATIONSHIPS AND TAXONOMY See *M. lateralis.*

DISTRIBUTION Endemic to the Gulf islands of Espíritu Santo and Partida Sur (Grismer 1990b,c; see map 6.12).

DESCRIPTION Same as *M. lateralis,* except for the following: adults reach 955 mm SVL; anterior loreal scale rectangular; 8 or 9 supralabials; 196 to 214 ventral scales; 122 to 150 divided subcaudals. **COLORATION** Dorsal ground color dull light green; continuous cream-colored preorbital stripe running from anterior margin of one eye to anterior margin of other eye; labials, gular region, and throat immaculate; union of dark ground color of head and body with white color of labials and gular region forms a straight border posterior to eye; top of head slightly mottled in large specimens; light dorsolateral stripes continuous to vent, and bordered in black; anterior portions of stripes greatly expanded at regular intervals with distinct cross-banding between expansions; 2 rows of faint, secondary ventrolateral stripes on body; secondary stripes fade one-third to half the way down the body; ventrum immaculate, orange to salmon-colored anteriorly on throat, blending posteriorly to cream-yellow on belly and faint pink in caudal region.

NATURAL HISTORY Islas Espíritu Santo and Partida Sur are highly eroded, rocky islands crossed with several large arroyos that drain the islands to the west. Although they lie within the Central Gulf Coast Region and are dominated by xerophilic shrubs, a significant portion of their flora is derived from Arid Tropical Region species because of the islands' association with the Cape Region. *Masticophis barbouri* is known from only eight specimens, and very little is known about its natural history. These specimens all come from the more mesic, vegetated, western sides of the islands. Like its closest relative, *M. aurigulus* from the Cape Region, *M. barbouri* is most commonly found near rocks. I observed an adult during early September at 1130 hours sunning itself on a large volcanic boulder on a south-facing hillside at Bahía Candelero. On my approach, it rapidly came down off the rock and escaped beneath an adjacent boulder. Francisco Reynoso informed me that he observed two other specimens at midday on the rocky hillsides. I also saw a specimen crawling through low-growing vegetation on the beach. Striped Racers have been collected from mid-March through early November.

FIG 6.25 *Masticophis bilineatus*

SONORAN WHIPSNAKE; CULEBRA LÁTIGO

Masticophis bilineatus Jan 1863:65. Type locality: None designated. Restricted to "Guaymas," Sonora, Mexico, by Smith and Taylor (1950:344). Holotype: MTKD 15523

IDENTIFICATION *Masticophis bilineatus* can be distinguished from all other snakes in the region of study by having a thin, elongate body; conspicuous dorsolateral stripes on dorsal scale rows 3 and 4, continuing the length of the body; paired light spots on the anterior corners of the dorsal scales; and dark spots on the ventral scales forming a median row extending 10 to 15 scales posterior to the head.

RELATIONSHIPS AND TAXONOMY *Masticophis bilineatus* is the sister species of *M. slevini* of Isla San Esteban; together these species form the sister lineage to a group containing *M. taeniatus* and *M. schotti* of the western United States and northern Mexico (Camper and Dixon 1994).

DISTRIBUTION *Masticophis bilineatus* ranges from central Arizona south through western Mexico to Colima (Camper and Dixon 1994; Camper 1996). It also occurs on Isla Tiburón, Sonora, in the Gulf of California (see map 6.12).

DESCRIPTION Body thin and elongate, subcircular in cross section, reaching 1,173 mm SVL; head large and elongate, moderately distinct from neck, somewhat pointed in dorsal profile; prominent orbital ridge; anterior loreal scale square; eyes large, pupils round; tail long, up to 31 percent of SVL; head scales large and platelike; 7 to 9 (usually 8) supralabials; genials contact medially; 8 to 11 (usually 9) infralabials; dorsal body scales smooth and imbricate; 17 scale rows at midbody; 182 to 221 ventrals, 120 to 167 divided

FIGURE 6.25 *Masticophis bilineatus* (Sonoran Whipsnake) from near Mocuzari, Sonora. (Photograph by Cecil R. Schwalbe)

subcaudals; anal plate single. **COLORATION** Dorsal ground color dark olive green fading to light green posteriorly and extending onto lateral margins of ventral scales; discontinuous, cream-colored preorbital stripe running from anterior margin of one eye to anterior margin of other eye; top of head dark and immaculate; paired cream-colored spots on anterior corners of dorsal scales; labials and gular region cream-colored; dark mottling on infralabials, chinshields, and gulars may be present; when present, it extends posteriorly as a median row on first 10 to 15 ventral scales; rest of belly usually immaculate; dark spotting may or may not occur in subcaudal region; union of dark ground color of head and body with white color of labials and gular region forms straight border posterior to eye; light dorsolateral body stripes on the edges of dorsal scale rows 3 and 4 and the edges of dorsal scale rows 1 and 2 fade posteriorly approximately halfway down body; often a third, light-colored stripe occurs on lower half of first dorsal scale row and lateral margins of ventral scales.

NATURAL HISTORY On Isla Tiburón, *M. bilineatus* is ubiquitous, and it seems to be more prevalent in the vicinity of vegetation. I have observed more Whipsnakes in wide arroyos with thick vegetation than in dry, rocky arroyos or on rocky hillsides. *Masticophis bilineatus* on Isla Tiburón has been collected from April through September, and adults are probably active from at least March through early November. Juveniles may be active year-round, depending on the weather. The Sonoran Whipsnake is an agile, diurnal, active forager that basks in the morning and searches for food among rocks and the lower branches of vegetation. It preys on lizards, birds, small mammals, and occasionally other snakes. Nothing has been reported on the reproductive biology of *M. bilineatus* from Isla Tiburón, but presumably it is not different from that of snakes from adjacent mainland Sonora, which have spring and early summer breeding seasons and a late summer to early fall hatching period (Goldberg 1998; Vitt 1975).

FOLKLORE AND HUMAN USE The Seri Indians believe that *Masticophis* stores venom in its teeth and that its bite is poisonous. Nonetheless, they consider it a food source (Malkin 1962).

FIG 6.26 *Masticophis slevini*

ISLA SAN ESTEBAN WHIPSNAKE; CULEBRA LÁTIGO

Masticophis bilineatus slevini Lowe and Norris 1955:93. Type locality: "San Estéban Island, Gulf of California, Sonora, Mexico." Holotype: SDSNH 3826

IDENTIFICATION *Masticophis slevini* can be distinguished from all other snakes in the region of study by having a thin, elongate body; conspicuous dorsolateral stripes on dorsal scale rows 3 and 4, continuing the length of the body; paired light spots on the anterior corners of the dorsal scales; and dark spots on the ventral surfaces forming a median row extending for at least 37 scales posterior to the head.

RELATIONSHIPS AND TAXONOMY See *M. bilineatus.* Camper and Dixon (1994) and Grismer (1999b) discuss taxonomy.

DISTRIBUTION Endemic to the Gulf island of San Esteban, Sonora (see map 6.12).

DESCRIPTION Same as *M. bilineatus,* except for the following: adults reach 900 mm SVL; 8 or 9 (usually 8) supralabials; 9 or 10 (usually 9) infralabials; 197 to 210 ventral scales; 132 to 142 divided subcaudals. **COLORATION** Dark mottling on infralabials, chinshields, and gulars, extending posteriorly as a median row on at least the first 37 ventral scales.

NATURAL HISTORY Isla San Esteban is a rocky, mountainous island with wide, sandy arroyos. It lies within the Central Gulf Coast Region and is dominated by a thorn scrub vegetation. *Masticophis slevini* is most commonly observed along the coast, near shrubs, at the mouths of arroyos. The interior of Isla San Esteban is well-vegetated, with many sizable shrubs and trees. However, toward the interior of the island, the lizard density (presumably this species' primary prey base) declines. Therefore, *M. slevini* are most abundant near the coastline, although one snake was found on one of the higher peaks of the island carrying a hatchling *Sauromalus varius* in its mouth (Sylber, personal communication, 1996). *Masticophis slevini* has been observed abroad from April through September. Its activity proba-

FIGURE 6.26 *Masticophis slevini* (Isla San Esteban Whipsnake) from Arroyo Limantur, Isla San Esteban, Sonora. (Photograph by Howard Lawler)

FIGURE 6.27 *Masticophis flagellum* (Coachwhip) from Sierra los Cucapás, BC

bly extends from March through early November, and, depending on the weather, juveniles may be active year-round. Nothing is known of its reproductive biology.

FOLKLORE AND HUMAN USE See *M. bilineatus.*

FIG 6.27 **Masticophis flagellum**

COACHWHIP; CHIRRIONERA

Coluber flagellum Shaw 1802:475. Type locality: "Carolina and Virginia," restricted to "Charleston, South Carolina" (Schmidt 1953). Holotype: None designated

IDENTIFICATION *Masticophis flagellum* can be distinguished from all other snakes in the region of study by having a thin, elongate body; relatively large head with a distinct orbital ridge; no stripes on the body; and no zigzag pattern of thin dark lines following the edges of the scales.

RELATIONSHIPS AND TAXONOMY Two subspecies are currently recognized in Baja California (Wilson 1971, 1973). *Masticophis flagellum fuliginosus* ranges from southern San Diego County, California, south throughout Baja California, but is replaced by *M. f. piceus* in the Lower Colorado Valley Region of northeastern Baja California (Wilson 1971, 1973). I have found both subspecies at Bahía San Luis Gonzaga and Paso de San Matías, and have not observed any signs of intergradation between them (Grismer 1994a). Wilson (1971) indicated that a similar situation may be occurring in San Diego County as well. Based on these observations, I consider *M. f. fuliginosus* a separate species.

DISTRIBUTION *Masticophis flagellum* ranges widely over the southern United States and northern Mexico (Wilson 1973). In northeastern Baja California, it extends from the foothills of the Sierra Juárez and Sierra San Pedro Mártir east to the Gulf coast and south to at least Bahía San Luis Gonzaga (Grismer 1994a; map 6.13). It is also known from Islas Tiburón and Dátil off the coast of Sonora in the Gulf of California.

DESCRIPTION Body thin and elongate, subcircular in cross section, reaching 1,290 mm SVL; head large and elongate, moderately distinct from neck, somewhat pointed in dorsal profile; prominent orbital ridge; anterior loreal scale rectangular; eyes large, pupils round; tail long, 22 to 28 percent of SVL; head scales large and platelike; 7 to 9 supralabials; genials contact medially; 8 to 12 infralabials; dorsal body scales smooth and imbricate; 17 scale rows at midbody; 183 to 205 ventral scales in males and females, 95 to 120 divided subcaudal scales in males and females; anal plate single. **COLORATION** Two very distinctive color patterns exist in the region of study. That of northeastern Baja California corresponds to the subspecies *Masticophis flagellum piceus,* and that of Islas Tiburón and Dátil corresponds to the subspecies *M. f. cingulum* (Wilson 1971). Both are used here as pattern classes. PICEUS: Dorsal ground color tan to dull red; discontinuous or continuous preorbital stripe; top of head dark, and may or may not have dark mottling; nasals, prefrontals, and frontals edged in white; labials and gular region cream-colored; dark mottling on infralabials; wide cross bands on neck, ranging from black to ground color; dark nape pattern less evident in young; large brownish blotches outlined in black on chinshields and gulars; blotches extend posteriorly on ventrum for approximately 65 ventral scales, then fade beyond recognition; rest of ventrum dull white and immaculate; union of dark ground color of head and body and white of labials and gular region extremely irregular. CINGULUM: Pattern extremely variable; dorsal ground color tan to dull white; preorbital stripe absent; top and sides of head reddish with large, irregularly shaped black blotches; labials and gular region cream-colored. Three color pattern phases exist (Wilson 1971): (1) dark mottling on infralabials; pattern of wide, dark red bands separated by light pink interspaces; (2) anterior one-third to half of body black, posterior red; (3) solid black dorsal coloration. Intermediates between these pattern phases also exist.

GEOGRAPHIC VARIATION See description of coloration.

NATURAL HISTORY The pattern class Piceus inhabits the Lower Colorado Valley Region of northeastern Baja California. These snakes are commonly found in the open Creosote Bush *(Larrea tridentata)* and Bursage *(Ambrosia dumosa)* flats, as well as in the rocky washes and foothills. I have also observed Coachwhip Snakes in the salt marshes near the coast at Bahía San Luis Gonzaga. Like other *Masticophis, M. flagellum* is diurnal and actively chases its prey, although Jones and Whitford (1989) report it being an effective ambush predator as well. I observed a large individual pursuing a Western Whiptail *(Cnemidophorus tigris)* in the Sierra San Felipe for approximately three minutes. The Coachwhip Snake crawled swiftly toward the Whiptail with the anterior fifth of its body elevated and its head parallel to the ground. When the Whiptail was motionless, the Coachwhip stopped crawling and moved its head from side to side. Eventually, the Coachwhip closed in on the Whiptail and came within lunging distance to capture it. Coachwhips are also known to prey on small birds, mammals, and other snakes, including Sidewinder Rattlesnakes *(Crotalus cerastes)*. Nabhan (forthcoming) reports finding a Coachwhip swallowing a Tiger Rattlesnake *(C. tigris)* at Ensenada de los Perros on Isla Tiburón.

I have observed gravid female Piceus in late April, May, and June in the Sierra los Cucapás, Paso de San Matías, and San Felipe, respectively. Hatchlings have been collected south of Puertecitos and at La Puerta during August and early September. These observations suggest that Piceus is a spring breeder and lays its eggs in the early summer.

Nothing has been reported concerning the natural history of Cingulum on Isla Tiburón. I have observed individuals abroad from late March through September. All appeared to be foraging for food in a wide, rocky arroyo among large shrubs. Zweifel and Norris (1955) recorded a mating pair in Sonora during early August.

FOLKLORE AND HUMAN USE See *M. bilineatus*.

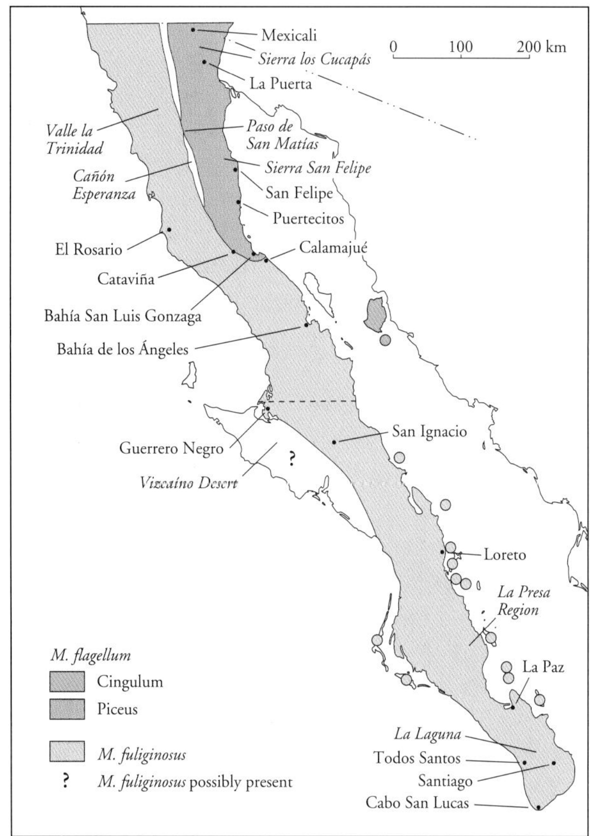

MAP 6.13 Distribution of *Masticophis fuliginosus* (Baja California Coachwhip) and the pattern classes of *M. flagellum* (Coachwhip)

FIGS. 6.28–30

Masticophis fuliginosus

BAJA CALIFORNIA COACHWHIP; CHIRRIONERA

Zamensis lateralis fuliginosus Cope 1895:679. Type locality: "Santa Margarita Island, Baja California del Sur, Mexico." Syntypes: USNM 15135 to 36

IDENTIFICATION *Masticophis fuliginosus* can be distinguished from all other snakes in the region of study by having a thin, elongate body; a relatively large head with a distinct orbital ridge; no stripes on the body; and wide, dark brown to black neck bands followed by a zigzag pattern of thin dark lines following the edges of the scales. Some populations may have a solid black or gray-brown to yellow dorsum.

RELATIONSHIPS AND TAXONOMY See *M. flagellum*.

DISTRIBUTION *Masticophis fuliginosus* ranges from extreme southwestern California south to Cabo San Lucas (Wilson 1973). In northern Baja California, *M. fuliginosus* ranges widely throughout the California Region but does not occur in the upper elevations of the Sierra San Pedro Mártir (Welsh 1988) or the Lower Colorado Valley Region of northeastern Baja California. It contacts the Gulf coast in the vicinity of Bahía Calamajué and continues south throughout the rest of the peninsula, but avoids the Vizcaíno Desert. Welsh (1988) reported observing a "black racer" along the lower east scarp of the Sierra San Pedro Mártir, at the mouth of Cañón Esperanza, which he could

FIGURE 6.28 Juvenile *Masticophis fuliginosus* (Baja California Coachwhip) from San Bruno, BCS

FIGURE 6.29 Adult *Masticophis fuliginosus* (Baja California Coachwhip) from Isla San José, BCS

not identify. Given that there are no black-phase *M. flagellum* known from Baja California (Welsh and Bury 1984; Wilson 1973), it is likely that the specimen was an *M. fuliginosus*. The black phase of this species is very common in northern Baja California (Wilson 1973). *Masticophis fuliginosus* is also known from the Pacific islands of Magdalena and Santa Margarita and the Gulf islands of Carmen, Cerralvo, Coronados, Danzante, Espíritu Santo, Monserrate, Partida Sur, San Ildefonso, San José, and San Marcos (see map 6.13). Its absence from the large continental island of Ángel de la Guarda is perplexing.

DESCRIPTION Same as *M. flagellum*, except for the following: adults reach 1,729 mm SVL; tail 20.9 to 29.5 percent of SVL; 8 to 11 infralabials; 175 to 205 ventral scales; 99 to 129 divided subcaudals. **COLORATION** Color pattern of this species is extremely variable. Wilson (1970) gave an excellent, detailed explanation that is summarized here and amended as necessary. Two color pattern phases occur in the region of study. DARK UNICOLORED PHASE: Body black, golden brown, or gray-brown with varying degrees of darker, wide neck bands; scales of anterolateral portion of dorsum have pale lateral edges, appearing as narrow lines along junction of scales; ventrolateral portion of body often lineate or stippled with dark brown; varying portions of ventrum cream-colored and often with brown spots. LIGHT PATTERNED PHASE: Dorsal ground color ivory white, tan, gray, or yellow with zigzag pattern of thin dark lines following dorsal scale edges; discontinuous or continuous preorbital stripe present; top of head dark with or without light brown mottling; nasals, prefrontals, and frontals not edged in white; labials and gular region light-colored; dark mottling on infralabials; wide cross bands on neck, ranging from black to ground color; large brownish or black blotches on chinshields and gulars; blotches extend posteriorly on neck; union of dark ground color of head and body with white of labials and gular region extremely irregular. JUVENILES: In both phases: head brown with cream-colored blotches laterally; body light to dark brown with varying degrees of mottling; dark brown neck bars separated by cream-colored lines; ventrum cream with dark spotting on neck; colors usually become darker and more intense with age.

GEOGRAPHIC VARIATION The dark unicolored phase is the most common throughout Baja California, except in the southernmost portion of the Cape Region, from Todos Santos in the west and Santiago in the east south to Cabo San Lucas. Here, only the light patterned phase occurs. Farther north, the light patterned phase occurs only sporadically, having been observed from San Ignacio, the vicinity of Guerrero Negro, and Valle la Trinidad. Wilson (1971) noted that Isla Carmen had the light patterned phase, although I have observed a dark unicolored phase specimen on this island as well. The nearby Isla Monserrate and the Pacific island of Magdalena also have both light and dark phases, whereas the other islands tend to have only a single phase. The light patterned phase is found on the Gulf islands of San Ildefonso, Coronados, and Espíritu Santo and the Pacific island of Santa Margarita. On Isla Cerralvo, only the dark unicolored phase is present, although some specimens have gray to orangish-pink throats and immaculate white ventra. The light patterned phase of whipsnakes from Isla San José is bright lemon yellow.

NATURAL HISTORY *Masticophis fuliginosus* is commonly found throughout all of Baja California, except in the Baja California Coniferous Forest and Lower Colorado Valley regions (Welsh 1988; Wilson 1973). It is common in all habitats, ranging from the banks of marshlands (Linsdale 1932)

FIGURE 6.30 Adult *Masticophis fuliginosus* (Baja California Coachwhip) from Isla Carmen, BCS

to coastal sand dunes (Bostic 1971; Leviton and Banta 1964). I have observed coachwhips in rocky arroyos, on rocky hillsides, and within thick thorn forests in the Vizcaíno, Magdalena, and Central Gulf Coast regions.

Masticophis fuliginosus is generally active year-round south of San Ignacio, with an activity peak from April through October. I have observed specimens crossing the road just south of Loreto on warm sunny days in late December and have found snakes hibernating in hollow Datilillo *(Yucca valida)* stumps on the Magdalena Plain on cold sunny days during early January. Leviton and Banta (1964) reported active specimens during December in the Cape Region, as well as dormant individuals within the cavities of decaying logs of Cardón Cactus *(Pachycereus pringlei).* North of San Ignacio, *M. fuliginosus* is generally inactive during the winter but begins to emerge around mid-March. By April it is usually the most common species of diurnal snake observed.

Coachwhips are highly active, aggressive, diurnal predators. As such, they are commonly observed foraging for prey with the anterior quarter of their bodies elevated off the ground. Bostic (1971) reported observing a Coachwhip foraging from shrub to shrub along open sand dunes, Cliff (1954a) reported specimens foraging in sandy flats, and Welsh (1988) observed Coachwhips in open areas among scrub vegetation. When abroad, *M. fuliginosus* is nearly always found close to brush (Alvarez et al. 1988; Bostic 1971; Linsdale 1932; Welsh 1988). I have often observed Coachwhips crawling among the lower branches of shrubs during both day and night. I found an adult at Juncalito 4 m above the ground in a salt cedar *(Tamarisk)* raiding the nest of a House Finch *(Caprodaucus mexicanus).* Cliff (1954a) has made similar observations on Isla Coronados, and I have observed this behavior many times on Islas Cerralvo and San José. *Masticophis fuliginosus* has been reported to use the hollow interiors of dead agave as retreats (Bostic 1971). I have observed this, as well as the use of hollow Datilillo, on numerous occasions. Linsdale (1932) reported finding a specimen beneath surface debris.

Masticophis fuliginosus preys on a wide variety of species but most commonly feeds on lizards, small snakes, birds, and small mammals. I observed a Coachwhip on Isla Cerralvo traveling from bush to bush catching small Isla Cerralvo Spiny Lizards *(Sceloporus grandaevus).* The snake's arrival in the bush would startle the lizards into moving, at which time the Coachwhip would locate them by sight and almost always successfully lunge for them. Near Santiago in the Cape Region, I watched one particularly large Coachwhip chase a Desert Iguana *(Dipsosaurus dorsalis).* The chase progressed for approximately 20 m at a surprisingly high speed before the snake gave up. On Isla Santa Margarita, I watched an adult Coachwhip chase an Orange-Throated Whiptail *(Cnemidophorus hyperythrus)* into a bush along the side of a hill, capture it, and eat it within seconds. Cliff (1954a) reported a specimen from Isla San José that had eaten a Ground Snake *(Sonora semiannulata).*

I have observed gravid *M. fuliginosus* in the Cape Region during early June and hatchlings from mid-July to mid-October. In northern Baja California, I saw a mating pair in late April and have seen hatchlings in early August. These observations suggest that *M. fuliginosus* has a typical spring breeding and summer to fall hatching period throughout its range.

Phyllorhynchus Cope

LEAF-NOSED SNAKES

The genus *Phyllorhynchus* is composed of two species of blunt-headed, terrestrial snakes (Klauber 1940b; McCleary and McDiarmid 1993). *Phyllorhynchus* are generally small, stout-bodied snakes that range throughout much of the Sonoran and Mojave deserts of the United States and northwestern Mexico. *Phyllorhynchus* frequents a variety of habitats but is most common in areas of loose soil where, during their nocturnal activities, snakes prey on lizards and their eggs. Leaf-nosed snakes derive their common name from the fact that the rostral scale is greatly enlarged and folds up over the front of the rostrum. The function of this enlarged scale is not entirely understood, although it may be related to uncovering the buried eggs of lizards.

FIG 6.31 ***Phyllorhynchus decurtatus***

SPOTTED LEAF-NOSED SNAKE; CULEBRITA

Phimothyra decurtata Cope 1868:310. Type locality: "Upper part of Lower California [BC]," Mexico. Holotype: ANSP 5489

IDENTIFICATION *Phyllorhynchus decurtatus* can be distinguished from all other snakes in the region of study by having a greatly enlarged triangular rostral scale and lacking a light middorsal stripe.

RELATIONSHIPS AND TAXONOMY *Phyllorhynchus decurtatus* is most closely related to *P. browni* of southern Arizona and extreme northwestern Mexico. Murphy and Ottley (1980), McCleary and McDiarmid (1993), and Grismer (1999b) discuss taxonomy.

DISTRIBUTION *Phyllorhynchus decurtatus* ranges widely throughout the Sonoran and Mojave deserts of the United States, Baja California, and northwestern mainland Mexico west of the Sierra Madre Occidental, south to central Sinaloa (McCleary and McDiarmid 1993). In Baja California, *P. decurtatus* ranges throughout the peninsula, except for the western portions of the cismontane California Region and the cool Pacific coastal areas from El Rosario south to Todos Santos (Grismer 1994d, 1996b). In the Cape Region, it is restricted to low-lying regions west of the Sierra la Laguna, north of Todos Santos. It also ranges along the east coast of the Cape Region from La Paz south to San Juan de los Planes, along the northeastern margin of the Sierra la Laguna. *Phyllorhynchus decurtatus* also occurs on the Gulf islands of Ángel de la Guarda, Cerralvo, Monserrate, San José, and San Marcos (map 6.14).

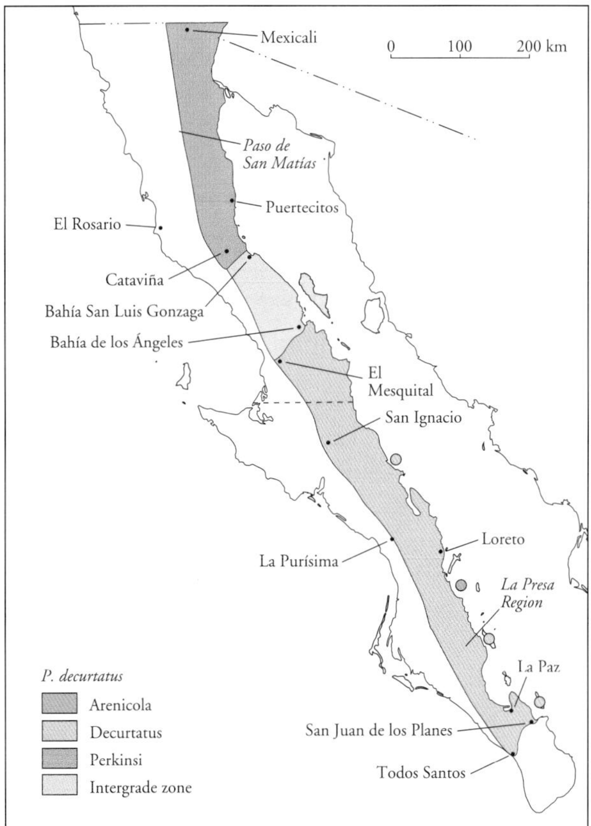

MAP 6.14 Distribution of the pattern classes of *Phyllorhynchus decurtatus* (Spotted Leaf-Nosed Snake)

FIGURE 6.31 Decurtatus pattern class of *Phyllorhynchus decurtatus* (Spotted Leaf-Nosed Snake) from La Paz, BCS

DESCRIPTION Body short and stout, cylindrical in cross section, reaching 435 mm SVL; head short, moderately distinct from neck, square in dorsal and lateral profile; greatly enlarged triangular rostral scale folding posteriorly over rostrum; eyes moderately protuberant, pupils elliptical; tail 11.2 to 16.4 percent of SVL in males, 8.0 to 10.8 percent of SVL in females; head scales large and platelike; 5 to 7 (usually 6) supralabials; genials contact medially; 7 to 10 (usually 8 or 9) infralabials; dorsal body scales imbricate and smooth anteriorly, usually keeled posteriorly in large males; 17 or 19 dorsal scale rows at midbody; 157 to 183 ventral scales in males, 167 to 196 in females; anal plate single; 32 to 42 divided subcaudals in males, 22 to 34 in females. **COLORATION** Dorsal ground color dull white to cream; varying amounts of dark mottling on top of head; dark, wide postorbital stripe may or may not continue through eye to varying degrees across top of head; 18 to 48 square to subcircular, brownish dorsal blotches; 2 to 9 brownish blotches on tail; 1 to 3 rows of brownish, irregularly shaped secondary rows of spots on sides of body; lateral body scales often densely stippled, rendering lateral spots inconspicuous; ventrum dull white to cream-colored and immaculate.

GEOGRAPHIC VARIATION The geographic variation of *P. decurtatus* in the region of study has been encompassed by the description of three subspecies: the peninsular *P. d. decurtatus* and *P. d. perkinsi* (Klauber 1935) and the insular *P. d. arenicola* (Savage and Cliff 1954; Murphy and Ottley 1980). All are treated here as pattern classes. Perkinsi range throughout the Lower Colorado Valley Region of northeastern Baja California (Welsh and Bury 1984) south to the vicinity of Bahía San Luis Gonzaga, where they begin to intergrade with Decurtatus of southern Baja California (Grismer 1996b). Perkinsi are characterized as having 157 to 167 ventral scales in males and 167 to 176 in females, and 33 to 36 subcaudals in males and 22 to 26 in females. They also have 18 to 33 narrow, transversely arranged, rectangular dorsal body blotches. The zone of intergradation between Perkinsi and Decurtatus is wide, extending from Bahía San Luis Gonzaga to at least 16 km southeast of El Mesquital, along the eastern edge of the Vizcaíno Desert (Grismer 1996b; Murray 1955; Soulé and Sloan 1966). Decurtatus have 164 to 183 ventral scales in males and 173 to 196 in females, 32 to 42 subcaudals in males and 24 to 34 in females, and 24 to 48 subcircular dorsal body blotches. In the north, where they begin to intergrade with Perkinsi, the dorsal blotches appear thinner and more rectangular. Individuals from the southern populations of Decurtatus tend to have fewer dorsal blotches and ventral scales. Leaf-Nosed Snakes from the vicinity of La Purísima, west of the Sierra la Giganta, have an orangish hue to their overall dorsal coloration, whereas those from the Gulf coast at that latitude are often exceptionally dark.

Phyllorhynchus decurtatus from Isla Monserrate was described as a distinct species, *P. arenicola* (Savage and Cliff 1954). Murphy and Ottley (1980) relegated it to a subspecies of *P. decurtatus.* As with other Decurtatus from central Baja California, Arenicola have distinctive lateral spots and wide dorsal blotches, many of which are spilt by a light medial area. In general, the other insular populations of *P. decurtatus* most closely resemble Leaf-Nosed Snakes of the adjacent peninsular regions. The only exception is the Isla Ángel de la Guarda population, which more closely resembles Perkinsi, having thin, transverse, rectangular dorsal blotches rather than the larger subcircular blotches of Decurtatus from the adjacent peninsula. As with *Coleonyx variegatus* of Isla Ángel de la Guarda, this coloration may have to do with the more northerly connection of the island to the peninsula near Bahía San Luis Gonzaga (Henyey and Bischoff 1973).

Leaf-Nosed Snakes from Isla San José have an overall yellow-orange hue. They have distinctive lateral spots, but the dorsal blotches are usually too irregularly shaped to be counted.

NATURAL HISTORY *Phyllorhynchus decurtatus* ranges throughout the Lower Colorado Valley Region south through the Vizcaíno, Magdalena, and Central Gulf Coast regions and into the northern portion of the Arid Tropical Region. In these areas, Leaf-Nosed Snakes are most often found in rocky habitats with sandy substrates, but they also can be common in the open flats of Creosote Bush *(Larrea tridentata)* and Bursage *(Ambrosia dumosa)* of the northeastern portion of the peninsula. Murphy and Ottley (1984) reported finding specimens beneath bushes and along arroyo edges. *Phyllorhynchus decurtatus* is also common in sand dune areas (Bostic 1971; Brattstrom 1953b; Klauber 1935) on islands and is often found in coastal dunes. During the day, Leaf-Nosed Snakes reside beneath rocks and decaying vegetation (Leviton and Banta 1964).

Phyllorhynchus decurtatus is generally active year-round in the Cape Region. In the northern portion of its distribution, however, it hibernates during the winter (Brattstrom 1953) and is active from March through October. *Phyllorhynchus decurtatus* is strictly nocturnal, is most active from dusk to midnight, and is often found abroad on relatively cold nights. It feeds almost exclusively on lizard eggs (Brattstrom 1953; Klauber 1935), most commonly those of Western Banded Geckos *(Coleonyx variegatus)* and Zebra-Tailed Lizards *(Callisaurus draconoides).* Klauber (1935) speculated that this species also feeds on the tails of *C. variegatus.* Brattstrom (1953) reported finding unidentified hymenopterans in the stomachs of Leaf-Nosed Snakes.

Klauber (1935) reported a *P. decurtatus* on Isla Monserrate being found during the evening with the anterior section of its body buried in a sand hummock; I have observed the same behavior on several occasions at Puerto Refugio on Isla Ángel de la Guarda. Additionally, at the foot of the Sierra la Giganta just south of Loreto, I found a specimen crawling along an exposed root of a Wild Fig *(Ficus palmeri)* on the bank of an arroyo. The snake moved slowly along the root, flicking its tongue and poking its head into shallow holes. It also made small excavations in the soft soil with sideways movements of its head. I suspect the snake was searching for buried squamate eggs.

I have found gravid Perkinsi most often in June and July, and occasionally in May. These observations are in accord with Brattstrom's (1953) experience in adjacent southern California of finding gravid individuals from April through August, but most commonly in June and July. In the Cape Region, I have found gravid Decurtatus only in July and hatchlings during September. Undoubtedly females are gravid in other months as well. Klauber (1935) and Brattstrom (1953) report that Perkinsi lays clutches of 2 to 4 eggs. Nothing is known of the number of eggs laid by Decurtatus. These observations suggest that *P. decurtatus* has a late spring to early summer breeding season.

Pituophis Holbrook

BULL, GOPHER, AND PINE SNAKES

The genus *Pituophis* is composed of five to seven species of large, heavy-bodied, terrestrial snakes (Duellman 1960;

Grismer 1994a, 1996c, 1997, 2001c; Kizirian 1987; Knight 1986; Reichling 1995; Rodríguez-Robles and Jesús-Escobar 2000; Sweet and Parker 1990). With additional systematic studies, however, this number may increase. *Pituophis* range through most of North America and Mexico. They are common diurnal species that prey primarily on small mammals (Rodríguez-Robles 1998). Gopher snakes are generally ubiquitous and are common in rural areas in association with humans. In fact, these snakes are often a conspicuous component of the wildlife in agricultural areas and are generally recognized by local inhabitants as helping to keep rodent populations under control. Three species of *Pituophis* occur in the region of study. Two *(P. vertebralis* and *P. insulanus)* are endemic to Baja California, and the third *(P. catenifer)* ranges into the northern portion of the peninsula (Grismer 1997, 2001c).

FIG 6.32

Pituophis catenifer

GOPHER SNAKE; TOPERA

Coluber catenifer Blainville 1835a:290. Type locality: "Californiae," restricted to "vicinity of San Francisco" by Schmidt (1953). Holotype: In MNHN of Paris.

IDENTIFICATION *Pituophis catenifer* can be distinguished from all other snakes in the region of study by having large head plates; subocular scales that do not prevent the supralabials from contacting the eye; keeled dorsal scales; a divided anal plate; a prominent black postorbital stripe; and no black subcaudal stripe.

RELATIONSHIPS AND TAXONOMY *Pituophis catenifer* contains several distinctive populations that show little or no sign of intergradation in zones of parapatry (Kizirian 1987; Knight 1986; Reichling 1995; Sweet and Parker 1990). A comprehensive reevaluation of this species is needed to formally delimit the number of species that actually occur under the name *Pituophis catenifer.* Grismer (1994a, 1996c, 1997, 2001c), Klauber (1946b), Rodríguez-Robles and Jesús-Escobar (2000), and Stull (1940) discuss taxonomy of Baja California populations.

DISTRIBUTION *Pituophis catenifer* ranges widely throughout North America and northern Mexico south into northern Baja California (Sweet and Parker 1990). In cismontane and montane northern Baja California, *P. catenifer* ranges to at least 42 km south of El Rosario (Grismer 1996c). In northeastern Baja California, it ranges as far south as Paso de San Matías and is known from the Pacific islands of Coronado Sur and San Martín and the Gulf island of Tiburón (map 6.15).

DESCRIPTION Body moderately stout, subcylindrical in cross section, reaching nearly 2 m SVL; head moderately distinct

FIGURE 6.32 Annectens pattern class of *Pituophis catenifer* (Gopher Snake) from El Rosario, BC

from neck, truncate to sharply rounded in dorsal and lateral profile; eyes moderately sized, pupils round; tail averages 13.6 to 17.7 percent of SVL in males and 12.5 to 16.0 percent of SVL in females; head scales large and platelike; usually 2 preoculars; fourth supralabial almost always in contact with eye; 7 to 10 (usually 8) supralabials; 11 to 16 (usually 12 or 13) infralabials; dorsal body scales imbricate, vertebral rows sharply keeled, dorsal scales less keeled laterally; usually 33, 35, or 37 dorsal scale rows at midbody, rarely 29; 223 to 251 ventral scales in males, 229 to 257 in females; anal plate single; 61 to 89 divided subcaudals in males, 50 to 82 in females. **COLORATION** Dorsal ground color dull white to cream; varying amounts of dark mottling on top of head; dark, vertical suborbital and oblique postorbital stripes; margins of labials edged in black; 34 to 106 black or brown dorsal body blotches; anterior body blotches often confluent anteriorly; 9 to 33 brownish blotches on tail; 2 or 3 secondary rows of brownish to black spots on sides of body; belly dull white to cream-colored and spotted in black; subcaudal region usually stippled in gray.

GEOGRAPHIC VARIATION The notable geographic variation in squamation and color pattern of *P. catenifer* from Baja California has prompted the designation of two peninsular and two insular subspecies (see Klauber 1946b; Sweet and Parker 1980). *Pituophis c. annectens,* recognized here as the pattern class Annectens, ranges throughout the California Region south to at least 42 km south of El Rosario. These snakes have a bluntly rounded snout, relatively long tail (average 17.7 percent of SVL in males and 15.5 percent in females); usually 31 or 33 dorsal scale rows at midbody; 223 to 242 ventrals in males and 230 to 253 in females, 75 to 89 subcaudals in males and 73 to 82 in females; 53 to 90 dark, very irregular dorsal body blotches that are confluent anteriorly and more rounded throughout the rest of the body; and 17 to 29 tail spots. Annectens from northernmost Baja

California, inland and south to the vicinity of Rancho San José, generally have black dorsal blotches and straw-colored sides. Near Ensenada, the dorsal blotches are dark brown, matching the color of the volcanic substrate. Inland, between Paso de San Matías and Mike's Sky Rancho, the color pattern of Annectens is generally more contrasted than in snakes from the coast, and some specimens have a slightly reddish cast. Gopher snakes from the Llano de San Quintín tend to have a lemon-yellow ventrum anteriorly. From approximately El Rosario south, adult Annectens have distinctly chocolate-colored dorsal blotches centrally, although the anterior and posterior blotches remain black. The juveniles resemble juveniles from the northern portion of the range.

Pituophis catenifer affinis, recognized here as the pattern class Affinis, occur in the Lower Colorado Valley Region of northeastern Baja California south to at least Paso de San Matías and on Isla Tiburón in the Gulf. They have a sharply rounded snout; relatively short tail (average 13.6 percent of SVL in males and 12.5 percent in females); usually 33 dorsal scale rows at midbody; 232 to 251 ventrals in males and 234 to 257 in females; 61 to 70 subcaudals in males and 50 to 63 in females; 34 to 63 brownish, biconcave dorsal body blotches; and 9 to 21 tail spots. Overall, they are much lighter in coloration than Annectens. The ground color of the head is tan and spotted with several conspicuous black markings. The dorsal ground color of the tail is nearly yellow, setting off the large, widely spaced brown caudal blotches.

Pituophis catenifer from Isla Coronado Sur has been described as the subspecies *P. c. coronalis* (Klauber 1946b) and is referred to here as the pattern class Coronalis. These snakes have large, platelike head scales that are irregularly fragmented; suborbital scales that prevent the supralabials from contacting the eye; 7 supralabials; 11 to 14 infralabials, with both series of labials often characterized by fused scales; 33 to 35 dorsal scale rows at midbody; 222 ventral scales in males, 229 to 333 in females; and 82 divided subcaudals in males, 69 to 71 in females. The dorsal ground color is buff or yellowish, suffused with brown punctations laterally; they have 64 to 70 narrow, circular, widely spaced dorsal body blotches; the anterior and posterior body blotches are black, with the anterior body blotches often confluent anteriorly; the central blotches are brownish, and there are 18 to 26 blotches on the tail. There are also 2 to 3 secondary rows of brownish to black spots on the sides of the body; the belly is dull white to cream-colored and spotted with black or gray; and the subcaudal region may be spotted.

Pituophis catenifer of Isla San Martín has been described as the subspecies *P. c. fuliginatus* (Klauber 1946b) and is referred to here as the pattern class Fuliginatus. These snakes have a single preocular; the fourth or fifth supralabials almost always contact the eye; there are 8 or 9 supralabials; 12 to 14 (usually 13) infralabials; 31 to 35 (usually 33) dorsal scale rows at midbody; 224 to 234 ventral scales in males and 236 to 245 in females; and 73 to 82 divided subcaudals in males and 67 to 72 in females. The dorsal ground color is dull white to cream. There is dark mottling on the top of the head; 55 to 70 irregular and confluent black to brown dorsal body blotches, with the anterior and posterior blotches black; central blotches are black, brown, or reddish-brown. There are 14 to 22 black blotches on the tail and 2 or 3 secondary rows of black spots on the sides of the body. The belly is dull white to cream-colored and spotted in black. Often a midventral light stripe in the subcaudal region is set off by the dark lateral margins of the subcaudals.

NATURAL HISTORY Annectens are ubiquitous in northwestern Baja California, ranging throughout the California and Baja California Coniferous Forest regions (Klauber 1946b; Welsh 1988). Although they are common in all habitats, they seem to prefer relatively flat, open areas. Annectens

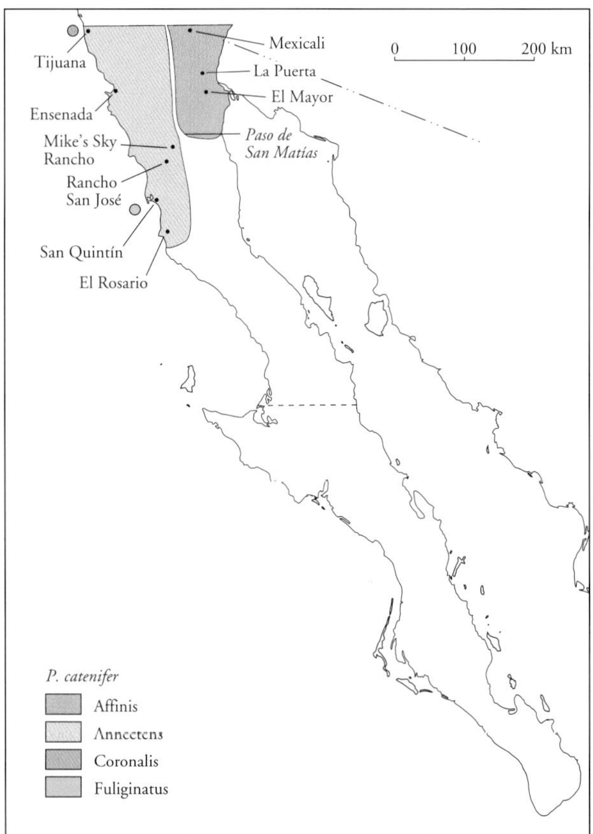

MAP 6.15 Distribution of the pattern classes of *Pituophis catenifer* (Gopher Snake)

are most active from March through September. Juveniles may be abroad on warm winter days in the California Region, but adults usually remain in hibernation. During most of the year, Annectens are diurnal. If daytime temperatures become too high, as is common in the southern portion of their range (Linsdale 1932) and in some of the inland valleys, they may shift their activities into the evening and night hours. Rodríguez-Robles (1998) indicates that *P. catenifer* primarily eat mammals. Annectens feed primarily on small mammals, including squirrels and rabbits, but will also take birds, lizards, and toads. I have observed gravid females from May to July and hatchlings from July to September. These observations suggest that Annectens have a spring breeding and summer laying season in Baja California, as in southern California (Klauber 1946b).

Affinis are common in the northern portion of the Lower Colorado Valley Region, especially near small towns and agricultural fields. Here, locals say they see these snakes catching "squirrels" and other small "animals" in the fields. During April and May, the number of dead Gopher Snakes on the road between El Mayor and La Puerta is staggering and probably corresponds to an activity peak. South of here, Affinis are uncommon, and in fact no specimens have been found south of the Sierra los Cucapás. I have observed active Affinis from March through October in northeastern Baja California, and Soulé and Sloan (1966) report on a specimen collected in mid-March from Isla Tiburón. I have found gravid road kills during May and early June, which suggests a spring breeding and summer laying season.

Isla Coronado Sur is a small, steep, rocky island lying within the California Region, 13 km off the coast of Tijuana. Virtually nothing has been reported concerning the natural history of Coronalis, and the snakes are known only from four preserved museum specimens, all collected before 1946. The only published natural history observation is that of Klauber (1946b), who noted that one of the juveniles he examined had eaten two fledgling birds.

Isla San Martín is a small island lying 5 km off the coast of San Quintín in northwestern Baja California. This island is the northernmost in a chain of six low volcanic cones bordering the western and northern edge of Bahía San Quintín. Isla San Martín lies offshore from a transition zone between the coastal scrub of the California Region and the Vizcaíno Region flora; although the island contains elements of both, it is dominated by dense stands of various cacti (*Opuntia* sp.). The island is covered with volcanic lava flows, tubes, caves, and pits, all of which provide refuge for Fuliginatus. These are common on the island and are often seen crawling within the lava flows, being much more saxicolous than their adjacent peninsular populations. Linsdale (1932) reported snakes being found within cactus patches, and all individuals I have observed have been within cactus stands on the volcanic flows. Klauber (1946b) reported two specimens being found beneath boards. I found a gravid female in late May. Fuliginatus have been reported to feed on rodents (Klauber 1946b; Linsdale 1932) and birds (Klauber 1946b).

REMARKS Further research will perhaps demonstrate that Annectens and Affinis are different species. Where their ranges come into contact in San Diego County, California, hybridization is very rare (Sweet, personal communication, 1992). Welsh and Bury (1984) reported a hybrid from 9.1 km east of Paso de San Matías, and I found one within Paso de San Matías. At the top of Paso de San Matías I have seen only Annectens.

The apparent rarity of Coronalis on the Islas Coronado is puzzling. I have spoken with military officials who live on the island year-round and are familiar with the herpetofauna. When shown pictures of living *P. catenifer* from the adjacent peninsula, they all said that they had never seen that snake on Isla Coronado Sur. Dr. Tomas A. Oberbauer, who has spent a number of years working on the flora of Isla Coronado Sur and even published a paper on its herpetofauna (Oberbauer 1992), has never seen a Gopher Snake there (personal communication, 1996). The possibility remains that this population has become extinct for unknown reasons. It may be that it has been outcompeted by the very common Islas Coronado Rattlesnake *(Crotalus caliginis)*.

FOLKLORE AND HUMAN USE The Seri Indians are well acquainted with Affinis. They know that it is harmless and use its vertebrae to construct necklaces. Their name for the Gopher Snake is *cocaznáacöl,* which is to them a humorous play on the snake's behavior of raising its head and shaking its tail during defensive displays. The name connotes that it "thinks it is a rattlesnake" (Nabhan, forthcoming).

FIGS. 6.33–34

Pituophis vertebralis

BAJA CALIFORNIA GOPHER SNAKE; CORALILLO

Coluber vertebralis Blainville 1835a:293. Type locality: "Californiae," restricted to "San Lucas, Cape [BCS]," Mexico, by Smith and Taylor (1950:108). Holotype: MNHN of Paris.

IDENTIFICATION *Pituophis vertebralis* can be distinguished from all other snakes in the region of study by having large head plates; keeled dorsal scales; a divided anal plate; 34 to 51 dorsal body blotches; a faded or absent rather than prominent black postorbital stripe; and usually a black subcaudal stripe.

FIGURE 6.33 Bimaris pattern class of *Pituophis vertebralis* (Baja California Gopher Snake) from San Ignacio, BCS

RELATIONSHIPS AND TAXONOMY *Pituophis vertebralis* is the sister species of *P. insulanus* of Isla Cedros (Grismer 1994a; Klauber 1946b). Grismer (1994a, 1996c, 1997, 2001c), Klauber (1946b), and Sweet and Parker (1990) discuss taxonomy. Rodríguez-Robles and Jesús-Escobar (2000) consider *P. vertebralis* a member of *P. catenifer* on the basis of DNA sequence data.

DISTRIBUTION *Pituophis vertebralis* is endemic to Baja California, ranging continuously throughout cismontane areas from at least 43 km south (by road) of El Rosario south to Cabo San Lucas (Sweet and Parker 1990; Grismer 1996c). One specimen has been reported much farther north in Valle la Trinidad (Klauber 1946b) in sympatry with *P. catenifer,* but it is not known whether the distribution between this specimen and the northern population south of El Rosario is continuous (Grismer 1997). *Pituophis vertebralis* comes close to the Gulf coast at Bahía de los Ángeles, but does not actually contact the coast there. It may do so somewhere immediately south of Bahía de los Ángeles in Valle San Rafael (map 6.16). *Pituophis vertebralis* is known from the Pacific islands of Magdalena and Santa Margarita as well as the Gulf island of San José (Grismer 1993, 1999a).

DESCRIPTION Body relatively thin, subcylindrical in cross section, reaching 1,530 mm SVL. (I found a road kill in the Vizcaíno Desert that was slightly over 2 m in length. However, because dead snakes relax, I cannot be sure of its actual size and do not include it here. Nonetheless, this species may grow considerably larger than reported here.) Head moderately distinct from neck, narrow and truncate in dorsal and lateral profile; eyes moderately protuberant, pupils round; tail averages 13.5 to 14.0 percent of SVL in males and 12.3 to 12.7 percent of SVL in females; head scales large and platelike; usually 3 preoculars, occasionally 4 or 5; fifth supralabial almost always in contact with eye; 8 to 10 (usually 9) supralabials; 11 to 15 (usually 13) infralabials; dorsal body scales imbricate and vertebral rows sharply keeled, dorsal scales less keeled laterally; usually 33, rarely 31 or 35 dorsal scale rows at midbody; 238 to 251 ventral scales in males, 246 to 257 in females; anal plate single; 60 to 72 divided subcaudals in males, 56 to 63 in females; occasionally some anterior subcaudals are undivided. **COLORATION** Dorsal ground color yellowish to light orange; varying amounts of dark mottling on top of head; suborbital and postorbital stripes usually absent or very faded; margins of labials not edged in black; 34 to 51 dorsal body blotches; 8 to 15 blotches on tail; color and shape of blotching geographically variable (see below); 2 or 3 secondary rows of black to orange spots on sides of body; belly dull white to light yellow and mottled in black posteriorly; black subcaudal stripe usually present.

GEOGRAPHIC VARIATION The notable geographic variation in the color pattern of *P. vertebralis* prompted Klauber (1946b) to recognize two subspecies, *P. catenifer* (= *vertebralis*) *vertebralis* and *P. c. bimaris,* which are used here as pattern classes. Bimaris range from Valle la Trinidad south to the Isthmus of La Paz, where they intergrade with populations

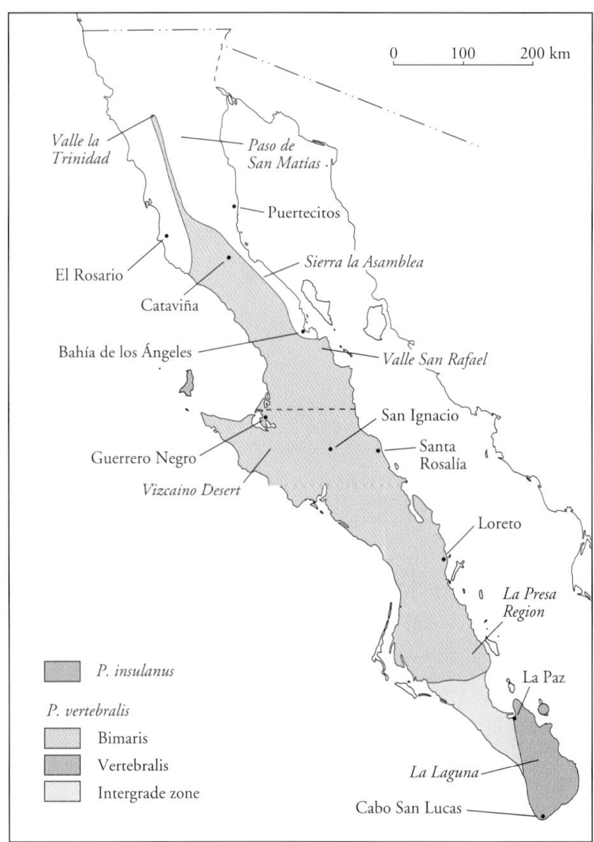

MAP 6.16 Distribution of *Pituophis insulanus* (Isla Cedros Gopher Snake) and the pattern classes of *P. vertebralis* (Baja California Gopher Snake)

of Vertebralis of the Cape Region (Sweet and Parker 1990). Bimaris have a dull yellow ground color and a darkly mottled head, with mottling increasing in more northerly locales. They have 34 to 46 biconcave dorsal body blotches that are black anteriorly and posteriorly and reddish-brown in the middle. In some northerly specimens, the blotches are black throughout, or at least the central blotches are streaked with black. The anterior blotches are widely confluent along their lateral margins, delimiting the light interspace as a large circular spot. North of the Vizcaíno Desert, the top of the head and the dorsal interspaces are usually dull yellow to light brownish-orange, although I have occasionally seen specimens (from the valley) with orange interspaces edging the western foothills of the Sierra la Asamblea. In the Vizcaíno Desert, Bimaris take on a distinctly red-orange dorsal coloration that is most pronounced at San Ignacio and south. Here, the top of the head and the circular dorsal body interspaces are a unicolored red-orange. The brilliance of the orange tends to fade along the Gulf coast, from at least Santa Rosalía to Loreto, but remains vibrant in populations on the western side of the Sierra la Giganta, especially in the Magdalena Plain. Here the neck blotching is a very rich orange. The color intensity also increases with age. The ventra of Bimaris are almost always immaculate. Toward the southern end of their distribution, near the Isthmus of La Paz, where they begin to intergrade with Vertebralis, the number of anterior black blotches decreases and the number of central reddish-brown blotches increases. Also, the lateral edges of the anterior blotches change from black to red-brown.

Immediately south of La Paz, Bimaris have transformed into Vertebralis, which are restricted to the Cape Region. Vertebralis have a yellow to light orange ground color and an immaculate orange head. I found a specimen 21 km north of La Paz in the isthmus that had no black in its color pattern. Vertebralis have 38 to 51 biconcave, red-orange dorsal body blotches that are occasionally streaked with black. The blotches remain red-orange until about two-thirds of the way down the body, where they turn to black. The anterior blotches are confluent along their lateral margins, which are marked with thin lines of orange, white, and black. As in Bimaris, the intensity of the orange increases with age. The ventra of Vertebralis are immaculate anteriorly but spotted with black posteriorly. Vertebralis from Isla Cerralvo tend to be somewhat faded overall, with a slight pinkish cast to their overall coloration and lineate tendencies in their dorsal blotching.

NATURAL HISTORY *Pituophis vertebralis* is fairly ubiquitous throughout most of the Vizcaíno, Central Gulf Coast, Magdalena, Arid Tropical, and Sierra la Laguna regions (Alvarez et al. 1988; Grismer 1994d) but does not range off the central highlands into the Lower Colorado Valley Region of northeastern Baja California. It occurs within some of the deeper, east-flowing arroyos draining higher regions around Cataviña, but it does not extend onto the desert floor. Where *P. vertebralis* is found, it appears to be a habitat generalist. I have commonly seen specimens in the cool, foggy areas dominated by Vizcaíno scrub in north-central Baja California; the open, flat, sandy Vizcaíno Desert; the rocky volcanic badlands of the Magdalena Region; the steep arroyos and coastal flats along the Central Gulf Coast Region; and throughout the Cape Region, including the pine-oak woodland of the upper elevations of the Sierra la Laguna. *Pituophis vertebralis* is commonly found beneath rocks in the Vizcaíno and Magdalena Regions, in brush piles in the Arid Tropical Region, and beneath pine logs in the Sierra la Laguna Region. Bostic (1971) reports finding snakes in rodent burrows and at the bases of agave leaves in north-central Baja California.

FIGURE 6.34 Vertebralis pattern class of *Pituophis vertebralis* (Baja California Gopher Snake) from El Chorro, BCS

Pituophis vertebralis is active year-round from at least the Vizcaíno Desert south, with an activity peak from March through October. North of the Vizcaíno Desert, adults are generally inactive from November through February. Juveniles may be active year-round, provided daytime temperatures are sufficiently high. I have found juveniles on warm evenings during mid-November near Cataviña. During the spring, *P. vertebralis* is usually diurnal. With the onset of higher summer temperatures, it becomes nocturnal.

Pituophis vertebralis preys primarily on small mammals (Bostic 1971) and birds, which it kills by constriction. I found two juveniles from San Ignacio that had eaten Pacific Treefrogs *(Hyla regilla)* and a juvenile from Mulegé that had eaten a Side-Blotched Lizard *(Uta stansburiana).*

Little is known about the reproductive biology of this species. I have found gravid females during May north of Guerrero Negro; very small juveniles during early July near Loreto and in the Cape Region; and hatchlings during early October near Cataviña. These observations suggest that *P. vertebralis* has a spring to early summer breeding season and a summer hatching period, but timing may vary with latitude.

FOLKLORE AND HUMAN USE The local inhabitants of Baja California are frightened by the *coralillo.* They believe that its bite is venomous during certain periods of the year (usually summer). Those in the Cape Region also believe that if one does not quickly wash one's skin where a *coralillo* has touched it, the skin will slough off.

FIG 6.35 ***Pituophis insulanus***

ISLA CEDROS GOPHER SNAKE; CORALILLO, RATONERA

Pituophis catenifer insulanus Klauber 1946b:11. Type locality: "Cedros (Cerros) Island off the west coast of Baja California," Mexico. Holotype: CAS 56353

IDENTIFICATION *Pituophis insulanus* can be distinguished from all other snakes in the region of study by having large head plates; keeled dorsal scales; a divided anal plate; 52 to 64 dorsal body blotches; and usually a black subcaudal stripe.

RELATIONSHIPS AND TAXONOMY See *P. vertebralis.*

DISTRIBUTION *Pituophis insulanus* is endemic to the Pacific island of Cedros (see map 6.16).

DESCRIPTION Same as *P. vertebralis,* except for the following: adults reach 1,162 mm SVL; tail averages 14.3 percent of SVL in males and 12.9 percent of SVL in females; 1 to 3 (usually 2) preoculars; 11 to 14 (usually 12 or 13) infralabials; usually 33, rarely 31 dorsal scale rows at midbody; 238 to 245 ventral scales in males, 247 to 248 in females; 58 to 65 divided subcaudals in males, 57 to 59 in females. **COLORATION** Dorsal ground color yellowish; ground color of head reddish; large dark spots on top of head; dark suborbital stripe present; postorbital stripe usually absent or very faded; margins of labials faintly edged in black; 52 to 64 large, somewhat irregularly shaped dorsal body blotches; anterior and posterior blotches black; central blotches slightly lighter because of a lightening of the scale centers; anterior blotches confluent laterally, forming isolated circular spots of ground color; 11 to 17 black blotches on tail; 2 or 3 secondary rows of black to brownish spots on sides of body; belly dull white to light yellow; subcaudal region heavily mottled.

FIGURE 6.35 *Pituophis insulanus* (Isla Cedros Gopher Snake) from Isla Cedros, BC. (Photograph by Chris Mattison)

NATURAL HISTORY Isla Cedros is a large landbridge island with diverse topography and habitat (Grismer 1993). Its lower elevations are dominated by coastal Vizcaíno Region flora (Oberbauer 1993). As elevation increases, many of the arroyos harbor relict chaparral scrub of the California Region, and the highest peaks maintain pine woodlands. *Pituophis insulanus* is ubiquitous on Isla Cedros. I have observed individuals in low-lying coastal arroyos along the east coast of the island as well as on the tops of some of the island's highest peaks. Residents say that this snake is also common in town and around the dump sites. *Pituophis insulanus* is active year-round, with an activity peak occurring from June through October. It is commonly found during the day, provided temperatures are moderate, which is usually the case during spring. With the onset of high temperatures, generally beginning in late June, *P. insulanus* becomes nocturnal. Nothing has been reported on its food habits, but they probably do not differ from those of *P. vertebralis,* which preys primarily on small mammals. Gravid females have been found in June. Residents of Isla Cedros say these snakes are very good rat eaters.

FOLKLORE AND HUMAN USE See *P. vertebralis.*

Rhinocheilus Baird and Girard

LONG-NOSED SNAKES

The genus *Rhinocheilus* contains two species of moderately sized, terrestrial, constricting snakes (Grismer 1999b). Long-nosed snakes are usually nocturnal and prey on small mammals, lizards, and reptile eggs. *Rhinocheilus* is fairly ubiquitous and frequents a number of brushy habitats, ranging from coastal chaparral areas to desert dunes. *Rhinocheilus* has long been considered closely related to kingsnakes and milksnakes *(Lampropeltis),* and recent work has shown that

Rhinocheilus is also closely related to scarlet *(Cnemophora)* and short-tailed *(Stilosoma)* snakes as well (Keogh 1996; Rodríguez-Robles and Jesús-Escobar 1999).

FIG 6.36 ***Rhinocheilus lecontei***

LONG-NOSED SNAKE; CORALILLO

Rhinocheilus lecontei Baird and Girard 1853:120. Type locality: "San Diego," California, USA. Holotype: MCZ 137

IDENTIFICATION *Rhinocheilus lecontei* can be distinguished from all other snakes in the region of study by having an enlarged, pointed, and slightly upturned rostral scale; elongate, rectangular loreal scales; fewer than 50 percent of the subcaudal scales undivided; black temporal regions; and a black tongue.

RELATIONSHIPS AND TAXONOMY *Rhinocheilus lecontei* is the sister species of *R. etheridgei* of Isla Cerralvo in the Gulf of California (Grismer 1990d, 1999b). Grismer (1990d) and Klauber (1941) discuss taxonomy.

DISTRIBUTION *Rhinocheilus lecontei* ranges through much of the American southwest and northern Mexico, extending into Baja California and along both sides of the Sierra Madre Occidental in mainland Mexico (Medica 1975). In Baja California, *R. lecontei* extends along both sides of the Peninsular Ranges to at least 45 km south (by road) of El Parador in the west and San Felipe in the east (Grismer 1994a; map 6.17). A rancher in the Sierra la Libertad, approximately 40 km south of El Parador, told me he has seen *R. lecontei* on his ranch.

FIGURE 6.36 *Rhinocheilus lecontei* (Long-Nosed Snake) from near Tecate, BC

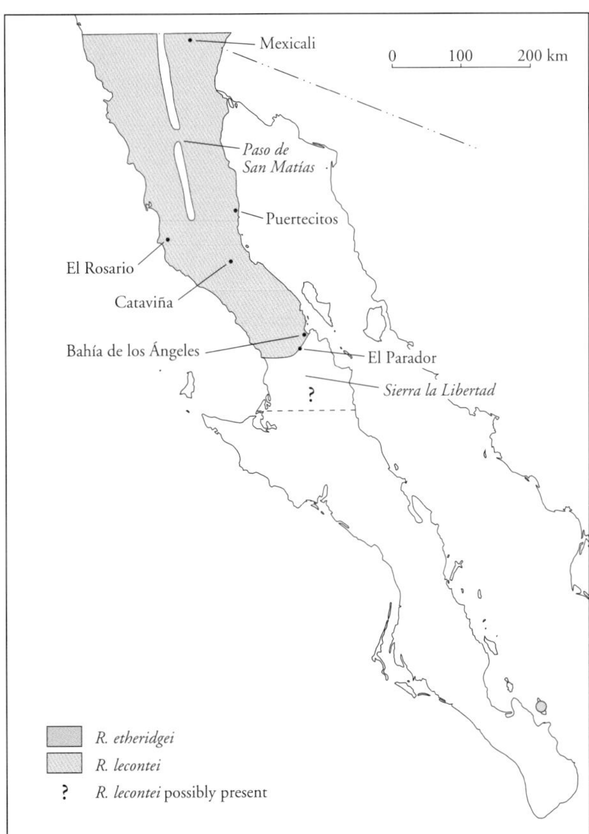

MAP 6.17 Distribution of *Rhinocheilus* (Long-Nosed Snakes)

DESCRIPTION Body relatively thin, subcylindrical in cross section, reaching 774 mm SVL; head weakly distinct from neck, narrow, and sharply pointed in dorsal and lateral profile; eyes moderately sized, pupils round; tail averages 13.2 to 14.9 percent of SVL in males and 12.1 to 14.0 percent of SVL in females; head scales large and platelike; rostral large, pointed, and slightly upturned; loreal rectangular, twice as long as it is high; frontal indents deeply into prefrontal; anterior temporals elongate, much larger than posterior temporals; 8 or 9 (usually 8) supralabials; 8 or 9 (usually 9) infralabials; dorsal body scales smooth, imbricate; 23 dorsal body scale rows at midbody; 202 to 216 ventral scales, with no significant sexual dimorphism; anal plate single; 48 to 60 usually undivided subcaudals in males, 41 to 51 in females. **COLORATION** Dorsal ground color dull white; top of head and temporal region usually black; scales of lower temporal region have white centers; rostral scale white, edged in black; margins of supralabials edged with wide black bands; 16 to 31 black dorsal blotches on body and 6 to 11 on tail, separated by white bands with extensive red infusions; scales of black blotches have white centers; black blotches extend ventrally to make contact with lateral margins of ventral scales; 1 secondary row of irregularly shaped black spots on sides of body; ground color of ventrum dull white and heavily mottled in black.

GEOGRAPHIC VARIATION *Rhinocheilus lecontei* tends to show an increase in black pigment from the southerly regions of its distribution. Individuals from the delta region of the Río Colorado in the northeast tend to have much narrower bands between the black dorsal blotches, with more red infusion. In fact, some Long-Nosed Snakes from these areas appear to have wide black bands separated by narrow red bands.

NATURAL HISTORY *Rhinocheilus lecontei* is common in the coastal sage scrub areas of the California Region of northwestern Baja California as well as the scrubland areas of the Vizcaíno Region farther south. It is absent from the Baja California Coniferous Forest Region (Welsh 1988). In northeastern Baja California, *R. lecontei* has become extremely common in the agricultural areas bordering the Río Colorado delta. It is less common in its natural habitat of Creosote Bush *(Larrea tridentata)* and Bursage *(Ambrosia dumosa)* scrub of the Lower Colorado Valley Region. *Rhinocheilus lecontei* is something of a burrower and is usually found in brushy areas with soft soils. It seems to prefer areas that have rocks, perhaps because such areas generally have more lizards, a favored prey item (Rodríguez-Robles, Bell, and Greene 1999b). *Rhinocheilus lecontei* is most abundant during May (Klauber 1941) but is active from at least March through October in Baja California. It is primarily nocturnal, and in Baja California has been found abroad on cold nights (20°C) when no other species were seen. Much of its nocturnal behavior may involve digging up and eating sleeping lizards (Tennant 1984) and reptile eggs. One night I observed a specimen in an arroyo along the western slopes of the Sierra la Asamblea that was eating a Western Whiptail *(Cnemidophorus tigris)*. The whiptail was dusty, and the Long-Nosed Snake had grains of sand on top of its head, presumably from unearthing the sleeping lizard. Rodríguez-Robles, Bell, and Greene (1999a) reported that *Cnemidophorus* may be a favored food item throughout this snake's range.

The reproductive biology of *R. lecontei* from Baja California appears to be the same as that reported by Klauber (1941) for Long-Nosed Snakes of adjacent San Diego County, California. Klauber reported that gravid females are found during June and July and that hatchlings first appear in August. I have observed gravid females in Baja California during late May and mid-July and have collected hatchlings during mid-September.

REMARKS Many authors have remarked on the possible transpeninsular distribution of *R. lecontei* in Baja California (Medica 1975; Soulé and Sloan 1966; Welsh 1988). This belief has been stimulated primarily by an erroneous record of this species in the vicinity of Bahía Magdalena (Lockington 1880) and the collection of a *Rhinocheilus* on Isla Cerralvo by Banks and Farmer (1963), which was properly identified by Soulé and Sloan (1966). Klauber (1941) discounted the validity of Lockington's specimen, although Welsh (1988) stated that the presence of *Rhinocheilus* in the "Central Desert Region" and on Isla Cerralvo gave credence to Lockington's report. Grismer (1990d) suggested that *R. lecontei* was not transpeninsular in distribution and did not range south into the Magdalena Region because it had never been reported from the heavily collected areas near San Ignacio. Additionally, Grismer demonstrated that the Isla Cerralvo *Rhinocheilus* was a lineage derived from Sonoran populations and not populations from Baja California (*contra* Murphy and Ottley 1984).

FIG 6.37

Rhinocheilus etheridgei

ISLA CERRALVO LONG-NOSED SNAKE; CORALILLO

Rhinocheilus lecontei etheridgei Grismer 1990d:4. Type locality: "Arroyo Viejos at 10 m in elevation on the southwestern end of Isla Cerralvo, Baja California Sur, Mexico." Holotype: SDSNH 66294

IDENTIFICATION *Rhinocheilus etheridgei* can be distinguished from all other snakes in the region of study by having an enlarged, pointed, and upturned rostral scale; square loreal scales; fewer than 50 percent of the subcaudal scales undivided; red temporal regions; and a red tongue with gray tips.

RELATIONSHIPS AND TAXONOMY See *R. lecontei.*

DISTRIBUTION *Rhinocheilus etheridgei* is endemic to Isla Cerralvo in the Gulf of California (see map 6.17).

DESCRIPTION Same as *R. lecontei,* except for the following: body stout, reaching 1,016 mm SVL; tail averages 14 percent of SVL in males (length in females unknown); rostral very large, pointed, and upturned; loreal scales square; frontal not indenting deeply into prefrontals; anterior and posterior temporals same size; 8 supralabials; 9 infralabials; 210 to 215 ventral scales; 50 to 53 subcaudals, usually single. **COLORATION** Dorsal ground color dull white; top of head usually black; black scales of temporal region have red centers, making temporal region appear mostly red; rostral scales white with red flecks and edged in black; margins of anterior supralabials not edged in black; 21 to 30 black dorsal blotches on body and 7 to 10 on tail, separated by predominantly red bands with black infusions; red bands on tail lack black infusions; scales of black blotches have white centers; blotches do not extend ventrally to make contact with lateral margins of ventral scales; 1 secondary row of

FIGURE 6.37 *Rhinocheilus etheridgei* (Isla Cerralvo Long-Nosed Snake) from Arroyo Viejos, Isla Cerralvo, BCS

irregularly shaped black spots on sides of body; ground color of ventrum dull white and immaculate. Juveniles may have an orangish rather than red hue.

NATURAL HISTORY Isla Cerralvo is a rocky, mountainous island within the Central Gulf Coast Region that is dominated by xerophilic vegetation. But because it lies off the coast of the Cape Region, it has marked wet and dry seasons, and a large component of its flora is derived from the Arid Tropical Region. *Rhinocheilus etheridgei* is a little-known snake that is represented by only four specimens and five field observations. It has been found only on the southern end of Isla Cerralvo, although it likely occurs throughout the island. It is most common in the coastal dune areas and in larger arroyos with sandy bottoms. Banks and Farmer (1963) reported finding a specimen during the day in late October, coiled beneath a bush and eating a Desert Iguana *(Dipsosaurus dorsalis).* I observed a juvenile one evening during late August foraging along the coastal dunes at the mouth of a large arroyo, and an adult farther inland, crawling on loose sand among the brush. Grismer (1990d) reported finding two specimens on an evening in mid-June. One was crawling across a gravelly arroyo, and the other was 2 m above the ground in an Elephant Tree *(Bursera microphylla).* The latter was presumably foraging for Isla Cerralvo Spiny Lizards *(Sator grandaevus),* which sleep in the trees at night.

Salvadora Baird and Girard

PATCH-NOSED SNAKES

The genus *Salvadora* is a small group of moderately sized, terrestrial snakes containing approximately seven species (Flores 1993). *Salvadora* ranges across much of the southwestern United States south to Chiapas, Mexico. Patch-nosed snakes are active, diurnal predators much like the whipsnakes, *Masticophis,* and they occur collectively in a variety of habitats ranging from arid deserts to tropical forests. Although *Salvadora* feeds primarily on lizards and snakes, its enlarged rostral scale (from which its name is derived) is used for digging up buried reptile eggs. The relationship of *Salvadora* to other colubrid genera is obscure, but this genus is likely allied to other North American whipsnakes and racers such as *Masticophis, Drymobius,* and *Dendrophidion.*

FIG 6.38 *Salvadora hexalepis*

WESTERN PATCH-NOSED SNAKE; CULEBRA CHATA

Phimothyra hexalepis Cope 1866:304. Type locality: "Fort Whipple, Arizona," USA. Holotype: USNM 7894

IDENTIFICATION *Salvadora hexalepis* can be distinguished from all other snakes in the region of study by having large head plates; a greatly enlarged rostral scale; and a light-colored vertebral stripe.

RELATIONSHIPS AND TAXONOMY The relationships of *S. hexalepis* to other species of *Salvadora* remain problematic. Bogert (1939, 1945) discusses taxonomy.

DISTRIBUTION *Salvadora hexalepis* ranges through much of the American southwest and most of northwestern Mexico. In Baja California, *S. hexalepis* ranges throughout the entire peninsula (Grismer 1994a) except for the upper elevations of the Sierra Juárez and Sierra San Pedro Mártir (Welsh 1988). It is known from the Pacific islands of San Gerónimo and Todos Santos and from the Gulf islands of Espíritu Santo, San José, and Tiburón (map 6.18).

DESCRIPTION Body long and thin, subcylindrical in cross section, reaching 883 mm SVL; head moderately distinct from body, snout square in dorsal and lateral profile; rostral scale greatly enlarged; eyes large, pupils round; tail averages 20 to 27 percent of total length in both males and females; head scales large and platelike; 8 to 10 (usually 9) supralabials; 8 to 12 (usually 10) infralabials; dorsal body scales smooth and imbricate, those above vent keeled in males; 15 or 17 dorsal body scale rows at midbody; 187 to 204 ventral scales in males and females; anal plate divided; 75 to 103 divided subcaudals in males and females. **COLORATION** Dorsal ground color brownish; top of head usually brown to olive drab; sides of body gray, brownish, or olive drab, leaving a prominent light-colored vertebral stripe running the length of the body; postorbital regions usually dark and same color as sides of body; faint light-colored, ventrolateral secondary stripes occasionally present; supralabials, gular region, and ventrum dull white and immaculate.

FIGURE 6.38 Klauberi pattern class of *Salvadora hexalepis* (Western Patch-Nosed Snake) from Miraflores, BCS

GEOGRAPHIC VARIATION *Salvadora hexalepis* shows a modest amount of geographic variation in squamation and color pattern, which has been indicated by the designation of three subspecies: *S. h. hexalepis, S. h. klauberi,* and *S. h. virgultea* (Bogert 1939, 1945). These are used here as pattern classes. Virgultea range throughout much of the California Region of northwestern Baja California. They are characterized by having dark olive drab sides; a light-colored vertebral stripe one scale wide; a dark brown to olive drab head; and usually one rather than two supralabial scales contacting the eye. Virgultea intergrade with Hexalepis of the Lower Colorado Valley Region of northeastern Baja California through Paso de San Matías and along the southern end of the Sierra San Pedro Mártir, and with Klauberi, of the southern three-fourths of Baja California, in the vicinity of El Mármol (Bogert 1945).

Hexalepis are characterized by being very light-colored overall, having gray sides, a nearly white head, a vertebral stripe that is three scales wide, secondary ventrolateral stripes, and usually one supralabial contacting the eye. They intergrade with Klauberi near Bahía San Luis Gonzaga.

Klauberi are characterized by their much larger adult size, dark brown to gray sides on the body, the top of the head usually being some shade of brown, a light vertebral stripe three scales wide, faint secondary ventrolateral stripes, and usually two supralabials (fifth and sixth) contacting the eye. Populations from low-lying regions north of the Isthmus of La Paz generally appear more faded than populations from the Cape Region and the upper elevations of the Sierra la Giganta. Secondary striping is most prominent in populations in the Cape Region. Specimens from the volcanic areas in the Magdalena Region tend to have a much darker overall grayish color than those found elsewhere.

NATURAL HISTORY *Salvadora hexalepis* is essentially ubiquitous in Baja California (Alvarez et al. 1988; Grismer 1994a), being absent only from the Baja California Coniferous Forest Region (Welsh 1988). Virgultea are most commonly associated with brushy foothill habitats in the California Region (Bogert 1945; Klauber 1931b; Welsh 1988), and I have often seen them in riparian areas south of Tecate. Hexalepis are most commonly encountered in the Creosote Bush *(Larrea tridentata)* and Bursage *(Ambrosia dumosa)* flats where loose soils prevail. Klauberi are ubiquitous and occur in habitats ranging from the low, flat, and cool windswept Vizcaíno Desert of west central Baja California to the endemic pine-oak woodland in the highest elevations of the Sierra la Laguna in the Cape Region.

Virgultea and Hexalepis are most abundant during May and June and remain active until October, although juveniles may be abroad in November. I have observed active adult Virgultea during mid-March at San Quintín. Klauberi are generally active year-round south of Misión San Borja, with a peak activity period from April through September.

Salvadora hexalepis is diurnal and occasionally ob-

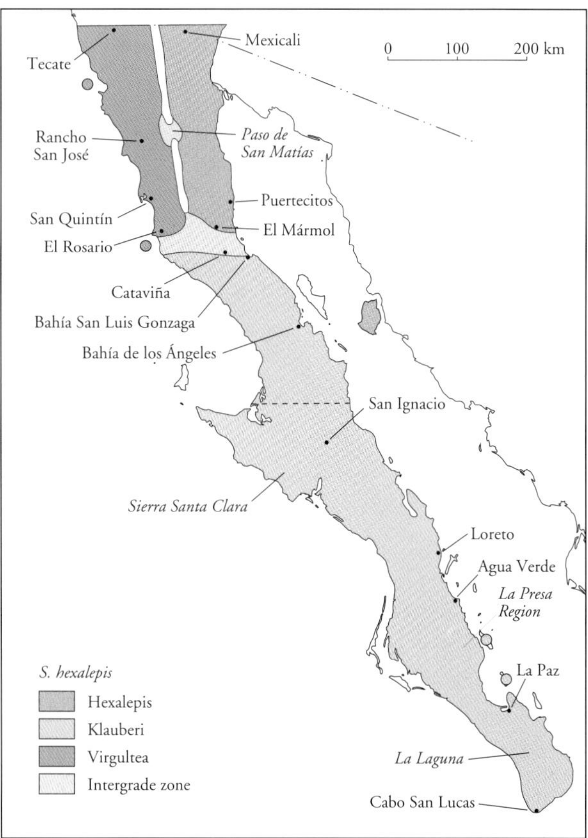

MAP 6.18 Distribution of the pattern classes of *Salvadora hexalepis* (Western Patch-Nosed Snake)

served lying motionless, basking in direct sunlight with its body pressed against the ground. More often, however, it is seen foraging through the brush and the lower branches of shrubs. Patch-Nosed Snakes also search burrows for lizards and other prey. I observed one individual exiting a burrow at midday in the Sierra Santa Clara with an Orange-Throated Whiptail *(Cnemidophorus hyperythrus)* in its mouth. *Salvadora hexalepis* is a swift, agile, active predator known to eat lizards, smaller snakes, and rodents. I found an adult Klauberi on the crest of the Sierra la Giganta, just north of Agua Verde, subduing a Western Whiptail *(Cnemidophorus tigris),* which it had grasped by the shoulders, crossways in its mouth. On my approach, the snake released the lizard but remained cautiously in position. The whiptail attempted to run but was obviously disabled, although the specific injury was not apparent. Eventually the snake left; I captured the lizard and noticed that its skin was torn above the left shoulder. Otherwise the lizard appeared fine, but it was unable to walk when prodded and died a few minutes later. *Salvadora hexalepis* is known to have enlarged, ungrooved rear teeth, but nothing has been reported concerning the toxicity of its saliva. Patch-Nosed Snakes prey on the eggs of other reptiles, and the enlarged rostral scale may be an adaptation for excavating buried clutches.

I observed a pair of Virgultea copulating near Rancho San José during late April and have seen hatchlings from mid-July through early September. Klauber (1931b) reports similar observations in southern California. I have found gravid Klauberi during late June throughout their distribution, hatchlings from July through early August in the Cape Region, and small juveniles during late December from San Ignacio south. These observations suggest that *S. hexalepis* has a spring breeding and summer hatching season.

Sonora Baird and Girard

GROUND SNAKES

The genus *Sonora* contains at least three species of small, secretive, terrestrial snakes (Frost 1983a) that collectively range from the central United States south to Guerrero, Mexico. Ground snakes are usually nocturnal and prey on arthropods and their larvae. Although they are probably very common, their secretive lifestyle makes them difficult to find. *Sonora* is part of a large natural group of relatively small terrestrial snakes containing 14 other genera (Savitzky 1995), among which sand snakes *(Chilomeniscus),* shovel-nosed snakes *(Chionactis),* and black-headed snakes *(Tantilla)* also occur in the region of study.

FIGURE 6.39 *Sonora semiannulata* (Ground Snake) from Cumbre de San Pedro, Sierra Guadalupe, BCS

FIG 6.39

Sonora semiannulata

GROUND SNAKE; CULEBRA

Sonora semiannulata Baird and Girard 1853:117. Type locality: "Sonora, Mex.," restricted to "vicinity of the Santa Rita Mountains of [Pima and Santa Cruz counties] Arizona" (Stickel 1943). Holotype: USNM 2109

IDENTIFICATION *Sonora semiannulata* can be distinguished from all other snakes in the region of study by having enlarged head plates and a normal rostral scale; lacking subocular scales; having loreal scales and a single preocular scale; the lower jaw not being countersunk; smooth dorsal scales; a divided anal plate; round pupils; and no orange neck ring.

RELATIONSHIPS AND TAXONOMY *Sonora semiannulata* is related to *S. aemula* and *S. michoacanensis* of mainland Mexico (Frost 1983b). Frost (1983b) and Frost and Van Devender (1979) discuss taxonomy.

DISTRIBUTION *Sonora semiannulata* ranges throughout much of the American southwest and northern Mexico, extending into Baja California and the Chihuahuan Desert of mainland Mexico (Frost 1983a). In Baja California, *S. semiannulata* ranges continuously throughout the peninsula into the Cape Region and is absent only from the extreme northwest (Frost 1983a; Grismer 1989e, 1994a). It is also known from Islas San José and San Marcos in the Gulf of California (map 6.19).

DESCRIPTION Body relatively thin, subcylindrical in cross section, reaching 371 mm SVL; head weakly distinct from neck, truncate in dorsal and lateral profile; eyes moderately sized, pupils round; tail averages 18.9 to 19.2 percent of SVL in males and 16.8 to 18.2 percent of SVL in females; head scales large and platelike; 7 supralabials; 7 or 8 (usually 7) infralabials; dorsal body scales smooth and imbricate; 13 to 15 (usually 15) dorsal body scale rows at midbody; 147 to 164 ventral scales in males, 158 to 171 in females; anal plate single; 43 to 57 divided subcaudals in males, 39 to 48 in females. **COLORATION** Dorsal ground color orangish, brown, or beige; top of head unicolored orange, brown, or beige, sometimes darker than body; dorsal body pattern banded, unicolored, or with a wide orange vertebral stripe (see below); ground color of ventrum dull white and immaculate. Hatchlings of striped phase may be unicolored or weakly banded.

GEOGRAPHIC VARIATION Snakes from northeastern Baja California in the Lower Colorado Valley Region have a dark brown head and a brownish dorsal ground color overlaid with a wide orange vertebral stripe extending the length of the body and nearly encompassing the tail. Snakes with this color pattern extend west and intergrade through Valle la Trinidad with snakes having a light-colored, banded pattern (Grismer 1989e). Banded snakes range throughout much of the California Region, from at least Cabo Colonet in the north to Rancho San Juan de Dios in the south. Ground Snakes from this area have a light brown ground color and well-defined grayish cross bands on the body and tail. From Rancho San Juan de Dios to just north of Bahía de los Ángeles, the banded pattern begins to fade as dark coloration invades the interspaces. In these areas, banded Ground Snakes intergrade with snakes from farther south that have the unicolored pattern. Ground Snakes with the unicolored phase extend from Bahía de los Ángeles south to Cabo San Lucas and are dull orange, dark brown, or gray north of the Isthmus of La Paz and light brown to beige in the Cape Region. Much of this color variation appears to be related to substrate matching. Intermediate specimens between the banded and unicolored phases have been found at San Ignacio (Frost 1983b), near San José de Magdalena, and near San José del Cabo.

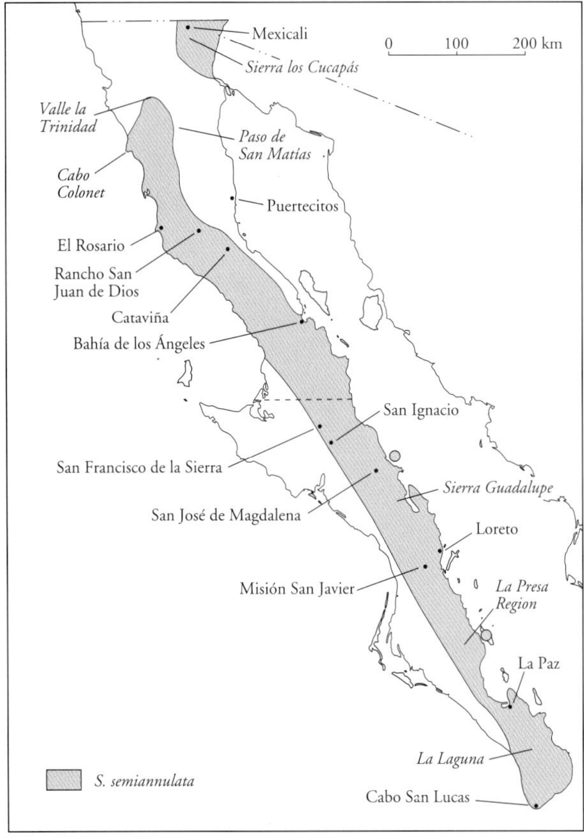

MAP 6.19 Distribution of *Sonora semiannulata* (Ground Snake)

NATURAL HISTORY Although quite a lot is known about the natural history of *S. semiannulata* outside Baja California (see Frost 1983a and citations therein), virtually nothing has been reported for Baja California populations. It is probably ubiquitous, but its secretive habits leave it seldom observed. *Sonora semiannulata* is most frequently found in rocky habitats, although it also occurs in flat, low-lying areas lacking rock. In the Lower Colorado Valley Region of northeastern Baja California, *S. semiannulata* is most often found in the agricultural areas of the Río Colorado flood plain. Ground Snakes also occur in coastal scrub localities in the California Region (Welsh 1988). In the Vizcaíno and Magdalena regions, *S. semiannulata* is found in rock piles and beneath rocks, and it is abundant in the vicinity of lava flows. I have found several specimens beneath various types of rock from Valle la Trinidad in the north to Misión San Javier in the south, and as high as 1800 m in the Sierra Guadalupe. In the Cape Region, Ground Snakes are most abundant in the low-lying, rocky areas of the Arid Tropical Region surrounding the Sierra la Laguna.

Sonora semiannulata is principally nocturnal, although I did observe one specimen crawling across open ground at midday in early August at San Francisco de la Sierra. These snakes feed on arthropods and their larvae, spiders, scorpions, and centipedes. They have grooved rear teeth, and their saliva may contain a toxin to help subdue their prey.

Gravid females found at the north end of the Sierra los Cucapás during late August had eggs that hatched in mid-October (Staedeli 1963). I have observed gravid females at San Ignacio during late July and in the Cape Region during early August.

Tantilla Baird and Girard

BLACK-HEADED SNAKES

The genus *Tantilla* is a large group of generally small, secretive, terrestrial snakes containing 53 species (Wilson

1999). *Tantilla* occurs in a variety of habitats and ranges from the southern United States south through Mexico, Central America, and South America as far south as northern Argentina (Wilson 1999). In North America, black-headed snakes are seldom seen above ground and apparently spend most of their time beneath rocks, within rodent burrows, in rock cracks, and within the fissures formed by expansive soils. Black-headed snakes are rear-fanged; in North America, they prey on arthropods, notably centipedes and beetle larvae. *Tantilla* is a member of a large natural group of other small New World terrestrial colubrids (Savitzky 1995), of which three, the sand snakes *(Chilomeniscus),* the shovel-nosed snakes *(Chionactis),* and the ground snakes *(Sonora),* occur in the region of study.

FIGURE 6.40 *Tantilla planiceps* (Western Black-Headed Snake) from Boca de la Sierra, BCS

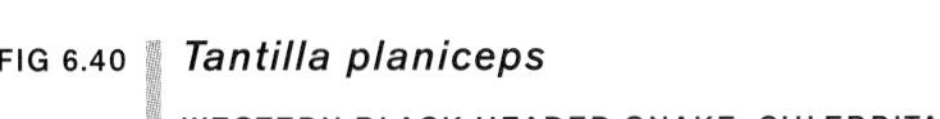

FIG 6.40 ***Tantilla planiceps***

WESTERN BLACK-HEADED SNAKE; CULEBRITA

Coluber planiceps Blainville 1835a:294, 295. Type locality: "Californiae" (Blainville 1835a:282). Restricted to "San Lucas, Cape [BCS, Mexico]" (Smith and Taylor 1950). Holotype: MNHN 818

IDENTIFICATION *Tantilla planiceps* can be distinguished from all other snakes in the region of study by being small, slender, and shiny and having a black head and brownish body.

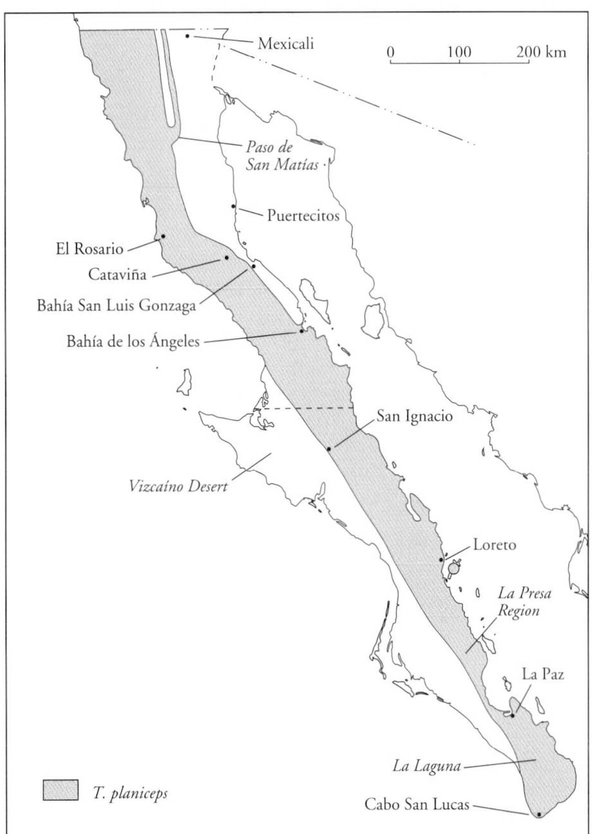

MAP 6.20 Distribution of *Tantilla planiceps* (Western Black-Headed Snake)

RELATIONSHIPS AND TAXONOMY *Tantilla planiceps* is the sister species of *T. yaquia* of southern Arizona and Sonora (Cole and Hardy 1981). Together they form a natural group with *T. gracilis, T. atriceps, T. hobartsmithi,* and *T. nigriceps* of the southwestern United States (Cole and Hardy 1981). Cole and Hardy (1981) discuss taxonomy.

DISTRIBUTION *Tantilla planiceps* ranges throughout much of cismontane California, south through Baja California to Cabo San Lucas (Cole and Hardy 1983). In Baja California, *T. planiceps* extends south through the California Region and along the eastern rocky slopes of the Sierra Juárez (Cole and Hardy 1981) and probably the Sierra San Pedro Mártir (Welsh 1988). It continues south to Cabo San Lucas but is not known to occur in the Vizcaíno Desert (Grismer et al. 1994). It is absent from the Lower Colorado Valley Region and is not known to contact the Gulf coast farther north than Bahía de los Ángeles; however, it probably extends northeast through the rocky Sierra Calamajué, contacting the Gulf coast along the southern border of Bahía San Luis Gonzaga. *Tantilla planiceps* is also known from Isla Carmen in the Gulf of California (map 6.20).

DESCRIPTION Body slender, subcylindrical in cross section, reaching 301 mm SVL; head not distinct from body, truncate in dorsal and lateral profile; eyes moderately sized, pupils round; tail averages 20.3 to 26.8 percent of SVL in males and 18.1 to 23.8 percent of SVL in females; head scales large and platelike; 7 supralabials; 6 infralabials; dorsal body scales smooth and imbricate; 15 dorsal body scale rows at midbody; 134 to 184 ventral scales in males, 148 to 197 in females; anal plate divided; 57 to 73 divided subcaudals in males, 49 to 70 in females. **COLORATION** Body unicolored, brownish to light olive green; top and sides of head black; white collar often present, bordered posteriorly by small black spots; labials and lower surfaces of head and gular region grayish; remainder of ventrum coral.

GEOGRAPHIC VARIATION *Tantilla planiceps* shows little geographic variation in Baja California. Generally, Western Black-Headed Snakes from the eastern edges of the Sierra Juárez tend to be lighter than those from cismontane populations in the California Region and lack the white neck collar. Snakes from the California Region to near San Ignacio tend to be more olive-colored and have a distinctive white collar. Some specimens from the Cape Region, especially around Miraflores, have grayish heads.

NATURAL HISTORY Because of its very secretive nature, little is known about the natural history of *T. planiceps* in Baja California or elsewhere. The majority of the specimens collected in Baja California have been found inadvertently, through excavation while searching for other species (e.g., Klauber 1943; Leviton and Banta 1964). *Tantilla planiceps* has been collected in the coastal scrub of the California Region (Klauber 1943; Tanner 1966b, Welsh 1988), scrub vegetation of the Vizcaíno Region (Bostic 1971), and vegetated areas farther south (Leviton and Banta 1964). In all these areas, Western Black-Headed Snakes are most common near rocks. Specimens I have observed in Baja California have been from the rocky eastern face of the Sierra Juárez, the lava flows in the vicinity of San Ignacio, or rocky areas along the eastern and western sides of the Sierra la Laguna in the Cape Region. Other specimens have been found beneath rocks at La Cumbre in the Sierra Guadalupe. Alvarez et al. (1988) report *T. planiceps* as being uncommon in the scrubland of the Arid Tropical Region and note that it extends into the oak woodland of the Sierra la Laguna.

Although *T. planiceps* has been collected throughout the year through excavations and rock-turning, it has been observed above ground and active only from mid-March through July. It is likely that it is occasionally active above ground through other warm months of the year as well, but it apparently spends the majority of its life underground or beneath rocks, surfacing infrequently on warm nights. The majority of specimens I have observed in the field have been beneath rocks and logs or in wells, mine shafts, and flumes where they had become trapped. Nothing is known of the reproductive biology of this snake in Baja California.

Thamnophis Fitzinger

GARTER SNAKES

The genus *Thamnophis* is a large group of moderately sized, active, semiaquatic snakes containing approximately 30 species (Rossman et al. 1996). *Thamnophis* is widely distributed throughout North and Central America and ranges from Canada to Costa Rica. Garter snakes occur in and around ponds and watercourses throughout a variety of habitats, ranging from extreme deserts to tropical forests; they tend to be locally abundant. Although they feed on a wide range of prey, they usually prefer fish and amphibians. *Thamnophis* is closely related to the New World water snakes *(Nerodia)*, which range through much of the eastern United States and Mexico. There are four species of *Thamnophis* in Baja California: one *(T. hammondii)* has a nearly transpeninsular distribution, two others *(T. elegans* and *T. marcianus)* are restricted to northern areas, and one *(T. validus)* is confined to the Cape Region.

FIG 6.41 ***Thamnophis elegans***

WESTERN TERRESTRIAL GARTER SNAKE; CULEBRA DEL AGUA

Eutaenia elegans Baird and Girard 1853:34, 295. Type locality: "El Dorado County, California," USA. Holotype: USNM 882

IDENTIFICATION *Thamnophis elegans* can be distinguished from all other snakes in the region of study by having large head plates; keeled dorsal scales; an entire anal plate; a wide yellow vertebral stripe; and no conspicuous dark blotches on the sides of the body.

RELATIONSHIPS AND TAXONOMY *Thamnophis elegans* is the sister species of a group of garter snakes composed of *T. brachystoma, T. butleri,* and *T. radix* of the northern United States and southern Canada (de Queiroz and Lawson 1994). Fitch (1940, 1983) and Rossman et al. (1996) discuss taxonomy.

DISTRIBUTION *Thamnophis elegans* ranges continuously from southwestern Canada, south through the western United States to central Nevada, Arizona, and New Mexico, nearly to the edges of the Mojave and Sonoran deserts. It is known from isolated populations in the Sierra Nevada and the San Bernardino Mountains of California as well as central New Mexico (Fitch 1983). In Baja California, *T. elegans* occurs in another isolated population from the Sierra San Pedro Mártir. This population occurs only on the main scarp above 1,820 m (Welsh 1988) and is separated from the nearest population of *T. elegans,* from the San Bernardino Mountains, by nearly 400 km (map 6.21).

DESCRIPTION Body moderate, subcylindrical in cross section, reaching 301 mm SVL; head moderately distinct from body, truncate in dorsal and lateral profile; eyes moderately sized, pupils round; tail averages 20.3 to 26.8 percent of SVL in males and 18.1 to 23.8 percent of SVL in females; head scales large and platelike; 7 or 8 supralabials; 9 or 10 infralabials; dorsal body scales keeled and imbricate; 21 dorsal body scale rows at midbody; 149 to 160 ventral scales in males, 151 to 156 in females; anal plate entire; 75 to 82 divided subcaudals in males, 66 to 71 in females. **COLORATION** Dorsal ground color dark olive drab; smaller individuals often have a yellowish cast; skin between scales lime green, often appearing as transversely oriented speckles; top of head slightly lighter, with dark-edged head scales; rostrum and side of head yellowish; small paired, yellow parietal spots present; postorbital regions dark; prominent yellow vertebral stripe extending from base of head to tip of tail and fading on tail; yellowish lateral stripe running along side of body; venter usually immaculate, although some specimens have dark markings; venter yellow anteriorly, darkening posteriorly.

FIGURE 6.41 *Thamnophis elegans* (Western Terrestrial Garter Snake) from Cerro Venado Blanco, Sierra San Pedro Mártir, BC

NATURAL HISTORY *Thamnophis elegans* is most commonly found in riparian areas within rocky, slow-moving watercourses (Fitch 1940; J. Grismer 1994; Welsh 1988). This species is by no means restricted to such habitats, however, and has been commonly observed on rocky hillsides (Fitch 1940; Linsdale 1932), in marshy meadows, in moist woodland areas (Welsh 1988), and in clumps of grass (Murray 1955). *Thamnophis elegans* is diurnal and actively forages in grassy areas along streamsides for Western Toads *(Bufo boreas)* and their larvae (Welsh 1988). I have seen *T. elegans* in small pools chasing the larvae of *B. boreas,* and J. Grismer (1994) noted a foraging individual on a stream bank that had captured a Southern Sagebrush Lizard *(Sceloporus vandenburgianus).* Nothing has been reported on the reproductive biology of *T. elegans* from Baja California, other than Welsh's (1988) noting a female with eight small ova during June, which suggests that this population has a spring to summer breeding season.

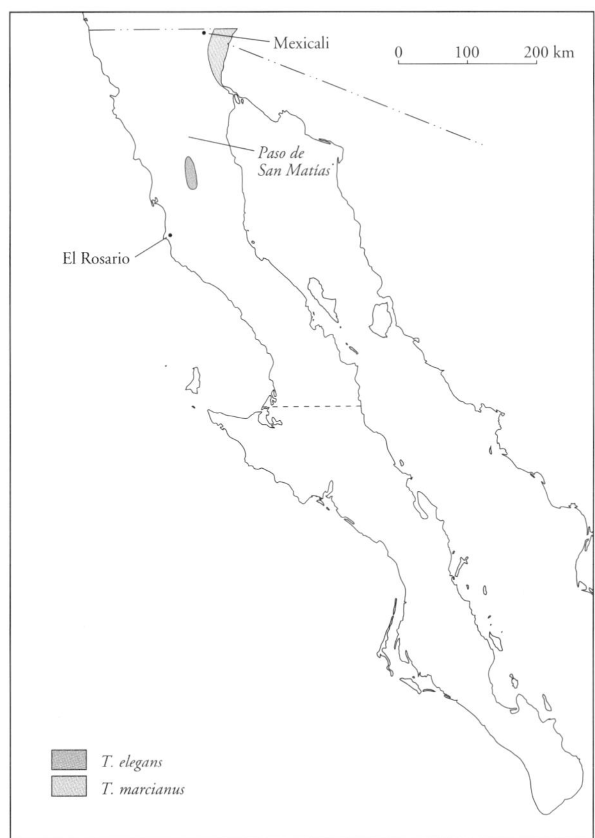

MAP 6.21 Distribution of *Thamnophis elegans* (Western Terrestrial Garter Snake) and *T. marcianus* (Checkered Garter Snake)

REMARKS The population of *T. elegans* from the Sierra San Pedro Mártir was originally described as the subspecies *T. ordinoides hueyi* (Van Denburgh and Slevin 1923). Fitch (1940) stated that *T. e. hueyi* was so similar to other *T. elegans* that it would be difficult to tell them apart if unlabeled specimens were mingled, although Rossman et al. (1996) provide a key to the different subspecies of *T. elegans.*

FIG 6.42 **_Thamnophis marcianus_**

CHECKERED GARTER SNAKE; CULEBRA DEL AGUA

Eutaenia marciana Baird and Girard 1853:36. Type locality: "Red River, Arkansas, [amended to include Cache Creek, Oklahoma, by Vandenburgh and Slevin (1918)]," USA. Cotype: USNM 844

IDENTIFICATION *Thamnophis marcianus* can be distinguished from all other snakes in the region of study by having large head plates; keeled dorsal scales; an entire anal plate; a wide yellow vertebral stripe; and two rows of large, conspicuous dark blotches on the sides of the body.

RELATIONSHIPS AND TAXONOMY See *T. hammondii.*

DISTRIBUTION *Thamnophis marcianus* ranges across the south-central United States and much of northern Mexico

FIGURE 6.42 *Thamnophis marcianus* (Checkered Garter Snake) from Mexicali, BC

west to central Arizona. It occurs again farther west in a disjunct population in the Lower Colorado River Basin of southeastern California and southwestern Arizona. From there, it extends south into northeastern Baja California in association with the Río Colorado and its tributaries (see map 6.21). In this region, its distribution has undoubtedly expanded because flood-plain agriculture has converted uninhabitable desert flats into well-watered fields with irrigation canals.

DESCRIPTION Same as *T. elegans,* except for the following: adults reach 1,000 mm SVL; tail averages 21.3 to 26.8 percent of total length (TL) in males and 18.9 to 26.9 percent of TL in females; 7 or 8 (usually 8) supralabials; 10 or 11 (usually 11) infralabials; dorsal body scales keeled and imbricate; 21 dorsal body scale rows at midbody; 157 to 162 ventral scales in males, 149 to 159 in females; anal plate entire; 77 to 79 divided subcaudals in males, 63 to 67 in females. **COLORATION** Dorsal ground color light brown; skin between scales dull white, appearing almost as speckles; top of head greenish brown; area immediately anterior to, posterior to, and below eye dull white and bordered by black vertical bars; small paired, yellow parietal spots present; wide, dark crescent markings on neck; prominent narrow, light-colored vertebral stripe running length of body; two rows of large, distinctive, black lateral body spots; lateral body stripes faint; venter dull white, often clouded with darker pigment.

NATURAL HISTORY *Thamnophis marcianus* is most commonly found along the banks of watercourses in northeastern Baja California, near vegetation. These snakes are often seen along irrigation ditches in the agricultural areas south of Mexicali, and I have even observed them on the outskirts of town in some of the open sewage ponds. From March through mid-May, *T. marcianus* is diurnal and actively forages for prey, which consists mostly of anuran larvae. During the summer, however, snakes forage during the cooler evening hours. Stebbins (1985) reported that 6 to 18 young are born from June to August. Nothing is known of the reproductive biology of the Baja California populations, but presumably it is no different from that of U.S. populations in the Lower Colorado Basin.

FIG 6.43 *Thamnophis hammondii*

TWO-STRIPED GARTER SNAKE; CULEBRA DEL AGUA

Eutaenia Hammondii Kennicott 1860:34, 332. Type locality: "San Diego, Cal. [California]," USA. Holotype: USNM 894

IDENTIFICATION *Thamnophis hammondii* can be distinguished from all other snakes in the region of study by having large head plates; keeled dorsal scales; an entire anal plate; and a unicolored olive drab dorsum with no vertebral stripe.

RELATIONSHIPS AND TAXONOMY *Thamnophis hammondii* is part of a natural group containing *T. couchii, T. marcianus,* and *T. eques* (de Queiroz and Lawson 1994) of the western United States. Fitch (1948), McGuire and Grismer (1993), and Rossman and Stewart (1987) discuss taxonomy.

DISTRIBUTION *Thamnophis hammondii* ranges through much of cismontane and montane southern California from San Benito County south to the La Presa Region of southern Baja California (McGuire and Grismer 1993; Rossman et al. 1996). Within Baja California, *T. hammondii* ranges continuously throughout the cismontane and montane areas south, possibly to Arroyo El Rosario (McGuire and Grismer 1993). Populations are also known from along the desert slopes of the Sierra Juárez and Sierra San Pedro Mártir, in well-watered canyons that drain the higher elevations. South of El Rosario, in the desert regions, *T. hammondii* is restricted to oases with relatively large amounts of permanent water (Grismer and McGuire 1993; McGuire and Grismer 1993; map 6.22).

DESCRIPTION Same as *T. elegans,* except for the following: adults reach 830 mm SVL; tail averages 22.2 to 25.2 percent of TL in males and 21.1 to 23.5 percent of TL in females; 13 to 18 supralabials; 18 to 22 infralabials; dorsal body scales keeled, imbricate; 17 to 21 (usually 19 or 20) dorsal body scale rows at midbody; 158 to 172 ventral scales in males, 150 to 169 in females; anal plate entire; 74 to 89 divided subcaudals in males, 67 to 75 in females. **COLORATION** Dorsal ground color dark olive drab; skin between scales lime green, most apparent laterally; top of head dark; sides of head yellow; supralabials edged with dark lines; small paired, yellow parietal spots present; dark crescent markings on neck; very faint vertebral stripe on anterior portion of neck; two rows of dark, relatively obscure lateral body spots; yellow lateral body stripe running along side of body and fading at tail; venter usually dusky orange and immaculate.

GEOGRAPHIC VARIATION McGuire and Grismer (1993) demonstrated that there were subtle differences in squamation and color pattern between the isolated populations of *T. hammondii* in the arid regions of central Baja California. These differences appear to be the accumulated effects of genetic drift, subsequent to the populations becoming fragmented with the formation of the peninsular deserts (McGuire and Grismer 1993). Populations from northwestern Baja California tend to be slightly darker in ground color than those from the desert foothills of the Sierra Juárez and Sierra San Pedro Mártir. Additionally, garter snakes from the latter populations have much brighter yellow lateral stripes. *Thamnophis hammondii* from Arroyo San Telmo in the California Region tend to be more brown than olive drab in dorsal coloration. Interestingly, garter snakes from San Ignacio are distinctive in having an overall darker color pattern. The lateral stripes, crescent neck markings, parietal spots, vestigial dorsal stripe, black labial sutures, and lateral body spots are obscure or absent. This coloration may result from substrate matching in a water system that courses through very dark, andesitic volcanic rock. However, the same variation is observed in some individuals from San Francisco de la Sierra, where the substrate is not nearly as dark. Some specimens from the La Presa region have a subtle mottling on their heads.

FIGURE 6.43 *Thamnophis hammondii* (Two-Striped Garter Snake) from San Francisco de la Sierra, BCS

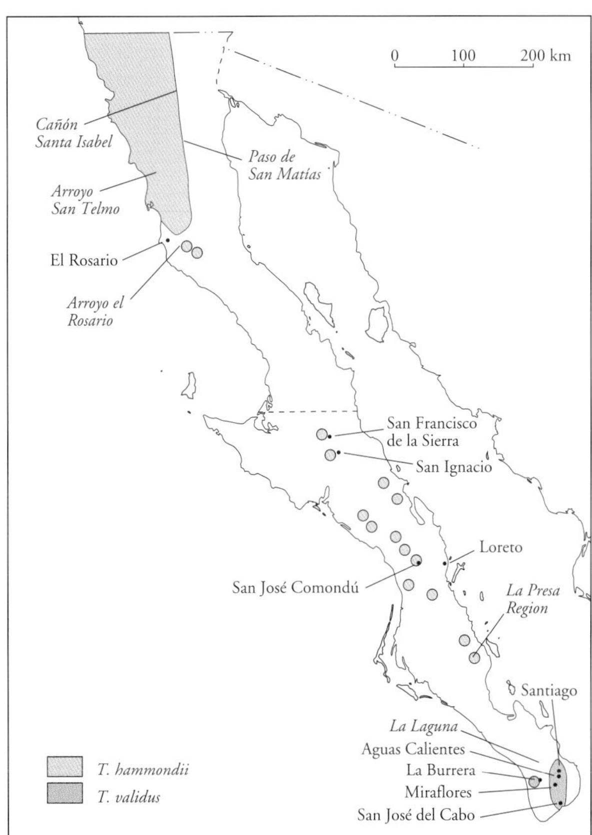

MAP 6.22 Distribution of *Thamnophis hammondii* (Two-Striped Garter Snake) and *T. validus* (West Coast Garter Snake)

NATURAL HISTORY *Thamnophis hammondii* is common throughout all types of riparian habitats in the California and Baja California Coniferous Forest regions (Welsh 1988). It is not uncommon along the western fringe of the Lower Colorado Valley Region, where large, well-watered canyons drain the eastern escarpments of the northern Peninsular Ranges. Farther south, in the Vizcaíno and Magdalena regions, *T. hammondii* is restricted to disjunct oases (McGuire and Grismer 1993). It is diurnal and commonly observed foraging along stream banks and among streamside rocks and vegetation, and it has even been seen 3.4 m above the ground in overstory riparian vegetation in northwestern Baja California (Welsh 1988). On several occasions, I have observed specimens up to 1 m above the ground in thorny vegetation along watercourses just west of San José Comondú and at La Presa. Individuals of the desert populations, from Arroyo Grande in the north to La Presa in the south, may be found foraging during warm spring and summer evenings.

Thamnophis hammondii is an active forager that, in Baja California, feeds primarily on frogs *(Bufo* and *Hyla)* and their larvae. I observed a hatchling in the La Presa region carrying a small fish in its mouth. It is common to see these garter snakes swimming underwater, foraging through the grass, or in clear open pools. When a snake comes into a body of water containing tadpoles, it chases them around somewhat erratically, lunging as it goes. While hunting, the garter snakes surface for air and apparently rest or bask by extending only the head above the water while the body floats just below the surface. Occasionally the head is placed on the stream bank or a rock and the body floats behind it. Snakes usually bask along the edge of the stream or on a rock in the stream within lunging distance of the water. Amazingly, at La Presa, Two-Striped Garter Snakes often bask on the rocky ledges above the water in direct sunlight, with daytime temperatures as high as 38°C. At San Francisco de la Sierra, these garter snakes crawl out onto floating stands of moss to "bask." They dive below the surface when startled, usually burrowing into the mud at the bottom of the pond.

Thamnophis hammondii is probably active year-round from San Francisco de la Sierra south. Ranchers at Rancho San Juanito, in the La Presa region, tell me they occasionally see these snakes during the winter, but they are far more common from March through October. In northwestern Baja California, Two-Striped Garter Snakes are rarely seen abroad during the winter, although juveniles are occasionally found on warm days. At this latitude, *T. hammondii* usually begins to emerge during February and remains active into early November. Throughout Baja California, I have observed pregnant females during May and June and newborns during July. These observations indicate that *T. hammondii* mates during the spring. I collected a female from Cañón Santa Isabel that gave birth to eight young during August. Stebbins (1954) reported a female from southern California giving birth to 25 young in October.

FIG 6.44 ***Thamnophis validus***

WEST COAST GARTER SNAKE; CULEBRA PRIETA

R. [Regina] valida Kennicott 1860:334. Type locality: "Durango, Mexico." Precise locality unknown (see Conant 1969). Holotype: USNM 1309

IDENTIFICATION *Thamnophis validus* can be distinguished from all other snakes in the region of study by having large head plates; keeled dorsal scales; a divided anal plate; and a black, brown, or pale gray body.

RELATIONSHIPS AND TAXONOMY *Thamnophis validus* is part of a large natural group containing *T. sauritus, T. proximus, T. errans, T. godmani, T. melanogaster,* and *T. scalaris* (de Queiroz and Lawson 1994) of the United States and Mexico. Conant (1946, 1969) and de Queiroz and Lawson (1994) discuss taxonomy.

DISTRIBUTION *Thamnophis validus* has a fragmented distribution along the west coast of southwestern Mexico from southern Sonora south to central Guerrero (McCranie and McAllister 1988). In Baja California, it also has a fragmented distribution and is known from water systems near La Burrera along the western face of the Sierra la Laguna and the watercourses and systems associated with Santiago, Aguas Calientes, and Miraflores along the eastern face of the Sierra la Laguna. The populations from the western and eastern sides of the mountains are probably allopatric. *Thamnophis validus* is also known from a disjunct population near San José del Cabo (Conant 1969) across the low-lying Valle San José (see map 6.22).

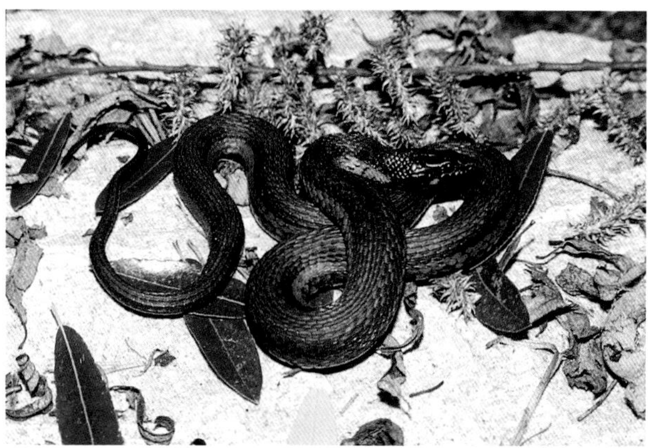

FIGURE 6.44 *Thamnophis validus* (West Coast Garter Snake) from Boca de la Sierra, BCS

DESCRIPTION Same as *T. elegans,* except for the following: adults reach 730 mm SVL; tail averages 23 to 28 percent of SVL in males and 22 to 27 percent of SVL in females; 7 to 9 (usually 8) supralabials; 9 or 10 (usually 10) infralabials; dorsal body scales keeled and imbricate; usually 17 dorsal body scale rows at midbody; 135 to 146 ventral scales in males, 134 to 145 in females; anal plate divided; 74 to 82 divided subcaudals in males, 67 to 75 in females. **COLORATION** *Thamnophis validus* is polymorphic in color pattern, with both dark and light color phases. DARK PHASE: Dorsal ground color uniformly black or very dark brown with indications of large dark blotches; a pale, very irregular lateral body stripe present; venter uniformly black or dark brown; top of head shiny black; pale spot on sixth supralabial. LIGHT PHASE: Dorsal ground color uniformly pale gray or brown or marked with numerous small, conspicuous dark spots arranged in four longitudinal rows; venter whitish to yellowish and immaculate; top of head shiny black.

GEOGRAPHIC VARIATION Both the dark and the light color phases occur in sympatry at Santiago and Aguas Calientes with no signs of intermediacy (Conant 1969). Only the dark phase occurs at La Burrera and Miraflores, although the light phase and intermediates occur at San José del Cabo (Conant 1969).

NATURAL HISTORY *Thamnophis validus* is locally abundant in the Arid Tropical Region of the Cape Region, along the foothills of the Sierra la Laguna, and in the vicinity of San José del Cabo (Conant 1946). This snake also ranges into the oak woodland of the Sierra la Laguna and borders the higher pine-oak woodland (Alvarez et al. 1988). Within these areas, *T. validus* is found in strict association with water, including marshy meadows such as those near San José del Cabo (Conant 1969); deep, rocky canyon pools and streams; and man-made concrete irrigation canals and dams such as those along the eastern face of the Sierra la Laguna. *Thamnophis validus* is active year-round, with a minimum of activity from November through February (Yarrow 1882b; Van Denburgh 1895; Banta and Leviton 1964; Conant 1969). This snake is diurnal, taking refuge and hunting prey within floating patches of moss and aquatic and streamside vegetation. I have often observed *T. validus* basking along the rocky ledges and shores and muddy stream banks in Cañón de San Bernardo at Boca de la Sierra.

Conant (1969) reported that the most important food source of *T. validus* is the Red-Spotted Toad *(Bufo punctatus)* and its tadpoles, but he noted that it also fed on Pacific Treefrogs *(Hyla regilla)* and possibly roaches. Van Denburgh (1985) reported small brackish-water fishes in the stomachs of specimens collected from San José del Cabo, and J. Grismer (2000) observed an adult eating a Couch's Spadefoot *(Scaphiopus couchii)* in Aguas Calientes. I watched adults and subadults hunting what I believed to be mosquito fish *(Gambusia)* in clear pools in Cañón San Bernardo. Nothing has been reported on the reproductive biology of *T. validus,* but I have noted an abundance of small juveniles in October, which suggests that reproduction may occur during the late spring and birth during the summer. I collected one female that gave birth to 10 young in August, and I have observed newborns during late September and early August.

Trimorphodon Cope

LYRE SNAKES

The genus *Trimorphodon* is composed of two species of moderately sized terrestrial snakes (Flores 1993; Gehlbach 1971; Scott and McDiarmid 1984) that collectively range across much of the southwestern United States south to Chiapas, Mexico. *Trimorphodon* occurs again in Guatemala and extends south along the west coast of Central America to northern Panama. Lyre snakes are secretive, nocturnal predators with large triangular heads and protuberant eyes containing vertical pupils. They have enlarged, grooved rear teeth that serve to channel a toxic saliva into their prey's body through a chewing action. Lyre snakes are moderately scansorial and saxicolous but occur in a variety of habitats ranging from arid, rocky deserts to moist tropical forests. The relationship of *Trimorphodon* to other colubrid genera is obscure at best. Cadle (1988), George and Dessauer (1970), Minton and Salanitro (1972), and Schwaner and Dessauer (1982) place it in a group that contains other North American colubrids such as *Lampropeltis, Elaphe (sensu stricto),* and *Pituophis.*

FIG 6.45 *Trimorphodon biscutatus*

LYRE SNAKE; VÍBORA SORDA

Dipsas bi-scutata Duméril, Bibron and Duméril 1854:1153. Type locality: "Mexique," restricted to "Tehuantepec (city, and environs)," Oaxaca, Mexico, by Smith and Taylor (1950:340). Holotype: MHNP 5900

IDENTIFICATION *Trimorphodon biscutatus* can be distinguished from all other snakes in the region of study by having a large triangular head; thin neck; large head plates; elliptical pupils; and a light-colored marking on the top of the head in the shape of a lyre.

RELATIONSHIPS AND TAXONOMY *Trimorphodon biscutatus* is the sister species of *T. tau* of central Mexico (Gehlbach 1971). Gehlbach (1971) and Grismer et al. (1994) discuss taxonomy.

DISTRIBUTION *Trimorphodon biscutatus* ranges disjunctly through much of the American southwest and extreme northern Mexico, south along the west coast of Mexico, and disjunctly through Central America to northern Panama (Scott and McDiarmid 1984). In Baja California, *T. biscutatus* ranges through most of the peninsula but is absent from the upper elevations of the Sierra San Pedro Mártir (Welsh 1988), Sierra Juárez, and Sierra la Laguna (Alvarez et al. 1988). Except for a population in the Sierra Vizcaíno and Sierra Santa Clara, from the southern portion of the Vizcaíno Peninsula, *T. biscutatus* does not occur in the Vizcaíno Desert (Grismer et al. 1994), and it has yet to be found along the western extremes of the Magdalena Plain. It is also known from the Gulf islands of Cerralvo, Danzante, San José, San Marcos, and Tiburón (map 6.23).

DESCRIPTION Body long and thin, subcylindrical in cross section, reaching 865 mm SVL; head triangular, distinct from neck; snout truncate in dorsal and lateral profile; eyes large and protuberant, pupils elliptical; tail averages 16 to 18 percent of total length in males and 14 to 15 percent in females; head scales large and platelike; 7 to 10 (usually 8 or 9) supralabials; 10 to 14 (usually 12 or 13) infralabials; dorsal body scales smooth, imbricate; 21 to 24 (usually 22

FIGURE 6.45 *Trimorphodon biscutatus* (Lyre Snake) from 50 km west of Santa Rosalía, BCS

or 23) dorsal body scale rows at midbody; 220 to 244 ventral scales in males and females; anal plate divided or single; 66 to 86 divided subcaudals in males and 58 to 76 in females. **COLORATION** Dorsal ground color gray; light band extending across rostrum; brown interocular bar across top of head bordered posteriorly by white band extending posterolaterally as wide, oblique postorbital stripes; light-colored, lyre-shaped marking on top of head; 23 to 35 dark, wide, saddle-shaped body blotches often extending to contact ventral scales; body blotches split by central, transversely oriented light area; dark, transversely elongate lateral body markings between lateral arms of dorsal blotches; ventrum dull white and immaculate. Juveniles tend to have a much more contrasting color pattern than adults.

GEOGRAPHIC VARIATION *Trimorphodon biscutatus* from Baja California and southwestern California shows little consistent geographic variation. Consequently, Grismer et al. (1994) synonymized *T. b. lyrophanes* of southern Baja California with *T. b. vandenburghi* of northern Baja California and southern California. Lyre snakes north of San Francisco de la Sierra usually have an undivided anal plate, whereas the anal plate of lyre snakes from the southern half of the peninsula is usually divided. Specimens of *T. biscutatus* from cismontane populations of northern Baja California are generally darker than individuals from desert populations. Lyre snakes from San Ignacio south generally have a more distinctive lyre marking and are darker than snakes farther north. Lyre snakes from areas dominated by light-colored granitic rocks have fewer small dark punctations in the interspaces between the dorsal saddles, resulting in an overall lightening and "clearing up" of the dorsal ground color. This patterning creates a bold contrast by highlighting the dark dorsal saddles; it is pronounced in lyre snakes from the Cape Region, Islas Cerralvo and San José, Cataviña, and the Sierra San Felipe. Extensive dark punctation occurs in specimens from habitats in central Baja California dominated by dark volcanic rocks, such as the Sierra la Giganta and Sierra Santa Clara, making the snakes appear more mottled and with far less contrast. Occasionally individuals from the vicinity of San Ignacio and just south of Loreto will have a unicolored slate gray dorsal ground color and narrow, widely separated dorsal saddles that do not extend nearly as far ventrally. These snakes look very similar to *T. b. vilkinsoni* (*sensu* Gehlbach 1971) of western Texas, southwestern New Mexico, and northern Chihuahua, Mexico.

NATURAL HISTORY *Trimorphodon biscutatus* occurs throughout most of the phytogeographic regions of Baja California and is almost always found near rocks. In the California Region, *T. biscutatus* is most common on boulder-covered hillsides along the western foothills and within the inland valleys of the Peninsular Ranges, although specimens also have been reported from the flat and rather featureless San Quintín Plain (Murray 1955). *Trimorphodon biscutatus* is very common along the rocky desert foothills of the Peninsular Ranges in the Lower Colorado Valley Region. In central Baja California, it is frequently found on lava flows, rocky mesas, bajadas, and rocky flats, and in rocky arroyos. In the Cape Region, Alvarez et al. (1988) reported lyre snakes as being common in rocky areas of the low-lying Arid Tropical Region surrounding the Sierra la Laguna and noted that it ranges up the Sierra la Laguna for a short distance in the rocky foothills but becomes scarce in the oak woodland and absent in the pine-oak woodland. Murphy and Ottley (1984) reported finding specimens on hillsides and among sparse low brush on Isla San Marcos, and on desert hardpan on Isla Danzante. I have found specimens on the eastern edges of the flat, somewhat featureless Magdalena Plain just east of Ciudad Constitución, where small to medium-sized rocks (ca. 0.5 m in diameter) provide ground cover. *Trimorphodon biscutatus* often takes refuge in deep rock cracks and fissures or beneath cap rocks. I have found small specimens in Valle la Trinidad beneath thick, vertically oriented exfoliations on granitic boulders. In southern Baja California, it often uses smaller rocks for refuge, and I have even found specimens beneath logs in a Date Palm *(Phoenix dactylifera)* grove at San Ignacio and beneath garbage near San Bartolo.

Trimorphodon biscutatus can be active in relatively low temperatures (Klauber 1940c) and thus has a longer activity period than some other snakes. Lyre snakes are generally active from March through October in northern Baja

California, although specimens can be found during February and November. I have seen lyre snakes during December and January basking in the sun at the openings of deep rock fissures near Cataviña and along the eastern foothills of the Sierra Juárez. I do not consider these individuals to have been active (or abroad). In northern Baja California, *T. biscutatus* is most active during April, May, and June. In the Cape Region, lyre snakes are somewhat active year-round, and I have often found individuals abroad at night during November and December. Peak activity in the Cape Region, however, is from March through October.

Trimorphodon biscutatus is nocturnal and scansorial. Although most lyre snakes are seen crossing roads at night, lantern-walking through their habitat reveals that they spend a modest amount of time above ground, crawling on large rocks. I observed one individual approximately 1.5 m above the ground crawling along a rock wall at La Presa. As it moved, it poked its head into spaces between the rocks while rapidly flicking its tongue. Earlier that day, I had seen several Black-Tailed Brush Lizards *(Urosaurus nigricaudus)* on that wall, and the lyre snake appeared to be systematically searching for sleeping lizards. Van Denburgh and Slevin (1921) reported finding a specimen crawling in a thatched roof in San José del Cabo, and the snake had mammal remains in its stomach.

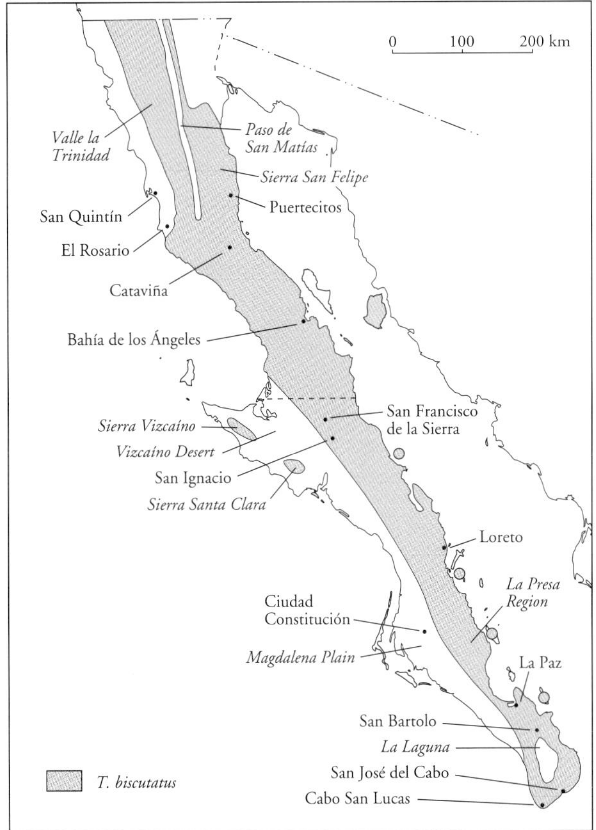

MAP 6.23 Distribution of *Trimorphodon biscutatus* (Lyre Snake)

Trimorphodon biscutatus feeds primarily on lizards and small mammals, including bats. Enlarged, grooved teeth on the backs of the maxillae facilitate the injection of a toxic saliva into the prey through a chewing action. This venom is more effective on lizards than it is on mammals, and with mammals snakes may actually use constriction as well as venom during prey abduction (Rodman 1939). In the Cape Region, I have found road-killed lyre snakes with Side-Blotched Lizards *(Uta stansburiana),* Black-Tailed Brush Lizards, Cape Spiny Lizards *(Sceloporus licki),* Hunsaker Spiny Lizards *(S. hunsakeri),* and Kangaroo Rats (*Dipodomys* sp.) in their stomachs. Klauber (1940c) reported a specimen that had eaten a juvenile Baja California Spiny-Tailed Iguana *(Ctenosaura hemilopha).*

Little is known about the reproductive biology of *T. biscutatus* in Baja California. Klauber (1940c) reported a specimen from southern Baja California that laid 13 eggs in mid-May, and I have seen gravid females in the Cape Region during June and hatchlings during late August. This suggests that lyre snakes in southern Baja California have a spring breeding season, as do lyre snakes in southern California and presumably northern Baja California.

FOLKLORE AND HUMAN USE The local inhabitants of Baja California fear lyre snakes, believing they are venomous and closely related to rattlesnakes. The Seri Indians also believe lyre snakes are venomous but maintain that they do not bite, although the venom is stored in the teeth as well as the flesh of the body (Malkin 1962).

Elapidae

CORAL SNAKES, SEA SNAKES, AND THEIR OLD WORLD RELATIVES

The Elapidae are a large, diverse group of advanced snakes containing approximately 62 genera and roughly 300 species (Pough et al. 2001). Elapids all have permanently erect, hollow fangs in the anterior portion of their immovable maxillae, which fit into a pair of slots on the floor of the buccal cavity when the mouth is closed. They have extremely diverse adaptive types, ranging from fossorial to marine species, and are found in virtually every kind of habitat. Elapids are found on every continent of the world except Antarctica, although they are generally absent from temperate zones and conspicuously absent from the Sahara Desert. Groups of marine-adapted elapids range along the continents in all the major oceans except the Atlantic, and one species is even transpacific. In the New World, elapids are represented by a generally tropical natural group known as the coral snakes, which contains the genera *Micrurus* (60+ species; Roze 1996) and the monotypic *Micruroides* (Slowinski 1995). Only the latter occurs in the region of study, and it is generally restricted to arid regions of the southwestern United States and northwestern Mexico (Roze 1974).

Micruroides Schmidt

WESTERN CORAL SNAKE

The genus *Micruroides* contains the single species *M. euryxanthus* (Roze 1974, 1996), which ranges throughout the American southwest and northwestern Mexico. It differs from other New World coral snakes *(Micrurus)* by occurring in arid as opposed to mesic regions. *Micruroides euryxanthus* is a secretive, somewhat fossorial species. Although highly toxic, it is not particularly dangerous to humans because its small mouth and short fangs are inefficient for injecting venom into large objects. Nonetheless, it should not be handled and should always be treated with extreme caution. *Micruroides euryxanthus* is the sister species to the remaining New World coral snakes of the genus *Micrurus* (Slowinski 1995).

FIG 6.46 *Micruroides euryxanthus*

WESTERN CORAL SNAKE; CORALILLO

Elaps euryxanthus Kennicott 1860:337. Type locality: Not given, but listed in the National Museum of Natural History as "Sonora, Mexico" (Roze 1974). Holotype: USNM 1122

IDENTIFICATION *Micruroides euryxanthus* can be distinguished from all other snakes in the region of study by having black, red, and white to cream bands, with the last two colors in contact.

RELATIONSHIPS AND TAXONOMY See above for *Micruroides*.

DISTRIBUTION *Micruroides euryxanthus* ranges from southeastern Arizona and southwestern New Mexico south along the west side of the Sierra Madre Occidental in northwestern Mexico to southern Sonora and southwestern Chihuahua. It occurs once again farther south, in southern Sinaloa (Roze 1974, 1996). *Micruroides euryxanthus* also occurs on Isla Tiburón in the Gulf of California (map 6.24).

DESCRIPTION Body long, thin, slightly flattened in cross section, reaching 540 mm SVL; head weakly distinct from neck; snout rounded in dorsal and lateral profile; eyes small, pupils rounded; head scales large and platelike; 7 supralabials; 7 infralabials; dorsal body scales smooth and imbricate; 15 dorsal body scale rows at midbody; 212 to 230 ventral scales in males and 219 to 245 in females; anal plate divided; 29 to 32 divided subcaudals in males and 19 to 27 in females. **COLORATION** Head black, followed by a wide, cream-colored nuchal ring; nuchal ring followed by red, black, and cream-colored rings in the sequence red, cream, black, cream, red; black pigment occurring at bases and tips of scales in red rings; red rings approximately same width as black rings, containing 4 to 9 vertebral dorsal scales.

FIGURE 6.46 *Micruroides euryxanthus* (Western Coral Snake) from Phoenix, Arizona

NATURAL HISTORY Although *M. euryxanthus* has been known to exist on Isla Tiburón for well over a century (Streets 1877), nothing has been reported on the natural history of this population. An excellent treatment of the natural history of *M. euryxanthus* in general is given in Lowe et al. (1986), from which most of this section is adapted.

Isla Tiburón is a large landbridge island within the Central Gulf Coast Region. Its large size and topographical complexity support a wide variety of desert habitats, such as rocky bajadas and desert slopes as well as wide, open valleys with Creosote Bush *(Larrea tridentata)* flats and sand dunes. Associated with these desert habitats are various floral compositions (Felger and Lowe 1976), which add to the habitat diversity of the island. *Micruroides euryxanthus* is most common in areas of loose soil in the vicinity of shrubs and trees on rocky slopes and bajadas, and in canyons. Western coral snakes are active from March through November, with peak activity occurring during the summer. During the spring, *M. euryxanthus* is diurnal, but as temperatures increase later in the year, activity is restricted to evening hours, especially following summer rains.

Micruroides euryxanthus feeds primarily on small snakes but will occasionally eat small lizards. The snake most heavily preyed on is the Western Blind Snake *(Leptotyphlops humilis)*. Roze (1996) reports that Sand Snakes *(Chilomeniscus cinctus)* are occasionally eaten. No blind snakes have been positively reported as occurring on Isla Tiburón. Although one specimen was discovered smashed between boards in a shipment of lumber brought from Sonora (R. Crombie, personal communication, 1995), it has not been considered a validating specimen for this species on Isla Tiburón. The presence of *Micruroides* on the island, however, suggests that *Leptotyphlops* may indeed occur there. Further, Lowe et al. (1986) noted that the distribution of *M. euryxanthus* may parallel that of Western Blind Snakes, and that Western Coral Snakes are often found in the burrow systems of blind snakes. *Micruroides euryxanthus* breeds in the spring and early summer, and eggs are laid in July and August, with hatching taking place in September. Hatchlings usually remain active into November.

FOLKLORE AND HUMAN USE Although Western Coral Snakes are relatively rare in coastal Sonora, the Seri Indians do recognize them as a potential danger (Malkin 1962). The Seri told Nabhan (forthcoming) that they were aware that the snake was poisonous but were not afraid of it because no Seri had ever been killed by one. They believe the snake will bite and that the venom is stored in the teeth as well as in the flesh. The bite is believed to be fatal to animals and Mexicans but not Seri men (Malkin 1962). Malkin reported that the Seri used the skin of specimens they found dead to make good-luck rings and hat bands.

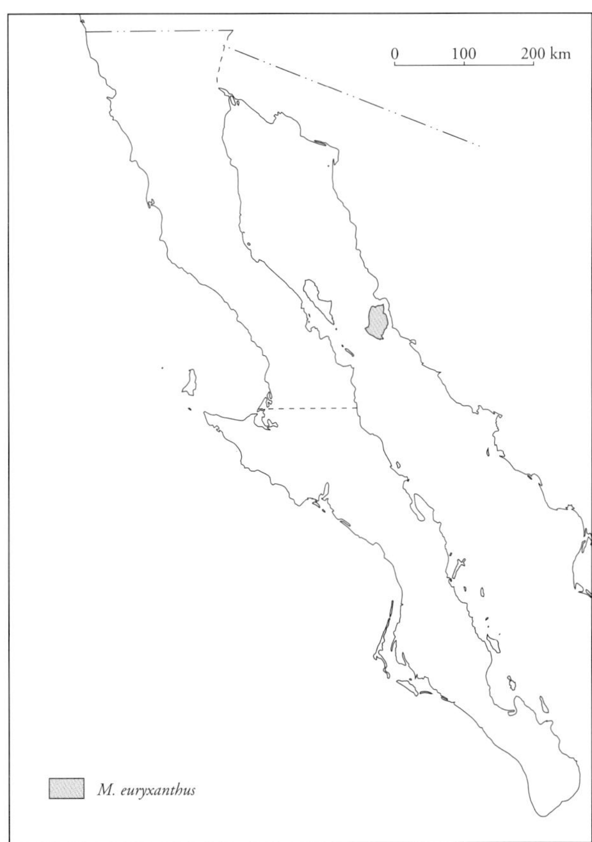

MAP 6.24 Distribution of *Micruroides euryxanthus* (Western Coral Snake)

Pelamis Daudin

PELAGIC SEA SNAKE

Pelamis is a monotypic genus that is part of a group of elapids known as the Hydrophiinae, or true sea snakes. All hydrophiines are highly adapted for a marine existence. They have laterally compressed bodies, paddlelike tails, valvular nostrils, and no transverse ventral scales, and they give live birth at sea. *Pelamis platurus* ranges continuously along the east coast of Africa, along the continental margins east through southeast Asia and Micronesia. It occurs again in the eastern Pacific along the west coast of Mexico and south to Ecuador (Pickwell and Culotta 1980; Reynolds 1984). Although *Pelamis* is the only truly pelagic sea snake, its occurrence in the open areas of the Indian and Pacific Oceans is sporadic and usually depends on ocean currents. *Pelamis* is absent from the Atlantic Ocean, although there are unconfirmed reports (Marinkelle 1966) of its presence in the Caribbean. The relationships of *Pelamis* to other hydrophiines is problematic, but Murphy (1988) indicates that it may be the sister species of the genus *Emydocephalus* of northern Australia and the South China Sea.

FIG 6.47 ***Pelamis platurus***

PELAGIC SEA SNAKE; ANGUILLA, CULEBRA DEL MAR

Anguis platura Linnaeus, 1766:391. Type locality and holotype unknown.

IDENTIFICATION *Pelamis platurus* can be distinguished from all other snakes in the region of study by having a laterally compressed body and tail and a black and yellow color pattern.

RELATIONSHIPS AND TAXONOMY See above for *Pelamis*.

DISTRIBUTION *Pelamis platurus* ranges throughout all areas of the Gulf of California (Pickwell et al. 1983). It has also been found along the west coast of Baja California at Bahía Magdalena and Bahía Blanca (Pickwell et al. 1983).

DESCRIPTION Body and tail laterally compressed; adults reach 500 mm SVL; head elongate, narrow, flattened, and tapering anteriorly; eyes moderate, pupils rounded; tail averages approximately 12 percent of total length in males and females; head scales large and platelike; 7 to 9 (usually 8) supralabials; 10 to 12 (usually 11) infralabials; dorsal body scales rectangular, juxtaposed, often tuberculate in adults;

FIGURE 6.47 *Pelamis platurus* (Pelagic Sea Snake) from the Gulf of Chiriquí, Panama. (Photograph by Paul Freed)

47 to 69 dorsal body scale rows at midbody. **COLORATION** Head and dorsum black; sides of body yellow, often with varying patterns of black; ground color of tail yellow or black; yellow tails are overlaid by black spots or thick black, sinuous lines; black tails are overlaid by yellow spots or thick yellow, sinuous lines.

NATURAL HISTORY Very little has been reported on the natural history of *P. platurus* from the Gulf of California, although Kropach (1975) gives an excellent account of this species' natural history in the Gulf of Panama. Although *P. platurus* ranges widely throughout the Gulf of California, it is considered a visitor and does not breed in its waters (Pickwell et al. 1983). Its occurrence in the Gulf may be passive as much as anything else. From June to November, northward currents prevail at the mouth of the Gulf and may bring sea snakes into its waters (Pickwell et al. 1983). *Pelamis platurus* breeds in tropical waters that maintain an isotherm of 18 to 20°C (Dunson and Ehlert 1971; Graham et al. 1971; Hect et al. 1974), and the farthest north that breeding populations have been found is Bahía Banderas of Jalisco, off the southwestern coast of Mexico (Pickwell et al. 1983). Pickwell et al. reported sightings in the Gulf of California at Isla Espíritu Santo, Guaymas, and San Felipe. Shaw (1961) reported a specimen from Bahía de los Ángeles. I have seen specimens at Puerto Escondido, Isla Santa Catalina, and Bahía San Luis Gonzaga.

Pelamis platurus may regularly occur along the west coast of Baja California as far north as Bahía Magdalena, the general northern extent of the northerly flowing, warm-water Davidson Current. Coastal areas north of here are dominated by the southerly flowing, cold-water California Current, whose cool temperatures probably exceed the physiological limits of *P. platurus* and restrict its northern distribution. During El Niño years, however, the Davidson Current remains on the surface and inshore to the California Current and has carried *P. platurus* with it as far north as San Clemente Island in southern California (Pickwell et al. 1983).

Pelamis platurus feeds almost exclusively on small fish. Sea snakes tend to float alongside vegetation or debris. This material attracts large schools of small fish, which the snake catches by violently lunging sideways into the school and moving its head back and forth with its mouth open (Kropach 1975). On a number of occasions I have seen two to five snakes feeding on small fish along the shore where fishermen were cleaning their catch. Smaller fish are attracted to the entrails thrown into the water, and sea snakes are attracted to the smaller fish.

FOLKLORE AND HUMAN USE Although the fishermen of the Gulf of California know that *P. platurus* is extremely poisonous and can be quite aggressive when taken out of the water, they pay little regard to this danger. Sea snakes are often found in the haul of fishing nets; the fishermen simply remove them by hand and throw them back into the water (Thomas and Scott 1997). None of the fishermen I have spoken with knows of anyone who has been bitten. Nabhan (forthcoming) reports on a fisherman in the northern Gulf near Puertecitos who was bitten. The wound bled and oozed a yellow liquid for three days, causing severe pain. In an interview two decades later, the man told Nabhan that he suffered permanent local nerve damage and still lacks a full range of motion in the foot.

Viperidae

VIPERS AND PIT VIPERS

The Viperidae are a large, diverse group of advanced snakes containing approximately 20 to 27 genera and approximately 227 species with a nearly cosmopolitan distribution (Pough et al. 2001). Viperids are highly venomous and possess long, hollow fangs anchored to a short, movable maxilla. There are three major lineages within the Viperidae: the viperines or vipers form a modestly large Old World radiation with approximately nine genera and 50 species. A peculiar montane species, *Azemiops feae,* restricted to the highlands of southern China, Tibet, and Burma, forms the second major viperid lineage; and the third lineage is known as the Crotalinae, or pit vipers. Crotalines are a New World and Asian group that contains approximately 14 genera and 120 species. Crotalines are unique among viperids in having a large pit organ between each eye and nostril that functions as an infrared heat receptor. This helps them to detect prey and direct their prey strikes during periods of low-intensity light. Crotalines have diverse lifestyles and habitat preferences, ranging from cryptically colored, forest floor species and brightly colored, arboreal, tropical forms to arid-adapted desert dwellers and aquatic swampland species. In Baja California and the Gulf of California, crotalines are represented by 14 species of rattlesnakes *(Crotalus)*, which in the region of study are usually terrestrial and generally arid-adapted.

Crotalus Linné

RATTLESNAKES

The genus *Crotalus* contains approximately 30 species that collectively range throughout most of North America south through Mexico and Central America and into South America to northern Argentina (Campbell and Lamar 1989). *Crotalus* is unique among snakes in having a rattle on the end of its tail, which is usually used in the snake's repertoire of defensive behavior. Rattlesnakes occupy a wide variety of habitats, ranging from barren, arid deserts and wet swamplands to tropical, alpine, and cloud forests. Some species are habitat generalists and range widely across broad geographic regions, whereas others are microhabitat specialists and may be restricted to hillsides containing a particular rock type. With the exception of some of the smaller, more secretive species, rattlesnakes are usually a conspicuous element of the environment, and on islands they are often among the top predators of the ecosystem. Rattlesnakes are common in the region of study, and 14 different species are recognized (Grismer 1999b; Campbell and Lamar 1989).

FIG 6.48 ***Crotalus atrox***

WESTERN DIAMONDBACK RATTLESNAKE; VÍBORA DE CASCABEL

Crotalus atrox Baird and Girard 1853:5. Type locality: "Indianola [Calhoun County], Texas," USA. Holotype: USNM 7761

IDENTIFICATION *Crotalus atrox* can be distinguished from all other snakes in the region of study by having a heat-sensing facial pit organ between the nostril and eye; an upper loreal scale; no hornlike processes on the supraocular scales; a well-developed inner rattle matrix; dorsal diamond markings composed of one internal row of light-colored scales bordered by one external row of dark scales; distinctive black and white rings on the tail; and fine black punctations in dorsal pattern.

RELATIONSHIPS AND TAXONOMY *Crotalus atrox* is the sister species of *C. tortugensis* of Isla Tortuga in the Gulf of California. Together they form the sister lineage of the *C. ruber* group, composed of *C. ruber, C. catalinensis,* and *C. lorenzoensis* (Murphy and Crabtree 1985). Amarál (1929) and Klauber (1930) discuss taxonomy.

DISTRIBUTION *Crotalus atrox* ranges through much of the southwestern United States and northern Mexico south to central Mexico on both sides of the Sierra Madre Occidental (Campbell and Lamar 1989). In Baja California, *C. atrox* occurs only in the extreme northeastern portion of the peninsula. It has been found as far west as the western side of the Sierra los Cucapás and as far south as 79 km below Mexicali. The southern limit of this species is the result of dispersal along Mexican Highway 1, south of the Sierra de los Cucapás. Here the Mexican government had to elevate the highway above the episodically flooding Laguna Salada. The construction created a narrow, rocky, and vegetated dispersal corridor in which *C. atrox* are occasionally found. *Crotalus atrox* also occurs on Islas Tiburón, Dátil, San Pedro Mártir, and Santa Cruz in the Gulf of California (map 6.25). Campbell and Lamar (1989) report *C. atrox* from the Gulf island of Santa María, but to my knowledge, no such island exists.

DESCRIPTION Body heavy and stout, reaching approximately 1,700 mm SVL; head triangular, distinct from neck; pupils elliptical; tail relatively short; 2 small internasals; prenasal usually in contact with first supralabial; loreal usually sin-

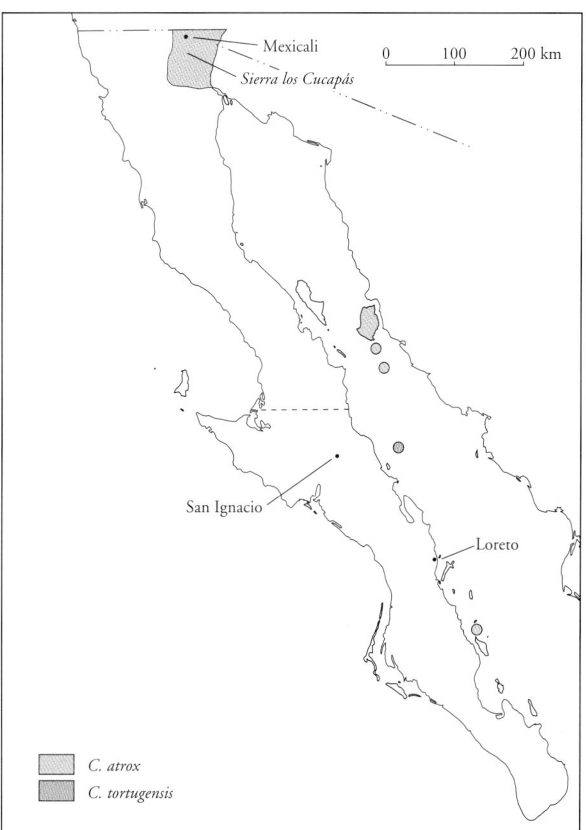

MAP 6.25 Distribution of *Crotalus atrox* (Western Diamondback Rattlesnake) and *C. tortugensis* (Isla Tortuga Rattlesnake)

FIGURE 6.48 *Crotalus atrox* (Western Diamondback Rattlesnake) from Isla Tiburón, Sonora

gle; head scales small and imbricate; 14 to 16 supralabials; 16 to 19 infralabials; dorsal body scales keeled and imbricate, 25 or 27 scale rows at midbody; 181 to 191 ventral scales in males, 185 to 190 in females; 21 to 28 subcaudals in males, 18 to 22 in females; subcaudals at beginning and end of series divided. **COLORATION** Dorsal ground color straw-colored to reddish-brown; distinctive oblique, light-colored pre- and postorbital stripes present; dorsal pattern composed of 24 to 45 diamond-shaped markings, most distinctive medially; diamonds composed of one internal row of light-colored scales bordered by single row of dark outer scales; diamonds fade posteriorly; small black punctations throughout dorsal pattern give an overall dusty appearance; tail composed of distinctive black and white caudal rings; black caudal rings often discontinuous; venter dull yellow, usually immaculate.

GEOGRAPHIC VARIATION *Crotalus atrox* blends in well with its habitat, and the majority of its geographic variation in the region of study is correlated with overall substrate matching. *Crotalus atrox* from the sandy, flat areas of northeastern Baja California are generally light gray in overall coloration, with a moderately distinct diamond pattern. Individuals from Isla Tiburón usually differ from specimens from the adjacent mainland coastline in that they lack the yellow highlighting in their scales and instead have a grayish hue. Western Diamondbacks from Isla San Pedro Mártir are yellowish to straw-colored, and the diamond pattern is occasionally faded. This fading is most evident in the posteriormost diamonds of larger specimens, which are often undiscernible. *Crotalus atrox* from Isla Santa Cruz are dark, having an overall coloration of rust-brown to reddish.

NATURAL HISTORY In Baja California, *C. atrox* is restricted to the Lower Colorado Valley Region of the extreme northeastern portion of the peninsula and is most common in the foothills of the Sierra los Cucapás. On the Gulf islands, however, Western diamondbacks occur within the Central Gulf Coast Region. Within all these habitats, *C. atrox* is most commonly found on rocky bajadas and flats and in arroyos in the vicinity of scrub vegetation. At night, individuals often crater themselves along rodent trails at the edge of vegetation, where they ambush prey. This behavior seems to be most common on Isla Tiburón, but I have also observed it on Isla San Pedro Mártir and occasionally on Isla Santa Cruz. *Crotalus atrox* is active from late March through late October. During the cooler spring months, *C. atrox* is both diurnal and nocturnal. During the hotter summer months, however, it is strictly nocturnal. On Isla San Pedro Mártir, Western Diamondbacks are abundant on summer nights, and I have seen as many as 10 snakes within the first few hours of darkness.

Crotalus atrox feeds primarily on small mammals; however, on Isla San Pedro Mártir, one specimen had a Blue-Footed Booby *(Sula nebouxi)* chick, an introduced rat (*Rattus* sp.), and an Isla San Pedro Mártir Side-Blotched Lizard *(Uta palmeri)* in its stomach (D. Hews, personal communication, 1993). *Crotalus atrox* outside the region of study apparently has two breeding periods, one in spring and another in late summer and early fall (see Ernst 1992). The mating season in the region of study has not been precisely determined. Cliff (1954a) reported on a presumably gravid female from Isla San Pedro Mártir on 14 May, and Bogert (1942) observed a copulating pair in Coachella Valley in southern California on 25 March. Lowe (1942) reported on a copulating pair from near Blythe in southern California on 8 October. The latter localities are just north of northeastern Baja California, and it is assumed that the *C. atrox* from these adjacent regions have similar reproductive strategies. These observations suggest that *C. atrox* in the region of study also has two different breeding periods during the year.

REMARKS Alvarez and Huerta (1974) report a specimen of *C. atrox* from Bahía San Francisquito in central Baja California. However, on examination of this specimen, I found it to be *C. ruber.*

FOLKLORE AND HUMAN USE The Seri Indians believe *C. atrox* live in packs of 20 to 25 after emerging from hibernation in March (Malkin 1962). Although they realize the snake is poisonous, it is a very common food item on the Seri table. They believe the poison resides in the head and the tail, and these areas are cut off and discarded before the animal is cooked (Malkin 1962). The vertebrae of *C. atrox* are common components of much Seri jewelry. The oil mixed with various ground-up portions of the snake is used for a number of medicinal purposes (see Nabhan, forthcoming).

FIG 6.49

Crotalus tortugensis

ISLA TORTUGA RATTLESNAKE; VÍBORA DE CASCABEL

Crotalus tortugensis Van Denburgh and Slevin 1921c:398. Type locality: "Tortuga Island, Gulf of California, Mexico." Holotype: CAS 50515

IDENTIFICATION *Crotalus tortugensis* can be distinguished from all other snakes in the region of study by having no upper loreal scale; no hornlike processes on the supraocular scales; a well-developed inner rattle matrix; dorsal diamond markings composed of one internal row of dark-colored scales bordered by one external row of light-colored scales; distinctive black and white rings on the tail; and fine black punctations in dorsal pattern.

RELATIONSHIPS AND TAXONOMY See *C. atrox.*

DISTRIBUTION *Crotalus tortugensis* is endemic to Isla Tortuga in the Gulf of California (see map 6.25).

DESCRIPTION Same as *C. atrox,* except for the following: adults reach 906 mm SVL; prenasal contacts first supralabial; 2 loreals; 14 to 18 supralabials; 14 to 19 infralabials; 180 to 190 ventral scales in males, 183 to 189 in females; 22 to 25 subcaudals in males, 16 to 25 in females; subcaudals at beginning and end of series divided. **COLORATION** Dorsal ground color gray to dark brown, often with reddish or purplish hue; distinctive oblique, light-colored pre- and postorbital stripes; top of head with irregular black speckling; dorsal pattern composed of 32 to 41 diamond-shaped or hexagonal markings, most distinctive medially and fading posteriorly; diamonds composed of one internal row of dark scales bordered by single row of white external scales; white scales of anterior border of diamonds reduced or absent; diamonds fade posteriorly; small black punctations throughout dorsal pattern give it an overall dusty appearance; tail composed of distinctive black and white caudal rings; black caudal rings sometimes discontinuous; venter dull white, usually immaculate. Juveniles tend to be more boldly marked.

NATURAL HISTORY Isla Tortuga lies within the Central Gulf Coast Region and is dominated by a depauperate thorn scrub vegetation. The island is little more than the top of an emergent volcano. It lacks beaches and coves, and its shoreline consists mostly of precipitous volcanic bluffs. The island is covered by irregularly shaped volcanic rocks within and beneath which *C. tortugensis* takes refuge. Van Denburgh and Slevin (1922) reported seeing several specimens in "cups" (wind-eroded indentations) beneath the rocky ledges. These are a common refuge for *C. tortugensis,* and I have found several specimens cratered in such areas. This species is extremely abundant, and as many as 26 individuals have been sighted in a few hours of nighttime observation during the summer. It is a very wary species that rattles while being approached from great distances; the sound gives away its location. Often snakes will rattle while buried deep in rock piles and completely out of view. *Crotalus tortugensis* is abundant throughout most parts of the island, but I have found no rattlesnakes inside the caldera of the volcano.

FIGURE 6.49 *Crotalus tortugensis* (Isla Tortuga Rattlesnake) from Isla Tortuga, BCS

This species is probably active year-round, depending on the weather. It appears to be most active from March through October. During the spring, specimens can be found abroad both day and night; during the summer, activity is restricted to the evening. At these times, rattlesnakes are seen foraging among the rocks or cratered along rodent trails. Van Denburgh and Slevin (1922) and Klauber (1972) reported finding deer mice *(Peromyscus)* in the stomachs of many of the specimens they examined. Nothing has been reported on the reproductive biology of this species, although I have seen gravid females during March and neonates during late May.

FIGS. 6.50–51

Crotalus ruber

RED DIAMOND RATTLESNAKE; VÍBORA DE CASCABEL

Crotalus adamanteus ruber Cope 1892b:690. Type locality: undesignated; restricted to "Dulzura," San Diego County, California, by Smith and Taylor (1950:356). Holotype: USNM 9209

IDENTIFICATION *Crotalus ruber* can be distinguished from all other snakes in the region of study by having a heat-sensing facial pit organ between the nostril and eye; no hornlike

FIGURE 6.50 Ruber pattern class of *Crotalus ruber* (Red Diamond Rattlesnake) from 63 km west of Santa Rosalía, BCS

processes on the supraocular scales; a transversely divided first infralabial scale; a well-developed inner rattle matrix; distinctive black and white caudal rings; and black caudal rings nearly as wide as the white caudal rings.

RELATIONSHIPS AND TAXONOMY *Crotalus ruber, C. lorenzoensis* of Isla San Lorenzo, and *C. catalinensis* on Isla Santa Catalina form a natural group (Murphy and Crabtree 1985). Grismer (1999b), Grismer et al. (1994), and Murphy et al. (1995) discuss taxonomy.

DISTRIBUTION Ranges from Los Angeles County, California, south throughout Baja California to Cabo San Lucas. *Crotalus ruber* is absent from the northern portion of the Lower Colorado Valley Region and first contacts the Gulf coast at San Felipe. It is also absent from the upper elevations of the Sierra San Pedro Mártir (Welsh 1988) and Sierra Juárez but is found along their eastern and western scarps. *Crotalus ruber* is known to occur on the Pacific islands of Cedros and Santa Margarita (Wong 1997) and the Gulf islands of Ángel de la Guarda, Danzante, Monserrate, Pond, San José, and San Marcos (map 6.26).

DESCRIPTION Body heavy and stout, reaching approximately 1,700 mm SVL; head triangular, distinct from neck; pupils elliptical; tail short, 6.9 to 7.3 percent of SVL in males and 5.2 to 5.3 percent in females; 2 small internasals; prenasal usually in contact with first supralabial; first infralabial usually transversely divided; 1 or 2 loreals; head scales small and imbricate; 12 to 19 supralabials; 13 to 21 infralabials; dorsal body scales keeled and imbricate, 25 to 33 scale rows at midbody; 181 to 203 ventral scales in males, 183 to 206 in females; anal undivided; 22 to 29 subcaudals in males, 16 to 25 in females; subcaudals at beginning and end of series divided; well-developed inner rattle matrix. **COLORATION** Dorsal ground color gray, tan, or reddish-brown; distinctive oblique, light-colored, pre- and postorbital stripes present; dorsal pattern composed of 29 to 42 diamond-shaped markings, most distinctive medially; diamonds composed of distinctive white scales, often bordered internally by a row of darker scales; diamonds fade posteriorly; tail composed of distinctive black and white caudal rings; black caudal rings often discontinuous and nearly as wide as white rings; venter dull yellow, usually immaculate.

GEOGRAPHIC VARIATION Three subspecies have been recognized in the region of study (Klauber 1945a; Murphy et al. 1995) and are used here as pattern classes. Ruber range south as far as Bahía Concepción, and Lucasensis extend from the vicinity of Loreto to Cabo San Lucas. Between Bahía Concepción and Loreto, Ruber and Lucasensis intergrade (Klauber 1949a; Murray 1955). Most Ruber have 27 or fewer dorsal scale rows, whereas most Lucasensis have 28 or more; most Ruber have a single loreal, whereas most Lucasensis have two or more; Ruber have 29 to 42 dorsal body blotches, whereas Lucasensis have 20 to 39; and Ruber are usually lighter and have a less contrasted diamond pattern than Lucasensis (Klauber 1949a; Radcliffe and Maslin 1975).

Ruber from cismontane populations in northwestern Baja California south to just below El Rosario are darker and much more reddish than the lighter, more tan-colored snakes from along the eastern slopes of the Sierra Juárez and Sierra San Pedro Mártir. Snakes from the upper elevations of the Sierra la Asamblea are dark reddish-brown. The southern populations of Ruber, from the vicinity of the Sierra la Asamblea south to Canipolé, tend to lack the upper portion of the thin, light-colored postorbital stripe. Ruber are fairly adept at substrate matching. A specimen collected from the dark volcanic flats at the foot of Cerro Afuera in the Sierra Santa Clara was relatively dark brown, with a typical diamond pattern. Just a few kilometers to the west, a specimen from Arroyo Destiladera, a deep gorge carved into light, sedimentary rock, was extremely pale in overall coloration and pattern. Ruber from Isla Ángel de la Guarda are very attractive, having a light red-orange tint to their dorsal ground color with a somewhat faded dorsal pattern. Specimens from Isla San Marcos are generally darker than those from the adjacent peninsula and more brown than red. A specimen collected at Laguna Chapala, just south of Calamajué, is anomalous in that the head lacks pigment and the body pattern is mottled (see Rubio 1999:45).

Crotalus ruber from Isla Cedros has most recently been considered the subspecies *C. r. exsul* (Murphy et al. 1995) and is treated here as a pattern class. Exsul usually have very faded anterior diamonds, many of the black caudal rings have lateral breaks, and most snakes have a pair of intergenials. Ruber from the Vizcaíno Peninsula are intermediate in these characteristics between Exsul and Ruber from the adjacent Peninsular Ranges (Grismer et al. 1994).

Adult Lucasensis from the vicinity of Agua Verde have a distinctive light green hue to their dorsal pattern, although juveniles are grayish. There is also a pattern anomaly that occurs in Lucasensis near Miraflores in the Cape Region: some specimens appear to be partly amelanistic. The ground color is beige and overlaid with varying degrees of irregular light brown blotches. The color pattern on the tail is usually normal, the eyes are black, and the tongue is often pink.

NATURAL HISTORY Much has been reported on the natural history of *C. ruber* from southern California (see Ernst 1992; Klauber 1972; and references therein); however, relatively little is known about the Baja California populations. Generally speaking, *C. ruber* occurs in a wide range of habitats and is absent only from the forests of the Baja California Coniferous Forest Region. In northern Baja California, Ruber range from cool ocean shoreline localities to hot inland valleys and mesas. They are most common in brushy areas with rocks, such as the boulder-strewn chaparral hillsides just east of Rancho San José. Ruber are also quite common along the eastern sides of the Sierra Juárez and Sierra San Pedro Mártir. Here they are most often observed on rocky hillsides and within riparian canyons (Klauber 1972; Linsdale 1932). In the latter habitats, I have found juveniles living within the cobblestone of dry streambeds. Further south, Ruber have been reported from coastal dunes, cobble beaches (Bostic 1971), and inland dunes (Mosauer 1936). Additionally, I have found Ruber to be very common on the brushy flats of the Vizcaíno Region, the moderately vegetated rocky canyons of the Magdalena Region, and along the volcanic lava flows near San Ignacio. In the Cape Region, Lucasensis are most common in rocky areas with Arid Tropical Region scrub, although they extend into the pine-oak woodland of the Sierra la Laguna (Alvarez et al. 1988).

MAP 6.26 Distribution of the pattern classes of *Crotalus ruber* (Red Diamond Rattlesnake), *C. catalinensis* (Isla Santa Catalina Rattlesnake), and *C. lorenzoensis* (Isla San Lorenzo Rattlesnake)

FIGURE 6.51 Exsul pattern class of *Crotalus ruber* (Red Diamond Rattlesnake) from Isla Cedros, BC

Insular populations are just as wide-ranging in their habitat tolerance. Cliff (1954a) reported specimens from coastal dunes and rocky canyons on Isla Monserrate. I have observed Ruber in rocky arroyos on Islas Ángel de la Guarda and San Marcos. I have observed Exsul in the same types of habitats on Isla Cedros, and Lucasensis on gravelly hillsides on Isla Santa Margarita. Murray (1955) reported finding Exsul at the mouth of a gravelly arroyo approximately 50 m from the shore on Isla Cedros.

Crotalus ruber is most active from March through May. North of the Sierra la Libertad, adult Ruber are not generally seen from November through February, although juveniles are occasionally observed during November and early December, provided the weather is sufficiently warm. From San Ignacio south, juvenile Lucasensis and Ruber may be active year-round. I have seen specimens abroad during the day in late December in the Cape Region and at Mulegé. Adults, however, appear to remain inactive throughout much of the winter. In early January, I found a pair of adult Ruber just south of La Poza Grande, on the Magdalena Plain, hibernating beneath a fallen Cardón Cactus *(Pachycereus pringlei)*. They were adjacent to one another, and both were tightly coiled and embedded in the soil.

During the spring, *Crotalus ruber* is generally diurnal, and snakes are commonly seen crawling among rocks or brush during the late morning. Earlier in the day, specimens may be seen coiled at the bases of large rocks or vegetation, basking in the sunlight. When afternoon temperatures begin to exceed the snakes' thermal capabilities, they often take refuge in the shade beneath brush, rocky ledges, or dead vegetation. On Isla Cedros, Exsul are commonly found beneath dead Agave (*Agave* sp.). During warmer months, *C. ruber* is nocturnal and commonly found along rocky ledges. Here, snakes crater themselves along rodent trails, waiting to ambush prey. I have even found individuals cratered beneath bushes and coiled on the tops of rocks. I observed a Ruber near San Ignacio on three consecutive nights during mid-June. It cratered in one of three places along rodent trails throughout the night, often switching from one location to another. All three locations were within 2 m of one another and criss-crossed by rodent trails. On another occasion during early October near Santa Rosalía, I noted that an adult Ruber had just cratered outside a rodent burrow at 2100 hours, and I observed it in the same spot repeatedly until 0200 hours. When I left the next morning at 0730 hours, it was still cratered in the same place even though the sun had risen. Klauber (1972) reported that adult *C. ruber* feed primarily on small rabbits, squirrels, kangaroo rats, and wood rats. Juveniles tend to feed more on mice and lizards, although Klauber noted that an adult Lucasensis from the Cape Region had eaten a Baja California Spiny-Tailed Iguana *(Ctenosaura hemilopha)*. I have removed Western Whiptails *(Cnemidophorus tigris)* from the stomachs of juvenile Ruber killed on the road just south of Bahía de los Ángeles.

In the northern portion of its range, Ruber mate from March through May. I observed courting pairs east of Tijuana during late April and just north of El Rosario during early May; and Linsdale (1932) reported a courting pair during early April from San Telmo. I have not observed courtship in this species south of El Rosario, but I have found newborn Lucasensis in the Cape Region during late September and juveniles during December. Newborn Ruber first appear in northern Baja California during August and September. These observations suggest that *C. ruber* breed in the spring and give birth from late summer to early fall.

FIG 6.52 *Crotalus lorenzoensis*

ISLA SAN LORENZO RATTLESNAKE; VÍBORA DE CASCABEL

Crotalus ruber lorenzoensis Radcliffe and Maslin 1975:490. Type locality: "San Lorenzo Sur Island in the Gulf of California, Baja California Norte [BC], Mexico." Holotype: SDNHS 46009

IDENTIFICATION *Crotalus lorenzoensis* can be distinguished from all other snakes in the region of study by having a heat-sensing facial pit organ between the nostril and eye; a transversely divided first infralabial scale; no hornlike processes on the supraocular scales; a poorly developed inner rattle matrix; distinctive black and white caudal rings, with the black rings much thinner than the white rings.

RELATIONSHIPS AND TAXONOMY See *C. ruber.*

DISTRIBUTION *Crotalus lorenzoensis* is endemic to Isla San Lorenzo Sur in the Gulf of California (see map 6.26).

DESCRIPTION Same as *C. ruber,* except for the following: adults reach approximately 810 mm SVL; 15 to 17 supralabials; 16 to 18 infralabials; 25 to 27 scale rows at midbody; 189 to 195 ventral scales in males, 186 to 193 in females; anal undivided; 21 to 23 subcaudals in males, 15 to 19 in females; subcaudals at beginning and end of series divided; poorly developed inner rattle matrix; outer matrix often absent. **COLORATION** Dorsal ground color reddish-brown; faded oblique, light-colored pre- and postorbital stripes; dorsal pattern composed of 31 to 39 diamond-shaped markings, most distinctive medially; diamonds composed of distinctive white scales, often bordered internally by a row of darker scales; diamonds fade posteriorly; tail composed of distinctive black and white caudal rings; black caudal rings often discontinuous and much thinner than white rings; ventrum dull yellow, usually immaculate.

NATURAL HISTORY Isla San Lorenzo Sur is a narrow, steep, rocky island within the Central Gulf Coast Region that is dominated by thorn scrub vegetation. *Crotalus lorenzoensis* is most commonly found in the rocky arroyos and bajadas along the west side of the island. I have also seen tracks in the dunelike areas along the coast. During the summer, rattlesnakes are nocturnal and are most often observed crawling along arroyo bottoms or cratered along rodent trails. During the day, they reside beneath large rocks or within shaded, cavelike retreats formed by adjacent large boulders. Hollingsworth and Mellink (1996) reported on one individual at night in a rocky arroyo that climbed to nearly 2 m above the ground in an Elephant Tree *(Bursera*

FIGURE 6.52 *Crotalus lorenzoensis* (Isla San Lorenzo Rattlesnake) from Isla San Lorenzo Sur, BC

FIGURE 6.53 *Crotalus catalinensis* (Isla Santa Catalina Rattlesnake) from Arroyo Mota, Isla Santa Catalina, BCS

microphylla) to escape. Presumably *C. lorenzoensis* feeds on Deer Mice (*Peromyscus* sp.), which are very common on the island. Nothing is known about the reproductive biology of *C. lorenzoensis,* but presumably it is not too different from that of *C. ruber* from the adjacent peninsula.

REMARKS Hollingsworth and Mellink (1996) suggest that the degenerate nature of the rattle of *C. lorenzoensis* may have something to do with its arboreal proclivities, as has been hypothesized for *C. catalinensis* (see below).

FIGS. 6.53–54

Crotalus catalinensis

ISLA SANTA CATALINA RATTLESNAKE; VÍBORA SORDA

Crotalus catalinensis Cliff 1954a:80. Type locality: "Isla Santa Catalina in the Gulf of California [BCS], Mexico." Holotype: CAS-SU 15631

IDENTIFICATION *Crotalus catalinensis* can be distinguished from all other snakes in the region of study by having a heat-sensing facial pit organ between the nostril and eye; no hornlike processes on the supraocular scales; no transversely divided infralabial scales; a very poorly developed inner rattle matrix, with the outer matrix usually absent; and black caudal rings nearly as wide as the white caudal rings.

RELATIONSHIPS AND TAXONOMY See *C. ruber.*

DISTRIBUTION *Crotalus catalinensis* is endemic to Isla Santa Catalina in the Gulf of California (Beaman and Wong 2001; see map 6.26).

DESCRIPTION Body relatively slender and short, reaching approximately 680 mm SVL; head triangular, distinct from neck; pupils elliptical; tail short, 7.9 to 9.3 percent of SVL in males and 5.2 to 7.3 percent in females; 2 kidney-shaped internasals; prenasal in contact with first supralabial; first infralabial not transversely divided; 1 or 2 loreals; prefrontal scales (and some interorbital scales) enlarged; 13 to 16 supralabials; 13 to 17 infralabials; dorsal body scales keeled and imbricate, 25 scale rows at midbody; 177 to 181 ventral scales in males, 182 to 189 in females; anal undivided; 24 to 28 divided subcaudals in males, 18 to 23 in females; poorly developed inner rattle matrix; outer rattle matrix occasionally present in juveniles, absent in adults. **COLORATION** *Crotalus catalinensis* is dimorphic in dorsal coloration, having an ashy gray and a brownish phase. BROWN PHASE: Dorsal ground color beige to brownish; oblique, light-colored pre- and postorbital stripes; postorbital stripe sometimes faded or absent; light supraocular crossbar continuous across top of head; dorsal pattern composed of 34 to 40 distinctive diamond-shaped blotches; blotches composed of distinctive white borders, edged internally by a row of darker scales; blotches fade posteriorly; large, brownish secondary blotches on sides of body; tail composed of distinctive black and white caudal rings; black caudal rings usually discontinuous and nearly as wide as white caudal rings; ventrum dull gray, usually immaculate. ASHY GRAY PHASE: Dorsal ground color light gray to dull white; pre- and postorbital stripes faint; supraocular crossbar faint; dorsal blotches faded, often indistinct posteriorly; large faded, gray secondary blotches on sides of body; tail composed of poorly defined black caudal rings; white caudal rings clouded with gray pigment; black caudal rings usually discontinuous and nearly as wide as white caudal rings; ventrum dull gray, usually immaculate.

NATURAL HISTORY Isla Santa Catalina is a granitic continental island lying within the Central Gulf Coast Region, dominated by thorn scrub vegetation and cacti. Its rugged topography consists of rocky hillsides separated by wide and narrow, sandy, heavily vegetated arroyos. *Crotalus catalinensis* is most commonly found in the arroyo bottoms and

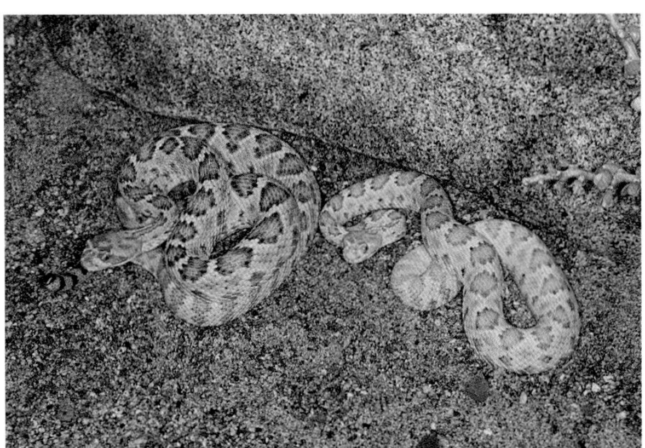

FIGURE 6.54 Brown and ashy gray color phases of *Crotalus catalinensis* (Isla Santa Catalina Rattlesnake) from Arroyo Mota, Isla Santa Catalina, BCS

has never been observed on the hillsides. Rattlesnakes are active from late March through mid-October. During the spring, *C. catalinensis* is both diurnal and nocturnal and can be observed abroad during mid-morning, crawling in direct sunlight. During the hotter months of May through August, *C. catalinensis* is nocturnal and appears to be most common just prior to and following storms. Jesús Sigala-R. (personal communication, 1999) observed a rattlesnake exiting a burrow approximately two hours after sundown. This may be where this species spends the day during the hotter months. Rattlesnakes have been found crawling in the arroyo bottoms, usually near vegetation, and I have occasionally found individuals cratered in the sand beneath the overhang of Elephant Tree *(Bursera microphylla)* branches and other shrubs, or in open areas in the middle of arroyos. More often, however, I have seen individuals crawling through the lower branches of Elephant Trees and Palo Verde trees *(Cercidium microphyllum)* between 0.3 m and 4.0 m above the ground. One evening during late July, I found two gravid females 1 m above the ground and approximately 1.5 m away from one another in a large Palo Verde tree. Four nights later, I found four individuals (three gravid females and one male) in the same tree, ranging from 0.5 m to 4 m above the ground. All were climbing along the outer edges of the tree rather than on the interior branches. To escape my lantern light, all climbed up rather than down toward the interior. I also noticed a Santa Catalina Island Deer Mouse *(Peromyscus slevini)* at the top of the tree in the outermost branches. I was amazed at the agility with which even the gravid female snakes climbed. There was nothing awkward or lumbering about them. They seemed to glide as swiftly through the branches as would any arboreal or semiarboreal snake.

Unlike most other rattlesnakes, *C. catalinensis* is thin, agile, and an adept climber. It is a short-tempered, aggressive species that often sidewinds when moving rapidly across the ground and climbs into vegetation to escape. The climbing ability of *C. catalinensis* may have to do with the fact that it eats birds. I have found the remains of Black-Throated Sparrows *(Amphispiza bilineata),* Isla Santa Catalina Spiny Lizards *(Sceloporus lineatulus),* and Isla Santa Catalina Deer Mice in the scat of freshly caught specimens. Juveniles probably eat small lizards, especially the nocturnal Isla Santa Catalina Leaf-Toed Gecko *(Phyllodactylus bugastrolepis).*

Very little is known about the reproductive biology of *C. catalinensis.* Cliff (1954a) reported an adult female collected in late March that had not ovulated. I have found gravid females from mid-July to early August. One of them had five early-developing embryos. I have found neonates during mid-August, and Jesús Sigala-R. (personal communication, 1999) told me that a female he collected in early August gave birth to two young. Armstrong and Murphy (1979) reported a female giving birth in September, and I have found juveniles during mid-March. These findings suggest that *C. catalinensis* has a spring to early summer breeding season and a late summer to early fall birthing period.

REMARKS The most notable characteristic of *C. catalinensis* is its degenerate, nonfunctional rattle. The inner matrix of the rattle is so poorly developed that the outer matrix falls off during shedding, leaving no loose structures. I have found shed skins with two or three buttons of an outer matrix still attached. It is not clear what the selection pressures (if any) may be for this degeneration. I believe, as others do, that it may be an adaptation involving stealth, for crawling through the lower branches of shrubs at night in search of sleeping birds. Additional support for this hypothesis may be provided by the fact that the teeth of *C. catalinensis* are much longer than those of other rattlesnakes and thus advantageous for biting through feathers. Also, their relatively slender bodies may facilitate maneuvering through branches.

Judging from the observation of numerous scats and skeletal material, there is a large population of feral cats on Isla Santa Catalina. Within the scats I have found the remains of the Isla Santa Catalina Spiny Lizard, the Night Snake *(Hypsiglena torquata),* and *C. catalinensis.*

FIG 6.55 *Crotalus cerastes*

SIDEWINDER; VÍBORA CORNUDA

Crotalus cerastes Hallowell 1854:95. Type locality: "Borders of the Mohave River, and in the desert of the Mohave," California, USA. Holotype: Designated but not by number, disposition unknown, but probably USNM 352, now PANS 7089 (Klauber 1972:33)

IDENTIFICATION *Crotalus cerastes* can be distinguished from all other snakes in the region of study by having a heat-sensing facial pit organ between the nostril and eye; an upper loreal scale; distinctive hornlike processes on the supraocular scales; a well-developed inner rattle matrix; and distinctive black and white caudal rings.

RELATIONSHIPS AND TAXONOMY *Crotalus cerastes* may be the sister species of *C. mitchellii* of the southwestern United States and Baja California (Foote and MacMahon 1977; Stille 1987).

DISTRIBUTION *Crotalus cerastes* ranges throughout most of the Great Basin and Mojave deserts as well as the Lower Colorado Valley Region of the Sonoran Desert. It extends from southern Nevada and eastern California south through southwestern Arizona to northwestern Sonora and northeastern Baja California (Klauber 1972). In Baja California, *C. cerastes* ranges south, east of the Peninsular Ranges, to at least Calamajué. The single specimen from "Bahía San Francisquito" (USNM 37565) reported by Klauber (1944) was collected by the explorers Charles Nelson and Alfonso Goldman on September 12, 1905. It is clear from Goldman (1951) and Nelson (1922), however, that what they referred

FIGURE 6.55 Laterorepens pattern class of *Crotalus cerastes* (Sidewinder) from near San Felipe, BC

to as Bahía San Francisquito is today known as Bahía Calamajué, much farther north. Campbell and Lamar (1989) report that *C. cerastes* extends as far south as the Llano de San Pedro, approximately 200 km south of Bahía Calamajué. Although there are no specimens to support this record, I have seen tracks of *C. cerastes* at San Francisquito east of Llano de San Pedro. Still, no specimens have been found between Calamajué and San Francisquito. *Crotalus cerastes* is also known from the Gulf island of Tiburón, Sonora (map 6.27).

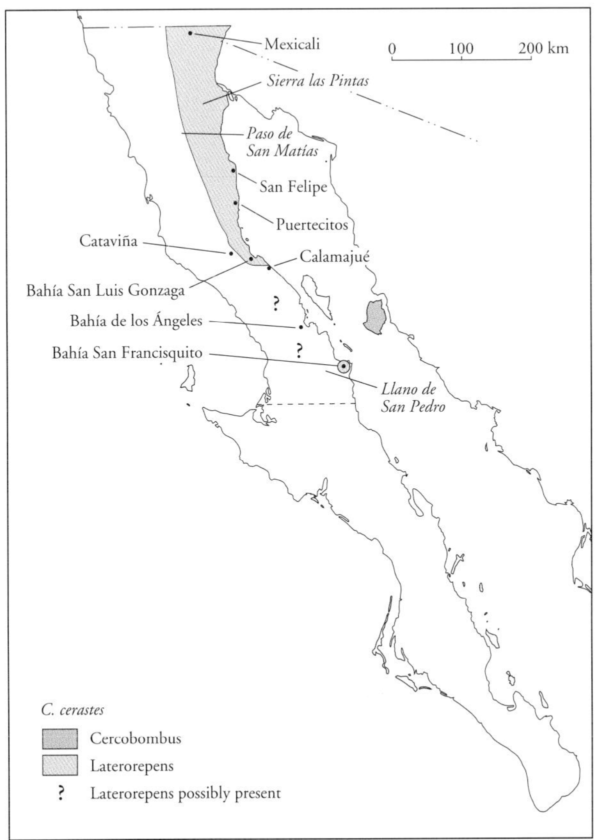

MAP 6.27 Distribution of the pattern classes of *Crotalus cerastes* (Sidewinder)

DESCRIPTION Body heavy and stout, reaching approximately 760 mm SVL; head triangular, distinct from neck; pupils elliptical; tail relatively short, averaging 8.6 percent of SVL in males and 6.2 percent in females; 2 moderately sized internasals; prenasal contacts first supralabial; loreal usually single; hornlike supraocular scales; head scales small and imbricate; 10 to 15 supralabials; 10 to 17 infralabials; dorsal body scales keeled and imbricate, 19 to 25 (usually 21 or 23) scale rows at midbody; middorsal rows pointed and tuberculate; 132 to 149 ventral scales in males, 136 to 154 in females; anal undivided; 20 to 26 subcaudals in males, 14 to 19 in females; most subcaudals undivided. **COLORATION** Dorsal ground color dull white to beige; distinctive wide, brown postorbital stripe; top of head with occasional brown spots; supraorbital bar bordered posteriorly in brown; dorsal pattern composed of 28 to 47 brown, squarish blotches, grading into bands posteriorly; interspaces between blotches much lighter, accentuating blotches; small black punctations throughout dorsal pattern give an overall dusty appearance; tail composed of distinctive black and white caudal rings; venter dull white, usually immaculate.

GEOGRAPHIC VARIATION *Crotalus cerastes* is composed of three weakly differentiated subspecies (Klauber 1944; Savage and Cliff 1953), two of which occur in the region of study and are recognized here as pattern classes. Laterorepens range throughout the northeastern portion of Baja California in the Lower Colorado Valley Region and possibly farther south into the Vizcaíno and Central Gulf Coast regions (see distribution section). They are generally lighter overall; males usually have more than 142 ventral scales and females more than 146. The supraoculars tend not to be as large as those in Cercobombus. In the southern portion of their range in Baja California, the posteriormost black caudal rings of Laterorepens tend to fuse, and often the entire posterior section of the tail is black. Cercobombus range from south-central Arizona to western Sonora and occur on Isla Tiburón in the region of study. This population generally has fewer than 141 ventral scales in males and fewer than 146 in females. They are generally darker overall as a result of an increase in the dark punctations. The supraocular horns tend to be larger than those of Laterorepens.

NATURAL HISTORY *Crotalus cerastes* is a common inhabitant of the Lower Colorado Valley Region of northeastern Baja California and the Central Gulf Coast Region of Isla Tiburón. In both regions, it is most often found in low-lying, flat areas and arroyo bottoms with loose sand, as well as along beaches. Sidewinders, however, are not restricted to such habitats. Murray (1955) reported finding a specimen beneath a rock on a talus slope at Punta San Felipe, and I observed a specimen during the day in early March crawling among the volcanic talus, 200 m up a hillside, in the Sierra las Pintas. I have observed active *C. cerastes* from late February to early November, with a peak from April through June. Klauber (1972) reported that sidewinders in San Diego County hibernate underground from December through February. *Crotalus cerastes* are generally nocturnal but are often observed abroad on spring days, provided that temperatures are not excessive. During the summer, sidewinders are strictly nocturnal but are often observed in the early morning cratered in the sand, perhaps basking to help digest the previous evening's meal (Ernst 1992). In one locality along the coastal dunes just south of the town of San Felipe, an area heavily frequented by campers, I observed seven cratered individuals one morning in early June. During the day, snakes usually take refuge in rodent burrows (Ernst 1992), although in the vicinity of Bahía San Luis Gonzaga, sidewinders are commonly found beneath fallen Ocotillo *(Fouquieria splendens).*

Funk (1965) presented a detailed diet analysis of *C. cerastes* from the vicinity of Yuma, just on the other side of the Río Colorado in northeastern Baja California. He noted that although sidewinders preyed on a number of small mammals, adults preferred kangaroo rats (*Dipodomys*) and pocket mice *(Perognathus).* Juvenile sidewinders preferred pocket mice and various lizards, of which Western Whiptails *(Cnemidophorus tigris)* and Fringe-Toed Lizards *(Uma notata)* were the most commonly eaten. He also noted that birds and small snakes were rarely taken. According to Nabhan (forthcoming), the Seri Indians say that *C. cerastes* is commonly found in rocky intertidal areas eating marine isopods.

Crotalus cerastes may have a dual mating season, from April to June and from September to October (Ernst 1992). Neonates are found from mid-August through November (Ernst 1992).

FOLKLORE AND HUMAN USE The Seri Indians believe *C. cerastes* live in packs of 15 to 20. As with *C. atrox,* they realize the snake is poisonous but consider it an important food item (Malkin 1962). Locals from around San Felipe believe that stepping on the horns of a sidewinder will result in an immediate, intense burning sensation. This is probably due to the subsequent bite and not the horns.

FIG 6.56 *Crotalus enyo*

BAJA CALIFORNIA RATTLESNAKE; VÍBORA DE CASCABEL

Caudisoma enyo Cope 1861:293. Type locality: "inhabits Lower California," collected at "Cape St. Lucas [Cabo San Lucas]," BCS, Mexico. Lectotype: ANSP 7761; designated by Beaman and Grismer (1994)

IDENTIFICATION *Crotalus enyo* can be distinguished from all other snakes in the region of study by having a heat-sensing facial pit organ between the nostril and eye; an upper loreal scale; no hornlike processes on the supraocular scales, although the outer edges of the supraocular scales are noticeably elevated; a well-developed inner rattle matrix; prominent white supraocular bar; and no distinctive black and white caudal rings.

RELATIONSHIPS AND TAXONOMY *Crotalus enyo* may be a member of a group containing *C. horridus* of the eastern United States, *C. durissus* of Mexico and Central and South America, *C. vegrandis* of Venezuela, and *C. unicolor* of the island of Aruba (Stille 1987). Beaman and Grismer (1994), Bostic (1971), Cliff (1954a,b), and Lowe and Norris (1954) discuss taxonomy.

DISTRIBUTION *Crotalus enyo* ranges throughout most of Baja California. In the north, its contact with the Pacific coast occurs in the vicinity of Cabo Colonet and with the Gulf coast near Bahía de los Ángeles. From here, *C. enyo* continues south throughout all of Baja California (Beaman and Grismer 1994). It is also known from the Pacific islands of Magdalena and Santa Margarita and the Gulf islands of Carmen, Cerralvo, Coronados, Espíritu Santo, Pardo, Partida Sur, San Francisco, San José, and San Marcos (Grismer 1999a; map 6.28).

DESCRIPTION Body heavy and stout, triangular in cross section, reaching approximately 820 mm SVL; head triangular, distinct from neck; eyes relatively large, pupils elliptical; tail relatively short, 8.4 to 9.8 percent of SVL in males and 6.1 to 7.2 percent in females; 2 small internasals; prenasal usually in contact with first supralabial; 1 to 5 loreal scales; head scales small and imbricate; 12 to 15 supralabials; 11 to 16 infralabials; dorsal body scales keeled and imbricate, 23 to 27 (usually 25) scale rows at midbody; 159 to 168 ventral scales in males, 161 to 177 in females; anal undivided; 22 to 28 subcaudals in males, 18 to 23 in females; subcaudals at beginning and end of series may be divided.

COLORATION Dorsal ground color tan, pale to dark brown, gray-brown, or silvery gray, becoming paler posteriorly; top of head has two pairs of large, dark brown, distinctive lineate blotches posteriorly; anterior pair usually smaller and more laterally located, posterior pair more medially located and usually longer, extending onto nape of neck; light supraocular bar distinct and bordered posteriorly by dark brown marking; wide, dark postorbital stripe present; dorsal pattern composed of 28 to 42 reddish to yellowish-brown dark-edged blotches; blotches subrectangular in shape anteriorly, hexagonal medially, and transversely elongate posteriorly; middorsal interspace between blotches usually much lighter than adjacent ground color; single row of dark secondary blotches occurs on sides of body; tail composed of 4 to 8 brownish rings; venter usually cream-colored and may be heavily mottled. The color pattern of newborns is very vivid.

FIGURE 6.56 Enyo pattern class of *Crotalus enyo* (Baja California Rattlesnake) from 23 km west of Santa Rosalía, BCS

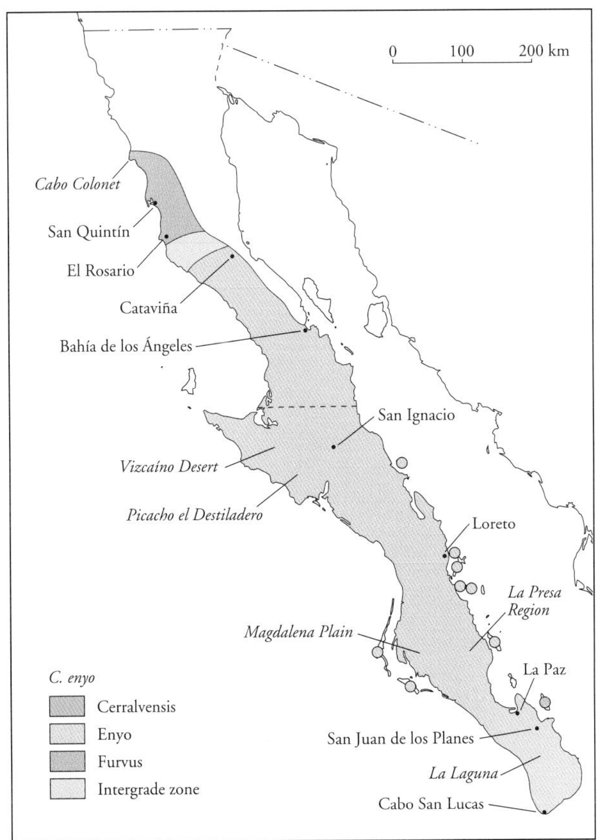

MAP 6.28 Distribution of the pattern classes of *Crotalus enyo* (Baja California Rattlesnake)

GEOGRAPHIC VARIATION *Crotalus enyo* exhibits clinal trends in both scalation and, to a lesser extent, color pattern. These trends have been delimited to some degree by the description of two peninsular subspecies, *C. e. furvus* and *C. e. enyo* (Lowe and Norris 1954), which are recognized here as pattern classes. A third subspecies, *C. e. cerralvensis* (Cliff 1954a) from Isla Cerralvo, is also considered a pattern class. Furvus occur in cismontane areas of northern Baja California from Cabo Colonet south to just beyond El Rosario. South of here, Furvus intergrade with Enyo for approximately 60 km (Beaman and Grismer 1994). Furvus tend to be darker than most Enyo in overall dorsal coloration, although Enyo from the Cape Region are the darkest of all. In Furvus, the third and fourth supralabials usually do not contact the lacunals as they usually do in Enyo. The subfoveals in Furvus are usually present, whereas they are usually absent in Enyo. Furvus also tend to have fewer ventral scales than Enyo (Beaman and Grismer 1994). Enyo from just north of Bahía de los Ángeles tend to have discontinuous lineate markings on the top of the head, and some individuals are very light in color. Snakes from the Vizcaíno Desert are often dark in overall coloration. Enyo from the Magdalena Plain are somewhat faded in coloration and tend to have a pinkish hue laterally. A similar pattern obtains in many specimens from San Juan de los Planes in the Cape Region.

Cerralvensis from Isla Cerralvo tend to have a more distinctive and vivid blotching pattern. In some individuals, the blotches are butterscotch-colored, and in others they are more gray-brown. Generally, adult Cerralvensis appear pedomorphic in coloration, tending to resemble juvenile snakes of the northern peninsular populations.

NATURAL HISTORY *Crotalus enyo* occurs in a number of habitats throughout most of Baja California. North of El Rosario, it inhabits the coastal sage scrub–Vizcaíno Region ecotone of the San Quintín and Santa María plains, where it is found among low-growing vegetation on the coastal plateaus (Lowe and Norris 1954). Although *C. enyo* is widespread throughout arid regions to the south, it is absent from the extremely arid Lower Colorado Valley Region of northeastern Baja California and does not contact the east coast of the peninsula until near Bahía de los Ángeles, which marks the southern extent of the Lower Colorado Valley Region. Throughout the rest of Baja California, *C. enyo* is widespread. It is a common inhabitant of the cooler Pacific coastal regions associated with the Vizcaíno Desert and the Magdalena Plain, where it is usually found in association with scrub vegetation and rocks (Bostic 1971; Grismer et al. 1994). It is also quite common in the Arid Tropical Region but does not extend into the oak or pine oak woodlands of the Sierra la Laguna and is scarce even along its lower slopes (Alvarez et al. 1988). *Crotalus enyo* is common in the vicinity of human habitations and is often found in trash piles (Grismer et al. 1994; Murphy and Ottley 1984; Soulé and Sloan 1966; Van Denburgh and Slevin 1921). Elsewhere, *C. enyo* has been observed under and within brush (Bostic 1971; Klauber 1972; Linsdale 1932; Lowe and Norris 1994; McGuire 1991b; Murphy and Armstrong 1978; Van Denburgh and Slevin 1921), under rocks (Grismer et al. 1994), in burrows (Lowe and Norris 1954), and among sand dunes on Isla Magdalena (Murray 1955). I have also seen specimens within the dunes along beaches on Isla Cerralvo. Klauber (1972) reported on a specimen near Cataviña that was nearly 1 m above the ground in a Bursage *(Ambrosia dumosa)*. McGuire (1991b) reported a similar observation on Isla Cerralvo and noted that the snake was approximately 1.5 m above the ground and moved from bush to bush.

Crotalus enyo is active year-round in the Cape Region, with minimal activity during December and January. North of the Vizcaíno Desert, it is generally inactive during the winter; however, I have observed active individuals just south of El Rosario on warm days in early December. In late December, I found an adult coiled beneath a small rock on the top of Picacho El Destiladero, one of the isolated peaks forming the Sierra Santa Clara. It had rained the day before, and the snake was wet. Judging from the poor cover offered by the rock, I assumed that this was not an overwinter retreat and that the snake had been active on recent warm days. The general activity period for *C. enyo* lasts from late March through early November. The period of greatest activity, especially from Bahía de los Ángeles south, extends from September through late October. During the spring, *C. enyo* is both diurnal and nocturnal. During the summer, activity becomes primarily nocturnal, although in the Cape Region from June through September, specimens may be abroad during the early morning and at dusk (Van Denburgh and Slevin 1921). I have noted that in central Baja California, *C. enyo* is quite active on cold nights during early fall, when more often than not it is the only active reptile.

Crotalus enyo feed primarily on small mammals and lizards (Klauber 1931c, 1972; Mocquard 1899; Taylor 2001; Van Denburgh and Slevin 1921) as juveniles and adults. Taylor reports that large adults (greater than 600 mm SVL) even eat centipedes *(Scolependra)*. *Crotalus enyo* seems to be a more active forager than the other rattlesnakes of Baja California. Where it is common to find *C. cerastes, C. ruber,* and *C. mitchellii* at night cratered along a rodent trail waiting to ambush prey, *C. enyo* are most often observed crawling about. The only cratered *C. enyo* I have come across were on Isla Cerralvo: one cratered in the sand at the base of a bush, and another coiled among some rocks.

Little is known about the reproductive biology of *C. enyo*. Klauber (1931c) reported on a specimen that had given birth in late March, but he did not indicate the origin of the specimen. I have found newborns in late July near Loreto, during early August just west of Bahía de los Ángeles, during mid-August near San Ignacio, and from early September to mid-October in the Cape Region. These observations suggest that mating generally takes place in the spring and birth in the summer and early fall. Based on an analysis of reproductive tracts of museum specimens, Taylor (1999) reported that 1 to 10 (mean 6.4) young are born primarily in the fall, but that spring births may also occur.

REMARKS Many species of reptiles are known to undergo daily changes in the intensity of their overall coloration. I have noted such a change in some specimens of *C. enyo*. Often, individuals collected during the night are noticeably darker the next morning.

FIGS. 6.57–59 **_Crotalus mitchellii_**

SPECKLED RATTLESNAKE; VÍBORA DE CASCABEL

Crotalus mitchellii Cope 1861:293. Type locality: "Cape St. Lucas, Lower California [Cabo San Lucas, BCS]," Mexico. Holotype: USNM 5219½

IDENTIFICATION *Crotalus mitchellii* can be distinguished from all other snakes in the region of study by having a heat-sensing facial pit organ between the nostril and eye; an upper loreal scale; no hornlike processes on the supraocular scales; prenasal usually separated from the rostral; upper preoculars usually divided; a well-developed inner rattle matrix; distinctive black and white rings on the tail; and fine black punctations in dorsal pattern.

RELATIONSHIPS AND TAXONOMY *Crotalus mitchellii* may be the sister species of *C. cerastes* of the southwestern United States and Baja California (Foote and MacMahon 1977; Stille 1987). Grismer (1999b) discusses taxonomy.

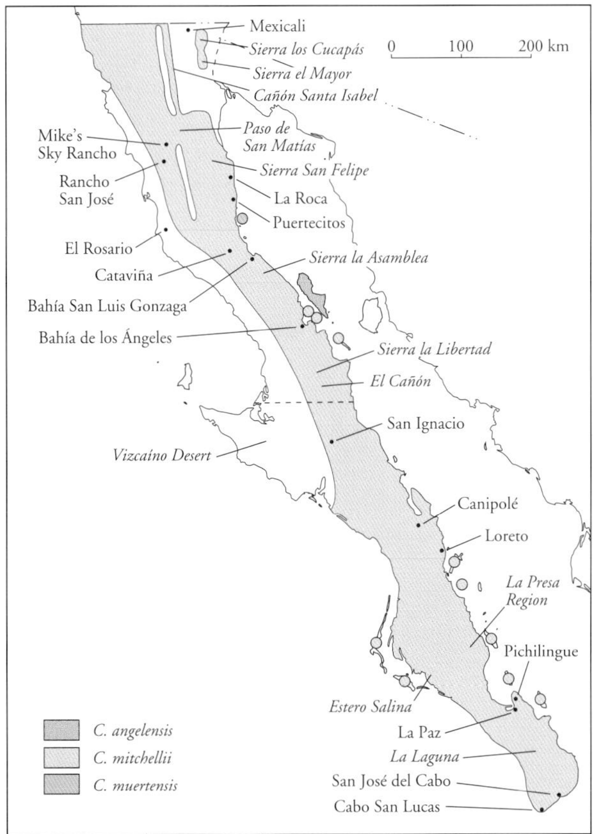

MAP 6.29 Distribution of *Crotalus mitchellii* (Speckled Rattlesnake), *C. angelensis* (Isla Ángel de la Guarda Rattlesnake), and *C. muertensis* (Isla el Muerto Rattlesnake)

FIGURE 6.57 *Crotalus mitchellii* (Speckled Rattlesnake) from sand dunes at Estero Salinas, BCS

DISTRIBUTION *Crotalus mitchellii* ranges throughout much of the Mojave and northern Sonoran deserts south into extreme northwestern Sonora, southern California, and nearly all of Baja California (McCrystal and McCoid 1986). In Baja California, *C. mitchellii* is transpeninsular in distribution, but it is largely restricted to rocky areas associated with the Peninsular Ranges and their foothills. An isolated population occurs in the Sierra los Cucapás and the Sierra el Mayor in northeastern Baja California. The only major region of the peninsula where *C. mitchellii* does not occur extends from coastal north-central Baja California (Bostic 1971) south through the Vizcaíno Desert (Grismer et al. 1994). It is widely distributed throughout the Magdalena Region, following the volcanic alluvia as it extends west toward the Pacific coast. *Crotalus mitchellii* is also known from the Pacific island of Santa Margarita (Wong 1997) and the Gulf islands of Carmen, Cerralvo, Espíritu Santo, Monserrate, Partida Sur, Piojo, Salsipuedes, San José, and Smith (map 6.29).

DESCRIPTION Body heavy and stout, reaching 1,114 mm SVL; head triangular, distinct from neck; pupils elliptical; tail relatively short, 6.7 to 8.0 percent of SVL in adult males and 5.3 to 6.3 percent in females; 2 to 4 small internasals; prenasal may or may not contact first supralabial; 0 to 5 (usually 2) loreals; prenasals may or may not contact rostral; upper preoculars usually divided; supraocular rarely divided; head scales small and imbricate; 13 to 19 supralabials and 13 to 17 infralabials; dorsal body scales keeled and imbricate, 21 to 27 (usually 25) scale rows at midbody; 163 to 187 ventral scales in males and females; anal undivided; 20 to 28 subcaudals in males, 16 to 24 in females; subcaudals usually entire, but posteriormost may be divided. **COLORATION** Dorsal ground color extremely variable, ranging from white, dull white, cream, brown, or tan to

FIGURE 6.58 *Crotalus mitchellii* (Speckled Rattlesnake) from 45 km west of Santa Rosalía, BCS

FIGURE 6.59 *Crotalus mitchellii* (Speckled Rattlesnake) from San Bartolo, BCS

orange, gray, blue-gray, pale gray, pinkish, light greenish, or yellowish; top of head speckled; supraocular bars indistinct or absent; dorsal pattern composed of 23 to 42 darker, variably shaped, transversely oriented dorsal blotches, transforming into bands posteriorly; lateral row of large, irregularly shaped, dark secondary blotches on sides of body; irregularly shaped and variably sized dark spots in dorsal and lateral blotches and interspaces give dorsal pattern a speckled appearance; dark spots usually absent middorsally; tail composed of 3 to 9 distinctive black and white caudal rings; most of black caudal rings offset and thus incomplete; venter dull yellow, usually faintly blotched or punctated.

GEOGRAPHIC VARIATION Although two peninsular subspecies (*C. mitchellii pyrrhus* from BC and *C. m. mitchellii* from BCS) have been recognized (Klauber 1936; McCrystal and McCoid 1996), the differences between them are so variable and subtle that they are not even useful as pattern classes. The vast majority of the geographic variation in *C. mitchellii* derives from its remarkable substrate-matching ability. Essentially, this species matches the color of any substrate on which it is found. For example, snakes from the light-colored granitic areas within the Sierra San Felipe are dull white in ground color, with a dark speckling pattern that matches the appearance of the dark tourmaline crystals in the rock; specimens from the reddish volcanic flows in the vicinity of El Cañón have a dull red ground color with light speckling; snakes from the lava flows near San Ignacio are usually dull gray with slightly darker flecks; snakes from the dark gray volcanic hills near Canipolé have a steel-blue ground color; and individuals from the volcanic area of Pichilingue are reddish. The most divergent population is that of Isla Salsipuedes. The substrate of the island is granitic and very light in color, and the *C. mitchellii* of this island have a white ground color with virtually no dorsal pattern except for a light peppering of black flecks. A similar color pattern obtains on Isla Santa Margarita, but here the ground color is a dull, light, nearly unicolored green. This same color pattern occurs in snakes from the adjacent peninsula, which occur in the sandy areas surrounding Estero Salinas.

NATURAL HISTORY *Crotalus mitchellii* ranges throughout all the phytogeographic regions of Baja California but is generally restricted to rocky areas. It does, however, extend short distances from rocky terrain onto surrounding flats and into sandy arroyos. In northwestern Baja California, *C. mitchellii* is a common inhabitant of rocky hillsides covered with foothill chaparral. In such areas, *C. mitchellii* takes refuge within rock fissures, beneath rocks, and within rodent burrows. Just east of Rancho San José, I encountered a juvenile beneath a vertically oriented granitic exfoliation approximately 1 m above the ground, on the side of a large boulder. Welsh (1988) noted that *C. mitchellii* occurred in rocky areas on both sides of the Sierra San Pedro Mártir in the chaparral–piñon pine ecotone on the west face and the creosote bush–piñon pine ecotone on the east. *Crotalus mitchellii* is common in the low-lying volcanic hills surrounding Bahía San Luis Gonzaga and the western slopes of the Sierra la Asamblea, as well as the volcanic hillsides and alluvia of the Magdalena and Central Gulf Coast regions of southern Baja California. In the Cape Region, *C. mitchellii* is common in rocky, low-lying areas but is relatively rare at higher elevations in the foothills and valleys of the Sierra la Laguna (Alvarez et al. 1988). On Isla Smith, *C. mitchellii* is often found foraging among rocks in the intertidal zone as well as through the vegetation along the beaches. On Isla Piojo,

C. mitchellii can be found beneath the large stick nests of Brown Pelicans *(Pelecanus occidentalis)*. The snakes may be attracted to the rodents that frequent these areas (E. Palacio, personal communication, 1996).

In the more northerly portions of its distribution, north of the Sierra la Libertad, *C. mitchellii* is active from at least mid-February through November and can be found abroad during the day through April. In the south, it may be active year-round depending on the weather. *Crotalus mitchellii* is most active from March through September. During the spring, it is both diurnal and nocturnal (Klauber 1936; Moore 1978). During morning hours in April and May I have often come across individuals along the eastern face of the Sierra Juárez, in the Sierra la Asamblea, and near Bahía San Luis Gonzaga, basking in the sun just outside rock crevices into which they retreated if startled. Welsh (1988) reported finding *C. mitchellii* crawling through the rocks during the morning hours in March and April in the Sierra San Pedro Mártir. I have made similar observations on Isla Smith during the same months. During the summer, activity is decidedly nocturnal, especially in southern Baja California. In the evening, *C. mitchellii* are most often found coiled along rodent trails to ambush prey. In Cañón Santa Isabel, I found an individual out on the flats, coiled outside a rodent burrow. At La Roca, I noticed two individuals coiled along the same trail during a full moon on three consecutive nights. They periodically adjusted their position through the night to remain within the shadow of a rocky overhang, to stay out of the moonlight. Often *C. mitchellii* can be found coiled at the bases of rocky ledges, and at La Presa, I found three individuals coiled on dead palm fronds in the palm grove. I have also found individuals coiled up on the lava flows near San Ignacio. One evening in El Cañón, I observed a juvenile crawling up the vertical face of a boulder. I observed two individuals 40 to 45 km from the nearest rocky area on the sand dunes of Estero Salinas in the Magdalena Plain. Both snakes were cratered into the sand along rodent trails in relatively open areas: one beneath a Leatherplant *(Jatropha cuneata)* and the other out in the open. The specimen beneath the Leatherplant remained in position throughout the night, even during brief periods of rain, after which it drank the water that had accumulated between the coils of its body. Just prior to daybreak, the snake crawled approximately 10 m to the nearest brushy area and coiled beneath the thicker vegetation for daytime cover.

Crotalus mitchellii feeds primarily on rodents but also takes lizards and occasionally birds (Klauber 1972). Klauber reported finding Western Whiptails *(Cnemidophorus tigris)*, Side-Blotched Lizards *(Uta stansburiana)*, Spiny Lizards (*Sceloporus* sp.), Zebra-Tailed Lizards *(Callisaurus draconoides)*, Western Skinks *(Eumeces skiltonianus)*, and Northern Chuckwallas *(Sauromalus obesus)* in the stomachs of *C. mitchellii.*

Crotalus mitchellii mates during the spring. I have observed copulating pairs at Mike's Sky Rancho during late April, in the Sierra la Asamblea during early May, and near Loreto during late May. I observed what I believed to be a gravid female during late June at Bahía San Luis Gonzaga, and Van Denburgh and Slevin (1921) reported on a female with three young during September from San José del Cabo. I have observed newborns on Isla Smith in late September. These observations indicate that *C. mitchellii* has an extended reproductive season lasting from April though October.

FIG 6.60 ***Crotalus angelensis***

ISLA ÁNGEL DE LA GUARDA RATTLESNAKE; VÍBORA DE CASCABEL

Crotalus mitchellii angelensis Klauber 1963:293. Type locality: "About 4 miles south of Refugio Bay, at 1500 feet elevation, Isla Ángel de la Guarda, Gulf of California [BC], Mexico (near 29° 29½' N, 113° 33' W)." Holotype: SDSNH 51994

IDENTIFICATION *Crotalus angelensis* can be distinguished from all other snakes in the region of study by having a heat-sensing facial pit organ between the nostril and eye; an upper loreal scale; no hornlike processes on the supraocular scales; prenasal usually separated from the rostral; upper preoculars usually divided; a well-developed inner rattle matrix; distinctive black and white rings on the tail; fine black punctations in dorsal pattern; and an SVL up to 1,410 mm.

RELATIONSHIPS AND TAXONOMY See *C. mitchellii.* Grismer (1999b) discusses taxonomy.

DISTRIBUTION *Crotalus angelensis* is endemic to Isla Ángel de la Guarda in the Gulf of California (see map 6.29).

DESCRIPTION Same as *C. mitchellii,* except for the following: body very heavy and stout, reaching 1,410 mm SVL; head very wide; supraoculars usually divided; tail relatively short, 7.7 percent of SVL in adult males and 5.7 percent in females; 12 to 15 supralabials and 13 to 17 infralabials; prenasal separated from rostral by a single scale; dorsal body scales keeled and imbricate; 25 or 27 (usually 27) scale rows at midbody; 175 to 184 ventral scales in males and 186 to 190 in females; anal undivided; 23 to 28 subcaudals in males, 19 to 21 in females; subcaudals usually entire, but posteri-

FIGURE 6.60 *Crotalus angelensis* (Isla Ángel de la Guarda Rattlesnake) from Puerto Refugio, Isla Ángel de la Guarda, BC

ormost may be divided. **COLORATION** Dorsal ground color tan, buff-orange, or pinkish; juveniles have a grayish ground color; indistinct spots on top of head; supraocular bars indistinct to absent; dorsal pattern composed of 36 to 46 darker, hexagonally shaped, transversely oriented dorsal blotches, transforming into bands posteriorly; lateral row of large, irregularly shaped, dark secondary blotches on sides of body; tail composed of 4 to 8 distinctive black and white caudal rings; most of black caudal rings offset and thus incomplete; ventrum pinkish to cinnamon, usually faintly blotched or punctated laterally.

GEOGRAPHIC VARIATION The dorsal ground color of adult *Crotalus angelensis* is usually salmon-pink. This coloration is remarkably consistent, in contrast to the extreme variability found in this snake's peninsular counterpart, *C. mitchellii.* Individuals from the northern portion of the island, in the vicinity of Puerto Refugio, tend to be more pinkish or salmon-colored than those from the southern end, opposite Isla Pond, where the ground color is generally more gray-brown.

NATURAL HISTORY Isla Ángel de la Guarda lies within the Central Gulf Coast Region and is dominated by a scant thorn scrub vegetation on its rocky hillsides. Virtually nothing has been reported on the natural history of *C. angelensis.* Its distribution on Isla Ángel de La Guarda is somewhat ubiquitous. Snakes are very common on the gravelly beaches along the shore as well as in the rocky arroyos, washes, and on the hillsides of the island's interior. Klauber (1963) reported finding specimens as high as 500 m elevation. *Crotalus angelensis* has been collected year-round but seems to have an activity peak from March through September. Although all specimens were observed after dark, it is likely that during the cooler spring and winter *C. angelensis* is diurnal. Juveniles are very common along the beaches and are often encountered coiled along the base of the precipitous bluffs at Puerto Refugio. Adults are occasionally found along the beaches but generally seem more abundant inland. On warm summer nights *C. angelensis* can be quite numerous.

Nothing has been reported on the food habits of *C. angelensis,* although they are probably not too different from *C. mitchellii* of the adjacent peninsula, given the abundance of lizards and rodents on the island. An adult *C. angelensis* was collected during late December with an adult Spiny Chuckwalla *(Sauromalus hispidus)* in its stomach. That this snake should eat a prey item of that size during December clearly indicates that it was an active individual.

The reproductive biology of *C. angelensis* is unknown. Small juveniles are common during the spring and early summer and are probably the progeny of broods from the previous year. Therefore birth probably occurs during the late summer and early fall.

REMARKS *Crotalus angelensis* is remarkable for its large adult size, which is nearly twice that of its peninsular counterpart, *C. mitchellii.* Although he did not give dimensions, Klauber (1963:77) noted that *C. angelensis* is proportionally thicker in body than *C. mitchellii:* "In appearance, the new subspecies *[C. m. angelensis]* is even more striking than indicated by length alone: a 1367 mm. snake is nearly twice as bulky or heavy as one *[C. mitchellii]* 1114 mm. long." *C. angelensis* also has smaller rattles than *C. mitchellii,* and Klauber stated that "*[C.] angelensis* never attains the rattle size of either *[C. m.] mitchellii* or *[C. m.] pyrrhus* until the snake has reached a body length which the mainland subspecies never attain" (1963:77).

FOLKLORE AND HUMAN USE The northern end of Isla Ángel de la Guarda is flanked by the two small islands Mejía and Granito. Together, these three islands form Puerto Refugio, which has provided a safe port for Mexican fishermen for years. The fishermen, however, are so frightened by the size and abundance of *Crotalus angelensis* that they camp only on Islas Mejía and Granito, which lack rattlesnakes. At the south end of Isla Ángel de la Guarda, the fishermen camp only on Isla Pond. They do, however, go to the main island occasionally to hunt rattlesnakes for food.

FIG 6.61

Crotalus muertensis

ISLA EL MUERTO RATTLESNAKE; VÍBORA DE CASCABEL

Crotalus mitchellii muertensis Klauber 1949b:97. Type locality: "El Muerto Island, Gulf of California [BC], Mexico." Holotype: SDSNH 37447

IDENTIFICATION *Crotalus muertensis* can be distinguished from all other snakes in the region of study by having a heat-sensing facial pit organ between the nostril and eye; an upper loreal scale; no hornlike processes on the supraocular scales; prenasal usually separated from the rostral; upper preoculars usually divided; a well-developed inner rattle matrix; distinctive black and white rings on the tail; fine black punctations in dorsal pattern; and an adult SVL never larger than 610 mm.

RELATIONSHIPS AND TAXONOMY See *C. mitchellii.*

DISTRIBUTION *Crotalus muertensis* is endemic to Isla El Muerto in the Gulf of California (see map 6.29).

DESCRIPTION Same as *C. mitchellii* except for the following: small, reaching 610 mm SVL; tail relatively short, 5.7 to 7.3 percent of SVL in adult males and 4.2 to 5.8 percent in females; 14 to 18 supralabials and 14 to 19 infralabials; prenasals usually separated from rostral by row of small scales; 23 to 25 (usually 23) scale rows at midbody; 175 to 184 ventral scales in males and 174 to 181 in females; 21 to 24 subcaudals in males, 16 to 18 in females. **COLORATION** Dorsal ground color grayish; top of head speckled; supraocular bars indistinct or absent; dorsal pattern composed of 32 to 39 blotches; bands reddish in juveniles; 2 to 6 black and white caudal rings; ventrum cream to buff, darker posteriorly, and mottled laterally.

NATURAL HISTORY Isla El Muerto lies within the Lower Colorado Valley Region and is dominated by Salt Bush *(Atriplex hymenelytra)* and Stipa Grass (*Stipa* sp.). The island is generally steep and extremely rocky. The volcanic rocks are angular, eroded, and very irregular in shape, and offer excellent refuge for rattlesnakes. *Crotalus muertensis* is ubiquitous on the island and has been reported from the intertidal zone to some of the island's highest crests (Klauber 1949b). Rattlesnakes are commonly seen coiled in deeply eroded contours of the rocks or in the rubble of the talus slopes (often their extremely cryptic coloration makes them difficult to see). *Crotalus muertensis* is diurnal and nocturnal from March through May but becomes nocturnal and crepuscular during the summer. According to Klauber (1949b), C. H. Lowe noted that during June, rattlesnakes "slept" during the morning, coiled on rocks, and retreated into the shade only after they were exposed to sunlight. I have found several coiled specimens during April, June, and July with portions of their bodies exposed to the sun during the morning and late afternoon. Full activity during the summer, however, has only been observed after dark. During these months, rattlesnakes are common in all habitats. I observed one very agile individual crawling up the vertical face of a rock. *Crotalus muertensis* is very wary and often rattles when approached from great distances.

FIGURE 6.61 *Crotalus muertensis* (Isla El Muerto Rattlesnake) from Isla El Muerto, BC

Crotalus muertensis is active from March through September and is commonly found coiled in the shade of the rocks. During the spring, individuals are often seen foraging during the day for lizards. In early April, I followed one set of tracks on the beach that led to three consecutive nests of Yellow-Footed Gulls *(Larus livens).* All three nests contained eggs. The small size of this rattlesnake would preclude it from eating anything but the smallest chicks. During the summer, snakes forage at night along the hillsides and in the intertidal zone. Klauber (1949b) reported that its diet consists of the Dead Side-Blotched Lizard *(Uta lowei),* Banded Rock Lizard *(Petrosaurus mearnsi),* and Deer Mouse (*Peromyscus* sp.). It probably also feeds on the Peninsular Leaf-Toed Gecko *(Phyllodactylus xanti),* which is common on the island. I have observed gravid females during May and newborns in late July and August. These observations suggest that reproduction takes place during the spring.

REMARKS Klauber (1949b) described *C. muertensis* on the basis of squamation, color pattern, and small body size. Only the last, however, is discretely diagnostic of this species. Klauber presented evidence that *C. muertensis* is a dwarfed race, based on maximum SVL and the significantly smaller size of gravid females. He noted that gravid females of *C. muertensis* range in size from 431 to 533 mm SVL, whereas the smallest gravid female from a noninsular population was 674 mm.

FIG 6.62 ***Crotalus molossus***

BLACK-TAILED RATTLESNAKE; VÍBORA DE CASCABEL

Crotalus molossus Baird and Girard 1853:10. Type locality: "Fort Webster, Santa Rita del Cobre [Grant County], New Mexico [USA]." Holotype: USNM 485

IDENTIFICATION *Crotalus molossus* can be distinguished from all other snakes in the region of study by a heat-sensing

FIGURE 6.62 *Crotalus molossus* (Black-Tailed Rattlesnake) from Isla Tiburón, Sonora. (Photograph by John Tashjian)

facial pit organ between the nostril and eye; no hornlike processes or sutures on the supraocular scales; two internasal scales; a solid dark brown to black tail; distinct dorsal blotches; and a prominently darkened internasal-prefrontal region.

RELATIONSHIPS AND TAXONOMY *Crotalus molossus* is most closely related to *C. estebanensis;* together these species may form a natural group with *C. horridus* of the eastern United States, Mexico, and Central America (Brattstrom 1964; Klauber 1972). Campbell and Lamar (1989) and Grismer (1999b) discuss taxonomy.

DISTRIBUTION *Crotalus molossus* ranges throughout the southwestern United States, from Arizona to central Texas and south to Oaxaca in southern Mexico. It occurs on the Gulf island of Tiburón (Campbell and Lamar 1989; Ernst 1992; map 6.30).

DESCRIPTION Body moderately sized, reaching less than 995 mm SVL; head triangular, distinct from neck; pupils elliptical; tail relatively short, 5.8 to 8.6 percent of SVL in adult males and 4.6 to 6.7 percent in females; 2 large internasals; prenasals rarely contact first supralabials; usually 2 to 4 loreals; head scales small and imbricate; 13 to 20 supralabials; 15 to 21 infralabials; dorsal body scales keeled and imbricate; 25 to 29 (usually 27) dorsal scale rows at midbody; 178 to 199 ventral scales in females; anal undivided; 22 to 29 subcaudals in males, 18 to 25 in females; subcaudals usually entire, but posteriormost may be divided; rattles normal, reaching parallelism at a length of approximately 14.5 mm. **COLORATION** Dorsal ground color brownish; top of head lightly mottled; internasal-prefrontal region dark; wide, dark postorbital stripe; dorsal pattern composed of 20 to 38 brownish, transversely oriented rhomboid dorsal blotches, transforming into bands posteriorly; first dorsal blotch usually lineate; anterior blotches tend to contain paler scales; interblotch scales on midline usually much lighter than surrounding ground color; lateral row of smaller, diamond-shaped dark secondary blotches on sides of body often confluent with dorsal blotches; tail solid brown to black, weakly banded in juveniles; ventrum dull yellow, usually faintly blotched or punctated.

NATURAL HISTORY Isla Tiburón is the largest island in the Gulf of California and is little more than an extension of mainland Sonora. It is topographically variable, having a series of large mountain ranges, wide arroyos, bajadas, and open flats. Isla Tiburón lies within the Central Gulf Coast Region and is dominated by a hot, arid climate and thorn scrub vegetation (Felger and Lowe 1976). *Crotalus molossus* is ubiquitous on the island and found in all habitats. Rattlesnakes are probably active year-round, with a minimum of activity occurring during the winter. Specimens have been collected or observed from at least April through November. During the spring, *C. molossus* is both diurnal and nocturnal, but it becomes decidedly nocturnal during the summer. Adult *C. molossus* prey primarily on rodents and occasionally small rabbits (Lowe et al. 1986). Ju-

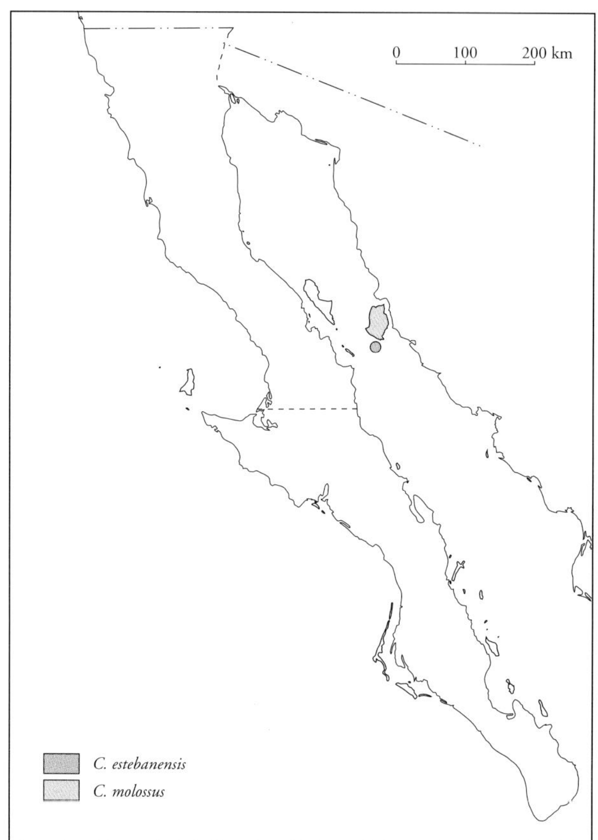

MAP 6.30 Distribution of *Crotalus molossus* (Black-Tailed Rattlesnake) and *C. estebanensis* (Isla San Esteban Rattlesnake)

veniles readily take lizards. Nothing has been reported concerning the reproductive biology of *C. molossus* from Isla Tiburón.

FOLKLORE AND HUMAN USE See *C. atrox.*

FIG 6.63 **_Crotalus estebanensis_**

ISLA SAN ESTEBAN RATTLESNAKE; VÍBORA DE CASCABEL

Crotalus molossus estebanensis Klauber 1949b:104. Type locality: "San Esteban Island, Gulf of California [Sonora], Mexico." Holotype: SDSNH 26792

IDENTIFICATION *Crotalus estebanensis* can be distinguished from all other snakes in the region of study by a heat-sensing facial pit organ between the nostril and eye; lacking hornlike processes or sutures on the supraocular scales; two internasal scales; a transversely and longitudinally compressed rattle (rattles usually degenerate); a solid dark brown to black tail; faint dorsal blotches; and a faintly darkened internasal-prefrontal region.

RELATIONSHIPS AND TAXONOMY See *C. molossus.*

DISTRIBUTION Endemic to the Gulf island of San Esteban (Grismer 1999a,b; see map 6.30).

DESCRIPTION Same as *C. molossus,* except for the following: adults reach 911 mm SVL; rattles usually degenerate, transversely and longitudinally compressed, reaching parallelism at a length of approximately 10 mm. **COLORATION** Dorsal ground color light olive green, rarely brownish; top of head unicolored to faintly mottled; internasal-prefrontal region same color as top of head or only slightly darker; wide, faint postorbital stripe; dorsal pattern composed of 38 to 43 indistinct, greenish, transversely oriented rhomboid dorsal blotches, transforming into faint bands posteriorly; posteriormost bands usually absent, anterior bands faint and difficult to discern; interblotch scales on midline usually much lighter than surrounding ground color; one lateral row of smaller diamond-shaped, dark secondary blotches on sides of body, often confluent with dorsal blotches; tail solid brown to black, weakly banded in juveniles; ventrum dull yellow, usually faintly blotched or punctated.

NATURAL HISTORY Isla San Esteban is a rocky, mountainous island within the Central Gulf Coast Region, crisscrossed by deep arroyos and steep hillsides and dominated by xerophilic thorn scrub. *Crotalus estebanensis* is commonly found in brushy, rocky arroyos and on talus hillsides, preferring these habitats to the more open arroyos. I have also observed specimens in rocky areas at the water's edge that were probably preying on introduced rats *(Rattus). Crotalus estebanensis* has been observed from early March to early October and is probably active year-round, with a minimum of activity during the winter. During April and May, *C. estebanensis* may be seen foraging during the day. As daytime temperatures rise later in the year, however, it becomes almost exclusively nocturnal. Nothing has been reported concerning the diet of *C. estebanensis,* but it is likely to include lizards (especially in juveniles) and rodents. Klauber (1949b) observed a gravid female during mid-April, and I have seen what I believe were gravid females during mid-June. These observations suggest that mating takes place in the spring, with newborns likely appearing during the summer.

FIGURE 6.63 *Crotalus estebanensis* (Isla San Esteban Rattlesnake) from Arroyo Limantur, Isla San Esteban, Sonora

FIG 6.64 **_Crotalus tigris_**

TIGER RATTLESNAKE; VÍBORA DE CASCABEL

Crotalus tigris Kennicott in Baird 1859b:14. Type locality: "Sierra Verde and Poza Verde," Sonora, Mexico. Holotype: USNM 471

IDENTIFICATION *Crotalus tigris* can be distinguished from all other snakes in the region of study by a heat-sensing facial pit organ between the nostril and eye; lacking hornlike processes or sutures on the supraocular scales; having two internasal scales; and having tail bands that are the same color and form as the bands on the body.

RELATIONSHIPS AND TAXONOMY The phylogenetic relationships of *C. tigris* remain problematic. Brattstrom (1964) indicated that *C. tigris* was the sister species of *C. mitchellii* and *C. viridis.* Stille (1987) removed *C. tigris* from being closely related to *C. mitchellii* and *C. viridis* and placed it in a very large, unresolved group that includes *C. viridis.*

DISTRIBUTION *Crotalus tigris* ranges from central southern Arizona south to southern Sonora, Mexico (Campbell and Lamar 1989; Ernst 1992). It is also known to occur on the Gulf island of Tiburón, Sonora (map 6.31).

DESCRIPTION *Crotalus tigris* from Isla Tiburón is known from only two specimens. The following description is based on those specimens and on data from Campbell and Lamar (1989) and Klauber (1972) concerning *C. tigris* in general. Body moderately built, reaching 805 mm SVL; head triangular, distinct from neck, relatively small; pupils elliptical; tail relatively short, averaging 8.4 percent of SVL in adult males and 6.4 percent in females; 2 small internasals; prenasals usually contact first supralabials; usually 1 loreal; head scales small and imbricate; 11 to 16 supralabials and infralabials; dorsal body scales keeled and imbricate; 20 to 27 (usually 23) dorsal scale rows at midbody; 156 to 172 ventral scales in males and 164 to 177 in females; anal undivided; 23 to 27 subcaudals in males, 16 to 21 in females; subcaudals usually entire, but posteriormost may be divided; rattles normal. **COLORATION** Dorsal ground color gray or blue-gray to brownish with buff to salmon-pinkish sides; top of head lightly speckled; wide, dark postorbital stripe; dorsal pattern composed of 35 to 52 brownish, transversely oriented bands composed of dense punctations that become slightly narrower and better defined on tail; interblotch scales on midline usually bluish; lateral row of smaller diamond-shaped, dark secondary blotches on sides of body, often confluent with dorsal blotches; ventrum yellowish to pink and moderately to heavily punctated.

FIGURE 6.64 *Crotalus tigris* (Tiger Rattlesnake) from near Mocuzari, Sonora. (Photograph by Cecil R. Schwalbe)

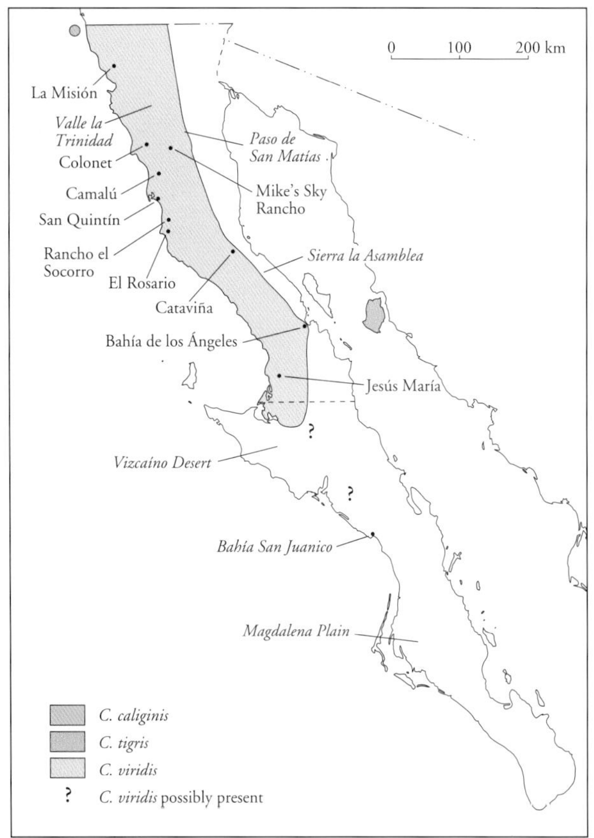

MAP 6.31 Distribution of *Crotalus tigris* (Tiger Rattlesnake), *C. viridis* (Western Rattlesnake), and *C. caliginis* (Islas Coronado Rattlesnake)

NATURAL HISTORY Isla Tiburón is a large landbridge island within the Central Gulf Coast Region. Its large size and topographical complexity support a wide variety of desert habitats, such as rocky bajadas and desert slopes, as well as wide, open valleys with Creosote Bush *(Larrea tridentata)* flats and sand dunes. Associated with these desert habitats are various floral compositions (Felger and Lowe 1976), which add to the island's habitat diversity. Nothing has been reported concerning the natural history of *C. tigris* from Isla Tiburón. Ernst (1992) and Lowe et al. (1986) give good summaries of the natural history of *C. tigris* from Arizona.

FIGS. 6.65–66

Crotalus viridis

WESTERN RATTLESNAKE; VÍBORA DE CASCABEL

Crotalinus viridis Rafinesque 1818:41. Type locality: "The Upper Missouri [Valley]." Restricted to "Gross, Boyd county, Nebraska" (Smith and Taylor 1950:358). Holotype: none designated

IDENTIFICATION *Crotalus viridis* can be distinguished from all other snakes in the region of study by having a heat-sensing facial pit organ between the nostril and eye; lacking hornlike processes or sutures on the supraocular scales; having the prenasal usually in contact with the rostral; usually three or more internasal scales; lacking distinct black and white caudal bands; and having an adult SVL greater than 675 mm.

RELATIONSHIPS AND TAXONOMY See *C. tigris*. Grismer (2001a) discusses taxonomy.

DISTRIBUTION *Crotalus viridis* ranges from southwestern Canada south through the less arid regions of the central and western United States to extreme north-central Mex-

FIGURE 6.65 *Crotalus viridis* (Western Rattlesnake) from Laguna Hanson, Sierra Juárez, BC

FIGURE 6.66 *Crotalus viridis* (Western Rattlesnake) from San Quintín, BC

ico and central Baja California (Campbell and Lamar 1989; Ernst 1992). In Baja California, *C. viridis* ranges south through cismontane and montane areas in the north to the central portion of the Vizcaíno Desert (Starrett 1993). Owing to the continuity of habitat and the consistency of climate along the west coast, this species probably extends much farther south throughout the Vizcaíno Desert, and may eventually be found to occur as far south as Bahía San Juanico in the Magdalena Plain. In north-central Baja California, its distribution extends east immediately south of the Sierra la Asamblea, where it approaches Bahía de los Ángeles. It has not been reported from the desert slopes of the northern Peninsular Ranges (see map 6.31).

DESCRIPTION Body moderately stout, reaching 1,110 mm SVL; head triangular, distinct from neck; pupils elliptical; tail relatively short, averaging 7.4 percent of SVL in adult males and 5.6 percent in females; usually 3 or more internasals; prenasals in contact with rostral; usually 1 loreal; supraoculars not elevated or sutured; head scales small and imbricate; 12 to 18 supralabials; 13 to 20 infralabials; dorsal body scales keeled and imbricate; 23 to 29 (usually 25) dorsal scale rows at midbody; 162 to 184 ventral scales in males, 166 to 189 in females; anal undivided; 19 to 29 subcaudals in males, 15 to 25 in females; subcaudals usually entire, but posteriormost may be divided; rattles normal, reaching parallelism at 15 to 17 mm. **COLORATION** ADULTS: Dorsal ground color olive green, gray to dark gray, or black; top of head unicolored; faint, thin, light-colored continuous supraocular band in adults, more prominent in juveniles; wide, light-colored postorbital stripe; wide, oblique, light-colored facial stripe beginning at the nostril and extending to corner of mouth; dorsal pattern composed of 27 to 43 dark hexagonal body blotches outlined by a single row of light-colored scales; blotches become narrow and more bandlike posteriorly; areas of ground color on sides of body appear as secondary rows of blotches; below these lie well-defined diamond-shaped secondary blotches; ventrum dull-colored, often weakly mottled. JUVENILES: Color pattern boldly marked with a nearly black ground color and white diamond markings; head markings particularly bold, especially supraocular band and postorbital stripe; facial stripe continues across frontal region and is confluent with white edging surrounding rostral scale; end of tail and first button of rattle bright yellow.

GEOGRAPHIC VARIATION Adult *C. viridis* from cismontane areas north of San Quintín usually have an olive green to dark gray ground color. The light-colored head and dorsal markings are not particularly striking, and the end of the tail is usually the same color as the ground color. Montane populations from the Sierra Juárez and the Sierra San Pedro Mártir are very dark, almost melanistic. All head and body markings are clouded with dark pigment, and some specimens are nearly solid black. There is a transition between the cismontane and montane color patterns in the foothill chaparral areas of the Sierra San Pedro Mártir in the vicinity of Mike's Sky Rancho. Presumably a similar transition is found in the Sierra Juárez. From just south of Colonet to approximately Rancho El Socorro, along the San Quintín Plain, *C. viridis* is lighter in overall ground color and becomes more light brown than dark gray. The head and body markings are better-defined and contrasted from the ground color, and the end of the tail takes on a buff coloration. From just northwest of Bahía de los Ángeles and south into the Vizcaíno Desert, the ground color becomes light gray, and the anteriormost blotches tend to connect and elongate. The light-colored head and body markings are even more prominent, and the base of the tail retains its yellow juvenile coloration into adulthood. This pattern reaches an extreme in populations from the Vizcaíno Desert: these apparently retain into adulthood the juvenile coloration of snakes from the northern cismontane populations.

NATURAL HISTORY *Crotalus viridis* is common in the California and Baja California Coniferous Forest regions as well as the western coastal portions of the Vizcaíno Region, including the Vizcaíno Desert. Within the first two regions, *C. viridis* is ubiquitous and found in virtually every habitat. I have found specimens in low-growing, sparse vegetation along the beaches near San Quintín, and Armstrong and Murphy (1979) have made similar observations from near Camalú. Klauber (1972) reported finding specimens in southern California along the wave wash of the beaches, and Bostic (1971) and Mosauer (1936) observed specimens beneath low-growing shrubs on coastal sand dunes near Jesús María and Rancho el Socorro, respectively. Welsh (1988) reported *C. viridis* as being common in riparian woodlands along the western lower slopes of the Sierra San Pedro Mártir. At higher elevations, in the Baja California Coniferous Forest above 2,000 m and in the coniferous forest–chaparral ecotones, *C. viridis* is most frequently observed in riparian edge habitats (Welsh 1988). *Crotalus viridis* is not a species of hot, arid deserts and has yet to be reported from the desert slopes of the northern Peninsular Ranges (Welsh 1988), although its distribution fringes the deserts farther north in southern California (Klauber 1972). In the cool Vizcaíno Desert, *C. viridis* is common in low-growing scrub vegetation (Bostic 1971; Jones 1981; Murray 1955; Ottley and Hunt 1981; Smith et al. 1971; Starrett 1993). This area is buffeted by onshore winds, and morning and evening fogs ameliorate the aridity and high temperatures. Where these weather patterns extend a significant distance inland, through coastal valleys, *C. viridis* follows. This accounts for its presence in the thick forests of Boojum Tree *(Fouquieria columnaris)* and Cardón Cactus *(Pachycereus pringlei)* just west of Bahía de los Ángeles. Wherever it occurs, *C. viridis* frequents a number of microhabitats and can be found beneath logs, rocks, and surface litter, within rock piles and cracks, and beneath brush (Klauber 1972). I have found these snakes to be most common beneath garbage at dump sites.

Crotalus viridis is active year-round, with a minimum of activity occurring during the winter. Throughout its distribution in Baja California I have observed adults abroad from February through October, and most commonly from April through July. In the winter juveniles are the most abundant age class found. *Crotalus viridis* is diurnal north of El Rosario. South of El Rosario, and especially in the Vizcaíno Desert, it is diurnal in the spring and crepuscular to nocturnal during the summer, when daytime temperatures are highest. Bostic (1971) observed an individual at 0920 hours during mid-July near Jesús María, coiled beneath a sparsely branched shrub.

Crotalus viridis eats a wide range of foods. Adults prey primarily on mammals, including various species of mice and rats, pocket gophers, small rabbits, and small ground squirrels. I observed a specimen near Valle la Trinidad that had entered the nest of a California Ground Squirrel *(Spermophilus beecheyi),* in the fallen branches of a tree, and was eating the young. Clark R. Mahrdt (personal communication, 1994) informed me that he removed an adult Arboreal Salamander *(Aneides lugubris)* from the stomach of a juvenile *C. viridis* from La Misión. Juvenile *C. viridis* eat small rodents but also prey heavily on whatever species of lizard may be in their area (Klauber 1972). Not much is known about the reproductive biology of *C. viridis* from Baja California, but it is probably not too different from that of southern California populations, in which mating occurs during the spring and early summer and newborns appear during the summer and fall.

REMARKS When threatened, juvenile and newborn *C. viridis* from the northern areas of Baja California often tuck their heads within their coils and display the yellow undersides of their tails by inverting them over the tops of their bodies. As the snakes get older, this behavior is lost, the yellow coloration on the tail fades to gray, and they become very defensive (which comes across as being very aggressive). As mentioned above, populations from the Vizcaíno Desert maintain the juvenile coloration into adulthood. Interestingly, the six living individuals I found from this area maintained into adulthood the juvenile behavior of displaying the yellow tail above the body and hiding the head when continually threatened. Snakes from this population are not particularly defensive when compared to the extremely defensive montane populations further north. With respect to the latter, I would have overlooked several individuals had they not rattled at me while I was still some distance away.

FIG 6.67

Crotalus caliginis

ISLAS CORONADO RATTLESNAKE; VÍBORA DE CASCABEL

Crotalinus viridis caliginis Klauber 1949b:90. Type locality: "South Coronado Island, off the northwest coast of Baja California, Mexico." Holotype: SDSNH 2800

IDENTIFICATION *Crotalus caliginis* can be distinguished from all other snakes in the region of study by having a heat-sensing facial pit organ between the nostril and eye; lacking hornlike processes or sutures on the supraocular scales; having the prenasal usually in contact with the rostral; usually having three or more internasal scales; lacking distinct black and white caudal bands; and having an adult SVL less than 675 mm.

RELATIONSHIPS AND TAXONOMY See *C. tigris.* Grismer (2001a) discusses taxonomy.

DISTRIBUTION *Crotalus caliginis* is known only from the Pacific island of Isla Coronado Sur, off the coast of northern Baja California (Grismer 2001a; see map 6.31).

DESCRIPTION Same as *C. viridis,* except for the following: body moderate, reaching 674 mm SVL; head triangular, distinct from neck; pupils elliptical; tail relatively short, averaging 7.9 percent of SVL in adult males and 6.7 percent in females; 2 to 6 (usually 4) internasals; 15 to 16 supralabials; 14 to 15 infralabials; 25 to 30 (usually 25) dorsal scale rows at midbody; 167 to 174 ventral scales in males, 171 to 179 in females; 22 to 28 subcaudals in males, 15 to 23 in females; subcaudals usually entire, but posteriormost may be divided; rattles normal, reaching parallelism at 10 to 11 mm. **COLORATION** ADULTS: Dorsal ground color olive green, gray to dark gray; top of head unicolored; faint, thin, light-colored, continuous supraocular band in adults, more prominent in juveniles; wide, light-colored postorbital stripe; wide, oblique, light-colored facial stripe beginning at the nostril and extending to corner of mouth; dorsal pattern composed of 28 to 37 dark, irregularly shaped hexagonal body blotches, darkened on the edges with lighter centers and outlined by a single row of light-colored scales; blotches become narrower and more bandlike posteriorly; areas of ground color on sides of body appear as secondary rows of blotches; below these lie well-defined, diamond-shaped secondary blotches; ventrum dull, often weakly mottled. JUVENILES: Color pattern boldly marked, with a nearly black ground color and white diamond markings; head markings particularly bold, especially supraocular band and postorbital stripe; facial stripe continues across frontal region and is confluent with white edging surrounding rostral scale; end of tail and first button of rattle bright yellow.

NATURAL HISTORY Isla Coronado Sur is a small, steep, rocky, fog-shrouded island lying 13 km off the coast of Tijuana, within the California Region. It is covered with low-growing brush and cactus that are generally wind-blown and stunted. *Crotalus caliginis* is abundant and ubiquitous on the island. It is most commonly found beneath the leaves of the Ice Plant *(Gasoul crystalinum)* that covers the island (Oberbauer 1992). I have found them within stands of *Opuntia* cactus. Klauber (1972) reported a specimen being found coiled on the top of a bush approximately 1 m off the ground. *Crotalus caliginis* is most active from March through August but may be found to some extent year-

FIGURE 6.67 *Crotalus caliginis* (Islas Coronado Rattlesnake) from Isla Coronado Sur, BC

round, provided the weather is sufficiently warm. Although the Islas Coronado are covered with fog for a large portion of the year, snakes remain active. They are principally diurnal and feed almost exclusively on lizards, although rodents are abundant (Klauber 1949b). Zweifel (1952a) noted that during periods of overcast, Islas Coronado Alligator Lizards *(Elgaria nana)* and Western Skinks *(Eumeces skiltonianus)* were abundant but that Side-Blotched Lizards *(Uta stansburiana)* and Western Whiptails *(Cnemidophorus tigris)* were not. This observation is consistent with the feeding records of Klauber (1949b) and Zweifel (1952a), indicating that the former two species are the most heavily preyed on. Mating takes place in the spring and early summer, and gravid females have been found in June and July. Armstrong and Murphy (1979) reported that a female with a single partial-term embryo was collected in early August.

REMARKS Klauber (1949b) described *C. caliginis* as a subspecies of *C. viridis.* Although he could find no differences in color pattern or squamation between it and the adjacent peninsular populations of *C. viridis (C. v. helleri),* he did observe significant differences in body proportions and SVL. He noted that no *C. caliginis* would ever grow to the size of an average adult *C. v. helleri* of the adjacent peninsula. The largest specimen he measured was 674 mm in total length, and he noted that *C. v. helleri* regularly reach or exceed 1,100 mm in length. Additional evidence that *C. v. caliginis* is a stunted population is that its rattles reach parallelism at a width of 10 to 11 mm, whereas in *C. v. helleri,* parallelism is reached at 15 to 17 mm (parallelism occurs at adulthood). Additionally, specimens of *C. caliginis* the same size as *C. v. helleri* have disproportionally smaller heads (Klauber 1938). Klauber (1949b) noted that a 1,100-mm *C. v. helleri* "would have 5 times the bulk" of a 650-mm *C. caliginis.*

APPENDIX A

INSULAR SPECIES CHECKLIST

The following is a checklist of the herpetofauna of the islands along the Pacific coast of Baja California and in the Gulf of California, Mexico. Less commonly used island names are given in parentheses. The species are listed alphabetically in the order salamanders, frogs, lizards, and snakes. An asterisk follows names of species endemic to one to three islands. Adapted from Grismer 1993, 1999b, and 2001a, with additions from Sánchez-Pacheco and Mellink 2001.

Pacific Islands

Isla Asunción
Uta stansburiana

Isla Cedros
Hyla regilla, Gambelia copeii, Phrynosoma coronatum, Sceloporus occidentalis, Sceloporus zosteromus, Uta stansburiana, Coleonyx variegatus, Cnemidophorus tigris, Elgaria cedrosensis, Leptotyphlops humilis, Lichanura trivirgata, Chilomeniscus stramineus, Hypsiglena torquata, Pituophis insulanus, Crotalus ruber*

Isla las Brosas (within Laguna Ojo de Liebre)
Uta stansburiana

Isla las Piedras (within Laguna Ojo de Liebre)
Uta stansburiana

Isla Magdalena
Dipsosaurus dorsalis, Gambelia copeii, Sceloporus zosteromus, Urosaurus nigricaudus, Uta stansburiana, Phyllodactylus xanti, Cnemidophorus hyperythrus, Cnemidophorus tigris, Chilomeniscus stramineus, Masticophis fuliginosus, Pituophis vertebralis, Crotalus enyo

Isla Natividad
Uta stansburiana, Cnemidophorus tigris, Lichanura trivirgata (?)

Isla Pata (within Laguna Ojo de Liebre)
Uta stansburiana

Isla San Geronimo
Uta stansburiana, Anniella geronimensis, Salvadora hexalepis

Isla San Martín
Batrachoseps major (?), Uta stansburiana, Elgaria multicarinata, Anniella geronimensis, Diadophis punctatus, Hypsiglena torquata, Pituophis catenifer

Isla San Roque
Uta stansburiana

Islas Coronado (includes all three islands unless noted otherwise)
Aneides lugubris (Norte), *Batrachoseps major, Uta stansburiana, Eumeces skiltonianus* (Norte and Sur), *Cnemidophorus tigris* (Norte and Sur), *Elgaria nana** (Norte and Sur), *Anniella pulchra* (Norte and Sur), *Hypsiglena torquata, Pituophis catenifer* (Sur), *Crotalus caliginis** (Sur)

Islas San Benito (three islands)
Uta stansburiana

Isla Santa Margarita
Bufo punctatus, Callisaurus draconoides, Dipsosaurus dorsalis, Gambelia copeii, Sceloporus zosteromus, Urosaurus nigricaudus, Uta stansburiana, Coleonyx variegatus, Phyllodactylus xanti, Cnemidophorus hyperythrus, Cnemidophorus tigris, Eridiphas slevini, Masticophis fuliginosus, Pituophis vertebralis, Crotalus enyo, Crotalus mitchellii, Crotalus ruber

Islas Todos Santos (Isla Sur only unless indicated otherwise)
Batrachoseps major, Sceloporus occidentalis, Uta stansburiana (Norte and Sur), *Eumeces skiltonianus* (Norte and Sur), *Anniella pulchra, Diadophis punctatus, Lampropeltis herrerae,* Salvadora hexalepis*

Gulf of California Islands

Isla Alcatraz (Isla Pelícano)
Sauromalus hispidus × obesus × varius, Uta stansburiana, Phyllodactylus xanti

Isla Ángel de la Guarda
Crotaphytus insularis, Dipsosaurus dorsalis, Sauromalus*

OPPOSITE *Crotalus angelensis* (Isla Ángel de la Guarda Rattlesnake) from Puerto Refugio, Isla Ángel de la Guarda, BC *hispidus,* Callisaurus draconoides, Petrosaurus slevini,* Uta stansburiana, Coleonyx variegatus, Phyllodactylus xanti, Cnemidophorus tigris, Lichanura trivirgata, Hypsiglena torquata, Lampropeltis getula, Phyllorhynchus decurtatus, Crotalus angelensis,* Crotalus ruber*

Isla Ballena (0.5 km west of Isla Espíritu Santo)
Sauromalus obesus, Sceloporus hunsakeri, Urosaurus nigricaudus, Uta stansburiana, Phyllodactylus unctus

Isla Bota (within Bahía de los Ángeles, BC)
Uta stansburiana

Isla Cabeza de Caballo (within Bahía de los Ángeles, BC)
Sauromalus hispidus, Uta stansburiana, Crotalus mitchellii

Isla Cardonosa Este (Isla Cardonosa; 0.5 km northeast of Isla Partida Norte)
Uta stansburiana, Phyllodactylus partidus, Cnemidophorus tigris*

Isla Carmen
Dipsosaurus dorsalis, Sauromalus slevini, Callisaurus draconoides, Sceloporus orcutti, Sceloporus zosteromus, Urosaurus nigricaudus, Uta stansburiana, Phyllodactylus xanti, Cnemidophorus carmenensis,* Cnemidophorus tigris, Leptotyphlops humilis, Hypsiglena torquata, Masticophis flagellum, Tantilla planiceps, Crotalus enyo, Crotalus mitchellii*

Isla Cayo (within Bahía Concepción, BCS)
Urosaurus nigricaudus, Phyllodactylus xanti

Isla Cerraja (within Bahía de Los Ángeles, BC)
Uta stansburiana

Isla Cerralvo
Bufo punctatus, Scaphiopus couchii, Ctenosaura hemilopha, Dipsosaurus dorsalis, Callisaurus draconoides, Sceloporus grandaevus, Phyllodactylus unctus; Cnemidophorus ceralbensis,* Leptotyphlops humilis, Lichanura trivirgata, Chilomeniscus savagei,* Eridiphas slevini, Hypsiglena torquata, Lampropeltis getula, Masticophis flagellum, Phyllorhynchus decurtatus, Rhinocheilus etheridgei,* Trimorphodon biscutatus, Crotalus enyo, Crotalus mitchellii*

Isla Cholludo (Isla Lobos, Son; Isla Roca Foca; between Islas Turners and Tiburón)
Ctenosaura conspicuosa, Phyllodactylus xanti*

Isla Coloradito (Isla Lobos, BC; Lobera)
*Uta tumidarostra**

Isla Coronados
Dipsosaurus dorsalis, Sauromalus slevini, Callisaurus draconoides, Sceloporus orcutti, Sceloporus zosteromus, Urosaurus nigricaudus, Uta stansburiana, Coleonyx variegatus, Phyllodactylus xanti, Cnemidophorus hyperythrus, Cnemidophorus tigris, Eridiphas slevini, Hypsiglena torquata, Masticophis flagellum, Crotalus enyo*

Isla Danzante
Sauromalus obesus, Callisaurus draconoides, Petrosaurus repens, Urosaurus nigricaudus, Uta stansburiana, Coleonyx variegatus, Phyllodactylus xanti, Cnemidophorus tigris, Leptotyphlops humilis, Bogertophis rosaliae, Chilomeniscus stramineus, Eridiphas slevini, Hypsiglena torquata, Masticophis flagellum, Trimorphodon biscutatus, Crotalus ruber

Isla Dátil (Isla Turners)
Gopherus agassizii (?), Uta stansburiana, Phyllodactylus xanti, Masticophis flagellum, Crotalus atrox

Isla El Coyote (within Bahía Concepción, BCS)
Sauromalus obesus, Urosaurus nigricaudus, Phyllodactylus xanti

Isla El Muerto (Isla Miramar; Isla Link)
Petrosaurus mearnsi, Phyllodactylus xanti, Uta lowei, Trimorphodon biscutatus, Crotalus muertensis**

Isla El Pardito (Isla Coyote)
Uta stansburiana, Phyllodactylus xanti

Isla El Requesón (within Bahía Concepción, BCS)
Urosaurus nigricaudus

Isla Encantada
*Uta encantadae**

Isla Espíritu Santo
Bufo punctatus, Scaphiopus couchii, Dipsosaurus dorsalis, Sauromalus obesus, Callisaurus draconoides, Petrosaurus thalassinus, Sceloporus hunsakeri, Sceloporus zosteromus, Urosaurus nigricaudus, Uta stansburiana, Coleonyx variegatus, Phyllodactylus unctus, Cnemidophorus espiritensis, Cnemidophorus tigris, Chilomeniscus stramineus, Masticophis barbouri,* Masticophis flagellum, Salvadora hexalepis, Crotalus enyo, Crotalus mitchellii*

Isla Flecha (within Bahía de los Ángeles, BC)
Sauromalus hispidus, Uta stansburiana

Isla Gallina (0.5 km west of Isla Espíritu Santo, BCS)
Urosaurus nigricaudus, Phyllodactylus unctus

Isla Gallo (0.5 km west of Isla Espíritu Santo)
Sauromalus obesus, Sceloporus hunsakeri, Urosaurus nigricaudus, Uta stansburiana, Phyllodactylus unctus

Isla Gaviota (0.5 km north of Pichilingue, BCS)
Urosaurus nigricaudus

Isla Granito
Sauromalus hispidus, Uta stansburiana

Isla Islitas (11.5 km south of Puerto Escondido, BCS)
Urosaurus nigricaudus

Isla Lagartija (0.25 km north of Isla Salsipuedes)
Uta stansburiana

Isla la Raza
Uta stansburiana, Phyllodactylus xanti

Isla la Ventana (Isla Nuevo Amor; within Bahía de los Ángeles, BC)
Sauromalus hispidus, Uta stansburiana, Phyllodactylus xanti

Isla las Ánimas (BCS)
Urosaurus nigricaudus, Phyllodactylus xanti

Isla las Galeras (two islands, east and west; 3 km north of Isla Monserrate)
Uta stansburiana

Isla Mejía
Sauromalus hispidus, Petrosaurus slevini, Uta stansburiana, Phyllodactylus xanti, Lichanura trivirgata, Hypsiglena torquata*

Isla Mitlán (within Bahía de los Ángeles, BC)
Sauromalus hispidus, Uta stansburiana

Isla Monserrate (Isla Monserrat; Isla Monserrato)
Dipsosaurus dorsalis, Sauromalus slevini, Sceloporus zosteromus, Uta stansburiana, Phyllodactylus xanti, Cnemidophorus pictus,* Chilomeniscus stramineus, Hypsiglena torquata, Lampropeltis getula, Masticophis flagellum, Phyllorhynchus decurtatus, Crotalus mitchellii, Crotalus ruber*

Isla Moscas (within Bahía Concepción, BCS; Dixon, 1966)
Phyllodactylus xanti

Isla Pardo (13.5 km south of Puerto Escondido, BCS)
Sauromalus obesus, Urosaurus nigricaudus, Phyllodactylus xanti, Crotalus enyo

Isla Partida Norte (Isla Partida; Isla Cardonosa)
*Uta stansburiana, Phyllodactylus partidus, Cnemidophorus tigris, Hypsiglena gularis**

Isla Partida Sur
Bufo punctatus, Scaphiopus couchii, Dipsosaurus dorsalis, Sauromalus obesus, Callisaurus draconoides, Petrosaurus thalassinus, Sceloporus hunsakeri, Sceloporus zosteromus, Urosaurus nigricaudus, Uta stansburiana, Coleonyx variegatus, Phyllodactylus unctus, Cnemidophorus espiritensis, Cnemidophorus tigris, Chilomeniscus stramineus, Hypsiglena torquata, Masticophis barbouri,* Masticophis flagellum, Crotalus enyo, Crotalus mitchellii*

Isla Pata (within Bahía de los Ángeles, BC)
Uta stansburiana

Isla Patos (16 km north of Isla Tiburón, Sonora)
Callisaurus draconoides, Uta stansburiana

Isla Piojo (within Bahía de los Ángeles, BC)
Sauromalus hispidus, Uta stansburiana, Phyllodactylus xanti, Crotalus mitchellii

Isla Pond (Isla Estanque; Isla La Víbora)
Sauromalus hispidus, Uta stansburiana, Phyllodactylus xanti, Cnemidophorus tigris, Crotalus ruber

Isla Roca Lobos (0.25 km southwest of north end of Isla Salsipuedes)
Sauromalus varius, Uta stansburiana*

Isla Salsipuedes
Uta stansburiana, Phyllodactylus xanti, Cnemidophorus canus, Hypsiglena torquata, Lampropeltis getula, Crotalus mitchellii*

Isla San Cosme (13 km W of Isla Monserrate)
Sauromalus obesus, Urosaurus nigricaudus, Cnemidophorus tigris

Isla San Damián (12 km west of Isla Monserrate)
Urosaurus nigricaudus

Isla San Diego
Sauromalus obesus, Sceloporus angustus, Phyllodactylus xanti*

Isla San Esteban
Ctenosaura conspicuosa, Sauromalus varius,* Uta stansburiana, Phyllodactylus xanti, Cnemidophorus tigris, Hypsiglena torquata, Lampropeltis getula, Masticophis slevini,* Crotalus estebanensis**

Isla San Francisco
Sauromalus obesus, Callisaurus draconoides, Sceloporus orcutti, Urosaurus nigricaudus, Uta stansburiana, Phyllodactylus xanti, Cnemidophorus celeripes, Cnemidophorus franciscensis,* Hypsiglena torquata, Crotalus enyo*

Isla San Ildefonso
Sceloporus orcutti, Uta stansburiana, Phyllodactylus xanti, Masticophis flagellum

Isla San José
Dipsosaurus dorsalis, Sauromalus obesus, Callisaurus draconoides, Sceloporus orcutti, Sceloporus zosteromus, Urosaurus nigricaudus, Uta stansburiana, Coleonyx variegatus, Phyllodactylus xanti, Cnemidophorus celeripes, Cnemidophorus danheimae,* Chilomeniscus stramineus, Hypsiglena torquata, Masticophis flagellum, Phyllorhynchus decurtatus, Pituophis vertebralis, Salvadora hexalepis, Sonora semiannulata, Trimorphodon biscutatus, Crotalus enyo, Crotalus mitchellii, Crotalus ruber*

Isla San Lorenzo Norte (Isla las Ánimas, BC)
Sauromalus hispidus, Uta stansburiana, Phyllodactylus xanti, Cnemidophorus canus, Lampropeltis getula*

Isla San Lorenzo Sur (Isla San Lorenzo)
Sauromalus hispidus, Uta stansburiana, Phyllodactylus xanti, Cnemidophorus canus, Hypsiglena torquata, Lampropeltis getula, Crotalus lorenzoensis**

Isla San Luis (Isla Encantada Grande; Isla Salvatierra; Isla San Luis Gonzaga)
Dipsosaurus dorsalis, Callisaurus draconoides, Uta stansburiana

Isla San Marcos
Dipsosaurus dorsalis, Sauromalus obesus, Callisaurus draconoides, Sceloporus orcutti, Urosaurus nigricaudus, Uta stansburiana, Coleonyx gypsicolus, Coleonyx variegatus, Phyllodactylus xanti, Cnemidophorus hyperythrus, Cnemidophorus tigris, Leptotyphlops humilis, Lichanura trivirgata, Bogertophis rosaliae, Chilomeniscus stramineus, Eridiphas marcosensis,* Hypsiglena torquata, Masticophis flagellum, Phyllorhynchus decurtatus, Sonora semiannulata, Trimorphodon biscutatus, Crotalus enyo, Crotalus ruber*

Isla San Pedro Mártir
Uta palmeri, Cnemidophorus martyris,* Lampropeltis getula, Crotalus atrox*

Isla San Pedro Nolasco
Ctenosaura nolascensis, Sceloporus clarkii, Uta nolascensis,* Phyllodactylus homolepidurus, Cnemidophorus bacatus,* Lampropeltis getula*

Isla Santa Catalina (Isla Catlán; Isla Catalana; Isla Catalano)
Dipsosaurus catalinensis, Sauromalus klauberi,* Sceloporus lineatulus,* Uta squamata,* Phyllodactylus bugastrolepis,* Cnemidophorus catalinensis,* Leptotyphlops humilis, Hypsiglena torquata, Lampropeltis catalinensis,* Crotalus catalinensis**

Isla Santa Cruz
Sauromalus obesus, Sceloporus angustus, Phyllodactylus xanti, Leptotyphlops humilis, Lampropeltis getula, Crotalus atrox*

Isla Santiago (0.25 km west of Isla Coronados)
Dipsosaurus dorsalis

Isla Smith (Isla Coronado; within Bahía de Los Ángeles, BC)
Sauromalus hispidus, Callisaurus draconoides, Uta stansburiana, Phyllodactylus xanti, Cnemidophorus tigris, Hypsiglena torquata, Crotalus mitchellii

Isla Tiburón
Bufo punctatus, Scaphiopus couchii, Gopherus agassizii, Crotaphytus dickersonae, Gambelia wislizenii, Sauromalus obesus, Callisaurus draconoides, Phrynosoma solare, Sceloporus clarkii, Sceloporus magister, Urosaurus ornatus, Uta stansburiana, Coleonyx variegatus, Phyllodactylus xanti, Cnemidophorus tigris, Lichanura trivirgata, Chilomeniscus stramineus, Hypsiglena torquata, Masticophis bilineatus, Masticophis flagellum, Pituophis melanoleucus, Salvadora hexalepis, Trimorphodon biscutatus, Micruroides euryxanthus, Crotalus atrox, Crotalus cerastes, Crotalus molossus, Crotalus tigris

Isla Tijeras (13 km south of Puerto Escondido, BCS)
Urosaurus nigricaudus

Isla Tortuga
*Sceloporus orcutti, Uta stansburiana, Hypsiglena torquata, Lampropeltis getula, Crotalus tortugensis**

Isla Willard (within Bahía San Luis Gonzaga, BC)
Sauromalus obesus, Uta stansburiana

Islas Santa Inez (three islands; records from southernmost island only)
Callisaurus draconoides, Coleonyx variegatus

Islotes Blancos (0.25 km southeast of Isla Encantada)
*Uta encantadae**

APPENDIX B

CLAVES TAXONÓMICAS

The following is a Spanish translation of the taxonomic keys given on pp. 45–53. I thank E. Mellink and P. Galina-T. for their advice and revisions.

SALAMANDRAS

1. Cuerpo muy alargado, semejante al de una lombriz; extremidades muy reducidas . . . . *Batrachoseps*
1. Cuerpo diferente 2
2. Cola con un estrechamiento en su base; dedos delgados y no cuadrados, sin membrana interdigital 3
2. Cola sin estrechamiento en su base; dedos cortos y cuadrados, con membrana interdigital . . . . *Aneides lugubris*
3. Con manchas anaranjadas grandes en el cuerpo . . . . *Ensatina klauberi*
3. Sin manchas anaranjadas en el cuerpo . . . . *Ensatina eschscholtzii*

RANAS Y SAPOS

1. Con un tubérculo negro conspicuo en el talón inmediatamente posterior al primer dedo; pupilas verticales 2
1. Talones sin tubérculos o con dos tubérculos grandes: el más grande en la parte interna del talón y el más pequeño en su parte externa; pupilas no verticales 3
2. Tubérculos en el talón rectangulares . . . . *Scaphiopus couchii*
2. Tubérculos semicirculares . . . . *Spea hammondii*
3. Extremidades posteriores muy rugosas y con verrugas 4
3. Extremidades posteriores lisas 9
4. Parte superior de las extremidades posteriores con verrugas grandes, conspicuas, y aisladas; una serie de 1–4 verrugas blancas conspicuas en la comisura de la boca . . . . *Bufo alvarius*
4. Verrugas en la parte superior de las extremidades posteriores pequeñas; sin verrugas conspicuas en la comisura de la boca 5
5. Una línea blanca delgada bien definida a lo largo del centro de la espalda 6
5. Espalda diferente 7
6. Con crestas craneales . . . . *Bufo woodhousii*
6. Sin crestas craneales . . . . *Bufo boreas*
7. Hocico comprimido lateralmente, casi puntiagudo; espalda cubierta con tubérculos o verrugas rojas . . . . *Bufo punctatus*
7. Hocico no comprimido lateralmente; espalda diferente 8
8. Manchas paravertebrales obscuras en pares y simétricas; glándulas parótidas generalmente de un solo color . . . . *Bufo cognatus*
8. Manchas obscuras en la espalda no simétricas; glándulas parótidas de color claro en sus partes anteriores y manchadas en su posterior . . . . *Bufo californicus*
9. Puntas de los dedos ensanchadas y en forma de discos 10
9. Dedos no ensanchados en sus puntas 11
10. Con una línea obscura extendiéndose a lo largo de la parte lateral de la cabeza y cruzando el ojo . . . . *Hyla regilla*
10. Sin línea obscura cruzando el ojo . . . . *Hyla cadaverina*
11. Sin pliegues dorsolaterales; con un gran pliegue semicircular detrás del tímpano . . . . *Rana catesbeiana*
11. Con pliegues dorsolaterales; sin pliegues detrás del tímpano 12

12. Pliegues dorsolaterales conspicuos ..... *Rana forreri*
12. Pliegues dorsolaterales poco desarrollados, presentes sólo posteriormente .... *Rana aurora*

TORTUGAS

1. Mano aplanada o en forma de remo; con una o dos uñas 2
1. Mano menos aplanada, nunca en forma de remo; con cinco uñas 6
2. Carapacho revestido de piel coriácea con siete quillas longitudinales .... *Dermochelys coriacea*
2. Carapacho diferente 3
3. Carapacho con cuatro escudos laterales 4
3. Carapacho con cinco escudos laterales 5
4. Un solo par de escamas prefrontales; escudos no imbricados .... *Chelonia mydas*
4. Dos pares de escamas prefrontales; escudos generalmente imbricado .... *Eretmochelys imbricata*
5. Puente con cuatro escudos inframarginales; carapacho ancho; un solo par de escudos abdominales .... *Lepidochelys olivacea*
5. Puente raramente con más de tres escudos inframarginales; carapacho no muy ancho ..... *Caretta caretta*
6. Sin escudos en el carapacho; hocico alargado y tubula .... *Apalone spinifera*
6. Carapacho con escudos; hocico no alargado 7
7. Extremidades posteriores elefantinas .... *Gopherus agassizii*
7. Extremidades posteriores no elefantinas 8
8. Con una mancha clara detrás del tímpano .... *Trachemys nebulosa*
8. Sin manchas detrás del tímpano .... *Clemmys marmorata*

LAGARTIJAS

CLAVE PARA LOS GÉNEROS

1. Sin párpado 2
1. Con párpado 6
2. Con cuatro patas 3
2. Sin patas .... *Anniella*
3. Laminillas subdigitales ensanchadas, ya sea a lo largo del dedo o solamente en la punta 4
3. Laminillas subdigitales no ensanchadas .... *Xantusia*
4. Laminillas subdigitales ensanchadas a lo largo del dedo 5
4. Laminillas subdigitales ensanchadas sólo en la punta del dedo .... *Phylodactylus*
5. Primer dedo de las manos sin uña ... *Gehyra mutilata*
5. Primer dedo de las manos con uña .... *Hemidactylus frenatus*
6. Sin pliegues gulares o postgulares, ni siquiera al lado del cuello 7
6. Con pliegues gulares o postgulares por lo menos al lado del cuello, aunque el pliegue gular está generalmente completo 9
7. Escamas del cuerpo pequeñas y granulares .... *Coleonyx*
7. Escamas del cuerpo imbricadas 8
8. Escamas del cuerpo aquilladas .... *Elgaria*
8. Escamas del cuerpo lisas .... *Eumeces*
9. Cuerpo y cabeza cubiertos de espinas grandes .... *Phrynosoma*
9. Cuerpo y cabeza sin espinas grandes 10
10. Una sola hilera de escamas vertebrales levantadas y agrandadas formando una cresta tenue o moderada en la espalda 11
10. Región vertebral diferente 12
11. Escamas caudales formando una espiral de espinas conspicuas alrededor de la cola .... *Ctenosaura*
11. Escamas caudales rectangulares; cola con una cresta conspicua a lo largo del dorso .... *Dipsosaurus*
12. Con una mancha en la axila (a veces muy tenue) ... . *Uta*
12. Sin mancha en la axila 13
13. Todas las escamas dorsales del cuerpo son aquilladas .... *Sceloporus*
13. Escamas dorsales del cuerpo no aquilladas o sólo las escamas vertebrales ligeramente aquilladas 14
14. Cabeza grande, notable y abruptamente más ancha que el cuello; escamas alargadas en el margen de los párpados inferiores, pero no en el de los superiores .... *Crotaphytus*
14. Cabeza diferente; escamas alargadas en el margen de ambos párpados o en ninguno 15
15. Escamas de la cabeza muy agrandadas, en forma de placa, y reducidas en número 16
15. Escamas de la cabeza numerosas y de tamaño más uniforme 18
16. Escamas ventrales del cuerpo agrandadas, rectangulares e imbricadas; escamas dorsales del cuerpo granulares .... *Cnemidophorus*
16. Escamas ventrales del cuerpo pequeñas y no rectangulares 17
17. Una sola banda dorsal negra y completa justo delante de las extremidades anteriores .... *Petrosaurus*
17. Sin una banda así .... *Urosaurus*

18. Supralabiales marcadamente aquilladas y oblicuas; mandíbula inferior ligeramente más chica que la superior, por lo que se inserta en ella 19
18. Supralabiales no aquilladas; mandíbula inferior del mismo tamaño que la superior 20
19. Una sola mancha obscura en cada lado del abdomen, equidistante entre las extremidades posteriores y anteriores; cuerpo muy ancho y aplanado .... *Uma notata*
19. Dos o tres manchas obscuras en cada costado, generalmente más cerca de las extremidades anteriores; cuerpo no ancho .... *Callisaurus draconoides*
20. Cuerpo aplanado y gordo con grandes pliegues de piel suelta a los lados; escamas muy agrandadas en la orilla anterior del oído .... *Sauromalus*
20. Diferentes .... *Gambelia*

CLAVE POR LAS ESPECIES

ANNIELLA

1. Rostro puntiagudo en perfil lateral; de las escamas supralabiales, la cuarta es la más grande .... *geronimensis*
1. Rostro redondo en perfil lateral; de las escamas supralabiales, la segunda es la más grande .... *pulchra*

CNEMIDOPHORUS

1. Escama frontoparietal entera 2
1. Escama frontoparietal dividida 8
2. Con líneas transversales claras en los costados .... *ceralbensis*
2. Sin líneas transversales claras 3
3. Vientre azulado; todas las escamas de la región gular pequeñas 5
3. Vientre grisáceo o anaranjado; algunas escamas de la región gular grandes 4
4. Vientre anaranjado en machos adultos ... *hyperythrus*
4. Vientre grisáceo en machos adultos .... *espiritensis*
5. Sin rayas en la cola 6
5. Con rayas en la cola 7
6. Rayas dorsales extremadamente tenues o ausentes ... . *pictus*
6. Rayas dorsales bien marcadas .... *carmenensis*
7. Extremidades posteriores con manchas claras .... *franciscensis*
7. Extremidades posteriores sin manchas claras .... *danheimae*
8. Todas las escamas de la región gular pequeñas .... *labialis*
8. Algunas escamas de la región gular grandes 9
9. Vientre gris con manchitas negras 10
9. Vientre negro 11
10. Rayas dorsales bien marcadas en adultos; color de fondo del dorso casi negro .... *celeripes*
10. Los adultos tienen rayas dorsales, pero poco marcadas y angostas; patrón dorsal claro y frecuentemente reticulado .... *tigris*
11. Dorso canelo-amarillento, casi unicolor, con reticulaciones tenues y finas .... *canus*
11. Dorso obscuro con reticulaciones finas o manchitas claras 12
12. Manchitas claras en el dorso 13
12. Dorso obscuro, con reticulaciones finas un poco más claras .... *martyris*
13. Parte anterior del cuerpo con manchitas claras .... *catalinensis*
13. Parte anterior del cuerpo sin manchitas claras .... *bacatus*

COLEONYX

1. Cuerpo con tubérculos 2
1. Cuerpo sin tubérculos .... *variegatus*
2. Más de 41 escamas en el margen de los párpados .... *gypsicolus*
2. Menos de 42 escamas en el margen de los párpados ... . *switaki*

CROTAPHYTUS

1. Collar negro posterior notablemente más ancho que el collar negro anterior .... *dickersonae*
1. Collar negro posterior un poco más angosto que el collar negro anterior o tan ancho como él 2
2. Dorso con bandas transversales blancas conspicuas ... . *vestigium*
2. Dorso con manchas, separadas o fusionadas, formando líneas sinuosas cortas 3
3. Collar negro posterior casi completo sobre el dorso; extremidades anteriores sin manchas .... *grismeri*
3. Collar negro posterior ausente o muy corto; extremidades anteriores manchadas .... *insularis*

CTENOSAURA

1. Bandas claras en las extremidades y en la región posterior al tórax presentes durante toda la vida .... *hemilopha*
1. En los adultos, la región posterior al tórax es unicolor o moteada tenuemente 2
2. Las manchas redondas y obscuras en la superficie ventral de las extremidades posteriores de los juveniles se mantienen durante la vida adulta .... *conspicuosa*
2. Lunares, pero no manchas redondas, en la superficie ventral de las extremidades posteriores de los adultos .... *nolascensis*

DIPSOSAURUS

1. Región gular clara con rayas longitudinales obscuras . . . . *dorsalis*
1. Región gular color café fuerte casi uniforme . . . . *catalinensis*

ELGARIA

1. Escamas del cuerpo, extremidades y cola marcadamente aquilladas; sin manchas blancas y negras conspicuas en las escamas labiales 2
1. Quillas poco desarrolladas; manchas blancas y negras conspicuas en las escamas labiales 3
2. Longitud hocico-cloaca (LHC) de adultos menor a 115 mm; en las Islas Coronado . . . . *nana*
2. LHC de adultos mayor a 116 mm . . . . . *multicarinata*
3. 14 hileras longitudinales de escamas dorsales . . . . *cedrosensis*
3. 16 hileras longitudinales de escamas dorsales 4
4. Ni las crías ni los adultos tienen bandas claras transversales . . . . *paucicarinata*
4. Las crías y casi siempre los adultos tienen bandas claras transversales . . . . *velazquezi*

EUMECES

1. Cola generalmente rojiza; cuerpo unicolor . . . *gilberti*
1. Cuerpo rayado 2
2. Cola rojiza o entre azul pálido y azul brillante . . . . *skiltonianus*
2. Cola rosa brillante o morada 3
3. Última escama supralabial en contacto amplio con la escama supratemporal secondaria superior . . . . *lagunensis*
3. Última escama supralabial no en contacto amplio con la escama supratemporal secondaria superior . . . . *gilberti*

GAMBELIA

1. Dorso pálido con muchos puntitos arreglados de manera asimétrica en la cabeza y el cuerpo . . . . *wislizenii*
1. Dorso obscuro con manchas paravertebrales en pares . . . . *copeii*

PETROSAURUS

1. Escamas dorsales de la cola aproximadamente del mismo tamaño que las escamas dorsales del cuerpo y no marcadamente aquilladas o espinosas 2
1. Escamas de la cola mucho mayores que las del cuerpo y marcadamente aquilladas y espinosa 3
2. Con una raya postorbital delgada conspicua; una escama entre la escama nasal y las escamas supralabiales . . . . *repens*
2. Sin raya postorbital; dos escamas entre la escama nasal y las escamas supralabiales . . . . *thalassinus*
3. Región gular con manchas redondas blancas ampliamente separadas . . . . *mearnsi*
3. Una retícula clara y densa sobrepuesta a la región gular . . . . *slevini*

PHRYNOSOMA

1. Bases de las espinas occipitales conectadas, formando una corona . . . . *solare*
1. Bases de las espinas occipitales separadas 2
2. Tímpano delgado y de apariencia normal; 6–8 hileras longitudinales de escamas gulares agrandadas . . . . *coronatum*
2. Tímpano grueso y cubierto de escamas que a veces no se diferencian de las escamas circundantes, por lo que el tímpano es difícil a ver; escamas gulares pequeñas, con solo 1–2 hileras de escamas agrandadas 3
3. Con un línea vertebral obscura y delgada en el dorso . . . . *mcallii*
3. Sin línea vertebral en el dorso . . . . *platyrhinos*

PHYLLODACTYLUS

1. Cuerpo con tubérculos 2
1. Cuerpo sin tubérculos . . . . *unctus*
2. Escamas alrededor del hocico al nivel de la tercera escama supralabial más numerosas que las escamas interorbitales 3
2. Escamas alrededor del hocico al nivel de la tercera escama supralabial menos numerosas que las escamas interorbitales 4
3. Dorso color café obscuro . . . . *partidus*
3. Dorso color grisáceo . . . . *homolepidurus*
4. Escamas del abdomen grandes . . . . *bugastrolepis*
4. Escamas del abdomen no grandes . . . . *xanti*

SAUROMALUS

1. El patrón de color es de manchas irregulares grandes entre rojizo-cafés y negras, sobre un fondo amarillo . . . . *varius*
1. Unicolor o con bandas 2
2. Escamas del cuello, el cuerpo, las extremidades y la cola muy espinosas . . . . *hispidus*
2. Las escamas del cuello pueden ser espinosas; las del cuerpo, las extremidades y la cola generalmente lisas 3
3. Cuerpo sin bandas 4
3. Cuerpo frecuentemente con bandas . . . . *obesus*

4. Escamas de la nuca homogéneas .... *slevini*
4. Escamas de la nuca heterogéneas, con algunas más pequeñas entre las grande .... *klauberi*

SCELOPORUS

1. Escamas de los costados mucho más pequeñas y marcadamente diferenciadas de las escamas dorsales 2
1. Escamas de los costados de casi el mismo tamaño que las dorsales 3
2. Escamas de los costados bien diferenciadas de las dorsales .... *grandaevus*
2. Las escamas de los costados se transforman gradualmente en dorsales .... *angustus*
3. Escamas dorsales relativamente pequeñas, con margen posterior espinoso no levantado, lo que hace que la espalda se vea lisa en perfil lateral; escamas de la cola de casi el doble del tamaño de las de la espalda .... *vandenburgianus*
3. Escamas dorsales grandes, con margen posterior espinoso levantado, lo que hace que la espalda se vea espinosa en perfil lateral; escamas de la cola del mismo tamaño de las de la espalda o sólo un poco más grandes 4
4. Con bandas blancas y negras conspicuas en las extremidades anteriores .... *clarki*
4. Sin bandas conspicuas en las extremidades anteriores 5
5. Con una mancha negra conspicua en el hombro .... *occidentalis*
5. Sin manchas negras conspicuas en el hombro 6
6. Escamas dorsales del cuerpo anchas, aplanadas y ligeramente aquilladas, con una espina corta proyectada posteriormente; mancha negra del hombro sin borde posterior claro .... *orcutti*
6. Escamas dorsales del cuerpo marcadamente aquilladas y más alargadas,debido a una espina prominente proyectada hacía atrás; mancha negra del hombro con borde posterior blanco 7
7. Primera escama sublabial en contacto con la escama del mentón, cuando menos en un lado del cuerpo 8
7. Primera escama sublabial no en contacto con la escama del mentón 10
8. Con rayas postorbitales negras delgadas .... *magister*
8. Sin rayas postorbitales negras delgadas 9
9. Dorso uniformemente dorado .... *lineatulus*
9. Dorso diferente .... *zosteromus*
10. Escama supraocular posterior en contacto con la escama superciliar .... *hunsakeri*
10. Escama supraocular posterior separada de la escama superciliar por pequeñas escamas supernumerarias ... . *licki*

UROSAURUS

1. Pliegue dorsolateral del cuerpo con escamas agrandadas o con tubérculos; escama frontal generalmente dividida 2
1. Pliegue dorsolateral del cuerpo sin escamas agrandadas ni tubérculos 3
2. Escamas supraoculares marcadamente convexas .... *ornatus*
2. Escamas supraoculares ligeramente convexas o planas .... *graciosus*
3. Con rayas paravertebrales longitudinales negras sinuosas .... *lahtelai*
3. Sin rayas paravertebrales .... *nigricaudus*

UTA

1. 75 o más escamas ventrales (contadas a medio vientre entre el pliegue gular y la apertura cloacal) .... *palmeri*
1. Menos de 75 escamas ventrales 2
2. Dorso casi uniformemente verde; vientre azulado .... *nolascensis*
2. Dorso diferente 3
3. Manchas de color de turquesa encima de la cabeza y del rostro .... *squamata*
3. Cabeza y rostro sin manchas de color de turquesa ... 4
4. Región prefrontal plana .... *stansburiana*
4. Región prefrontal marcadamente convexa en perfil lateral 5
5. Dos escamas prefrontales entre las escamas frontonasal y frontal anterior, por lo menos en un lado de la cara .... *tumidarostra*
5. Una escama prefrontal entre las escamas frontonasal y frontal anterior 6
6. Escamas frontoparietales no en contacto en su parte media, o solo ligeramente en contacto .... *lowei*
6. Escamas frontoparietales ampliamente en contacto en su parte media .... *encantadae*

XANTUSIA

1. Cuerpo muy plano; extremidades no cortas; dorso cubierto de manchas obscuras grandes de forma irregular .... *henshawi*
1. Cuerpo cilíndrico; extremidades cortas; dorso sin manchas grandes .... *vigilis*

SERPIENTES Y ANFISBÉNIDOS

CLAVE PARA LOS GÉNEROS

1. Con dos extremidades anteriores .... *Bipes biporus*
1. Sin extremidades 2

2. Cuerpo con bandas rojo-naranjas, negras y blancas, estando las rojas y blancas en contacto .... *Micruroides euryxanthus*
2. Cuerpo diferente 3
3. Cola comprimida lateralmente y con un patrón reticulado de amarillo y negro .... *Pelamis platurus*
3. Cola cilíndrica en corte transversal 4
4. Escamas ventrales del mismo tamaño y forma que las dorsales .... *Leptotyphlops humilis*
4. Escamas ventrales alargadas transversalmente 5
5. Con una foseta profunda conspicua en la región loreal, entre el ojo y la fosa nasal (o nostrilo) .... *Crotalus*
5. Sin foseta loreal 6
6. Sin placas grandes en la cabeza; escamas ventrales pequeñas .... *Lichanura trivirgata*
6. Con placas grandes sobre la cabeza; escamas ventrales grandes 7
7. Escama anal entera 8
7. Escama anal dividida 14
8. Todas o algunas de las escamas dorsales bien aquilladas 9
8. Todas las escamas dorsales lisas, o no más de una serie de escamas vertebrales ligeramente aquilladas 10
9. 27 o más hileras de escamas dorsales en la mitad del cuerpo; dorso con manchas obscuras .... *Pituophis*
9. Menos de 27 hileras de escamas dorsales en la mitad del cuerpo; dorso rayado, por lo menos en los costados .... *Thamnophis*
10. Escama rostral muy agrandada y triangular extendiéndose ampliamente en la región prefrontal; parte anterior de la cabeza cuadrada en perfil lateral .... *Phyllorhynchus decurtatus*
10. Escama rostral diferente; cabeza de forma normal en perfil lateral 11
11. Ojos grandes y protuberantes; cabeza triangular, mucho más ancha que el cuello; pupilas elípticas .... *Trimorphodon biscutatus*
11. Ojos normales; cabeza no triangular; pupilas redondas 12
12. Escama rostral agrandada y proyectada hacía adelante; 50 por ciento o menos de las escamas subcaudales divididas .... *Rhinocheilus*
12. Escama rostral normal; 50–100 por ciento de las escamas subcaudales divididas 13
13. Banda obscura sobre la cabeza conectando los ojos; dorso manchado; vientre sin manchas obscuras .... *Arizona*
13. Sin banda obscura sobre la cabeza; dorso bandeado o rayado; vientre con manchas obscuras ... *Lampropeltis*
14. Pupilas elípticas 15
14. Pupilas redondas 17
15. Collar prominente en tres partes; la raya obscura en el lado de la cabeza cruza sobre el ojo ...... *Hypsiglena*
15. Collar reducido o ausente; sin raya en el lado de la cabeza 16
16. Mancha clara en forma de lira encima de la cabeza detrás de los ojos .... *Trimorphodon biscutatus*
16. Cabeza moteada .... *Eridiphas*
17. Escama rostral muy agrandada; con una raya a lo largo del dorso .... *Salvadora hexalepis*
17. Escama rostral no agrandada 18
18. Con escamas suboculares; cuerpo rojizo-café .... *Bogertophis rosaliae*
18. Sin escamas suboculares 19
19. Sin escama loreal 20
19. Con escama loreal 21
20. Mandíbula inferior más corta que la superior; la mitad delantera de la cabeza no negra .... *Chilomeniscus*
20. Mandíbula inferior del mismo tamaño que la superior; cabeza enteramente negra .... *Tantilla planiceps*
21. Hocico aplanado, parecido a una pala en perfil lateral; mandíbula inferior más corta que la superior .... *Chionactis occipitalis*
21. Hocico de forma normal; mandíbula inferior del mismo tamaño que la superior 22
22. Anillo anaranjado en el cuello .... *Diadophis punctatus*
22. Sin anillo en el cuello 23
23. Cabeza completamente negra .... *Thamnophis validus*
23. Cabeza no negra 24
24. Con una cresta conspicua en el ángulo del ojo .... *Masticophis*
24. Sin cresta en el ángulo del ojo .... *Sonora semiannulata*

ARIZONA

1. 51–83 manchas transversales cortas en el cuerpo .... *elegans*
1. 36–41 manchas subcirculares alargadas en el cuerpo .... *pacata*

CHILOMENISCUS

1. Escamas prefrontales en contacto en su línea media .... *stramineus*
1. Escamas prefrontales no en contacto .... *savagei*

CROTALUS

1. Cascabel reducido, sin segmentos sueltos (o sólo uno, raramente dos) .... *catalinensis*
1. Cascabel normal o solamente un poco reducido, con

al menos un segmento suelto después de la etapa juvenil 2
2. Color de fondo del dorso rojizo; primer par de escamas infralabiales divididas transversalmente; bandas blancas y negras bien definidas en la cola 3
2. Color de fondo del dorso cenizo o café claro; primer par de escamas infralabiales no divididas transversalmente; bandas negras y blancas en la cola con márgenes difusos 4
3. Anillos negros de la cola casi del mismo ancho o un poco más delgados que los anillos blancos de la cola . . . . . *ruber*
3. Anillos negros de la cola mucho más angostos que los anillos blancos de la cola . . . . . *lorenzoensis*
4. Orillas laterales de las escamas supraoculares visiblemente elevadas 5
4. Orillas laterales de las escamas supraoculares no elevadas 6
5. Orillas laterales de las escamas supraoculares en forma de cuerno . . . . . *cerastes*
5. Orillas laterales de las escamas supraoculares no en forma de cuerno . . . . . *enyo*
6. Muchos puntos obscuros en el dorso, que le dan una apariencia jaspeada; escamas prenasales generalmente separadas de la escama rostral 9
6. Dorso no jaspeado; escamas prenasales generalmente en contacto con la escama rostral 7
7. LHC de adultos menor a 637 mm; de Isla El Muerto . . . . . *muertensis*
7. LHC de adultos mayor a 636 mm; no de Isla El Muerto 8
8. LHC de adultos mayor a 1,114 mm; de Isla Ángel de la Guarda . . . . . *angelensis*
8. LHC de adultos menor a 1,113 mm; no de Isla Ángel de la Guarda . . . . . *mitchellii*
9. Generalmente 3 o más escamas internasales 10
9. Generalmente 2 escamas internasales 11
10. LHC de adultos menor a 675 mm; de Isla Coronado Sur . . . . . *caliginis*
10. LHC de adultos mayor a 674 mm . . . . . *viridis*
11. Bandas caudales blancas y negras muy contrastantes 12
11. Bandas caudales no muy contrastantes 13
12. Con escama loreal superior . . . . . *atrox*
12. Sin escama loreal superior . . . . . *tortugensis*
13. Cola negra o café 14
13. Cola diferente . . . . . *tigris*
14. Región prefrontal muy obscura . . . . . *molossus*
14. Región prefrontal ligeramente obscurecida . . . . . *estebanensis*

ERIDIPHAS

1. Más de 196 escamas ventrales . . . . . *marcosensis*
1. Menos de 197 escamas ventrales . . . . . *slevini*

HYPSIGLENA

1. Por lo menos un par de escamas gulares de casi igual tamaño que los escudos de la barbilla . . . . . *gularis*
1. Escamas gulares más pequeñas que los escudos de la barbilla . . . . . *torquata*

LAMPROPELTIS

1. Con una banda blanca prominente en la parte posterior de la cabeza en contacto con las escamas parietales 2
1. Sin banda blanca en la parte posterior de la cabeza 3
2. Con bandas rojas; con rojo en el rostro . . . . . *zonata*
2. Sin bandas rojas; rostro sin rojo . . . . . *herrerae*
3. Dorso con bandas café-negras y blanco-amarillas alternadas; o con algún grado de rayado . . . . . *getula*
3. Dorso diferente; costados cubiertos de manchas amarillas pequeñas; color de fondo del dorso desde morado hasta negro . . . . . *catalinensis*

MASTICOPHIS

1. Con una raya dorsolateral conspicua en las hileras 3 y 4 de escamas dorsales; las rayas pueden o no extenderse a lo largo del cuerpo 2
1. Sin rayas en las hileras 3 y 4 de escamas dorsales . . . 6
2. En el tercio anterior la raya dorsolateral tiene anchura uniforme; con manchas obscuras en el mentón y la garganta . . . . . *lateralis*
2. En el tercio anterior la raya dorsolateral se ensancha a intervalos regulares de cada 2–7 hileras de escamas; barbilla y garganta generalmente sin manchas 4
3. Manchas obscuras en las primeras 10–15 escamas ventrales . . . . . *bilineatus*
3. Manchas obscuras en por lo menos las primeras 40 escamas ventrales . . . . . *slevini*
4. Las rayas dorsolaterales se extienden a lo largo del cuerpo 5
4. Las rayas dorsolaterales se decoloran en la parte posterior del cuerpo . . . . . *aurigulus*
5. En su porción anterior las rayas dorsolaterales tienen anchura uniforme; sin bandas transversales en la nuca . . . . . *lateralis*
5. En su porción anterior las rayas dorsolaterales se ensanchan a intervalos regulares; con bandas transversales en la nuca . . . . . *barbouri*

6. Con bandas obscuras delgadas en patrón de zigzag siguiendo la orilla de las escamas, o cabeza y cuerpo enteramente negros .... *fuliginosa*
6. Color de cuerpo diferente .... *flagellum*

PITUOPHIS

1. Con una raya subcaudal negra 2
1. Sin una raya subcaudal .... *catenifer*
2. Más de 52 manchas obscuras dorsales en el cuerpo ... . *insulana*
2. Menos de 53 manchas dorsales en el cuerpo .... *vertebralis*

RHINOCHEILUS

1. Escama loreal cuadrada; manchas dorsales negras del cuerpo no en contacto con las escamas ventrales .... *etheridgei*
1. Escama loreal rectangular; manchas dorsales negras del cuerpo en contacto con las escamas ventrales .... *lecontei*

THAMNOPHIS

1. Con raya vertebral 2
1. Sin raya vertebral 3
2. Raya dorsal de 2–3 hileras de escamas de anchura; sin manchas negras conspicuas en los costados ... *elegans*
2. Raya dorsal de 1 hilera de escamas de anchura; con manchas negras conspicuas en los costados .... *marcianus*
3. Color de fondo del dorso verdoso obscuro; cabeza sin negro .... *hammondii*
3. Color de fondo del dorso negro; cabeza negra .... *validus*

GLOSSARY

The definitions offered here are specific to this book. Many of these terms are used differently by different authors. For a more comprehensive list of terms, see Peters 1964.

abiotic *adj* Referring to the nonliving components of the environment.

adpress *v* To press closely against the body: refers to measurements and counts made with the limbs pressed against the body.

allopatric *adj* Pertaining to allopatry; nonoverlapping.

allopatry *n* The condition in which the geographic distributions of two or more populations do not overlap.

alveolar surface *n* Area of the jawbone usually occupied by teeth. In turtles, the chewing surfaces of the jaws.

ambient *adj* Referring to the surrounding environmental conditions.

ambush feeders *n* Predators that remain motionless and wait for prey to come within the range of a strike or lunge, in contrast to active foragers, which move through the environment searching for prey.

amplexus *n* The sexual embrace of a female anuran by a male using his forelimbs.

anal plate *n* The large scale of a snake that forms the anterior margin of the anus or vent.

annulus (pl. annuli) *n* The body segment of an amphisbaenian. Primary annuli are defined by integumentary grooves that encircle the body. Secondary annuli are defined by grooves that do not completely encircle the body.

antefemoral *adj* Anterior to the femoral region.

antehumeral *adj* Anterior to the humeral region.

anterior *adj* Of or pertaining to the front.

anteroventral *adj* Of or pertaining to the anterior region of the ventral surface of a body structure.

anticyclonic *adj* Referring to hurricanes north of the equator in the eastern Pacific that spin counterclockwise.

Anura *n* The taxonomic order that contains frogs and toads.

apical *adj* Of, at, or constituting the apex of a structure (e.g., scales).

apical pit *n* A pit or depression occurring at the posterior tip of a scale.

arboreal *adj* Referring to an organism that spends a large portion of its time off the ground in bushes or trees and usually manifests some adaptations for doing so.

arenicolous *adj* Referring to an organism that spends a large portion of its time living in or on sand and usually manifests some adaptations for doing so.

auricular scales *n* A series of enlarged scales usually forming the anterior border of the ear opening in lizards.

axilla *n* Posterior angle formed by the body and the forelimb (armpit).

axillary *adj* Of or pertaining to the axilla.

azygous *adj* Referring to unpaired scales on the midline of a structure and thus lacking a counterpart.

basal *adj* Referring to the earliest lineage of a group to have evolved.

beak *n* A structure projecting or ending in a point that is associated with the rostrum (e.g., the snout of turtles).

bicuspid *adj* Of or pertaining to having two toothlike projections; referring to the paired terminal toothlike projections on the tomial surfaces of some turtles.

Bidder's organ *n* The upper cortical lobe of the progonad in the Bufonidae, which retains its ovarial nature in males even though the rest of the gonad differentiates into a testis.

biogeography *n* The science that studies the geographic distribution of organisms.

biotic *adj* Referring to the living components of the environment.

body fields *n* The regions between the body stripes on the dorsum of an organism; referring to the different regions between the body stripes in the whiptails *(Cnemidophorus).*

brachium *n* The upper portion of the forelimb above the elbow.

bridge *n* The portion of a turtle's shell that connects the carapace and the plastron.

buccal cavity *n* The oral cavity.

canthal *adj* Of or pertaining to the canthal ridge or the region of the head between the tip of the snout and the eye.

canthal ridge *n* The angle between the top of the head and the side of the snout.

canthus rostralis *n* *See* canthal ridge.

carapace *n* The upper portion of a turtle's shell.

carapace length *n* A measurement of a turtle shell extending in a straight line from the anterior margin of the nuchal plate to the posterior margin of the supracaudal plate.

carnivorous *adj* Eating or living on flesh; referring to species that eat other vertebrates.

caudal *adj* Of or pertaining to the tail.

caudal whorls *n* A series of differentiated caudal scales arranged in whorls encircling the tail, such as those in the spiny-tailed iguanas *(Ctenosaura).*

cephalic *adj* Of or pertaining to the head.

chemosensory *adj* Of or pertaining to taste or olfaction (smell).

chinshield *n* The paired elongate scales on the lower jaws of snakes.

cismontane *adj* Referring to geographic regions west of the Peninsular Ranges.

clinal *adj* Referring to the gradual nature of the geographic variation of a species' characteristic (e.g., color pattern or a particular scale count) from one region to the next.

cloaca *n* The terminal portion of the gastrointestinal tract that receives the contents of digestion, excretion, and, in males, semen. It opens to the environment through the vent or anus.

cloacal spur *n* One of a pair of upward-curving projections at the base of the tail in male banded geckos *(Coleonyx).*

coccyx *n* The postsacral fusion of the vertebrae.

conspecific *adj* Of the same species.

continental *adj* Of or pertaining to a continent or to an island that was separated from a continental margin because of geological faulting.

coracoid *n* One of the bones of the pectoral girdle (the series of bones in the thoracic region that anchor the forelimbs).

coracoid fenestra *n* An emarginated region of the coracoid.

cornified *adj* Covered by or impregnated with keratin.

copora lutea *n* The remnants of a ruptured egg follicle following ovulation.

cosmopolitan *adj* Occurring worldwide.

costal *adj* Of or pertaining to a rib or the sides of the body.

costal groove *n* Vertical integumentary groove along the side of the body in salamanders.

cotype *n* One of a pair of specimens that constitutes the type series of a taxon.

countershading *adj* Referring to the highlighting of the dorsal bands in lizards, usually by the scales immediately posterior to the bands being white or black, as in alligator lizards *(Elgaria).*

countersunk *adj* Referring to the lower jaw being enclosed within the upper jaw when the mouth is shut, as in sand snakes *(Chilomeniscus).*

cranial *adj* Of or pertaining to the skull.

cranial crest *n* An elevated, bony ridge on the dorsal surface of the skull in some toads.

crepuscular *adj* Active at dawn and dusk.

crypsis *adj* Visual concealment by means of camouflage.

cutaneous *adj* Of or pertaining to the skin.

cycloid *adj* Referring to the semicircular shape of the scales in skinks *(Eumeces).*

dentary *n* A bone in the lower jaw.

denticle *n* One of the toothlike structures surrounding the mouth of some tadpoles.

denticulate *adj* Of or pertaining to the toothlike projections along the edge of a structure.

depauperate *adj* Of or pertaining to a region with a low number of species.

depositional filling *n* The process by which a geographical basin is filled with earthen deposits over a long period.

dermal platelet *n* A series of small dermal ossifications in the carapace of hatchling and juvenile Leatherback Sea Turtles *(Dermochelys coriacea).*

diphasic *n* Having two phases: for example, the mating call of the male Pacific Treefrog *(Hyla regilla)* has two parts.

direct development *n* Complete embryonic development within the egg of some amphibians, with no free-living larval stage.

discrete diagnosability *n* The ability to differentiate between populations by nonoverlapping morphological characteristics within their range of variation. For example, one population of lizards may have 20 to 25 subdigital lamellae on the fourth toe, and another may have 28 to 32 subdigital lamellae.

disjunct *adj* Referring to geographically separated distributions (*see* allopatry).

disk *n* The adhesive, flat circular structure on the tip of the digits of some frogs.

distal *adj* Situated away from the base or point of attachment of a structure.

diurnal *adj* Active during the day.

diversity *n* The number of species in an area.

dorsal *adj* Of or pertaining to the top of a body structure.

dorsolateral *adj* Of or pertaining to the upper portion of the side of a structure.

dorsolateral fields *n* The area on the body between the paravertebral and lateral body stripes. *See* body fields.

dorsolateral fold *n* A fold of skin along the upper portion of the side of the body. Commonly found in frogs of the group Ranidae.

dorsoposterior *adj* Of or pertaining to the upper portion of the rear of a body structure.

dorsoventrally compressed *adj* Of or pertaining to a body structure flattened from top to bottom.

dorsum *n* The upper surface of the body or a body structure.

ectopterygoid *n* A dermal bone in the palate of most reptiles.

edentate *adj* Of or pertaining to an absence of teeth.

elephantine *adj* Elephant-like. Used herein to describe the shape of the hind limbs of desert tortoises *(Gopherus).*

emarginate *adj* Of or pertaining to the margin or edge of a structure that is indented or notched.

endemic 1. *adj* Found only in a particular

area or region. 2. *n* A population or taxon that is restricted to a particular region or area, as on an island.

Eocene *n* A geological era beginning 35 million years ago and ending 55 million years ago.

epidermal *adj* Of or pertaining to the outermost layer of skin (the epidermis). Many structures, such as scales, claws, feathers, and tubercles, are epidermal derivatives.

evolutionary species concept *n* A philosophy of systematics that recognizes genetically isolated populations (which have no gene flow in or out) as species.

exostosed *adj* Having undergone exostosis, which refers here to the roughening of the dorsal surface of the skull in some toads, with subsequent fusion to the dermis.

extirpated *adj* Having undergone extirpation, the local extinction of a population of a species from a single area but not extinction of the species itself.

eyelid fringe scales *n* The series of outermost scales on the margins of the eyelids.

femoral *adj* Of or pertaining to the femur or femora.

femoral pore *n* One of a series of pores on the posteroventral surface of the femora in some species of lizards. They are usually better developed in males than in females.

femur *n* The bone of the upper hind limb; thigh bone.

foramen *n* An opening (usually in bone) through which pass various elements, such as blood vessels and nerves.

fossorial *adj* Pertaining to burrowing or digging; refers here to species that spend a great deal of time underground.

frontal *adj* Of or pertaining to the front of the head, anterior and dorsal to the eyes.

frontal scale *n* A large unpaired scale on the dorsal surface of the head.

frontonasal scale *n* A median scale on the dorsal surface of the head in lizards and turtles that usually lies between the prefrontals, loreals, and internasal.

frontoparietal scale *n* A median scale (or paired scales) lying between the frontal and parietal scales of some lizards.

fusiform *adj* Tapering from the middle toward each end.

generic *adj* Of or pertaining to genera.

genial scale *n* *See* chinshield.

geomorphological *adj* Of or pertaining to the development, configuration, and distribution of the surface features of the earth.

glandular *adj* Of or pertaining to glands. Often used to describe the nature of the skin in amphibians.

granular *adj* Possessing granules. Used to describe the nature of the epidermal scales in many reptiles.

granules *n* Very small flat scales in reptiles or small round elevations in the skin of some species of frogs.

grass-swimming *n* A type of locomotion in limbless or nearly limbless reptiles wherein sinuous curves of the body are used to propel the body through thick grass. In lizards, the reduced limbs are pressed against the sides of the body during locomotion.

gravid *adj* Of or pertaining to a female with eggs.

groin *n* The angle formed by the body and the anterior surface of the hind limb at its insertion.

gular *adj* Of or pertaining to the ventral surface of the head between the lower jaws, throat, and neck.

gular fold *n* A transverse fold of skin found in the gular region of some salamanders and lizards.

hatchling *n* Newly hatched animal.

hemipenis (pl. hemipenes) *n* One of the paired copulatory organs of male snakes and lizards.

herbivorous *adj* Feeding only on plant material.

heterogeneous *adj* Consisting of parts or elements that are different or unrelated.

holotype *n* The single specimen designated by the author (describer) of a taxon as the exemplar of that taxon.

homogeneous *adj* Consisting of the same, similar, or related elements.

imbricate *adj* Overlapping; referring to the distal portion of one scale overlapping the proximal portion of another.

immaculate *adj* Without mark; referring to color patterns or portions of color patterns composed of a single color.

infrahumeral *adj* Of or pertaining to the ventral surface of the upper portion of the forelimb.

infralabial scale *n* One of a series of epidermal scales bordering the lower lips of lizards and snakes.

inframarginal plate *n* The epidermal plates or scutes lying between the marginal plates of the carapace and the adjacent plastral plates in turtles.

inguinal *adj* Of or pertaining to the groin.

inguinal fold *n* A fold of skin in the groin region of lizards of the lineage Xantusiidae.

inner tarsal tubercle *n* A tubercle on the inner or medial side of the palm in some species of anurans.

insular *adj* Of or pertaining to islands. Organisms occurring on islands are said to be insular.

insectivorous *adj* Feeding only on insects, which here includes all arthropods.

intercalary cartilage *n* A small cartilage between the last and second-to-last phalanges in the digits of some frogs.

intergrade 1. *n* An individual from an intermediate population (intergradient). 2. *v* To gradually merge or blend into one another. Pertains here to the intermediate morphology and color pattern of geographically intermediate populations formed from different pattern classes in areas where their distributions come into contact.

internasal *n* Of or pertaining to the region between the external nares on the dorsal surface of the skull.

internasal scale *n* One of the two or more scales lying between the nasal scales in lizards and snakes.

interoccipital scale *n* A median scale on the dorsal surface of the head posterior to the interparietal in some lizards and snakes.

interoccipital spine *n* A median bony spine lying between the occipital spines in horned lizards *(Phrynosoma)*.

interorbital *adj* Of or pertaining to the dorsal region of the skull between the eyes.

interorbital bar *n* A dark transverse band across the top of the head that contacts the eyes in some snakes.

interorbital scale *n* One of a series of scales on the dorsal surface of the head between the eyes.

interspaces *n* The spaces between the dorsal body bands in some lizards and snakes.

interstitial granules *n* Very small, round scales found between larger scales or tubercles in some lizards.

iridophore *n* A type of chromatophore (cell with color pigments) that contains the iridescent pigment guanine and contributes to the overall coloration of an animal.

isthmus *n* A narrow strip of land connecting two larger landmasses.

jowls *n* The fleshy part beneath the lower jaw.

jugal *n* A skull bone forming part of the lower border of the orbit (the portion of the head in which the eye is located).

juxtaposed *adj* Placed side by side.

keel *n* A raised median ridge running down the long axis of a scale.

labial *adj* Of or pertaining to the lips.

labial tooth row *n* Small, keratinized toothlike structures surrounding the mouth in tadpoles. The length and number of labial tooth rows vary from species to species, and each combination is known as a different formula.

lamellae (sing. lamella) *n* The scales on the underside of the digits in lizards (i.e., subdigital lamellae).

landbridge *n* A young island (less than ten thousand years old) whose origin was the result of a rise in sea level, erosion, or a marine inundation.

lateral *adj* Of or pertaining to the side of a body or structure.

lateral field *n* The area between the paravertebral

and lateral body stripes in some lizards (e.g., *Cnemidophorus*).

lateral profile *n* A side view of a body or structure.

lectotype *n* One of a series of syntypes designated by an author, subsequent to the publication of the original description, to serve as the type specimen.

lineage *n* 1. A group of populations or taxa that share a most recent common ancestor (a clade). 2. A population of genetically cohesive individual organisms lacking external gene flow.

longitudinal dorsal scale row *n* A row of scales running parallel to the long axis of the body.

loreal *adj* Of or pertaining to the region on the side of the head between the eye and the nostril.

loreal pit *n* The external opening to the infrared heat receptors (pit organs) in the loreal region of pit vipers.

loreal scale *n* One of the scales on the lateral side of the head between the eye and the nostril.

marginal lamina *n* One of the outer epidermal plates along the edge of the carapace in turtles.

maxillae *n* In amphibians and reptiles, the two principal bones in the upper jaw, which usually bear teeth.

maxillary *adj* Of or pertaining to the maxillae.

medial *adj* Of or pertaining to the middle or midline of a body or structure.

mental *adj* Of or pertaining to the chin or the chin region.

mental gland *n* A gland in the chin region of some male salamanders.

mental groove *n* A medial groove lying between the chinshields in most snakes.

mental scale *n* An unpaired median scale at the anterior margin of the lower jaw.

mesic *adj* Characterized by a moderate supply of moisture. Used herein to denote nondesert regions.

mesophilic *adj* Of or pertaining to plants and animals that require a moderate amount of environmental moisture to survive.

mesoptychial scale *n* One of a series of scales that border the anterior margin of the gular fold in lizards.

metatarsal tubercle *n* An elevated epidermal thickening on the posterior portion of the bottom of the foot in anurans.

microhabitat *n* The immediate surroundings in which an organism lives.

middorsal *adj* Referring to the portion of the dorsal surface of a body or structure that lies along the midline.

midsagittal fold *n* A medial fold of skin in the gular region of some lizards.

midventral *adj* Referring to the portion of the ventral surface of a body or structure that lies along the midline.

Miocene *n* A geological era that began 25 million years ago and ended 5 million years ago.

monophyletic *adj* Having one origin: refers to a group of populations, taxa, or lineages that exclusively share a most recent common ancestor.

monotypic *adj* Having only one type: refers to taxa above the species level that contain only one species.

monsoon *n* A seasonal summer storm characterized by a high amount of precipitation in a relatively short period.

montane *adj* Of or pertaining to mountains.

morphology *n* The form of an organism, especially considered as a whole.

mucronate *adj* Having a spine projecting off the trailing edge, as in the scales of some lizards.

multicuspid *adj* Having more than one cusp, as on the occlusal margin of a tooth (the surface that contacts the opposing surface of another tooth).

myomere *n* A segment of trunk musculature.

nape *n* The back of the neck.

naris (pl. nares) *n* A nasal opening. Those that open to the outside of the body are referred to as nostrils or external nares.

nasal scale *n* A scale perforated by or bordering the external nares.

nasolabial groove *n* A thin, chemical-conducting depression or trough running from the margin of the upper lip to the external nares.

natural group *n* A group of taxa that are monophyletic.

neonate *n* Newly hatched or born organism.

nuchal *adj* Of or pertaining to the dorsal surface of the neck.

nuchal loop *adj* A light-colored band extending around the nuchal region, connecting one eye to the other in eublepharid geckos.

nuptial excrescence *n* A thick, dark, roughened epidermal pad usually found on the thumb of sexually active anurans.

obtuse *adj* Blunt or rounded at an extremity.

occipital *adj* Of or pertaining to the posterior portion of the skull.

oceanic *adj* Of or pertaining to the ocean; or to an island that has never had a continental connection.

ocellus (pl. ocelli) *n* A spot of color or a structure often resembling an eye.

ocular *adj* Of or pertaining to the eye.

ocular scale *n* A scale covering the eye in some snakes (e.g., *Leptotyphlops*).

olfaction *n* The process of smell.

olfactory *adj* Of or pertaining to the sense of smell.

omnivorous *n* Feeding on both plants and animals.

ontogenetic *adj* Of or pertaining to ontogeny; refers here to changes in an organism as it matures.

ontogeny *n* The developmental history of an individual from fertilization until death.

oral disk *n* The area in tadpoles that surrounds the mouth and supports various mouth parts.

oral papilla *n* One of the small elongate epidermal projections surrounding the mouth in tadpoles.

ossified *adj* Transformed into bone.

osteoderm *n* A small dermal ossification found beneath the scales in some reptiles (e.g., *Eumeces*).

oviductal egg *n* An egg within the oviduct, sometimes referred to as an oviducal egg.

palatine teeth *n* Teeth on the palatine bone of the skull (a bone that in part forms the roof of the mouth).

parallelism *n* In rattlesnakes, the point at which all rattle segments attain the same width.

parapatric *adj* Of or pertaining to geographic distributions that are adjacent or juxtaposed but not overlapping.

parasagittal fold *n* A longitudinal fold of skin lateral to the midgular region in some lizards (e.g., *Sauromalus*).

parasphenoid tooth patch *n* A patch of teeth on the parasphenoid bones of the skull (bones that in part form the roof of the mouth in amphibians), found only in plethodontid salamanders.

paratype *n* Any specimen or specimens of a type series not designated as the holotype.

paravertebral *adj* Of or pertaining to an area immediately lateral to the dorsal midline.

paravertebral field *n* The area between the vertebral and paravertebral body stripes in some lizards (e.g., *Cnemidophorus*).

paravertebral fold *n* A fold of skin extending along the paravertebral region of the body in some lizards (e.g., *Urosaurus*).

paravertebral stripe *n* A fold or stripe extending along the paravertebral region of the body in some lizards (e.g., *Cnemidophorus*).

parietal *adj* Of or pertaining to the posteromedial portion of the dorsal surface of the head.

parietal eye *n* A sensory structure in many lizards capable of light reception, located on the dorsal surface of the brain and opening to the outside through the parietal foramen of the skull.

parietal scale *n* 1. In lizards, a scale on the dorsal surface of the head posterior to the frontoparietal scales. 2. In snakes, one of a pair of large scales on the dorsal surface of the head immediately posterior to the frontal scale.

parietal sulcus *n* A narrow but prominent depression or furrow in the parietal region of some lizards (e.g., *Uta tumidarostra*).

parotid gland *n* A large glandular structure posterior to the eye and often elongate and extending onto the neck and/or shoulder. Best developed in some anurans (e.g., *Bufo*).

pattern class *n* A putatively genetically nonexclusive population of individual organisms from the same geographic region with similar distinguishing characteristics.

pectoral *adj* Of or pertaining to the chest.

pelagic *adj* Of or pertaining to the open ocean.

phalanx *n* One of the bones in the digits.

phylogeny *n* The history of the evolution of a species or monophyletic group of taxa.

phylogenetic *adj* Of or pertaining to phylogeny.

physiography *n* The study of the physical geography of a region.

phytogeography *n* The study of the geographical distribution and relationships of plants.

pit viper *n* Any venomous snake of the lineage Crotalinae (e.g., *Crotalus*).

plantar scale *n* A scale on the ventral surface of the foot.

plastral *adj* Of or pertaining to the plastron.

plastron *n* The ventral portion of the shell of turtles.

plate *n* Generally, a large epidermal scale.

Pleistocene *n* A geological era that began 1.9 million years ago and ended 100,000 years ago.

pleural scale *n* One of a series of epidermal laminae covering the ribs in the carapace.

Pliocene *n* A geological era that began 5 million years ago and ended 1.9 million years ago.

polymorphism *n* The presence of more than one morphological type in a population. Often refers to differences in anatomy or coloration between males and females of the same species.

postanal *adj* Of or pertaining to the region immediately posterior to the anus.

postanal scale *n* One of the enlarged scales immediately posterior to the anus in the males of some lizards (e.g., *Callisaurus draconoides*).

posterior *adj* Pertaining to the rear of a body or structure.

posterolateral *adj* Pertaining to the side of the rear of a body or structure.

posteromedial *adj* Pertaining to the midline or middle region of the rear of a body or structure.

postfemoral *adj* Pertaining to the posterior side of the femoral region.

postfemoral mite pocket *n* An invagination in the skin on the posterior surface of the femur at its insertion point on the body.

postgular fold *n* A transverse fold of skin posterior to the gular fold and immediately anterior to the forelimb insertion points on the body. In many lizards (e.g., *Sceloporus*), the postgular fold is incomplete medially.

postmental scale *n* One or a series of scales immediately posterior to and in contact with the mental scale.

postoccipital *adj* Of or pertaining to the area immediately posterior to the occipital region.

postocular *adj* Of or pertaining to the region immediately behind the eye.

postocular scale *n* One of the scales immediately posterior to the margin of the eye in lizards and snakes.

postparietal scale *n* One of the scales immediately behind and in contact with the parietal scale in lizards and snakes.

postrictal spine *n* An epidermal spine immediately posterior to the angle of the mouth in horned lizards (*Phrynosoma*).

postrostral scale *n* One of the scales immediately posterior to and in contact with the rostral scale.

post-thoracic *adj* Referring to the area immediately behind the thoracic region.

preanal *adj* Referring to the area immediately in front of the anus.

preanal pore *n* One of a series of openings of exocrine glands anterior to the anus or vent in the males of some species of lizards (e.g., *Coleonyx*).

prefrontal scale *n* One of a pair (or more) of epidermal scales on the dorsal surface of the head immediately in front of the frontal scale.

prehallux *n* An extra digit on the inner margin of the foot, usually appearing as a bony process on the hallux (the first or innermost digit on the hind foot), in many burrowing anurans.

prehensile *adj* Adapted for grasping or holding, especially by coiling around an object.

prenasal *adj* Of or pertaining to the area immediately in front of the nasal region.

prenasal scale *n* One of a number of scales immediately anterior to the nostrils.

preocular *adj* Of or pertaining to the region immediately in front of the eye.

protuberant *adj* Swollen beyond the surrounding surface.

proximal *adj* Situated near or toward the point of attachment.

proximate *adj* In immediate relation with something else.

pustule *n* A small wart, tubercle, or excrescence on the skin of amphibians.

rattle matrix *n* The fleshy terminus of the tail in rattlesnakes *(Crotalus)* on which new rattles are formed.

rear-fanged *adj* Referring to snakes in which one or more teeth at or near the posterior end of the maxillary bone are enlarged and grooved.

reticulate *adj* Referring to a net or meshlike color pattern.

retroarticular process *n* The projection on the posterior end of the jaw in reptiles to which jaw musculature attaches.

rostral *adj* Of or pertaining to the snout.

rostral scale *n* An unpaired scale at the tip of the snout that separates the left and right supralabial series.

rostrum *n* The snout.

rugose *adj* Rough, wrinkled, or folded.

sacral diapophysis *n* One of a pair of transverse processes of the sacral vertebrae that articulates with the pelvic girdle.

sacrum *n* The last single vertebra in the anuran vertebral column.

saxicolous *adj* Dwelling on or among rocks.

scale organ *n* Microscopic structure on the dorsal surface of the scales in reptiles.

scansorial *adj* Climbing.

scapular *adj* Of or pertaining to the scapula (shoulder blade).

secondary palate *n* A shelf of bone within the buccal or oral cavity serving to place the internal nares immediately anterior to the pharynx.

secondary upper supratemporal *n* The uppermost scale of a series of supratemporal scales, located immediately posterior to the primary supratemporals in skinks *(Eumeces)*.

serrated *adj* Sawtoothed in appearance.

sexual dichromatism *n* A color pattern difference between males and females of the same species.

sexual dimorphism *n* A morphological difference between males and females of the same species.

sinistral *adj* Of or pertaining to the left side. Often refers to the location of the spiracle in tadpoles.

sister group *n* The group with which another group shares a most recent common ancestor.

sister species *n* The species with which another species shares a most recent common ancestor.

snout-vent length (SVL) *n* The straight-line distance from the the tip of the snout to the posterior margin of the vent or anus.

spade *n* The dark, keratinized protuberance on the outer margin of the foot in burrowing anurans.

spectacle *n* The immovable transparent covering of the eye of snakes and some lizards, formed by the fusion of the eyelids.

spines *n* The elongated pointed projections extending off the posterior scale margin in some lizards (e.g., *Sceloporus*).

spinose *adj* Possessing spines.

spiracle *n* The opening of the gill chamber to the outside in tadpoles.

subarticular tubercle *n* A tubercle lying at the base of the digits in anurans.

subcaudal *adj* Of or pertaining to the ventral surface of the tail.

subcaudal scale *n* One of a series of scales on the ventral surface of the tail.

subdigital *adj* Of or pertaining to the ventral surface of the digit.

sublabial scale *n* One of a series of scales adjacent to and below the infralabial scales.

subocular *adj* Of or pertaining to the area immediately below the eye.

subocular scale *n* One of the scales bordering the lower margin of the eye.

subrictal spine *n* The spine located below the corner of the mouth and the rictal spine in horned lizards *(Phrynosoma).*

subspecies *n* A geographic subdivision within a species that is given taxonomic recognition.

subspecific epithet *n* The third name of a trinomial.

sulcus *n* A narrow channel or furrow in brain tissue.

superciliary scale *n* One of a series of scales lying along the outer margin of the supraocular scales in lizards; in snakes, a synonym of supraocular scale or superorbital scale.

supracaudal *adj* Of or pertaining to the dorsal surface of the tail.

suprahumeral *adj* Of or pertaining to the dorsal surface of the upper portion of the forelimb.

supralabial scale *n* One of the scales bordering the upper lip.

supranasal scale *n* A scale bordering the dorsal margin of the nostrils.

supranumerary tubercle *n* Any tubercle on the bottom of the hand or foot of an anuran that is not a subarticular tubercle.

supraocular *adj* Of or pertaining to the area immediately above the eye.

supraocular scale *n* 1. One of the scales lying along the outer margin of the eye in lizards. 2. One of a pair of enlarged scales lying immediately above the eye in some snakes (also known as supraorbital).

supraorbital semicircle *n* An arc of small scales bordering the larger supraocular scales in some lizards (e.g., *Sceloporus*).

supratibial *adj* Of or pertaining to the dorsal surface of the lower leg.

supratympanic fold *n* A fold of skin lying above the tympanum in some anurans.

suture *n* The region of contact between different scales and different bones.

SVL *See* snout-vent length

sympatric *adj* Having overlapping geographic distributions.

sympatry *n* An area of overlapping geographic distributions.

symphysis *n* The union of two opposing halves of a structure.

syntopic *adj* Occurring at the same place or in the same microhabitat.

syntype *n* Any member of the type series of a taxon whose author did not designate a holotype.

tarsal fold *n* A fold of skin running along the tarsus in some species of anurans.

tarsus *n* The hind limb segment lying between the tibia and fibula and the metatarsal bones; the ankle.

taxon *n* Any hierarchical category in a classification.

taxonomy *n* The area of systematic biology concerned with classification and the application of scientific names.

tectonic *adj* Relating to the deformation of the structure of the earth's crust.

temporal *adj* Of or pertaining to the region of the skull behind the eye.

temporal scale *n* One of a series of scales in the temporal region of the skull.

terminal *adj* An anatomical position pertaining to the end of a structure.

terminal nares *n* Used herein to refer to the position of the nostrils at the end of the elongated snout in soft-shell turtles *(Apalone).*

terminal phalanx *n* The distalmost phalanx of a digit.

terrestrial *adj* Living on the ground.

tomium *n* The cutting edge of the keratinized jaw sheaths of turtles.

transpeninsular *adj* Ranging throughout the length of a peninsula.

transverse process *n* A laterally projecting knob of the centrum of a vertebra.

tripartite *adj* Having three parts.

truncate *adj* Flattened or appearing cut off; usually refers to the blunt appearance of extremities.

tubercle *n* A small, rounded epidermal bump.

tuberculate *adj* Bearing tubercles.

tympanum *n* The membrane covering the external opening of the middle ear; the eardrum.

type locality *n* The place where the holotype or lectotype originated.

type series *n* The specimens on which a description of a species is based.

ubiquitous *adj* Existing or seeming to exist everywhere.

undulate *adj* Having a wavy margin.

vent *n* The cloacal aperture; the anus.

ventral *adj* Of or pertaining to the lower surface of a body or structure.

ventrolateral *adj* Of or pertaining to the side of the lower surface of a body or structure.

ventrolateral fold *n* A fold of skin running along the outside edge of the ventral surface of the body in some lizards (e.g., *Elgaria*).

ventrum *n* The lower surface of the body or body structures; the venter.

vertebral *adj* Of or pertaining to the vertebral region of the dorsal surface of the body.

vertebral crest *n* A single row of raised, enlarged scales running along the vertebral region in some lizards (e.g., *Ctenosaura*).

vertebral stripe *n* A longitudinal middorsal stripe.

vocal sac *n* A ventral outpocketing of the buccal cavity in most male anurans that expands under the pressure of exhaled air and amplifies sounds emanating from the vocal cords.

wart *n* A rounded, elevated bump on the skin of certain anurans.

whorl *n* A ring of scales encircling the tail of some lizards (e.g., *Ctenosaura*) and crocodilians.

xeric *adj* Arid, lacking in moisture.

xerophilic *adj* Of or pertaining to plants and animals that require little or no environmental moisture to survive.

LITERATURE CITED

Adest, G. A. 1987. Genetic differentiation among populations of the zebra tailed lizard, *Callisaurus draconoides* (Sauria: Iguanidae). Copeia 1987:854–859.

Agassiz, L. 1857. Contributions to the natural history of the United States of America. Vol. 1. Boston: Little, Brown and Co.

Aguirre-L., G., D. J. Morafka, and R. W. Murphy. 1999. The peninsular archipelago of Baja California: A thousand kilometers of tree lizard genetics. Herpetologica 55:369–381.

Allen, M. J. 1933. Report on a collection of amphibians and reptiles from Sonora, Mexico, with the description of a new lizard. Occ. Pap. Mus. Zool. Univ. of Michigan. 259:1–15.

Altig, R., and P. C. Dumas. 1972. *Rana aurora.* Cat. Amer. Amph. Rept. 160.1–4.

Alvarez, C. S., P. Galina-T., and A. González-R. 1988. Herpetofauna. In L. Arriaga and A. Ortega, eds., La Sierra de la Laguna de Baja California Sur, 167–184. La Paz, Baja California Sur: Centro de Investigaciones Biológicas de Baja California Sur A.C.

Alvarez, T., and P. Huerta. 1974. Nuevo registro de *Crotalus atrox* para la peninsula de Baja California. Rev. Soc. Mex. Hist. Nat. 53:113–115.

Amarál, A. do. 1929. On *Crotalus tortugensis* Van Denburgh and Slevin, 1921, *Crotalus atrox elegans* Schmidt, 1922, and *Crotalus atrox lucasensis* Van Denburgh, 1920. Bull. Antivenin Inst. Amer. 2:85–86.

Armstrong, B. L., and J. B. Murphy. 1979. The natural history of Mexican rattlesnakes. Univ. of Kansas Mus. Nat. Hist. Spec. Pub. 5:1–88.

Aschmann, H. 1959. The central desert of Baja California: Demography and ecology. Ibero-Americana 42:1–315. Berkeley: Univ. of California Press.

Asplund, K. K. 1967. Ecology of lizards in the relictual cape flora, Baja California. Amer. Midl. Nat. 77:462–475.

———. 1974. Body size and habitat utilization in whiptail lizards *(Cnemidophorus).* Copeia 1974:695–703.

Atsatt, S. R. 1952. Observations on the early natural history of the lizards *Sceloporus graciosus vandenburgianus* and *Gerrhonotus multicarinatus webbi.* Copeia 1952:276.

Auffenberg, W., and R. Franz. 1978. *Gopherus agassizii.* Cat. Amer. Amph. Rept. 212.12.

Axelrod, D. I. 1979. Age and origin of the Sonoran Desert vegetation. Occ. Pap. California Acad. Sci. 132:1–74.

Baegert, J. J. 1772. Nachrichten von der Amerikanischen Halbinsel Californien: Mit einen zweyfachen Anhang falscher Nachrichten. Mannheim. Translated, with an introduction and notes, as Observations in Lower California, by M. M. Brandenburg and Carl L. Baumann, 1952. Berkeley: Univ. of California Press.

Baharav, D. 1975. Movement of the horned lizard *Phrynosoma solare.* Copeia 1975:649–657.

Baird, S. F. 1854. Descriptions of new genera and species of North American frogs. Proc. Acad. Nat. Sci. Philadelphia 7:59–62.

———. 1859a. Reptiles: Report on reptiles collected on the survey. Vol. 10, pt. 3, no. 3, 11–16. Washington, D.C.: 33rd Congress, 2nd sess., Senate Exec. Doc. No. 78.

———. 1859b. Reptiles: Report on reptiles collected on the survey. Vol. 10, pt. 3, no. 3, 17–20. Washington, D.C.: 33rd Congress, 2nd sess., Senate Exec. Doc. No. 78.

———. 1859c. Reptiles: Report on reptiles collected on the survey. Vol. 10, pt. 4, no. 4, pp. 9–13. Washington, D.C.: 33rd Congress, 2nd sess., Senate Exec. Doc. No. 78.

———. 1859d. Description of new genera and species of North American lizards in the museum of the Smith-

OPPOSITE *Pituophis vertebralis* (Baja California Gopher Snake) from near Santa Rosalía, BCS

sonian Institution. [1858]. Proc. Acad. Nat. Sci. Philadelphia 10:253–256.

Baird, S. F., and C. Girard. 1852a. Characteristics of some new reptiles in the museum of the Smithsonian Institution. Proc. Acad. Nat. Sci. Philadelphia. 6:68–70.

———. 1852b. Appendix C. Reptiles. In Howard Stansbury, An expedition to the valley of the Great Salt Lake of Utah, 336–365. Philadelphia: Lippincott, Grambo and Co.

———. 1852c. Description of new species of reptiles, collected by the U.S. Exploring Expedition under the command of Capt. Charles Wilkes, U.S.N. Part I. Proc. Acad. Nat. Sci. Philadelphia 6:174–177.

———. 1852d. Characteristics of some new reptiles in the museum of the Smithsonian Institution. Proc. Acad. Nat. Sci. Philadelphia 6:125–129.

———. 1853. Catalogue of North American reptiles in the museum of the Smithsonian Institution, part 1: Serpentes. Smithsonian Misc. Coll. 2:1–172.

Ballinger, R. E. 1983. Life-history variation. In R. B. Huey, E. R. Pianka, and T. W. Schoener, eds., Lizard ecology: Studies of a model organism, 241–260. Cambridge: Harvard Univ. Press.

Ballinger, R. E., and D. W. Tinkle. 1972. Systematics and evolution of the genus *Uta* (Sauria: Iguanidae). Misc. Pub. Mus. Zool. Univ. of Michigan 145:1–83.

Banks, R. C. 1962. A history of explorations for vertebrates on Cerralvo Island, Baja California. Proc. California Acad. Sci., 4th ser., 30:117–125.

Banks, R. C., and W. M. Farmer. 1963. Observations on reptiles of Cerralvo Island, Mexico. Herpetologica 18:246–250.

Banta, B. H., and A. E. Leviton. 1963. Remarks upon *Arizona elegans pacata.* Herpetologica 18:277–279.

Banta, B. H., and W. W. Tanner. 1968. The systematics of *Crotaphytus wislizeni,* the leopard lizards (Sauria: Iguanidae), part II: A review of the status of the Baja California peninsular populations and a description of a new subspecies from Cedros Island. Great Basin Nat. 38:183–194.

Barnard, F. L. 1970. Structural geology of the Sierra de los Cucupas, Northeastern Baja California, Mexico, and Imperial County, California. Ph.D. diss., Univ. of Colorado.

Bauer, A. M. 1991. An annotated type catalogue of the geckos (Reptilia: Gekkonidae) in the Zoological Museum, Berlin. Mitt. zool. Mus. Berlin 67:279–310.

———. 1994. Familia Gekkonidae (Reptilia, Sauria), part I: Australia and Oceania. In H. Wermuth and M. Fischer, eds., Das Tierreich (The animal kingdom), 1–306. New York: Walter de Gruyter.

Beaman, K. R., and L. L. Grismer. 1994. *Crotalus enyo.* Cat. Amer. Amph. Rept. 589.1–6.

Beaman, K. R., B. D. Hollingsworth, H. E. Lawler, and C. H. Lowe. 1997. Bibliography of *Sauromalus* (Duméril 1856), the chuckwallas. Smithsonian Herpetol. Info. Serv. 116:1–44.

Beaman, K. R., and N. Wong. 2001. *Crotalus catalinensis.* Cat. Amer. Amph. Rept. 733.1–4.

Belding, L. 1887. Reptiles of the Cape Region of Lower California. West Amer. Sci. 3:93–97.

Bell, E. L. 1954. A preliminary report on the subspecies of the western fence lizard, *Sceloporus occidentalis,* and its relationships to the eastern fence lizard, *Sceloporus undulatus.* Herpetologica 10:31–36.

Bell, E. L., and A. H. Price. 1996. *Sceloporus occidentalis.* Cat. Amer. Amph. Rept. 631.1–17.

Berry, K. H. 1974. The ecology and social behavior of the chuckwalla, *Sauromalus obesus.* Univ. of California Pub. Zool. 101:1–60.

Bezy, R. L. 1972. Karyotypic variation and the evolution of the lizards in the family Xantusiidae. Contrib. Sci. Los Angeles Co. Mus. Nat. Hist. 227:1–29.

———. 1982. *Xantusia vigilis.* Cat. Amer. Amph. Rept. 302.1–4.

Bezy, R. L., G. C. Gorman, Y. J. Kim, and J. W. Wright. 1977. Chromosomal and genetic divergence in the fossorial lizards of the family Anniellidae. Syst. Zool. 26:57–71.

Bezy, R. L., and J. W. Sites Jr. 1987. A preliminary study of allozyme evolution in the lizard family Xantusiidae. Herpetologica 43:280–292.

Bickham, J. W., T. Lamb, P. Minx, and J. C. Patton. 1996. Molecular systematics of the genus *Clemmys* and the intergeneric relationships of emydid turtles. Herpetologica 52:89–97.

Bishop, S. C. 1943. Handbook of salamanders: The salamanders of the United States, Canada, and Lower California. Ithaca, N.Y.: Comstock Publishing Associates.

Blainville, H. M. de D. 1835a. Description de quelques espèces de reptiles de la Californie précédée de l'analyse d'un système général d'herpétologie et d'amphibologie. Nouv. Ann. Mus. Nat. Hist. Nat. Paris 4:232–296.

———. 1835b. Description de quelques espèces de reptiles de la Californie. Nouv. Ann. Mus. Nat. Hist. Nat. Paris 4:1–64.

Blair, W. F. 1972. Evolution in the genus *Bufo.* Austin: Univ. of Texas Press.

Blanchard, F. N. 1924. A new snake of the genus *Arizona.* Occ. Pap. Mus. Zool. Univ. of Michigan 150:1–5.

———. 1942. The ring-neck snakes, genus *Diadophis.* Bull. Chicago Acad. Sci. 7:1–144.

Blaney, R. M. 1977. Systematics of the common kingsnake, *Lampropeltis getulus* (Linnaeus). Tulane Stud. Zool. Bot. 19:47–103.

Blázquez, M. C., and A. Ortega-Rubio. 1996. Lizard winter activity at Baja California Sur, Mexico. J. Arid Environ. 33:247–253.

Blázquez, M. C., and R. Rodríguez-Estrella. 1997. Factors

influencing the selection of basking perches on cardon cacti by spiny-tailed iguanas *(Ctenosaura hemilopha).* Biotropica 29:344–348.

Blázquez, M. C., R. Rodríguez-Estrella, and Miguel Delibes. 1997. Escape behavior and predation risk of mainland and island spiny-tailed iguanas *(Ctenosaura hemilopha).* Ethology 103:990–998.

Bleakney, J. S. 1965. Reports of marine turtles from New England and eastern Canada. Can. Fld. Nat. 79:120–128.

Bogert, C. M. 1939. A study of the genus *Salvadora,* the patch-nosed snakes. Univ. of California Pub. Biol. Sci. 1:177–236.

———. 1942. Field note on the copulation of *Crotalus atrox* in California. Copeia 1942:262.

———. 1945. Two additional races of the patch-nosed snake, *Salvadora hexalepis.* Amer. Mus. Novitates 1285:1–14.

Bogert, C. M., and J. A. Oliver. 1945. A preliminary analysis of the herpetofauna of Sonora. Bull. Amer. Mus. Nat. Hist. 83:297–426.

Bostic, D. L. 1965. Home range of the teiid lizard *Cnemidophorus hyperythrus beldingi.* Southwest. Nat. 10:278–281.

———. 1966a. Food and feeding behavior of the teiid lizard *Cnemidophorus hyperythrus beldingi.* Herpetologica 22:12–31.

———. 1966b. Threat display in the lizards *Cnemidophorus hyperythrus* and *Cnemidophorus labialis.* Herpetologica 22:77–79.

———. 1966c. Preliminary report on reproduction in the teiid lizard *Cnemidophorus hyperythrus beldingi.* Herpetologica 22:81–90.

———. 1966d. Thermoregulation and hibernation of the teiid lizard, *Cnemidophorus hyperythrus beldingi* (Sauria: Teiidae). Southwest. Nat. 11:275–289.

———. 1968. Thermal relations, distribution, and habitat of *Cnemidophorus labialis* (Sauria: Teiidae). Trans. San Diego Soc. Nat. Hist. 15:21–30.

———. 1971. Herpetofauna of the Pacific coast of north central Baja California, Mexico, with a description of a new subspecies of *Phyllodactylus xanti.* Trans. San Diego Soc. Nat. Hist. 16:237–264.

———. 1975. A natural history guide to the Pacific coast of north central Baja California and adjacent islands. Vista, Calif.: Biological Education Expeditions.

Boulenger, G. A. 1882. Catalogue of the Batrachia Gradientia S. Caudata and Batrachia Apoda in the collection of the British Museum. 2nd ed. London: Taylor and Francis.

———. 1883. Description of new species of lizards and frogs collected by Herr A. Forrer in Mexico. Ann. Mag. Nat. Hist. 5:342–344.

Bowen, B. W., F. A. Abreu-Grobois, G. H. Balazs, N. Kamezaki, C. J. Limpus, and R. J. Ferl. 1995. Trans-pacific migrations of the loggerhead sea turtle demonstrated with mitochondrial DNA markers. Proc. National Acad. Sci. U.S. Amer. 92:3731–3734.

Bowen, B. W., N. Kamezaki, C. J. Limpus, G. H. Hughes, A. B. Meylan, and J. C. Avise. 1994. Global phylogeography of the loggerhead turtle *(Caretta caretta)* as indicated by mitochondrial DNA haplotypes. Evolution 48:1820–1828.

Bowen, B. W., and S. A. Karl. 1997. Population genetics, phylogeography, and molecular evolution. In P. L. Lutz and J. A. Musik, eds., The biology of sea turtles, 29–50. New York: CRC Press.

Bowen, B. W., A. B. Meylan, J. P. Ross, C. L. Limpus, G. H. Balazs, and J. C. Avise. 1992. Global population structure and natural history of the green turtle *(Chelonia mydas)* in terms of matriarchal phylogeny. Evolution 46:865–881.

Brattstrom, B. H. 1951. Two additions to the herpetofauna of Baja California, Mexico. Herpetologica 7:196.

———. 1952. Diurnal activities of a nocturnal animal. Herpetologica 8:61–63.

———. 1953a. An ecological restriction of the type locality of the western worm snake, *Leptotyphlops h. humilis.* Herpetologica 8:180–181.

———. 1953b. Notes on a population of leaf-nosed snakes, *Phyllorhynchus decurtatus perkinsi.* Herpetologica 9:57–64.

———. 1955. Notes on the herpetology of the Revillagigedo Islands, Mexico. Amer. Midl. Nat. 54:219–229.

———. 1964. Evolution of the pit vipers. Trans. San Diego Soc. Nat. Hist. 13:185–286.

———. 1965. Body temperatures of reptiles. Amer. Midl. Nat. 73:376–422.

———. 1997. Status of the subspecies of the coast horned lizard, *Phrynosoma coronatum.* J. Herpetol. 31:434–436.

Brown, C. W. 1974. Hybridization among the subspecies of the plethodontid salamander *Ensatina eschscholtzi.* Univ. of California Pub. Zool. 98:1–57.

Buckley, L. J., and R. W. Axtell. 1997. Evidence for the specific status of the Honduran lizards formerly referred to as *Ctenosaura palearis* (Reptilia: Squamata: Iguanidae). Copeia 1997:138–150.

Burrage, B. R. 1964. The eggs and young of *Gerrhonotus multicarinatus nanus* Fitch. Herpetologica 20:133.

———. 1965. Notes on the eggs and young of the lizards *Gerrhonotus multicarinatus webbi* and *G. m. nanus.* Copeia 1965:512.

Burt, C. E. 1929. The genus of teiid lizards, *Verticaria* Cope, 1869, considered as a synonym of *Cnemidophorus* Wagler, 1830, with a key to the primitive genera of the Teiidae. Proc. Biol. Soc. Washington 42:153–156.

———. 1931. A study of the teiid lizards of the genus *Cnemidophorus,* with special reference to their phylogenetic relationships. Bull. Amer. Mus. Nat. Hist. 154:1–286.

Bury, R. B. 1970. *Clemmys marmorata.* Cat. Amer. Amph. Rept. 100.1–3.

———. 1983. *Anniella nigra argentea:* Geographic distribution. Herpetol. Rev. 14:83–84.

Bury, R. B., and P. S. Corn. 1995. Have desert tortoises undergone a long-term decline in abundance? Wildlife Soc. Bull. 23:41–47.

Bury, R. B., R. A. Luckenbach, and L. R. Muñoz. 1978. Observations on *Gopherus agassizii* from Isla Tiburón, Sonora, Mexico. Proc. Desert Tortoise Council 1978:69–79.

Cadle, J. E. 1984a. Molecular systematics of neotropical xenodontine snakes, III: Overview of xenodontine phylogeny and the history of New World snakes. Copeia 1984:641–652.

———. 1984b. Molecular systematics of neotropical xenodontine snakes, II: Central American xenodontines. Herpetologica 40:21–30.

———. 1988. Phylogenetic relationships among advanced snakes: A molecular perspective. Univ. of California Pub. Zool. 119:1–77.

Caldwell, D. K. 1962a. Sea turtles in Baja California waters (with special reference to those of the Gulf of California), and the description of a new subspecies of northeastern Pacific green turtle. Contrib. Sci. Los Angeles Co. Mus. Nat. Hist. 61:1–31.

———. 1962b. Carapace length–body weight relationship and size and sex ratio of the northeastern Pacific green sea turtle, *Chelonia mydas carrinegra.* Contrib. Sci. Los Angeles Co. Mus. Nat. Hist. 62:1–10.

Camp, C. L. 1915. *Batrachoseps major* and *Bufo cognatus californicus,* new Amphibia from southern California. Univ. of California Pub. Zool. 12:327–334.

Campbell, J. A., and W. W. Lamar. 1989. The venomous reptiles of Latin America. Ithaca, N.Y.: Cornell Univ. Press.

Camper, J. D. 1996. *Masticophis bilineatus.* Cat. Amer. Amph. Rept. 637.1–4.

Camper, J. D., and J. R. Dixon. 1994. Geographic variation and systematics of the striped whipsnakes (*Masticophis taeniatus* complex; Reptilia: Serpentes: Colubridae). Ann. Carnegie Mus. 63:1–48.

Cannatella, D. C. 1985. A phylogeny of primitive frogs (Archaeobatrachians). Ph.D. diss., Univ. of Kansas.

Carl, G., and J. P. Jones. 1979. Spiny chuckwallas bred at the Fort Worth Zoo. Herpetol. Rev. 9:57.

Carothers, J. H. 1986. An experimental confirmation of morphological adaptation: Toe fringes in the sand-dwelling lizard *Uma scoparia.* Evolution 40:871–874.

Carr, A. 1952. Handbook of turtles of the United States, Canada, and Baja California. Ithaca, N.Y.: Comstock Publishing Associates.

Case, T. J. 1982. Ecology and evolution of the insular giant chuckawallas, *Sauromalus hispidus* and *S. varius.* In G. M. Burghardt and A. S. Rand, eds., Iguanas of the world: Their behavior, ecology, and conservation, 184–212. Park Ridge, N.J.: Noyes Publications.

———. 1983. The reptiles: Ecology. In T. J. Case and M. L. Cody, eds., Island biogeography in the Sea of Cortéz, 159–209. Berkeley: Univ. of California Press.

Castro-Franco, R., and M. G. Bustos-Zagal. 1994. List of reptiles of Morelos, Mexico, and their distribution in relation to vegetation types. Southwest. Nat. 39:171–175.

Censky, E. J. 1986. *Sceloporus graciosus.* Cat. Amer. Amph. Rept. 386.1–4.

Clarkson, R. W., and J. C. deVos Jr. 1986. The bullfrog, *Rana catesbeiana* Shaw, in the Lower Colorado River, Arizona-California. J. Herpetol. 20:42–49.

Clavigero, F. J. 1789. Storia della California. Venice. Translated and annotated as The history of [lower] California, by S. E. Lake and A. A. Gray, 1937. Stanford: Stanford Univ. Press.

Cliff, F. S. 1954a. Snakes of the islands in the Gulf of California, Mexico. Trans. San Diego Soc. Nat. Hist. 12:67–98.

———. 1954b. Variation and evolution of the reptiles inhabiting the islands in the Gulf of California, Mexico. Ph.D. diss., Stanford Univ.

Cliffton, K., D. O. Cornejo, and R. S. Felger. 1982. Sea turtles of the Pacific coast of Mexico. In K. A. Bjorndal, ed., Biology and conservation of sea turtles, 199–209. Washington, D.C.: Smithsonian Institution.

Cochran, D. M. 1961. Type specimens of reptiles and amphibians in the United States National Museum. Bull. U.S. National Mus. 220:1–199.

Cocroft, R. B. 1994. A cladistic analysis of chorus frog phylogeny (Hylidae: *Pseudacris*). Herpetologica 50:420–437.

Cody, M. L., R. Moran, and H. Thompson. 1983. The plants. In T. J. Case and M. L. Cody, eds., Island biogeography in the Sea of Cortéz, 49–97. Berkeley: Univ. of California Press.

Cogger, H. G., E. E. Cameron, and H. M. Cogger. 1983. Zoological catalogue of Australia. Vol. 1. Amphibia and Reptilia. Canberra: Australian Govt. Printing Serv.

Cole, C. J., and C. Gans. 1987. Chromosomes of *Bipes, Mesobaena,* and other amphisbaenians (Reptilia), with comments on their evolution. Amer. Mus. Novitates 2869:1–9.

Cole, C. J., and L. M. Hardy. 1981. Systematics of North American colubrid snakes related to *Tantilla planiceps* (Blainville). Bull. Amer. Mus. Nat. Hist. 171:199–284.

———. 1983. *Tantilla planiceps.* Cat. Amer. Amph. Rept. 319.1–2.

Conant, R. 1946. Studies on North American water snakes, II: The subspecies of *Natrix valida.* Amer. Midl. Nat. 35:250–275.

———. 1958. A field guide to reptiles and amphibians of eastern North America. Boston: Houghton Mifflin Co.

———. 1969. A review of the water snakes of the genus *Natrix* in Mexico. Bull. Amer. Mus. Nat. Hist. 142:1–140.

Cooper, J. G. 1863. Description of *Xerobates agassizii*. Proc. California Acad. Sci. 2:120.

Cope, E. D. 1860. Notes and descriptions of new and little known species of American reptiles. Proc. Acad. Nat. Sci. 12:339–345.

———. 1861. Contributions to the ophiology of Lower California, Mexico, and Central America. Proc. Acad. Nat. Sci. Philadelphia. 13:292–306.

———. 1863. Descriptions of new American Squamata in the museum of the Smithsonian Institution, Washington. Proc. Acad. Nat. Sci. Philadelphia 15:100–106.

———. 1864. Contributions to the herpetology of tropical America. Proc. Acad. Nat. Sci. Philadelphia 16:166–181.

———. 1865. Third contribution to the herpetology of tropical America. Proc. Acad. Nat. Sci. Philadelphia 17:185–198.

———. 1866. Fifth contribution to the herpetology of tropical America. Proc. Acad. Nat. Sci. Philadelphia 18:317–323.

———. 1868. Sixth contribution to the herpetology of tropical America. Proc. Acad. Nat. Sci. Philadelphia 20:305–313.

———. 1889. The Batrachia of North America. Bull. U.S. National Mus. 34:5–525.

———. 1892a. A synopsis of the species of the teiid genus *Cnemidophorus*. Trans. Amer. Phil. Soc. 17:27–52.

———. 1892b. A critical review of the characters and variations of the snakes of North America. Proc. U.S. National Mus. 14:589–694.

———. 1894. On the genera and species of Euchirotidae. Amer. Nat. 28:436–437.

———. 1895. On some new North American snakes. Amer. Nat. 29:676–680.

———. 1896a. On the genus *Callisaurus*. Amer. Nat. 30:1049–1050.

———. 1896b. On two new species of lizards from southern California. Amer. Nat. 30:883–836.

Cornejo, D. 1987. The enchanted islands. In L. Lindblad and S. Lindblad, eds., Baja California, 81–112. New York: Rizzoli.

Cowles, R. B., and C. M. Bogert. 1944. A preliminary study of the thermal requirements of desert reptiles. Bull. Amer. Mus. Nat. Hist. 83:261–296.

Cozens, T. H. 1974. Behavioral ecology of the banded rock lizard *Petrosaurus mearnsi* (Sauria: Iguanidae). M.S. thesis, San Diego State Univ.

Crosby, H. W. 1994. Antigua California: Mission and colony on the peninsular frontier, 1679–1768. Albuquerque: Univ. of New Mexico Press.

Cross, J. K. 1979. Multivariate and univariate character geography in *Chionactis* (Reptilia: Serpentes). Ph.D. diss., Univ. of Arizona.

Crumly, C. R. 1994. Phylogenetic systematics of North American tortoises (Genus *Gopherus*): Evidence for their classification. In R. B. Bury and D. J. Germano, eds., Biology of North American tortoises, 7–31. Fish and Wildlife Res. 13.

Crumly, C. R., and L. L. Grismer. 1994. Validity of the tortoise *Xerobates lepidocephalus* Ottley and Velazquez in Baja California. In R. B. Bury and D. J. Germano, eds., Biology of North American tortoises, 32–36. Fish and Wildlife Res. 13.

Cryder, M. R. 1999. Molecular systematics and evolution of the *Ctenosaura hemilopha* complex (Squamata: Iguanidae). M.S. thesis, Loma Linda Univ.

Cunningham, J. D. 1956. Food habits of the San Diego alligator lizard. Herpetologica 12:225–230.

Davis, J. I., and K. C. Nixon. 1992. Populations, variation, and the delimitation of phylogenetic species. Syst. Biol. 41:421–435.

Dawson, E. Y. 1944. Some ethnobotanical notes on the Seri Indians. Desert Plant Life 16:133–138.

De Lisle, H. F. 1991. Behavioral ecology of the banded rock lizard *(Petrosaurus mearnsi)*. Bull. So. Cal. Acad. Sci. 90:102–117.

de Queiroz, A., and R. Lawson. 1994. Phylogenetic relationships of the garter snakes based on DNA sequence and allozyme variation. Biol. J. Linn. Soc. 53:209–229.

de Queiroz, K. 1987. Phylogenetic systematics of iguanine lizards: A comparative osteological study. Univ. of California Pub. Zool. 118:1–203.

———. 1989. Morphological and biochemical evolution in the sand lizards. Ph.D. diss., Univ. of California, Berkeley.

———. 1992. Phylogenetic relationships and rates of allozyme evolution among the lineages of sceloporine sand lizards. Biol. J. Linn. Soc. 45:333–362.

———. 1995. Checklist and key to the extant species of Mexican iguanas (Reptilia: Iguanidae). Publicaciones Especiales del Museo de Zoología 9:1995.

Dial, B. E., R. E. Gatten Jr., and S. Kamel. 1987. Energetics of concertina locomotion in *Bipes biporus* (Reptilia: Amphisbaenia). Copeia 1987:470–477.

Dickerson, M. C. 1919. Diagnoses of twenty-three new species and a new genus of lizards from Lower California. Bull. Amer. Mus. Nat. Hist. 61:461–477.

Diller, L. V., and R. L. Wallace. 1986. Aspects of the life history of the desert night snake, *Hypsiglena torquata deserticola:* Colubridae, in southwestern Idaho. Southwestern Nat. 31:56–64.

Dixon, J. R. 1964. The systematics and distributions of the lizards of the genus *Phyllodactylus* in North and Central America. New Mexico State Univ. Res. Center Sci. Bull. 64–1:1–139.

———. 1966. Speciation and systematics of the gekkonid lizard genus *Phyllodactylus* of the islands in the Gulf of California. Proc. California Acad. Sci., 4th ser., 33:415–452.

———. 1970. *Coleonyx variegatus.* Cat. Amer. Amph. Rept. 96.1–4.

Dixon, J. R., and R. H. Dean. 1986. Status of the southern populations of the night snake (*Hypsiglena:* Colubridae) exclusive of California and Baja California. Southwest. Nat. 31:307–318.

Dixon, J. R., and R. R. Fleet. 1976. *Arizona elegans.* Cat. Amer. Amph. Rept. 179.1–.4.

Dodd, C. K. 1988. Synopsis of the biological data on the loggerhead sea turtle, *Caretta caretta* (Linnaeus 1758). U.S. Fish and Wildl. Serv. Biol. Rep. 88:1–110.

———. 1990. *Caretta caretta.* Cat. Amer. Amph. Rept. 483.1–7.

Dowling, H. G. 1957. A taxonomic study of the ratsnakes, genus *Elaphe* Fitzinger, IV: A checklist of the American forms. Occ. Pap. Mus. Zool. Univ. of Michigan 541:1–12.

Dowling, H. G., and J. V. Jenner. 1987. Taxonomy of American xenodontine snakes, II: The status of and relationships of *Pseudoleptodeira.* Herpetologica 43:190–200.

Dowling, H. G., and R. M. Price. 1988. A proposed new genus for *Elaphe subocularis* and *E. rosaliae.* The Snake 20:52–63.

Duellman, W. E. 1958. A monographic study of the colubrid snake genus *Leptodeira.* Bull. Amer. Mus. Nat. Hist. 114:1–152.

———. 1960. A taxonomic study of the middle American snake, *Pituophis deppei.* Lawrence, Kans.: Univ. of Kansas Pub. Mus. Nat. Hist. 10:599–610.

———. 1970. The hylid frogs of Middle America. Lawrence, Kans.: Monograph of the Museum of Natural History Univ. of Kansas, 1:1–753.

———. 1993. Amphibian species of the world: Additions and corrections. Spec. Pub. No. 21. Lawrence, Kans.: Univ. of Kansas Nat. Hist. Pub.

Duellman, W. E., and R. G. Zweifel. 1962. A synopsis of the lizards of the *sexlineatus* group (genus *Cnemidophorus*). Bull. Amer. Mus. Nat. Hist. 123:155–210.

Duméril, A. 1856. Descriptions des reptiles nouveaux et imparfaitement connus de la collection du Museum d'Histoire Naturelle et remarques sur la classification et les caractères des reptiles. Arch. Mus. Hist. Nat. 8:437–589.

Duméril, A., and G. Bibron. 1836. Erpétologie générale, ou histoire naturelle complète des reptiles. Vol. 3. Paris: Libraire Encyclopédique de Roret.

Duméril, A. M. C., G. Bibron, and A. Duméril. 1854. Erpétologie générale, ou histoire naturelle complète des reptiles. Vol. 7, part 2. Paris: Libraire Encyclopédique de Roret.

Dunham, A. E. 1982. Demographic and natural history variation among populations of the iguanid lizard *Urosaurus ornatus:* Implication for the study of life-history phenomena in lizards. Herpetologica 38:208–221.

———. 1983. Realized niche overlap, resource abundance, and intensity of interspecific competition. In R. B. Huey, E. R. Pianka, and T. W. Schoener, eds., Lizard ecology: Studies of a model organism, 261–280. Cambridge: Harvard Univ. Press.

Dunham, A. E., D. B. Miles, and D. N. Reznick. 1994. Natural history patterns in squamate reptiles. In C. Gans and R. B. Huey, eds., Biology of the Reptilia, Vol. 16: Ecology B, Defense and natural history, 441–522. Ann Arbor: Branta Books.

Dunn, E. R. 1926. The salamanders of the family Plethodontidae. Northampton, Mass.: Smith College.

———. 1929. A new salamander from southern California. Proc. U.S. National Mus. 74:1–3.

Dunson, W. A. 1976. Salt glands in reptiles. In C. Gans and W. R. Dawson, eds., Biology of the Reptilia, Vol. 5: Physiology A, 413–415. New York: Academic Press.

Dunson, W. A., and G. W. Ehlert. 1971. Effects of temperature, salinity, and surface water flow on distribution of the sea snake *Pelamis.* Limnol. Oceanogr. 16:845–853.

Dutton, P. H., S. K. Davis, T. Guerra, and D. Owens. 1996. Molecular phylogeny for marine turtles based on sequences of the ND4-leucine tRNA and control regions of the mitochondrial DNA. Molec. Phylo. Evol. 5:511–533.

Eckert, K. L. 1991. Leatherback sea turtles: A declining species of the global commons. In N. G. E. M. Borgese and J. R. Morgan, eds., Ocean yearbook, Vol. 9, 73–90. Chicago: Univ. of Chicago Press.

Eckert, S. A., and L. M. Sarti. 1997. Distant fisheries implicated in the loss of the world's largest leatherback nesting population. Marine Sci. Newslt. 78:2–7.

Ernst, C. H. 1992. Venomous reptiles of North America. Washington, D.C.: Smithsonian Institution Press.

Ernst, C. H., and R. W. Barbour. 1989. Turtles of the world. Washington, D.C.: Smithsonian Institution Press.

Ernst, C. H., J. E. Lovich, and R. W. Barbour. 1994. Turtles of the United States and Canada. Washington, D.C.: Smithsonian Institution Press.

Eschscholtz, J. F. 1829–1833. Zoologischer Atlas, enthaltend Abbildungen und Beschreibungen neuer Thierarten, während des Flottcapitains von Kotzebue zweiter Reise um die Welt, auf der Russisch-Kaiserlichen Kriegsschlupp *Predpriaetië* in den Jahren 1823–1826, 1:1–28. Berlin: G. Reimer.

Estes, R., K. de Queiroz, and J. Gauthier. 1988. Phylogenetic relationships within Squamata. In R. Estes and G. Pregill, eds., Phylogenetic relationships of the lizard families, 119–281. Stanford: Stanford Univ. Press.

Etheridge, R. E. 1961. Additions to the herpetofauna of Isla Cerralvo in the Gulf of California, México. Herpetologica 17:57–60.

———. 1982. Checklist of the iguanine and Malagasy iguanid lizards. In G. M. Burghardt and A. S. Rand, eds., Iguanas of the world: Their behavior, ecology, and conservation, 7–37. Park Ridge, N.J.: Noyes Publications.

Felger, R. S. 1966. Ecology of the islands and the Gulf Coast of Sonora, Mexico. Ph.D. diss., Univ. of Arizona.

———. 1990. Seri Indians and their herpetofauna. Sonoran Herpetol. Newsl. 3:41–44.

Felger, R. S., K. Cliffton, and J. P. Regal. 1976. Winter dormancy in sea turtles: Independent discovery and exploitation in the Gulf of California by two cultures. Science 191:283–285.

Felger, R. S., and C. H. Lowe. 1976. The island and coastal vegetation and flora of the northern part of the Gulf of California. Contrib. Sci. Los Angeles Co. Mus. Nat. Hist. 285:1–159.

Felger, R. S., and M. B. Moser. 1985. People of the desert and sea: Ethnobotany of the Seri Indians. Tucson: Univ. of Arizona Press.

Felger, R. S., M. B. Moser, and E. W. Moser. 1981. The desert tortoise in Seri Indian culture. Proc. Desert Tortoise Council 1981:113–120.

Ferguson, J. H., and C. H. Lowe. 1969. Evolutionary relationships in the *Bufo punctatus* group. Amer. Midl. Nat. 81:435–466.

Fitch, H. S. 1934a. New alligator lizards from the Pacific coast. Copeia 1934:6–7.

———. 1934b. A shift in specific names in the genus *Gerrhonotus.* Copeia 1934:172–173.

———. 1938. Systematic account of the alligator lizards *(Gerrhonotus)* in the western United States and Lower California. Amer. Midl. Nat. 20:381–424.

———. 1940. A biogeographical study of the ordinoides Artenkreis of garter snakes (genus *Thamnophis*). Univ. of California Pub. Zool. 44:1–150.

———. 1970. Reproductive cycles in lizards and snakes. Lawrence, Kans.: Misc. Pub. Univ. of Kansas Mus. Nat. Hist. 70:1–72.

———. 1983. *Thamnophis elegans.* Cat. Amer. Amph. Rept. 320.1–4.

Flores-Villela, O. 1993. Herpetofauna Mexicana. Carnegie Mus. Nat. Hist. Spec. Pub. 17:1–73.

Foote, R., and J. H. MacMahon. 1977. Electrophoretic studies of rattlesnake *(Crotalus* and *Sistrurus)* venom: Taxonomic implications. Comp. Biochem. Physiol. 57B:235–241.

Ford, L. S., and D. C. Cannatella. 1993. The major clades of frogs. Herpetol. Monogr. 7:94–117.

Fouquette, M. J., Jr. 1968. Remarks on the type specimen of *Bufo alvarius* Girard. Great Basin Nat. 28:70–72.

———. 1970. *Bufo alvarius.* Cat. Amer. Amph. Rept. 93.1–4.

Frazier, J. 1985. Misidentifications of sea turtles in the East Pacific: *Caretta caretta* and *Lepidochelys olivacea.* J. Herpetol. 19:1–11.

Fritts, T. H., and R. D. Jennings. 1994. Distribution, habitat use, and status of the desert tortoise in Mexico. In R. B. Bury and D. J. Germano, eds., Biology of North American tortoises, 49–56. Fish and Wildlife Res. 13.

Fritts, T. H., H. L. Snell, and R. L. Martin. 1982. *Anarbylus switaki* Murphy: An addition to the herpetofauna of the United States, with comments on relationships with *Coleonyx.* J. Herpetol. 16:39–52.

Fritts, T. H., M. L. Stinson, and R. Márquez-M. 1982. Status of sea turtle nesting in southern Baja California, México. Bull. So. California Acad. Sci. 81:51–60.

Frost, D. R. 1983a. Relationships of the Baja California ground snakes, genus *Sonora.* Trans. Kansas Acad. Sci. 86:31–37.

———. 1983b. *Sonora semiannulata.* Cat. Amer. Amph. Rept. 333.1–4.

Frost, D. R., ed. 1985. Amphibian species of the world. A taxonomic and geographical reference. Lawrence, Kans.: Allen Press Inc., and the Association of Systematic Collections.

Frost, D. R., and R. Etheridge. 1989. A phylogenetic analysis and taxonomy of iguanian lizards (Reptilia: Squamata). Univ. of Kansas Mus. Nat. Hist. Misc. Pub. 81:1–65.

Frost, D. R., and D. M. Hillis. 1990. Species concepts and practice: Herpetological applications. Herpetologica 46:87–104.

Frost, D. R., and A. G. Kluge. 1994. A consideration of epistemology in systematic biology, with special reference to species. Cladistics 10:259–94.

Frost, D. R., A. G. Kluge, and D. M. Hillis. 1992. Species in contemporary herpetology: Comments on phylogenetic inference and taxonomy. Herpetol. Rev. 23:46–54.

Frost, D. R., and T. R. Van Devender. 1979. The relationships of the ground snakes *Sonora semiannulata* and *S. episcopa* (Serpentes: Colubridae). Occ. Pap. Mus. Zool. Louisiana State Univ. 52:1–9.

Funk, R. S. 1965. Food of *Crotalus cerastes laterorepens* in Yuma County, Arizona. Herpetologica 21:15–17.

———. 1981. *Phrynosoma mcallii.* Cat. Amer. Amph. Rept. 281.1–2.

Gaffney, E. S., and P. A. Meylan. 1988. A phylogeny of turtles. In Benton, M. J., ed., The phylogeny and classification of the tetrapods, Vol. 1, 157–219. Oxford: Clarendon Press.

Galina-Tessaro, P. 1994. Estudio comparativo de tres especies de lacertilios en un matorral desértico de la región del cabo Baja California Sur, México. M.S. thesis, Universidad Nacional Autónoma de México.

Galina-Tessaro, P., A. Ortega-Rubio, and S. Alvarez-Cardenas. 2000. Diet of the Black-Tailed Brush Lizard *Urosaurus nigricaudus* of the Cape Region, Baja California Sur, México. Herpetol. Nat. Hist. 7:35–40.

Gartska, W. R. 1982. Systematics of the *mexicana* species group of the colubrid genus *Lampropeltis,* with a hypothesis of mimicry. Breviora 466:1–35.

Gastil, R. G., and D. Krummenacher. 1977. Reconnaissance geology of coastal Sonora between Puerto Lobos and Bahía Kino. Geol. Soc. Amer. Bull. 88:189–198.

Gastil, R. G., J. Minch, and R. P. Philips. 1983. The geology and ages of islands. In T. J. Case and M. L. Cody, eds., Island biogeography in the Sea of Cortéz, 13–25. Berkeley: Univ. of California Press.

Gaudin, A. J. 1979. *Hyla cadaverina.* Cat. Amer. Amph. Rept. 225.1–2.

Gehlbach, F. R. 1970. Death-feigning and erratic behavior in leptotyphlopid, colubrid, and elapid snakes. Herpetologica 26:24–34.

———. 1971. Lyre snakes of the *Trimorphodon biscutatus* complex: A taxonomic résumé. Herpetologica 27:200–211.

George, D. W., and H. C. Dessauer. 1970. Immunological correspondence of transferrins and the relationships of colubrid snakes. Comp. Biochem. Physiol. 33:617–627.

Gergus, E. W. A. 1994. Systematics and biogeography of *Bufo microscaphus* (Anura: Bufonidae), with a preliminary report on *americanus* group phylogeny. M.S. thesis, San Diego State Univ.

———. 1998. Systematics of the *Bufo microscaphus* complex: Allozyme evidence. Herpetologica 54:317–325.

Gergus, E. W. A., L. L. Grismer, and K. Beaman. 1997. *Bufo californicus:* Geographic distribution. Herpetol. Rev. 28:47–48.

Gergus, E. W. A., B. K. Sullivan, and K. B. Malmos. 1997. Call variation in the *Bufo microscaphus* complex: Implications for species boundaries and the evolution of mate recognition. Ethology 103:979–989.

Girard, C. 1854. A list of North American bufonids with diagnoses of new species. Proc. Acad. Nat. Sci. Philadelphia 7:86–88.

———. 1859. Reptiles of the boundary. In S. F. Baird, United States and Mexican boundary survey. Washington, D.C.

Good, D. A. 1988a. Allozyme variation and phylogenetic relationships among the species of *Elgaria* (Squamata: Anguidae). Herpetologica 44:154–162.

———. 1988b. Phylogenetic relationships among gerrhonotine lizards: An analysis of external morphology. Univ. of California Pub. Zool. 121:1–139.

Goldberg, S. R. 1997. Reproduction in the Western Shovelnose Snake, *Chionactis occipitalis* (Colubridae), from California. Great Basin Nat. 57:85–87.

———. 1998. Reproduction in the Sonoran whipsnake, *Masticophis bilineatus* (Serpentes: Colubridae). Southwest. Nat. 43:412–415.

Goldberg, S. R., and P. C. Rosen. 1999. Reproduction in the Sonoran Shovelnose Snake *(Chionactis palarostris)* and the Western Shovelnose Snake *(Chionactis occipitalis)* (Serpentes: Colubridae). Texas J. Sci. 51:153–158.

Goldman, E. A. 1951. Biological investigations in México. Smithsonian Misc. Coll. 115:1–476.

Graham, J. B., I. Rubinoff, and M. K. Hect. 1971. Temperature physiology of the sea snake *Pelamis platurus:* An index of its colonization potential in the Atlantic Ocean. Proc. National Acad. Sci. 68:1360–1363.

Gray, A. A. 1937 (ed.). The History of [Lower] California. Bryn Mawr, California: Manessier Pub. Co.

Gray, J. E. 1845. Catalogue of the specimens of lizards in the collection of the British Museum. London: Edward Newman.

———. 1850. Catalogue of the specimens of Amphibia in the collection of the British Museum, Part II: Batrachia Gradientia, etc. London: Taylor and Francis.

———. 1852. Description of several new genera of reptiles, principally from the collection of the H. M. S. *Herald.* Ann. Mag. Nat. Hist. 10:437–440.

Graybeal, A. 1997. The phylogenetic utility of cytochrome b: Lessons from bufonid frogs. Molec. Phylo. Evol. 2:256–269.

Green, D. M. 1985a. Differentiation in heterochromatin amount between subspecies of the red-legged frog, *Rana aurora.* Copeia 1985:1071–1074.

———. 1985b. Biochemical identification of red-legged frogs, *Rana aurora draytoni* (Ranidae), at Duckwater, Nevada. Southwest. Nat. 30:614–616.

———. 1986a. Systematics and evolution of western North American frogs allied to *Rana aurora* and *Rana boylii:* Karyological evidence. Syst. Zool. 35:273–282.

———. 1986b. Systematics and evolution of western North American frogs allied to *Rana aurora* and *Rana boylii:* Electrophoretic evidence. Syst. Zool. 35:283–296.

Greene, H. W. 1994. Antipredator mechanisms in reptiles. In C. Gans and R. B. Huey, eds., Biology of the Reptilia, Vol. 16: Ecology B, defense and natural history, 1–152. Ann Arbor: Branta Books.

———. 1997. Snakes: The evolution of mystery in nature. Berkeley: Univ. of California Press.

Greer, A. E. 1970. A subfamilial classification of scincid lizards. Bull. Mus. Comp. Zool. 139:151–184.

Grinnell, J., and C. L. Camp. 1917. A distribution list of the amphibians and reptiles of California. Univ. of California Pub. Zool. 17:127–208.

Grismer, J. L. 1994. Food observations on the endemic Sierra San Pedro Mártir Garter Snake *(Thamnophis elegans hueyi)* from Baja California, México. Herpetol. Nat. Hist. 2:107–108.

———. 2000. *Thamnophis validus:* Natural history notes—diet. Herpetol. Rev. 31:106.

Grismer, L. L. 1982. A new population of slender salamander *(Batrachoseps)* from northern Baja California, Mexico. San Diego Herpetological Soc. Newsl. 4:3–4.

———. 1988a. Geographic variation, taxonomy, and biogeography of the anguid genus *Elgaria* (Reptilia: Squamata) in Baja California, México. Herpetologica 44:431–439.

———. 1988b. Phylogeny, taxonomy, classification, and

biogeography of eublepharid geckos. In R. Estes and G. Pregill, eds., Phylogenetic relationships of the lizard families, 369–469. Stanford: Stanford Univ. Press.
———. 1989a. *Sceloporus occidentalis biseriatus:* Geographic distribution. Herpetol. Rev. 20:75.
———. 1989b. *Sceloporus licki:* Geographic distribution. Herpetol. Rev. 20:13.
———. 1989c. *Urosaurus g. graciosus:* Geographic distribution. Herpetol. Rev. 20:13.
———. 1989d. *Chionactis occipitalis:* Geographic distribution. Herpetol. Rev. 20:13.
———. 1989e. *Sonora semiannulata:* Geographic distribution. Herpetol. Rev. 20:13.
———. 1990a. *Coleonyx switaki.* Cat. Amer. Amph. Rept. 464.1–2.
———. 1990b. The relationships, taxonomy, and geographic variation of the *Masticophis lateralis* complex (Serpentes: Colubridae) of Baja California, México. Herpetologica 46:66–77.
———. 1990c. *Masticophis aurigulus.* Cat. Amer. Amph. Rept. 499.1–2.
———. 1990d. A new long nosed snake *(Rhinocheilus lecontei)* from Isla Cerralvo, Baja California Sur, México. Proc. San Diego Soc. Nat. Hist. 4:1–7.
———. 1993. The insular herpetofauna of the Pacific coast of Baja California, México. Herpetol. Nat. Hist. 1:1–10.
———. 1994a. The origin and evolution of the peninsular herpetofauna of Baja California, México. Herpetol. Nat. Hist. 2:51–106.
———. 1994b. Ecogeography of the peninsular herpetofauna of Baja California, México, and its utility in historical biogeography. In J. W. Wright and P. Brown, eds., Proc. Conf. Herpetology of North American Deserts, 89–125. Van Nuys, Calif.: Southwest. Herpetol. Soc., Spec. Pub. No. 5.
———. 1994c. Geographic origins for the reptiles on islands in the Gulf of California, México. Herpetol. Nat. Hist. 2(2):17–40.
———. 1994d. The evolutionary and ecological biogeography of the herpetofauna of Baja California and the Sea of Cortés, México. Ph.D. diss., Loma Linda Univ.
———. 1994e. Three new side-blotched lizards (genus *Uta*) from the Gulf of California. Herpetologica 50:451–474.
———. 1996a. Geographic variation, taxonomy, and distribution of *Eumeces skiltonianus* and *E. lagunensis* (Squamata: Scincidae) in Baja California, México. Amphibia-Reptilia 17:361–375.
———. 1996b. *Phyllorhynchus decurtatus:* Geographic distribution. Herpetol. Rev. 27:35.
———. 1996c. *Pituophis melanoleucus annectens:* Geographic distribution. Herpetol. Rev. 27:35.
———. 1996d. *Eridiphas slevini slevini:* Geographic distribution. Herpetol. Rev. 32–33.
———. 1997. The distribution of *Pituophis melanoleucus* and *P. vertebralis* in northern Baja California, México. Herpetol. Rev. 27:28:68–70.
———. 1999a. Checklist of amphibians and reptiles on islands in the Gulf of California, México. Bull. So. California Acad. Sci. 98:45–56.
———. 1999b. An evolutionary classification of reptiles on islands in the Gulf of California, México. Herpetologica 55:446–469.
———. 1999c. Phylogeny, taxonomy, and biogeography of *Cnemidophorus hyperythrus* and *C. ceralbensis* (Squamata: Teiidae) in Baja California, México. Herpetologica 55:28–42.
———. 2000a. Evolutionary biogeography on Mexico's Baja California peninsula: A synthesis of molecules and historical geology. Proc. National Acad. Sci. 97:14017–14018.
———. 2000b. *Goniurosaurus murphyi* Orlov and Darevsky: A junior synonym for *Goniurosaurus lichtenfelderi* Mocquard. J. Herpetol. 34:486–488.
———. 2001a. An evolutionary classification and checklist of amphibians and reptiles on islands along the Pacific coast of Baja California, México. Bull. So. California Acad. Sci. 100:12–23.
———. 2001b. Geographic variation of color pattern in *Coleonyx switaki* (Squamata: Eublepharidae) from Baja California, México, and southern California. Gekko 2:4–19.
———. 2001c. Comments on the taxonomy of Gopher Snakes from Baja California, México: A reply to Rodríguez-Robles and de Jesús-Escobar. Herpetol. Rev. 32:81–83.
Grismer, L. L., K. R. Beaman, and H. E. Lawler. 1995. *Sauromalus hispidus.* Cat. Amer. Amph. Rept. 615.1–4.
Grismer, L. L., and D. D. Edwards. 1988. Notes on the natural history of the barefoot banded gecko *Coleonyx switaki* (Squamata: Eublepharidae). In H. F. De Lisle, P. R. Brown, B. Kaufman, and B. M. McGurty, eds., Proc. Conf. California Herpetology, 43–50. Van Nuys, Calif.: Southwest. Herpetol. Soc. Spec. Pub. No. 4.
Grismer, L. L., and M. A. Galvan. 1986. A new night lizard *(Xantusia henshawi)* from a sandstone habitat in San Diego County, California. Trans. San Diego Soc. Nat. Hist. 21:155–165.
Grismer, L. L., and B. D. Hollingsworth. 1996. *Cnemidophorus tigris* does not occur on Islas San Benito, Baja California, México. Herpetol. Rev. 27:69–70.
———. 2000. *Lampropeltis getula:* Geographic distribution. Herpetol. Rev. 31:56.
———. 2001. A taxonomic review of the endemic alligator lizard *Elgaria paucicarinata* (Anguidae: Squamata) of Baja California, México, with a description of a new species. Herpetologica 57:488–496.
Grismer, L. L., and C. R. Mahrdt. 1996a. *Petrosaurus repens:* Geographic distribution. Herpetol. Rev. 27:153.

———. 1996b. *Masticophis lateralis:* Geographic distribution. Herpetol. Rev. 27:34

Grismer, L. L., and J. A. McGuire. 1993. The oases of central Baja California, México, part I: A preliminary account of the relict mesophilic herpetofauna and the status of the oases. Bull. So. California Acad. Sci. 92:2–24.

———. 1996. Taxonomy and biogeography of the *Sceloporus magister* complex (Squamata: Phrynosomatidae) in Baja California, México. Herpetologica 52:416–427.

Grismer, L. L., J. A. McGuire, and B. D. Hollingsworth. 1994. A report on the herpetofauna of the Vizcaíno Peninsula, Baja California, México, with a discussion of its biogeographic and taxonomic implications. Bull. So. California Acad. Sci. 93:45–80.

Grismer, L. L., and E. Mellink. 1994. The addition of *Sceloporus occidentalis* to the herpetofauna of Isla de Cedros, Baja California, México, and its historical and taxonomic implications. J. Herpetol. 28:120–126.

Grismer, L. L., and J. R. Ottley. 1988. A preliminary analysis of geographic variation in *Coleonyx switaki* (Squamata: Eublepharidae), with a description of a new subspecies. Herpetologica 44:143–154.

Grismer, L. L., B. E. Viets, and L. J. Boyle. 1999. Two new species of *Goniurosaurus* (Squamata: Eublepharidae) with a phylogeny and evolutionary classification of the genus. J. Herpetol. 333:82–93.

Grismer, L. L., H. Wong, and P. Galina-T. 2002. Geographic variation and taxonomy of the Sand Snakes, *Chilomeniscus* (Squamata: Colubridae). Herpetologica 58:18–31.

Groombridge, B. 1982. The IUCN Amphibia-Reptilia red data book: Testudines, Crocodylia, Rhynchocephalia, part 1. Gland, Switzerland: IUCN.

Günther, A. C. L. G. 1900. Biologia Centrali-Americana: Reptilia and Batrachia. London: Porter.

Hager, S. B. 1992. Surface activity, movements and home range of the San Diego horned lizard, *Phrynosoma coronatum blainvillii.* M.S. thesis, California State Univ., Fullerton.

Hager, S. B., and B. H. Brattstrom. 1997. Surface activity of the San Diego horned lizard, *Phrynosoma coronatum blainvillii.* Bull. So. California Acad. Sci. 42:339–344.

Hahn, D. E. 1979a. *Leptotyphlops.* Cat. Amer. Amph. Rept. 230.1–4

———. 1979b. *Leptotyphlops humilis.* Cat. Amer. Amph. Rept. 232.1–4.

Hall, W. P., and H. M. Smith. 1979. Lizards of the *Sceloporus orcutti* complex of the Cape Region of Baja California. Breviora 452:1–26.

Hallowell, E. 1852. Descriptions of new species of reptiles inhabiting North America. Proc. Acad. Nat. Sci. Philadelphia 6:177–184.

———. 1853. Reptiles. In Lorenzo Sitgreaves, Report of an expedition down the Zuni and Colorado Rivers, 106–147. Washington, D.C.: 32nd Congress, 2nd sess., Senate Exec. Doc. No. 78.

———. 1854. Descriptions of new reptiles from California. Proc. Acad. Nat. Sci. Philadelphia 7:91–97.

Hanson, J. A., and J. L. Vial. 1956. Defensive behavior and effects of toxins in *Bufo alvarius.* Herpetologica 12:141–149.

Hardy, L. M., and R. W. McDiarmid. 1969. The amphibians and reptiles of Sinaloa, México. Lawrence, Kans.: Univ. of Kansas Pub. Mus. Nat. Hist. 18:39–252.

Hastings, J. R., and R. R. Humphrey. 1969. Climatological data and statistics for Baja California. Univ. of Arizona Inst. Atmos. Phys. Tech. Rep. Meterol. Climatol. Arid Regions 18:1–96.

Hastings, J. R., and R. M. Turner. 1965. Seasonal precipitation regimes in Baja California, Mexico. Geogr. Ann. 47 (ser. A):204–223.

Hayes, M. P. 1975. The taxonomy and evolution of *Lampropeltis zonata.* M.S. thesis, California State Univ., Chico.

Hayes, M. P., and M. M. Miyamoto. 1984. Biochemical, behavioral, and body size differences between the red-legged frogs *Rana aurora aurora* and *Rana aurora draytoni.* Copeia 1984:1018–1022.

Hazard, L. C., V. H. Shoemaker, and L. L. Grismer. 1998. Salt gland secretion by an intertidal lizard, *Uta tumidarostra.* Copeia 1998:231–234.

Heath, J. E. 1965. Temperature regulation and diurnal activity in horned lizards. Univ. of California Pub. Zool. 64:97–136.

Hect, M. K., C. Kropach, and B. M. Hect. 1974. Distribution of the yellow-bellied sea snake, *Pelamis platurus,* and its significance in relation to the fossil record. Herpetologica 30:387–396.

Hedges, S. B. 1986. An electrophoretic analysis of Holarctic hylid frog evolution. Syst. Zool. 35:1–21.

Henyey, T. L., and J. L. Bischoff. 1973. Tectonic elements of the northern part of the Gulf of California. Geol. Soc. Amer. Bull. 84:315–330.

Hews, D. K. 1990. Resource, defense, sexual selection, and sexual dimorphism in the lizard *Uta palmeri.* Ph.D. diss., Univ. of Texas, Austin.

Hews, D. K., and J. C. Dickhaut. 1989. *Uta palmeri:* Natural history notes—cannibalism. Herpetol. Rev. 20:71.

Higgins, E. B. 1959. Type localities of vascular plants in San Diego County, California. Trans. San Diego Soc. Nat. Hist. 12:347–406.

Highton, R. 1998. Is *Ensatina eschscholtzii* a ring species? Herpetologica 54:254–278.

Hirth, H. F. 1980. *Chelonia mydas.* Cat. Amer. Amph. Rept. 249.1–4.

Holland, D.C. 1992. Level and pattern in morphological

variation: A phylogenetic study of the western pond turtle *(Clemmys marmorata)*. Ph.D. diss., Univ. of Southwestern Louisiana.

———. 1994. The western pond turtle: Habitat and history. Portland, Or.: U.S. Dept. of Energy, Bonneville Power Administration.

Hollingsworth, B. D. 1998. The systematics of chuckwallas *(Sauromalus),* with a phylogenetic analysis of other iguanid lizards. Herpetol. Monogr. 11:38–191.

———. 1999. The molecular systematics of the side-blotched lizards (Phrynosomatidae: *Uta*). Ph.D. diss., Loma Linda Univ.

Hollingsworth, B. D., C. R. Mahrdt, L. L. Grismer, B. H. Banta, and C. Sylber. 1997. An additional population of *Sauromalus varius* on a satellite island in the Gulf of California, México. Herpetol. Rev. 28:26–28.

Hollingsworth, B. D., and E. Mellink. 1996. *Crotalus exsul lorenzoensis:* Natural history notes—arboreal behavior. Herpetol. Rev. 27:143–144.

Hulse, A. C. 1992. *Dipsosaurus.* Cat. Amer. Amph. Rept. 542.1–6.

Humphrey, R. R. 1974. The Boojum and its home. Tucson: Univ. of Arizona Press.

Hunsaker, D. 1965. The ratsnake *Elaphe rosaliae* in Baja California. Herpetologica 21:71–72.

Hunt, L. E. 1983. A nomenclatural rearrangement of the genus *Anniella* (Sauria: Anniellidae). Copeia 1983:79–89.

Ineich, I. 1987. Recherches sur le peuplement et l'évolution des reptiles terrestres de la Polynésie française. Ph.D. diss., Université des Sciences et Techniques du Languedoc.

Iverson, J. B. 1992. A revised checklist with distribution maps of the turtles of the world. Richmond, Indiana: privately printed.

Jackman, T. R., and D. B. Wake. 1994. Evolutionary and historical analysis of protein variation in the blotched forms of salamanders of the *Ensatina* complex. Evolution 48:876–897.

———. 1998. Molecular and historical evidence for the introduction of clouded salamanders (genus *Aneides*) to Vancouver Island, British Columbia, Canada, from California. Can. J. Zool 76:1570–1580.

James, E. 1823. Account of an expedition from Pittsburgh to the Rocky Mountains, performed in the years 1819 and 1820, by order of the Hon. J. C. Calhoun, Sec'y of War: Under the command of Major Stephen H. Long, 2:1–442. Philadelphia: H. C. Varey and I. Lea.

Jameson, D. L., J. P. Mackey, and R. C. Richmond. 1966. The systematics of the Pacific treefrog, *Hyla regilla.* Proc. California Acad. Sci. 33:551–620.

Jan, G. 1863. Elenco sistematico degli ofidi descritti e disegnati per l'iconografia generale. Milan: Lombardi.

Janzen, F. J., S. L. Hoover, and H. B. Shaffer. 1997. Molecular phylogeny of the western pond turtle *(Clemmys marmorata):* Preliminary results. Linnaeus Fund research report. Chelonian Conservation and Biology 2(4):623–626.

Jennings, M. R. 1983. *Masticophis lateralis.* Cat. Amer. Amph. Rept. 343.1–2.

———. 1988a. *Phrynosoma coronatum.* Cat. Amer. Amph. Rept. 428.1–5.

———. 1988b. *Phrynosoma cerroense.* Cat. Amer. Amph. Rept. 427.1–2.

———. 1990a. *Petrosaurus mearnsi.* Cat. Amer. Amph. Rept. 495.1–3.

———. 1990b. *Petrosaurus thalassinus.* Cat. Amer. Amph. Rept. 496.1–3.

Jennings, M. R., and M. P. Hayes. 1985. Pre-1900 overharvest of California red-legged frogs *(Rana aurora draytoni):* the inducement for bullfrog *(Rana catesbeiana)* introduction. Herpetologica 41:94–103.

———. 1994. Amphibian and reptile species of special concern in California. Rancho Cordova: California Department of Fish and Game.

Jockusch, E. L., D. B. Wake, and K. P. Yanev. 1998. New species of slender salamanders, *Batrachoseps* (Amphibia: Plethodontidae), from the Sierra Nevada of California. Contrib. Sci. Los Angeles Co. Mus. Nat. Hist. 472:1–17.

Johnson, J. D. 1977. The taxonomy and distribution of the neotropical whipsnake *Masticophis mentovarius* (Reptilia, Serpentes, Colubridae). J. Herpetol. 11:287–309.

Jones, K. B. 1985. *Eumeces gilberti.* Cat Amer. Amph. Rept. 372.1–3.

Jones, K. B., and W. G. Whitford. 1989. Feeding behavior of free-roaming *Masticophis flagellum:* An efficient ambush predator. Southwestern Nat. 34:460–467.

Kamezaki, N., and M. Matsui. 1995. Geographic variation in the skull morphology of the green turtle, *Chelonia mydas,* with a taxonomic discussion. J. Herpetol. 29:51–60.

Karl, S. A., B. W. Bowen, and J. C. Avise. 1992. Global population, genetic structure, and male-mediated gene flow in the green turtle *(Chelonia mydas):* RFLP analyses of anonymous nuclear loci. Genetics 131:163–173.

Kellogg, R. 1932. Mexican tailless amphibians in the United States National Museum. Bull. U.S. National Mus. 160:1–224.

Kennicott, R. 1860. Descriptions of new species of North American serpents in the Museum of the Smithsonian Institution, Washington. Proc. Acad. Nat. Sci. Philadelphia 12:328–338.

Keogh, S. J. 1996. Evolution of the colubrid snake tribe Lampropeltini: A morphological perspective. Herpetologica 52:406–416.

Kim, Y. J., G. C. Gorman, T. J. Papenfuss, and A. K. Roychoudhury. 1976. Genetic relationships and genetic variation in the amphisbaenian genus *Bipes.* Copeia 1976:120–124.

King, F. E. 1932. Herpetological records and notes from the vicinity of Tucson, Arizona, June and July, 1930. Copeia 1932:175–177.

Kingsbury, B. A. 1989. Factors influencing activity patterns in *Coleonyx variegatus.* J. Herpetol. 23:399–404.

———. 1993. Thermoregulatory set points of the eurythermic lizard *Elgaria multicarinata.* J. Herpetol. 27:241–247.

———. 1994. Thermal constraints and eurythermy in the lizard *Elgaria multicarinata.* Herpetologica 50:266–273.

———. 1995. Field metabolic rates of a eurythermic lizard. Herpetologica 51:155–159.

Kizirian, D. A. 1987. Geographic variation in the zone of contact between two subspecies of gopher snakes *(Pituophis melanoleucus).* M.S. thesis, Univ. of Texas, El Paso.

Klauber, L. M. 1927. Notes on the salamanders of San Diego County, California. Bull. Zool. Soc. San Diego 3:1–4.

———. 1928. A list of the amphibians and reptiles of San Diego County, California. Bull. Zool. Soc. San Diego 4:1–8.

———. 1930. Differential characteristics of southwestern rattlesnakes allied to *Crotalus atrox.* Bull. Zool. Soc. San Diego 6:1–72.

———. 1931a. Notes on the worm snakes of the southwest, with descriptions of two new subspecies. Trans. San Diego Soc. Nat. Hist. 6:333–352.

———. 1931b. A statistical survey of the snakes of the southern border of California. Bull. Zool. Soc. San Diego. 8:1–93.

———. 1931c. *Crotalus tigris* and *Crotalus enyo,* two little known rattlesnakes of the southwest. Trans. San Diego Soc. Nat. Hist. 6:353–370.

———. 1932. Notes on the silvery footless lizard, *Anniella pulchra.* Copeia 1932:4–6.

———. 1933. Notes on *Lichanura.* Copeia 1933:214–215.

———. 1934. Annotated list of the amphibians and reptiles of the southern border of California. Bull. Zool. Soc. San Diego 11:2–28.

———. 1935. *Phyllorhynchus,* the leaf-nosed snake. Bull. Zool. Soc. San Diego 14:1–31.

———. 1936. *Crotalus mitchellii,* the speckled rattlesnake. Trans. San Diego Nat. Hist. 8:149–184.

———. 1939. Studies of reptile life in the arid southwest. Bull. Zool. Soc. San Diego 14:1–100.

———. 1940a. The worm snakes of the genus *Leptotyphlops* in the United States and northern Mexico. Trans. San Diego Soc. Nat. Hist. 9:87–162.

———. 1940b. Two new sub-species of *Phyllorhynchus,* the leaf-nosed snake, with notes on the genus. Trans. San Diego Soc. Nat. Hist. 9:195–214.

———. 1940c. The lyre snakes (genus *Trimorphodon*) of the United States. Trans. San Diego Soc. Nat. Hist. 9:163–194.

———. 1941. The long-nosed snakes of the genus *Rhinocheilus.* Trans. San Diego Soc. Nat. Hist. 9:289–332.

———. 1943. A desert subspecies of the snake *Tantilla eiseni.* Trans. San Diego Soc. Nat. Hist. 10:71–74.

———. 1944. The sidewinder, *Crotalus cerastes,* with description of a new subspecies. Trans. San Diego Soc. Nat. Hist. 10:91–126.

———. 1945. The geckos of the genus *Coleonyx,* with descriptions of new subspecies. Trans. San Diego Soc. Nat. Hist. 10:135–215.

———. 1946a. The glossy snake, *Arizona,* with descriptions of new subspecies. Trans. San Diego Soc. Nat. Hist. 10:311–398.

———. 1946b. The gopher snakes of Baja California, with descriptions of new subspecies of *Pituophis catenifer.* Trans. San Diego Soc. Nat. Hist. 11:1–40.

———. 1948. Some misapplications of the Linnean names applied to American snakes. Copeia 1948:1–14.

———. 1949a. The relationship of *Crotalus ruber* and *Crotalus lucasensis.* Trans. San Diego Soc. Nat. Hist. 11:57–60.

———. 1949b. Some new and revived subspecies of rattlesnakes. Trans. San Diego Soc. Nat. Hist. 11:61–116.

———. 1951. The shovel-nosed snake, *Chionactis,* with descriptions of two new subspecies. Trans. San Diego Soc. Nat. Hist. 11:141–204.

———. 1963. A new insular subspecies of the speckled rattlesnake. Trans. San Diego Soc. Nat. Hist. 13:73–80.

———. 1972. Rattlesnakes: Their habits, life histories, and influence on mankind. Berkeley: Univ. of California Press.

Kluge, A. G. 1987. Cladistic relationships in the Gekkonoidae (Squamata, Sauria). Misc. Pub. Mus. Zool. Univ. of Michigan 173:1–54.

———. 1989. A concern for evidence and a phylogenetic hypothesis of relationships among *Epicrates* (Boidae, Serpentes). Syst. Zool. 38:7–25.

———. 1991. Boine phylogeny and research cycles. Misc. Pub. Mus. Zool. Univ. of Michigan 178:1–58.

———. 1993a. *Calabaria* and the phylogeny of erycine snakes. Zool. J. Linn. Soc. 107:293–351.

———. 1993b. *Aspidites* and the phylogeny of pythonine snakes. Rec. Australian Mus. 19:1–77.

———. 2001. Gekkonid lizard taxonomy. Hamadryad 26:1–209.

Knight, J. L. 1986. Variation in snout morphology in the North American snake *Pituophis melanoleucus.* J. Herpetol. 20:77–79.

Kropach, C. 1975. The yellow-bellied sea snake, *Pelamis,* in the eastern Pacific. In W. A. Dunson, ed., The biology of sea snakes, 185–213. Baltimore: University Park Press.

Krupa, J. J. 1990. *Bufo cognatus.* Cat. Amer. Amph. Rept. 457.1–8.

Lais, M. P. 1976a. *Gerrhonotus cedrosensis.* Cat. Amer. Amph. Rept. 177.1.

———. 1976b. *Gerrhonotus multicarinatus.* Cat. Amer. Amph. Rept. 187.1–4.

———. 1976c. *Gerrhonotus paucicarinata.* Cat. Amer. Amph. Rept. 188.1–2.

Lamb, T., J. C. Avise, and J. W. Gibbons. 1989. Phylogenetic patterns in mitochondrial DNA of the desert tortoise *(Xerobates agassizi),* and evolutionary relationships among North American gopher tortoises. Evolution 43:76–87.

Lamb, T. N. 1992. The geology of the Coronado Islands. In L. Perry, ed., Natural history of the Coronado Islands, Baja California, Mexico, 32–83. San Diego: San Diego Assoc. Geol.

Larsen, K. R., and W. W. Tanner. 1975. Evolution of the sceloporine lizards (Iguanidae). Great Basin Nat. 35:1–20.

Larson, A., D. B. Wake, L. R. Maxon, and R. Highton. 1981. A molecular phylogenetic perspective on the origins of morphological novelties in the salamanders of the tribe Plethodontini (Amphibia, Plethodontidae). Evolution 35:405–422.

Laurenti, J. N. 1768. Specimen medicum, exhibens synopsin reptilium emendatum cum experimentis circa venena et antidota reptilium austriacorum. Vienna.

Lawler, H. E., K. R. Beaman, and L. L. Grismer. 1995. *Sauromalus varius.* Cat. Amer. Amph. Rept. 616.1–4.

Lawlor, T. E. 1983. The mammals. In T. J. Case and M. L. Cody, eds., Island biogeography in the Sea of Cortéz, 265–289. Berkeley: Univ. of California Press.

Laylander, D. P. 1987. Sources and strategies for the prehistory of Baja California. M.S. thesis, San Diego State Univ.

Lee, J. C. 1974. The diel activity cycle of the lizard, *Xantusia henshawi.* Copeia 1974:934–940.

———. 1975. The autecology of *Xantusia henshawi* (Sauria: Xantusiidae). Trans. San Diego Soc. Nat. Hist. 17:259–277.

———. 1976. *Xantusia henshawi.* Cat. Amer. Amph. Rept. 189.1–2.

Legler, J. M. 1990. The genus *Pseudemys* in Mesoamerica: Taxonomy, distribution, and origin. In J. W. Gibbons, ed., Life history and ecology of the slider turtle, 82–105. Washington, D.C.: Smithsonian Institution Press.

LeSueur, C. A. 1827. Note sur deux espèces de tortues du genre *Trionyx* Geoffrey Saint-Hilaire. Mém. Mus. Nat. Hist. Nat. Paris 15:257–268.

Leviton, A. E., and B. H. Banta. 1964. Midwinter reconnaissance of the herpetofauna of the Cape Region of Baja California, Mexico. Proc. California Acad. Sci., 4th ser., 30:127–156.

Leviton, A. E., R. H. Gibbs, E. Heal, and C. E. Dawson. 1985. Standards in herpetology and ichthyology, part 1: Standard symbolic codes for institutional resource collections in herpetology and ichthyology. Copeia 1985:802–832.

Leviton, A. E., and W. W. Tanner. 1960. The generic allocation of *Hypsiglena slevini* Tanner (Serpentes: Colubridae). Occ. Pap. California Acad. Sci. 27:1–7.

Lichtenstein, H. 1939. Beitrag zur ornithologischen Fauna von California. Abh. Akad. Wiss. Berlin, 1838:417–451.

Limpus, C. J., E. Gyuris, and J. D. Miller. 1988. Reassessment of the taxonomic status of the sea turtle genus *Natator* McCulloch 1908, with a redescription of the genus and species. Trans. Royal Soc. So. Australia 112:1–10.

Liner, E. A. 1994. Scientific and common names for the amphibians and reptiles of Mexico in English and Spanish. Society for the Study of Amphibians and Reptiles, Herpetological Circular No. 23.

Linnaeus, C. 1758. Systema naturae per regna tria naturae, secundum classes, ordines, genera, species cum characteribus, differentiis, synonymis, locis. Vol. 1. 10th ed. Stockhom: L. Salvius.

———. 1766. Systema naturae per regna tria naturae, secundum classes, ordines, genera, species cum characteribus, differentiis, synonymis, locis. Vol. 1. 12th ed. Stockholm: L. Salvius.

Linsdale, J. M. 1932. Amphibians and reptiles from Lower California. Univ. of California Pub. Zool. 38:345–386.

List, J. C. 1966. Comparative osteology of the snake families Typhlopidae and Leptotyphlopidae. Illinois Biol. Monogr. 36:1–112.

Lockington, W. N. 1880. List of Californian reptiles and batrachia collected by Mr. Dunn and Mr. W. J. Fisher in 1876. Amer. Nat. 1880:295–96.

Logan, R. F. 1968. Causes, climates, and distributions of deserts. In G. W. Brown Jr., ed., Desert biology, vol. 1, 21–50. New York: Academic Press.

Lonsdale, P. 1989. Geology and tectonic history of the Gulf of California. In E. L. Winterer, D. M. Hussong, and R. W. Decker, eds., The eastern Pacific Ocean and Hawaii: The geology of North America, 123–146. Boulder: Geological Society of America.

Loomis, R. B. 1965. The yellow-legged frog, *Rana boylei,* from the Sierra San Pedro Mártir, Baja California Norte, México. Herpetologica 21:78–80.

Loveridge, A. 1947. Revision of the African lizards of the family Gekkonidae. Bull. Mus. Comp. Zool. 98:1–469.

Lovich, J. E., A. F. Laemmerzahl, C. H. Ernst, and J. F. McBreen. 1991. Relationships among turtles of the genus *Clemmys* (Reptilia, Testudines, Emydidae) as suggested by plastron scute morphology. Zool. Script. 20:425–429.

Lowe, C. H., Jr. 1942. Notes on the mating of desert rattlesnakes. Copeia 1942:261–262.

Lowe, C. H., and K. S. Norris. 1954. Analysis of the herpetofauna of Baja California, Mexico. Trans. San Diego Soc. Nat. Hist. 12:47–64.

———. 1955. Analysis of the herpetofauna of Baja California, Mexico, III: New and revived reptilian subspecies of Isla San Esteban, Gulf of California, Sonora, Mexico, with notes on other islands in the Gulf of California. Herpetologica 11:89–96.

Lowe, C. H., Jr., C. R. Schwalbe, and T. B. Johnson. 1986. The venomous reptiles of Arizona. Phoenix: Arizona Game and Fish Department.

Lowe, C. H., J. W. Wright, C. J. Cole, and R. L. Bezy. 1970. Chromosomes and evolution of the species groups of *Cnemidophorus* (Reptilia: Teiidae). Syst. Zool. 19:128–141.

Lynch, J. F., and D. B. Wake. 1974. *Aneides lugubris.* Cat. Amer. Amph. Rept. 159.1–2.

MacArthur, R. H., and E. O. Wilson. 1967. The theory of island biogeography. Princeton: Princeton Univ. Press.

Macey, J. R., J. A. Schulte II, A. Larson, B. S. Tuniyev, N. Orlov, and T. J. Papenfuss. 1999. Molecular phylogenetics, tRNA evolution, and historical biogeography in anguid lizards and related taxonomic families. Molec. Phylo. Evol. 12:250–272.

Mahrdt, C. R. 1975. The occurrence of *Ensatina eschscholtzii eschscholtzii* in Baja California, México. J. Herpetol. 9:240–242.

Mahrdt, C. R., and B. H. Banta. 1996. *Chionactis occipitalis annulata:* Natural history notes—predation and diurnal activity. Herpetol. Rev. 27:81.

Mahrdt, C. R., K. R. Beaman, P. C. Rosen, and P. A. Holm. 2001. *Chionactis occipitalis.* Cat. Amer. Amph. Rept. 731.1–12.

Mahrdt, C. R., R. H. McPeak, and L. L. Grismer. 1998. The discovery of *Ensatina eschscholtzii klauberi* (Plethodontidae) in the Sierra San Pedro Mártir, Baja California, México. Herpetol. Nat. Hist. 6:73–76.

Malkin, B. 1962. Seri ethnozoology. Occ. Pap. Idaho State. Coll. Mus. 7:1–68.

Marinkelle, C. J. 1966. Accidents by venomous animals in Colombia. Indust. Med. Surg. 1966:988–992.

Markham, C. G. 1972. Baja California's climate. Weatherwise 25:66–101.

Marlow, R. W., J. M. Brodie, and D. B. Wake. 1979. A new slender salamander, genus *Batrachoseps,* from the Inyo Mountains of California, with a discussion of relationships in the genus. Contrib. Sci. Los Angeles Co. Mus. Nat. Hist. 308:1–17.

Márquez, M. R., A. Villanueva, and C. Peñaflores. 1976. Sinopsis de datos biológicos sobre la tortuga golfina, *Lepidochelys olivacea* (Eschscholtz, 1829). Inst. Nac. Pesca, INP Sinop. Pesca 2:1–61.

Maslin, T. P., and D. E. Secoy. 1986. A checklist of the lizard genus *Cnemidophorus* (Teiidae). Contrib. Zool. Univ. of Colorado Mus. 1:1–60.

Mautz, W. J., and T. J. Case. 1974. The diurnal activity cycle in the granite night lizard, *Xantusia henshawi.* Copeia 1974:243–251.

Mayhew, W. W. 1963a. Reproduction in the granite spiny lizard, *Sceloporus orcutti.* Copeia 1963:144–152.

———. 1963b. Biology of the granite spiny lizard, *Sceloporus orcutti.* Amer. Midl. Nat. 69:310–327.

———. 1963c. Temperature preferences of *Sceloporus orcutti.* Herpetologica 18:217–233.

———. 1965. Hibernation in the horned lizard, *Phrynosoma m'calli.* Comp. Biochem. Physiol. 16:103–119.

McClanahan, L., Jr. 1967. Adaptations of the spadefoot toad, *Scaphiopus couchi,* to desert environments. Comp. Biochem. Physiol. 20:73–99.

McClanahan, L., Jr., R. Ruibal, and V. H. Shoemaker. 1994. Frogs and toads in deserts. Sci. Amer. 270:64–70.

McCleary, R. J. R., and R. W. McDiarmid. 1993. *Phyllorhynchus decurtatus.* Cat. Amer. Amph. Rept. 580.1–7.

McCranie, J. R., and C. T. McAllister. 1988. *Nerodia valida.* Cat. Amer. Amph. Rept. 431.1–3.

McCrystal, H. K., and M. J. McCoid. 1986. *Crotalus mitchellii.* Cat. Amer. Amph. Rept. 388.1–4.

McGee, W. J. 1896. The Seri Indians. Smithsonian Inst. Bur. Amer. Ethnol. Ann. Rept. 17:1–344.

McGuire, J. A. 1991a. *Crotaphytus insularis vestigium:* Geographic distribution. Herpetol. Rev. 22:135.

———. 1991b. *Crotalus enyo cerralvensis:* Behavior. Herpetol. Rev. 22:100.

———. 1994. A new species of collared lizard (Iguania: Crotaphytidae) from northeastern Baja California. Herpetologica 50:438–450.

———. 1996. Phylogenetic systematics of crotaphytid lizards (Reptilia: Iguania: Crotaphytidae). Bull. Carnegie Mus. Nat. Hist. 32:1–143.

McGuire, J. A., and L. L. Grismer. 1993. The taxonomy and biogeography of *Thamnophis hammondii* and *T. digueti* (Reptilia: Squamata: Colubridae) in Baja California, México. Herpetologica 49:354–365.

Medica, P. A. 1975. *Rhinocheilus lecontei.* Cat. Amer. Amph. Rept. 175.1–4.

Meek, S. E. 1905. An annotated list of a collection of reptiles from southern California and northern Lower California. Field Columbian Mus. Zool. Ser. 7:1–19.

Meigs, P. 1953. World distribution of arid and semi-arid homoclimates. In Reviews of research on arid zone hydrology: Arid zone prog. 1, 202–210. Paris: UNESCO.

———. 1966. Geography of coastal deserts. Arid Zone Research 28:1–140. Paris: UNESCO.

Mellink, E. 1995. The potential effect of commercialization of reptiles from Mexico's Baja California peninsula and its associated islands. Herpetol. Nat. Hist. 3:95–99.

———. 1996. *Eridiphas slevini:* Geographic distribution. Herpetol. Rev. 27:88.

Mellink, E., and V. Ferreira-Bartrina. 2000. On the wildlife of wetlands of the Mexican portion of the Río Colorado delta. Bull. So. California Acad. Sci. 99:115–127.

Merker, G., and C. Merker. 1995. The allure of rosy boas. Reptiles 2:48–63.

Merkle, D. A. 1975. A taxonomic analysis of the *Clemmys*

complex (Reptilia: Testudinidae). Bull. Mus. Comp. Zool. 123:113–127.

Mertens, R., and L. Müller. 1928. Liste der Amphibien und Reptilien Europas. Abh. Senck. Naturf. Gesell. 41:1–62.

Meylan, P. A. 1987. The phylogenetic relationships of soft-shelled turtles (Family Trionychidae). Bull. Amer. Mus. Nat. Hist. 186:1–101.

Middendorf, G. A., III, and W. C. Sherbrooke. 1992. Canid elicitation of blood-squirting in a horned lizard *(Phrynosoma cornutum).* Copeia 1992:519–527.

Miller, C. M. 1944. Ecologic relations and adaptations of the limbless lizards of the genus *Anniella.* Ecol. Monogr. 14:271–289.

Miller, M. R. 1951. Some aspects of the natural history of the yucca night lizard, *Xantusia vigilis.* Copeia 1951:114–120.

———. 1954. Further observations on reproduction in the lizard *Xantusia vigilis.* Copeia 1954:38–40.

Miller, R. R. 1946. The probable origin of the soft-shelled turtle in the Colorado River Basin. Copeia 1946:46.

Miller, W. E. 1978. Pleistocene terrestrial vertebrates from southern Baja California. Geol. Soc. Amer., Abstracts with Programs 9:468.

———. 1980. The Late Pliocene Las Tunas local fauna from southernmost Baja California, Mexico. J. Paleo. 54:762–805.

Minton, S. A., and S. K. Salanitro. 1972. Serological relationships among some colubrid snakes. Copeia 1972:246–252.

Mittleman, M. B. 1942. A summary of the iguanid genus *Urosaurus.* Bull. Mus. Comp. Zool. 91:103–181.

Mocquard, F. 1899. Contribution à la faune herpétologique de la Basse Californie. Nouv. Arch. Mus. Nat. Hist. Nat. 4:297–344.

Montanucci, R. R. 1978. Dorsal pattern polymorphism and adaptation in *Gambelia wislizenii* (Reptilia, Lacertilia, Iguanidae). J. Herpetol. 12:73–81.

———. 1987. A phylogenetic study of the horned lizards, genus *Phrynosoma,* based on skeletal and external morphology. Contrib. Sci. Los Angeles Co. Mus. Nat. Hist. 390:1–36.

Montanucci, R. R., R. W. Axtell, and H. C. Dessauer. 1975. Evolutionary divergence among collared lizards *(Crotaphytus),* with comments on the status of *Gambelia.* Herpetologica 31:336–347.

Montanucci, R. R., H. M. Smith, K. Adler, A. L. Auth, R. W. Axtell, T. J. Case, D. Chizar, J. T. Collins, R. Conant, R. Murphy, K. Petren, and R. C. Stebbins. 2001. *Euphryne obesus* Baird, 1858: Proposed precedence of the specific name over that of *Sauromalus ater* Duméril, 1856. Bull. Zool. Nomen. 58:3740.

Moore, R. G. 1978. Seasonal and daily activity patterns and thermoregulation in the southwestern speckled rattlesnake *(Crotalus mitchellii pyrrhus)* and the Colorado Desert sidewinder *(Crotalus cerastes laterorepens).* Copeia 1978:439–442.

Morafka, D. J., and B. H. Banta. 1976. Biogeographical implications of pattern variation in the salamander *Aneides lugubris.* Copeia 1976:580–586.

Moritz, C., T. J. Case, D. T. Bolger, and S. Donnellan. 1993. Genetic diversity and the history of Pacific island house geckos *(Hemidactylus* and *Lepidodactylus).* Biol. J. Linn. Soc. 48:113–133.

Moritz, C., C. J. Schneider, and D. B. Wake. 1992. Evolutionary relationships within the *Ensatina eschscholtzii* complex confirm the ring species interpretation. Syst. Zool. 41:273–291.

Mosauer, W. 1936. The reptilian fauna of the sand dune areas of the Vizcaino Desert and of northwestern Lower California. Occ. Pap. Mus. Zool. Univ. of Michigan 329:1–22.

Murphy, J. B., and B. L. Armstrong. 1978. Maintenance of rattlesnakes in captivity. Spec. Pub. Univ. of Kansas Mus. Nat. Hist. 3:1–40.

Murphy, R. W. 1974. A new genus and species of eublepharine gecko (Sauria: Gekkonidae) from Baja California, Mexico. Proc. California Acad. Sci., 4th ser., 60:87–92.

———. 1975. Two new blind snakes (Serpentes: Leptotyphlopidae) from Baja California, Mexico, with a contribution to the biogeography of peninsular and insular herpetofauna. Proc. California Acad. Sci., 4th ser., 40:87–92.

———. 1983a. Paleobiogeography and genetic differentiation of the Baja California herpetofauna. Occ. Pap. California Acad. Sci. 137:1–48.

———. 1983b. The reptiles: origin and evolution. In T. J. Case and M. L. Cody, eds., Island biogeography in the Sea of Cortéz, 130–158. Berkeley: Univ. of California Press.

———. 1988. The problematic phylogenetic analysis of interlocus heteropolymer isozyme characters: A case study from sea snakes and cobras. Can. J. Zool. 66:2628–2633.

Murphy, R. W., and B. Crabtree. 1985. Genetic relationships of the Santa Catalina Island rattleless rattlesnake, *Crotalus catalinensis* (Serpentes: Viperidae). Acta Zool. Mex. 9:1–16.

Murphy, R. W., V. Kovac, O. Haddrath, G. S. Allen, A. Fishbein, and N. E. Mandrak. 1995. mtDNA gene sequence, allozyme, and morphological uniformity among red diamond rattlesnakes, *Crotalus ruber* and *Crotalus exsul.* Can. J. Zool. 73:270–281.

Murphy, R. W., and J. R. Ottley. 1980. A genetic evaluation of the leafnose snake, *Phyllorhynchus arenicolus.* J. Herpetol. 14:263–268.

———. 1984. Distribution of amphibians and reptiles on islands in the Gulf of California. Ann. Carnegie Mus. 53:207–230.

Murphy, R. W., and H. M. Smith. 1985. Conservation of the name *Anniella pulchra* for the California legless lizard. Herpetol. Rev. 16:68.

———. 1991. *Anniella pulchra* Gray, 1852 (Reptilia, Squamata): Proposed designation of a neotype. Bull. Zool. Nomen. 48:316–318.

Murray, K. F. 1955. Herpetological collections from Baja California. Herpetologica 11:33–48.

Nabhan, G. P. Forthcoming. Singing the turtles to sea: The Comcáac (Seri) art and science of reptiles. Berkeley: Univ. of California Press.

Nelson, E. W. 1922. Lower California and its natural resources. Mem. National Acad. Sci. Washington 16:1–194.

Nichols, W. J., A. Resendiz, and J. A. Seminoff. 2000. Transpacific loggerhead turtle migration monitored with satellite telemetry. Marine Turtle Newsl. 89:4–7.

Norris, K. S. 1953. The ecology of the desert iguana *Dipsosaurus dorsalis.* Ecology 34:265–287.

———. 1958. The evolution and systematics of the iguanid genus *Uma* and its relation to the evolution of other North American desert reptiles. Bull. Amer. Mus. Nat. Hist. 114:251–326.

———. 1967. Color adaptation in desert reptiles and its thermal relationships. In W. W. Milstead, ed., Lizard ecology: A symposium, 162–229. Columbia: Univ. of Missouri Press.

Oberbauer, T. A. 1992. Herpetology of Islas Coronados. In L. Perry, ed., Natural history of the Coronado Islands, Baja California, Mexico, 24–26. San Diego: San Diego Assoc. Geol.

———. 1993. Floristic analysis of vegetation communities on Isla Cedros, Baja California, Mexico. In F. G. Hochberg, ed., Third California island symposium: Recent advances in research on the California islands, 115–131. Santa Barbara: Santa Barbara Museum of Natural History.

Oliver, J. A. 1946. An aggregation of Pacific sea turtles. Copeia 1946:103.

Ortenburger, A. I. 1928. The whipsnakes and racers: Genera *Masticophis* and *Coluber.* Mem. Univ. of Michigan Mus. 1:1–127.

Ortenburger, A. I., and R. D. Ortenburger. 1926. Field observations on some amphibians and reptiles of Pima County, Arizona. Proc. Oklahoma Acad. Sci. 6:101–121.

Ottley, J. R. 1978. A new subspecies of the snake *Lichanura trivirgata* from Cedros Island, Mexico. Great Basin Nat. 38:411–416.

Ottley, J. R., and L. E. Hunt. 1981. *Crotalus viridis helleri:* Geographic distribution. Herpetol Rev. 12:65.

Ottley, J. R., and E. E. Jacobsen. 1983. Pattern and coloration of juvenile *Elaphe rosaliae,* with notes on natural history. J. Herpetol. 17:189–190.

Ottley, J. R., and R. W. Murphy. 1981. *Petrosaurus thalassinus repens:* Geographic distribution. Herpetol. Rev. 12:65.

———. 1983. *Gerrhonotus paucicarinatus:* Geographic distribution. Herpetol. Rev. 14:27.

Ottley, J. R., R. W. Murphy, and G. V. Smith. 1980. The taxonomic status of the rosy boa *Lichanura roseofusca* (Serpentes: Boidae). Great Basin Nat. 40:59–62.

Ottley, J. R., and W. W. Tanner. 1978. New range and a new subspecies for the snake *Eridiphas slevini.* Great Basin Nat. 38:406–410.

Ottley, J. R., and V. M. Velazques. 1989. An extant, indigenous tortoise population in Baja California Sur, Mexico, with the description of a new species of *Xerobates* (Testudines: Testudinidae). Great Basin Nat. 49:496–502.

Papenfuss, T. J. 1982. The ecology and systematics of the amphisbaenian genus *Bipes.* Occ. Pap. California Acad. Sci. 136:1–42.

Papenfuss, T. J., J. R. Macey, and J. A. Schulte II. 2001. A new lizard species in the genus *Xantusia* from Arizona. Sci. Pap. Nat. Hist. Mus. Univ. of Kansas. 23:1–9.

Parker, W. S. 1972. Aspects of the ecology of Sonoran Desert population of the western banded gecko, *Coleonyx variegatus.* Amer. Nat. 88:20–24.

———. 1974. *Phrynosoma solare.* Cat. Amer. Amph. Rept. 162.1–2.

———. 1982. *Sceloporus magister.* Cat. Amer. Amph. Rept. 290.1–4.

Parker, W. S., and E. R. Pianka. 1974. Further ecological observations on the western banded gecko, *Coleonyx variegatus.* Copeia 1974:528–531.

Pase, C. P. 1982. Sierran montane conifer forest. In D. E. Brown, ed., Biotic communities of the American southwest: United States and Mexico, 49–51. Desert Plants 4:1–342.

Peters, J. A. 1964. Dictionary of herpetology: A brief and meaningful definition of words and terms used in herpetology. New York: Hafner Publishing Co.

Peterson, J. A., and R. L. Bezy. 1985. The microstructure and evolution of scale surfaces in xantusiid lizards. Herpetologica 41:298–324.

Petron, K., and T. J. Case. 1997. A phylogenetic analysis of body size evolution and biogeography in chuckwallas *(Sauromalus)* and other iguanines. Evolution 51:206–219.

Phelan, R. L., and B. H. Brattstrom. 1955. Geographic variation in *Sceloporus magister.* Herpetologica 11:1–14.

Pianka, E. R. 1991. *Phrynosoma platyrhinos.* Cat. Amer. Amph. Rept. 517.1–4.

Pianka, E. R., and W. S. Parker. 1972. Ecology of the iguanid lizard *Callisaurus draconoides.* Copeia 1972:493–508.

———. 1975. Ecology of the horned lizards: Review with special reference to *Phrynosoma platyrhinos.* Copeia 1975:141–162.

Pickwell, G. V., R. L. Bezy, and J. E. Fitch. 1983. Northern occurrences of the sea snake, *Pelamis platurus,* in the eastern Pacific, with a record of predation on the species. California Fish and Game 69:172–177.

Pickwell, G. V., and W. A. Culotta. 1980. *Pelamis platurus.* Cat. Amer. Amph. Rept. 255.1–4.

Platz, J. E., R. W. Clarkson, J. C. Rorabaugh, and D. M. Hillis. 1990. *Rana berlandieri:* recently introduced populations in Arizona and southeastern California. Copeia 1990:324–333.

Platz, J. E., and J. S. Frost. 1984. *Rana yavapaiensis,* a new species of leopard frog (*Rana pipiens* complex). Copeia 1984:940–948.

Pough, F. H. 1969. Physiological aspects of the burrowing sand lizards (*Uma,* Iguanidae) and other lizards. Comp. Biochem. Physiol. 31:869–884.

———. 1977. *Uma notata.* Cat. Amer. Amph. Rept. 197.1–1.

Pough, F. H., R. M. Andrews, J. E. Cadle, M. L. Crump, A. H. Savitzky, and K. D. Wells. 2001. Herpetology. 2nd ed. Upper Saddle River, N.J: Prentice-Hall.

Powers, A. L., and B. H. Banta. 1974. *Chilomeniscus straminius* (Cope) recorded from Cerralvo Island, Gulf of California, Mexico. J. Herpetol. 8:386–387.

Price, A. H., and B. K. Sullivan. 1988. *Bufo microscaphus.* Cat. Amer. Amph. Rept. 415.1–3.

Price, R. M. 1990. *Bogertophis rosaliae.* Cat. Amer. Amph. Rept. 498.1–3.

Pritchard, P. C. H. 1971. The leatherback or leathery luth. IUCN Monogr. 1:1–39.

———. 1979. Encyclopedia of turtles. Jersey City, N.J.: T. F. H. Publications.

———. 1980. *Dermochelys, D. coriacea.* Cat. Amer. Amph. Rept. 238.1–4.

———. 1983. Review of Conserving sea turtles, by N. Mrosovsky. Copeia 1983:1108–1111.

———. 1997. Evolution, phylogeny, and current status. In P. L. Lutz and J. A. Musik, eds., The biology of sea turtles, 1–28. New York: CRC Press.

Pritchard, P. C. H., and P. Trebbau. 1984. The turtle of Venezuela. Soc. Stud. Amph. Rept. Contrib. Herpetol. 2:1–403.

Quijada-M., A. 1992. Feeding and foraging of *Callisaurus draconoides* (Sauria: Phrynosomatidae) in the intertidal zone of coastal Sonora. Southwest. Nat. 37:311–314.

Radcliffe, C. W., and T. P. Maslin. 1975. A new subspecies of the red rattlesnake, *Crotalus ruber,* San Lorenzo Sur Island, Baja California Norte, Mexico. Copeia 1975:490–493.

Radtkey, R. R., S. M. Fallon, and T. J. Case. 1997. Character displacement in some *Cnemidophorus* lizards revisited: A phylogenetic analysis. Proc. National Acad. Sci. 94:9740–9745.

Rathbun, G. B., N. Siepel, and D. Holland. 1992. Nesting behavior and movements of Western Pond Turtles, *Clemmys marmorata.* Southwest Nat. 37:319–329.

Rau, C. S. 1980. The genus *Urosaurus* (Reptilia, Lacertilia, Iguanidae) of Baja California, Mexico. M.S. thesis, California State Univ., Long Beach.

Rau, C. S., and R. B. Loomis. 1977. A new species of *Urosaurus* (Reptilia, Lacertilia, Iguanidae) from Baja California, Mexico. J. Herpetol. 11:25–29.

Reeder, T. W. 1995. Phylogenetic relationships among phrynosomatid lizards as inferred from mitochondrial ribosomal DNA sequences: Substitutional bias and information content of transitions relative to transversions. Molec. Phylo. Evol. 4:203–222.

Reeder T. W., and R. R. Montanucci. 2001. Phylogenetic analysis of the horned lizards (Phrynosomatidae: *Phrynosoma*): Evidence from mitochondrial DNA and morphology. Copeia 2001:309–323.

Reeder, T. W., and J. J. Wiens. 1996. Evolution of the lizard family Phrynosomatidae as inferred from diverse types of data. Herpetol. Monogr. 10:43–84.

Reeve, W. L. 1952. Taxonomy and distribution of the horned lizard genus *Phrynosoma.* Univ. of Kansas Sci. Bull. 34:817–960.

Reichling, S. B. 1995. The taxonomic status of the Louisiana pine snake *(Pituophis melanoleucus ruthveni)* and its relevance to the evolutionary species concept. J. Herpetol. 29:186–198.

Resendiz, A., W. J. Nichols, J. A. Seminoff, and N. Kamezaki. 1998. One-way transpacific migration of Loggerhead sea turtles *(Caretta caretta)* as determined through flipper tag recovery and satellite tracking. In S. P. Epperly and J. Braun, compilers, Proc. 17th Ann. Sea Turtle Symposium, 253. NOAA Tech. Memo. NMFS-SEFC-415.

Resendiz, A., B. Resendiz, W. J. Nichols, J. A. Seminoff, and N. Kamezaki. 1998. First confirmed east-west transpacific movement of loggerhead sea turtle, *Caretta caretta,* released in Baja California, Mexico. Pacific Sci. 52:151–153.

Reyes-Osario, S., and R. B. Bury. 1982. Ecology of the desert tortoise *(Gopherus agassizii)* on Tiburon Island. In R. B. Bury, ed., North American tortoises: Conservation and ecology, 39–49 . U.S. Fish and Wildlife Service, Wildlife Research Report 12.

Reynolds, F. A. 1943. Notes on the western glossy snake in captivity. Copeia 1943:196.

Reynolds, R. P. 1984. Records of the yellow-bellied sea snake, *Pelamis platurus,* from the Galápagos Islands. Copeia 1984:786–789.

Reynoso, F. 1990a. *Gehyra mutilata:* Geographic distribution. Herpetol. Rev. 21:22.

———. 1990b. *Hemidactylus frenatus:* Geographic distribution. Herpetol. Rev. 21:22.

———. 1990c. *Lichanura trivirgata trivirgata:* Geographic distribution. Herpetol. Rev. 21:23.

———. 1990d. *Arizona elegans pacata:* Geographic distribution. Herpetol. Rev. 21:23.

———. 1990e. *Eridiphas slevini slevini:* Geographic distribution. Herpetol. Rev. 21:23.

Richmond, N. D. 1965. Distribution of *Gerrhonotus paucicarinatus* Fitch. Copeia 1965:375.

Roberts, N. 1982. A preliminary report on the status of Chelydridae, Trionychidae, and Testudinidae in the region of Baja California, Mexico. Proc. Desert Tortoise Council 1982:154–161.

Robinson, M. D. 1972. Chromosomes, protein polymorphism, and systematics of insular chuckwalla lizards (genus *Sauromalus*) in the Gulf of California, Mexico. Ph.D. diss., Univ. of Arizona.

———. 1973. Chromosomes and systematics of the Baja California whiptail lizards *Cnemidophorus hyperythrus* and *C. ceralbensis* (Reptilia: Teiidae). Syst. Zool. 22:30–35.

———. 1974. Chromosomes of the insular species of chuckwalla lizards (genus *Sauromalus*) in the Gulf of California, Mexico. Herpetologica 30:162–167.

Rodgers, T. L., and H. S. Fitch. 1947. Variation in the skinks (Reptilia: Lacertilia) of the *skiltonianus* group. Univ. of California Pub. Zool. 48:169–220.

Rodman, G. B., Jr. 1939. Habits of *Trimorphodon vandenburghi* in captivity. Copeia 1939:50.

Rodríguez-Robles, J. A. 1998. Alternative perspectives on the diet of gopher snakes (*Pituophis catenifer,* Colubridae): Literature records versus stomach contents of wild and museum specimens. Copeia 1998:463–466.

Rodríguez-Robles, J. A., C. J. Bell, and H. W. Greene. 1999a. Gape size and evolution of diet in snakes: Feeding ecology of erycine boas. J. Zool. Lond. 248:49–58.

———. 1999b. Food habits of the glossy snake, *Arizona elegans,* with comparisons to the diet of sympatric long-nosed snakes, *Rhinocheilus lecontei.* J. Herp. 33:87–92.

Rodríguez-Robles, J. A., D. F. DeNardo, and R. E. Staub. 1999. Phylogeography of the California mountain kingsnake, *Lampropeltis zonata* (Colubridae). Molec. Ecol. 8:1923–1934.

Rodríguez-Robles, J. A., and H. W. Greene. 1999. Food habits of the long-nosed snake *(Rhinocheilus lecontei),* a "specialist" predator. J. Zool. London 248:489–499.

Rodríguez-Robles, J. A., and J. M. de Jesús-Escobar. 1999. Molecular systematics of New World lampropeltine snakes (Colubridae): Implications for biogeography and food habits. Biol. J. Linn. Soc. 68:355–385.

———. 2000. Molecular systematics of New World gopher, bull, and pinesnakes (*Pituophis:* Colubridae), a transcontinental species complex. Molec. Phylo. Evol. 14:35–50.

Rodríguez-Robles, J. A., D. G. Mulcahy, and H. W. Greene. 1999. Feeding ecology of the desert nightsnake, *Hypsiglena torquata* (Colubridae). Copeia 1999:93–98.

Rogers, J. S. 1972. Discriminant function analysis of morphological relationships within *Bufo cognatus* species group. Copeia 1972:381–382.

Romero-Schmidt, H. L., and A. Ortega-Rubio. 2000. Reproduction of the Cape Orange-Throat Whiptail *Cnemidophorus hyperythrus hyperythrus.* Herpetol. Nat. Hist. 7:1–8.

Romero-Schmidt, H. L., A. Ortega-Rubio, and M. Acevedo-Beltrán. 1999. Reproductive characteristics of the black-tailed brush lizard, *Urosaurus nigricaudus* (Phrynosomatidae). Rev. Biol. Trop. 16:21–38.

Rosenberg, J. 1997. Documenting and revitalizing traditional ecological knowledge: the curriculum development component of the Seri Ethnozoology Education Project. M.S. thesis, Univ. of Arizona.

Rossman, D. A., N. B. Ford, and R. A. Seigel. 1996. The garter snakes: Evolution and ecology. Norman: Univ. of Oklahoma Press.

Rossman, D. A., and G. R. Stewart. 1987. Taxonomic reevaluation of *Thamnophis couchii* (Serpentes: Colubridae). Occ. Pap. Mus. Zool. Louisiana State Univ. 63:1–25.

Roze, J. A. 1974. *Micruroides euryxanthus.* Cat. Amer. Amph. Rept. 163.1–4

———. 1996. Coral snakes of the Americas: Biology, identification, and venoms. Malabar, Fla.: Krieger Publishing Co.

Rubio, M. 1999. Rattlesnake: Portrait of a predator. Washington, D.C.: Smithsonian Institution Press.

Ruth, S. B. 1974. A kingsnake from Isla Tortuga in the Gulf of California, Mexico. Herpetologica 30:97–98.

Sánchez-Pacheco, J. A., and E. Mellink. 2001. *Anniella geronimensis:* Geographic distribution. Herpetol. Rev. 32:192.

Sarti, L. M., S. A. Eckert, N. T. Garcia, and A. R. Barragan. 1996. Decline of the world's largest nesting assemblage of leatherback turtles. Marine Turtle Newslt. 74:2–5.

Savage, J. M. 1952. Studies on the lizard family Xantusiidae, I: The systematic status of the Baja California night lizards allied to *Xantusia vigilis,* with the description of a new subspecies. Amer. Midl. Nat. 48:467–479.

———. 1954. Notulae herpetologicae 1–7. 1. Range extensions for amphibians in Sonora, Mexico. 2. *Sphaerodactylus argus argus* from Cuba and Key West. 3. The status of *Sphaerodactylus samanensis* Cochran, 1932. 4. On the validity of *Cnemidophorus gadaovi* Burger, 1950. 5. The origin of the types of *Cnemidophorus labialis* Stejneger, 1894. 6. The distribution of *Phrynosoma solare* Gray, 1845. 7. The generic type of *Zonosaurus* Boulenger, 1887, with notes on a method of citing type species. Trans. Kansas Acad. Sci. 57:326–334.

———. 1960. Evolution of a peninsular herpetofauna. Syst. Zool. 9:184–212.

———. 1967. Evolution of the insular herpetofauna. Proc. Symp. Biology of the California Islands, 219–227. Santa Barbara: Santa Barbara Botanic Garden.

Savage, J. M., and F. S. Cliff. 1953. A new subspecies of sidewinder, *Crotalus cerastes,* from Arizona. Chicago Acad. Sci. Nat. Hist. Misc. Pub. 119:1–7.

———. 1954. A new snake, *Phyllorhynchus arenicola,* from the Gulf of California, Mexico. Proc. Biol. Soc. Washington 67:69-76.

Savitzky, A. H. 1995. An osteological synapomorphy uniting a lineage of North American colubrid snakes. Program and Abstracts, ASIH-HL Meetings, 175.

Schmidt, K. P. 1922. The amphibians and reptiles of Lower California and the neighboring islands. Bull. Amer. Mus. Nat. Hist. 46:607–707.

———. 1953. Checklist of North American amphibians and reptiles. Chicago: Pub. Amer. Soc. Ichthyol. Herpetol.

Schwaner, T. D., and H. C. Dessauer. 1982. Comparative immunodiffusion survey of snake transferrins focused on the relationships of the natricines. Copeia 1982:541–549.

Scott, N. J., Jr., and R. W. McDiarmid. 1984. *Trimorphodon biscutatus.* Cat. Amer. Amph. Rept. 353.1–4.

Scudder, K. M., A. L. Powers, and H. M. Smith. 1983. Comparisons of the desert iguanas *(Dipsosaurus)* from Cerralvo Island and adjacent Baja California, Mexico. Trans. Kansas Acad. Sci. 86:149–153.

Seeliger, L. M. 1945. Variation in the Pacific mud turtle. Copeia 1945:150–162.

Seifert, W. 1980. *Arizona elegans:* Geographic distribution. Herpetol. Rev. 11:39.

Seminoff, J. A., W. J. Nichols, and A. Resendiz. 1998. Diet composition of the black sea turtle, *Chelonia mydas agassizii,* in the central Gulf of California. In S. P. Epperly and J. Braun, compilers, Proc. 17th Ann. Sea Turtle Symposium, 89–91. NOAA Tech. Memo. NMFS-SEFC.

Seminoff, J. A., W. A. Nichols, A. Resendiz, and A. Galvan. 1999. Diet composition of the black sea turtle, *Chelonia mydas agassizii,* near Baja California, Mexico. In F. A. Abreu-Grobois, R. Briseño, R. Márquez, and L. Sarti, compilers, Proc. 18th Ann. Symp. Sea Turtle Biology and Conservation, 166–168. NOAA Tech. Memo NMFS-SEFC.

Seminoff, J. A., W. A. Nichols, A. Resendiz, and S. Hidalgo. 2000. *Chelonia mydas agassizii* (east Pacific green turtle): Diet. Herpetol. Rev.31:103.

Shaw, C. E. 1940. A new species of legless lizard from San Geronimo Island, Lower California. Trans. San Diego Soc. Nat. Hist. 9:225–228.

———. 1941. A new chuckwalla from Santa Catalina Island, Gulf of California, Mexico. Trans. San Diego Soc. Nat. Hist. 9:285–288.

———. 1943. Hatching of the eggs of the San Diego alligator lizard. Copeia 1943:194.

———. 1945. The chuckwallas, genus *Sauromalus.* Trans. San Diego Soc. Nat. Hist. 10:296–306.

———. 1947. First records of the red-brown loggerhead turtle from the eastern Pacific. Herpetologica 4:55–56.

———. 1949. *Anniella geronimensis* on the Baja California peninsula. Herpetologica 5:27–28.

———. 1952. Notes on the eggs and young of some United States and Mexican lizards, I. Herpetologica 8:71–79.

———. 1953. *Anniella pulchra* and *Anniella geronimensis,* sympatric species. Herpetologica 8:167–170.

———. 1961. Snakes of the sea. Zoonooz 34:3–5.

Shaw, G. 1802. General zoology or systematic natural history. Vol. 3. London: Thomas Davidson.

Shreve, F. 1936. The transition from desert to chaparral in Baja California. Madroño 3:257–320.

———. 1951. Vegetation and flora of the Sonoran Desert, Vol. 1: Vegetation. Carnegie Inst. Washington Pub. 591:1–192.

Shreve, F., and I. R. Wiggins. 1964. Vegetation and flora of the Sonoran Desert. 2 vols. Stanford: Stanford Univ. Press.

Sites, J. W., Jr., J. W. Archie, C. J. Cole, and O. Flores-V. 1992. A review of phylogenetic hypotheses for lizards of the genus *Sceloporus* (Phrynosomatidae): Implications for ecological and evolutionary studies. Bull. Amer. Mus. Nat. Hist. 213:1–110.

Sites, J. W., Jr., S. K. Davis, T. Guerra, J. B. Iverson, and H. L. Snell. 1996. Character congruence and phylogenetic signal in molecular and morphological data sets: A case study in the living iguanas (Squamata, Iguanidae). Molec. Biol. Evol. 13:1087–1105.

Slowinski, J. B. 1995. A phylogenetic analysis of New World coral snakes (Elapidae: *Leptomicrurus, Micruroides,* and *Micrurus*) based on allozymic and morphological characters. J. Herpetol. 29:325–338.

Smith, D.C. 1977. Interspecific competition and the demography of two lizards. Ph.D. diss., Univ. of Michigan.

Smith, H. M. 1935. Miscellaneous notes on Mexican lizards. Univ. of Kansas Sci. Bull. 22:119–155.

———. 1939. The Mexican and Central American lizards of the genus *Sceloporus.* Field Mus. Nat. Hist. Zool. Ser. 26:1–397.

———. 1940. Descriptions of new lizards from Mexico and Guatemala. Proc. Biol. Soc. Washington 53:55–64.

———. 1972. The Sonoran subspecies of the lizard *Ctenosaura hemilopha.* Great Basin Nat. 32:104–111.

Smith, H. M., and R. L. Holland. 1971. Noteworthy snakes and lizards from Baja California. J. Herpetol. 5:56–59.

Smith, H. M., R. L. Holland, and R. L. Brown. 1971. The prairie rattlesnake in Baja California. J. Herpetol. 5:200.

Smith, H. M., and R. B. Smith. 1976. Synopsis of the herpetofauna of Mexico, vol. III: Source analysis and index for Mexican reptiles. North Bennington, Vt.: John Johnson.

———. 1979. Synopsis of the herpetofauna of Mexico, vol.

VI: Guide to Mexican turtles, bibliographic addendum III. North Bennington, Vt.: John Johnson.

Smith, H. M., and W. W. Tanner. 1972. Two new subspecies of *Crotaphytus* (Sauria: Iguanidae). Great Basin Nat. 32:25–34.

Smith, H. M., and E. H. Taylor. 1945. An annotated checklist and key to the snakes of Mexico. Bull. U.S. Nat. Mus. 187:1–239.

———. 1948. An annotated checklist and key to the amphibia of Mexico. Bull. U.S. National Mus. 194:1–118.

———. 1950. An annotated checklist and key to the reptiles of Mexico, exclusive of the snakes. Bull. U.S. National Mus. 199:1–253.

Soulé, M., and A. J. Sloan. 1966. Biogeography and distribution of the reptiles and amphibians on islands in the Gulf of California, Mexico. Trans. San Diego Soc. Nat. Hist. 14:137–156.

Spiteri, D. E. 1986. Taxonomy of the rosy boa *(Lichanura),* with a description of a new subspecies. M.S. thesis, California State Univ., Fullerton.

———. 1988. Geographic variability of the species *Lichanura trivirgata* and a description of a new subspecies. In H. F. De Lisle, P. R. Brown, B. Kaufman, and B. M. McGurty, eds., Proc. Conf. California Herpetology, 113–130. Van Nuys, California: Southwestern Herpetologists Society Spec. Pub. No. 4.

———. 1992. The questionable status of *Lichanura trivirgata bostici,* the Cedros Island boa. Bull. Chicago Herpetol. Soc. 27:181.

———. 1994. Further analysis of the Cedros Island boa, *Lichanura trivirgata bostici.* Bull. Chicago Herpetol. Soc. 29:248–250.

Staedeli, J. H. 1963. Island reptile prizes. Zoonooz 36:11–15.

Starrett, B. L. 1993. *Crotalus viridis helleri:* Geographic distribution. Herpetol Rev. 24:109.

Stebbins, R. C. 1944. Some aspects of the ecology of the iguanid genus *Uma.* Ecol. Monogr. 14:311–322.

———. 1954. Amphibians and Reptiles of western North America. New York: McGraw-Hill Book Co.

———. 1962. Amphibians of western North America. Berkeley: Univ. of California Press.

———. 1985. A field guide to western reptiles and amphibians. 2nd rev. ed. Boston: Houghton Mifflin Co.

Stejneger, L. 1890a. Annotated list of the reptiles and batrachians collected by Dr. C. Hart Merriam and Vernon Bailey on the San Francisco Mountain plateau and the desert of the little Colorado, Arizona, with descriptions of new species. N. Amer. Fauna 3:103–118.

———. 1890b. Description of a new lizard from Lower California. Proc. U.S. National Mus. 12:643–644.

———. 1891a. Description of a new North American lizard of the genus *Sauromalus.* Proc. U.S. National Mus. 14:409–411.

———. 1891b. Description of a new species of lizard from the island of San Pedro Mártir, Gulf of California. Proc. U.S. National Mus. 14:407–408.

———. 1893a. Annotated list of reptiles and batrachians collected by the Death Valley expedition in 1891, with descriptions of new species. N. Amer. Fauna 7:159–228.

———. 1893b. Identification of a new California lizard. Proc. U.S. National Mus. 16:467.

———. 1894. Description of *Uta mearnsi,* a new lizard from California. Proc. U.S. National Mus. 17:589–591.

Stejneger, L., and T. Barbour. 1943. A check list of North American amphibians and reptiles. 5th ed. Bull. Mus. Comp. Zool. 93:1–260.

Stickel, W. H. 1943. The Mexican snakes of the genera *Sonora* and *Chionactis,* with notes on the status of other colubrid genera. Proc. Biol. Soc. Washington 56:109–123.

Stille, B. 1987. Dorsal scale microdermatoglyphics and rattlesnake *(Crotalus* and *Sistrurus)* phylogeny (Reptilia: Viperidae: Crotalinae). Herpetologica 43:98–104.

Stock, J. M., and K. V. Hodges. 1989. Pre-Pliocene extension around the Gulf of California and the transfer of Baja California to the Pacific Plate. Tectonics 8:99–115.

Storer, T. I. 1925. A synopsis of the amphibia of California. Univ. of California Pub. Zool. 27:1–342.

Streets, T. H. 1877. Contributions to the natural history of the Hawaiian and Fanning islands and Lower California. Bull. U.S. National Mus. 7:1–172.

Stull, O. G. 1940. Variations and relationships in the snakes of the genus *Pituophis.* Bull. U.S. National Mus. 175:1–225.

Sullivan, B. K. 1992. Calling behavior of the southwestern toad *(Bufo microscaphus).* Herpetologica 48:383–389.

Sullivan, B. K., and K. B. Malmos. 1994. Call variation in the Colorado River toad *(Bufo alvarius):* Behavioral and phylogenetic implications. Herpetologica 50:146–156.

Sullivan, B. K., K. B. Malmos, and M. F. Given. 1996. Systematics of the *Bufo woodhousii* complex (Anura: Bufonidae): Advertisement call variation. Copeia 1996:274–280.

Sweet, S. S., and W. S. Parker. 1990. *Pituophis melanoleucus.* Cat. Amer. Amph. Rept. 474.1–8.

Sylber, C. K. 1985a. Feeding habits, reproduction, and relocation of the insular giant chuckwallas. Ph.D. diss., Colorado State Univ.

———. 1985b. Eggs and hatchlings of the yellow giant chuckwalla and the black giant chuckwalla in captivity. Herpetol. Rev. 16:18–21.

———. 1988. Feeding habits of the lizards *Sauromalus varius* and *S. hispidus* in the Gulf of California. J. Herpetol. 22:413–424.

Tanner, W. W. 1943. Two new species of *Hypsiglena* from western North America. Great Basin Nat. 4:49–54.

———. 1944. A taxonomic study of the genus *Hypsiglena.* Great Basin Nat. 5:25–92.

———. 1954. Additional note on the genus *Hypsiglena* with a description of a new subspecies. Herpetologica 10:54–56.

———. 1955. A new *Sceloporus magister* from eastern Utah. Great Basin Nat. 15:32–34.

———. 1966a. The night snakes of Baja California. Trans. San Diego Soc. Nat. Hist. 14:189–196.

———. 1966b. A re-evaluation of the genus *Tantilla* in the southwestern United States and northwestern Mexico. Herpetologica 22:134–152.

———. 1981. A new *Hypsiglena* from Tiburon Island, Sonora, Mexico. Great Basin Nat. 14:189–196.

———. 1985. Snakes of western Chihuahua. Great Basin Nat. 45:615–676.

———. 1988. *Eumeces skiltonianus.* Cat. Amer. Amph. Rept. 447.1–4.

Tanner, W. W., and B. H. Banta. 1962. Description of a new *Hypsiglena* from San Mártin Island, México, with a résumé of the reptile fauna of the island. Herpetologica 18:21–25.

Taylor, E. H. 1922. The lizards of the Philippine Islands. Philippine Bur. Sci. Pub. 17:1–269.

———. 1935. A taxonomic study of the cosmopolitan scincoid lizards of the genus *Eumeces,* with an account of the distribution and relationships of its species. Univ. of Kansas Sci. Bull. 23:1–643.

———. 1950. Fourth contribution to the herpetology of San Luis Potosí. Univ. of Kansas Sci. Bull. 35:1587–1614.

Taylor, E. N. 1999. Diet and reproductive biology of the Baja California rattlesnake, *Crotalus enyo:* Program Book and Abstracts, Joint Meetings of ASIH, HL, and SSAR, 217.

Taylor, E. N. 2001. Diet of the Baja California rattlesnake, *Crotalus enyo* (Viperidae). Copeia 2001:553–555.

Taylor, H. L., and J. M. Walker. 1987. Reproductive characteristics of the Cerralvo Island whiptail lizard, *Cnemidophorus ceralbensis* (Teiidae). Southwest. Nat. 32:391–413.

———. 1996. Application of the names *Cnemidophorus tigris disparilis* and *C. t. punctilinealis* to valid taxa (Sauria: Teiidae) and relegation of the names *C. t. gracilis* and *C. t. dickersonae* to appropriate synonymies. Copeia 1996:140–148.

Tennant, A. 1984. The snakes of Texas. Austin: Texas Monthly Press.

Terron, C. C. 1920. Datos para una monografía sobre la fauna erpetológica de la península de la Baja California. Boln. Dir. Estud. Biol. Mex. 2:398–402.

———. 1921. Datos para una monografía sobre la fauna erpetológica de la península de la Baja California. Mems. Revta. Soc. Cient. "Antonio Alzate" 39:161–171.

Tevis, L., Jr. 1944. Herpetological notes from Lower California. Copeia 1944:6–18.

Thomas, C., and S. Scott. 1997. All Stings Considered. Honolulu: Univ. of Hawaii Press.

Thompson, J. S., B. I. Crother, and A. H. Price. 1998. *Cnemidophorus hyperythrus.* Cat. Amer. Amph. Rept. 655.1–6.

Thorne, R. F. 1976. The vascular plant communities of California. In J. Latting, ed., Plant communities of Southern California, 1–31. California Native Plant Soc. Spec. Pub. No. 2.

Tinkle, D. W., and A. E. Dunham. 1986. Comparative life histories of two syntopic sceloporine lizards. Copeia 1986:1–18.

Townsend, C. H. 1916. Voyage of the *Albatross* to the Gulf of California. Bull. Amer. Mus. Nat. Hist. 35:399–476.

Turner, F. B., and P. A. Medica. 1982. The distribution and abundance of the flat-tailed horned lizard *(Phrynosoma mcallii).* Copeia 1982:815–823.

Turner, R. M., and D. E. Brown. 1982. Tropical-subtropical desertlands. In D. E. Brown, ed., Biotic communities of the American Southwest United States and Mexico, 180–221. Desert Plants 4:1–342.

Upton, D. E., and R. W. Murphy. 1997. Phylogeny of the side-blotched lizards (Phrynosomatidae: *Uta*), based on mtDNA sequences: Support for a midpeninsular seaway in California. Molec. Phylo. Evol. 8:109–13.

Van Denburgh, J. 1895. Review of the herpetology of Lower California. Proc. California Acad. Sci. 2:77–163.

———. 1896. Description of a new lizard *(Eumeces gilberti)* from the Sierra Nevada of California. Proc. California Acad. Sci. 6:350–352.

———. 1905. The reptiles and amphibians of the islands of the Pacific coast of North America from the Farallons to Cape San Lucas and the Revilla Gigedos. Proc. California Acad. Sci. 4:1–41.

———. 1916. Four species of salamanders new to the state of California, with a description of *Plethodon elongatus,* a new species, and notes on other salamanders. Proc. California Acad. Sci., 4th ser., 6:215–221.

———. 1920. Description of a new lizard *(Dipsosaurus dorsalis lucasensis)* from Lower California. Proc. California Acad. Sci., 4th ser., 10:33–34.

———. 1922. The reptiles of western North America. Proc. California Acad. Sci., Vols. 1–2:1–1028.

Van Denburgh, J., and J. R. Slevin. 1914. Reptiles and amphibians of islands of the west coast of North America. Proc. California Acad. Sci. 4:129–152.

———. 1918. The garter-snakes of western North America. Proc. California Acad. Sci. 8:181–270.

———. 1921a. A list of amphibians and reptiles of the peninsula of Lower California, with notes on the species in the California Academy. Proc. California Acad. Sci. 11:27–72.

———. 1921b. Preliminary diagnoses of new species of reptiles from islands in the Gulf of California, Mexico. Proc. California Acad. Sci. 11:95–98.

———. 1921c. Preliminary diagnoses of more new species of reptiles from islands in the Gulf of California, Mexico. Proc. California Acad. Sci. 11:395–398.

———. 1923. Preliminary diagnoses of four new snakes from Lower California, Mexico. Proc. California Acad. Sci. 13:1–2.

Vitt, L. J. 1975. Observations on reproduction in five species of Arizona snakes. Herpetologica 31:83–84.

Vitt, L. J., J. D. Congdon, A. C. Hulse, and J. R. Platz. 1974. Territorial aggressive encounters and tail breaks in the lizard *Sceloporus magister.* Copeia 1974:990–992.

Vitt, L. J., and R. D. Ohmart. 1978. Herpetofauna of the lower Colorado River: Davis dam to the Mexican border. Western Found. Vert. Zool. 2:35–72.

Wake, D. B. 1966. Comparative osteology and evolution of the lungless salamanders, family Plethodontidae. Mem. So. California Acad. Sci. 4:1–111.

———. 1996. A new species of *Batrachoseps* (Amphibia: Plethodontidae) from the San Gabriel Mountains, southern California. Contrib. Sci. Los Angeles Co. Mus. Nat. Hist. 463:1–12.

Wake, D. B., and E. L. Jockusch. 2000. Detecting species borders using diverse data sets: Examples from the plethodontid salamanders in California. In B. Bruce, ed., The Biology of Plethodontid Salamanders, 95–119. New York: Plenum Publishers.

Wake, D. B., and C. J. Schneider. 1998. Taxonomy of the plethodontid salamander genus *Ensatina.* Herpetologica 54:279–298.

Walker, J. M. 1966. On the status of the teiid lizard *Cnemidophorus celeripes* Dickerson. Copeia 1966:373–376.

———. 1980. Reproductive characteristics of the San Pedro Martir whiptail, *Cnemidophorus martyris.* J. Herpetol. 14:431–432.

———. 1981a. A new subspecies of *Cnemidophorus tigris* from south Coronado Island, Mexico. J. Herpetol. 15:193–197.

———. 1981b. On the status of the lizard *Cnemidophorus tigris dickersonae* Van Denburgh and Slevin. J. Herpetol. 15:199–206.

Walker, J. M., and T. P. Maslin. 1969. A review of the San Pedro Nolasco whiptail lizard (*Cnemidophorus bacatus* Van Denburgh and Slevin). Amer. Midl. Nat. 82:127–139.

———. 1981. Systematics of the Santa Catalina whiptail *(Cnemidophorus catalinensis)* with reference to the superspecies *Cnemidophorus tigris.* Amer. Midl. Nat. 105:84–92.

Walker, J. M., and H. L. Taylor. 1968. Geographical variation in the teiid lizard *Cnemidophorus hyperythrus,* I: The *caeruleus*-like subspecies. Am. Midl. Nat. 80:1–27.

Walker, J. M., H. L. Taylor, and T. P. Maslin. 1966a. Evidence for specific recognition of the San Esteban whiptail lizard *(Cnemidophorus estebanensis).* Copeia 1966:498–505.

———. 1966b. Morphology and relations of the teiid lizard, *Cnemidophorus ceralbensis.* Copeia 1966:585–588.

Wasserman, A. O. 1970. *Scaphiopus couchii.* Cat. Amer. Amph. Rept. 85.1–4.

Webb, R. G. 1962. North American recent soft-shelled turtles (Family Trionychidae). Univ. of Kansas Pub. Mus. Nat. Hist. 13:429–611.

Weintraub, J. D. 1968. Winter behavior of the granite spiny lizard, *Sceloporus orcutti* Stejneger. Copeia 1968: 708–712.

———. 1969. Size relationships of the granite spiny lizard, *Sceloporus orcutti.* Herpetologica 25:25–29.

———. 1980. *Sceloporus orcutti.* Cat. Amer. Amph. Rept. 265.1–2.

Wells, R., and C. R. Wellington. 1985. A classification of the amphibia and reptilia of Australia. Austral. J. Herpetol. Suppl. Ser. 1:1–61.

Welsh, H. H., Jr. 1988. An ecogeographic analysis of the herpetofauna of the Sierra San Pedro Mártir Region, Baja California, with a contribution to the biogeography of the Baja California herpetofauna. Proc. California Acad. Sci., 4th ser., 46:1–72.

Welsh, H. H., Jr., and R. B. Bury. 1984. Additions to the herpetofauna of the south Colorado Desert, Baja California, with comments on the relationships of *Lichanura trivirgata.* Herpetol. Rev. 15:53–56.

Wetherall, J. A., G. H. Balazs, R. A. Tokunaga, and M. Y. Y. Yong. 1993. Bycatch of marine turtles in North Pacific high-seas driftnet fisheries and impacts on the stocks. In J. Ito, W. Shaw, and R. L. Burgner, eds., International North Pacific Fisheries Commission Symp. Biology, Distribution and Stock Assessment of Species Caught in the High-Seas Driftnet Fisheries in the North Pacific Ocean, 519–538. Vancouver, Canada: International North Pacific Fisheries Commission, Bull. 53 (III).

Wiegmann, A. F. A. 1835. Beiträge zur Zoologie, gesammelt auf einer Reise um die Erde, von Dr. F. J. F. Meyen, M.d.A.d.N. Amphibien. Acta Acad. caes. leop. carol. Nat. Cur. 17:184–268.

Wiens, J. J. 1993a. Phylogenetic relationships among phrynosomatid lizards and monophyly of the *Sceloporus* group. Copeia 1993:287–299.

———. 1993b. Phylogenetic systematics of the tree lizards (genus *Urosaurus*). Herpetologica 49:399–420.

Wiens, J. J., and T. W. Reeder. 1997. Phylogeny of the spiny lizards *(Sceloporus)* based on molecular and morphological evidence. Herpetol. Monogr. 11:1–101.

Wiens, J. J., and T. A. Titus. 1991. A phylogenetic analysis of *Spea* (Anura: Pelobatidae). Herpetologica 47:21–28.

Wiggins, I. L. 1980. Flora of Baja California. Stanford: Stanford Univ. Press.

Wilcox, B. A. 1980. Species number, stability, and equilibrium status of reptile faunas on the California islands. In D. M. Power, ed., The California Islands: Proc. Multidisciplinary Symp., 551–564. Santa Barbara: Santa Barbara Museum of Natural History.

Wiley, E. O. 1978. The evolutionary species concept reconsidered. Syst. Zool. 29:76–80.

Wilgenbusch, J., and K. de Queiroz. 2000. Phylogenetic

relationships among the phrynosomatid sand lizards inferred from mitochondrial DNA sequences generated by heterogeneous evolutionary processes. Syst. Biol. 49:592–612.

Williams, K. L. 1988. Systematics and natural history of the American milk snake, *Lampropeltis triangulum.* Milwaukee: Milwaukee Public Museum.

Wilson, L. D. 1971. The coachwhip snake *Masticophis flagellum* (Shaw): Taxonomy and distribution. Tulane Stud. Zool. Bot. 16:31–99.

———. 1973. *Masticophis.* Cat. Amer. Amph. Rept. 145.1–4.

———. 1999. Checklist and key to the species of the genus *Tantilla* (Serpentes: Colubridae), with some commentary on distribution. Smithsonian Herpetol. Info. Service No. 122:1–36

Winker, C. D., and S. M. Kidwell. 1986. Paleocurrent evidence for lateral displacement of the Pliocene Colorado River delta by the San Andreas fault system, southeastern California. Geology 14:788–791.

Wong, H. 1997. Comments on the snake records of *Chilomeniscus cinctus, Crotalus exsul,* and *C. mitchellii* from islas Magdalena and Santa Margarita. Herpetol. Rev. 28:188–189.

Wright, J. W. 1993. Evolution of the whiptail lizards (genus *Cnemidophorus*). In J. W. Wright and L. J. Vitt, eds., Biology of whiptail lizards (Genus *Cnemidophorus*), 27–81. Norman: Oklahoma Mus. Nat. Hist.

Wyles, J. S., and G. C. Gorman. 1978. Close relationship between the lizard genus *Sator* and *Sceloporus utiformis* (Reptilia, Lacertilia, Iguanidae): Electrophoretic and immunological evidence. J. Herpetol. 12:343–350.

Yanev, K. P. 1980. Biogeography and distribution of three parapatric salamander species in coastal and borderland California. In D. M. Power, ed., The California islands: Proc. Multidisciplinary Symp., 531–555 . Santa Barbara: Santa Barbara Mus. of Nat. Hist.

Yarrow, H. C. 1882a. Checklist of North American reptilia and batrachia. Bull. U.S. Nat. Mus. 24:1–249.

———. 1882b. Description of new species of reptiles and amphibians in the U.S. National Museum. Proc. U.S. National Mus. 5:438–443.

———. 1883. Check list of North American reptilia and batrachia, with catalogue of specimens in U.S. National Museum. U.S. National Mus. Bull. 24:1–249.

Yingling, R. P. 1982. *Lichanura trivirgata.* Cat. Amer. Amph. Rept. 294.1–2.

Zangerl, R., L. P. Hendrickson, and J. R. Hendrickson. 1988. A redescription of the Australian flatback sea turtle, *Natator depressus.* Bishop Mus. Bull. Zool. 1:1–69.

Zug, G. R. 1996. *Chelonia agassizii:* Valid or not? Mar. Sci. News. 72:2.

Zug, G. R., C. H. Ernst, and R. V. Wilson. 1998. *Lepidochelys olivacea.* Cat. Amer. Amph. Rept. 653.1–13.

Zweifel, R. G. 1952a. Notes on the lizards of the Coronados Islands, Baja California, Mexico. Herpetologica 8:9–11.

———. 1952b. Pattern variation and evolution of the mountain kingsnake, *Lampropeltis zonata.* Copeia 1952:152–168.

———. 1958. Results of the Puritan-American Museum of Natural History expedition to western Mexico. 2. Notes on the reptiles and amphibians from the Pacific coastal islands of Baja California. Amer. Mus. Nat. Hist. 1895:1–17.

———. 1975. *Lampropeltis zonata.* Cat. Amer. Amph. Rept. 174.1–4.

———. 1981. Genetics of color pattern polymorphism in the California kingsnake. J. Hered. 72:238–244.

———. 1982. Color pattern morphs of the kingsnake *(Lampropeltis getulus)* in southern California: Distribution and evolutionary status. Bull. So. California Acad. Sci. 80:70–81.

Zweifel, R. G., and C. H. Lowe. 1966. The ecology of a population of *Xantusia vigilis,* the desert night lizard. Amer. Mus. Novitates 2247:1–57.

Zweifel, R. G., and K. S. Norris. 1955. Contributions to the herpetology of Sonora, Mexico: Descriptions of new subspecies of snakes *(Micruroides euryxanthus* and *Lampropeltis getulus)* and miscellaneous collecting notes. Amer. Midl. Nat. 54:230–249.

INDEX

Boldface page numbers refer to detailed species accounts, which include photographs and distribution maps for the given species. Other illustrations are indicated by page numbers in italics.

Ctenosaura conspicuosa (Isla San Esteban Spiny-Tailed Iguana)
at sunrise on Isla San Esteban, Sonora

DESIGN Nicole Hayward TEXT 10.25/13 Adobe Garamond DISPLAY Adobe Garamond Titling, Akzidenz Grotesk
COMPOSITION & COLOR SEPARATION Integrated Composition Systems PRINTING & BINDING Tien Wah Press
CARTOGRAPHY Bill Nelson INDEX Beaver Wood Associates